D0948200

# Preface

The purpose of this book is to describe methods for solving problems in applied electromagnetic theory using basic concepts from functional analysis and the theory of operators. Although the book focuses on certain mathematical fundamentals, it is written from an applications perspective for engineers and applied scientists working in this area.

Part I is intended to be a somewhat self-contained introduction to operator theory and functional analysis, especially those elements necessary for application to problems in electromagnetics. The goal of Part I is to explain and synthesize these topics in a logical manner. Examples principally geared toward electromagnetics are provided.

With the exception of Chapter 1, which serves as a review of basic electromagnetic theory, Part I presents definitions and theorems along with associated discussion and examples. This style was chosen because it allows one to readily identify the main concepts in a particular section. A proof is provided for all theorems whose proof is simple and straightforward. A proof is also provided for theorems that require a slightly more elaborate proof, yet one that is especially enlightening, being either constructive or illustrative. Generally, theorems are stated but not proved in cases where either the proof is too involved or the details of the proof would take one too far afield of the topic at hand, such as requiring additional lemmas that are not clearly useful in applications.

The material introduced in Part I is subsequently applied in Part II to problems in classical electromagnetics. A variety of problems are discussed, with some emphasis given to spectral formulations. Although the problem formulations and solution procedures are largely taken from the applied electromagnetics literature, the intrinsic connection to mathematical operator theory is highlighted, and the benefits of abstracting problems to an operator level are emphasized. For example, the completeness property of the eigenfunctions associated with many differential waveguide operators (justifying associated modal expansions) follows immediately after identifying the operators as being of the regular Sturm–Liouville type. Similarly, the fact that discrete, real-valued resonance frequencies occur in perfectly conducting electromagnetic cavities follows from spectral properties of the Laplacian operator. Many other examples that illustrate related concepts

George W. Hanson    Alexander B. Yakovlev

# Operator Theory for Electromagnetics

## An Introduction

With 77 Illustrations

George W. Hanson
Department of Electrical Engineering
and Computer Science
University of Wisconsin
Milwaukee, WI 53211
USA
george@uwm.edu

Alexander B. Yakovlev
Department of Electrical Engineering
University of Mississippi
University, MS 38677
USA
yakovlev@olemiss.edu

Library of Congress Cataloging-in-Publication Data
Hanson, George W., 1963–
Operator theory for electromagnetics : an introduction / George W. Hanson, Alexander B. Yakovlev.
p. cm.
Includes bibliographical references and index.
ISBN 0-387-95278-0 (alk. paper)
1. Electromagnetic theory. 2. Operator theory. I. Yakovlev, Alexander B. II. Title.
QC670.7.H36 2001
530.14′1—dc21 2001020438

Printed on acid-free paper.



Production managed by Lesley Poliner; manufacturing supervised by Joe Quatela.
Photocomposed copy prepared from the authors' LaTeX files by Macrotex, Savoy, IL.
Printed and bound by Edwards Brothers, Inc., Ann Arbor, MI.
Printed in the United States of America.

9 8 7 6 5 4 3 2 1

ISBN 0-387-95278-0 SPIN 10833374

Springer-Verlag New York Berlin Heidelberg
*A member of BertelsmannSpringer Science+Business Media GmbH*

# Operator Theory for Electromagnetics

**Springer**
*New York*
*Berlin*
*Heidelberg*
*Barcelona*
*Hong Kong*
*London*
*Milan*
*Paris*
*Singapore*
*Tokyo*

To
Sydney, Eric, and Anastasia

# Contents

**Preface** vii

**List of Symbols and Notation** xvii

**Part I: Basic Theory** 1

**1 Electromagnetic Fundamentals** 3

1.1 Maxwell's Equations . . . . . . . . . . 4
1.1.1 Maxwell's Equations—Differential Form . . . . . . . 4
1.1.2 Maxwell's Equations—Integral Form . . . . . . . . . 6
1.1.3 Boundary Conditions . . . . . . . . . . 7
1.1.4 Time-Harmonic and Transform-Domain Fields . . . 7
1.2 Field Theorems . . . . . . . . . . 13
1.2.1 Duality . . . . . . . . . . 13
1.2.2 Poynting's Theorem . . . . . . . . . . 14
1.2.3 The Lorentz Reciprocity Theorem . . . . . . . . . . 15
1.3 Wave Equations . . . . . . . . . . 17
1.3.1 Vector Wave and Vector Helmholtz Equations for Electric and Magnetic Fields . . . . . . . . . . 17
1.3.2 Vector and Scalar Potentials and Associated Helmholtz Equations . . . . . . . . . . 18
1.3.3 Solution of the Scalar Helmholtz Equations and Scalar Green's Functions . . . . . . . . . . 20
1.3.4 Improper Integrals Associated with Scalar Green's Functions . . . . . . . . . . 26
1.3.5 Solution of the Vector Wave and Vector Helmholtz Equations—Dyadic Green's Functions and Associated Improper Integrals . . . . . . . . . . 38
1.4 Integral Equations . . . . . . . . . . 48
1.4.1 Domain Integral Equations . . . . . . . . . . 48
1.4.2 Surface Integral Equations . . . . . . . . . . 52
1.5 Uniqueness Theorem and Radiation Conditions . . . . . . . 55
Bibliography . . . . . . . . . . 59

**2 Introductory Functional Analysis** **63**
2.1 Sets . . . . . 64
2.1.1 Sets of Numbers . . . . . 64
2.1.2 Sets of Functions—Function Spaces . . . . . 65
2.1.3 Sets of Sequences—Sequence Spaces . . . . . 73
2.2 Metric Space . . . . . 83
2.2.1 Continuity and Convergence . . . . . 88
2.2.2 Denseness, Closure, and Compactness . . . . . 94
2.2.3 Completeness . . . . . 97
2.3 Linear Space . . . . . 100
2.3.1 Linear Subspace . . . . . 101
2.3.2 Dimension of a Linear Space, Basis . . . . . 103
2.4 Normed Linear Space . . . . . 105
2.4.1 Completeness, Banach Space . . . . . 109
2.5 Inner Product Space . . . . . 110
2.5.1 Inner Product . . . . . 110
2.5.2 Pseudo Inner Product, Reaction . . . . . 113
2.5.3 Strong and Weak Convergence . . . . . 114
2.5.4 Completeness, Hilbert Space . . . . . 115
2.5.5 Orthogonality . . . . . 116
2.5.6 Expansions and Projections . . . . . 117
2.5.7 Basis of a Hilbert Space . . . . . 121
Bibliography . . . . . 125

**3 Introductory Linear Operator Theory** **129**
3.1 Linear Operators . . . . . 130
3.1.1 General Concepts . . . . . 130
3.1.2 Examples Related to Linear Operators . . . . . 134
3.1.3 Isomorphisms . . . . . 141
3.2 Linear Functionals . . . . . 143
3.2.1 General Concepts . . . . . 143
3.2.2 Examples Related to Linear Functionals . . . . . 144
3.3 Adjoint Operators . . . . . 146
3.3.1 General Concepts: Bounded Operators . . . . . 146
3.3.2 Examples of Operator Adjoints:
Bounded Operators . . . . . 147
3.3.3 General Concepts: Unbounded Operators . . . . . 149
3.3.4 Examples of Operator Adjoints:
Unbounded Operators . . . . . 150
3.4 Self-Adjoint, Symmetric, Normal, and Unitary Operators . 154
3.4.1 General Concepts . . . . . 154
3.4.2 Examples Relating to Self-Adjointness . . . . . 158
3.4.3 Pseudo Adjoints, Pseudo Symmetry,
and Reciprocity . . . . . 161
3.5 Definiteness and Convergence in Energy . . . . . 164

3.5.1 General Concepts . . . . . . . . . . . . . . . . . . . . . . 164
3.5.2 Examples Relating to Definiteness . . . . . . . . . . 166
3.5.3 Convergence in Energy . . . . . . . . . . . . . . . . . . 167
3.6 Compact Operators . . . . . . . . . . . . . . . . . . . . . . . . 169
3.6.1 General Concepts . . . . . . . . . . . . . . . . . . . . . . 169
3.6.2 Examples Relating to Compactness . . . . . . . . . . 171
3.7 Continuity and Compactness of Matrix Operators . . . . . . 175
3.7.1 Matrix Operators in $\mathbf{l}^2$ . . . . . . . . . . . . . . . . . 176
3.7.2 Gribanov's Theorems . . . . . . . . . . . . . . . . . . . 177
3.8 Closed and Closable Operators . . . . . . . . . . . . . . . . . 180
3.9 Invertible Operators . . . . . . . . . . . . . . . . . . . . . . . . 181
3.9.1 General Concepts . . . . . . . . . . . . . . . . . . . . . . 181
3.9.2 Examples Relating to Inverses . . . . . . . . . . . . . 189
3.9.3 Green's Functions and Green's Operators . . . . . . 190
3.10 Projection Operators . . . . . . . . . . . . . . . . . . . . . . . 202
3.11 Solution of Operator Equations . . . . . . . . . . . . . . . . . 204
3.11.1 Existence of Solutions . . . . . . . . . . . . . . . . . . . 204
3.11.2 Convergence of Solutions . . . . . . . . . . . . . . . . 207
Bibliography . . . . . . . . . . . . . . . . . . . . . . . . . . . . . . 213

**4 Spectral Theory of Linear Operators 219**
4.1 General Concepts . . . . . . . . . . . . . . . . . . . . . . . . . 220
4.1.1 Operators on Finite-Dimensional Spaces . . . . . . . 222
4.1.2 Operators on Infinite-Dimensional Spaces . . . . . . 223
4.2 Spectral Properties of Operators . . . . . . . . . . . . . . . . 228
4.2.1 General Properties . . . . . . . . . . . . . . . . . . . . . 228
4.2.2 Bounded Operators . . . . . . . . . . . . . . . . . . . . . 232
4.2.3 Invertible Operators . . . . . . . . . . . . . . . . . . . . 233
4.2.4 Compact Operators . . . . . . . . . . . . . . . . . . . . . 234
4.2.5 Self-Adjoint Operators . . . . . . . . . . . . . . . . . . . 234
4.2.6 Nonnegative and Positive Operators . . . . . . . . . 236
4.2.7 Compact Self-Adjoint Operators . . . . . . . . . . . 236
4.2.8 Normal and Unitary Operators . . . . . . . . . . . . 237
4.2.9 Generalized Eigenvalue Problems . . . . . . . . . . . 239
4.2.10 Eigenvalue Problems under a Pseudo Inner Product . . . . . . . . . . . . . . . . . . . . . . . . 241
4.2.11 Pseudo Self-Adjoint, Nonstandard Eigenvalue Problems . . . . . . . . . . . . . . . . . . . 242
4.2.12 Steinberg's Theorems for Compact Operators . . . . 244
4.3 Expansions and Representations, Spectral Theorems . . . . 247
4.3.1 Operators on Finite-Dimensional Spaces . . . . . . . 248
4.3.2 Operators on Infinite-Dimensional Spaces . . . . . . 257
4.3.3 Spectral Expansions Associated with Boundary Value Problems . . . . . . . . . . . . . . . . . 260

4.3.4 Spectral Expansions Associated with Integral Operators . . . 263
4.3.5 Generalized Eigenvectors and the Root System . . . 269
4.3.6 Spectral Representations . . . 271
4.4 Functions of Operators . . . 273
4.4.1 Functions of Operators via Spectral Representations . . . 273
4.4.2 Functions of Operators via Series Expansions . . . 276
4.4.3 Functions of Operators via the Dunford Integral Representation . . . 277
4.5 Spectral Methods in the Solution of Operator Equations . . . 278
4.5.1 First- and Second-Kind Operator Equations . . . 279
4.5.2 Spectral Methods and Green's Functions . . . 282
4.5.3 Convergence of Nonstandard Eigenvalues in Projection Techniques . . . 285
Bibliography . . . 287

**5 Sturm–Liouville Operators 291**
5.1 Regular Sturm–Liouville Problems . . . 292
5.1.1 Spectral Properties of the Regular Sturm–Liouville Operator . . . 294
5.1.2 Solution of the Regular Sturm–Liouville Problem . . . 297
5.1.3 Eigenfunction Expansion Solution of the Regular Sturm–Liouville Problem . . . 303
5.1.4 Completeness Relations for the Regular Sturm–Liouville Problem . . . 305
5.2 Singular Sturm–Liouville Problems . . . 307
5.2.1 Classification of Singular Points . . . 311
5.2.2 Identification of the Continuous Spectrum and Improper Eigenfunctions . . . 314
5.2.3 Continuous Expansions and Associated Integral Transforms . . . 321
5.2.4 Simultaneous Occurrence of Discrete and Continuous Spectra . . . 326
5.3 Nonself-Adjoint Sturm–Liouville Problems . . . 327
5.3.1 Green's Function Methods in the Nonself-Adjoint Case . . . 327
5.3.2 Spectral Methods in the Nonself-Adjoint Case . . . 331
5.4 Special Functions Associated with Singular Sturm–Liouville Problems . . . 340
5.4.1 Cylindrical Coordinate Problems . . . 341
5.4.2 Spherical Coordinate Problems . . . 350
5.5 Classical Orthogonal Polynomials and Associated Bases . . . 358

5.5.1 Laguerre Polynomials . . . . . . . . . . . . . . . . . 359
5.5.2 Hermite Polynomials . . . . . . . . . . . . . . . . . . 359
5.5.3 Chebyshev Polynomials . . . . . . . . . . . . . . . . 360
Bibliography . . . . . . . . . . . . . . . . . . . . . . . . . . 363

**Part II: Applications in Electromagnetics** **365**

**6 Poisson's and Laplace's Boundary Value Problems: Potential Theory** **367**
6.1 Problem Formulation . . . . . . . . . . . . . . . . . . . . 368
6.2 Operator Properties of the Negative Laplacian . . . . . . . 372
6.3 Spectral Properties of the Negative Laplacian . . . . . . . . 376
6.4 Solution Techniques for Self-Adjoint Problems . . . . . . . . 378
6.5 Integral Methods and Separation of Variables Solutions for Nonself-Adjoint Problems . . . . . . . . . . . . . . . . . . 384
6.6 Integral Equation Techniques for Potential Theory . . . . . 391

**7 Transmission-Line Analysis** **403**
7.1 General Analysis . . . . . . . . . . . . . . . . . . . . . . . 404
7.2 Transmission-Line Resonators . . . . . . . . . . . . . . . . 412
7.3 Semi-Infinite and Infinite Transmission Lines . . . . . . . . 415
7.4 Multiconductor Transmission Lines . . . . . . . . . . . . . . 417

**8 Planarly Layered Media Problems** **421**
8.1 Two-Dimensional Problems . . . . . . . . . . . . . . . . . . 422
8.1.1 General Analysis . . . . . . . . . . . . . . . . . . . . 422
8.1.2 Homogeneously Filled Parallel-Plates . . . . . . . . . 434
8.1.3 Two-Dimensional Resonator . . . . . . . . . . . . . . 444
8.1.4 Semi-Open Structure . . . . . . . . . . . . . . . . . . 446
8.1.5 Free Space . . . . . . . . . . . . . . . . . . . . . . . . 446
8.1.6 Impedance Plane Structure . . . . . . . . . . . . . . 447
8.1.7 Grounded Dielectric Layer . . . . . . . . . . . . . . . 462
8.1.8 Comments on General Multilayered Media Problems, Completeness, and Associated Eigenfunctions . . . . . . . . . . . . . . 480
8.2 Two-Dimensional Scattering Problems in Planar Media . . 482
8.2.1 TE Scattering from Inhomogeneous Dielectric Cylinders in Layered Media . . . . . . . . . . . . . . 482
8.2.2 TE Scattering from Thin Conducting Strips in Layered Media . . . . . . . . . . . . . . . . . . . . 486
8.3 Three-Dimensional Planar Problems . . . . . . . . . . . . . 489
8.3.1 General Analysis . . . . . . . . . . . . . . . . . . . . 490
8.3.2 Parallel-Plate Structure . . . . . . . . . . . . . . . . 499

**9 Cylindrical Waveguide Problems** **503**
9.1 Scattering Problems for Waveguides Filled with a Homogeneous Medium . . . 504
9.1.1 Green's Dyadics—General Concepts and Definitions . . . 504
9.1.2 Infinite Waveguide Containing Metal Obstacles . . . 507
9.1.3 Infinite Waveguide with Apertures . . . 512
9.1.4 Semi-Infinite Waveguides Coupled through Apertures in a Common Ground Plane . . . 515
9.1.5 Semi-Infinite Waveguides with Metal Obstacles Coupled through Apertures in a Common Ground Plane . . . 516
9.2 Green's Dyadics for Waveguides Filled with a Homogeneous Medium . . . 518
9.2.1 Magnetic Potential Green's Dyadics . . . 519
9.2.2 Electric Green's Dyadics . . . 526
9.3 Scattering Problems for Waveguides Filled with a Planarly Layered Medium . . . 532
9.3.1 Interacting Electric- and Magnetic-Type Discontinuities in a Layered-Medium Waveguide . . . 532
9.3.2 A Waveguide-Based Aperture-Coupled Patch Array . . . 546
9.4 Electric Green's Dyadics for Waveguides Filled with a Planarly Layered Medium . . . 553
9.4.1 Electric Green's Dyadics of the Third Kind . . . 554
9.5 The Method of Overlapping Regions . . . 560
9.5.1 Integral Equation Formulations for Overlapping Regions . . . 561
9.5.2 Shielded Microstrip Line . . . 567

**10 Electromagnetic Cavities** **575**
10.1 Problem Formulation . . . 575
10.2 Operator Properties and Eigenfunction Methods . . . 577
10.3 Integral Equation Methods for Cavities . . . 586
10.4 Cavity Resonances from Vector Potentials . . . 589

**Bibliography for Part II** **597**

**A Vector, Dyadic, and Integral Relations** **605**
A.1 Vector Identities . . . 605
A.2 Dyadic Identities . . . 606
A.3 Dyadic Analysis . . . 607
A.4 Integral Identities . . . 608
A.5 Useful Formulas Involving the Position Vector and Scalar Green's Functions . . . 611

A.6 Scalar and Vector Differential Operators in the Three Principal Coordinate Systems . . . . . . . . . . . . . 612

**B Derivation of Second-Derivative Formula (1.59) 615**

**C Gram–Schmidt Orthogonalization Procedure 619**

**D Coefficients of Planar-Media Green's Dyadics 621**

**E Additional Function Spaces 625**

**Index 629**

# List of Symbols and Notation

$\mathbf{A}, \mathbf{a}$ — vector

$\underline{\mathbf{A}}, \underline{\mathbf{a}}$ — dyadic

$\underline{\mathbf{G}}^{\wedge}(\mathbf{r}, \mathbf{r}')$ — form of a dyadic Green's function that includes a depolarizing dyadic contribution, 44

$[A]$ — a matrix

$[A]^{\top}$ — transpose of matrix $A$

$\mathbf{n}$ — unit vector normal to a surface

$\widehat{\alpha}$ — unit vector in $\alpha$-direction

$\overline{a}$ — complex conjugate of $a$

$i$ — imaginary unit, $i = \sqrt{-1}$

$\|\cdot\|$ — norm, 106

$d(x, y)$ — metric, 83

$\langle \cdot, \cdot \rangle$ — inner product, 110

$\langle \cdot, \cdot \rangle_p$ — pseudo inner product, 113

$[A, B]$ — commutator of operators $A, B$, 157

$x_n \to x$ — strong convergence, 114

$x_n \overset{w}{\to} x$ — weak convergence, 114

$M^{\perp}$ — orthogonal complement, 116

$D^{\alpha}$ — generalized partial derivative, 69

$L^{\alpha}(\nabla)$ — generalized vector derivative operator of order $\alpha$, 71

$l$ — linear functional, 143

$D_A$ — domain of operator $A$, 130

$R_A$ — range of operator $A$, 131

$N_A$ — null space of operator $A$, 131

$A^*$ — adjoint of operator $A$, 146

$A^{-1}$ — inverse of operator $A$, 181

$A_L^{-1}$ ——— left inverse of operator $A$, 188

$A_R^{-1}$ ——— right inverse of operator $A$, 188

$P$ ——— projection operator, 202

$\widetilde{S}$ ——— closure of set $S$, 95

$(a, b)$ ——— open interval in one dimension, e.g., $a < x < b$

$[a, b] = \widetilde{(a, b)}$ closed interval in one dimension, e.g., $a \leq x \leq b$

$\Omega$ ——— open region in $m$ dimensions

$\mathbf{\Gamma}$ ——— boundary of region $\Omega$

$\widetilde{\Omega} = \Omega \cup \Gamma$ — closed region in $m$ dimensions

$d\Omega$ ——— differential element of $\Omega$

$\in$ ——— "in" or "belongs to," e.g., $x \in [a, b]$ means $a \leq x \leq b$

$\subseteq$ ——— subset, e.g., $X \subseteq Y$ means the set $X$ is contained in the set $Y$ ($\subset$ indicates a proper subset)

$\cup$ ——— union, e.g., $X \cup Y$ is the set of all elements in $X$ and in $Y$

$\cap$ ——— intersection, e.g., $X \cap Y$ is the set of all elements that are in both $X$ and in $Y$

$Y \setminus X$ ——— the set $Y$ excluding the set $X$; alternately,

$Y - X$ ——— the set $Y$ excluding the set $X$

$x_>, x_<$ ——— indicates the greater and lesser of the pair $(x, x')$, respectively, e.g., if $x > x'$, then $x_> = x$ and $x_< = x'$, 301

$\forall$ ——— "for all," e.g., $\forall\ y \in H$ means "for all $y$ in $H$"

$\exists$ ——— "there exists," e.g., $\exists\ y \in H$ means "there exists a $y$ in $H$"

$\varepsilon_n$ ——— Neumann's number, 307

$\delta_{ij}$ ——— Kronecker delta function, 116

$H$ ——— a general Hilbert space, 115

$\mathbf{C}$ ——— the space of all complex numbers, 64

$\mathbf{R}$ ——— the space of all real numbers, 64

$\mathbf{Z}$ ——— the space of all integers, 64

$\mathbf{N}$ ——— the space of all positive integers, 64

$\mathbf{C}^n (\mathbf{R}^n)$ ——— the space of all complex- (real-) valued $n$-tuples, 65

$\mathbf{C}^k (\Omega)$ ——— the space of all continuous functions on $\Omega$ with continuous derivatives up to order $k$, 65

$\mathbf{M} (\Omega)$ ——— the space of bounded functions on $\Omega$, 66

$\mathbf{C} (\Omega)$ ——— the space of continuous functions on $\Omega$, 66

$\mathbf{L}^p (\Omega)$ ——— the space of all functions Lebesgue-integrable to the $p$th power on $\Omega$, 66

$\mathbf{L}^p(\Omega)^m$ —— the space of all $m$-tuple vector functions on $\Omega \subseteq \mathbf{R}^n$ Lebesgue-integrable to the $p$th power, 68

$\mathbf{H}_{k,\alpha}(\Omega)$ —— Hölder space, 70

$\mathbf{H}^*_\beta(\Omega)$ —— Hölder space, 625

$\mathbf{W}^k_p$, $\mathbf{H}^k(\Omega)$ Sobolev space of order $k$, 72

$H(\mathrm{curl}, \Omega)$ — Sobolev space, 73

$H(\mathrm{div}, \Omega)$ — Sobolev space, 73

$\mathbf{l}^p$ —— the space of all sequences summable to the $p$th power, 73

$\mathbf{l}^\infty$ —— the space of all bounded sequences, 74

$O, o$ —— order symbols, 22

# Part I
# Basic Theory

The basic theory part of the text consists of five chapters. Chapter 1 presents some fundamental concepts from classical electromagnetic theory. The chapter begins with a presentation of the governing (Maxwell's) equations for macroscopic electromagnetic phenomena, relevant constitutive parameters and boundary conditions, and time-domain and transform-domain formulations. Wave equations are formulated for both fields and potentials, and methods for solving the various wave equations are discussed in some detail. Green's functions are developed, reading to improper integrals and a careful treatment of the source-point singularity. The volume equivalence principle is presented, leading to the formulation of domain integral equations, and surface integral equations are developed for perfectly conducting scatterers. The chapter concludes with conditions under which solutions of the wave equations are unique.

The second chapter covers some basic concepts from functional analysis, including metric spaces, linear spaces, normed spaces, and inner product spaces. The material in this chapter is intended to present, from an applied perspective, the concepts necessary for understanding the notion of function spaces, particularly Hilbert and Banach spaces. Much of the presented material is associated with properties of operators encountered in electromagnetic problems.

Chapter 3 introduces linear operator theory, primarily for Hilbert spaces. In some sense this and the next chapter form the core of the first part of the text, in that the basic properties of various types of operators are presented. The main properties of linear operators and, as a subclass, linear functionals are described, as are operator adjoints. Properties of various classes of operators, including self-adjoint, symmetric, normal, unitary, and compact, are discussed. The important concepts of Green's functions and Green's operators are introduced, and the chapter concludes with a partial discussion of solvability conditions for operator equations.

The fourth chapter covers the spectral theory of linear operators in Hilbert spaces, with special emphasis on the eigenvalue problem. The spectral properties of operators on both finite- and infinite-dimensional spaces are described for a variety of operator types. Spectral theorems are presented which describe classes of operators that admit relatively simple spectral representations. Functions of operators are discussed, followed by spectral methods for solving operator equations.

Chapter 5 is devoted to the important Sturm–Liouville operator. The regular Sturm–Liouville problem is considered in detail, and salient points from the theory of singular Sturm–Liouville problems are presented. Spectral properties of Sturm–Liouville operators are discussed, both in the self-adjoint and nonself-adjoint cases, and many examples directly applicable to the waveguiding problems considered in Part II are presented. The chapter concludes with a discussion of special functions and classical orthogonal polynomials that arise from certain singular Sturm–Liouville problems.

# 1

# Electromagnetic Fundamentals

The coverage of electromagnetics in this chapter is somewhat brief, especially the physical aspects of the theory, since it is assumed that the reader is familiar with basic field theory at an undergraduate or beginning graduate level. For a more extensive introduction to electromagnetic theory, the references at the end of this chapter may be consulted.

The chapter begins with a presentation of Maxwell's equations for macroscopic electromagnetic phenomena, relevant constitutive parameters and boundary conditions, and time-domain and transform-domain formulations. Then some important field theorems are presented. They lead to insight into the field theory and are necessary for formulating and solving many problems in electromagnetics. Next, wave equations are formulated for both fields and potentials in the form of vector Helmholtz equations and vector wave equations. Methods for solving the various wave equations are discussed in some detail. The topic of differentiating the weakly singular volume integrals associated with solutions of the wave equations are considered for both the scalar and vector cases, leading to a discussion of the depolarizing dyadic contribution for the electric dyadic Green's function. The volume equivalence principle is described, leading to the formulation of domain integral equations. Surface integral equations are then developed for perfectly conducting scatterers. The chapter concludes with conditions under which solutions of the wave equations are unique.

# 1.1 Maxwell's Equations

## 1.1.1 Maxwell's Equations—Differential Form

Classical macroscopic electromagnetic phenomena are governed by a set of vector equations known collectively as *Maxwell's equations*. Maxwell's equations in differential form are

$$\begin{aligned}
\nabla \cdot \mathbf{D}(\mathbf{r},t) &= \rho_e(\mathbf{r},t), \\
\nabla \cdot \mathbf{B}(\mathbf{r},t) &= \rho_m(\mathbf{r},t), \\
\nabla \times \mathbf{E}(\mathbf{r},t) &= -\frac{\partial}{\partial t}\mathbf{B}(\mathbf{r},t) - \mathbf{J}_m(\mathbf{r},t), \\
\nabla \times \mathbf{H}(\mathbf{r},t) &= \frac{\partial}{\partial t}\mathbf{D}(\mathbf{r},t) + \mathbf{J}_e(\mathbf{r},t),
\end{aligned} \tag{1.1}$$

where $\mathbf{E}$ is the electric field intensity (V/m), $\mathbf{D}$ is the electric flux density (C/m$^2$), $\mathbf{B}$ is the magnetic flux density (Wb/m$^2$), $\mathbf{H}$ is the magnetic field intensity (A/m), $\rho_e$ is the electric charge density (C/m$^3$), $\mathbf{J}_e$ is the electric current density (A/m$^2$), $\rho_m$ is the magnetic charge density (Wb/m$^3$), and $\mathbf{J}_m$ is the magnetic current density (V/m$^2$), and where V stands for volts, C for coulombs, Wb for webers, A for amperes, and m for meters.[1]

The equations are known, respectively, as *Gauss' law*, the *magnetic-source law* or *magnetic Gauss' law*, *Faraday's law*, and *Ampère's law*. The magnetic charge and magnetic current density have not been shown to physically exist, and so often those terms are set to zero. However, their inclusion provides a nice mathematical symmetry to Maxwell's equations. More importantly, they are useful as sources in equivalence problems, such as in problems concerning aperture radiation (see, e.g., Chapter 9).

The constitutive equations

$$\begin{aligned}
\mathbf{D}(\mathbf{r},t) &= \epsilon_0\mathbf{E}(\mathbf{r},t) + \mathbf{P}(\mathbf{r},t), \\
\mathbf{B}(\mathbf{r},t) &= \mu_0\mathbf{H}(\mathbf{r},t) + \mu_0\mathbf{M}(\mathbf{r},t),
\end{aligned} \tag{1.2}$$

provide relations between the four field vectors in a material medium, where $\mathbf{P}$ is the polarization density (C/m$^2$), $\mathbf{M}$ is the magnetization density (A/m), $\epsilon_0$ is the permittivity of free space ($\simeq 8.85 \times 10^{-12}$ F/m), and $\mu_0$ is the permeability of free space ($\simeq 4\pi \times 10^{-7}$ H/m), and where F stands for farads and H for henrys. For dimensional analysis, C $=$ A $\cdot$ s $=$ F $\cdot$ V and Wb $=$ V $\cdot$ s $=$ H $\cdot$ A, where s stands for seconds.

The form (1.2) notwithstanding, often the field quantities $\mathbf{E}, \mathbf{B}$ are considered the fundamental fields because these are implicated by the fundamental Lorentz force law

$$\mathbf{F} = q\left(\mathbf{E} + \mathbf{v} \times \mathbf{B}\right)$$

[1] Rationalized mksA units are used throughout. For a detailed discussion of units, see [38].

where $\mathbf{F}$ is the force on charge $q$, which has velocity $\mathbf{v}$. However, there is some debate on this issue [1].

The polarization and magnetization densities are associated with electric and magnetic dipole moments, respectively, in a given material. These dipole moments include both induced effects and permanent dipole moments. In free space these quantities vanish.

In general, we will assume, unless otherwise noted, that all of the relevant electromagnetic quantities are continuous and continuously differentiable[2] at *most* points in space. If we apply multiple partial derivative operators to quantities, it will be implicitly assumed that those quantities are sufficiently differentiable, with the resulting differentiated quantities being continuous. This allows, among other things, the free interchange of the order of partial derivative operators[3] and the application of various vector calculus theorems such as the divergence theorem. Additional smoothness conditions are suggested by the time-harmonic or temporal transform-domain quantities, as will be discussed later. Certain continuity requirements are not necessarily expected at the location of sources or at material boundaries.

In the preceding equations $\mathbf{r}$ is the "field-point" position vector $\mathbf{r} = \sum_{i=1}^{n} \widehat{\mathbf{x}}_i x_i$ in $n$-dimensional space, e.g., for $n = 3$ in rectangular coordinates, $\mathbf{r} = \widehat{\mathbf{x}}x + \widehat{\mathbf{y}}y + \widehat{\mathbf{z}}z$. Generally, a primed coordinate system will denote the "source-point" position vector, e.g., $\mathbf{r}' = \widehat{\mathbf{x}}x' + \widehat{\mathbf{y}}y' + \widehat{\mathbf{z}}z'$. The vector that points from the source point to the field point is denoted by

$$\mathbf{R}(\mathbf{r}, \mathbf{r}') \equiv \mathbf{r} - \mathbf{r}' = \widehat{\mathbf{R}}(\mathbf{r}, \mathbf{r}')\, R$$

with $R(\mathbf{r}, \mathbf{r}') = |\mathbf{r} - \mathbf{r}'| = R(\mathbf{r}', \mathbf{r})$ and

$$\widehat{\mathbf{R}}(\mathbf{r}, \mathbf{r}') = (\mathbf{r} - \mathbf{r}')/\,|\mathbf{r} - \mathbf{r}'| = -\widehat{\mathbf{R}}(\mathbf{r}', \mathbf{r}).$$

An important equation that demonstrates that charge conservation is embedded in (1.1) is known as the *continuity equation*. Taking the divergence of Ampère's law we have

$$0 = \nabla \cdot \nabla \times \mathbf{H} = \nabla \cdot \mathbf{J}_e + \nabla \cdot \frac{\partial \mathbf{D}}{\partial t}$$

[2] A function of one variable is said to be continuously differentiable in an open region $(a, b)$ if the first derivative is continuous in $(a, b)$. For a function of $n$ variables we consider the open region $\Omega \subseteq \mathbf{R}^n$ and the generalized partial derivative described on p. 69.

[3] If $\partial^2 f(x, y)/\partial x\, \partial y$ and $\partial^2 f(x, y)/\partial y\, \partial x$ are both continuous in a region $\Omega$, then

$$\frac{\partial^2 f(x, y)}{\partial x\, \partial y} = \frac{\partial^2 f(x, y)}{\partial y\, \partial x}$$

throughout that region [11, p. 36]. In fact, it is sufficient that $\partial f(x, y)/\partial x$, $\partial f(x, y)/\partial y$, and either $\partial^2 f(x, y)/\partial x\, \partial y$ or $\partial^2 f(x, y)/\partial y\, \partial x$ are continuous in $\Omega$ for the equality of the two second partial derivatives to hold.

and, upon interchanging the spatial and temporal derivatives and invoking Gauss' law, we obtain the continuity equation

$$\nabla \cdot \mathbf{J}_e + \frac{\partial \rho_e}{\partial t} = 0.$$

Similarly, starting with Faraday's law we obtain

$$\nabla \cdot \mathbf{J}_m + \frac{\partial \rho_m}{\partial t} = 0.$$

Conversely, the two divergence equations are not independent equations within the set (1.1), in the sense that they are embedded in the two curl equations and the continuity equation. Therefore, in macroscopic electromagnetics for $\omega \neq 0$, one may consider the relevant set of equations to be solved as

$$\begin{aligned} \nabla \times \mathbf{E}(\mathbf{r},t) &= -\frac{\partial}{\partial t}\mathbf{B}(\mathbf{r},t) - \mathbf{J}_m(\mathbf{r},t), \\ \nabla \times \mathbf{H}(\mathbf{r},t) &= \frac{\partial}{\partial t}\mathbf{D}(\mathbf{r},t) + \mathbf{J}_e(\mathbf{r},t), \\ \nabla \cdot \mathbf{J}_{e(m)}(\mathbf{r},t) &= -\frac{\partial \rho_{e(m)}(\mathbf{r},t)}{\partial t}, \end{aligned} \tag{1.3}$$

subject to appropriate boundary conditions. For $\omega = 0$, the divergence equations must also be included in (1.3).

### 1.1.2 Maxwell's Equations—Integral Form

Starting with the differential (point) form of Maxwell's equations, an integral (large-scale) form may be derived. Applying the divergence theorem

$$\int_V \nabla \cdot \mathbf{A}\, dV = \oint_S \mathbf{n} \cdot \mathbf{A}\, dS = \oint_S \mathbf{A} \cdot d\mathbf{S}$$

(see Appendix A.4) to the divergence and continuity equations, and Stokes' theorem

$$\int_S \nabla \times \mathbf{A} \cdot d\mathbf{S} = \oint_l \mathbf{A} \cdot d\mathbf{l}$$

to the curl equations, leads to the integral form

$$\begin{aligned} \oint_S \mathbf{D}(\mathbf{r},t) \cdot d\mathbf{S} &= \int_V \rho_e(\mathbf{r},t)\, dV, \\ \oint_S \mathbf{B}(\mathbf{r},t) \cdot d\mathbf{S} &= \int_V \rho_m(\mathbf{r},t)\, dV, \\ \oint_l \mathbf{E}(\mathbf{r},t) \cdot d\mathbf{l} &= -\frac{d}{dt}\int_S \mathbf{B}(\mathbf{r},t) \cdot d\mathbf{S} - \int_S \mathbf{J}_m(\mathbf{r},t) \cdot d\mathbf{S}, \\ \oint_l \mathbf{H}(\mathbf{r},t) \cdot d\mathbf{l} &= \frac{d}{dt}\int_S \mathbf{D}(\mathbf{r},t) \cdot d\mathbf{S} + \int_S \mathbf{J}_e(\mathbf{r},t) \cdot d\mathbf{S}, \\ \oint_S \mathbf{J}_{e(m)}(\mathbf{r},t) \cdot d\mathbf{S} &= -\frac{d}{dt}\int_V \rho_{e(m)}(\mathbf{r},t)\, dV, \end{aligned} \tag{1.4}$$

assuming that the conditions implied by the divergence and Stokes' theorems are satisfied and that the derivative and integral operators may be interchanged. Throughout this chapter $V$ is an open region bounded by a closed surface $S$.

### 1.1.3 Boundary Conditions

Knowledge of the variation of a field quantity across a regular smooth (nonsingular) boundary is often necessary in solving or formulating electromagnetic problems. These boundary conditions are usually deduced from the integral form of Maxwell's equations (1.4) and can be given as [2]

$$\begin{aligned} \mathbf{n} \times (\mathbf{H}_2 - \mathbf{H}_1) &= \mathbf{J}_{e,s}, \\ \mathbf{n} \times (\mathbf{E}_2 - \mathbf{E}_1) &= -\mathbf{J}_{m,s}, \\ \mathbf{n} \cdot (\mathbf{D}_2 - \mathbf{D}_1) &= \rho_{e,s}, \\ \mathbf{n} \cdot (\mathbf{B}_2 - \mathbf{B}_1) &= \rho_{m,s}, \end{aligned} \tag{1.5}$$

where $\mathbf{J}_{e,s}$, $\mathbf{J}_{m,s}$ ($\rho_{e,s}$, $\rho_{m,s}$) are surface currents (charges) on the boundary, $\mathbf{E}_2, \mathbf{H}_2$ are the fields infinitely close to the boundary on the side into which $\mathbf{n}$, the unit vector normal to the surface, is directed, and $\mathbf{E}_1, \mathbf{H}_1$ are the corresponding fields infinitely close to the boundary on the opposite side. The tangential boundary conditions arise from the curl equations, whereas the normal boundary conditions are deduced from the divergence equations. Thus, the normal boundary conditions are not independent from the tangential boundary conditions.

When $\mathbf{J}_{e,s} = 0$ ($\mathbf{J}_{m,s} = 0$) the tangential magnetic (electric) fields are continuous across the interface. This is usually the case of physical interest, true surface currents being a theoretical construct. However, as a practical matter, one often considers an idealized material having infinite electrical (magnetic) conductivity, such that electric (magnetic) surface currents will occur. An illuminating discussion of electromagnetic boundary conditions can be found in [3].

### 1.1.4 Time-Harmonic and Transform-Domain Fields

#### Time-Harmonic Fields

Often the fields of interest vary harmonically (sinusoidally) with time. Because Maxwell's equations are linear, if we assume linear constitutive equations (which is often the case, with a notable exception being certain optical materials), then time-harmonic sources $\rho, \mathbf{J}$ will maintain time-harmonic fields $\mathbf{E}, \mathbf{D}, \mathbf{B}, \mathbf{H}$. Assuming we can separate the space and time dependence of the field and source quantities,[4] e.g.,

$$\mathbf{E}(\mathbf{r},t) = \mathbf{E}_0(\mathbf{r})\cos(\omega t + \phi_E),$$
$$\mathbf{B}(\mathbf{r},t) = \mathbf{B}_0(\mathbf{r})\cos(\omega t + \phi_B),$$

etc., we can eliminate the time derivatives in the following manner. Considering the electric field, we form

$$\begin{aligned}\mathbf{E}(\mathbf{r},t) &= \mathbf{E}_0(\mathbf{r})\,\mathrm{Re}\left\{e^{i(\omega t+\phi_E)}\right\} = \mathrm{Re}\left\{\mathbf{E}_0(\mathbf{r})e^{i\omega t}e^{i\phi_E}\right\} \\ &= \mathrm{Re}\left\{\mathbf{E}(\mathbf{r})e^{i\omega t}\right\},\end{aligned}$$

where $\mathbf{E}(\mathbf{r}) \equiv \mathbf{E}_0(\mathbf{r})e^{i\phi_E}$ is a complex phasor. We have assumed that $\mathbf{E}_0(\mathbf{r})$ is a real-valued vector, consistent with the usual assumption that the time-domain field is real-valued. The other field quantities follow similarly.

Next, consider, for instance, Faraday's law,

$$\nabla \times \mathbf{E}(\mathbf{r},t) = -\frac{\partial}{\partial t}\mathbf{B}(\mathbf{r},t) - \mathbf{J}_m(\mathbf{r},t).$$

This can be rewritten as[5]

$$\mathrm{Re}\left\{\left[\nabla \times \mathbf{E}(\mathbf{r}) + i\omega\mathbf{B}(\mathbf{r}) + \mathbf{J}_m(\mathbf{r})\right]e^{i\omega t}\right\} = \mathbf{0},$$

which must be true for all $t$. When $\omega t = 0$, $\mathrm{Re}\{\nabla\times\mathbf{E}(\mathbf{r})+i\omega\mathbf{B}(\mathbf{r})+\mathbf{J}_m(\mathbf{r})\} = \mathbf{0}$, and when $\omega t = \pi/2$, $\mathrm{Im}\{\nabla \times \mathbf{E}(\mathbf{r}) + i\omega\mathbf{B}(\mathbf{r}) + \mathbf{J}_m(\mathbf{r})\} = \mathbf{0}$. If both the real part and the imaginary part of a complex number are zero, then the number itself is zero, and thus $\nabla\times\mathbf{E}(\mathbf{r}) = -i\omega\mathbf{B}(\mathbf{r}) - \mathbf{J}_m(\mathbf{r})$. Repeating for all of (1.1) and the continuity equations, we get the time-harmonic forms

$$\begin{aligned}\nabla\cdot\mathbf{D}(\mathbf{r}) &= \rho_e(\mathbf{r}), \\ \nabla\cdot\mathbf{B}(\mathbf{r}) &= \rho_m(\mathbf{r}), \\ \nabla\times\mathbf{E}(\mathbf{r}) &= -i\omega\mathbf{B}(\mathbf{r}) - \mathbf{J}_m(\mathbf{r}), \\ \nabla\times\mathbf{H}(\mathbf{r}) &= \ i\omega\mathbf{D}(\mathbf{r}) + \mathbf{J}_e(\mathbf{r}), \\ \nabla\cdot\mathbf{J}_{e(m)}(\mathbf{r}) &= -i\omega\,\rho_{e(m)}(\mathbf{r}),\end{aligned} \tag{1.6}$$

[4] To be general, we should allow the various scalar components of fields to have different phases. However, for simplicity, here we assume that all scalar components of a vector field have the same phase.

[5] For the real-part operator [25, pp. 16–17]

$$\mathrm{Re}\,(f_1) + \mathrm{Re}\,(f_2) = \mathrm{Re}\,(f_1 + f_2), \quad \frac{\partial}{\partial x_i}\mathrm{Re}\,(f_1(\mathbf{r})) = \mathrm{Re}\left(\frac{\partial}{\partial x_i}f_1(\mathbf{r})\right),$$

$$\mathrm{Re}\,(\alpha f_1) = \alpha\,\mathrm{Re}\,(f_1), \quad \int \mathrm{Re}\,(f_1(\mathbf{r}))\,dx_i = \mathrm{Re}\int f_1(\mathbf{r})\,dx_i,$$

where $f_1, f_2 \in C$, $\alpha \in R$. This is similar for the operator $\mathrm{Im}\,(\cdot)$.

where all quantities are time-harmonic phasors.[6] Time-domain quantities can be recovered from the phasor quantities as, for instance, $\mathbf{E}(\mathbf{r},t) = \operatorname{Re}\left\{\mathbf{E}(\mathbf{r})e^{i\omega t}\right\}$. Due to the properties of the operator $\operatorname{Re}(\cdot)$, other than the condition that the quantities in (1.1) are time-harmonic, no additional constraints need to be placed on the quantities appearing in (1.1) for (1.1) to imply (1.6).

### Transform-Domain Fields

If the electromagnetic sources vary arbitrarily with time (time-harmonic variation being a special case), it is often convenient to work in the temporal Fourier transform domain. The Fourier transform pair is given as[7]

$$\mathbf{K}(\mathbf{r},\omega) = F\{\mathbf{K}(\mathbf{r},t)\} = \int_{-\infty}^{\infty} \mathbf{K}(\mathbf{r},t)e^{-i\omega t}dt, \tag{1.7}$$

$$\mathbf{K}(\mathbf{r},t) = F^{-1}\{\mathbf{K}(\mathbf{r},\omega)\} = \frac{1}{2\pi}\int_{-\infty}^{\infty} \mathbf{K}(\mathbf{r},\omega)e^{i\omega t}d\omega. \tag{1.8}$$

To see how Maxwell's equations are Fourier-transformed, consider again Faraday's law. Assuming each field quantity $\mathbf{E}, \mathbf{B}$ is Fourier transformable[8] and substituting the various transforms into Faraday's law leads to

$$\begin{aligned}\nabla \times \frac{1}{2\pi}\int_{-\infty}^{\infty} \mathbf{E}(\mathbf{r},\omega)e^{i\omega t}d\omega = &-\frac{\partial}{\partial t}\frac{1}{2\pi}\int_{-\infty}^{\infty} \mathbf{B}(\mathbf{r},\omega)e^{i\omega t}d\omega \\ &-\frac{1}{2\pi}\int_{-\infty}^{\infty} \mathbf{J}_m(\mathbf{r},\omega)e^{i\omega t}d\omega.\end{aligned}$$

Assuming the differential and integral operators can be interchanged,[9] we

[6]Note that in the time-harmonic case we obtain the convenient correspondence $\partial/\partial t \leftrightarrow i\omega$ and $\int(\cdot)\,dt \leftrightarrow (i\omega)^{-1}$.

[7]Note that we use the same symbol for a function and its transform. The correct interpretation should be clear from the context of the problem.

[8]Classically, a function $f(t)$ is Fourier transformable if it is *absolutely integrable*, i.e., $\int_{-\infty}^{\infty}|f(t)|\,dt < \infty$. Sufficient conditions for a function $f$ to be absolutely integrable are that $f(t)$ is piecewise smooth and that $f(t) \to 0$ sufficiently fast (e.g., $f(t) = O\left(t^{-1-\varepsilon}\right)$ for $\varepsilon > 0$) as $t \to \pm\infty$. Integrals for which $\int_{-\infty}^{\infty} f(t)\,dt$ exists yet $\int_{-\infty}^{\infty}|f(t)|\,dt$ does not exist are called *conditionally convergent*. Classical functions leading to conditionally convergent integrals, as well as generalized functions, usually can be Fourier-transformed within generalized function theory.

[9]Leibnitz's rule covers the interchange of differential and integral operators for proper integrals having finite, possibly variable limits of integration (see note on p. 31). For improper integrals of the type $\int_a^{\infty} f(x,y)dy$, it can be shown that [44, p. 243], [35, p. 392]

$$\frac{d}{dx}\int_a^{\infty} f(x,y)dy = \int_a^{\infty}\frac{\partial}{\partial x}f(x,y)dy$$

if $f, \partial f/\partial x$ are continuous on $[c,d] \times [a,\infty)$, $\int_a^{\infty} f(x,y)dy$ converges on $[c,d]$, and

have

$$\frac{1}{2\pi}\int_{-\infty}^{\infty}\{\nabla\times\mathbf{E}(\mathbf{r},\omega)+i\omega\mathbf{B}(\mathbf{r},\omega)+\mathbf{J}_m(\mathbf{r},\omega)\}\,e^{i\omega t}d\omega=\mathbf{0}.$$

Because $\mathbf{K}(\mathbf{r},t) = F^{-1}\{\mathbf{K}(\mathbf{r},\omega)\} = \mathbf{0}$ for all $t$ implies that $\mathbf{K}(\mathbf{r},\omega) = F\{\mathbf{K}(\mathbf{r},t)\} = F\{\mathbf{0}\} = \mathbf{0}$, we obtain $\nabla\times\mathbf{E}(\mathbf{r},\omega) = -i\omega\mathbf{B}(\mathbf{r},\omega) - \mathbf{J}_m(\mathbf{r},\omega)$. Repeating for all equations yields

$$\begin{aligned}\nabla\cdot\mathbf{D}(\mathbf{r},\omega) &= \rho_e(\mathbf{r},\omega),\\ \nabla\cdot\mathbf{B}(\mathbf{r},\omega) &= \rho_m(\mathbf{r},\omega),\\ \nabla\times\mathbf{E}(\mathbf{r},\omega) &= -i\omega\mathbf{B}(\mathbf{r},\omega)-\mathbf{J}_m(\mathbf{r},\omega),\\ \nabla\times\mathbf{H}(\mathbf{r},\omega) &= \ i\omega\mathbf{D}(\mathbf{r},\omega)+\mathbf{J}_e(\mathbf{r},\omega),\\ \nabla\cdot\mathbf{J}_{e(m)}(\mathbf{r},\omega) &= -i\omega\,\rho_{e(m)}(\mathbf{r},\omega).\end{aligned} \tag{1.9}$$

The frequency-domain quantities have the units of the corresponding time-domain quantities multiplied by time units (seconds). Elimination of the temporal derivative allows for the electric (magnetic) field to be easily found once the corresponding magnetic (electric) field is determined. Note also that assuming the time-domain fields are real-valued, from (1.7) we have

$$\overline{\mathbf{E}(\mathbf{r},\omega)} = \mathbf{E}(\mathbf{r},-\overline{\omega})$$

and similarly for the other quantities appearing in (1.9), where the overbar indicates complex conjugation.

For general linear anisotropic media, the constitutive equations (1.2) become

$$\begin{aligned}\mathbf{D}(\mathbf{r},\omega) &= \underline{\widetilde{\epsilon}}(\mathbf{r},\omega)\cdot\mathbf{E}(\mathbf{r},\omega),\\ \mathbf{B}(\mathbf{r},\omega) &= \underline{\widetilde{\mu}}(\mathbf{r},\omega)\cdot\mathbf{H}(\mathbf{r},\omega),\end{aligned}$$

where $\underline{\widetilde{\epsilon}}$, $\underline{\widetilde{\mu}}$ are the dyadic permittivity (F/m componentwise) and permeability (H/m componentwise), respectively, of the medium, although we'll often work with the simpler form

$$\begin{aligned}\mathbf{D}(\mathbf{r},\omega) &= \widetilde{\epsilon}\,\mathbf{E}(\mathbf{r},\omega),\\ \mathbf{B}(\mathbf{r},\omega) &= \widetilde{\mu}\,\mathbf{H}(\mathbf{r},\omega),\end{aligned} \tag{1.10}$$

valid for homogeneous, isotropic media.

Another relationship that is often useful for lossy media is the point form of Ohm's law,

$$\begin{aligned}\mathbf{J}_e(\mathbf{r},\omega) &= \underline{\sigma}_e(\mathbf{r},\omega)\cdot\mathbf{E}(\mathbf{r},\omega),\\ \mathbf{J}_m(\mathbf{r},\omega) &= \underline{\sigma}_m(\mathbf{r},\omega)\cdot\mathbf{H}(\mathbf{r},\omega),\end{aligned} \tag{1.11}$$

---

$\int_a^\infty \partial f(x,y)/\partial x\ dy$ converges uniformly on $[c,d]$. The same relation holds for improper integrals of the form $\int_a^b f(x,y)dy$ with $f(x,b)$ unbounded.

where $\underline{\sigma}_e$ ($\text{ohms}^{-1}/\text{m}$ componentwise) is the dyadic electrical conductivity of the medium and $\underline{\sigma}_m$ (ohms/m componentwise) is the dyadic magnetic conductivity of the medium, with V = A·ohms.

If we assume at some point in space the source quantities $(\rho, \mathbf{J})$ and constitutive quantities $(\underline{\widetilde{\epsilon}}, \underline{\widetilde{\mu}}, \underline{\sigma}_{e(m)})$ are infinitely differentiable, and the constitutive quantities invertible, then (1.9) implies at that point the field quantities are infinitely differentiable, at least in the sense of the curl. To see this, consider that for the curl equations in (1.9) to make sense $\mathbf{E}, \mathbf{H}$ must be differentiable in the sense of the curl. However, from (1.10), and assuming that the constitutive quantities are differentiable and invertible, then $\mathbf{D}, \mathbf{B}$ must be differentiable. If that is the case, and assuming the differentiability constraints on the source terms hold, then $\nabla \times \mathbf{E}, \nabla \times \mathbf{H}$ must be differentiable. Therefore, we are allowed to apply another curl operator to the curl equations in (1.9), resulting in $\nabla \times \nabla \times$ applied to the left-side terms and a single curl operator acting on the right-side terms. But from previous arguments the single curl terms on the right side are differentiable, and therefore so are the double curl terms. Continuing in this way proves the statement.

At most points in space the constitutive quantities will be differentiable and invertible; in fact they are often modeled as nonzero piecewise constant functions. It is at discontinuity points of the constitutive quantities or source densities that field components may experience a discontinuity.

Equations (1.9) seem very similar to (1.6), and indeed they are identical under formal time-independent operations. The difference is in interpretation: In (1.6) the quantities are time-harmonic phasors, having the same units as the time-domain quantities, with time dependence recovered as $\mathbf{E}(\mathbf{r}, t) = \text{Re}\left\{\mathbf{E}(\mathbf{r})e^{i\omega t}\right\}$. However, in (1.9) the quantities reside in the Fourier transform domain and carry units of the corresponding time-domain field multiplied by units of time, with recovery of the time dependence only through Fourier inversion. In many cases the inversion contour in the $\omega$-plane needs to be carefully considered.

Allowing for generalized functions, in the special case of time-harmonic fields in the Fourier transform domain,

$$F\left\{\mathbf{E}_0(\mathbf{r})\cos(\omega_0 t)\right\} = \mathbf{E}_0(\mathbf{r})\pi\left\{\delta(\omega - \omega_0) + \delta(\omega + \omega_0)\right\},$$

and recovery of the time-dependent fields through Fourier inversion is trivial. In the following we work with equations most generally of the form (1.9), although the $\omega$ dependence is typically not shown.

### Complex Constitutive Parameters

In the transform domain it becomes particularly easy to separate applied quantities from induced effects. In (1.9) the field quantities represent the total fields at a point in space. Assume that an impressed current density $\mathbf{J}_e^i(\mathbf{r}) \neq \mathbf{0}$ maintains $\mathbf{E}, \mathbf{D}, \mathbf{H}, \mathbf{B} \neq \mathbf{0}$, which, in turn, results in $\mathbf{J}_e^c(\mathbf{r}) =$

$\underline{\sigma}_e \cdot \mathbf{E}(\mathbf{r}) \neq \mathbf{0}$, where $\mathbf{J}_e^c$ is an induced conduction current density. The total electric current is

$$\mathbf{J}_e(\mathbf{r}) = \mathbf{J}_e^i(\mathbf{r}) + \mathbf{J}_e^c,$$

and Faraday's law becomes

$$\begin{aligned} \nabla \times \mathbf{H}(\mathbf{r}) &= i\omega\underline{\widetilde{\varepsilon}} \cdot \mathbf{E}(\mathbf{r}) + \mathbf{J}_e^i(\mathbf{r}) + \underline{\sigma}_e \cdot \mathbf{E}(\mathbf{r}) \\ &= i\omega \left( \underline{\widetilde{\varepsilon}} - \frac{i}{\omega}\underline{\sigma}_e \right) \cdot \mathbf{E}(\mathbf{r}) + \mathbf{J}_e^i(\mathbf{r}). \end{aligned}$$

Defining a new complex permittivity tensor as

$$\underline{\varepsilon} \equiv \left( \underline{\widetilde{\varepsilon}} - \frac{i}{\omega}\underline{\sigma}_e \right)$$

leads to

$$\nabla \times \mathbf{H}(\mathbf{r}) = i\omega\underline{\varepsilon} \cdot \mathbf{E}(\mathbf{r}) + \mathbf{J}_e^i(\mathbf{r}),$$

where we have separated the induced effects from the applied source. Repeating for $\mathbf{J}_m(\mathbf{r}) = \mathbf{J}_m^i(\mathbf{r}) + \underline{\sigma}_m \cdot \mathbf{H}(\mathbf{r}) = \mathbf{J}_m^i(\mathbf{r}) + \mathbf{J}_m^c$ and noting that

$$\nabla \cdot \mathbf{J}_{e(m)}^i + i\omega\rho_{e(m)}^i = 0,$$

we have

$$\begin{aligned} \nabla \cdot (\underline{\varepsilon} \cdot \mathbf{E}(\mathbf{r})) &= \rho_e^i(\mathbf{r}), \\ \nabla \cdot (\underline{\mu} \cdot \mathbf{H}(\mathbf{r})) &= \rho_m^i(\mathbf{r}), \\ \nabla \times \mathbf{E}(\mathbf{r}) &= -i\omega\underline{\mu} \cdot \mathbf{H}(\mathbf{r}) - \mathbf{J}_m^i(\mathbf{r}), \\ \nabla \times \mathbf{H}(\mathbf{r}) &= i\omega\underline{\varepsilon} \cdot \mathbf{E}(\mathbf{r}) + \mathbf{J}_e^i(\mathbf{r}), \end{aligned} \tag{1.12}$$

where

$$\underline{\mu} \equiv \left( \underline{\widetilde{\mu}} - \frac{i}{\omega}\underline{\sigma}_m \right).$$

Assuming $\underline{\widetilde{\varepsilon}}$, $\underline{\widetilde{\mu}}$ are real, the imaginary parts of $\underline{\varepsilon}$, $\underline{\mu}$ account for conduction loss. In general, other loss mechanisms may also be present [4]. In this case $\underline{\widetilde{\varepsilon}}$, $\underline{\widetilde{\mu}}$ will themselves be complex, with imaginary parts relating to those other loss mechanisms. Even more generally, in other media (such as gyrotropic) $\underline{\widetilde{\varepsilon}}$, $\underline{\widetilde{\mu}}$ may have imaginary components that are not related to loss (see Section 1.2.2). In any case we can write the above quantities as

$$\begin{aligned} \underline{\varepsilon} &= \left( \underline{\varepsilon}' - i\underline{\varepsilon}'' \right), \\ \underline{\mu} &= \left( \underline{\mu}' - i\underline{\mu}'' \right), \end{aligned}$$

where, in general, each component $\underline{\varepsilon}'$, $\underline{\varepsilon}''$, $\underline{\mu}'$, $\underline{\mu}''$ is a function of $(\mathbf{r}, \omega)$. The material dyadics will usually be invertible at most points of space, i.e., $\underline{\varepsilon}^{-1}, \underline{\mu}^{-1}$ will exist such that $\underline{\mu} \cdot \underline{\mu}^{-1} = \underline{\mu}^{-1} \cdot \underline{\mu} = \underline{\mathbf{I}}$, with $\underline{\mathbf{I}}$ being the identity dyadic, and similarly for the other constitutive quantities.

For later convenience it is useful to relax our notation in (1.12) and simply work with

$$\begin{aligned} \nabla \cdot (\underline{\varepsilon} \cdot \mathbf{E}(\mathbf{r})) &= \rho_e(\mathbf{r}), \\ \nabla \cdot (\underline{\mu} \cdot \mathbf{H}(\mathbf{r})) &= \rho_m(\mathbf{r}), \\ \nabla \times \mathbf{E}(\mathbf{r}) &= -i\omega\underline{\mu} \cdot \mathbf{H}(\mathbf{r}) - \mathbf{J}_m(\mathbf{r}), \\ \nabla \times \mathbf{H}(\mathbf{r}) &= i\omega\underline{\varepsilon} \cdot \mathbf{E}(\mathbf{r}) + \mathbf{J}_e(\mathbf{r}), \end{aligned} \tag{1.13}$$

where it should be kept in mind that if $\underline{\mu}$, $\underline{\varepsilon}$ are complex dyadics, generalized conduction currents may be already accounted for, and $\mathbf{J}_e, \mathbf{J}_m$ need to be interpreted accordingly.

# 1.2 Field Theorems

In this section we consider several field theorems that are important in electromagnetics and that are utilized in later sections.

## 1.2.1 Duality

Maxwell's equations (1.13) are symmetric among electric and magnetic quantities, except for a sign change. This symmetry can be utilized to simplify some electromagnetic problems. Considering the set of equations comprising Maxwell's equations and the continuity equations, the substitutions

$$\begin{gathered} \mathbf{E} \to \mathbf{H}, \mathbf{H} \to -\mathbf{E}, \mathbf{B} \to -\mathbf{D}, \mathbf{D} \to \mathbf{B}, \mathbf{J}_e \to \mathbf{J}_m, \\ \mathbf{J}_m \to -\mathbf{J}_e,\ \rho_e \to \rho_m,\ \rho_m \to -\rho_e,\ \underline{\varepsilon} \to \underline{\mu},\ \underline{\mu} \to \underline{\varepsilon} \end{gathered} \tag{1.14}$$

leave the set unchanged.[10] This *duality* is often used when a solution $(\mathbf{E}_e, \mathbf{H}_e)$ is obtained for the fields caused by electric sources $\mathbf{J}_e, \rho_e$, with magnetic sources set to zero. Then, upon the replacements

$$\begin{gathered} \mathbf{E} \to \mathbf{H}, \mathbf{H} \to -\mathbf{E}, \mathbf{J}_e \to \mathbf{J}_m, \\ \rho_e \to \rho_m,\ \underline{\varepsilon} \to \underline{\mu},\ \underline{\mu} \to \underline{\varepsilon}, \end{gathered}$$

one has the solution for the electric and magnetic fields $(\mathbf{E}_m, \mathbf{H}_m)$ maintained by magnetic sources.

[10] The substitution is not unique;

$$\begin{gathered} \mathbf{E} \to \mathbf{H},\ \mathbf{H} \to \mathbf{E},\ \mathbf{B} \to -\mathbf{D},\ \mathbf{D} \to -\mathbf{B},\ \mathbf{J}_e \to -\mathbf{J}_m, \\ \mathbf{J}_m \to -\mathbf{J}_e,\ \rho_e \to -\rho_m,\ \rho_m \to -\rho_e,\ \underline{\varepsilon} \to -\underline{\mu},\ \underline{\mu} \to -\underline{\varepsilon} \end{gathered}$$

is also valid.

### 1.2.2 Poynting's Theorem

In analogy with electrical circuit theory where $V \cdot I$ represents power in watts ($\mathrm{W} = \mathrm{V} \cdot \mathrm{A}$), it is well known that $\mathbf{E} \times \mathbf{H}$ represents electromagnetic power density ($\mathrm{W/m^2}$). The quantity $\mathbf{E} \times \mathbf{H}$ is called the *Poynting vector*. Assuming the fields are time-harmonic with period $2\pi$, the time-average Poynting's vector (directed time-average complex power density) can be written as

$$\begin{aligned}[\mathbf{E}(\mathbf{r},t) \times \mathbf{H}(\mathbf{r},t)]_{\text{avg}} &= \frac{1}{2\pi}\int_0^{2\pi} \mathbf{E}(\mathbf{r},t) \times \mathbf{H}(\mathbf{r},t)\, d(\omega t) \\ &= \frac{1}{2}\,\mathrm{Re}\left\{\mathbf{E}(\mathbf{r}) \times \overline{\mathbf{H}}(\mathbf{r})\right\}.\end{aligned}$$

We can study energy conservation by forming the complex Poynting theorem,

$$\begin{aligned}\nabla \cdot \left[\frac{1}{2}\left(\mathbf{E} \times \overline{\mathbf{H}}\right)\right] &= \frac{1}{2}\left[\overline{\mathbf{H}} \cdot \nabla \times \mathbf{E} - \mathbf{E} \cdot \nabla \times \overline{\mathbf{H}}\right] \\ &= 2i\omega\frac{1}{4}\left[\mathbf{E} \cdot \overline{\underline{\varepsilon}} \cdot \overline{\mathbf{E}} - \overline{\mathbf{H}} \cdot \underline{\mu} \cdot \mathbf{H}\right] - \frac{1}{2}\left[\overline{\mathbf{H}} \cdot \mathbf{J}_m(\mathbf{r}) + \mathbf{E} \cdot \overline{\mathbf{J}_e}(\mathbf{r})\right],\end{aligned}$$

which can be written in integral form by taking the volume integral and invoking the divergence theorem (outward unit normal), leading to

$$\begin{aligned}\oint_S \frac{1}{2}\left(\mathbf{E} \times \overline{\mathbf{H}}\right) \cdot d\mathbf{S} = &- 2i\omega \int_V \frac{1}{4}\left(\overline{\mathbf{H}} \cdot \underline{\mu} \cdot \mathbf{H} - \mathbf{E} \cdot \overline{\underline{\varepsilon}} \cdot \overline{\mathbf{E}}\right) dV \\ &- \int_V \frac{1}{2}\left(\overline{\mathbf{H}} \cdot \mathbf{J}_m(\mathbf{r}) + \mathbf{E} \cdot \overline{\mathbf{J}_e}\right) dV.\end{aligned} \tag{1.15}$$

The left side above can be interpreted as being the net time-average complex power flow out of the region $V$ bounded by $S$. In (1.15) the material dyadics are the complex quantities defined in the last section, accounting for material loss as well as polarization effects.

If we separate the total current density into induced and applied terms $\mathbf{J}_e(\mathbf{r}) = \mathbf{J}_e^i(\mathbf{r}) + \underline{\sigma}_e \cdot \mathbf{E}(\mathbf{r})$ and $\mathbf{J}_m(\mathbf{r}) = \mathbf{J}_m^i(\mathbf{r}) + \underline{\sigma}_m \cdot \mathbf{H}(\mathbf{r})$, where the $\underline{\sigma}_{e(m)}$ account for all loss mechanisms (and to be consistent we now consider the material dyadics to only account for polarization), we have

$$\begin{aligned}\oint_S \frac{1}{2}\left(\mathbf{E} \times \overline{\mathbf{H}}\right) \cdot d\mathbf{S} = &- 2i\omega \int_V \frac{1}{4}\left(\overline{\mathbf{H}} \cdot \underline{\widetilde{\mu}} \cdot \mathbf{H} - \mathbf{E} \cdot \overline{\underline{\widetilde{\varepsilon}}} \cdot \overline{\mathbf{E}}\right) dV \\ &- \int_V \frac{1}{2}\left(\overline{\mathbf{H}} \cdot \mathbf{J}_m^i + \mathbf{E} \cdot \overline{\mathbf{J}_e^i}\right) dV \\ &- \int_V \frac{1}{2}\left(\overline{\mathbf{H}} \cdot \underline{\sigma}_m \cdot \mathbf{H} + \mathbf{E} \cdot \overline{\underline{\sigma}}_e \cdot \overline{\mathbf{E}}\right) dV.\end{aligned} \tag{1.16}$$

The first term on the right side represents stored power, the second term (including the negative sign) is interpreted as the complex power supplied by the sources $\mathbf{J}^i_{e(m)}$, and the last term represents the complex power dissipated by the medium characterized by $\underline{\sigma}_{e(m)}$.

By setting the impressed sources to zero and considering real power flow we can gain insight into the allowed properties of the constitutive parameters of the region. Because $\mathrm{Re}\left(i\,\mathbf{E}\cdot\overline{\mathbf{D}}\right) = (i/2)\left(\mathbf{E}\cdot\overline{\mathbf{D}} - \overline{\mathbf{E}}\cdot\mathbf{D}\right)$,

$$
\begin{aligned}
\mathrm{Re} \oint_S \frac{1}{2} \left(\mathbf{E}\times\overline{\mathbf{H}}\right) \cdot d\mathbf{S} &= \mathrm{Re}\, \frac{i\omega}{2} \int_V \left(\mathbf{E}\cdot\overline{\underline{\varepsilon}} \cdot \overline{\mathbf{E}} - \overline{\mathbf{H}} \cdot \underline{\mu} \cdot \mathbf{H}\right) dV \\
&= \frac{i\omega}{4} \int_V \left[\left(\mathbf{E}\cdot\overline{\underline{\varepsilon}} \cdot \overline{\mathbf{E}} - \overline{\mathbf{E}}\cdot\underline{\varepsilon} \cdot \mathbf{E}\right)\right. \\
&\quad \left. - \left(\overline{\mathbf{H}} \cdot \underline{\mu} \cdot \mathbf{H} - \mathbf{H} \cdot \overline{\underline{\mu}} \cdot \overline{\mathbf{H}}\right)\right] dV \\
&= \frac{i\omega}{4} \int_V \left[\mathbf{E}\cdot \left(\overline{\underline{\varepsilon}} - \underline{\varepsilon}^\top\right) \cdot \overline{\mathbf{E}} + \mathbf{H} \cdot \left(\overline{\underline{\mu}} - \underline{\mu}^\top\right) \cdot \overline{\mathbf{H}}\right] dV,
\end{aligned}
\tag{1.17}
$$

where $\underline{\varepsilon}^\top$ is the transpose of $\underline{\varepsilon}$.

For a passive lossless medium containing no sources, the left side of (1.17) must equal zero because there is no net time-average real power flow into or out of the region. Therefore, assuming $\omega > 0$, unless the expression on the right side happens to integrate to zero, which will not occur for $V$ arbitrary, or $\mathbf{E}\cdot \left(\overline{\underline{\varepsilon}} - \underline{\varepsilon}^\top\right) \cdot \overline{\mathbf{E}} = -\mathbf{H} \cdot \left(\overline{\underline{\mu}} - \underline{\mu}^\top\right) \cdot \overline{\mathbf{H}}$, which will generally not occur, then $i\left(\overline{\underline{\varepsilon}} - \underline{\varepsilon}^\top\right)$ and $i\left(\overline{\underline{\mu}} - \underline{\mu}^\top\right)$ must vanish. This leads to the conclusion that for a passive lossless medium

$$
\underline{\varepsilon} = \overline{\underline{\varepsilon}^\top}, \qquad \underline{\mu} = \overline{\underline{\mu}^\top}, \tag{1.18}
$$

requiring the permittivity dyadics to be Hermitian. This condition does not require the material dyadics to have real-valued entries, although for an isotropic lossless medium, $\mathrm{Im}\,(\varepsilon) = \mathrm{Im}(\mu) = 0$. For most lossless materials, the material dyadics will be real-valued and hence symmetric. As a counterexample, gyrotropic media, a category that includes biased electron plasmas and biased ferrites, do not have real-valued material parameters in the lossless case [5, p. 91].

If we now consider a passive, lossy medium without sources, the left side of Poynting's theorem must be negative, representing a net inward flow of real time-average power to compensate for power dissipated in the medium. Then $i\left(\overline{\underline{\varepsilon}} - \underline{\varepsilon}^\top\right)$ and $i\left(\overline{\underline{\mu}} - \underline{\mu}^\top\right)$ must be negative-definite (see Definition 3.19) quantities.

### 1.2.3 The Lorentz Reciprocity Theorem

The Lorentz reciprocity theorem describes the mutual interaction between two groups of sources. Here we follow the method outlined in [2]. Consider an inhomogeneous, anisotropic medium characterized by $\underline{\varepsilon}$, $\underline{\mu}$, which will be considered to be symmetric dyadics, $\underline{\varepsilon} = \underline{\varepsilon}^\top, \underline{\mu} = \underline{\mu}^\top$. Such media are called *reciprocal*[11] (this implies that lossless reciprocal media have real-valued material dyadics). Let $\mathbf{E}_1, \mathbf{H}_1$ and $\mathbf{E}_2, \mathbf{H}_2$ be the fields caused by sources $\mathbf{J}_{e1}, \mathbf{J}_{m1}$ and $\mathbf{J}_{e2}, \mathbf{J}_{m2}$, respectively. Starting with Maxwell's curl equations

$$\begin{aligned} \nabla \times \mathbf{E}_1(\mathbf{r}) &= -i\omega\underline{\mu} \cdot \mathbf{H}_1(\mathbf{r}) - \mathbf{J}_{m1}(\mathbf{r}), \\ \nabla \times \mathbf{H}_1(\mathbf{r}) &= \ i\omega\underline{\varepsilon} \cdot \mathbf{E}_1(\mathbf{r}) + \mathbf{J}_{e1}(\mathbf{r}), \\ \nabla \times \mathbf{E}_2(\mathbf{r}) &= -i\omega\underline{\mu} \cdot \mathbf{H}_2(\mathbf{r}) - \mathbf{J}_{m2}(\mathbf{r}), \\ \nabla \times \mathbf{H}_2(\mathbf{r}) &= \ i\omega\underline{\varepsilon} \cdot \mathbf{E}_2(\mathbf{r}) + \mathbf{J}_{e2}(\mathbf{r}), \end{aligned} \tag{1.19}$$

we form

$$\begin{aligned} &\nabla \cdot (\mathbf{E}_1 \times \mathbf{H}_2 - \mathbf{E}_2 \times \mathbf{H}_1) \\ &\qquad = \mathbf{H}_2 \cdot \nabla \times \mathbf{E}_1 - \mathbf{E}_1 \cdot \nabla \times \mathbf{H}_2 - \mathbf{H}_1 \cdot \nabla \times \mathbf{E}_2 + \mathbf{E}_2 \cdot \nabla \times \mathbf{H}_1. \end{aligned}$$

Substituting for $\nabla \times \mathbf{E}_{1,2}, \nabla \times \mathbf{H}_{1,2}$ and using symmetry of the material dyadics lead to

$$\nabla \cdot (\mathbf{E}_1 \times \mathbf{H}_2 - \mathbf{E}_2 \times \mathbf{H}_1) = -\mathbf{H}_2 \cdot \mathbf{J}_{m1} - \mathbf{E}_1 \cdot \mathbf{J}_{e2} + \mathbf{H}_1 \cdot \mathbf{J}_{m2} + \mathbf{E}_2 \cdot \mathbf{J}_{e1}.$$

Taking the volume integral and invoking the divergence theorem lead to

$$\begin{aligned} &\oint_S (\mathbf{E}_1 \times \mathbf{H}_2 - \mathbf{E}_2 \times \mathbf{H}_1) \cdot d\mathbf{S} \\ &\quad = \int_V (-\mathbf{H}_2 \cdot \mathbf{J}_{m1} - \mathbf{E}_1 \cdot \mathbf{J}_{e2} + \mathbf{H}_1 \cdot \mathbf{J}_{m2} + \mathbf{E}_2 \cdot \mathbf{J}_{e1}) \, dV \\ &\quad = -\langle \mathbf{H}_2, \mathbf{J}_{m1} \rangle_p + \langle \mathbf{E}_2, \mathbf{J}_{e1} \rangle_p - \langle \mathbf{E}_1, \mathbf{J}_{e2} \rangle_p + \langle \mathbf{H}_1, \mathbf{J}_{m2} \rangle_p, \end{aligned} \tag{1.20}$$

where we use the reaction (pseudo) inner product (see Definition 2.34),

$$\langle \mathbf{A}, \mathbf{B} \rangle_p = \int_V \mathbf{A} \cdot \mathbf{B} \, dV.$$

Equation (1.20) is known as the *Lorentz reciprocity theorem.*

Various special cases are often useful. For instance, assume $\mathbf{n} \times \mathbf{E}_1 = \mathbf{n} \times \mathbf{E}_2 = \mathbf{0}$ on surface $S$, such as in the case of a perfectly conducting surface. Because $(\mathbf{E}_1 \times \mathbf{H}_2) \cdot \mathbf{n} = \mathbf{H}_2 \cdot (\mathbf{n} \times \mathbf{E}_1)$, the surface terms vanish and we obtain

$$\langle \mathbf{E}_2, \mathbf{J}_{e1} \rangle_p - \langle \mathbf{H}_2, \mathbf{J}_{m1} \rangle_p = \langle \mathbf{E}_1, \mathbf{J}_{e2} \rangle_p - \langle \mathbf{H}_1, \mathbf{J}_{m2} \rangle_p.$$

We obtain the same result if we consider an infinite region containing sources having compact support. Then the surface integral over the infinite surface $S$ vanishes since the fields will satisfy a radiation condition (see Sections 1.5 and 3.4.3).

For a region without sources we have

$$\oint_S (\mathbf{E}_1 \times \mathbf{H}_2 - \mathbf{E}_2 \times \mathbf{H}_1) \cdot d\mathbf{S} = 0.$$

[11] In the case of a nonreciprocal medium, the derivation of the reciprocity theorem is more involved. See [5, p. 408] for details.

## 1.3 Wave Equations

Although there are many techniques for solving Maxwell's equations, in this text we are primarily interested in solving the differential Maxwell equations in the temporal transform domain. Even with the time derivatives eliminated from (1.1), the two curl equations in (1.13) still represent a complicated set of two first-order, vector, coupled partial differential equations. These equations can be decoupled in several ways, as considered in the following.

### 1.3.1 Vector Wave and Vector Helmholtz Equations for Electric and Magnetic Fields

Repeating from (1.13), we start with

$$\begin{aligned} \nabla \times \mathbf{E}(\mathbf{r}) &= -i\omega\underline{\mu}(\mathbf{r}) \cdot \mathbf{H}(\mathbf{r}) - \mathbf{J}_m(\mathbf{r}), \\ \nabla \times \mathbf{H}(\mathbf{r}) &= \ i\omega\underline{\varepsilon}(\mathbf{r}) \cdot \mathbf{E}(\mathbf{r}) + \mathbf{J}_e(\mathbf{r}). \end{aligned}$$

Taking the curl of $\underline{\mu}(\mathbf{r})^{-1} \cdot \nabla \times \mathbf{E}(\mathbf{r})$ and of $\underline{\varepsilon}(\mathbf{r})^{-1} \cdot \nabla \times \mathbf{H}(\mathbf{r})$ leads to

$$\begin{aligned} \nabla\times\underline{\mu}(\mathbf{r})^{-1} \cdot \nabla \times \mathbf{E}(\mathbf{r}) - \omega^2\underline{\varepsilon}(\mathbf{r}) \cdot \mathbf{E}(\mathbf{r}) & \qquad (1.21) \\ = -i\omega\mathbf{J}_e(\mathbf{r}) - \nabla \times \underline{\mu}(\mathbf{r})^{-\mathbf{1}} \cdot \mathbf{J}_m(\mathbf{r}), & \end{aligned}$$

$$\begin{aligned} \nabla\times\underline{\varepsilon}(\mathbf{r})^{-1} \cdot \nabla \times \mathbf{H}(\mathbf{r}) - \omega^2\underline{\mu}(\mathbf{r}) \cdot \mathbf{H}(\mathbf{r}) & \qquad (1.22) \\ = -i\omega\mathbf{J}_m(\mathbf{r}) + \nabla \times \underline{\varepsilon}(\mathbf{r})^{-\mathbf{1}} \cdot \mathbf{J}_e(\mathbf{r}), & \end{aligned}$$

where (1.22) could also be obtained from (1.21) using duality. These are the *vector wave equations* for the fields. Either (1.21) or (1.22) may be solved, with the undetermined field quantity found via the curl equations.

Various simplifications to the above can be found. For instance, if the medium is isotropic, we have

$$\begin{aligned} \nabla\times\mu(\mathbf{r})^{-1}\nabla \times \mathbf{E}(\mathbf{r}) - \omega^2\varepsilon(\mathbf{r})\mathbf{E}(\mathbf{r}) & \\ = -i\omega\mathbf{J}_e(\mathbf{r}) - \nabla\times\ \mu(\mathbf{r})^{-\mathbf{1}}\mathbf{J}_m(\mathbf{r}), & \\ \nabla\times\varepsilon(\mathbf{r})^{-1}\nabla \times \mathbf{H}(\mathbf{r}) - \omega^2\mu(\mathbf{r})\mathbf{H}(\mathbf{r}) & \\ = -i\omega\mathbf{J}_m(\mathbf{r}) + \nabla\times\ \varepsilon(\mathbf{r})^{-\mathbf{1}}\mathbf{J}_e(\mathbf{r}). & \end{aligned} \qquad (1.23)$$

Furthermore, if the medium is isotropic and homogeneous,

$$\begin{aligned} \nabla \times \nabla \times \mathbf{E}(\mathbf{r}) - \omega^2\varepsilon\mu\mathbf{E}(\mathbf{r}) &= -i\omega\mu\mathbf{J}_e(\mathbf{r}) - \nabla \times \mathbf{J}_m(\mathbf{r}), \\ \nabla \times \nabla \times \mathbf{H}(\mathbf{r}) - \omega^2\varepsilon\mu\mathbf{H}(\mathbf{r}) &= -i\omega\varepsilon\mathbf{J}_m(\mathbf{r}) + \nabla \times \mathbf{J}_e(\mathbf{r}). \end{aligned} \qquad (1.24)$$

Of course (1.24) also applies to individual homogeneous subregions within an isotropic inhomogeneous region. The equivalence of the wave equations (1.24) and Maxwell's equations (1.13) (for isotropic homogeneous space) is discussed in [6, pp. 74–75], assuming sufficient differentiability requirements.

Noting that $\nabla \times \nabla \times \mathbf{A} = \nabla(\nabla \cdot \mathbf{A}) - \nabla^2 \mathbf{A}$, we also have for isotropic homogeneous media

$$\begin{aligned} \nabla^2 \mathbf{E}(\mathbf{r}) + \omega^2 \varepsilon\mu \mathbf{E}(\mathbf{r}) &= i\omega\mu \mathbf{J}_e(\mathbf{r}) + \nabla \times \mathbf{J}_m(\mathbf{r}) + \frac{\nabla \rho_e}{\varepsilon}, \\ \nabla^2 \mathbf{H}(\mathbf{r}) + \omega^2 \varepsilon\mu \mathbf{H}(\mathbf{r}) &= i\omega\varepsilon \mathbf{J}_m(\mathbf{r}) - \nabla \times \mathbf{J}_e(\mathbf{r}) + \frac{\nabla \rho_m}{\mu}. \end{aligned} \tag{1.25}$$

These are known as *vector Helmholtz equations.*[12] Yet another form can be obtained using the continuity equations, leading to

$$\begin{aligned} \nabla^2 \mathbf{E}(\mathbf{r}) + \omega^2 \varepsilon\mu \mathbf{E}(\mathbf{r}) &= i\omega\mu \left[\underline{\mathbf{I}} + \frac{\nabla\nabla}{\omega^2 \mu\varepsilon}\right] \cdot \mathbf{J}_e(\mathbf{r}) + \nabla \times \mathbf{J}_m(\mathbf{r}), \\ \nabla^2 \mathbf{H}(\mathbf{r}) + \omega^2 \varepsilon\mu \mathbf{H}(\mathbf{r}) &= i\omega\varepsilon \left[\underline{\mathbf{I}} + \frac{\nabla\nabla}{\omega^2 \mu\varepsilon}\right] \cdot \mathbf{J}_m(\mathbf{r}) - \nabla \times \mathbf{J}_e(\mathbf{r}). \end{aligned} \tag{1.26}$$

### 1.3.2 Vector and Scalar Potentials and Associated Helmholtz Equations

The source terms on the right side of (1.25) and (1.26) are quite complicated. Introducing a potential function can simplify the form of the source term, which in turn leads to a reduction of many vector problems to scalar ones. Another benefit of the potential approach is that the integrals providing the potentials from the sources are less singular than those relating the electric and magnetic fields to the sources. For simplicity we proceed assuming homogeneous isotropic media.

Consider first only the case of electric sources in (1.13). By virtue of the identity $\nabla \cdot \nabla \times \mathbf{A} = 0$, Maxwell's equation $\nabla \cdot \mathbf{B} = 0$ leads to the relationship

$$\mathbf{B} = \nabla \times \mathbf{A}, \tag{1.27}$$

where $\mathbf{A}$ is known as the *magnetic vector potential* (Wb/m). Substitution of this into Faraday's law results in $\nabla \times (\mathbf{E} + i\omega\mathbf{A}) = \mathbf{0}$. From the vector identity $\nabla \times \nabla\phi = \mathbf{0}$ we can obtain

$$\mathbf{E} = -\, i\omega\mathbf{A} - \nabla\phi_e, \tag{1.28}$$

where $\phi_e$ is known as the *electric scalar potential* (V). Ampère's law then becomes

$$\begin{aligned} \nabla \times \mathbf{H}(\mathbf{r}) &= \frac{1}{\mu}\nabla \times \nabla \times \mathbf{A} \;=\; \frac{1}{\mu}\left(\nabla\nabla \cdot \mathbf{A} - \nabla^2 \mathbf{A}\right) \\ &= i\omega\varepsilon \mathbf{E}(\mathbf{r}) + \mathbf{J}_e(\mathbf{r}) = i\omega\varepsilon\left(-i\omega\mathbf{A} - \nabla\phi_e\right) + \mathbf{J}_e(\mathbf{r}), \end{aligned}$$

[12]Note, however, that we cannot obtain Maxwell's divergence equations directly from (1.25). This implies that we need to solve (1.25) subject to $\nabla \cdot \mathbf{E}(\mathbf{r}) = \rho_e(\mathbf{r})/\varepsilon$, $\nabla \cdot \mathbf{H}(\mathbf{r}) = \rho_m(\mathbf{r})/\mu$.

leading to

$$\nabla^2 \mathbf{A} + k^2 \mathbf{A} = \nabla \left( \nabla \cdot \mathbf{A} + i\omega\mu\varepsilon\phi_e \right) - \mu \mathbf{J}_e(\mathbf{r}), \tag{1.29}$$

where $k^2 = \omega^2 \mu\varepsilon$.

So far only the curl of **A** has been specified. According to the Helmholtz theorem a vector field is determined by specifying both its curl and its divergence. We are at liberty to set $\nabla \cdot \mathbf{A}$ such that the right-side of (1.29) is simplified. Accordingly, we let $\nabla \cdot \mathbf{A} = -i\omega\mu\varepsilon\phi_e$, which is known as the *Lorenz gauge*, resulting in

$$\nabla^2 \mathbf{A} + k^2 \mathbf{A} = -\mu \mathbf{J}_e(\mathbf{r}). \tag{1.30}$$

Because we also have $\nabla \cdot \mathbf{E} = - i\omega \nabla \cdot \mathbf{A} - \nabla^2 \phi_e = \rho_e / \varepsilon$, then

$$\nabla^2 \phi_e + k^2 \phi_e = -\rho_e / \varepsilon. \tag{1.31}$$

Now consider only magnetic sources. Maxwell's equation $\nabla \cdot \mathbf{E} = 0$ leads to

$$\mathbf{E} = -\nabla \times \mathbf{F}, \tag{1.32}$$

where **F** is known as the *electric vector potential* (V). Substituting into Ampère's law leads to $\nabla \times (\mathbf{H} + i\omega\varepsilon \mathbf{F}) = \mathbf{0}$, while the vector identity $\nabla \times \nabla\phi = \mathbf{0}$ results in

$$\mathbf{H} = - i\omega\varepsilon \mathbf{F} - \nabla \phi_m, \tag{1.33}$$

where $\phi_m$ is known as the *magnetic scalar potential* (A). Faraday's law is then

$$\begin{aligned} \nabla \times \mathbf{E}(\mathbf{r}) &= -\nabla \times \nabla \times \mathbf{F} = - \left( \nabla\nabla \cdot \mathbf{F} - \nabla^2 \mathbf{F} \right) \\ &= - i\omega\mu \mathbf{H}(\mathbf{r}) - \mathbf{J}_m(\mathbf{r}) = -i\omega\mu \left( -i\omega\varepsilon \mathbf{F} - \nabla \phi_m \right) - \mathbf{J}_m(\mathbf{r}), \end{aligned}$$

leading to

$$\nabla^2 \mathbf{F} + k^2 \mathbf{F} = \nabla \left( \nabla \cdot \mathbf{F} + i\omega\mu\phi_m \right) - \mathbf{J}_m(\mathbf{r}). \tag{1.34}$$

Accordingly, let $\nabla \cdot \mathbf{F} = -i\omega\mu\phi_m$, resulting in

$$\nabla^2 \mathbf{F} + k^2 \mathbf{F} = -\mathbf{J}_m(\mathbf{r}). \tag{1.35}$$

Because we also have $\nabla \cdot \mathbf{H} = - i\omega\varepsilon \nabla \cdot \mathbf{F} - \nabla^2 \phi_m = \rho_m / \mu$, then

$$\nabla^2 \phi_m + k^2 \phi_m = -\rho_m / \mu. \tag{1.36}$$

To summarize, for the various potentials in the Lorenz gauge the Helmholtz equations are

$$\left( \nabla^2 + k^2 \right) \begin{pmatrix} \mathbf{A} \\ \mathbf{F} \\ \phi_e \\ \phi_m \end{pmatrix} = - \begin{pmatrix} \mu \mathbf{J}_e \\ \mathbf{J}_m \\ \rho_e / \varepsilon \\ \rho_m / \mu \end{pmatrix}. \tag{1.37}$$

The Lorenz gauge, and the corresponding gauge for static fields ($\nabla \cdot \mathbf{A} = 0$, $\nabla \cdot \mathbf{F} = 0$), known as the *Coulomb gauge*, are the usual ones of interest. However, for problems involving spherical coordinates, other choices are sometimes preferable (see Section 10.4).

Note that the Helmholtz equations for the potentials have much simpler source terms than those for the fields. In particular, in a homogeneous space the vectors $\mathbf{A}$, $\mathbf{F}$ will be colinear with the source terms $\mathbf{J}_{e(m)}$, respectively, often reducing the vector problem to a simpler scalar one.

Using superposition we obtain the fields from the Lorenz-gauge potentials as

$$\begin{aligned} \mathbf{B} &= \nabla \times \mathbf{A} - i\omega\mu\varepsilon\mathbf{F} + \frac{1}{i\omega}\nabla\nabla \cdot \mathbf{F}, \\ \mathbf{E} &= -i\omega\mathbf{A} + \frac{1}{i\omega\mu\varepsilon}\nabla\nabla \cdot \mathbf{A} - \nabla \times \mathbf{F}. \end{aligned} \tag{1.38}$$

Boundary conditions for the potentials follow from those of the fields, (1.5).

### 1.3.3 Solution of the Scalar Helmholtz Equations and Scalar Green's Functions

So far we have considered vector wave and vector Helmholtz equations for the fields, vector Helmholtz equations for the vector potentials, and scalar Helmholtz equations for the scalar potentials. To begin the discussion of solving these equations, we first study the solution of the scalar Helmholtz equation. Because[13] $\nabla^2\mathbf{A} = \widehat{\mathbf{x}}\,\nabla^2 A_x + \widehat{\mathbf{y}}\,\nabla^2 A_y + \widehat{\mathbf{z}}\,\nabla^2 A_z$, it is worthwhile to study

$$\left(\nabla^2 + k^2\right)\psi(\mathbf{r}) = -s(\mathbf{r}), \tag{1.39}$$

where $\psi$ represents any of the terms $A_\alpha$, $F_\alpha$, $\phi_e$, $\phi_m$ ($A_\alpha$, $F_\alpha$ being the $\alpha$th component of $\mathbf{A}$, $\mathbf{F}$, respectively) and $s$ represents the various source terms $\mu J_{e,\alpha}$, $J_{m,\alpha}$, $\rho_e/\varepsilon$, $\rho_m/\mu$ associated with $A_\alpha$, $F_\alpha$, $\phi_e$, $\phi_m$, respectively.

We now consider the solution of (1.39), using the well known method of Green's functions (see also Section 3.9.3).

#### Green's Functions for the Scalar Helmholtz Equation

The Green's function for a particular linear problem is defined as the solution to the equation governing the problem for a unit point-source excitation. By linearity one can then obtain the solution for an arbitrary source as a superposition of point-source responses. We therefore define the scalar Green's function for the scalar Helmholtz equation to be the solution of

$$\left(\nabla^2 + k^2\right)g(\mathbf{r}, \mathbf{r}') = -\delta(\mathbf{r} - \mathbf{r}'), \tag{1.40}$$

where it is customary to explicitly denote the source-point and field-point coordinates, $\mathbf{r}', \mathbf{r}$, respectively. Of course, (1.40) must be understood in the

[13]Note that this convenient decomposition occurs only in rectangular coordinates.

sense of generalized functions [7] (see [8] for a nice presentation of distribution theory in electromagnetics). In this chapter we use predominantly classical analysis rather than distributional theory, although we retain the delta-function notation and make use of the properties

$$\delta(\mathbf{r}-\mathbf{r}') = 0, \quad \mathbf{r} \neq \mathbf{r}',$$

$$\int_V \delta(\mathbf{r}-\mathbf{r}')\, dV = 1,$$

$$\int_V \delta(\mathbf{r}-\mathbf{r}') f(\mathbf{r})\, dV = f(\mathbf{r}'),$$

where in the integrals we assume the point $\mathbf{r} = \mathbf{r}'$ occurs in $V$. We will first determine the Green's function itself and then solve for the response $\psi(\mathbf{r})$ in terms of the Green's function.

Consider an unbounded medium with origin at $\mathbf{r}' = \mathbf{0}$. From (1.40), and noting the spherical symmetry of the point source $\delta(\mathbf{r})$, we see that in spherical coordinates with radial distance from the origin $r > 0$ we expect $g(\mathbf{r}) = g(r)$ to be well behaved since it satisfies

$$\left(\nabla^2 - k^2\right) g(r) = 0,$$

which can be written as

$$\left(\frac{\partial^2}{\partial r^2} + k^2\right) r\, g(r) = 0.$$

The solution of the above is obviously

$$rg(r) = A\, e^{-ikr} + B\, e^{+ikr},$$

and, since $r \neq 0$, we have

$$g(r) = A\,\frac{e^{-ikr}}{r} + B\,\frac{e^{+ikr}}{r}.$$

The first term represents an outward-traveling spherical wave (for an $e^{i\omega t}$ time dependence, or by consideration of (1.8)), and the second term an inward-traveling wave. Because we assume no sources at infinity, $B = 0$, and the constant of integration $A$ can be determined by enforcing the proper singular behavior of the Green's function at $r = 0$. Let

$$g_a(r) = A\,\frac{e^{-ikr}\left(1 - e^{-\frac{r}{a}}\right)}{r}$$

for some $a > 0$, where $g_a \to g$ as $a \to 0$. Note that $g_a$ is finite at $r = 0$. Substituting $g_a$ for $g$ in (1.40) and integrating over a small spherical volume $V_\delta$ centered at the origin lead to

$$\int_{V_\delta} \nabla^2 \frac{e^{-ikr}\left(1 - e^{-\frac{r}{a}}\right)}{r}\, dV + k^2 \int_{V_\delta} \frac{e^{-ikr}\left(1 - e^{-\frac{r}{a}}\right)}{r}\, dV = -1/A.$$

We then note that as $V_\delta \to 0$ $(r \to 0)$ with $a$ fixed, the second integral on the left side vanishes since the integrand is[14] $O(1/r)$ while the volume element is $O(r^2)$. The first term is

$$\begin{aligned}
&\int_{V_\delta} \nabla\cdot\nabla \frac{e^{-ikr}\left(1-e^{-\frac{r}{a}}\right)}{r}\, dV \\
&= \oint_{S_\delta} \mathbf{n}\cdot\nabla \frac{e^{-ikr}\left(1-e^{-\frac{r}{a}}\right)}{r}\, dS \\
&= \int_0^{2\pi}\int_0^{\pi} \widehat{\mathbf{r}}\cdot\widehat{\mathbf{r}}\, \frac{e^{-ikr}\left[r\left(-ik+\left(ik+\frac{1}{a}\right)e^{-\frac{r}{a}}\right)-\left(1-e^{-\frac{r}{a}}\right)\right]}{r^2}\, r^2 \sin\theta\, d\theta\, d\phi \\
&= -4\pi
\end{aligned}$$

for $a \to 0$ (with $r$ initially fixed, then $r \to 0$), where the divergence theorem has been used. Therefore, $A = 1/\,4\pi$ and, restoring the dependence $\mathbf{r}-\mathbf{r}'$, we have

$$g(\mathbf{r},\mathbf{r}') = \frac{e^{-ik\left|\mathbf{r}-\mathbf{r}'\right|}}{4\pi\left|\mathbf{r}-\mathbf{r}'\right|}, \tag{1.41}$$

which is recognized as the usual free-space scalar Green's function for the Helmholtz equation. This represents a particular solution to (1.40) and behaves as a classical function everywhere except at $\mathbf{r}=\mathbf{r}'$, where it must be regarded in a distributional sense. If desired, we can add homogeneous solutions of (1.40) to (1.41) to satisfy some desired boundary condition, leading to

$$g(\mathbf{r},\mathbf{r}') = \frac{e^{-ik\left|\mathbf{r}-\mathbf{r}'\right|}}{4\pi\left|\mathbf{r}-\mathbf{r}'\right|} + g^h(\mathbf{r},\mathbf{r}'), \tag{1.42}$$

where $g^h$ satisfies $\left(\nabla^2+k^2\right)g^h(\mathbf{r},\mathbf{r}')=0$ (and so is well behaved at $\mathbf{r}=\mathbf{r}'$). Similar methods lead to

$$g(\mathbf{r},\mathbf{r}') = \frac{1}{4i}H_0^{(2)}(k\left|\mathbf{r}-\mathbf{r}'\right|) + g^h(\mathbf{r},\mathbf{r}') \tag{1.43}$$

in two dimensions ($H_0^{(1)}$ for an $e^{-i\omega t}$ time dependence), $\mathbf{r}=\widehat{\mathbf{x}}x+\widehat{\mathbf{y}}y$ so that $\left|\mathbf{r}-\mathbf{r}'\right| = \sqrt{(x-x')^2+(y-y')^2}$, and

$$g(x,x') = \frac{e^{-ik\left|x-x'\right|}}{2ik} + g^h(x,x') \tag{1.44}$$

---

[14] The expression "$f$ is $O(\nu)$" or "$f = O(\nu)$," expressed as "$f$ is of the order of $\nu$," means there exists $k>0$ such that $|f| \le k\,|\nu|$. If $f$, $\nu$ are functions of $\mathbf{r}$, then $f = O(\nu)$ could apply in the neighborhood of a point or in a region. In typical engineering usage one means that $f$ is of the same order of magnitude as $\nu$, although $f$ could be much smaller in magnitude as well.

For later use note that the expression "$f$ is $o(\nu)$" or "$f = o(\nu)$," as $\varepsilon \to 0$ means $|f\,/\,\nu| \to 0$. That is, $f$ has a lower order of magnitude than $\nu$. If $\nu \to 0$ as $\varepsilon \to 0$, then $f$ goes to zero faster than $\nu$.

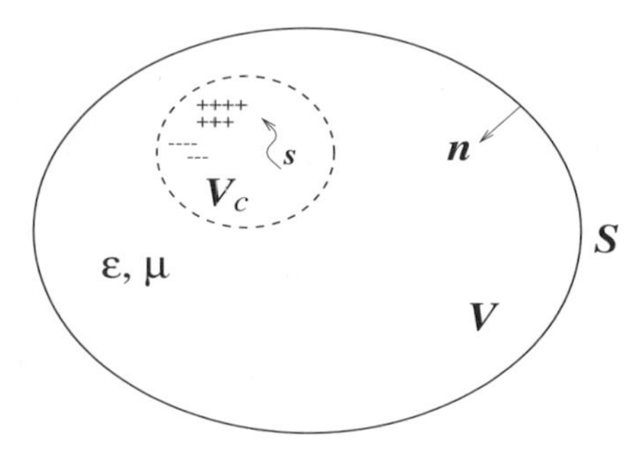


Figure 1.1: Electromagnetic sources $s$ in region $V_c \subseteq V$ bounded by a smooth surface $S$ containing a medium characterized by constants $\varepsilon, \mu$.

in one dimension.

For static problems ($k = 0$) the small argument limits of $e^{-ik|\mathbf{r}-\mathbf{r}'|}$ and $H_0^2(k|\mathbf{r}-\mathbf{r}'|)$ lead to, respectively,

$$g(\mathbf{r},\mathbf{r}') = \frac{1}{4\pi|\mathbf{r}-\mathbf{r}'|} + g^h(\mathbf{r},\mathbf{r}'), \tag{1.45}$$

$$g(\mathbf{r},\mathbf{r}') = -\frac{1}{2\pi}\ln(|\mathbf{r}-\mathbf{r}'|) + g^h(\mathbf{r},\mathbf{r}') \tag{1.46}$$

where for the one-dimensional static problem we have

$$g(x,x') = -\frac{1}{2}|x-x'| + g^h(x,x'). \tag{1.47}$$

It is important to note that the source-point singularity at $\mathbf{r} = \mathbf{r}'$ encountered in $n \geq 2$ dimensions arises from the static case $k = 0$.

### Solution of the Scalar Helmholtz Equation

We now turn to solving (1.39), where we will concentrate on the three-dimensional case. Consider the geometry depicted in Figure 1.1 showing a volume $V$ containing a medium characterized by constants $\varepsilon, \mu$ and bounded by a smooth surface $S$ (normal unit vector $\mathbf{n}$ points inward). All sources reside within $S$, and we are concerned with determining field values inside $S$.

We use Green's second theorem,

$$\int_V \left[\psi_1\nabla^2\psi_2 - \psi_2\nabla^2\psi_1\right] dV = \oint_S (\psi_2\nabla\psi_1 - \psi_1\nabla\psi_2)\cdot d\mathbf{S} \tag{1.48}$$

and let $\psi_1 = \psi(\mathbf{r})$ and $\psi_2 = g(\mathbf{r},\mathbf{r}')$, leading to

$$\psi(\mathbf{r}') = \int_V g(\mathbf{r},\mathbf{r}')\, s(\mathbf{r})\, dV - \oint_S [g(\mathbf{r},\mathbf{r}')\nabla\psi(\mathbf{r}) - \psi(\mathbf{r})\nabla g(\mathbf{r},\mathbf{r}')]\cdot d\mathbf{S}. \tag{1.49}$$

The volume integral extends over all parts of $V$ where $s(\mathbf{r}) \neq 0$, shown as $V_c$ ($\subseteq V$) in the figure, and the right side vanishes for $\mathbf{r}' \notin V$. In (1.49) $\mathbf{r}'$ now represents the field point and $\mathbf{r}$ the source point.

The above equation represents a general solution to (1.39) in the sense that no boundary conditions have been specified on $\psi$ or $g$. All we have required is that $\psi, g$ satisfy (1.39) and (1.40), respectively. Although we are not concerned here with the uniqueness of the Green's function, we must specify boundary conditions on $\psi$ for (1.39) to have a unique solution (see Section 1.5). Often these conditions will simplify the surface term in (1.49). For instance, if $\psi(\mathbf{r})|_S = 0$ (homogeneous Dirichlet condition) or $\left.\frac{\partial \psi}{\partial n}\right|_S = 0$ (homogeneous Neumann condition), the surface integral in (1.49) is simplified.

We are at liberty to take $g$ as any solution to (1.40). For example, we may take $g$ to be the free-space Green's function (1.41), even in the presence of boundaries. If $g$ in (1.49) contains homogeneous solutions to (1.40) as well, desirable behavior of $g$ on $S$ may be achievable. For instance, we may be able to choose $g^h$ such that $g(\mathbf{r})|_S = 0$, further simplifying the evaluation of the surface integral in (1.49). If one of the surface terms in (1.49) vanishes because of the boundary condition on $\psi$, then if the remaining surface term can be made to vanish by choice of the Green's function, (1.49) reduces to the volume term, which represents an explicit solution of (1.39). Also, for a surface with apertures, by proper choice of the Green's function it may be possible to reduce the surface integral term to one only over the apertures. However, often the Green's function for a particular space is difficult to obtain, and one utilizes the free-space Green's function. In this case (1.49) represents an integral relation for $\psi$ rather than an explicit solution of (1.39).

We may also use (1.49) to obtain an expression for the solution due to a source with compact support in free space. Suppose $s(\mathbf{r})$ has compact support $V_c \subset V$. Then, with $g$ being the free-space Green's function and letting the boundary $S$ of $V$ recede to infinity (and assuming the existence of appropriate radiation conditions as discussed in Section 1.5), we obtain

$$\psi(\mathbf{r}') = \int_{V_c} g(\mathbf{r}, \mathbf{r}') \, s(\mathbf{r}) \, dV, \tag{1.50}$$

valid for $\mathbf{r} \in V$ (both interior and exterior to $V_c$).

One can criticize the derivation of (1.49) since Green's theorem (1.48) requires $\psi_{1,2}$, $\partial \psi_{1,2}/\partial n$ to be continuous in the closed region $V \cup S$ with $\psi_{1,2}$ having piecewise continuous second derivatives in $V$. With the source density $s$ at least piecewise continuous, from (1.39) one would expect $\psi$ to have piecewise continuous second derivatives in $V$. However, the substitution $\psi_2 = g(\mathbf{r}, \mathbf{r}')$ violates the conditions of Green's theorem. We can alleviate this difficulty by following the usual procedure of excluding the point $\mathbf{r}' = \mathbf{r}$ from our integration. Considering Figure 1.2 (note again that

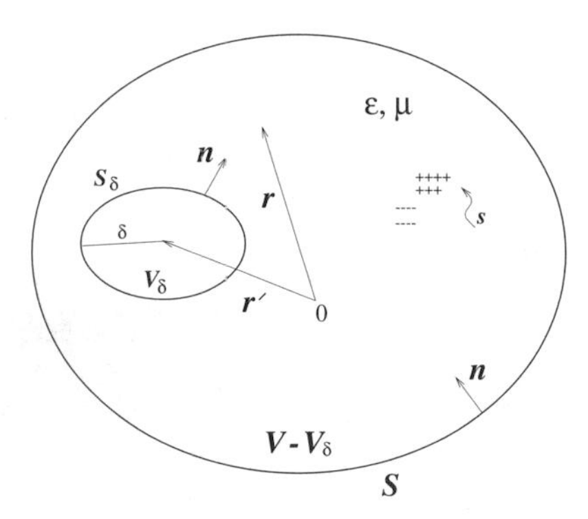


Figure 1.2: Region $V$ containing sources $s$. Observation point $\mathbf{r}'$ is excluded from $V$ by arbitrary exclusion volume $V_\delta$.

the unit normal vector $\mathbf{n}$ points into $V$), we exclude the point $\mathbf{r}'$ from the volume $V$ by containing it within an arbitrary volume $V_\delta$ bounded by the smooth surface $S_\delta$. The application of Green's second theorem to the region $V - V_\delta$, with the substitutions $\psi_1 = \psi(\mathbf{r})$ and $\psi_2 = g(\mathbf{r}, \mathbf{r}')$, is now rigorously valid, leading to

$$\int_{V-V_\delta} \left[\psi(\mathbf{r})\nabla^2 g(\mathbf{r},\mathbf{r}') - g(\mathbf{r},\mathbf{r}')\nabla^2\psi(\mathbf{r})\right] dV$$
$$= \oint_{S+S_\delta} \left(g(\mathbf{r},\mathbf{r}')\nabla\psi(\mathbf{r}) - \psi(\mathbf{r})\nabla g(\mathbf{r},\mathbf{r}')\right) \cdot d\mathbf{S}.$$

We write this as

$$\oint_{S_\delta} \left(g(\mathbf{r},\mathbf{r}')\nabla\psi(\mathbf{r}) - \psi(\mathbf{r})\nabla g(\mathbf{r},\mathbf{r}')\right) \cdot d\mathbf{S} = \int_{V-V_\delta} g(\mathbf{r},\mathbf{r}')\, s(\mathbf{r})\, dV$$
$$- \oint_{S} \left(g(\mathbf{r},\mathbf{r}')\nabla\psi(\mathbf{r}) - \psi(\mathbf{r})\nabla g(\mathbf{r},\mathbf{r}')\right) \cdot d\mathbf{S},$$

where, taking the limit as $\delta \to 0$, the left side becomes[15]

$$\lim_{\delta\to 0} \oint_{S_\delta} \frac{e^{-ikR}}{4\pi R} \nabla\psi(\mathbf{r}) \cdot \mathbf{n}\, dS - \lim_{\delta\to 0} \oint_{S_\delta} \psi(\mathbf{r}) \frac{e^{-ikR}}{4\pi R} (ik) \left(-\widehat{\mathbf{R}}\right) \cdot \mathbf{n}\, dS$$
$$- \lim_{\delta\to 0} \oint_{S_\delta} \psi(\mathbf{r}) \frac{e^{-ikR}}{4\pi} \frac{\left(-\widehat{\mathbf{R}} \cdot \mathbf{n}\right)}{R^2}\, dS.$$

The first term vanishes since $R \leq \delta$ ($\delta$ is the maximum chord of $V_\delta$) so that the integrand is $O(1/\delta)$, while the surface element is $O\left(\delta^2\right)$. The second

[15]The possible inclusion of $g^h$ in $g$ does not affect the result since $g^h$ is regular at $\mathbf{r} = \mathbf{r}'$, and so we use the free-space Green's function.

term vanishes for the same reason, while the third term leads to $\psi(\mathbf{r}')$. In evaluating the third term we assume $\psi(\mathbf{r})$ is well behaved for $\mathbf{r}$ near $\mathbf{r}'$, so that it can be brought outside the integral as a constant on $S_\delta$. The solid-angle formula [4, p. 9]

$$\oint_{S_\delta} \frac{\widehat{\mathbf{R}} \cdot \mathbf{n}}{R^2}\, dS = 4\pi, \tag{1.51}$$

where both unit vectors point outward from $S_\delta$ and $R = 0$ is contained inside $S_\delta$ then leads to the desired result. For $\mathbf{r}' \in V$ we therefore get

$$\psi(\mathbf{r}') = \lim_{\delta \to 0} \int_{V - V_\delta} g(\mathbf{r}, \mathbf{r}')\, s(\mathbf{r})\, dV - \oint_S [g(\mathbf{r}, \mathbf{r}')\nabla\psi(\mathbf{r}) - \psi(\mathbf{r})\nabla g(\mathbf{r}, \mathbf{r}')] \cdot d\mathbf{S}, \tag{1.52}$$

where for $\mathbf{r}' \notin V$ the right side of (1.52) is zero.

In order to reconcile (1.49) with (1.52), it is useful to recall some details from the theory of improper integrals and potential theory.

### 1.3.4 Improper Integrals Associated with Scalar Green's Functions

Classically, if a function $f(\mathbf{r}, \mathbf{r}')$ is piecewise continuous everywhere in a region $V$, except at $\mathbf{r} = \mathbf{r}'$ where it becomes unbounded, then the *improper integral* $\int_V f(\mathbf{r}, \mathbf{r}')\, dV$ is said to exist (converge to a unique function of $\mathbf{r}'$) and is equal to

$$\lim_{\delta \to 0} \int_{V - V_\delta} f(\mathbf{r}, \mathbf{r}')\, dV$$

if the latter integral exists [9, p. 147], [10, p. 412], [11].[16] In the latter expression $V_\delta$ is a small volume containing the singular point $\mathbf{r}'$, and so $V_\delta = V_\delta(\mathbf{r}')$. The only restrictions on $V_\delta$ are that the point $\mathbf{r}'$ is interior to $V_\delta$ and that the maximum chord of $V_\delta$ does not exceed $\delta$. As the limit is taken, the shape, position, and orientation with respect to $\mathbf{r}'$ are maintained. The integral is said to exist (converge) if the limiting integral converges to a finite value independent of the shape of the exclusion region. For our purposes we will broaden the concept of convergence and say that an integral is *conditionally convergent*[17] (exists in a conditional sense) if the limiting integral converges to a finite value that is dependent of the shape of the exclusion region.

If the integral does not converge, it is said to *diverge*. This general topic can also be considered as a regularization problem in the theory of

---

[16]By definition, in the Riemann integral theory $\int_V f(\mathbf{r}, \mathbf{r}')\, dV$ is understood as $\lim_{\delta \to 0} \int_{V - V_\delta} f(\mathbf{r}, \mathbf{r}')\, dV$, while for the Lebesgue theory these quantities are quite different [46, p. 8]. The following remarks apply in the Riemann sense.

[17]This notion of conditional convergence is different from that described in the footnote on p. 9.

generalized functions [7, p. 10], although that method will not be pursued here.

A useful check for the existence of an improper integral is that if the improper integral $\int_V f(\mathbf{r},\mathbf{r}')\,dV$ exists, then [9, p. 147]

$$\lim_{\delta\to 0}\int_{V_\delta} f(\mathbf{r},\mathbf{r}')\,dV = 0.$$

It can also be observed that if we regard the improper integral $\int_V f(\mathbf{r},\mathbf{r}')\,dV$ as $\lim_{\delta\to 0}\int_{V-V_\delta} f(\mathbf{r},\mathbf{r}')\,dV$, the tempting decomposition

$$\int_V f(\mathbf{r},\mathbf{r}')\,dV = \lim_{\delta\to 0}\int_{V-V_\delta} f(\mathbf{r},\mathbf{r}')\,dV + \lim_{\delta\to 0}\int_{V_\delta} f(\mathbf{r},\mathbf{r}')\,dV$$

does not make sense unless we know a priori that the left volume integral converges, in which case $\lim_{\delta\to 0}\int_{V_\delta} f(\mathbf{r},\mathbf{r}')\,dV = 0$. On the other hand, for $V_S$ *finite* and containing the singular point $\mathbf{r}'$, the decomposition

$$\int_V f(\mathbf{r},\mathbf{r}')\,dV = \int_{V-V_S} f(\mathbf{r},\mathbf{r}')\,dV + \int_{V_S} f(\mathbf{r},\mathbf{r}')\,dV$$

is meaningful in the sense that the term $\int_{V_S} f(\mathbf{r},\mathbf{r}')\,dV$ is then regarded as $\lim_{\delta\to 0}\int_{V_S-V_\delta} f(\mathbf{r},\mathbf{r}')\,dV$, which will exist if the original integral exists.

It is also useful for later purposes to state that the relation

$$\int_V [f_1(\mathbf{r},\mathbf{r}') + f_2(\mathbf{r},\mathbf{r}')]\,dV = \int_V f_1(\mathbf{r},\mathbf{r}')\,dV + \int_V f_2(\mathbf{r},\mathbf{r}')\,dV,$$

where we assume the left-side integral exists, is only strictly valid when both right-side integrals independently exist. The existence of the two right-side integrals does not necessarily follow from the existence of the left-side integral.[18]

As an example of an improper integral in one dimension,

$$I = \int_{-a}^{a}\frac{1}{x}\,dx = \lim_{\left(\substack{\varepsilon\to 0\\ \delta\to 0}\right)}\left[\int_{-a}^{-\varepsilon}\frac{1}{x}\,dx + \int_{\delta}^{a}\frac{1}{x}\,dx\right] = \lim_{\left(\substack{\varepsilon\to 0\\ \delta\to 0}\right)}\left[\ln(\varepsilon) - \ln(\delta)\right].$$

If the limit variables are related, say $\varepsilon = \alpha\delta$, then $I = \ln(\alpha)$ and the integral converges (conditionally) to a number for a given $\alpha$, but that number is

---

[18] As a one-dimensional example, $\int_0^1 [(e^x/x) - (1/x)]\,dx$ can be evaluated in terms of an exponential integral to yield $\simeq 1.3179$, while neither $\int_0^1 (e^x/x)\,dx$ nor $\int_0^1 (1/x)\,dx$ exists individually. It is also true that

$$\lim_{\delta\to 0^+}\int_\delta^1 [(e^x/x) - (1/x)]\,dx = \lim_{\delta\to 0^+}\left[\int_\delta^1 (e^x/x)\,dx - \int_\delta^1 [(1/x)\,dx\right]$$

exists, whereas $\left[\lim_{\delta\to 0^+}\int_\delta^1 (e^x/x)\,dx - \lim_{\delta\to 0^+}\int_\delta^1 (1/x)\,dx\right]$ does not exist.

not unique.[19] Note that if $\varepsilon, \delta$ are unrelated then the integral is not even finite. So it is seen that in this instance the value of the integral depends on the "shape" of the exclusion region. For this one-dimensional case, the integral $\int_{-a}^{a}(1/x^n)\,dx$ converges for $n > 1$, exists as a Cauchy principal value for $n = 1$, and experiences no singularity for $n < 1$.

As a related example,

$$\int_0^a \frac{1}{x^n}\,dx = \lim_{\delta\to 0+} \int_\delta^a \frac{1}{x^n}\,dx = \lim_{\delta\to 0+} -\frac{a^{1-n} - \delta^{1-n}}{-1+n},$$

which is not finite for $n \geq 1$ but exists for $0 \leq n < 1$.

We find a similar situation for integrals over $\mathbf{R}^3$; the improper integral $\int_V (1/R^n)\,dV$ (with $R = |\mathbf{r} - \mathbf{r}'|$) converges, meaning

$$\lim_{\delta\to 0} \int_{V-V_\delta} \frac{1}{R^n}\,dV$$

exists in the sense described above, for $0 \leq n < 3$ [9, p. 148]. The integrand $(1/R^n)$ with $0 < n < 3$ is called *weakly singular* or is said to have an *integrable singularity*, and for $n \geq 3$ the integral is divergent. For a more general type of integral given as

$$\int_V \frac{k(\mathbf{r}, \mathbf{r}')}{R^n} u(\mathbf{r})\,dV$$

similar conditions on $n$ hold as long as $k$ and $u$ are reasonably well behaved at $\mathbf{r} = \mathbf{r}'$ (see Section 3.6.2). In fact, the nature of $k$ may cause integrals that would otherwise be divergent to converge (see [12] for an in-depth discussion of the singular case $n = 3$).

The above discussion also applies to $N$-fold integrals in $N$-dimensional space: For $\Omega \subset \mathbf{R}^N$, the improper integral $\int_\Omega (1/R^n)\,d\Omega$ converges for $0 \leq n < N$ and diverges for $n \geq N$ [10, p. 417]. The same holds for integrals on an $N$-dimensional surface in an $(N+1)$ dimensional space. Note for $N = 1$ that the "polar" integral $\int_0^a (1/x^n)\,dx$ is the one-dimensional analog of $\int_\Omega (1/R^n)\,d\Omega$.

A very useful sufficiency test for convergence is the following [9, p. 147], [10, p. 419].[20]

---

[19] The case $\varepsilon = \delta$ for this conditionally convergent integral is called the *Cauchy principal value* of the integral, i.e.,

$$\text{P.V.} \int_{-a}^{a} f(x)\,dx \equiv \lim_{\delta\to 0} \left[ \int_{-a}^{-\delta} f(x)\,dx + \int_\delta^a f(x)\,dx \right].$$

Sometimes a divergent improper integral has a Cauchy principal value due to cancellation effects on either side of the singularity. In $N$ dimensions the Cauchy principal value is defined using $N$-dimensional balls [10, p. 413].

[20] Comparison tests of this nature provide sufficient conditions for convergence of improper integrals. Their usefulness lies in the fact that they are very easy to apply. Cauchy tests, which are not discussed here, are necessary and sufficient yet are more difficult to apply in practice.

**Theorem 1.1.** *If $\int_\Omega f(\mathbf{r})\, d\Omega$ is an $N$-fold improper integral as described above, and there is a function $g(\mathbf{r})$ such that $|f(\mathbf{r})| \le g(\mathbf{r})$ for all $\mathbf{r} \in \Omega \subset R^N$ and such that $\int_\Omega g(\mathbf{r})\, d\Omega$ is convergent, then $\int_\Omega f(\mathbf{r})\, d\Omega$ is convergent.*[21] *Conversely, if there is a function $g(\mathbf{r})$ such that $|f(\mathbf{r})| \ge g(\mathbf{r})$ for all $\mathbf{r} \in \Omega$ and such that $\int_\Omega g(\mathbf{r})\, d\Omega$ diverges, then $\int_\Omega f(\mathbf{r})\, d\Omega$ diverges.*

We now move on to integrals of the form

$$\phi(\mathbf{r}) = \int_V s(\mathbf{r}') \frac{e^{-ik|\mathbf{r}-\mathbf{r}'|}}{4\pi |\mathbf{r}-\mathbf{r}'|}\, dV' = \int_V s(\mathbf{r}')\, g(\mathbf{r},\mathbf{r}')\, dV', \qquad (1.53)$$

where again the volume integral is to be interpreted as $\lim_{\delta\to 0} \int_{V-V_\delta} (\cdot)\, dV'$ and we assume the source density $s(\mathbf{r})$ is at least piecewise continuous.

This expression is found to occur for $\mathbf{r} \in V$ as well as for $\mathbf{r} \notin V$ (see, e.g., (1.50)). Because for small $k\,|\mathbf{r}-\mathbf{r}'|$ we can expand the exponential term in (1.53) in a Taylor's series,

$$e^{-ik|\mathbf{r}-\mathbf{r}'|} = 1 + (-ik\,|\mathbf{r}-\mathbf{r}'|) + \frac{1}{2}(-ik\,|\mathbf{r}-\mathbf{r}'|)^2 + \cdots,$$

the Green's function can be written as

$$\frac{e^{-ik|\mathbf{r}-\mathbf{r}'|}}{4\pi |\mathbf{r}-\mathbf{r}'|} = \frac{1}{4\pi}\left[\frac{1}{|\mathbf{r}-\mathbf{r}'|} - ik - \frac{1}{2}k^2\,|\mathbf{r}-\mathbf{r}'| + \cdots\right]$$

near the source-point singularity. Regarding the singularity of (1.53) it is then useful to consider the electrostatic scalar potential

$$\phi_p(\mathbf{r}) = \int_V \frac{s(\mathbf{r}')}{4\pi |\mathbf{r}-\mathbf{r}'|}\, dV' = \lim_{\delta\to 0} \int_{V-V_\delta} \frac{s(\mathbf{r}')}{4\pi |\mathbf{r}-\mathbf{r}'|}\, dV', \qquad (1.54)$$

which is the solution to Poisson's equation, $\nabla^2 \phi_p(\mathbf{r}) = -s(\mathbf{r})$.

In the following we will take $g(\mathbf{r},\mathbf{r}') = (e^{-ik|\mathbf{r}-\mathbf{r}'|}) / (4\pi |\mathbf{r}-\mathbf{r}'|)$, where $k$ may be zero for the static case or nonzero for the dynamic case. The singular nature of the volume integral is not substantially affected by the presence or absence of the term $e^{-ik|\mathbf{r}-\mathbf{r}'|}$(which does, of course, affect the form of the differentiated integrand), since the source-point singularity arises from the static case. The solution (1.53) will be called a *fundamental solution* of the Helmholtz equation.[22]

---

[21] In general, absolute convergence, which means convergence of $\int_\Omega |f(\mathbf{r})|\, d\Omega$, implies ordinary convergence. In ($N \ge 2$) dimensional space, ordinary convergence implies absolute convergence as well [10, p. 421].

[22] Note that often $g$ itself is called a fundamental solution.

**First Derivatives of Fundamental Solutions**

For the case $\mathbf{r} \notin V$, $\mathbf{r} = \mathbf{r}'$ cannot occur so that (1.53) represents a proper convergent integral over fixed limits. As such it can be differentiated arbitrarily often, with derivatives brought under the integral sign[23] (see, e.g., [9, pp. 121–122] for the case $k = 0$). We now consider the case $\mathbf{r} \in V$.

It is well known that if $s(\mathbf{r})$ is piecewise continuous in $V$ the volume integral (1.53) exists for any $\mathbf{r}$ (i.e., converges independent of the shape of $V_\delta$) as $\delta \to 0$. This is easy to see from Theorem 1.1. Indeed,

$$|\phi(\mathbf{r})| = \left| \int_V s(\mathbf{r}') \frac{e^{-ik|\mathbf{r}-\mathbf{r}'|}}{4\pi |\mathbf{r}-\mathbf{r}'|} \, dV' \right| \leq \int_V \frac{|s(\mathbf{r}')|}{|\mathbf{r}-\mathbf{r}'|} \, dV',$$

assuming $k$ is a real constant. Because $s(\mathbf{r})$ is piecewise continuous in the finite region $V$, and hence bounded in $V$, $|s(\mathbf{r})| \leq B$. Then, for $\mathbf{r} \simeq \mathbf{r}'$

$$\frac{|s(\mathbf{r}')|}{|\mathbf{r}-\mathbf{r}'|} \leq \frac{B}{|\mathbf{r}-\mathbf{r}'|} < \frac{B}{|\mathbf{r}-\mathbf{r}'|^n},$$

$1 < n < 3$, and so the integral converges.[24]

Moreover, the volume integral (1.53) is *uniformly convergent* to a continuous function $\phi(\mathbf{r})$ and, in fact, is differentiable with derivatives allowed to be taken under the integral sign[25] (see [9, pp. 150–152], [10, pp. 471–475] for the case $k = 0$, and [13, pp. 28–32] for the case of $k$ a real-valued constant),

$$\frac{\partial}{\partial x_i} \lim_{\delta \to 0} \int_{V-V_\delta} s(\mathbf{r}') g(\mathbf{r}, \mathbf{r}') \, dV' = \lim_{\delta \to 0} \int_{V-V_\delta} s(\mathbf{r}') \frac{\partial}{\partial x_i} g(\mathbf{r}, \mathbf{r}') \, dV'. \tag{1.55}$$

---

[23] By an extension of Leibnitz's theorem (see p. 31) for multidimensional space, for a proper integral over fixed finite limits a sufficient condition for this to occur is that both the original and the differentiated integrands are continuous. As an example, for $\mathbf{r} \in \Omega$, $\mathbf{r}' \in \Lambda$, where $\Omega, \Lambda$ are bounded regions in $N, M$-dimensional space, respectively, we have

$$\frac{\partial}{\partial x_i'} \int_\Omega f(\mathbf{r}, \mathbf{r}') \, d\Omega = \int_\Omega \frac{\partial}{\partial x_i'} f(\mathbf{r}, \mathbf{r}') \, d\Omega$$

if $f, \partial f / \partial x$ are continuous on the closed region $\widetilde{\Omega} \times \widetilde{\Lambda}$ [10, p. 469]. Note that we also have

$$\int_\Lambda \int_\Omega f(\mathbf{r}, \mathbf{r}') \, d\Omega d\tau = \int_\Omega \int_\Lambda f(\mathbf{r}, \mathbf{r}') \, d\tau d\Omega$$

if $f$ is continuous on $\widetilde{\Omega} \times \widetilde{\Lambda}$.

[24] In fact, it can be shown that $|\phi(\mathbf{r})| \leq B \, 2\pi \, (3V / 4\pi)^{2/3}$.

[25] For an $N$-dimensional improper integral of this type ($\Omega$ finite, $f$ unbounded at $\mathbf{r} = \mathbf{r}'$) the expression

$$\frac{\partial}{\partial x_i'} \int_\Omega f(\mathbf{r}, \mathbf{r}') \, d\Omega = \int_\Omega \frac{\partial}{\partial x_i'} f(\mathbf{r}, \mathbf{r}') \, d\Omega$$

is valid if $f$ and the resulting differentiated integrand are continuous over the regions of interest and the right side integral is uniformly convergent.

The resulting differentiated integrals are everywhere continuous.

Because $V_\delta = V_\delta(\mathbf{r})$ (see, e.g., [14], [15]), the validity of passing $\partial / \partial x_i$ through the limiting integral needs to be established carefully, as can be seen from the cited references. In some situations this interchange of operations can also be accomplished through the use of Leibnitz's theorem [16], [17].[26] When (1.55) holds, one can see that the "extra terms" given in Leibnitz's theorem, generated by the rigorous interchange of operators, vanish.

Using (1.55), one can see that for first derivatives (usually $\partial\phi / \partial x_i$ and $\nabla\phi$ in the scalar potential case, and $\nabla \cdot \mathbf{A}$ and $\nabla \times \mathbf{A}$ in the case of the vector potential) the final result is the same as if the derivative was formally passed through the integral without regard for either the limiting operation or the integration limits depending on the differentiation variable (see also [15], [16] for the proof in the vector potential case). Thus we obtain

$$\begin{aligned}
\nabla \lim_{\delta\to 0} \int_{V-V_\delta} s(\mathbf{r}')\, g(\mathbf{r},\mathbf{r}')\, dV' &= \lim_{\delta\to 0} \int_{V-V_\delta} s(\mathbf{r}')\, \nabla g(\mathbf{r},\mathbf{r}')\, dV', \\
\frac{\partial}{\partial x_i} \lim_{\delta\to 0} \int_{V-V_\delta} \mathbf{s}(\mathbf{r}')\, g(\mathbf{r},\mathbf{r}')\, dV' &= \lim_{\delta\to 0} \int_{V-V_\delta} \mathbf{s}(\mathbf{r}')\, \frac{\partial}{\partial x_i} g(\mathbf{r},\mathbf{r}')\, dV', \\
\nabla \cdot \lim_{\delta\to 0} \int_{V-V_\delta} \mathbf{s}(\mathbf{r}')\, g(\mathbf{r},\mathbf{r}')\, dV' &= \lim_{\delta\to 0} \int_{V-V_\delta} \mathbf{s}(\mathbf{r}') \cdot \nabla g(\mathbf{r},\mathbf{r}')\, dV', \\
\nabla \times \lim_{\delta\to 0} \int_{V-V_\delta} \mathbf{s}(\mathbf{r}')\, g(\mathbf{r},\mathbf{r}')\, dV' &= \lim_{\delta\to 0} \int_{V-V_\delta} -\mathbf{s}(\mathbf{r}') \times \nabla g(\mathbf{r},\mathbf{r}')\, dV',
\end{aligned} \tag{1.56}$$

where the right-side integrals are uniformly convergent [10, p. 472].

### Second Derivatives of the Fundamental Solution

Second derivatives of the fundamental solution (1.53) may not necessarily exist when $\mathbf{r} \in V$, but if the source density is piecewise continuous in $V$,

[26] In the one-dimensional case, for $g_1(x), g_2(x)$ continuously differentiable functions on $x_1 < x < x_2$ and $f(x,y)$, $f'_x(x,y)$ continuous on $g_1(x) < y < g_2(x)$, $x_1 < x < x_2$, Leibnitz's theorem states

$$\frac{\partial}{\partial x} \int_{g_1(x)}^{g_2(x)} f(x,y)\, dy = \int_{g_1(x)}^{g_2(x)} \frac{\partial}{\partial x} f(x,y)\, dy + f(x, g_2(x))\, \frac{\partial g_2(x)}{\partial x} - f(x, g_1(x))\, \frac{\partial g_1(x)}{\partial x}.$$

In the case of interest here one has to be careful with this procedure since Leibnitz's theorem and its generalization to three dimensions [16] apply to proper integrals. If, however, it can be shown that

$$\frac{\partial}{\partial x_i} \lim_{\delta\to 0} \int_{V-V_\delta} s(\mathbf{r}') g(\mathbf{r},\mathbf{r}')\, dV' = \lim_{\delta\to 0} \frac{\partial}{\partial x_i} \int_{V-V_\delta} s(\mathbf{r}') g(\mathbf{r},\mathbf{r}')\, dV'$$

holds, then Leibnitz's theorem may be rigorously applied. The validity of this interchange for the curl operator and the type of integrand of interest here is described in [16].

then at any point in $V$ (the bounding surface $S$ is not part of $V$) where the source density $s(\mathbf{r})$ satisfies a Hölder condition (see Definition 2.5)

$$|s(\mathbf{r}) - s(\mathbf{r}')| \leq k|\mathbf{r} - \mathbf{r}'|^{\alpha}, \tag{1.57}$$

where $k > 0$ and $0 < \alpha \leq 1$, then the second-order partial derivative

$$\frac{\partial^2}{\partial x_j \, \partial x_i} \phi(\mathbf{r}) = \frac{\partial^2}{\partial x_j \, \partial x_i} \lim_{\delta \to 0} \int_{V - V_\delta} s(\mathbf{r}') \, g(\mathbf{r}, \mathbf{r}') \, dV' \tag{1.58}$$

exists as well [9, pp. 153–156], [13, p. 41]. If the source density satisfies the same Hölder condition everywhere in $V$, then second partial derivatives are (Hölder) continuous in $V$, although they will not, in general, be continuous on the boundary $S$.

Even if existence of the second-derivative is established, second-derivative operators may not generally be brought under the integral sign without careful consideration of the source-point singularity. If the integral and second-derivative operator are formally interchanged, the integral of the resulting differentiated integrand is often no longer convergent, let alone uniformly convergent (the differentiated integrand being singular, $O(1/R^3)$). One procedure to evaluate (1.58) is motivated by a method presented in [18, p. 248] for the case of $i = j$ with $g$ being the static Green's function. Generalizing this method for any $i, j$, and for $g$ being the free-space Green's function (static or dynamic), we obtain

$$\begin{aligned} \frac{\partial^2}{\partial x_j \, \partial x_i} \lim_{\delta \to 0} \int_{V - V_\delta} & s(\mathbf{r}') \, g(\mathbf{r}, \mathbf{r}') \, dV' \\ &= -s(\mathbf{r}) \oint_S \frac{\partial}{\partial x_j} g(\mathbf{r}, \mathbf{r}') \, \widehat{\mathbf{x}}_i \cdot \mathbf{n} \, dS' \\ &\quad + \lim_{\delta \to 0} \int_{V - V_\delta} [s(\mathbf{r}') - s(\mathbf{r})] \frac{\partial^2}{\partial x_i' \, \partial x_j'} g(\mathbf{r}, \mathbf{r}') \, dV', \end{aligned} \tag{1.59}$$

where $S$ is the boundary surface of $V$, $\mathbf{n}$ is an outward unit normal vector on $S$, and $\mathbf{r}, \mathbf{r}' \in V$. This equation holds for $s$ being Hölder continuous, with the derivation of (1.59) presented in Appendix B.

The form (1.59) can be used to pass various second-order derivative operators ($\nabla^2$, $\nabla \times \nabla \times$, $\nabla \nabla \cdot$, etc.) through integrals of the form (1.53), scalar or vector case as appropriate. For instance, it is a simple matter to use (1.59), with $g(\mathbf{r}, \mathbf{r}') = 1 / 4\pi \, |\mathbf{r} - \mathbf{r}'|$, to verify that (1.54) is a solution to Poisson's equation and that in the dynamic case with $g(\mathbf{r}, \mathbf{r}') = e^{-ik\,|\mathbf{r} - \mathbf{r}'|} / 4\pi \, |\mathbf{r} - \mathbf{r}'|$, (1.53) satisfies the Helmholtz equation (1.39). Indeed, for the dynamic case

$$\left(\nabla^2 + k^2\right) \lim_{\delta \to 0} \int_{V - V_\delta} s(\mathbf{r}') \, g(\mathbf{r}, \mathbf{r}') \, dV' = -s(\mathbf{r}) \oint_S \sum_{i=1}^{3} \frac{\partial}{\partial x_i} g(\mathbf{r}, \mathbf{r}') \, \widehat{\mathbf{x}}_i \cdot \mathbf{n} \, dS'$$

$$+\lim_{\delta\to 0}\int_{V-V_\delta}\left[[s(\mathbf{r}')-s(\mathbf{r})]\sum_{i=1}^{3}\frac{\partial^2}{\partial x_i'\,\partial x_i'}\,g(\mathbf{r},\mathbf{r}')+s(\mathbf{r}')\,k^2\,g(\mathbf{r},\mathbf{r}')\right]dV'$$

$$=-s(\mathbf{r})\oint_S(\nabla\,g(\mathbf{r},\mathbf{r}'))\cdot\mathbf{n}\,dS'$$

$$+\lim_{\delta\to 0}\int_{V-V_\delta}\left[s(\mathbf{r}')\left[\nabla^2+k^2\right]g(\mathbf{r},\mathbf{r}')-s(\mathbf{r})\,\nabla^2 g(\mathbf{r},\mathbf{r}')\right]dV' \qquad (1.60)$$

where the first term in the integrand of the volume integral vanishes since the point $\mathbf{r}=\mathbf{r}'$ is excluded from the domain of integration. Note that $\nabla^2 g=\nabla\cdot\nabla g$; then the remaining volume integral is converted to a surface integral using the divergence theorem, such that the right side of (1.60) becomes

$$s(\mathbf{r})\left[-\oint_S\mathbf{n}\cdot\nabla\,g(\mathbf{r},\mathbf{r}')\,dS'+\lim_{\delta\to 0}\oint_{S+S_\delta}\mathbf{n}\cdot\nabla\,g(\mathbf{r},\mathbf{r}')\,dS'\right]. \qquad (1.61)$$

The integrals over $S$ cancel, leaving the integral over the exclusion surface. Because $\nabla g(\mathbf{r},\mathbf{r}')=-\widehat{\mathbf{R}}\,/\,4\pi\,R^2$ on the limitingly small surface $S_\delta$, and using the solid-angle formula $\oint_{S_\delta}(\mathbf{n}\cdot\widehat{\mathbf{R}}\,/\,4\pi\,R^2)\,dS'=1$, (1.61) becomes $-s(\mathbf{r})$, proving the statement that (1.53) is a solution of (1.39). For statics ($k=0$), it turns out to be the integral over $S$ that provides the source contribution.

It is important to note that the surface integral terms have combined together such that the solid-angle formula can be applied. This occurs for the Laplacian operator, but not generally for other second-order derivative operators. The usual procedure of formally passing the operator $\nabla^2$ through the volume integral (1.53), often justified by noting that the integral is over primed coordinates while $\nabla^2$ is with respect to unprimed coordinates, is not rigorously valid since the limits of integration are in fact functions of the variable $\mathbf{r}$ through the limiting process.[27] This interchange happens to work for the Laplacian operator, which has led to this widespread practice, but does not generally yield correct results for other second-order derivative operators. This fact is obvious for $\partial^2\,/\,\partial x_j\,\partial x_i$ from (1.59) and will be shown later for the important second-order differential operator $(\nabla\nabla\cdot)$ acting on integrals of the form (1.53) with vector source terms $\mathbf{s}$.

To see what effect the assumption of Hölder continuity has on the second-derivative formula (1.59) for the volume integral (1.53), it is sufficient to study the static Green's function and consider the term for $i=j$,

[27] This can be remedied by formally interchanging the derivative and integral operators and considering the resulting differentiated integrand as a generalized function. It may then be possible to perform a regularization of the resulting divergent integral [7, p. 10] leading to the correct result (see [19], [20] for the operator $\partial^2\,/\,\partial x_j\,\partial x_i$). Note that although the consideration of generalized functions allows for many convenient operations, like interchanging the order of operators, termwise differentiation of infinite series, etc., one often needs to apply generalized function theory carefully to obtain correct results.

noting that

$$\frac{\partial^2}{\partial x_j \partial x_i} \frac{1}{|\mathbf{r}-\mathbf{r}'|} = \frac{3(x_i - x_i')^2}{|\mathbf{r}-\mathbf{r}'|^5} - \frac{1}{|\mathbf{r}-\mathbf{r}'|^3}.$$

Because $|x_i - x_i'| \leq |\mathbf{r}-\mathbf{r}'|$, the first term is less than or equal to $3/|\mathbf{r}-\mathbf{r}'|^3$, and so we'll consider the term

$$\int_V [s(\mathbf{r}') - s(\mathbf{r})] \frac{1}{|\mathbf{r}-\mathbf{r}'|^3} dV'. \tag{1.62}$$

If the source density $s$ is Hölder continuous, satisfying (1.57), we conclude that the above term exists because

$$\begin{aligned}\left| \int_V [s(\mathbf{r}') - s(\mathbf{r})] \frac{1}{|\mathbf{r}-\mathbf{r}'|^3} dV' \right| &\leq \int_V \frac{|s(\mathbf{r}') - s(\mathbf{r})|}{|\mathbf{r}-\mathbf{r}'|^3} dV' \\ &\leq k \int_V \frac{|\mathbf{r}-\mathbf{r}'|^\alpha}{|\mathbf{r}-\mathbf{r}'|^3} dV' \\ &= k \int_V \frac{1}{|\mathbf{r}-\mathbf{r}'|^{3-\alpha}} dV',\end{aligned}$$

where $0 < \alpha \leq 1$, and by Theorem 1.1 the integral converges.

### Alternative Method for Evaluating Second Derivatives of the Fundamental Solution

Another method of evaluating the second partial derivatives of the fundamental solution (1.53) was developed in [19], [20] with regards to the electric dyadic Green's function singularity. Using concepts from generalized function theory [7], it is shown that

$$\begin{aligned}&\frac{\partial^2}{\partial x_j \partial x_i} \int_V s(\mathbf{r}') \frac{e^{-ik|\mathbf{r}-\mathbf{r}'|}}{4\pi |\mathbf{r}-\mathbf{r}'|} dV' \\ &= \int_{V-V_\varepsilon} s(\mathbf{r}') \frac{\partial^2}{\partial x_j' \partial x_i'} \frac{e^{-ik|\mathbf{r}-\mathbf{r}'|}}{4\pi |\mathbf{r}-\mathbf{r}'|} dV' \\ &\quad + \int_{V\varepsilon} \left[ s(\mathbf{r}') \frac{\partial^2}{\partial x_j' \partial x_i'} \frac{e^{-ik|\mathbf{r}-\mathbf{r}'|}}{4\pi |\mathbf{r}-\mathbf{r}'|} - s(\mathbf{r}) \frac{\partial^2}{\partial x_j' \partial x_i'} \frac{1}{4\pi |\mathbf{r}-\mathbf{r}'|} \right] dV' \\ &\quad - s(\mathbf{r}) \oint_{S_\varepsilon} \frac{(\widehat{\mathbf{x}}_j \cdot \mathbf{n}) \left( \widehat{\mathbf{x}}_i \cdot \widehat{\mathbf{R}} \right)}{4\pi R^2} dS' \\ &= A_{ji} + B_{ji} + C_{ji},\end{aligned} \tag{1.63}$$

where $V_\varepsilon$ is an arbitrary *finite* region within $V$ containing the point $\mathbf{r}$, and $\mathbf{n}$, $\widehat{\mathbf{R}}$ both point inward to $V_\varepsilon$ (as in Figure B.1 if $V_\varepsilon$ replaces $V_\delta$). The right side of (1.63) is independent of $V_\varepsilon$ [19], but it is not quite proper to take the limit as $V_\varepsilon$ vanishes, at least in a termwise fashion. The fact that $V_\varepsilon$ should remain finite is pointed out in [21] for a related expression. The problem is that if

$$\lim_{\varepsilon\to 0}\int_{V-V_\varepsilon} s(\mathbf{r}')\frac{\partial^2}{\partial x'_j\,\partial x'_i}\frac{e^{-i\,k\,|\mathbf{r}-\mathbf{r}'|}}{4\pi\,|\mathbf{r}-\mathbf{r}'|}\,dV'$$

exists, then

$$\lim_{\varepsilon\to 0}\int_{V_\varepsilon} s(\mathbf{r}')\frac{\partial^2}{\partial x'_j\,\partial x'_i}\frac{e^{-i\,k\,|\mathbf{r}-\mathbf{r}'|}}{4\pi\,|\mathbf{r}-\mathbf{r}'|}\,dV'$$

must vanish. Because the second-derivative operator introduces a singularity $O\left(1\,/\,R^3\right)$, this clearly will not occur for most source densities of interest. Therefore, taken individually, the limiting procedure is meaningless.

In general, though, as long as the three terms are kept together, the limiting procedure leads to the correct answer. Therefore, *operationally,* we can use the result

$$\begin{aligned}\frac{\partial^2}{\partial x_j\,\partial x_i}&\int_V s(\mathbf{r}')\frac{e^{-i\,k\,|\mathbf{r}-\mathbf{r}'|}}{4\pi\,|\mathbf{r}-\mathbf{r}'|}\,dV'\\ &=\lim_{\varepsilon\to 0}\int_{V-V_\varepsilon} s(\mathbf{r}')\,\frac{\partial^2}{\partial x'_j\,\partial x'_i}\frac{e^{-i\,k\,|\mathbf{r}-\mathbf{r}'|}}{4\pi\,|\mathbf{r}-\mathbf{r}'|}\,dV'\\ &\quad- s(\mathbf{r})\lim_{\varepsilon\to 0}\oint_{S_\varepsilon}\frac{(\widehat{\mathbf{x}}_j\cdot\mathbf{n})\left(\widehat{\mathbf{x}}_i\cdot\widehat{\mathbf{R}}\right)}{4\pi\,|\mathbf{r}-\mathbf{r}'|^2}\,dS'\end{aligned}\tag{1.64}$$

where it can be seen that the middle term on the right side of (1.63) has vanished for $\varepsilon\to 0$.

This can also be written in the operational form

$$\frac{\partial^2}{\partial x_j\,\partial x_i}\int_V s(\mathbf{r}')\frac{e^{-i\,k\,|\mathbf{r}-\mathbf{r}'|}}{4\pi\,|\mathbf{r}-\mathbf{r}'|}\,dV'=\int_V s(\mathbf{r}')\,g^{\wedge}_{j\,i}(\mathbf{r},\mathbf{r}')\,dV',$$

where

$$g^{\wedge}_{j\,i}(\mathbf{r},\mathbf{r}')\equiv \text{P.V.}\,\frac{\partial^2}{\partial x'_j\,\partial x'_i}\,g(\mathbf{r},\mathbf{r}')-L_{j\,i}(\mathbf{r})\delta(\mathbf{r}-\mathbf{r}'),\tag{1.65}$$

with

$$L_{j\,i}(\mathbf{r})=\lim_{\varepsilon\to 0}\oint_{S_\varepsilon}\frac{(\widehat{\mathbf{x}}_j\cdot\mathbf{n})\left(\widehat{\mathbf{x}}_i\cdot\widehat{\mathbf{R}}\right)}{4\pi\,|\mathbf{r}-\mathbf{r}'|^2}\,dS'\tag{1.66}$$

and P.V. indicates the integral for that term should be performed in the principal value sense.[28]

[28] This concept is slightly different than the usual principal value integral, since the shape of the exclusion volume must correlate with the shape of $S_\delta$ in $L_{j\,i}$.

Both terms on the right side of (1.64) are generally dependent on the shape of the exclusion volume but together form a unique, shape-independent result. To see this, consider an exclusion sphere centered at $\mathbf{r} = \mathbf{0}$ with outward normal vector $\mathbf{n} = \widehat{\mathbf{R}} = -\widehat{\mathbf{r}}'$, leading to $L_{ij} = 1/(3\delta_{ij})$. Alternately, for a pillbox with normal along the $x_3$-axis, $L_{ij} = \delta_{ij}\delta_{i3}$. Therefore, the second term on the right side of (1.64) is finite and shape-dependent. Because we know the left side exists, the first term on the right side of (1.64) must be finite and shape-dependent such that the sum of the two terms is finite and shape-independent.

It is interesting to see the origin of the problem with the limiting procedure from another perspective. Starting from the second-derivative formula (1.59), one can derive the generalized function result (1.63). Indeed, for the static case we have from (1.59)

$$\begin{aligned}
-s(\mathbf{r}) \oint_S \frac{\partial}{\partial x_j} \frac{1}{4\pi |\mathbf{r}-\mathbf{r}'|} \widehat{\mathbf{x}}_i \cdot \mathbf{n}\, dS' + \int_V [s(\mathbf{r}') - s(\mathbf{r})] \frac{\partial^2}{\partial x_i \partial x_j} \frac{1}{4\pi |\mathbf{r}-\mathbf{r}'|} dV' \\
= -s(\mathbf{r}) \oint_S \frac{\partial}{\partial x_j} \frac{1}{4\pi |\mathbf{r}-\mathbf{r}'|} \widehat{\mathbf{x}}_i \cdot \mathbf{n}\, dS' \\
+ \int_{V-V_\varepsilon} [s(\mathbf{r}') - s(\mathbf{r})] \frac{\partial^2}{\partial x_i \partial x_j} \frac{1}{4\pi |\mathbf{r}-\mathbf{r}'|} dV' \\
+ \int_{V_\varepsilon} [s(\mathbf{r}') - s(\mathbf{r})] \frac{\partial^2}{\partial x_i \partial x_j} \frac{1}{4\pi |\mathbf{r}-\mathbf{r}'|} dV',
\end{aligned}$$

where the decomposition $\int_V (\cdot)\, dV' = \int_{V-V_\varepsilon} (\cdot)\, dV' + \int_{V_\varepsilon} (\cdot)\, dV'$ is clearly permissible for $V_\varepsilon$ finite and, in fact, is allowed for the case of limiting $V_\varepsilon$ since $\lim_{\varepsilon\to 0} \int_{V_\varepsilon} (\cdot)\, dV' = 0$ (because we have kept the term $[s(\mathbf{r}') - s(\mathbf{r})]$ together; if we first split the corresponding integral into two integrals, then $V_\varepsilon$ must remain finite). The last term on the right side is recognized as $B_{ji}$ in (1.63) (by direct calculation one can see that the order of the partial derivatives may be interchanged in this case, even though the term is discontinuous at $\mathbf{r} = \mathbf{r}'$), and so we will consider the remaining terms. The second term on the right side is then written as

$$\begin{aligned}
\int_{V-V_\varepsilon} [s(\mathbf{r}') - s(\mathbf{r})] \frac{\partial^2}{\partial x_i \partial x_j} \frac{1}{4\pi |\mathbf{r}-\mathbf{r}'|} dV' \\
= \int_{V-V_\varepsilon} s(\mathbf{r}') \frac{\partial^2}{\partial x_i \partial x_j} \frac{1}{4\pi |\mathbf{r}-\mathbf{r}'|} dV' \\
- s(\mathbf{r}) \int_{V-V_\varepsilon} \frac{\partial^2}{\partial x_i \partial x_j} \frac{1}{4\pi |\mathbf{r}-\mathbf{r}'|} dV'
\end{aligned} \tag{1.67}$$

where the first term is recognized as $A_{ji}$. We therefore have the term

$$-s(\mathbf{r}) \oint_S \frac{\partial}{\partial x_j} \frac{1}{4\pi |\mathbf{r}-\mathbf{r}'|} \widehat{\mathbf{x}}_i \cdot \mathbf{n}\, dS' - s(\mathbf{r}) \int_{V-V_\varepsilon} \frac{\partial^2}{\partial x_i \partial x_j} \frac{1}{4\pi |\mathbf{r}-\mathbf{r}'|} dV' \tag{1.68}$$

remaining. To show this is equivalent to $C_{ji}$ in (1.63) we again use (1.48) to express the volume term as

$$\int_{V-V_\varepsilon} \frac{\partial^2}{\partial x_i \partial x_j} \frac{1}{4\pi |\mathbf{r}-\mathbf{r}'|} dV' = -\int_{V-V_\varepsilon} \frac{\partial^2}{\partial x_i' \partial x_j} \frac{1}{4\pi |\mathbf{r}-\mathbf{r}'|} dV'$$
$$= -\oint_{S+S_\varepsilon} \frac{\partial}{\partial x_j} \frac{1}{4\pi |\mathbf{r}-\mathbf{r}'|} \widehat{\mathbf{x}}_i \cdot \mathbf{n}\, dS'. \tag{1.69}$$

Substituting (1.69) into (1.68) leads to

$$s(\mathbf{r}) \oint_{S_\varepsilon} \frac{\partial}{\partial x_j} \frac{1}{4\pi |\mathbf{r}-\mathbf{r}'|} \widehat{\mathbf{x}}_i \cdot \mathbf{n}\, dS' = -s(\mathbf{r}) \oint_{S_\varepsilon} \frac{(\widehat{\mathbf{x}}_j \cdot \mathbf{n}) \left(\widehat{\mathbf{x}}_i \cdot \widehat{\mathbf{R}}\right)}{4\pi R^2} dS',$$

which is the term $C_{ji}$ in (1.63). A similar technique recovers the result (1.63) for the dynamic case.

To get an idea of why the limiting procedure is not strictly valid in (1.63), we can examine the steps in deriving (1.63) from (1.59). An important point is that the splitting of the integral into two integrals described in (1.67) is valid as long as $V_\varepsilon$ remains finite, since the source-point singularity is avoided. As $\lim_{\varepsilon\to 0}$ is taken, that decomposition is no longer strictly valid since individually the terms

$$\lim_{\varepsilon\to 0} \int_{V-V_\varepsilon} s(\mathbf{r}') \frac{\partial^2}{\partial x_i \partial x_j} \frac{1}{4\pi |\mathbf{r}-\mathbf{r}'|} dV',$$
$$s(\mathbf{r}) \lim_{\varepsilon\to 0} \int_{V-V_\varepsilon} \frac{\partial^2}{\partial x_i \partial x_j} \frac{1}{4\pi |\mathbf{r}-\mathbf{r}'|} dV'$$

do not exist. Therefore, it is clear that the volume $V_\varepsilon$ in (1.63) must remain finite. If one recognizes that the two terms on the right side of (1.64) are both finite yet shape-dependent (and therefore conditionally convergent), then the limiting procedure is acceptable within that framework.

Returning to the solutions (1.49) and (1.52), it is concluded that the volume integral in (1.49) needs to be regarded in the same limiting form as (1.52) to make sense. If this is then recognized, the formal manipulations leading to (1.49) indeed produce the correct answer. The proof that (1.52) satisfies the original Helmholtz equation is complicated by the necessity of careful consideration of the second derivative of the fundamental solution (1.53). In the remainder of the text we take the improper integral $\int_V f(\mathbf{r},\mathbf{r}')\, dV$ to mean $\lim_{\delta\to 0} \int_{V-V_\delta} f(\mathbf{r},\mathbf{r}')\, dV$ unless otherwise specified.

### 1.3.5 Solution of the Vector Wave and Vector Helmholtz Equations—Dyadic Green's Functions and Associated Improper Integrals

#### Dyadic Green's Functions and Solutions of the Vector Helmholtz Equations

As an example of solving (1.26), consider again the situation depicted in Figure 1.1, which shows a volume $V$ containing a medium characterized by constants $\varepsilon, \mu$ bounded by a smooth surface $S$. For convenience let $k^2 = \omega^2 \mu\varepsilon$, such that we can write (1.26) as

$$\begin{aligned} \nabla^2 \mathbf{E}(\mathbf{r}) + k^2 \mathbf{E}(\mathbf{r}) &= -\mathbf{i}_e, \\ \nabla^2 \mathbf{H}(\mathbf{r}) + k^2 \mathbf{H}(\mathbf{r}) &= -\mathbf{i}_m \end{aligned} \tag{1.70}$$

at any point $\mathbf{r} \in V$, where

$$\begin{aligned} \mathbf{i}_e &\equiv -i\omega\mu \left[\underline{\mathbf{I}} + \frac{\nabla\nabla}{k^2}\right] \cdot \mathbf{J}_e(\mathbf{r}) - \nabla \times \mathbf{J}_m(\mathbf{r}), \\ \mathbf{i}_m &\equiv -i\omega\varepsilon \left[\underline{\mathbf{I}} + \frac{\nabla\nabla}{k^2}\right] \cdot \mathbf{J}_m(\mathbf{r}) + \nabla \times \mathbf{J}_e(\mathbf{r}). \end{aligned}$$

We can use the result (1.52) for each rectangular component of (1.70) and add the components together to form the vector solution, but we'll consider the problem using another approach.

Consider the dyadic Green's function defined by

$$\nabla^2 \underline{\mathbf{G}}(\mathbf{r}, \mathbf{r}') + k^2 \underline{\mathbf{G}}(\mathbf{r}, \mathbf{r}') = -\underline{\mathbf{I}}\, \delta(\mathbf{r} - \mathbf{r}'), \tag{1.71}$$

where $\mathbf{r}, \mathbf{r}' \in V$. We can solve for the Green's dyadic using the method of Levine and Schwinger [22], [2, p. 96]. From (1.71), and using the fact that $\left(\nabla^2 + k^2\right) g(\mathbf{r}, \mathbf{r}') = -\delta(\mathbf{r} - \mathbf{r}')$, we have

$$\left(\nabla^2 + k^2\right) \underline{\mathbf{G}}(\mathbf{r}, \mathbf{r}') = \underline{\mathbf{I}} \left(\nabla^2 + k^2\right) g(\mathbf{r}, \mathbf{r}')$$

or

$$\left(\nabla^2 + k^2\right) \left(\underline{\mathbf{G}}(\mathbf{r}, \mathbf{r}') - \underline{\mathbf{I}}\, g(\mathbf{r}, \mathbf{r}')\right) = \underline{\mathbf{0}}.$$

A particular solution to the above is

$$\underline{\mathbf{G}}(\mathbf{r}, \mathbf{r}') = \underline{\mathbf{I}}\, g(\mathbf{r}, \mathbf{r}') = \underline{\mathbf{I}}\, \frac{e^{-ikR}}{4\pi R}, \tag{1.72}$$

which is a free-space dyadic Green's function.

Using the vector-dyadic Green's second theorem ($\mathbf{n}$ points into $V$ from $S$)

$$\begin{aligned} \int_V \left[(\nabla^2 \mathbf{A}) \cdot \underline{\mathbf{B}} - \mathbf{A} \cdot \nabla^2 \underline{\mathbf{B}}\right] dV = \oint_S \{&(\mathbf{n} \times \mathbf{A}) \cdot (\nabla \times \underline{\mathbf{B}}) + (\mathbf{n} \times \nabla \times \mathbf{A}) \cdot \underline{\mathbf{B}} \\ &+ [\mathbf{n} \cdot \mathbf{A}\ \nabla \cdot \underline{\mathbf{B}} - \mathbf{n} \cdot \underline{\mathbf{B}}\ \nabla \cdot \mathbf{A}]\}\, dS, \end{aligned}$$

with $\underline{\mathbf{B}} = \underline{\mathbf{G}}(\mathbf{r}, \mathbf{r}')$ and $\mathbf{A} = \mathbf{E}(\mathbf{r})$ or $\mathbf{H}(\mathbf{r})$, we have

$$
\begin{aligned}
\mathbf{E}(\mathbf{r}') = & \int_V \mathbf{i}_e(\mathbf{r}) \cdot \underline{\mathbf{G}}(\mathbf{r}, \mathbf{r}') \, dV + \oint_S (\mathbf{n} \times \mathbf{E}(\mathbf{r})) \cdot (\nabla \times \underline{\mathbf{G}}(\mathbf{r}, \mathbf{r}')) \, dS \\
& + \oint_S \{(\mathbf{n} \times \nabla \times \mathbf{E}(\mathbf{r})) \cdot \underline{\mathbf{G}}(\mathbf{r}, \mathbf{r}') \\
& + [\mathbf{n} \cdot \mathbf{E}(\mathbf{r}) \, \nabla \cdot \underline{\mathbf{G}}(\mathbf{r}, \mathbf{r}') - \mathbf{n} \cdot \underline{\mathbf{G}}(\mathbf{r}, \mathbf{r}') \, \nabla \cdot \mathbf{E}(\mathbf{r})]\} \, dS, \\
\mathbf{H}(\mathbf{r}') = & \int_V \mathbf{i}_m(\mathbf{r}) \cdot \underline{\mathbf{G}}(\mathbf{r}, \mathbf{r}') \, dV + \oint_S (\mathbf{n} \times \mathbf{H}(\mathbf{r})) \cdot (\nabla \times \underline{\mathbf{G}}(\mathbf{r}, \mathbf{r}')) \, dS \\
& + \oint_S \{(\mathbf{n} \times \nabla \times \mathbf{H}(\mathbf{r})) \cdot \underline{\mathbf{G}}(\mathbf{r}, \mathbf{r}') \\
& + [\mathbf{n} \cdot \mathbf{H}(\mathbf{r}) \, \nabla \cdot \underline{\mathbf{G}}(\mathbf{r}, \mathbf{r}') - \mathbf{n} \cdot \underline{\mathbf{G}}(\mathbf{r}, \mathbf{r}') \, \nabla \cdot \mathbf{H}(\mathbf{r})]\} \, dS.
\end{aligned}
\tag{1.73}
$$

It is easy to see that the integrals on the right side of (1.73) vanish identically for $\mathbf{r}$ external to $S$ (since the delta-function is never encountered during the volume integration). Of course the same procedure applies to the vector Helmholtz equations for the vector potentials $\mathbf{A}, \mathbf{F}$ in (1.37), yielding

$$
\begin{aligned}
\mathbf{A}(\mathbf{r}') = & \int_V \mu \mathbf{J}_e(\mathbf{r}) \cdot \underline{\mathbf{G}}(\mathbf{r}, \mathbf{r}') \, dV + \oint_S (\mathbf{n} \times \mathbf{A}(\mathbf{r})) \cdot (\nabla \times \underline{\mathbf{G}}(\mathbf{r}, \mathbf{r}')) \, dS \\
& + \oint_S \{(\mathbf{n} \times \nabla \times \mathbf{A}(\mathbf{r})) \cdot \underline{\mathbf{G}}(\mathbf{r}, \mathbf{r}') \\
& + [\mathbf{n} \cdot \mathbf{A}(\mathbf{r}) \, \nabla \cdot \underline{\mathbf{G}}(\mathbf{r}, \mathbf{r}') - \mathbf{n} \cdot \underline{\mathbf{G}}(\mathbf{r}, \mathbf{r}') \, \nabla \cdot \mathbf{A}(\mathbf{r})]\} \, dS, \qquad (1.74) \\
\mathbf{F}(\mathbf{r}') = & \int_V \mathbf{J}_m(\mathbf{r}) \cdot \underline{\mathbf{G}}(\mathbf{r}, \mathbf{r}') \, dV + \oint_S (\mathbf{n} \times \mathbf{F}(\mathbf{r})) \cdot (\nabla \times \underline{\mathbf{G}}(\mathbf{r}, \mathbf{r}')) \, dS \\
& + \oint_S \{(\mathbf{n} \times \nabla \times \mathbf{F}(\mathbf{r})) \cdot \underline{\mathbf{G}}(\mathbf{r}, \mathbf{r}') \\
& + [\mathbf{n} \cdot \mathbf{F}(\mathbf{r}) \, \nabla \cdot \underline{\mathbf{G}}(\mathbf{r}, \mathbf{r}') - \mathbf{n} \cdot \underline{\mathbf{G}}(\mathbf{r}, \mathbf{r}') \, \nabla \cdot \mathbf{F}(\mathbf{r})]\} \, dS,
\end{aligned}
$$

with fields $\mathbf{E}, \mathbf{H}$ recovered from the potentials using (1.38). From the previous discussion it is clear that for $\mathbf{r}' \in V$ the volume integrals are improper and we need to interpret them in the limiting sense. Because the singularity is the same as in the scalar case (1.53), the volume integrals all converge, first partial derivatives $\partial / \partial x_i'$ can be brought under the integral sign, and second partial derivatives exist assuming the vector source terms $(\mathbf{i}_{e(m)}, \mathbf{J}_{e(m)})$ obey a Hölder condition. Homogeneous solutions can also be added to (1.72) so that the total Green's dyadic satisfies specified boundary conditions.

As with the derivations in the scalar case, the formal manipulations leading to (1.73) and (1.74) can be criticized since the continuity requirements on the functions in the vector-dyadic Green's second theorem were violated. A limiting procedure similar to that used to derive (1.52) can be used to overcome this difficulty. Indeed, for the electric field, applying the Green's theorem to the volume $V - V_\delta$ bounded by the smooth surface $S + S_\delta$ as depicted in Figure 1.2 leads to

$$\begin{aligned}
\int_{V-V_\delta} & \left[(\nabla^{2}\mathbf{E}(\mathbf{r})) \cdot \underline{\mathbf{G}}(\mathbf{r},\mathbf{r}') - \mathbf{E}(\mathbf{r}) \cdot \nabla^{2}\underline{\mathbf{G}}(\mathbf{r},\mathbf{r}')\right] dV \\
&= \oint_{S+S_\delta} (\mathbf{n} \times \mathbf{E}(\mathbf{r})) \cdot (\nabla \times \underline{\mathbf{G}}(\mathbf{r},\mathbf{r}'))\, dS \\
&\quad + \oint_{S+S_\delta} (\mathbf{n} \times \nabla \times \mathbf{E}(\mathbf{r})) \cdot \underline{\mathbf{G}}(\mathbf{r},\mathbf{r}')\, dS \\
&\quad + \oint_{S+S_\delta} [\mathbf{n} \cdot \mathbf{E}(\mathbf{r})\ \nabla \cdot \underline{\mathbf{G}}(\mathbf{r},\mathbf{r}') - \mathbf{n} \cdot \underline{\mathbf{G}}(\mathbf{r},\mathbf{r}') \nabla \cdot \mathbf{E}(\mathbf{r})]\, dS.
\end{aligned} \tag{1.75}$$

Consider the integration over $S_\delta$ in the limit $\delta \to 0$. Because $\underline{\mathbf{G}}(\mathbf{r},\mathbf{r}') = \underline{\mathbf{I}}\ g(\mathbf{r},\mathbf{r}')$, we see that the second integral on the right side, and the second term in the third integral on the right side, will vanish as $\delta \to 0$, assuming $\mathbf{E}(\mathbf{r})$ is well behaved at $\mathbf{r} = \mathbf{r}'$. Using $\nabla \times \underline{\mathbf{I}}\, g(\mathbf{r},\mathbf{r}') = \underline{\mathbf{I}} \times \nabla g(\mathbf{r},\mathbf{r}')$ and $\nabla \cdot \underline{\mathbf{I}}\ g(\mathbf{r},\mathbf{r}') = \nabla g(\mathbf{r},\mathbf{r}')$, we obtain

$$\begin{aligned}
\lim_{\delta \to 0} \oint_{S_\delta} & \left[(\mathbf{n} \times \mathbf{E}(\mathbf{r})) \cdot (\nabla \times \underline{\mathbf{G}}(\mathbf{r},\mathbf{r}')) + \mathbf{n} \cdot \mathbf{E}(\mathbf{r})\ \nabla \cdot \underline{\mathbf{G}}(\mathbf{r},\mathbf{r}')\right] dS \\
&= \lim_{\delta \to 0} \oint_{S_\delta} \big[\mathbf{E}(\mathbf{r})(\mathbf{n} \cdot \nabla g(\mathbf{r},\mathbf{r}')) + (\mathbf{n} \cdot \mathbf{E}(\mathbf{r})) \nabla g(\mathbf{r},\mathbf{r}') \\
&\qquad\qquad - \mathbf{n}(\nabla g(\mathbf{r},\mathbf{r}') \cdot \mathbf{E}(\mathbf{r}))\big] dS.
\end{aligned}$$

For simplicity, assume the exclusion volume is spherical and use

$$\nabla g(\mathbf{r},\mathbf{r}') = -\widehat{\mathbf{R}}(\mathbf{r},\mathbf{r}') \frac{e^{-ik\delta}}{4\pi\delta} \left(\frac{1}{\delta} + ik\right)$$

and $\mathbf{n} = \widehat{\mathbf{R}}$, such that the last two terms cancel and the above term becomes

$$\begin{aligned}
\lim_{\delta \to 0} \oint_{S_\delta} (\cdot)\ dS &= -\mathbf{E}(\mathbf{r}') \lim_{\delta \to 0} \int_0^{2\pi} \int_0^{\pi} \frac{e^{-ik\delta}}{4\pi\delta} \left(\frac{1}{\delta} + ik\right) \delta^2 \sin\theta\, d\theta\, d\phi \\
&= -\mathbf{E}(\mathbf{r}').
\end{aligned}$$

We therefore obtain from (1.75), after substituting $\nabla^{2}\mathbf{E}(\mathbf{r}) = -\mathbf{i}_e - k^2\mathbf{E}(\mathbf{r})$, $\nabla^2\underline{\mathbf{G}}(\mathbf{r},\mathbf{r}') = -k^2\underline{\mathbf{G}}(\mathbf{r},\mathbf{r}')$,

$$\begin{aligned}
\mathbf{E}(\mathbf{r}') = \lim_{\delta \to 0} \int_{V-V_\delta} & \mathbf{i}_e(\mathbf{r}) \cdot \underline{\mathbf{G}}(\mathbf{r},\mathbf{r}')\ dV + \oint_S (\mathbf{n} \times \mathbf{E}(\mathbf{r}) \cdot (\nabla \times \underline{\mathbf{G}}(\mathbf{r},\mathbf{r}'))\ dS \\
&+ \oint_S \{(\mathbf{n} \times \nabla \times \mathbf{E}(\mathbf{r})) \cdot \underline{\mathbf{G}}(\mathbf{r},\mathbf{r}') \\
&+ [\mathbf{n} \cdot \mathbf{E}(\mathbf{r})\ \nabla \cdot \underline{\mathbf{G}}(\mathbf{r},\mathbf{r}') - \mathbf{n} \cdot \underline{\mathbf{G}}(\mathbf{r},\mathbf{r}')\ \nabla \cdot \mathbf{E}(\mathbf{r})]\}\, dS
\end{aligned} \tag{1.76}$$

in agreement with the formal method.[29] Similar manipulations, or duality, lead to the magnetic field equation and the equations for the potentials.

We now further consider the forms of the solutions (1.76) to the vector Helmholtz equations. The equations for $\mathbf{E}, \mathbf{H}$ are somewhat unsatisfactory due to the complicated form of the physical source densities $\mathbf{J}_{e(m)}$ appearing in $\mathbf{i}_{e(m)}$. Actually, if the source densities are given functions with sufficiently nice properties (in this case perhaps $\mathbf{J}_{e(m)}$ being sufficiently differentiable), then (1.76) and similar equations provide numerically stable formulations that are quite useful since the integrand singularity is fairly weak, being $O(1/R)$. On the other hand, these source densities may be numerically determined or approximated such that subsequent differentiation can introduce large errors. In this case we would like to move the derivative operators onto the known Green's function. Also, from an analytical standpoint it is very convenient to work with terms involving a Green's dyadic and an undifferentiated current density. Both situations lead us to look for an integral of the form $\lim_{\delta\to 0}\int_{V-V_\delta}\mathbf{J}_e(\mathbf{r})\cdot\underline{\mathbf{G}}_{ee}(\mathbf{r},\mathbf{r}')\,dV$.

### Free Space Electric Dyadic Green's Function and Solution of the Vector Helmholtz Equation

For convenience we will consider a current density having compact support residing in an infinite, homogeneous space characterized by $\mu, \varepsilon$. We can obtain this case if we let the surface $S$ recede to infinity, and both $\mathbf{E}$ and $\underline{\mathbf{G}}$ satisfy a radiation condition,

$$\lim_{r\to\infty} r\left(\nabla\times\mathbf{E}+ik\,\widehat{\mathbf{r}}\times\mathbf{E}\right)=\mathbf{0},$$
$$\lim_{r\to\infty} r\left(\nabla\times\underline{\mathbf{G}}+ik\,\widehat{\mathbf{r}}\times\underline{\mathbf{G}}\right)=\underline{\mathbf{0}},$$

in which case the surface integrals in (1.76) vanish. But rather than manipulate the resulting volume integrals in (1.73), it is more convenient to obtain the desired integral relation directly as discussed in the following.

A rigorous method to achieve the desired form is described in considerable detail (for the case of electric sources, the magnetic-source case follows

[29] As an aside, using various vector and dyadic identities one can show that (1.76) can be put in the classic Stratton–Chu form [41, pp. 464–470] (see also [48, pp. 17–21])

$$\begin{aligned}\mathbf{E}(\mathbf{r}') = &\int_V \left[-i\omega\mu\,\mathbf{J}_e(\mathbf{r})\,g(\mathbf{r},\mathbf{r}') + \frac{\rho_e(\mathbf{r})}{\varepsilon}\nabla g(\mathbf{r},\mathbf{r}') + \nabla g(\mathbf{r},\mathbf{r}')\times\mathbf{J}_m(\mathbf{r})\right] dV \\ &+ \oint_S \Big\{(\mathbf{n}\times\mathbf{E}(\mathbf{r}))\times\nabla g(\mathbf{r},\mathbf{r}') - i\omega\mu\,(\mathbf{n}\times\mathbf{H}(\mathbf{r}))\,g(\mathbf{r},\mathbf{r}') \\ &+ (\mathbf{n}\cdot\mathbf{E}(\mathbf{r}))\,\nabla g(\mathbf{r},\mathbf{r}')\Big\}\,dS.\end{aligned} \tag{1.77}$$

easily) in [14],[30] yielding

$$\begin{aligned}\mathbf{E}(\mathbf{r}') = &- i\omega\mu \lim_{\delta\to 0}\int_{V-V_\delta} \mathbf{J}_e(\mathbf{r})\cdot\underline{\mathbf{G}}_{ee}(\mathbf{r},\mathbf{r}')\,dV - \frac{\underline{\mathbf{L}}\,(\mathbf{r}')\cdot\mathbf{J}_e(\mathbf{r}')}{i\omega\varepsilon}\\ &+ \lim_{\delta\to 0}\int_{V-V_\delta} \mathbf{J}_m(\mathbf{r})\cdot\underline{\mathbf{G}}_{em}(\mathbf{r},\mathbf{r}')\,dV,\end{aligned} \tag{1.78}$$

where

$$\begin{aligned}\underline{\mathbf{G}}_{ee}(\mathbf{r},\mathbf{r}') &= \left[\underline{\mathbf{I}} + \frac{\nabla\nabla}{k^2}\right] g(\mathbf{r},\mathbf{r}'),\\ \underline{\mathbf{G}}_{em}(\mathbf{r},\mathbf{r}') &= -\nabla g(\mathbf{r},\mathbf{r}')\times\underline{\mathbf{I}},\\ \underline{\mathbf{L}}\,(\mathbf{r}') &= -\oint_{S_\delta} \mathbf{n}\,\nabla g(\mathbf{r},\mathbf{r}')\,dS = \frac{1}{4\pi}\oint_{S_\delta}\frac{\mathbf{n}\,\widehat{\mathbf{R}}(\mathbf{r},\mathbf{r}')}{R^2}\,dS,\end{aligned} \tag{1.79}$$

and both $\mathbf{n}$, $\widehat{\mathbf{R}}$ point outward from $S_\delta$ into $V - V_\delta$. In general, $\underline{\mathbf{G}}_{\alpha\beta}$ provides the field $\alpha$ caused by the source $\beta$.

In connection with the scalar problem considered previously, from (1.64) and (1.66) we see that

$$L_{j\,i}(\mathbf{r}') = \widehat{\mathbf{x}}_j\cdot\underline{\mathbf{L}}\,(\mathbf{r}')\cdot\widehat{\mathbf{x}}_i, \tag{1.80}$$

$$\frac{\partial^2}{\partial x_j\,\partial x_i} g(\mathbf{r},\mathbf{r}') = \widehat{\mathbf{x}}_j\cdot\nabla\nabla\, g(\mathbf{r},\mathbf{r}')\cdot\widehat{\mathbf{x}}_i, \tag{1.81}$$

so that operationally

$$\begin{aligned}&\frac{\partial^2}{\partial x_j\,\partial x_i}\int_V s(\mathbf{r}')\frac{e^{-i\,k\,|\mathbf{r}-\mathbf{r}'|}}{4\pi\,|\mathbf{r}-\mathbf{r}'|}\,dV'\\ &\quad = \lim_{\delta\to 0}\int_{V-V_\delta} s(\mathbf{r}')\,\widehat{\mathbf{x}}_j\cdot\nabla\nabla\, g(\mathbf{r},\mathbf{r}')\cdot\widehat{\mathbf{x}}_i\,dV'\\ &\quad\quad - s(\mathbf{r})\,\widehat{\mathbf{x}}_j\cdot\underline{\mathbf{L}}\,(\mathbf{r}')\cdot\widehat{\mathbf{x}}_i.\end{aligned} \tag{1.82}$$

The dyadic $\underline{\mathbf{G}}_{ee}$ is the free-space electric dyadic Green's function, and $\underline{\mathbf{L}}$ is known as the *depolarizing dyadic* [14]. The depolarizing dyadic arises from careful consideration of the strong source-point singularity in $\underline{\mathbf{G}}_{ee}$. The limit $\lim_{\delta\to 0}$ can be omitted in the integral for $\underline{\mathbf{L}}$ since the integral is independent of the size of $V_\delta$ and only dependent on its shape. Furthermore, it is easy to see that the singularity associated with $\underline{\mathbf{G}}_{em}$ is $O(1/R^2)$,

[30] In this important paper it is the vector Helmholtz equations

$$\nabla^2\left[\mathbf{E}(\mathbf{r}) + \frac{\mathbf{J}_e^i(\mathbf{r})}{i\omega\varepsilon}\right] + k^2\left[\mathbf{E}(\mathbf{r}) + \frac{\mathbf{J}_e^i(\mathbf{r})}{i\omega\varepsilon}\right] = -\frac{\nabla\times\nabla\times\mathbf{J}_e^i(\mathbf{r})}{i\omega\varepsilon},$$
$$\nabla^2\mathbf{H}(\mathbf{r}) + k^2\mathbf{H}(\mathbf{r}) = -\nabla\times\mathbf{J}_e^i(\mathbf{r})$$

that are actually solved.

such that the associated volume integrals converge independently of the shape of the exclusion volume. It is otherwise with the term $\underline{\mathbf{G}}_{ee}$, which is $O(1/R^3)$.[31]

As explained in [14], the volume integral associated with $\underline{\mathbf{G}}_{ee}$ is dependent on the shape of $V_\delta$ in just the right way to cancel the shape dependence of the depolarizing dyadic term, resulting in a unique value for $\mathbf{E}$ independent of the shape of the exclusion volume. This is the vector/dyadic version of the same phenomena encountered in the scalar case (1.64). A discussion of the depolarizing dyadic in the time domain may be found in [6].

Using duality we obtain

$$\begin{aligned}\mathbf{H}(\mathbf{r}') = &- i\omega\varepsilon \lim_{\delta\to 0}\int_{V-V_\delta} \mathbf{J}_m(\mathbf{r})\cdot\underline{\mathbf{G}}_{mm}(\mathbf{r},\mathbf{r}')\,dV - \frac{\underline{\mathbf{L}}\cdot\mathbf{J}_m(\mathbf{r}')}{i\omega\mu} \\ &+ \lim_{\delta\to 0}\int_{V-V_\delta} \mathbf{J}_e(\mathbf{r})\cdot\underline{\mathbf{G}}_{me}(\mathbf{r},\mathbf{r}')\,dV,\end{aligned} \tag{1.83}$$

where

$$\begin{aligned}\underline{\mathbf{G}}_{mm}(\mathbf{r},\mathbf{r}') &= \left[\underline{\mathbf{I}} + \frac{\nabla\nabla}{k^2}\right] g(\mathbf{r},\mathbf{r}') = \underline{\mathbf{G}}_{ee}(\mathbf{r},\mathbf{r}'), \\ \underline{\mathbf{G}}_{me}(\mathbf{r},\mathbf{r}') &= \nabla g(\mathbf{r},\mathbf{r}')\times\underline{\mathbf{I}} = -\underline{\mathbf{G}}_{em}(\mathbf{r},\mathbf{r}').\end{aligned} \tag{1.84}$$

Using the symmetry property of the free-space Green's dyadics,[32]

$$\begin{aligned}\left[\underline{\mathbf{G}}_{ee}(\mathbf{r},\mathbf{r}')\right]^\top &= \underline{\mathbf{G}}_{ee}(\mathbf{r}',\mathbf{r}), \\ \left[\underline{\mathbf{G}}_{em}(\mathbf{r},\mathbf{r}')\right]^\top &= \underline{\mathbf{G}}_{em}(\mathbf{r}',\mathbf{r}) = -\underline{\mathbf{G}}_{em}(\mathbf{r},\mathbf{r}'),\end{aligned} \tag{1.85}$$

the expressions (1.78) and (1.83) take on the conventional form

$$\begin{aligned}\mathbf{E}(\mathbf{r}) = &- i\omega\mu \lim_{\delta\to 0}\int_{V-V_\delta} \underline{\mathbf{G}}_{ee}(\mathbf{r},\mathbf{r}')\cdot\mathbf{J}_e(\mathbf{r}')\,dV' - \frac{\underline{\mathbf{L}}\cdot\mathbf{J}_e(\mathbf{r})}{i\omega\varepsilon} \\ &+ \lim_{\delta\to 0}\int_{V-V_\delta} \underline{\mathbf{G}}_{em}(\mathbf{r},\mathbf{r}')\cdot\mathbf{J}_m(\mathbf{r}')\,dV', \\ \mathbf{H}(\mathbf{r}) = &- i\omega\varepsilon \lim_{\delta\to 0}\int_{V-V_\delta} \underline{\mathbf{G}}_{mm}(\mathbf{r},\mathbf{r}')\cdot\mathbf{J}_m(\mathbf{r}')\,dV' - \frac{\underline{\mathbf{L}}\cdot\mathbf{J}_m(\mathbf{r})}{i\omega\mu} \\ &+ \lim_{\delta\to 0}\int_{V-V_\delta} \underline{\mathbf{G}}_{me}(\mathbf{r},\mathbf{r}')\cdot\mathbf{J}_e(\mathbf{r}')\,dV'.\end{aligned} \tag{1.86}$$

---

[31] Note that the operations involved in the definition of $\underline{\mathbf{G}}_{ee}(\mathbf{r},\mathbf{r}')$ are readily carried out, leading to

$$\underline{\mathbf{G}}_{ee}(\mathbf{r},\mathbf{r}') = g(\mathbf{r},\mathbf{r}')\left\{\left(3\widehat{\mathbf{R}}\widehat{\mathbf{R}} - \underline{\mathbf{I}}\right)\left(\frac{1}{k^2R^2} - \frac{1}{ikR}\right) - \left(\widehat{\mathbf{R}}\widehat{\mathbf{R}} - \underline{\mathbf{I}}\right)\right\},$$

where $\widehat{\mathbf{R}} = \widehat{\mathbf{R}}(\mathbf{r},\mathbf{r}')$. One can see that the source-point singularity is much stronger, $O\left(1/R^3\right)$, for the electric dyadic Green's function than for the Green's function $\underline{\mathbf{I}}g(\mathbf{r},\mathbf{r}')$, which has an $O\left(1/R\right)$ singularity.

[32] These are easily shown using the explicit forms for $\underline{\mathbf{G}}_{ee}$ and $\underline{\mathbf{G}}_{em}$. A more general method for determining symmetry properties of other (than free-space) dyadic Green's functions is shown in [3].

The form of the symmetric dyadic $\underline{\mathbf{L}}$ for various exclusion volumes is presented in [14]. For a sphere, $\underline{\mathbf{L}} = \underline{\mathbf{I}}\,/\,3$ is independent of the position of the origin within the sphere. For a cube with origin at the center of the cube, $\underline{\mathbf{L}} = \underline{\mathbf{I}}\,/\,3$ as well. For a pillbox of arbitrary cross-section $\underline{\mathbf{L}} = \widehat{\mathbf{x}}_i\,\widehat{\mathbf{x}}_i$, where $\widehat{\mathbf{x}}_i$ is a unit vector in the direction of the axis of the pillbox. The same dyadic is found for the "slice" exclusion volume [17], which is the natural form of the pillbox for laterally infinite layered-media geometries. Note the similarity with the values for the scalar term $L_{j\,i}$ (1.66).

For theoretical developments, it is often useful to absorb the depolarizing dyadic term into the volume integral. For instance, we may write

$$
\begin{aligned}
\underline{\mathbf{G}}^{\wedge}_{ee}(\mathbf{r},\mathbf{r}') &\equiv \text{P.V.}\left\{-i\omega\mu\underline{\mathbf{G}}_{ee}(\mathbf{r},\mathbf{r}')\right\} - \frac{\underline{\mathbf{L}}\delta\left(\mathbf{r}-\mathbf{r}'\right)}{i\omega\varepsilon} \\
\underline{\mathbf{G}}^{\wedge}_{mm}(\mathbf{r},\mathbf{r}') &\equiv \text{P.V.}\left\{-i\omega\varepsilon\underline{\mathbf{G}}_{mm}(\mathbf{r},\mathbf{r}')\right\} - \frac{\underline{\mathbf{L}}\delta\left(\mathbf{r}-\mathbf{r}'\right)}{i\omega\mu}
\end{aligned}
\tag{1.87}
$$

where P.V. indicates that the associated term is to be integrated in the principal value sense. This leads to the simpler form

$$
\begin{aligned}
\mathbf{E}(\mathbf{r}) &= \int_V \underline{\mathbf{G}}^{\wedge}_{ee}(\mathbf{r},\mathbf{r}')\cdot\mathbf{J}_e(\mathbf{r}')\,dV' + \int_V \underline{\mathbf{G}}_{em}(\mathbf{r},\mathbf{r}')\cdot\mathbf{J}_m(\mathbf{r}')\,dV', \\
\mathbf{H}(\mathbf{r}) &= \int_V \underline{\mathbf{G}}^{\wedge}_{mm}(\mathbf{r},\mathbf{r}')\cdot\mathbf{J}_m(\mathbf{r}')\,dV' + \int_V \underline{\mathbf{G}}_{me}(\mathbf{r},\mathbf{r}')\cdot\mathbf{J}_e(\mathbf{r}')\,dV'.
\end{aligned}
\tag{1.88}
$$

### Origin of the Depolarizing Dyadic

Aside from the method of [14], the origin of the depolarizing dyadic is also clear when one obtains the fields from the potentials. Consider the case of electric sources. We will use (1.59) to evaluate the electric field from the magnetic vector potential.[33] Starting from

$$
\begin{aligned}
\mathbf{E}(\mathbf{r}) &= -i\omega\,\mathbf{A}(\mathbf{r}) + \frac{1}{i\omega\varepsilon\mu}\nabla\nabla\cdot\mathbf{A}(\mathbf{r}) \\
&= \left(-i\omega + \frac{1}{i\omega\varepsilon\mu}\nabla\nabla\cdot\right)\lim_{\delta\to 0}\int_{V-V_\delta}\mu\, g(\mathbf{r},\mathbf{r}')\,\mathbf{J}_e(\mathbf{r}')\,dV'
\end{aligned}
\tag{1.89}
$$

and using (1.59), we obtain

$$
\begin{aligned}
\mathbf{E}(\mathbf{r}) = &-i\omega\lim_{\delta\to 0}\int_{V-V_\delta}\mu\, g(\mathbf{r},\mathbf{r}')\,\mathbf{J}_e(\mathbf{r}')\,dV' \\
&+ \frac{1}{i\omega\varepsilon}\sum_{i=1}^{3}\widehat{\mathbf{x}}_i\lim_{\delta\to 0}\int_{V-V_\delta}\sum_{j=1}^{3}\left[J_{j\,e}(\mathbf{r}') - J_{j\,e}(\mathbf{r})\right]\frac{\partial^2}{\partial x_i\,\partial x_j}g(\mathbf{r},\mathbf{r}')\,dV' \\
&- \frac{1}{i\omega\varepsilon}\sum_{i=1}^{3}\widehat{\mathbf{x}}_i\sum_{j=1}^{3}J_{j\,e}(\mathbf{r})\oint_S\frac{\partial}{\partial x_i}g(\mathbf{r},\mathbf{r}')\,\widehat{\mathbf{x}}_j\cdot\mathbf{n}\,dS',
\end{aligned}
$$

[33] See also [15], [16] for the case where $\mathbf{E}$ is obtained from $\mathbf{A}$ via the magnetic field.

which can be shown to be equal to

$$\begin{aligned}\mathbf{E}(\mathbf{r}) = &- i\omega\mu \lim_{\delta\to 0}\int_{V-V_\delta} g(\mathbf{r},\mathbf{r}')\,\mathbf{J}_e(\mathbf{r}')\,dV' \\ &+ \frac{1}{i\omega\varepsilon}\lim_{\delta\to 0}\int_{V-V_\delta} \nabla\nabla\, g(\mathbf{r},\mathbf{r}')\cdot[\mathbf{J}_e(\mathbf{r}')-\mathbf{J}_e(\mathbf{r})]\,dV' \\ &- \frac{1}{i\omega\varepsilon}\mathbf{J}_e(\mathbf{r})\cdot\oint_S \mathbf{n}\,\nabla g(\mathbf{r},\mathbf{r}')\,dS'.\end{aligned} \tag{1.90}$$

Recall that $S$ is the boundary of $V$, not the boundary of $V_\delta$. All terms in the last expression are convergent, and so the limiting operation may be suppressed according to convention. Of course, this means that there is no depolarizing dyadic term associated with (1.90). To obtain (1.86) from (1.90), we need to decompose the middle term in (1.90) into two terms as

$$\begin{aligned}&\lim_{\delta\to 0}\int_{V-V_\delta} \nabla\nabla\, g(\mathbf{r},\mathbf{r}')\cdot[\mathbf{J}_e(\mathbf{r}')-\mathbf{J}_e(\mathbf{r})]\,dV' \\ &\qquad = \lim_{\delta\to 0}\int_{V-V_\delta} \nabla\nabla\, g(\mathbf{r},\mathbf{r}')\cdot\mathbf{J}_e(\mathbf{r}')\,dV' \\ &\qquad\quad - \lim_{\delta\to 0}\int_{V-V_\delta} \nabla\nabla\, g(\mathbf{r},\mathbf{r}')\,dV' \cdot\mathbf{J}_e(\mathbf{r}).\end{aligned} \tag{1.91}$$

As before in obtaining (1.67), the individual integrals on the right side of (1.91) are not convergent, and so the decomposition is not strictly allowed in the sense of obtaining two "independent" convergent integrals. Taken together, though, the two terms lead to a shape-independent, finite result, just as in the scalar case. Proceeding in this manner, the second term may be rigorously converted into a surface integral since the point $\mathbf{r}=\mathbf{r}'$ is excluded from the domain of integration. Substituting $\nabla\nabla g(\mathbf{r},\mathbf{r}') = -\nabla'\,\nabla\, g(\mathbf{r},\mathbf{r}')$ and using the dyadic gradient theorem (see Appendix A.4), we get a surface term that cancels with the surface integral in (1.90), and a surface integral over the exclusion surface, which provides the depolarizing dyadic contribution.

It can be seen then that one view of the origin of the shape dependence of the volume integral in (1.86) is the improper decomposition of the (convergent) second term on the right side of (1.91) into two finite, shape-dependent terms.

It is interesting to compare the process of rigorously bringing the operators $(-i\omega + (1/i\omega\varepsilon\mu)\nabla\,\nabla\cdot)$ and $\left(\nabla^2+k^2\right)$ through the integral operator in

$$\lim_{\delta\to 0}\int_{V-V_\delta} \mu\, g(\mathbf{r},\mathbf{r}')\mathbf{J}_e(\mathbf{r}')\,dV'.$$

In the latter case the various second derivatives for the Laplacian combine to "naturally" decompose (since one term in the integrand vanishes) the volume integral on the right side of (1.60). Thus, in this case the step analogous to the decomposition in (1.91) presents no difficulties. The remaining volume integral produces[34] $\mathbf{J}(\mathbf{r})\int_{S+S_\delta}\mathbf{n}\cdot\nabla g(\mathbf{r},\mathbf{r}')\,dS'$ such that the integral over $S$ cancels the surface integral in (1.60) and the integral over $S_\delta$ provides the quantity $\mathbf{J}(\mathbf{r})$ because of the resulting solid-angle integration.

The important differences in obtaining (1.90) are that

- we obtain a surface term $\mathbf{J}(\mathbf{r})\cdot\oint_{S+S_\delta}\mathbf{n}\,\nabla g(\mathbf{r},\mathbf{r}')\,dS'$ (without a dot product between $\mathbf{n}$ and $\nabla g$), and
- the decomposition of the term involving $[\mathbf{J}_e(\mathbf{r}')-\mathbf{J}_e(\mathbf{r})]$ does not occur naturally and is in fact somewhat problematic, leading to two shape-dependent terms (in the scalar case for the operator $\partial^2/\partial x_j\,\partial x_i$ as well).

Even though the integral over $S$ cancels with the surface integral in (1.90), the integral over $S_\delta$ results in a shape-dependent result, $\mathbf{J}(\mathbf{r})\cdot\underline{\mathbf{L}}$. This is due to the way in which the partial derivatives associated with $(\nabla\,\nabla\cdot)$ combine, as opposed to those of the Laplacian operator $\left(\nabla^2\right)$.

### Solution of the Vector Wave Equation in Terms of the Electric Dyadic Green's Function

In analogy with the scalar case, we can derive the solution to the vector wave equation (1.24) from a vector-dyadic Green's theorem. Assume the geometry depicted in Figure 1.1. The medium is isotropic and homogeneous, so the vector wave equation for the electric field, considering electric sources, is

$$\nabla\times\nabla\times\mathbf{E}(\mathbf{r})-k^2\mathbf{E}(\mathbf{r})=-i\omega\mu\mathbf{J}_e(\mathbf{r}). \tag{1.92}$$

Defining an electric dyadic Green's function as

$$\nabla\times\nabla\times\underline{\mathbf{G}}_{ee}(\mathbf{r},\mathbf{r}')-k^2\underline{\mathbf{G}}_{ee}(\mathbf{r},\mathbf{r}')=\underline{\mathbf{I}}\,\delta(\mathbf{r}-\mathbf{r}') \tag{1.93}$$

and using

$$\begin{aligned}&\int_V\left[(\nabla\times\nabla\times\mathbf{A})\cdot\underline{\mathbf{B}}-\mathbf{A}\cdot(\nabla\times\nabla\times\underline{\mathbf{B}})\right]dV\\&\quad=-\oint_S\left((\mathbf{n}\times\mathbf{A})\cdot(\nabla\times\underline{\mathbf{B}})+(\mathbf{n}\times\nabla\times\mathbf{A})\cdot\underline{\mathbf{B}}\right)dS,\end{aligned} \tag{1.94}$$

with $\mathbf{A}=\mathbf{E}$ and $\underline{\mathbf{B}}=\underline{\mathbf{G}}_{ee}(\mathbf{r},\mathbf{r}')$, we have

$$\begin{aligned}\mathbf{E}(\mathbf{r}')&=-i\omega\mu\int_V\mathbf{J}_e(\mathbf{r})\cdot\underline{\mathbf{G}}_{ee}(\mathbf{r},\mathbf{r}')\,dV\\&\quad+\oint_S\left[(\mathbf{n}\times\mathbf{E}(\mathbf{r}))\cdot\nabla\times\underline{\mathbf{G}}_{ee}(\mathbf{r},\mathbf{r}')+(\mathbf{n}\times\nabla\times\mathbf{E}(\mathbf{r}))\cdot\underline{\mathbf{G}}_{ee}(\mathbf{r},\mathbf{r}')\right]dS.\end{aligned} \tag{1.95}$$

[34] In (1.60) we considered the scalar source $s(\mathbf{r})$ rather than $\mathbf{J}(\mathbf{r})$; because $\nabla^2\mathbf{J}=\widehat{\mathbf{x}}\,\nabla^2J_x+\widehat{\mathbf{y}}\,\nabla^2J_y+\widehat{\mathbf{z}}\,\nabla^2J_z$, we can combine the scalar components to form $\mathbf{J}$.

Equation (1.95) is not quite correct in the sense that the improper volume integral must be carefully defined. Unlike in the scalar case (1.49) and the vector case (1.73) and (1.74), where the volume integrals are unambiguously defined as convergent improper integrals, the improper volume integral in (1.95) is not convergent (without placing severe, nonphysical restrictions on $\mathbf{J}_e$). This is due to the fact that we have terms like $\int_V (\mathbf{J}_e(\mathbf{r}) \,/\, R^3)\, dV$, which do not exist in the classical sense. Although the correct expression as in (1.78) may be obtained by writing the volume integral as $\int_V (\cdot)\, dV = \lim_{\delta\to 0} \int_{V-V_\delta} (\cdot)\, dV + \lim_{\delta\to 0} \int_{V_\delta} (\cdot)\, dV$ and performing some simple manipulations, the decomposition of the integral into two terms is not valid since $\lim_{\delta\to 0} \int_{V_\delta} (\cdot)\, dV \neq 0$. It is therefore seen that, unlike in the scalar and vector cases, in the dyadic case the formal Green's theorem procedure *does not* yield the correct result since the volume integral does not exist, at least in the classical sense.

If we rigorously apply a limiting procedure as was done for the scalar and vector cases, we obtain

$$
\begin{aligned}
-i\omega\mu \lim_{\delta\to 0} \int_{V-V_\delta} & \mathbf{J}_e(\mathbf{r}) \cdot \underline{\mathbf{G}}_{ee}(\mathbf{r},\mathbf{r}')\, dV \qquad (1.96) \\
= - \lim_{\delta\to 0} \oint_{S+S_\delta} & \left[ (\mathbf{n} \times \mathbf{E}(\mathbf{r})) \cdot \nabla \times \underline{\mathbf{G}}_{ee}(\mathbf{r},\mathbf{r}') \right. \\
& \left. + (\mathbf{n} \times \nabla \times \mathbf{E}(\mathbf{r})) \cdot \underline{\mathbf{G}}_{ee}(\mathbf{r},\mathbf{r}') \right] dS \\
= - \lim_{\delta\to 0} \oint_{S+S_\delta} & \mathbf{n} \cdot \left[ \mathbf{E}(\mathbf{r}) \times \nabla \times \underline{\mathbf{G}}_{ee}(\mathbf{r},\mathbf{r}') + (\nabla \times \mathbf{E}(\mathbf{r})) \times \underline{\mathbf{G}}_{ee}(\mathbf{r},\mathbf{r}') \right] dS.
\end{aligned}
$$

By manipulations detailed in [42], the integral over the exclusion surface becomes

$$
\begin{aligned}
\lim_{\delta\to 0} \oint_{S_\delta} & \mathbf{n} \cdot \left[ \mathbf{E}(\mathbf{r}) \times \nabla \times \underline{\mathbf{G}}_{ee}(\mathbf{r},\mathbf{r}') + (\nabla \times \mathbf{E}(\mathbf{r})) \times \underline{\mathbf{G}}_{ee}(\mathbf{r},\mathbf{r}') \right] dS \\
& = - \left[ \frac{1}{k^2} \nabla' \times \nabla' \times \mathbf{E}(\mathbf{r}') - \mathbf{E}(\mathbf{r}') \right] \cdot \underline{\mathbf{L}}\,(\mathbf{r}') - \mathbf{E}(\mathbf{r}'),
\end{aligned}
$$

where $\underline{\mathbf{L}}$ is defined in (1.79). We therefore get

$$
\begin{aligned}
\mathbf{E}(\mathbf{r}') = & - i\omega\mu \lim_{\delta\to 0} \int_{V-V_\delta} \mathbf{J}_e(\mathbf{r}) \cdot \underline{\mathbf{G}}_{ee}(\mathbf{r},\mathbf{r}')\, dV \\
& - \left[ \frac{1}{k^2} \nabla' \times \nabla' \times \mathbf{E}(\mathbf{r}') - \mathbf{E}(\mathbf{r}') \right] \cdot \underline{\mathbf{L}}\,(\mathbf{r}') \qquad (1.97) \\
& + \oint_S \left[ (\mathbf{n} \times \mathbf{E}(\mathbf{r})) \cdot \nabla \times \underline{\mathbf{G}}_{ee}(\mathbf{r},\mathbf{r}') + (\mathbf{n} \times \nabla \times \mathbf{E}(\mathbf{r})) \cdot \underline{\mathbf{G}}_{ee}(\mathbf{r},\mathbf{r}') \right] dS,
\end{aligned}
$$

which is not an explicit solution for the electric field. If, however, we again note (1.92), we obtain

$$\mathbf{E}(\mathbf{r}') = -\, i\omega\mu \lim_{\delta\to 0} \int_{V-V_\delta} \mathbf{J}_e(\mathbf{r}) \cdot \underline{\mathbf{G}}_{ee}(\mathbf{r},\mathbf{r}')\, dV - \frac{\underline{\mathbf{L}}\,(\mathbf{r}') \cdot \mathbf{J}_e(\mathbf{r}')}{i\omega\varepsilon} \qquad (1.98)$$
$$+ \oint_S \left[ (\mathbf{n} \times \mathbf{E}(\mathbf{r})) \cdot \nabla \times \underline{\mathbf{G}}_{ee}(\mathbf{r},\mathbf{r}') + (\mathbf{n} \times \nabla \times \mathbf{E}(\mathbf{r})) \cdot \underline{\mathbf{G}}_{ee}(\mathbf{r},\mathbf{r}') \right] dS$$

for $\mathbf{r}' \in V$, which is the desired result.

The Green's function $\underline{\mathbf{G}}_{ee}$ is any solution of (1.93), where (1.98) can be written using the form (1.87) as

$$\mathbf{E}(\mathbf{r}) = \lim_{\delta\to 0} \int_{V-V_\delta} \underline{\mathbf{G}}_{ee}^{\wedge}(\mathbf{r},\mathbf{r}') \cdot \mathbf{J}_e(\mathbf{r})\, dV' \qquad (1.99)$$
$$+ \oint_S \left[ (\mathbf{n} \times \mathbf{E}(\mathbf{r})) \cdot \nabla \times \underline{\mathbf{G}}_{ee}^{\wedge}(\mathbf{r},\mathbf{r}') + (\mathbf{n} \times \nabla \times \mathbf{E}(\mathbf{r})) \cdot \underline{\mathbf{G}}_{ee}^{\wedge}(\mathbf{r},\mathbf{r}') \right] dS.$$

If $\underline{\mathbf{G}}_{ee}^{\wedge}$ satisfies some boundary conditions on $S$, the surface term in (1.99) may be further simplified; in free space the solution reduces to (1.88).

The preceding derivation is presented because of its simplicity. A method is presented in [23] to obtain (1.98) from (1.96) directly,[35] without reinvoking (1.92) in (1.97), as was necessary above. That direct procedure should be viewed as the correct method, although the details will be omitted here.

## 1.4 Integral Equations

Equations of the form (1.88), along with appropriate boundary conditions, lead to a variety of integral equations commonly used in electromagnetics. In this section we formulate several integral equations that are subsequently used in Part II to analyze scattering and waveguiding problems. For a general discussion of integral equations arising in electromagnetics, see [24, Ch. 8].

### 1.4.1 Domain Integral Equations

We first consider *domain integral equations*, which are particularly appropriate for solving problems involving inhomogeneous dielectric regions.

#### Volume Equivalence Principle

Consider the scattering problem depicted in Figure 1.3. Electromagnetic sources $\mathbf{J}^i_{e(m)}$ (shown located in the region $\mathbf{r} \notin V$, but which may, in fact, be within $V$) create a field $\mathbf{E}^i, \mathbf{H}^i$ that interacts with an inhomogeneous,

[35] The surface term is not considered in the analysis [23] and is incidental to the question of the origin of the depolarizing dyadic.

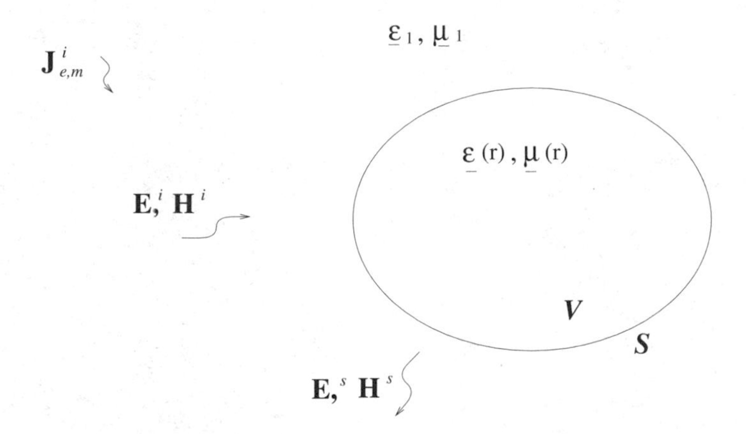


Figure 1.3: Formulation of the volume equivalence principle: scattering of an electromagnetic wave $\mathbf{E}^i, \mathbf{H}^i$ by an inhomogeneous anisotropic region $V$.

anisotropic region $V$, producing scattered fields $\mathbf{E}^s, \mathbf{H}^s$. For simplicity, $\underline{\mu}_1$ and $\underline{\varepsilon}_1$ are assumed to be constants, although $V$ may be multiply-connected, $V = \cup_{n=1}^{N} V_n$, implying $N$ scatterers. From (1.12), Maxwell's curl equations are given as

$$\begin{aligned} \nabla \times \mathbf{E}(\mathbf{r}) &= -i\omega \underline{\mu}(\mathbf{r}) \cdot \mathbf{H}(\mathbf{r}) - \mathbf{J}_m^i(\mathbf{r}), \\ \nabla \times \mathbf{H}(\mathbf{r}) &= \ i\omega \underline{\varepsilon}(\mathbf{r}) \cdot \mathbf{E}(\mathbf{r}) + \mathbf{J}_e^i(\mathbf{r}), \end{aligned}$$

valid at any point in space. By adding and subtracting the term $-i\omega \underline{\mu}_1 \cdot \mathbf{H}(\mathbf{r})$ on the right side of the first equation, and $i\omega \underline{\varepsilon}_1 \cdot \mathbf{E}(\mathbf{r})$ on the right side of the second equation, we obtain

$$\begin{aligned} \nabla \times \mathbf{E}(\mathbf{r}) &= -i\omega \left\{ \left[ \underline{\mu}(\mathbf{r}) - \underline{\mu}_1 \right] \cdot \mathbf{H}(\mathbf{r}) + \underline{\mu}_1 \cdot \mathbf{H}(\mathbf{r}) \right\} - \mathbf{J}_m^i(\mathbf{r}), \\ \nabla \times \mathbf{H}(\mathbf{r}) &= \ i\omega \left\{ \left[ \underline{\varepsilon}(\mathbf{r}) - \underline{\varepsilon}_1 \right] \cdot \mathbf{E}(\mathbf{r}) + \underline{\varepsilon}_1 \cdot \mathbf{E}(\mathbf{r}) \right\} + \mathbf{J}_e^i(\mathbf{r}) \end{aligned} \tag{1.100}$$

for all $\mathbf{r}$. We define the scattered fields as the difference between the total fields $\mathbf{E}, \mathbf{H}$ and the incident fields, $\mathbf{E}^i, \mathbf{H}^i$,

$$\begin{aligned} \mathbf{E}^s(\mathbf{r}) &\equiv \mathbf{E}(\mathbf{r}) - \mathbf{E}^i(\mathbf{r}), \\ \mathbf{H}^s(\mathbf{r}) &\equiv \mathbf{H}(\mathbf{r}) - \mathbf{H}^i(\mathbf{r}), \end{aligned} \tag{1.101}$$

and note that the incident fields satisfy

$$\begin{aligned} \nabla \times \mathbf{E}^i(\mathbf{r}) &= -i\omega \underline{\mu}_1 \cdot \mathbf{H}^i(\mathbf{r}) - \mathbf{J}_m^i(\mathbf{r}), \\ \nabla \times \mathbf{H}^i(\mathbf{r}) &= \ i\omega \underline{\varepsilon}_1 \cdot \mathbf{E}^i(\mathbf{r}) + \mathbf{J}_e^i(\mathbf{r}), \end{aligned} \tag{1.102}$$

valid for all $\mathbf{r}$; these are the fields caused by $\mathbf{J}_{e(m)}^i$ in a homogeneous space characterized by $\underline{\mu}_1, \underline{\varepsilon}_1$, i.e., in the *absence* of region $V$. Subtracting (1.102)

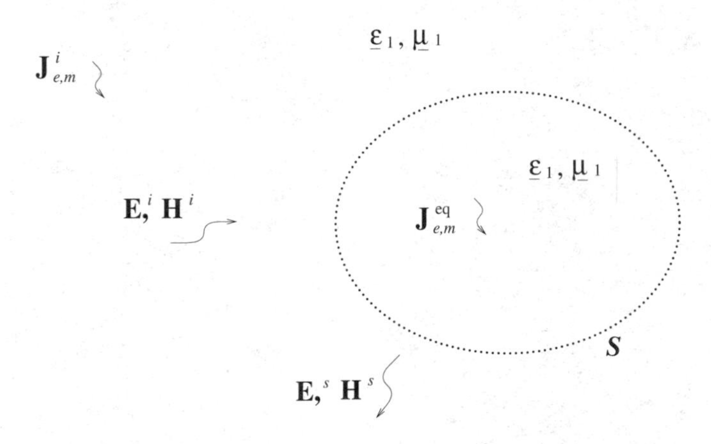


Figure 1.4: Formulation of the volume equivalence principle: problem equivalent to Figure 1.3 in terms of volume polarization sources $\mathbf{J}^{\text{eq}}_{e,m}$.

from (1.100) leads to

$$\begin{aligned} \nabla \times \mathbf{E}^s(\mathbf{r}) &= -i\omega \left[ \underline{\mu}(\mathbf{r}) - \underline{\mu}_1 \right] \cdot \mathbf{H}(\mathbf{r}) - i\omega \underline{\mu}_1 \cdot \mathbf{H}^s(\mathbf{r}), \\ \nabla \times \mathbf{H}^s(\mathbf{r}) &= \ i\omega \left[ \underline{\varepsilon}(\mathbf{r}) - \underline{\varepsilon}_1 \right] \cdot \mathbf{E}(\mathbf{r}) + i\omega \underline{\varepsilon}_1 \cdot \mathbf{E}^s(\mathbf{r}). \end{aligned} \tag{1.103}$$

If we define equivalent polarization currents as

$$\begin{aligned} \mathbf{J}^{\text{eq}}_e &\equiv i\omega \left[ \underline{\varepsilon}(\mathbf{r}) - \underline{\varepsilon}_1 \right] \cdot \mathbf{E}(\mathbf{r}), \\ \mathbf{J}^{\text{eq}}_m &\equiv i\omega \left[ \underline{\mu}(\mathbf{r}) - \underline{\mu}_1 \right] \cdot \mathbf{H}(\mathbf{r}), \end{aligned} \tag{1.104}$$

then (1.103) becomes

$$\begin{aligned} \nabla \times \mathbf{E}^s(\mathbf{r}) &= -i\omega \underline{\mu}_1 \cdot \mathbf{H}^s(\mathbf{r}) - \mathbf{J}^{\text{eq}}_m(\mathbf{r}), \\ \nabla \times \mathbf{H}^s(\mathbf{r}) &= i\omega \underline{\varepsilon}_1 \cdot \mathbf{E}^s(\mathbf{r}) + \mathbf{J}^{\text{eq}}_e(\mathbf{r}), \end{aligned} \tag{1.105}$$

valid for all $\mathbf{r}$. Thus, the scattered fields appear to be produced by the localized equivalent electric and magnetic volume polarization currents $\mathbf{J}^{\text{eq}}_{e(m)}$ residing in a homogeneous medium characterized by $\underline{\mu}_1$, $\underline{\varepsilon}_1$. Therefore, the inhomogeneous medium situation, depicted in Figure 1.3, is equivalent to the homogeneous medium problem depicted in Figure 1.4 [25, p. 126]. This is known as the *volume equivalence principle*, which is easily extended to general bianisotropic media [26]. For $\mathbf{r} \notin V$, $\mathbf{J}^{\text{eq}}_{e(m)} = \mathbf{0}$.

### Domain Integral Equations

Because $\mathbf{J}^{\text{eq}}_{e(m)}$ are undetermined, it may seem that we have not really gained anything in arriving at the equivalent problem depicted in Figure 1.4. However, from (1.105), if the relationship between currents and fields has the form (1.88), then

$$\mathbf{E}^s(\mathbf{r}) = \int_V \underline{\mathbf{G}}_{ee}^{\wedge}(\mathbf{r},\mathbf{r}') \cdot \mathbf{J}_e^{\mathrm{eq}}(\mathbf{r}')\, dV' + \int_V \underline{\mathbf{G}}_{em}(\mathbf{r},\mathbf{r}') \cdot \mathbf{J}_m^{\mathrm{eq}}(\mathbf{r}')\, dV',$$
$$\mathbf{H}^s(\mathbf{r}) = \int_V \underline{\mathbf{G}}_{mm}^{\wedge}(\mathbf{r},\mathbf{r}') \cdot \mathbf{J}_m^{\mathrm{eq}}(\mathbf{r}')\, dV' + \int_V \underline{\mathbf{G}}_{me}(\mathbf{r},\mathbf{r}') \cdot \mathbf{J}_e^{\mathrm{eq}}(\mathbf{r}')\, dV', \tag{1.106}$$

where the Green's dyadics are those that are appropriate for a homogeneous region characterized by $\underline{\mu}_1$, $\underline{\varepsilon}_1$. With the equivalent polarization currents given as (1.104), and the scattered fields by (1.101), we then obtain

$$\begin{aligned}
\mathbf{E}(\mathbf{r}) - \int_V \underline{\mathbf{G}}_{ee}^{\wedge}(\mathbf{r},\mathbf{r}') \cdot \mathbf{J}_e^{\mathrm{eq}}(\mathbf{r}')\, dV' \\
- \int_V \underline{\mathbf{G}}_{em}(\mathbf{r},\mathbf{r}') \cdot \mathbf{J}_m^{\mathrm{eq}}(\mathbf{r}')\, dV' = \mathbf{E}^i(\mathbf{r}), \\
\mathbf{H}(\mathbf{r}) - \int_V \underline{\mathbf{G}}_{mm}^{\wedge}(\mathbf{r},\mathbf{r}') \cdot \mathbf{J}_m^{\mathrm{eq}}(\mathbf{r}')\, dV' \\
- \int_V \underline{\mathbf{G}}_{me}(\mathbf{r},\mathbf{r}') \cdot \mathbf{J}_e^{\mathrm{eq}}(\mathbf{r}')\, dV' = \mathbf{H}^i(\mathbf{r}),
\end{aligned} \tag{1.107}$$

valid for all $\mathbf{r}$, although $\mathbf{r} \in V$ is enforced in solving the above equations. Upon defining

$$\begin{aligned}
\underline{\mathbf{G}}_{e,ea}(\mathbf{r},\mathbf{r}') &\equiv i\omega \underline{\mathbf{G}}_{ee}^{\wedge}(\mathbf{r},\mathbf{r}') \cdot [\underline{\varepsilon}(\mathbf{r}') - \underline{\varepsilon}_1], \\
\underline{\mathbf{G}}_{e,ma}(\mathbf{r},\mathbf{r}') &\equiv i\omega \underline{\mathbf{G}}_{em}(\mathbf{r},\mathbf{r}') \cdot \left[\underline{\mu}(\mathbf{r}') - \underline{\mu}_1\right], \\
\underline{\mathbf{G}}_{m,ea}(\mathbf{r},\mathbf{r}') &\equiv i\omega \underline{\mathbf{G}}_{me}(\mathbf{r},\mathbf{r}') \cdot [\underline{\varepsilon}(\mathbf{r}') - \underline{\varepsilon}_1], \\
\underline{\mathbf{G}}_{m,ma}(\mathbf{r},\mathbf{r}') &\equiv i\omega \underline{\mathbf{G}}_{mm}^{\wedge}(\mathbf{r},\mathbf{r}') \cdot \left[\underline{\mu}(\mathbf{r}') - \underline{\mu}_1\right],
\end{aligned} \tag{1.108}$$

we obtain

$$\begin{aligned}
\mathbf{E}(\mathbf{r}) - \int_V \underline{\mathbf{G}}_{e,ea}(\mathbf{r},\mathbf{r}') \cdot \mathbf{E}(\mathbf{r}')\, dV' \\
- \int_V \underline{\mathbf{G}}_{e,ma}(\mathbf{r},\mathbf{r}') \cdot \mathbf{H}(\mathbf{r}')\, dV' = \mathbf{E}^i(\mathbf{r}), \\
\mathbf{H}(\mathbf{r}) - \int_V \underline{\mathbf{G}}_{m,ma}(\mathbf{r},\mathbf{r}') \cdot \mathbf{H}(\mathbf{r}')\, dV' \\
- \int_V \underline{\mathbf{G}}_{m,ea}(\mathbf{r},\mathbf{r}') \cdot \mathbf{E}(\mathbf{r}')\, dV' = \mathbf{H}^i(\mathbf{r}),
\end{aligned} \tag{1.109}$$

which are coupled second-kind integral equations for the unknown fields $\mathbf{E}, \mathbf{H}$ (note again that $V$ may be multiply connected). Mathematical properties of domain integral equations for objects in free space have been explored in a series of papers [27]–[30] (see also [31, Ch. 9] and [13]), although the analysis makes use of a less singular relation between fields and currents, equivalent to (1.89). In the scalar (acoustic) case, integral equations (1.109) are known as *Lippmann–Schwinger* equations.

If, for instance, the magnetic contrast vanishes, then (1.109) become uncoupled and we only need to solve

$$\mathbf{E}(\mathbf{r}) - \int_V \underline{\mathbf{G}}_{e,ea}(\mathbf{r},\mathbf{r}') \cdot \mathbf{E}(\mathbf{r}')\, dV' = \mathbf{E}^i(\mathbf{r}), \tag{1.110}$$

and similarly for the case where electric contrast vanishes.

Generally, solving (1.110) requires a complicated numerical procedure. However, in the case of small contrast $\underline{\varepsilon}(\mathbf{r}') \simeq \underline{\varepsilon}_1$, the second term on the left side of (1.110) is small compared to the first term, and $\mathbf{E} \simeq \mathbf{E}^i$. In this case the field $\mathbf{E} \in V$ is approximately determined as

$$\mathbf{E}(\mathbf{r}) = \mathbf{E}^i(\mathbf{r}) + \int_V \underline{\mathbf{G}}_{e,ea}(\mathbf{r},\mathbf{r}') \cdot \mathbf{E}^i(\mathbf{r}')\, dV', \tag{1.111}$$

which is known as the *Born approximation* [32].

The solution of (1.109) leads to the fields $\mathbf{E}, \mathbf{H}$ for $\mathbf{r} \in V$. By (1.104) and (1.106), one may obtain the scattered field for $\mathbf{r} \notin V$ as

$$\begin{aligned}\mathbf{E}^s(\mathbf{r}) &= \int_V \underline{\mathbf{G}}_{e,ea}(\mathbf{r},\mathbf{r}') \cdot \mathbf{E}(\mathbf{r}')\, dV' + \int_V \underline{\mathbf{G}}_{e,ma}(\mathbf{r},\mathbf{r}') \cdot \mathbf{H}(\mathbf{r}')\, dV', \\ \mathbf{H}^s(\mathbf{r}) &= \int_V \underline{\mathbf{G}}_{m,ma}(\mathbf{r},\mathbf{r}') \cdot \mathbf{H}(\mathbf{r}')\, dV' + \int_V \underline{\mathbf{G}}_{m,ea}(\mathbf{r},\mathbf{r}') \cdot \mathbf{E}(\mathbf{r}')\, dV'.\end{aligned} \tag{1.112}$$

(1.109) are applicable to a variety of other problems, not only for the analysis of scattering. For example, natural resonances and waveguiding problems may be solved from the homogeneous form of (1.109), i.e., by setting $\mathbf{E}^i = \mathbf{H}^i = \mathbf{0}$.

### 1.4.2 Surface Integral Equations

In solving the domain integral equations (1.109), the entire scatterer region $V$ must be considered. If the region $V$ in Figure 1.3 is homogeneous, then integral equations over the surface $S$ of $V$ can be derived [24]. Rather than considering the general case, in this section we formulate *surface integral equations* for perfectly conducting closed scatterers.

In Figure 1.5, sources $\mathbf{J}^i_{e(m)}$ create a field $\mathbf{E}^i, \mathbf{H}^i$ that interacts with a perfectly conducting scatterer having a surface $S$ with surface normal $\mathbf{n}$, inducing an electric surface current $\mathbf{J}^s_e$ on $S$ ($\mathbf{J}^s_m = \mathbf{0}$ on a perfect electric conductor). From (1.88), the field maintained by the sources in the absence of the scatterer is

$$\begin{aligned}\mathbf{E}^i(\mathbf{r}) &= \int_V \underline{\mathbf{G}}^{\wedge}_{ee}(\mathbf{r},\mathbf{r}') \cdot \mathbf{J}^i_e(\mathbf{r}')\, dV' + \int_V \underline{\mathbf{G}}_{em}(\mathbf{r},\mathbf{r}') \cdot \mathbf{J}^i_m(\mathbf{r}')\, dV', \\ \mathbf{H}^i(\mathbf{r}) &= \int_V \underline{\mathbf{G}}^{\wedge}_{mm}(\mathbf{r},\mathbf{r}') \cdot \mathbf{J}^i_m(\mathbf{r}')\, dV' + \int_V \underline{\mathbf{G}}_{me}(\mathbf{r},\mathbf{r}') \cdot \mathbf{J}^i_e(\mathbf{r}')\, dV',\end{aligned} \tag{1.113}$$

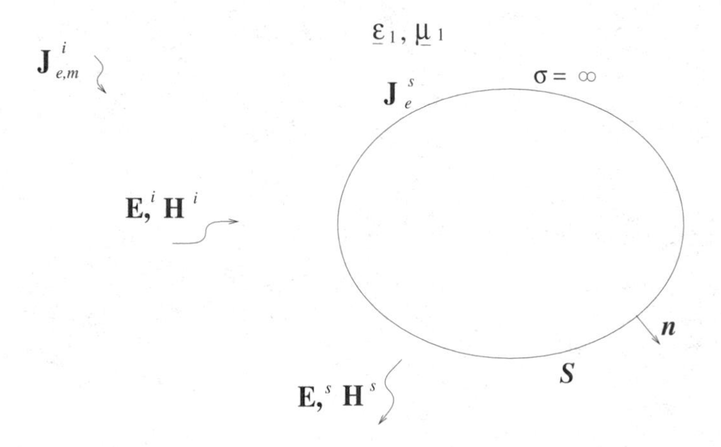


Figure 1.5: Scattering of an electromagnetic wave $\mathbf{E}^i, \mathbf{H}^i$ by a perfectly conducting surface $S$.

and the field maintained by the induced surface current is

$$\begin{aligned} \mathbf{E}^s(\mathbf{r}) &= \int_S \underline{\mathbf{G}}_{ee}^{\wedge}(\mathbf{r}, \mathbf{r}') \cdot \mathbf{J}_e^s(\mathbf{r}')\, dS', \\ \mathbf{H}^s(\mathbf{r}) &= \int_S \underline{\mathbf{G}}_{me}(\mathbf{r}, \mathbf{r}') \cdot \mathbf{J}_e^s(\mathbf{r}')\, dS'. \end{aligned} \tag{1.114}$$

At this point two formulations are possible.

### Electric Field Integral Equation

An *electric field integral equation* (EFIE) is obtained by enforcing the boundary condition for the electric field at a perfectly conducting surface. From (1.5), the electric field at the surface of a perfect conductor is

$$\mathbf{n} \times \mathbf{E}|_S = \mathbf{0}.$$

Upon noting that the total electric field on the surface $S$ (and more generally for all $\mathbf{r} \notin V$) is the sum of the incident and scattered fields,

$$\mathbf{E}(\mathbf{r}) = \mathbf{E}^s(\mathbf{r}) + \mathbf{E}^i(\mathbf{r}),$$

then

$$\mathbf{n}\times \left[ \int_S \underline{\mathbf{G}}_{ee}^{\wedge}(\mathbf{r}, \mathbf{r}') \cdot \mathbf{J}_e^s(\mathbf{r}')\, dS' + \mathbf{E}^i(\mathbf{r}) \right] = \mathbf{0}$$

for $\mathbf{r} \in S$. The above is usually rewritten as

$$\mathbf{n}\times \int_S \underline{\mathbf{G}}_{ee}^{\wedge}(\mathbf{r}, \mathbf{r}') \cdot \mathbf{J}_e^s(\mathbf{r}')\, dS' = -\mathbf{n} \times \mathbf{E}^i(\mathbf{r}), \qquad \mathbf{r} \in S, \tag{1.115}$$

which is called the electric field integral equation. This is a *first-kind integral equation* and is extensively applied to both scattering and antenna (radiation) problems. For scatterers forming closed bodies, this formulation suffers from spurious solutions associated with cavity resonances [24], [2], [33], although this topic is not discussed here.

### Magnetic Field Integral Equation

A *magnetic field integral equation* (MFIE) is obtained by enforcing the boundary condition for the magnetic field at a perfectly conducting scatterer. From (1.5), the magnetic field at the surface of a perfect conductor is

$$\mathbf{n} \times \mathbf{H}|_S = \mathbf{J}_e^s. \tag{1.116}$$

With

$$\mathbf{H}(\mathbf{r}) = \mathbf{H}^s(\mathbf{r}) + \mathbf{H}^i(\mathbf{r}),$$

then

$$\mathbf{n}\times \left[\int_S \underline{\mathbf{G}}_{me}(\mathbf{r},\mathbf{r}') \cdot \mathbf{J}_e^s(\mathbf{r}')\, dS' + \mathbf{H}^i(\mathbf{r})\right] = \mathbf{J}_e^s$$

for $\mathbf{r} \in S$. Actually, the above equation is correct only when $\mathbf{r} \in S^+$, where $S^+$ is a surface an infinitesimal distance outside the surface $S$. This limiting procedure is necessary since the tangential components of the surface integral are discontinuous across the surface $S$ (see Section 6.6), whereas this is not the case for the surface integral in (1.115). The above second-kind integral equation can be rewritten as

$$\mathbf{n}\times \int_{S^+} \underline{\mathbf{G}}_{me}(\mathbf{r},\mathbf{r}') \cdot \mathbf{J}_e^s(\mathbf{r}')\, dS' - \mathbf{J}_e^s(\mathbf{r}) = -\mathbf{n} \times \mathbf{H}^i(\mathbf{r}), \qquad \mathbf{r} \in S^+, \tag{1.117}$$

and is called the magnetic field integral equation. It is common to perform the limiting procedure and express the result as [2, pp. 141–144]

$$\int_S \left[\underline{\mathbf{G}}_{me}(\mathbf{r},\mathbf{r}') \cdot \mathbf{J}_e^s(\mathbf{r}')\right] \times \mathbf{n}\, dS' + \frac{1}{2}\mathbf{J}_e^s(\mathbf{r}) = \mathbf{n} \times \mathbf{H}^i(\mathbf{r}), \qquad \mathbf{r} \in S. \tag{1.118}$$

As with the EFIE, this formulation suffers from spurious solutions associated with cavity resonances.

Note that the Green's function singularity is weaker in the MFIE compared to the EFIE. The MFIE is somewhat less general though, since it cannot be applied to open structures with infinitely thin shells, because the boundary condition (1.116) applies only when the magnetic field on one side of the surface vanishes.

It should be noted that for all the integral equations derived above, the appropriate Green's dyadics are those for the space obtained by removing the scatterer. For instance, if this is simply a homogeneous, isotropic space characterized by $\mu, \varepsilon$, then the Green's dyadics are explicitly given by (1.79), (1.84), with (1.87).

### Integro-Differential Equations

Rather than representing the fields as in (1.88), which involve highly singular Green's dyadics, the fields may be given in terms of potentials, leading to *integro-differential equations.* For instance, for a current $\mathbf{J}_e$ in a homogeneous, isotropic space characterized by $\mu, \varepsilon$, from (1.74) we have

$$\mathbf{A}(\mathbf{r}) = \mu \int_V \underline{\mathbf{G}}(\mathbf{r}, \mathbf{r}') \cdot \mathbf{J}_e(\mathbf{r}')\, dV',$$

where simply $\underline{\mathbf{G}}(\mathbf{r}, \mathbf{r}') = \underline{\mathbf{I}}\, g\,(\mathbf{r}, \mathbf{r}') = \underline{\mathbf{I}}\,(e^{-ikR}/\,4\pi R)$ from (1.72). With the electric field as (1.38),

$$\mathbf{E}\,(\mathbf{r}) = -i\omega \mathbf{A}\,(\mathbf{r}) + \frac{1}{i\omega\mu\varepsilon} \nabla\,\nabla \cdot \mathbf{A}\,(\mathbf{r})\,, \tag{1.119}$$

then the incident electric field is given by (1.119) with $\mathbf{A} = \mathbf{A}^i$, where

$$\mathbf{A}^i(\mathbf{r}) = \mu \int_V g(\mathbf{r}, \mathbf{r}') \mathbf{J}_e^i(\mathbf{r}')\, dV',$$

and the scattered electric field is given by (1.119) with $\mathbf{A} = \mathbf{A}^s$, where

$$\mathbf{A}^s(\mathbf{r}) = \mu \int_S g(\mathbf{r}, \mathbf{r}') \mathbf{J}_e^s(\mathbf{r}')\, dS'.$$

The integral equation analogous to (1.115) is

$$\mathbf{n} \times \left[ -i\omega \mathbf{A}^s\,(\mathbf{r}) + \frac{1}{i\omega\mu\varepsilon} \nabla\,\nabla \cdot \mathbf{A}^s\,(\mathbf{r}) + \mathbf{E}^i\,(\mathbf{r}) \right] = \mathbf{0},$$

valid for $\mathbf{r} \in S$. This can be rewritten as

$$\mathbf{n} \times \left[ k^2 + \nabla\,\nabla \cdot \right] \int_S g(\mathbf{r}, \mathbf{r}') \mathbf{J}_e^s(\mathbf{r}')\, dS' = -i\omega\varepsilon\, \mathbf{n} \times \mathbf{E}^i\,(\mathbf{r})\,, \qquad \mathbf{r} \in S, \tag{1.120}$$

where $k^2 = \omega^2 \mu\varepsilon$. The singularity of the surface integral in (1.120) is much weaker than in (1.115). However, subsequent to performing the integration, the differentiation operation may cause difficulties in numerical solution procedures. The benefit of the form (1.115) is that derivatives are transferred to the Green's function and can be performed analytically.

## 1.5 Uniqueness Theorem and Radiation Conditions

In order for fields obtained from differential equation formulations to be unique, one must impose boundary conditions on finite surfaces and radiation conditions at infinity. This is analogous to the case of a scalar one-dimensional differential equation, where a sufficient number of conditions must be specified to uniquely determine the "constants of integration" in the general solution. For scalar and vector Helmholtz equations, the topic of appropriate boundary conditions is discussed in detail in many books, and so here we simply summarize the results.

For fields satisfying Maxwell's equations to be unique inside a finite volume $V$ bounded by surface $S$, it is necessary and sufficient that [5, pp. 372–373]

- the tangential electric field $\mathbf{n} \times \mathbf{E}$ is specified on $S$, or
- the tangential magnetic field $\mathbf{n} \times \mathbf{H}$ is specified on $S$, or
- the tangential electric field $\mathbf{n} \times \mathbf{E}$ is specified on part of $S$ and the tangential magnetic field $\mathbf{n} \times \mathbf{H}$ is specified on the remainder of $S$.

If multiple surfaces $S_1, \ldots, S_n$ bounding volumes $V_1, \ldots, V_n$ are contained within $S$, then $V$ is the region interior to $S$ and exterior to $S_1, \ldots, S_n$, and the above conditions apply to the total surface $S_T = S \cup (\cup_i S_i)$.

If the "outer" surface $S$ has radius $r$ with $r \to \infty$, then we no longer need to specify boundary conditions on $S$, but rather we need to enforce radiation conditions at infinity. These are found to be [5, pp. 332–333]

$$\lim_{r \to \infty} r \left[ \nabla \times \begin{pmatrix} \mathbf{E} \\ \mathbf{H} \\ \underline{\mathbf{G}}_{ee} \end{pmatrix} + ik\widehat{\mathbf{r}} \times \begin{pmatrix} \mathbf{E} \\ \mathbf{H} \\ \underline{\mathbf{G}}_{ee} \end{pmatrix} \right] = \begin{pmatrix} \mathbf{0} \\ \mathbf{0} \\ \underline{\mathbf{0}} \end{pmatrix}, \tag{1.121}$$

where, using Maxwell's equations, one obtains

$$\lim_{r \to \infty} r \left[ \begin{pmatrix} -i\omega\mu\mathbf{H} \\ i\omega\varepsilon\mathbf{E} \\ -\underline{\mathbf{G}}_{em} \end{pmatrix} + ik\widehat{\mathbf{r}} \times \begin{pmatrix} \mathbf{E} \\ \mathbf{H} \\ \underline{\mathbf{G}}_{ee} \end{pmatrix} \right] = \begin{pmatrix} \mathbf{0} \\ \mathbf{0} \\ \underline{\mathbf{0}} \end{pmatrix}. \tag{1.122}$$

For scalar fields we have

$$\lim_{r \to \infty} r \left[ \frac{\partial \psi}{\partial r} + ik\psi \right] = 0.$$

Generally these radiation conditions are called *Sommerfeld radiation conditions*, although sometimes the vector ones are called *Silver–Müller radiation conditions*.[36]

Without considering in detail the topic of radiation conditions, it is worthwhile to illustrate the fact that they are at least reasonable. Consider

[36] For a lossy medium we replace the Sommerfeld radiation condition with the much simpler requirement that the fields vanish at infinity. In this case we can work within an $\mathbf{L}^2$ space as described in the next chapter.

the Stratton–Chu form (1.77), repeated here as

$$\mathbf{E}(\mathbf{r}') = \int_V \left\{ -i\omega\mu\, \mathbf{J}_e(\mathbf{r})\, g(\mathbf{r},\mathbf{r}') + \frac{\rho_e(\mathbf{r})}{\varepsilon} \nabla g(\mathbf{r},\mathbf{r}') + \nabla g(\mathbf{r},\mathbf{r}') \times \mathbf{J}_m(\mathbf{r}) \right\} dV$$
$$+ \oint_S \{(\mathbf{n} \times \mathbf{E}(\mathbf{r})) \times \nabla g(\mathbf{r},\mathbf{r}') - i\omega\mu\, (\mathbf{n} \times \mathbf{H}(\mathbf{r}))\, g(\mathbf{r},\mathbf{r}') \tag{1.123}$$
$$+ (\mathbf{n} \cdot \mathbf{E}(\mathbf{r}))\, \nabla g(\mathbf{r},\mathbf{r}')\}\, dS.$$

Let $S$ be a sphere of radius $R$. If $R \to \infty$, we expect that the contribution from the surface terms should vanish. Using

$$\nabla g(\mathbf{r},\mathbf{r}') = -g(\mathbf{r},\mathbf{r}') \left[ jk + \frac{1}{R(\mathbf{r},\mathbf{r}')} \right] \widehat{\mathbf{R}}(\mathbf{r},\mathbf{r}') \underset{R\to\infty}{\to} -jkg\widehat{\mathbf{R}}$$

with $\mathbf{n} = \widehat{\mathbf{R}}$, we have for the surface term

$$-\oint_S \left\{ jkg \left( \widehat{\mathbf{R}} \times \mathbf{E} \right) \times \widehat{\mathbf{R}} + i\omega\mu g \left( \widehat{\mathbf{R}} \times \mathbf{H} \right) + jkg \left( \widehat{\mathbf{R}} \cdot \mathbf{E} \right) \widehat{\mathbf{R}} \right\} dS.$$

With $dS = O\left(R^2\right)$ and $g = O\,(1/R)$ the surface integral will vanish if

$$\mathbf{R} \left[ i\omega\varepsilon \mathbf{E} + ik\widehat{\mathbf{R}} \times \mathbf{H} \right] \to \mathbf{0},$$

in agreement with (1.122).

# Bibliography


[1] Rothwell, E.J. and Cloud, M.J. (2001). *Electromagnetics*, Boca Raton, FL: CRC Press.

[2] Collin, R.E. (1991). *Field Theory of Guided Waves*, 2nd ed., New York: IEEE Press.

[3] Tai, C.T. (1994). *Dyadic Green Functions in Electromagnetic Theory*, 2nd ed., New York: IEEE Press.

[4] Portis, A.M. (1978). *Electromagnetic Fields: Sources and Media,* New York: Wiley.

[5] Kong, J.A. (1990). *Electromagnetic Wave Theory*, 2nd ed., New York: Wiley.

[6] Hansen, T.B. and Yaghjian, A.D. (1999). *Plane-Wave Theory of Time-Domain Fields*, New York: IEEE Press.

[7] Gel'fand, I.M. and Shilov, G.E. (1964). *Generalized Functions,* Vol. I, New York: Academic Press.

[8] Van Bladel, J. (1991). *Singular Electromagnetic Fields and Sources*, Oxford: Clarendon Press.

[9] Kellogg, O.D. (1929). *Foundations of Potential Theory*, New York: Dover.

[10] Budak, B.M. and Fomin, S.V. (1973). *Multiple Integrals, Field Theory and Series*, Moscow: MIR Publishers.

[11] Courant, R. (1974). *Introduction to Calculus and Analysis*, Vol. II, New York: Wiley.

[12] Mikhlin, S.G. (1965). *Multidimensional Singular Integrals and Integral Equations*, Oxford: Pergamon Press.

[13] Müller, C. (1969). *Foundations of the Mathematical Theory of Electromagnetic Waves*, New York: Springer-Verlag.

[14] Yaghjian, A.D. (1980). Electric dyadic Green's functions in the source region, *IEEE Proc.*, Vol. 68, pp. 248–263, Feb.

[15] Yaghjian, A.D. (1985). Maxwellian and cavity electromagnetic fields within continuous sources, *Am. J. Phys.*, Vol. 53, pp. 859–863, Sept.

[16] Silberstein, M. (1991). Application of a generalized Leibnitz rule for calculating electromagnetic fields within continuous source regions, *Radio Science*, Vol. 26, pp. 183–190, Jan.–Feb.

[17] Viola, M.S. and Nyquist, D.P. (1988). An observation on the Sommerfeld-integral representation of the electric dyadic Green's function for layered media, *IEEE Trans. Microwave Theory Tech.*, Vol. 36, pp. 1289–1292, Aug.

[18] Courant, R. and Hilbert, D. (1953). *Methods of Mathematical Physics*, Vol. II, New York: Interscience.

[19] Lee, S.W., Boersma, J., Law, C.L., and Deschamps, G.A. (1980). Singularity in Green's function and its numerical evaluation, *IEEE Trans. Ant. Prop.*, Vol. AP-28, pp. 311–317, May.

[20] Asvestas, J.S. (1983). Comments on "Singularity in Green's function and its numerical evaluation," *IEEE Trans. Ant. Prop.*, Vol. AP-31, pp. 174–177, Jan.

[21] Fikioris, J.G. (1965). Electromagnetic field inside a current-carrying region, *J. Math. Phys.*, Vol. 6, pp. 1617–1620, Nov.

[22] Levine, H. and Schwinger, J. (1950). On the theory of electromagnetic wave diffraction by an aperture in an infinite plane conducting screen, *Comm. Pure and Appl. Math.*, Vol. III, pp. 355–391.

[23] Ball, J.A.R. and Khan, P.J. (1980). Source region electric field derivation by a dyadic Green's function approach, *IEE Proc.*, Vol. 127, Part H, no. 5, pp. 301–304, Oct.

[24] Chew, W.C. (1990). *Waves and Fields in Inhomogeneous Media*, New York: IEEE Press.

[25] Harrington, R.F. (1961). *Time-Harmonic Electromagnetic Fields*, New York: McGraw-Hill.

[26] Hanson, G.W. (1996). A numerical formulation of dyadic Green's functions for planar bianisotropic media with applications to printed transmission lines, *IEEE Trans. Microwave Theory Tech.*, Vol. 44, pp. 144–151, Jan.

[27] Samokhin, A.B. (1990). Investigation of problems of the diffraction of electromagnetic waves in locally non-uniform media, *Comput. Maths. Math. Phys.*, Vol. 30, no. 1, pp. 80–90.

[28] Samokhin, A.B. (1992). Diffraction of electromagnetic waves by a locally non-homogeneous body and singular integral equations, *Comput. Maths. Math. Phys.*, Vol. 32, no. 5, pp. 673–686.

[29] Samokhin, A.B. (1993). Integral equations of electrodynamics for three-dimensional structures and iteration methods for solving them (a review)." *J. Comm. Tech. Electron.*, Vol. 38, no. 15, pp. 15–34.

[30] Samokhin, A.B. (1994). An iterative method for the solution of integral equations applied to a scattering problem on a three-dimensional transparent body, " *Diff. Equations*, Vol. 30, no. 12, pp. 1987–1997.

[31] Colton, D. and Kress, R. (1998). *Inverse Acoustic and Electromagnetic Scattering Theory*, New York: Springer.

[32] Born, N. and Wolf, E. (1980). *Principles of Optics*, 6th ed., New York: Pergamon Press.

[33] Peterson, A.F., Ray, S.L., and Mittra, R. (1998). *Computational Methods for Electromagnetics*, New York: IEEE Press.

[34] Cessenat, M. (1996). *Mathematical Methods in Electromagnetism*, Singapore: World Scientific.

[35] Mattuck, A. (1999). *Introduction to Analysis*, Englewood Cliffs, NJ: Prentice-Hall.

[36] Jones, D.S. (1986). *Acoustic and Electromagnetic Waves*, Oxford: Clarendon Press.

[37] Jones, D.S. (1964). *The Theory of Electromagnetism*, Oxford: Pergamon Press.

[38] Jackson, J.D. (1975). *Classical Electrodynamics*, 2nd ed., New York: John Wiley.

[39] Van Bladel, J. (1985). *Electromagnetic Fields*, Washington: Hemisphere.

[40] Lindell, I.V. (1992). *Methods for Electromagnetic Field Analysis*, Oxford: Clarendon Press.

[41] Stratton, J. (1941). *Electromagnetic Theory*, New York: McGraw-Hill.

[42] Collin, R.E. (1986). The dyadic Green's functions as an inverse operator, *Radio Science*, Vol. 21, pp. 883–890, Nov.–Dec.

[43] Pandofsky, W.K.H. and Phillips, M. (1962). *Classical Electricity and Magnetism*, Reading, MA: Addison-Wesley.

[44] Amazigo, J.C. and Rubenfeld, L.A. (1980). *Advanced Calculus*, New York: Wiley.

[45] Sternberg, W.J. and Smith, T.L. (1946). *The Theory of Potential and Spherical Harmonics*, Toronto: University of Toronto Press.

[46] Champeney, D.C. (1987). *A Handbook of Fourier Theorems*, Cambridge: Cambridge University Press.

[47] Mikhlin, S.G. (1970). *Mathematical Physics, An Advanced Course*, Amsterdam: North-Holland.

[48] Elliott, R.S. (1981). *Antenna Theory and Design*, Englewood Cliffs, NJ: Prentice-Hall.

# 2

# Introductory Functional Analysis

The purpose of this chapter is to introduce basic elements of functional analysis, especially those concepts necessary for a study of the operators arising in electromagnetics. The main idea is to start with the simple concept of a set and then introduce increasing levels of mathematical structure. The goal is to gradually build the foundation for understanding the concept and utility of formulating problems in an appropriate function space, usually a Hilbert or Banach space.

Where possible, working within such a space is very desirable. Within a Hilbert space we have the idea of a collection of elements (sets), the distance between elements (a metric), the algebraic concepts of addition and scalar multiplication (a linear space), the size of the space (the dimension), the size of the elements themselves (the norm), and an extension of the idea of perpendicularity (orthogonality) among elements using inner products. Thus, Hilbert spaces come equipped with abstractions, most importantly to collections of functions, of many familiar geometrical concepts. The same can be said for Banach spaces, with the exception of the inner product. This correspondence is very satisfying to applied scientists, who are of course comfortable with geometric structure. Most importantly, though, Hilbert and Banach spaces have completeness properties that differentiate them from other spaces with seemingly similar structures. The concept of completeness allows one to represent an element (function, vector, etc.) in terms of basis elements, i.e., as a generalized Fourier series, which often have desirable properties. Working within this framework also allows one to draw many conclusions about physical problems cast into operator form, which are discussed in the next chapter.

Many of the concepts discussed in this chapter are based on properties of sequences. It is also true that a great many problems in applied electromagnetics make extensive use of the idea of a discretization (for example, Galerkin-type projection techniques) of function-level equations to the level of an infinite system of linear equations. The corresponding correlation (an isomorphism) between function spaces and sequence spaces plays a significant role in understanding questions of existence and solvability of such problems. Sequence spaces determine the fundamental basis for properties (continuity, compactness, etc.) of matrix operators and provide important criteria governing infinite matrix equations of the first and second kinds. The mathematical apparatus of sequence spaces, applied within the framework of linear operator theory, can then be applied to fundamental boundary value problems in mathematical physics.

## 2.1 Sets

We will start with the concept of a set, the elements of which satisfy the *axiomatics* (postulates) of the set. Due to the many possible sets of interest, we will introduce sets of numbers (real and complex), sets of functions (function spaces), and sets of sequences (sequence spaces). Some of the sets of primary importance in applied electromagnetics are listed below; they are extensively referred to in the text. We use the terms "set" and "space" interchangeably.

Throughout the text certain problems are developed with some generality as to spatial dimension $m$. Unless it is otherwise clear from the context, $\Omega$ represents an open region of space, with $\Gamma$ its sufficiently smooth boundary. The closure $\widetilde{\Omega} = \Omega \cup \Gamma$ indicates all points within $\Omega$ and on its boundary $\Gamma$. Integrations over $\Omega$ are represented as $\int_\Omega (\cdot)\, d\Omega$, and integrations over the boundary as $\int_\Gamma (\cdot)\, d\Gamma$. Obviously for $m = 3$, $\Omega$ is a volume and $\Gamma$ a closed surface. For $m = 2$, $\Omega$ is an open surface and $\Gamma$ a closed line segment, and for $m = 1$, $\Omega = (a, b)$ with $\widetilde{\Omega} = [a, b]$. For $m > 3$, the presented formulations also usually hold, but these cases are generally not of physical interest in electromagnetics.

### 2.1.1 Sets of Numbers

1. $\mathbf{C}$ ($\mathbf{R}$) is the set of all complex (real) numbers.

2. $\mathbf{Z}$ is the set of all integers, including 0.

3. $\mathbf{N}$ is the set of all positive integers (e.g., $0 \notin \mathbf{N}$).

4. $\mathbf{R}^n$ is the set of all ordered $n$-tuples ($n$-component vectors) of real numbers, $\mathbf{x} = (x_1, x_2, \ldots, x_n)$, $x_i \in \mathbf{R}$.

5. $\mathbf{C}^n$ is the set of all ordered $n$-tuples of complex numbers, $\mathbf{z} = (z_1, z_2, \ldots z_n)$, $z_i \in \mathbf{C}$.

   For instance,

   (a) $\mathbf{x} = (1,3,5) \in \mathbf{R}^3$, $\mathbf{z} = (1+i, 1-i) \in \mathbf{C}^2$.

   (b) $\mathbf{R}^2 = \mathbf{R} \times \mathbf{R}$, $\mathbf{R}^3 = \mathbf{R} \times \mathbf{R} \times \mathbf{R}$, $\mathbf{C}^3 = \mathbf{C} \times \mathbf{C} \times \mathbf{C}$, etc., where "$\times$" indicates the *Cartesian product.*[1]

## 2.1.2 Sets of Functions—Function Spaces

### Continuous Functions

$\mathbf{C}^k(a,b)$ is the set of all continuous functions on $(a,b)$ whose derivatives up to order $k$ are continuous. We denote the set of continuous functions with no restrictions on the continuity of its derivatives as $\mathbf{C}(a,b)$ $(\equiv \mathbf{C}^0(a,b))$. Note that

$$\mathbf{C}^\infty(a,b) \subset \cdots \subset \mathbf{C}^k(a,b) \subset \mathbf{C}^{k-1}(a,b) \subset \cdots \subset \mathbf{C}^1(a,b) \subset \mathbf{C}(a,b)$$

and that $k$ determines the degree of smoothness of the function, in the sense that larger values of $k$ correspond to smoother functions, compared to smaller $k$ values. The space $\mathbf{C}^\infty(a,b)$ is the set of all infinitely differentiable functions defined on $(a,b)$. Clearly, the space $\mathbf{C}^k(a,b)$, and its generalization to accommodate multidimensional scalar and vector functions as described later, are of considerable interest in electromagnetic problems, where one often specifies continuity of certain field components within a region or on a boundary. For a closed interval we use $[a,b]$ and consider continuity from the left or right.

Another useful set is $\mathbf{C}_0^\infty(a,b)$, defined as the set of all continuous functions with continuous derivatives of all orders, and having *compact support*[2] on $(a,b)$.

### Examples

1. $\sin(t)$, $\sin(t)/t \in \mathbf{C}^\infty(-\infty,\infty)$.

2. $\left\{ \begin{array}{ll} 0, & -a \le t < 0, \\ t^3, & 0 \le t \le b \end{array} \right\} \in \mathbf{C}^2[-a,b]$.

3. $|t| \in \mathbf{C}(-\infty,\infty)$.

4. Heaviside function (unit step function) $H(t-t_0)$;

$$H(t-t_0) = \left\{ \begin{array}{ll} 0, & t \le t_0, \\ 1, & t > t_0 \end{array} \right\} \notin \mathbf{C}^k[a,b] \text{ for } t_0 \in [a,b],\ k = 0,1,\ldots,\infty.$$

[1] If $X_1, X_2, \ldots, X_n$ are sets, then the Cartesian product $X_1 \times X_2 \times \cdots \times X_n$ is the set of all ordered $n$-tuples $(x_1, x_2, \ldots, x_n)$, where $x_i \in X_i$.

[2] A function has compact support if it vanishes outside some compact set.

5. Ramp function $R(t-t_0)$;

$$R(t-t_0)=\left\{\begin{array}{ll} 0, & t\le t_0, \\ t, & t>t_0 \end{array}\right\}\in \mathbf{C}[a,b], \text{ for } t_0\in[a,b].$$

### Bounded and Bounded Continuous Functions

$\mathbf{M}\,(a,b)$ is the set of all bounded functions on $(a,b)$, i.e.,

$$\mathbf{M}\,(a,b)\equiv\{x(t):|x(t)|<\infty \text{ for all } t\in(a,b)\}.$$

If $x(t)$ is a continuous function defined on a closed bounded interval $[a,b]$, then $x(t)$ is bounded on $[a,b]$, and, furthermore,[3] $\sup|x(t)| = \max|x(t)|$ and $\inf|x(t)| = \min|x(t)|$ for $t\in[a,b]$. Note that $\mathbf{C}[a,b]\subset \mathbf{M}[a,b]$, but not every bounded function is continuous.

On an infinite interval not every continuous function is bounded, so we distinguish $\mathbf{BC}\,(a,b)$ as the set of all bounded continuous functions on $(a,b)$.

### Examples

1. $x(t)=t^n\in\mathbf{C}[a,b]\subset\mathbf{M}[a,b]$ for $n\ge 1$ and $0\le a<b<\infty$, with $\sup|x(t)|=b^n$ and $\inf|x(t)|=a^n$.

2. $x(t)=\left\{\begin{array}{ll} t, & -1\le t<0, \\ e^t, & 0\le t\le 1 \end{array}\right\}\notin\mathbf{C}\,[-1,1]$; however, $x(t)\in\mathbf{M}\,[-1,1]$ with $\sup|x(t)|=e$ and $\inf|x(t)|=-1$.

3. $\sin(t),\cos(t)\in\mathbf{BC}[a,b]$.

4. $e^t, t^n\in\mathbf{BC}[a,b]$, $0<a<b<\infty$.

5. $t^{-n}\notin\mathbf{BC}\,[-1,1]$ for $n>0$.

### Lebesgue-Integrable Functions

$\mathbf{L}^p(a,b)$ is the set of all functions $x(t)$ Lebesgue-integrable[4] on $(a,b)$ such that

$$\int_a^b |x(t)|^p\,dt<\infty$$

---

[3] The supremum of a set $X$, denoted as $\sup(X)$, is the least upper bound of the set. The infimum of a set $X$, denoted as $\inf(X)$, is the greatest lower bound of the set. For all sets bounded above (below), a sup (inf) exists. Further, if the set $X$ possesses a maximum (minimum), then $\sup(X)=\max(X)$ $(\inf(X)=\min(X))$.

[4] $\mathbf{L}$ indicates the integral is taken in the Lebesgue sense. For functions arising in applied problems the Lebesgue integral of a function is the same as the usual Riemann integral. All Riemann integrable functions leading to proper integrals are Lebesgue-integrable (also called Lebesgue measurable), in which case the results of Riemann and Lebesgue integration coincide.

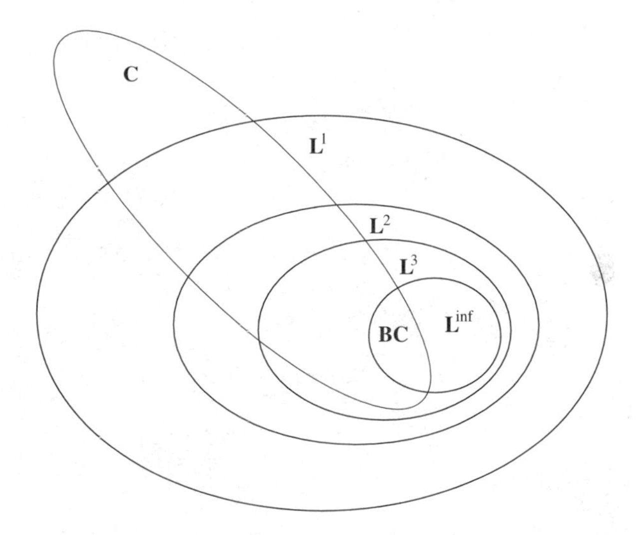


Figure 2.1: Illustration of some interrelationships among the function spaces **C**, **BC**, and $\mathbf{L}^p$ for finite regions $\Omega \subset \mathbf{R}^n$.

for $1 \leq p < \infty$. Note that $x(t) \in \mathbf{L}^1(a, b)$ implies that $x$ is *absolutely integrable.*

The most important set corresponds to $p = 2$. Such functions are said to be *square integrable*. For most functions arising in applications, the condition $\int_a^b |x(t)|^2 dt < \infty$, where $a, b$ are finite, indicates that the function is associated with finite energy.

It can be shown that for finite intervals[5] $(a, b)$ the following inclusion is true;[6]

$$\mathbf{L}^\infty(a, b) \subset \cdots \subset \mathbf{L}^p \subset \mathbf{L}^{p-1} \subset \cdots \subset \mathbf{L}^2 \subset \mathbf{L}^1(a, b).$$

For infinite regions ($m(E) = \infty$), say $(0, \infty)$, neither $\mathbf{L}^p \subset \mathbf{L}^{p-1}$ nor $\mathbf{L}^{p-1} \subset \mathbf{L}^p$ is generally true. Continuous functions do not, in general, form a subset of any space $\mathbf{L}^p$, but bounded continuous functions on finite intervals are a subset of $\mathbf{L}^\infty$ and hence of all $\mathbf{L}^p$ spaces. These relationships are depicted in Figure 2.1.

[5]Finite intervals are said to have finite Lebesgue measure $m(E) < \infty$ of the measurable set of real numbers $E$, defined by $m(E) = \int_a^b \Lambda_E(t)\, dt$ with $\Lambda_E(t) = 1$ for $t \in E$ and $\Lambda_E(t) = 0$ for $t \notin E$.

[6]$L^\infty(a, b)$ is the space of all Lebesgue-measurable functions $x(t)$ on $(a, b)$ and *bounded almost everywhere* (a.e.) on $(a, b)$ defined by

$$L^\infty(a, b) \equiv \{x : |x(t)| \leq \alpha \text{ a.e.}, \ t \in (a, b), \ \alpha > 0\}.$$

**Examples**

1. $\sin(x) \in \mathbf{L}^p(-\pi, \pi)$ for $p \geq 1$.

2. $t^{-1/\alpha} \in \mathbf{L}^p(a, b)$ for $\alpha > 0$, $p \geq 1$ and $0 \leq a < b < \infty$, where $a \neq 0$ for $\alpha = p$.

3. $x(t) = 1/t \notin \mathbf{L}^1(a, \infty)$, although $x(t) \in \mathbf{L}^p(a, \infty)$ for $p > 1$ if $a > 0$.

4. $t^{-1/p} \notin \mathbf{L}^p(a, \infty)$ for $p \geq 1$.

5. If $x(t) \in \mathbf{L}^2(-\infty, \infty)$, then Parseval's relation [1]

$$\infty > \int_{-\infty}^{\infty} |x(t)|^2 \, dt = \int_{-\infty}^{\infty} |X(\omega)|^2 \, d\omega$$

   states that $X(\omega) \in \mathbf{L}^2(-\infty, \infty)$, where $X(\omega)$ is the Fourier transform.

Weighted Lebesgue spaces $\mathbf{L}_w^p(a, b)$ are also of interest in many applications, where $x \in \mathbf{L}_w^p(a, b)$ if

$$\int_a^b |x(t)|^p \, w(t) \, dt < \infty$$

for some weight $w$.

**Lebesgue Spaces of Multivariable and Vector Functions**

$\mathbf{L}^p(\Omega)$ and $\mathbf{L}^p(\Omega)^m$: The space $\mathbf{L}^p(a, b)$ defined above for functions of one variable can be generalized for functions of several variables in a natural way, e.g., $\mathbf{L}^p(\Omega)$ is the set of all functions Lebesgue integrable to the $p$th power on $\mathbf{t} = (t_1, t_2, \ldots, t_n) \in \Omega \subseteq \mathbf{R}^n$ such that

$$\int_\Omega |x(\mathbf{t})|^p \, d\Omega < \infty$$

for $1 \leq p < \infty$, where $d\Omega = dt_1 \, dt_2 \cdots dt_n$.

Further generalizing, the elements could be complex-valued $m$-tuples of $n$-dimensional functions, i.e., $\mathbf{x}(\mathbf{t}) = (x_1(\mathbf{t}), x_2(\mathbf{t}), \ldots, x_m(\mathbf{t}))$, where $\mathbf{t} = (t_1, t_2, \ldots, t_n) \in \Omega \subseteq \mathbf{R}^n$. In electromagnetics this is a common occurrence, where we consider a complex-valued three-component vector field, with each component a function of three spatial coordinates and possibly a time coordinate, e.g., $\mathbf{E} = (E_x(x, y, z, t), E_y(x, y, z, t), E_z(x, y, z, t))$. We will make a notational distinction for this case, such that most generally $\mathbf{L}^p(\Omega)^m$ is the set of all $m$-tuples of $n$-dimensional functions $\mathbf{x}(\mathbf{t}) =$

$(x_1(\mathbf{t}), x_2(\mathbf{t}), \ldots, x_m(\mathbf{t}))$, Lebesgue integrable to the $p$th power such that[7]

$$\int_\Omega |\mathbf{x}(\mathbf{t})|^p \, d\Omega < \infty, \tag{2.1}$$

where $1 \leq p \leq \infty$. Note that

$$|\mathbf{x}(\mathbf{t})|^p = (\mathbf{x}(\mathbf{t}) \cdot \overline{\mathbf{x}}(\mathbf{t}))^{p/2},$$

which naturally reduces to the scalar multivariable ($\mathbf{L}^p(\Omega)^1 \equiv \mathbf{L}^p(\Omega)$) and single-variable cases (see also [2], [3] for the important case $p = 2$).

As an example, in a three-dimensional space $\Omega \subset \mathbf{R}^3$, $\mathbf{L}^2(\Omega)^3$ is the set of vector fields $\mathbf{E}(\mathbf{r}) = (E_x(\mathbf{r}), E_y(\mathbf{r}), E_z(\mathbf{r}))$ such that

$$\int_\Omega |\mathbf{E}(\mathbf{r})|^2 \, d\Omega < \infty.$$

Clearly this indicates that the energy associated with the field in $\Omega$ is finite.

The spaces $\mathbf{C}^k(a,b)$, $\mathbf{M}(a,b)$, and $\mathbf{BC}(a,b)$ are similarly generalized for $n$-dimensional scalar functions on $\Omega \subset \mathbf{R}^n$ as $\mathbf{C}(\Omega)$, $\mathbf{C}^k(\Omega)$, $\mathbf{M}(\Omega)$, and $\mathbf{BC}(\Omega)$, respectively, and for $m$-tuple $n$-variable vectors as $\mathbf{C}(\Omega)^m$, $\mathbf{C}^k(\Omega)^m$, $\mathbf{M}(\Omega)^m$, and $\mathbf{BC}(\Omega)^m$, respectively.[8]

For $\mathbf{C}^k(\Omega)$ we need to consider all possible combinations of various partial derivatives, the total order of which is $k$, such that $D^k x(\mathbf{t}) \in \mathbf{C}(\Omega)$ where $D^k$ is the *generalized partial derivative operator*, defined for all $|\alpha| \leq k$ as[9]

$$D^\alpha x(\mathbf{t}) = \frac{\partial^{|\alpha|} x(\mathbf{t})}{\partial t_1^{\alpha_1} \partial t_2^{\alpha_2} \cdots \partial t_n^{\alpha_n}},$$

where $|\alpha| = \alpha_1 + \alpha_2 + \cdots + \alpha_n$, with $\alpha_1, \alpha_2, \ldots, \alpha_n \geq 0$.

As an example, if $n = 2$ and $\alpha = (3,1)$,

$$D^\alpha x(t_1, t_2) = \frac{\partial^4 x(\mathbf{t})}{\partial t_1^3 \partial t_2}.$$

For $\alpha = 1$, $D^1 = d/dt$ and for $\alpha = 0$, $D^0 = 1$.

In relation to adjoints discussed in Chapter 3, it is useful to note that for $x, y \in \mathbf{C}^k(\Omega)$,

$$\int_\Omega D^\alpha x(\mathbf{t}) y(\mathbf{t}) \, d\Omega = (-1)^{|\alpha|} \int_\Omega x(\mathbf{t}) D^\alpha y(\mathbf{t}) \, d\Omega.$$

---

[7] Note that $\mathbf{L}^p(\Omega)^m$ for $m$-tuple vectors represents a direct sum (see Section 2.3.1, Definition 2.25) of spaces $\mathbf{L}^p(\Omega)$ for scalar $n$-dimensional components such that $\mathbf{L}^p(\Omega)^m = (\oplus_i \mathbf{L}^p(\Omega))_{i=1}^m = \mathbf{L}^p(\Omega) \oplus \mathbf{L}^p(\Omega) \oplus \cdots \oplus \mathbf{L}^p(\Omega)$. Therefore, one can see that if $\mathbf{x} \in \mathbf{L}^p(\Omega)^m$, then scalar components of $\mathbf{x}$ are in $\mathbf{L}^2(\Omega)$.

[8] $\mathbf{C}(\Omega)^m, \mathbf{C}^k(\Omega)^m, \mathbf{M}(\Omega)^m$, and $\mathbf{BC}(\Omega)^m$ represent a direct sum of $\mathbf{C}(\Omega)$, $\mathbf{C}^k(\Omega)$, $\mathbf{M}(\Omega)$, and $\mathbf{BC}(\Omega)$, respectively.

[9] Most generally, $D^\alpha x(\mathbf{t})$ is understood in the distributional sense [35] as the $\alpha$th *distributional partial derivative* of a distribution $x(\mathbf{t})$ defined on $\Omega$. Within this framework, $D^\alpha x(\mathbf{t})$ represents the $\alpha$th *weak* partial derivative. If $|\alpha| \leq k$, this definition coincides with the conventional $\alpha th$ derivative.

### Hölder Space

The Hölder space $\mathbf{H}_{0,\alpha}(a,b)$ is the set of all functions $x(t)$ defined on a bounded interval $(a,b)$ such that for any two values $t$ and $s$ from this interval there is a constant $k>0$ such that

$$|x(t)-x(s)| \leq k\,|t-s|^{\alpha},$$

where $0<\alpha\leq 1$. The case of $\alpha=1$ is also known as a *Lipschitz* space. Functions $x(t)\in\mathbf{H}_{0,\alpha}(a,b)$ are said to be Hölder (Lipschitz) continuous, as discussed in Section 2.2.1 (see Definition 2.5).

The one-dimensional Hölder space $\mathbf{H}_{0,\alpha}(a,b)$ can be generalized for functions $x(\mathbf{t})$ of $n$ variables ($\mathbf{t}\in\Omega$) such that $x(\mathbf{t})\in\mathbf{H}_{0,\alpha_1\ldots\alpha_n}(\Omega)$ if for any two points $\mathbf{t}$ and $\mathbf{s}$ from $\Omega$ there exist constants $k_i>0$ such that

$$|x(\mathbf{t})-x(\mathbf{s})| \leq \sum_{i=1}^{n} k_i\,|t_i-s_i|^{\alpha_i},$$

where $0<\alpha_i\leq 1$ [4, p. 12]. Note that if $x(\mathbf{t})\in\mathbf{H}_{0,\alpha_1\ldots\alpha_n}(\Omega)$, then it satisfies the Hölder condition uniformly with respect to each variable $t_j$,

$$|x(t_1,\ldots,t_j,\ldots,t_n)-x(t_1,\ldots,s_j,\ldots,t_n)| \leq k_j\,|t_j-s_j|^{\alpha_j}$$

for $j=1,\ldots,n$. It can be shown that for any $\mathbf{t}$ from the local neighborhood of $\mathbf{s}\in\Omega$ (i.e., $|\mathbf{t}-\mathbf{s}|=o(\mathbf{1})$ as $\mathbf{t}\to\mathbf{s}$), if $1>\alpha_p>\alpha_r>0$, then

$$\mathbf{H}_{0,1}(\Omega)\subset\mathbf{H}_{0,\alpha_p}\subset\mathbf{H}_{0,\alpha_r}(\Omega).$$

Often in electromagnetics one is interested in the space of $m$-tuple complex vector functions of $n$-variables $\mathbf{x}(\mathbf{t})\in\mathbf{H}_{0,\alpha}(\Omega)^m$, defined by the condition

$$|\mathbf{x}(\mathbf{t})-\mathbf{x}(\mathbf{s})| \leq k\,|\mathbf{t}-\mathbf{s}|^{\alpha},$$

where $0<\alpha\leq 1$ and $k>0$. This type of condition is often imposed on current densities, as discussed in Section 1.3.4, and on field quantities.

### Examples

1. $x(t)=\left\{\begin{array}{cc} 1/\ln t, & 0<t\leq\frac{1}{2}, \\ 0, & t=0 \end{array}\right\}.$

   This function is continuous on the closed interval $t\in[0,1/2]$, i.e., $x(t)\in\mathbf{C}[0,1/2]$. However, one can observe that the function $t^{\alpha}$ approaches the origin faster than $\ln t$ approaches $-\infty$ for *any* $\alpha>0$, and $\lim_{t\to 0}t^{\alpha}\ln t=0$. This means that we can always find a value of $t$ such that

$$|x(t)-x(0)|=\left|\frac{1}{\ln t}\right|>kt^{\alpha}$$

for any $k$ and $\alpha$ $(0 < \alpha \leq 1)$. Obviously, $x(t) \notin \mathbf{H}_{0,\alpha}[\alpha, 1/2]$, where $\alpha \in (0, 1/2)$. However, $x(t)$ satisfies the Hölder condition on the semi-open interval $t \in (\alpha, 1/2]$ such that $x(t) \in \mathbf{H}_{0,\alpha}(\alpha, 1/2]$ [5, p. 6].

2. $x(t) = t^{1/n} \in \mathbf{H}_{0,\alpha}[0, \infty)$ for $\alpha = 1/n,\ n \in \mathbf{N}$ and $t^{1/n} \in \mathbf{H}_{0,1}(0, \infty)$ (Lipschitz continuous on the open interval $t \in (0, \infty)$) for any $n \in \mathbf{N}$. Note that the function $x(t)$ is not analytic at the origin.

3. The singular integral $\int_a^b x(t) / (t - \tau)\, dt$, where $x(t)$ satisfies a Hölder condition, is equal to

$$\text{P.V.} \int_a^b \frac{x(t)}{t - \tau}\, dt = \int_a^b \frac{x(t) - x(\tau)}{t - \tau}\, dt + x(\tau) \ln \frac{b - \tau}{\tau - a}$$

in the sense of the Cauchy principal value [4], [5]. The first integral on the right side exists as an improper one where the integrand is considered in view of the Hölder condition

$$\left| \frac{x(t) - x(\tau)}{t - \tau} \right| \leq \frac{k}{|t - \tau|^{1-\alpha}}$$

with $0 < \alpha \leq 1$ (see the discussion of Equation (1.62)). The second term involves the singular integral $\int_a^b 1 / (t - \tau)\, dt$ taken in the sense of the Cauchy principal value

$$\text{P.V.} \int_a^b \frac{dt}{t - \tau} = \lim_{\varepsilon \to 0} \left[ \int_a^{\tau - \varepsilon} \frac{dt}{t - \tau} + \int_{\tau + \varepsilon}^b \frac{dt}{t - \tau} \right] = \ln \frac{b - \tau}{\tau - a}.$$

We can also introduce a Hölder space $\mathbf{H}_{1,\alpha}(a, b)$ of continuously differentiable functions $x(t)$ such that $dx(t)/dt$ belongs to the Hölder space $\mathbf{H}_{0,\alpha}(a, b)$, where $0 < \alpha \leq 1$. Further generalizing, we can define the Hölder space $\mathbf{H}_{k,\alpha}(\Omega)$ of functions $x(\mathbf{t})$ which are $k$-times Hölder continuously differentiable, i.e., $D^k x(\mathbf{t}) \in \mathbf{H}_{0,\alpha}(\Omega)$, where

$$\mathbf{H}_{\infty,\alpha}(\Omega) \subset \cdots \subset \mathbf{H}_{k,\alpha} \subset \mathbf{H}_{k-1,\alpha} \subset \cdots \subset \mathbf{H}_{1,\alpha} \subset \mathbf{H}_{0,\alpha} \subset \mathbf{C}(\Omega).$$

For $m$-tuple vectors of $n$ variables, $\mathbf{x}(\mathbf{t})$, one can define Hölder spaces $\mathbf{H}_{k,\alpha}(\Omega)^m$ similarly, i.e., $\mathbf{x}(\mathbf{t}) \in \mathbf{H}_{k,\alpha}(\Omega)^m$ if $L^k(\nabla)\, \mathbf{x}(\mathbf{t}) \in \mathbf{H}_{0,\alpha}(\Omega)^m$, where $L^k(\nabla)$ is an appropriately defined $k$th-order operator involving $\nabla$ such that $L^0 = 1$. For example, $\mathbf{H}_{k,\alpha}(\Omega)^3$ could be defined such that $\mathbf{x}(\mathbf{t}) \in \mathbf{H}_{k,\alpha}(\Omega)^3$ if

$$\underbrace{(\nabla \times \nabla \times \nabla \times \cdots)}_{k \text{ times}} \mathbf{x} \in \mathbf{H}_{0,\alpha}(\Omega)^3.$$

As another example, $\mathbf{H}_{1,\alpha}(\Omega)^3$ could be defined such that $\mathbf{x}(\mathbf{t}) \in \mathbf{H}_{1,\alpha}(\Omega)^3$ if $\nabla \cdot \mathbf{x} \in \mathbf{H}_{0,\alpha}(\Omega)$, and $\mathbf{H}_{1,\alpha}(\Omega)$ could be defined such that $x(\mathbf{t}) \in \mathbf{H}_{1,\alpha}(\Omega)$ if $\nabla x \in \mathbf{H}_{0,\alpha}(\Omega)^3$. These types of spaces are naturally of interest when $L^k(\nabla)\mathbf{x}$ represents a source term for a field (see, e.g., (1.76), wherein $\nabla \times \mathbf{J}_{e(m)}(\mathbf{r})$ represent source terms in the volume integral leading to the electric field).

### Sobolev Space

A notion relevant to $\mathbf{L}^2$ spaces is that of a finite energy condition, both for field components and for fields. For vectors $\mathbf{x}(\mathbf{t}) \in \mathbf{L}^2(\Omega)^3$ representing an electric- or magnetic field quantity in the time-harmonic or transform domain, if $\nabla \times \mathbf{x}(\mathbf{t}) \in \mathbf{L}^2(\Omega)^3$, then the total energy in a bounded domain $\Omega$ containing a homogeneous isotropic medium should be finite, i.e.,[10]

$$\iiint_\Omega \left(|\mathbf{x}(\mathbf{t})|^2 + |\nabla \times \mathbf{x}(\mathbf{t})|^2\right) d\Omega < \infty.$$

For $n$-dimensional scalars with $x(\mathbf{t}) \in \mathbf{L}^2(\Omega)$ and $\nabla x(\mathbf{t}) \in \mathbf{L}^2(\Omega)^3$, we obtain

$$\int_\Omega \left(|x(\mathbf{t})|^2 + |\nabla x(\mathbf{t})|^2\right) d\Omega < \infty,$$

and in one dimension if $x(t) \in \mathbf{L}^2(a,b)$ and $dx(t)/dt \in \mathbf{L}^2(a,b)$, then

$$\int_a^b \left(|x(t)|^2 + |dx(t)/dt|^2\right) dt < \infty.$$

The idea of enforcing a finite energy condition on the total field leads to the concept that subspaces of $\mathbf{L}^p$ accommodating a finite energy condition should be distinguished. For $n$-dimensional scalars the *Sobolev space* $\mathbf{W}_p^k(\Omega)$ of order $k$ ($k \geq 0$, $k$ an integer) defined in an $n$-dimensional domain $\Omega \subset \mathbf{R}^n$ is the space of all functions $x(\mathbf{t})$ of $n$ variables in $\mathbf{L}^p(\Omega)$ $(1 \leq p < \infty)$ that, together with all of their partial derivatives up to order $k$, belong to $\mathbf{L}^p(\Omega)$;

$$\mathbf{W}_p^k(\Omega) \equiv \{x(\mathbf{t}) : \ x(\mathbf{t}) \in \mathbf{L}^p(\Omega),\ D^\alpha x(\mathbf{t}) \in \mathbf{L}^p(\Omega),\ |\alpha| \leq k\}.$$

[10]From Section 1.2.2 we have $\int_\Omega \left(\varepsilon|\mathbf{E}|^2 + \mu|\mathbf{H}|^2\right) d\Omega < \infty$, which can also be represented in terms of $\mathbf{E}$ ($\mathbf{H}$) and $\nabla \times \mathbf{E}$ ($\nabla \times \mathbf{H}$). The finite energy condition can often be reduced to a scalar one dealing with scalar field components $E_\alpha$, $H_\beta$, which are connected via a derivative operator.

In practical applications of problems with edges we require that

$$\lim_{V \to 0} \int_V \left(\varepsilon|\mathbf{E}|^2 + \mu|\mathbf{H}|^2\right) dV = 0,$$

which guarantees that there is no source in $V$ and imposes certain restrictions on the field behavior in the vicinity of the edge, namely that components of $\mathbf{E}$ and $\mathbf{H}$ normal to the edge will not grow faster than $O(\delta^{-1+\tau})$ as $\delta \to 0$, where $\tau > 0$ ($\tau$ depends on the curvature of the edge) and $\delta$ is the distance to the edge.

Obviously, $\mathbf{W}_p^k(\Omega) \subset \mathbf{L}^p(\Omega)$ and $\mathbf{W}_p^0(\Omega) = \mathbf{L}^p(\Omega)$. It can be shown that if $l \geq k$ then

$$\mathbf{W}_p^l(\Omega) \subseteq \mathbf{W}_p^k(\Omega)$$

for any $1 \leq p < \infty$.

The Sobolev space for $p = 2$, $\mathbf{W}_2^k(\Omega)$, is a Hilbert space (see Section 2.5.4), often denoted by $\mathbf{H}^k(\Omega)$, which determines a set of Lebesgue square-integrable functions $x(\mathbf{t})$ that, together with their partial derivatives, satisfy a finite energy condition in any bounded domain $\Omega$. The order $k$ determines the degree of smoothness of the function $x(\mathbf{t}) \in \mathbf{W}_p^k(\Omega)$ in the sense of the existence of a set of $k$th-order partial derivatives that are Lebesgue integrable to the $p$th power on $\Omega$.

For the case of $m$-tuple vectors of $n$ variables we can also define various Sobolev spaces $\mathbf{W}_p^k(\Omega)^m$ in terms of the general operator $L^k(\nabla)$. For example, several different spaces $\mathbf{W}_2^1(\Omega)^3$ (with $L^1(\nabla) = \nabla\times$ and $L^1(\nabla) = \nabla\cdot$) are particularly important in electromagnetics [6], [2],

$$\mathbf{H}(\text{curl}, \Omega) \equiv \{\mathbf{x}(\mathbf{t}) : \ \mathbf{x}(\mathbf{t}) \in \mathbf{L}^2(\Omega)^3, \ \nabla \times \mathbf{x}(\mathbf{t}) \in \mathbf{L}^2(\Omega)^3\},$$
$$\mathbf{H}(\text{div}, \Omega) \equiv \{\mathbf{x}(\mathbf{t}) : \ \mathbf{x}(\mathbf{t}) \in \mathbf{L}^2(\Omega)^3, \ \nabla \cdot \mathbf{x}(\mathbf{t}) \in \mathbf{L}^2(\Omega)\},$$

and

$$\mathbf{H}(\text{curl}, \text{div}, \Omega) \equiv \{\mathbf{x}(\mathbf{t}) : \ \mathbf{x}(\mathbf{t}) \in \mathbf{L}^2(\Omega)^3, \nabla \times \mathbf{x}(\mathbf{t}) \in \mathbf{L}^2(\Omega)^3, \nabla \cdot \mathbf{x}(\mathbf{t}) \in \mathbf{L}^2(\Omega)\},$$

where $\mathbf{H}(\text{curl}, \text{div}, \Omega) = \mathbf{H}(\text{curl}, \Omega) \cap \mathbf{H}(\text{div}, \Omega)$. Note that even if $\mathbf{x}$ is such that $\mathbf{x} \in \mathbf{H}(\text{curl}, \text{div}, \Omega)$, derivatives of individual components of $\mathbf{x}$ are not necessarily in $\mathbf{L}^2(\Omega)$ [6].

### 2.1.3 Sets of Sequences—Sequence Spaces

1. $\mathbf{l}^p$ is the set of all (real or complex) sequences $\mathbf{x} = \{x_1, x_2, \ldots\}$ such that

$$\sum_{i=1}^{\infty} |x_i|^p < \infty,$$

where $1 \leq p < \infty$. If $\infty > p_2 > p_1 \geq 1$, then the strict inclusion $\mathbf{l}^{p_1} \subset \mathbf{l}^{p_2}$ can be shown to be true. We can also consider $\mathbf{l}^p$ to be defined for sequences $\{\ldots, x_{-1}, x_0, x_1, \ldots\}$ by $\sum_{i=-\infty}^{\infty} |x_i|^p < \infty$.

**Examples**

(a) $\{1/n^{\alpha/p}\} \in \mathbf{l}^p$ for $\alpha > 1$ and any $1 \leq p < \infty$.
(b) $\{1/n\} \in \mathbf{l}^2$, although $\{\frac{1}{n}\} \notin \mathbf{l}^1$.
(c) $\{\ln(\alpha n)/n\} \in \mathbf{l}^2$ for $\alpha > 0$.

We will see that in many function spaces the expansion $\sum_{n=1}^{\infty} a_n x_n$ converges (in the norm sense) if and only if $\{a_n\} \in \mathbf{l}^2$ (Theorem 2.11), making this sequence space one of primary importance. The strong relationship between $\mathbf{l}^2$ and $\mathbf{L}^2$ (more generally, any separable Hilbert space) is discussed in Section 3.1.3. We can also define sequences of $m$-tuple vectors $\{\mathbf{x}_1, \mathbf{x}_2, \ldots\}$ in $\mathbf{l}^p_m$ in a similar manner, where $\mathbf{x}_i = \{x_1^i, x_2^i, \ldots, x_m^i\}$.

2. $\mathbf{l}^\infty$ is the set of all bounded sequences $\mathbf{x} = \{x_1, x_2, \ldots\}$ (real or complex) such that

$$\sup |x_i| < \infty$$

for $i \in \mathbf{N}$. It can be shown that for any $p \geq 1$, $\mathbf{l}^p \subset \mathbf{l}^\infty$. We can also define a space $\mathbf{c} \subset \mathbf{l}^\infty$ that contains all convergent sequences (it is obvious that each convergent sequence is bounded, but the converse is not necessarily true). The space $\mathbf{c}_0 \subset \mathbf{c}$ contains only those sequences that converge to zero, i.e., $\lim_{i \to \infty} x_i = 0$.

3. $\mathbf{l}^2(\mu)$ is the set of all (real or complex) sequences $\mathbf{x} = \{x_1, x_2, \ldots\}$ such that

$$\sum_{i=1}^{\infty} |x_i|^2 \, i^{\mu} < \infty$$

for $\mu \geq 0$. Note that $\mathbf{l}^2(0) = \mathbf{l}^2$, and $\mathbf{l}^2(1) = \widetilde{\mathbf{l}^2}$, where $\widetilde{\mathbf{l}^2}$ is the set of all (real or complex) sequences $\mathbf{x} = \{x_1, x_2, \ldots\}$ such that

$$\sum_{i=1}^{\infty} |x_i|^2 \, i < \infty.$$

It is obvious that $\mathbf{l}^1 \subset \widetilde{\mathbf{l}^2} \subset \mathbf{l}^2$ and, in general, $\mathbf{l}^2(\mu_1) \subset \mathbf{l}^2(\mu_2)$ for $\mu_1 > \mu_2$.

Figure 2.2 indicates interrelationships among several sequence spaces.

Sequence spaces represent the analog of function spaces in the sense of an isomorphism, as discussed in Section 3.1.3, correlating functional (e.g., integral, differential) equations with matrix equations. The examples below demonstrate some correlations between function and sequence spaces occurring in boundary value problems of applied electromagnetics. By necessity, some details concerning the problem formulations are omitted, although references are made to Part II of the text where possible. Some related material is discussed in Sections 3.7 and 9.5.

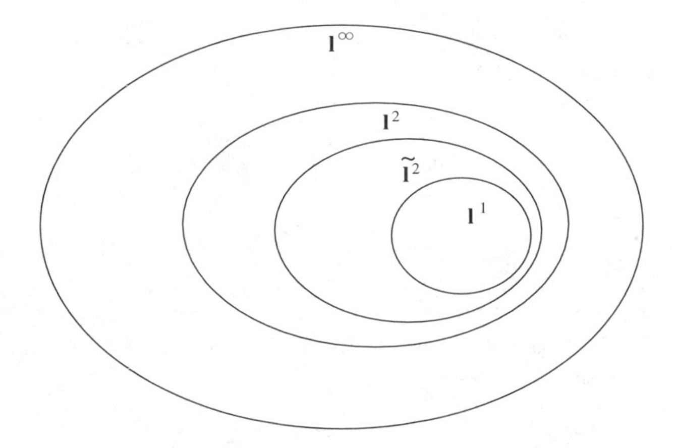


Figure 2.2: Illustration of some interrelationships among the sequences spaces $\mathbf{l}^p$ and $\widetilde{\mathbf{l}}^2$.

**Examples**

1. In Example 5.3 in Section 5.1.3, we seek a solution of the differential equation

$$\left(-\frac{d^2}{dx^2}+\gamma^2\right)u(x)=f(x)$$

in the form of an eigenfunction expansion, where $u,f\in\mathbf{L}^2(0,a)$, with $\operatorname{Re}\gamma>0$ and $u(0)=u(a)=0$. The solution is

$$u(x)=\sum_{n=1}^{\infty}a_n\sin\frac{n\pi}{a}x, \tag{2.2}$$

where

$$a_n=\frac{\frac{2}{a}\int_0^a f(x)\sin\frac{n\pi}{a}x\,dx}{\left(\frac{n\pi}{a}\right)^2+\gamma^2}. \tag{2.3}$$

For (2.2) to converge in the $\mathbf{L}^2$ norm, we must have $\{a_n\}\in\mathbf{l}^2$ (see Theorem 2.11), i.e., $\int_0^a f(x)\sin(n\pi x/a)\,dx\sim O(n)$. In this case, if we make the weak assumption that $f$ is integrable (i.e., $f\in\mathbf{L}^1(0,a)$), then the coefficients are in $\mathbf{l}^1\subset\mathbf{l}^2$. Indeed, as $n\to\infty$

$$|a_n|\sim\left|\frac{\frac{2}{a}\int_0^a f(x)\sin\frac{n\pi}{a}x\,dx}{\left(\frac{n\pi}{a}\right)^2}\right|\leq\frac{2a}{n^2\pi^2}\int_0^a|f(x)|\,dx.$$

2. In this example we show that the imposition of a finite energy condition leads to the sequence space $\widetilde{\mathbf{l}}^2$.

Consider a section of two-dimensional parallel-plate waveguide, within which an electric field can be represented as[11]

$$E_y(x,z) = \sqrt{\frac{2}{a}} \sum_{m=1}^{\infty} a_m \sin \frac{m\pi}{a} x e^{-\gamma_m z} \tag{2.4}$$

with the propagation constant $\gamma_m = \sqrt{(m\pi/a)^2 - k_0^2}$. If $E_y(x,z) \in \mathbf{L}^2(\Omega)$ (or any separable Hilbert space), then for (2.4) to converge we must have $\{a_m\} \in \mathbf{l}^2$ by Theorem 2.11, as discussed later. We can obtain a sharper condition on the coefficients by imposing the structure of a Sobolev space (finite energy condition) on $E_y$.

Let the solution $E_y(x,z)$ and its gradient $\nabla E_y(x,z)$ be defined within $\mathbf{L}^2(\Omega)$, where $\Omega \subset \mathbf{R}^2$ is the closed domain $\Omega = \{(x,z) : x \in [0,a], z \in [0,c]\}$. Therefore, $E_y$ is defined in the Sobolev space $\mathbf{W}_2^1(\Omega)$ such that

$$\int_0^a \int_0^c \left( |E_y(x,z)|^2 + |\nabla E_y(x,z)|^2 \right) dz\, dx < \infty. \tag{2.5}$$

The gradient $\nabla E_y(x,z)$ is written as

$$\begin{aligned} \nabla E_y(x,z) = &\, \hat{\mathbf{x}} \sum_{m=1}^{\infty} a_m \sqrt{\frac{2}{a}} \frac{m\pi}{a} \cos \frac{m\pi}{a} x e^{-\gamma_m z} \\ & - \hat{\mathbf{z}} \sum_{m=1}^{\infty} a_m \sqrt{\frac{2}{a}} \gamma_m \sin \frac{m\pi}{a} x e^{-\gamma_m z}, \end{aligned}$$

where, using orthogonality of the eigenfunctions $\phi_m$,

$$\begin{aligned} & \int_0^a \int_0^c |\nabla E_y(x,z)|^2 \, dz\, dx \\ & \quad = \int_0^a \int_0^c \sum_{m=1}^{\infty} |a_m|^2 \frac{2}{a} \left( \frac{m\pi}{a} \right)^2 \cos^2 \frac{m\pi}{a} x e^{-2\,\mathrm{Re}\,\gamma_m z} dz\, dx \\ & \quad\quad + \int_0^a \int_0^c \sum_{m=1}^{\infty} |a_m|^2 \frac{2}{a} |\gamma_m|^2 \sin^2 \frac{m\pi}{a} x e^{-2\,\mathrm{Re}\,\gamma_m z} dz\, dx \\ & \quad \sim \sum_{m=1}^{\infty} |a_m|^2 m. \end{aligned}$$

Therefore, the finite energy condition (2.5) associated with the definition of the Sobolev space $\mathbf{W}_2^1$ leads to

$$\sum_{m=1}^{\infty} |a_m|^2 m < \infty$$

[11]Complete details leading to (2.4) can be found in Section 8.1.2 (see (8.70)).

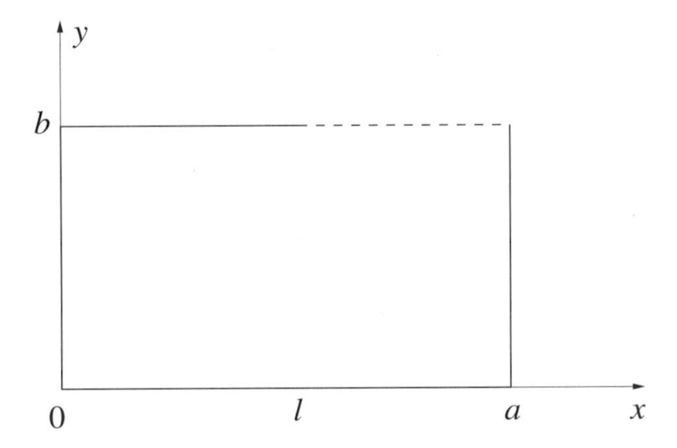


Figure 2.3: Rectangular waveguide with an aperture to demonstrate that the Hölder space $\mathbf{H}_{0,\alpha}$ for a solution in the vicinity of an edge determines the sequence space $\mathbf{l}^1$ for the coefficients in the associated eigenmode expansion.

such that $\{a_m\} \in \widetilde{\mathbf{l}^2} \subset \mathbf{l}^2$. Note that $E_y(x,z) \in \mathbf{W}_2^1(\Omega) \subset \mathbf{L}^2(\Omega)$, such that (2.5) further narrows the function space $\mathbf{L}^2(\Omega)$ in which we seek the solution $E_y(x,z)$.

3. In this example we show that if we seek the solution to a boundary value problem in the Hölder space $\mathbf{H}_{0,\alpha}$ on a boundary, then coefficients in its expansion necessarily belong to the sequence space $\mathbf{l}^1$.

   Consider a uniform rectangular waveguide containing an aperture, as shown in Figure 2.3. This geometry may be encountered in the decomposition of a complex structure into simple substructures (e.g., a mode-matching analysis of shielded transmission lines and waveguides). A boundary value problem can be formulated for a scalar field component $E(x,y)$ (see (1.37)) defined on the cross-section $S = \{(x,y) : x \in [0,a], y \in [0,b]\}$,

$$\left(\frac{\partial^2}{\partial x^2} + \frac{\partial^2}{\partial y^2} + \kappa^2\right) E(x,y) = 0, \qquad (x,y) \in S,$$

   where $\kappa^2 = k_0^2 + \gamma^2$ and $k_0 = \omega\sqrt{\varepsilon_0\mu_0}$, with $\gamma$ the propagation constant of the waveguide ($e^{j\omega t - \gamma z}$ dependence assumed and suppressed). We enforce homogeneous Dirichlet boundary conditions on the conducting part of the closed contour $\Gamma$ (excluding the aperture $\Gamma_a = \{(x,y) : x \in [l,a], y = b\}$),

$$E(x,y) = 0, \qquad (x,y) \in \Gamma - \Gamma_a.$$

Using Green's second theorem, the solution of the above problem can be obtained as an integral over the aperture $\Gamma_a$,

$$E(x,y) = -\int_l^a E(x',b)\frac{\partial g(x,y;x',b)}{\partial y'}\,dx', \tag{2.6}$$

where the Green's function $g$ is expressed as a series expansion over the complete system of eigenfunctions $\Phi_n$ of the one-dimensional Laplacian (see Chapter 9),

$$g(x,y;x',y') = \sum_{n=1}^{\infty} \Phi_n(y)\Phi_n(y')g_x(x,x',k_{xn}). \tag{2.7}$$

The eigenfunctions are

$$\Phi_n(y) = \sqrt{\frac{2}{b}}\sin\frac{n\pi}{b}y,$$

and the characteristic Green's function is

$$g_x(x,x',k_{xn}) = \frac{-1}{k_{xn}\sin k_{xn}a}\begin{cases} \sin k_{xn}x'\sin k_{xn}(x-a), & x \geq x', \\ \sin k_{xn}x\sin k_{xn}(x'-a), & x < x', \end{cases}$$

where $k_{xn} = \sqrt{\kappa^2 - (n\pi/b)^2}$. Using (2.7) in (2.6) we obtain along the line $x = l$, $y \in [0,b]$,

$$E(l,y) = \sum_{n=1}^{\infty} a_n\Phi_n(y), \tag{2.8}$$

where

$$a_n = (-1)^n\sqrt{\frac{2}{b}}\frac{n\pi}{b}\frac{\sin k_{xn}l}{k_{xn}\sin k_{xn}a}\int_l^a E(x',b)\sin k_{xn}(x'-a)\,dx'.$$

For (2.8) to converge in $\mathbf{L}^2(0,b)$ we must have $\{a_n\} \in \mathbf{l}^2$ by Theorem 2.11. Taking absolute values and noting that

$$k_{xn} \sim jO(n), \qquad n \to \infty,$$
$$\frac{\sin k_{xn}l}{\sin k_{xn}a} \sim O\left(e^{-\frac{n\pi}{b}(a-l)}\right), \qquad n \to \infty,$$

then the weak assumption that $E(x',b) \in \mathbf{L}^1(\Gamma_a)$ leads to $|a_n| \sim O(1)$. However, if we require the much stronger condition that the solution $E(x',b)$ belongs to the Hölder space $\mathbf{H}_{0,\alpha}(\Gamma_a)$, then we obtain the result $\{a_n\} \in \mathbf{l}^1 \subset \mathbf{l}^2$.

Indeed, if $E(x',b)$ is Hölder continuous, there exist $k>0$ and $0<\alpha\le 1$ such that

$$|E(x',b)-E(l,b)|\le k\,|x'-l|^{\alpha}.$$

Due to the boundary value $E(l,b)=0$, this inequality becomes

$$E(x',b)\le k\,|x'-l|^{\alpha},$$

leading to

$$a_n\le(-1)^n\sqrt{\frac{2}{b}}\frac{n\pi}{b}\frac{\sin k_{xn}l}{k_{xn}\sin k_{xn}a}kI_n,$$

where

$$I_n=\int_l^a |x'-l|^{\alpha}\sin k_{xn}(x'-a)\,dx'. \tag{2.9}$$

The integral (2.9) can be evaluated in terms of hypergeometric Kummer's functions [7],

$$\begin{aligned}I_n=&\frac{1}{2}j(1+\alpha)^{-1}(a-l)^{1+\alpha}\{M(1,2+\alpha,\frac{n\pi}{b}(a-l))\\&-M(1,2+\alpha,-\frac{n\pi}{b}(a-l))\},\end{aligned}$$

and, using representations (13.1.4) and (13.1.5) in [8, p. 504], we estimate the integral $I_n$ as

$$I_n\sim jC(\alpha)O\left(\frac{1}{n^{1+\alpha}}e^{\frac{n\pi}{b}(a-l)}\right),\qquad n\to\infty.$$

Therefore, the final result is

$$|a_n|\le O\left(\frac{1}{n^{1+\alpha}}\right),\qquad n\to\infty,$$

and by the example on p. 73 we see that $\{a_n\}\in\mathbf{l}^1$.

4. In this example the method of overlapping regions (see Section 9.5) is applied to the analysis of a rectangular coaxial line. As in the previous example we show the correlation between the Hölder space $\mathbf{H}_{0,\alpha}$ and the sequence space $\mathbf{l}^1$.

   Consider propagation of TM$^z$-odd modes (with respect to the $E_z$ component) in a rectangular coaxial line (because of the symmetry by electric walls only a quarter of the structure will be analyzed, as shown in Figure 2.4). We can decompose the domain surrounding the inner conductor ($x\in[0,l]$, $y\in[0,t]$) into two overlapping regions $S_1$ and $S_2$,

$$\begin{aligned}S_1&=\{(x,y):x\in[0,a],\ y\in[t,b]\},\\S_2&=\{(x,y):x\in[l,a],\ y\in[0,b]\}.\end{aligned}$$

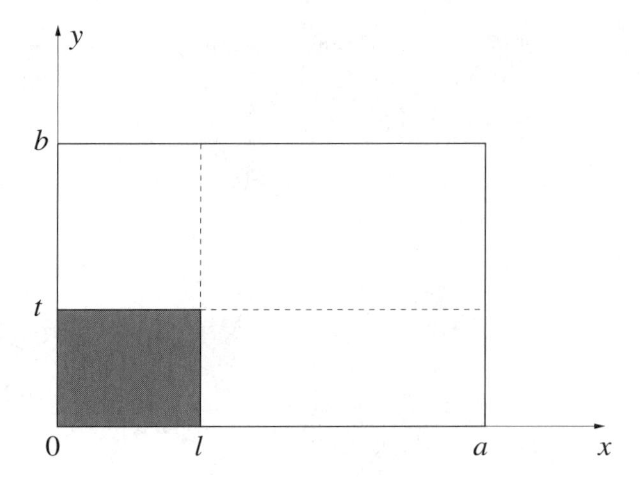


Figure 2.4: A rectangular coaxial line for the problem of the propagation of $TM^z$-modes (one quarter of the structure is shown; symmetry by two electric walls). Solution of the Fredholm-type integral equation of the second kind in the Hölder space $\mathbf{H}_{0,\alpha}$ correlates with $\mathbf{l}^1$ space for the coefficients in the associated eigenmode expansion.

The solution of boundary value problems formulated in regions $S_1$ and $S_2$ (similar to the problem shown in Example 3 above) can be obtained using Green's second theorem, leading to[12]

$$E_{z1}(x,y) = \int_l^a E_{z2}(x',t)\frac{\partial G_1(x,y;x',t)}{\partial y'}\,dx', \quad (x,y) \in \partial\sigma_2, \tag{2.10}$$

$$E_{z2}(x',y') = \int_t^b E_{z1}(l,y'')\frac{\partial G_2(x',y';l,y'')}{\partial x''}\,dy'', \quad (x',y') \in \partial\sigma_1, \tag{2.11}$$

where $\partial\sigma_2 = \{x = l,\ y \in [t,b]\}$, $\partial\sigma_1 = \{x' \in [l,a],\ y' = t\}$, and $G_1$ and $G_2$ are Green's functions for regions $S_1$ and $S_2$, respectively, obtained in the form of partial eigenfunction expansions (similar to that shown in Example 3),

$$G_1(x,y;x',y') = \sum_{n=1}^{\infty} \Phi_n(y)\Phi_n(y')f_n(x,x'),$$

$$G_2(x',y';x'',y'') = \sum_{m=1}^{\infty} \Theta_m(x')\Theta_m(x'')g_m(y',y'').$$

[12]Similar integral representations in the vector-dyadic form (9.199) and (9.200) are given in Section 9.5.1.

Here, the eigenfunctions are

$$\Phi_n(y) = \sqrt{\frac{2}{b-t}} \sin \frac{n\pi}{b-t}(y-t),$$

$$\Theta_m(x') = \sqrt{\frac{2}{a-l}} \sin \frac{m\pi}{a-l}(x'-l),$$

and the characteristic Green's functions are

$$f_n(x,x') = \frac{-1}{k_{xn} \sin k_{xn} a} \begin{cases} \sin k_{xn} x' \sin k_{xn}(x-a), & x \geq x', \\ \sin k_{xn} x \sin k_{xn}(x'-a), & x < x', \end{cases}$$

$$g_m(y',y'') = \frac{-1}{k_{ym} \sin k_{ym} b} \begin{cases} \sin k_{ym} y'' \sin k_{ym}(y'-b), & y' \geq y'', \\ \sin k_{ym} y' \sin k_{ym}(y''-b), & y' < y'', \end{cases}$$

where $k_{xn} = \sqrt{\kappa^2 - (n\pi/(b-t))^2}$ and $k_{ym} = \sqrt{\kappa^2 - (m\pi/(a-l))^2}$.

According to the method of overlapping regions, a Fredholm-type integral equation of the second kind can be obtained by substituting the representation (2.11) into (2.10), resulting in

$$E_{z1}(x,y) = \int_l^a \int_t^b E_{z1}(l,y'') \frac{\partial G_1(x,y;x',t)}{\partial y'} \frac{\partial G_2(x',y';l,y'')}{\partial x''} \, dy'' \, dx'.$$

Substituting representations of the Green's functions in the above equation, we can express the solution $E_{z1}(x,y)$ along the line $x = l$, $y \in [t,b]$ as an eigenfunction expansion,

$$E_{z1}(l,y) = \sum_{n=1}^{\infty} a_n \Phi_n(y),$$

where

$$\begin{aligned} a_n \;=\; & \frac{2}{(a-l)} \sqrt{\frac{2}{b-t}} \frac{n\pi}{(b-t)} \frac{\sin k_{xn} l \sin k_{xn}(a-l)}{k_{xn} \sin k_{xn} a} \\ & \cdot \sum_{m=1}^{\infty} \left(\frac{m\pi}{a-l}\right)^2 \frac{\sin k_{ym} t}{(k_{xn}^2 - (\frac{m\pi}{a-l})^2) k_{ym} \sin k_{ym} b} \\ & \cdot \int_t^b E_{z1}(l,y'') \sin k_{ym}(y''-b) \, dy''. \end{aligned}$$

If $E_{z1}(l,y'') \in \mathbf{H}_{0,\alpha}(\partial\sigma_2)$, then[13]

$$|E_{z1}(l,y'') - E_{z1}(l,t)| \leq k \left|y'' - t\right|^{\alpha},$$

[13]Note that the integral in Green's second theorem should be considered in the Lebesgue sense and, therefore, it defines the $\mathbf{L}$ space for $E_z(x,y)$. In this example, as well as in the previous one, the solution should be sought in the intersection of Lebesgue and Hölder spaces, $E_z(x,y) \in \mathbf{L}^2 \cap \mathbf{H}_{0,\alpha}$.

where $k > 0$, $0 < \alpha \leq 1$, and, because of the boundary condition $E_{z1}(l,t) = 0$ the inequality is reduced to

$$E_{z1}(l, y'') \leq k\, |y'' - t|^{\alpha}\,.$$

The integral is estimated using the result of Example 3,

$$I_m = \int_t^b |y'' - t|^{\alpha} \sin k_{ym}(y'' - b)\, dy'' \sim jC_1(\alpha) O\left(\frac{1}{m^{1+\alpha}} e^{\frac{m\pi}{a-l}(b-t)}\right),$$

$m \to \infty$, and the asymptotic behavior for the coefficients $a_n$ can be obtained as

$$a_n \leq jC_2(\alpha) \sum_{m=1}^{\infty} \frac{m}{m^{1+\alpha}(m^2 + \eta^2 n^2)}, \qquad n \to \infty,$$

where $\eta = (a-l)/(b-t)$. It can be shown [7] that for $0 < \alpha < 1$ the series is evaluated as

$$\sum_{m=1}^{\infty} \frac{m}{m^{1+\alpha}(m^2 + \eta^2 n^2)} \sim \frac{\pi \sin(\alpha \frac{\pi}{2})}{\sin(1-\alpha)\pi} \frac{1}{(\eta n)^{1+\alpha}}$$

and for $\alpha = 1$,

$$\begin{aligned} \sum_{m=1}^{\infty} \frac{1}{m(m^2 + \eta^2 n^2)} &= \frac{1}{2(\eta n)^2} \left(\psi(1 + j\eta n) + \psi(1 - j\eta n) - 2\psi(1)\right) \\ &\sim \frac{\ln \eta n - \psi(1)}{(\eta n)^2}, \end{aligned}$$

where $\psi(z)$ is the psi-function [8]. This leads to the asymptotics for $a_n$ as

$$\begin{aligned} |a_n| &\leq O\left(\frac{1}{(\eta n)^{1+\alpha}}\right), \qquad 0 < \alpha < 1, \quad n \to \infty, \\ |a_n| &\leq O\left(\frac{\ln \eta n}{(\eta n)^2}\right), \qquad \alpha = 1, \quad n \to \infty, \end{aligned}$$

and to the conclusion that $\{a_n\} \in \mathbf{l}^1$.

Furthermore, we obtain

$$E_{z1}(x, y) = \sum_{n=1}^{\infty} a_n \frac{n\pi}{b-t} \sin \frac{n\pi}{b-t}(y-t) \frac{\sin k_{xn}x \sin k_{xn}(a-l)}{k_{nx} \sin k_{xn}a}$$

for $x \in [0, l]$ and $y \in [t, b]$, and the derivative of $E_{z1}$ with respect to $y$ at $y = t$ is

$$\frac{\partial E_{z1}(x,t)}{\partial y} = \sum_{n=1}^{\infty} a_n \left(\frac{n\pi}{b-t}\right)^2 \frac{\sin k_{xn}x \sin k_{xn}(a-l)}{k_{nx} \sin k_{xn}a}.$$

Note that $\partial E_{z1}/\partial y \sim H_{x1}$, and that the asymptotic behavior of the normal $H$ component in the vicinity of a rectangular edge should be $H_{x1} = O(\rho^{-1/3})$ as $\rho \to 0$ [9]. If we assume $a_n = O(n^\tau)$ as $n \to \infty$, then

$$\frac{\partial E_{z1}(x,t)}{\partial y} \sim \sum_{n=1}^{\infty} n^{\tau+1} e^{-n(l-x)}, \qquad x \to l,$$

and, using the result of [9, p. 11], we obtain

$$\sum_{n=1}^{\infty} n^{\tau+1} e^{-n(l-x)} \sim \Gamma(\tau+1)(l-x)^{-\tau-2}, \quad l-x > 0, \quad -1 < \tau+1 < 0,$$

where $\Gamma(z)$ is the gamma-function. A comparison of the above asymptotics for $\partial E_{z1}/\partial y$, and the requirement $H_{x1} = O(\rho^{-1/3})$ lead to the condition $-\tau - 2 = -1/3$, so that $\tau = -5/3$. Therefore, $a_n = O(n^{-5/3})$, $n \to \infty$, and $\{a_n\} \in \mathbf{l}^1$.

The requirement that the solution belongs to the Hölder space $\mathbf{H}_{0,\alpha}$ is consistent with assuming $a_n = O(n^\tau)$ as $n \to \infty$ and imposing the correct asymptotic behavior on the coefficients $a_n$. Indeed, from the previous example we obtained $a_n = O(n^{-1-\alpha})$, where $\alpha = 2/3$ for longitudinal components in the vicinity of a rectangular edge, which is consistent with the result $\tau = -5/3$. In summary, the Hölder space imposes an additional condition (edge condition) on the solution, which is associated with the condition $\{a_n\} \in \mathbf{l}^1$.

The material related to edge problems in connection with appropriate function and sequence spaces has been extensively covered in the Russian and Ukrainian literature; see, for example, [10]–[20], among others.

## 2.2 Metric Space

A metric specifies a distance between any two elements of a set. A metric space arises naturally from the set concept by imposing the structure of a metric on the set. Without a metric or related concept (e.g., a norm) one cannot have a sense that an element of a space is "close" to another element of the space, which is necessary for the notion of convergence and continuity.

**Definition 2.1.** *A set $S$ of elements $x, y, z, \ldots$ is a metric space if for each pair of elements $x, y \in S$ there is associated a number $d(x,y) \in \mathbf{R}$ satisfying*

a. $d(x,y) \geq 0$ *and* $d(x,x) = 0$

b. *If* $d(x,y) = 0$, *then* $x = y$

c. $d(x,y) = d(y,x)$

d. $d(x,y) \leq d(x,z) + d(z,y)$ *(triangle inequality)*, $z \in S$.

The number $d$ is called the *metric*. For any given nonempty set with at least two elements, an infinite number of metrics may be defined.

## Examples

1. For the set of complex numbers $S = \mathbf{C}$ the usual metric is $d(x,y) = |x-y|$ (absolute value), which is easily seen to satisfy the above axioms.

2. For the set of complex-valued $n$-tuples of numbers $S = \mathbf{C}^n$ with $\mathbf{x} = (x_1, x_2, \ldots, x_n)$ and $\mathbf{y} = (y_1, y_2, \ldots, y_n)$ the following metrics are common:

$$d_1(\mathbf{x},\mathbf{y}) = \sum_{i=1}^{n} |x_i - y_i| \quad (\textit{one-metric}),$$

$$d_2(\mathbf{x},\mathbf{y}) = \left( \sum_{i=1}^{n} |x_i - y_i|^2 \right)^{1/2} \quad (\textit{two-metric}),$$

$$d_\infty(\mathbf{x},\mathbf{y}) = \max_i |x_i - y_i| \quad (\textit{infinity- or max-metric}),$$

all of which are specific cases of the *p-metric*

$$d_p(\mathbf{x},\mathbf{y}) = \left( \sum_{i=1}^{n} |x_i - y_i|^p \right)^{1/p},$$

where $1 \leq p < \infty$. The metric $d_2$ is also known as the *Euclidean metric.*

For sequences of numbers in $\mathbf{l}^p$ (or $m$-tuple vectors in $\mathbf{l}^p_m$) the preceding metrics hold if the upper summation limit is replaced by infinity, and for $d_\infty$ "max" is replaced by "sup" since infinite-dimensional vectors need not possess a maximum element, although we'll assume the set is at least bounded. The notion of sup naturally reduces to max if a maximum exists. The metric in $\mathbf{c}$ and $\mathbf{c}_0$ spaces is the same as in $\mathbf{l}^\infty$.

3. For all sequences of complex or real numbers in $\mathbf{l}^2(\mu)$, $\mu \geq 0$, the usual metric is

$$d_2^\mu(\mathbf{x},\mathbf{y}) = \left( \sum_{i=1}^{\infty} |x_i - y_i|^2 \, i^\mu \right)^{1/2}.$$

For $\mu = 0$ the metric $d_2^0$ is the Euclidean metric $d_2$, and for $\mu = 1$ the metric $d_2^1$ represents the usual metric in $\widetilde{\mathbf{l}^2}$ space.

4. For the spaces $\mathbf{M}(\Omega)^m$, $\mathbf{C}(\Omega)^m$, and $\mathbf{BC}(\Omega)^m$ the *sup-metric* is defined as

$$d_\infty(\mathbf{x},\mathbf{y}) = \sup_{\mathbf{t}\in\Omega} |\mathbf{x}(\mathbf{t}) - \mathbf{y}(\mathbf{t})| \,.$$

For vector-valued functions,

$$|\mathbf{x}(\mathbf{t})| \equiv (\mathbf{x}(\mathbf{t}) \cdot \overline{\mathbf{x}}(\mathbf{t}))^{1/2}$$

and in this sense the absolute value provides the Euclidean metric.

5. For functions in $\mathbf{L}^p(\Omega)^m$ the usual ($p$) metric is

$$d_p(\mathbf{x},\mathbf{y}) = \left( \int_\Omega |\mathbf{x}(\mathbf{t}) - \mathbf{y}(\mathbf{t})|^p \, d\Omega \right)^{1/p},$$

where $1 \leq p \leq \infty$ with $d_\infty(\mathbf{x},\mathbf{y}) = \sup_{\mathbf{t}\in\Omega} |\mathbf{x}(\mathbf{t}) - \mathbf{y}(\mathbf{t})|$. The above integral metric satisfies the second axiom if we are willing to say that $\mathbf{x} = \mathbf{y}$ means $\mathbf{x}(\mathbf{t}) = \mathbf{y}(\mathbf{t})$ *almost everywhere* (a.e.), i.e., for all $\mathbf{t} \in \Omega$ except possibly on a set of measure zero.[14] This metric can also be applied to the space $\mathbf{C}(\Omega)^m$.

6. For sequences of $m$-tuple vectors of functions in $\mathbf{L}^p(\Omega)^m$, i.e., $\{\mathbf{x}_1(\mathbf{t}), \mathbf{x}_2(\mathbf{t}), \mathbf{x}_3(\mathbf{t}), \dots\}$ where $\mathbf{x}_i(\mathbf{t}) = \{x_1^i(\mathbf{t}), x_2^i(\mathbf{t}), \dots, x_m^i(\mathbf{t})\}$, the usual ($p$) metric is

$$d_p(\mathbf{x},\mathbf{y}) = \left( \sum_{i=1}^{\infty} \int_\Omega |\mathbf{x}_i(\mathbf{t}) - \mathbf{y}_i(\mathbf{t})|^p \, d\Omega \right)^{1/p},$$

which can also be applied for $m$-tuple vectors of functions in $\mathbf{C}(\Omega)^m$.

7. For functions in $\mathbf{C}^k(\Omega)$ $(0 \leq k < \infty)$ a particularly strong statement concerning the closeness of two functions is obtained by the metric

$$d_\infty^k(x,y) = \sum_{|\alpha| \leq k} \sup_{\mathbf{t}\in\Omega} \{|D^\alpha x(\mathbf{t}) - D^\alpha y(\mathbf{t})|\}$$

for $|\alpha| \leq k$, which states that the function together with all of its first $k$ derivatives differ by no more than $d_\infty^k$ at any point $\mathbf{t} \in \Omega$.

---

[14] The concept of a "set of measure zero" comes from Lebesgue measure (integral) theory. $x = y$ a.e. means that $x$ and $y$ are the same except possibly at a countable number of points; for engineering applications in such cases we can assume that $x$ and $y$ are the same.

8. In Hölder spaces $\mathbf{H}_{0,\alpha}(\Omega)^m$ the metric is defined for any $\mathbf{t}, \mathbf{s} \in \Omega$, $\mathbf{t} \neq \mathbf{s}$, such that

$$d^{0,\alpha}(\mathbf{x}, \mathbf{y}) = \sup_{\mathbf{t}\in\Omega}\{|\mathbf{x}(\mathbf{t}) - \mathbf{y}(\mathbf{t})|\} + \sup_{\mathbf{t},\mathbf{s}\in\Omega}\left\{\left|\frac{|\mathbf{x}(\mathbf{t}) - \mathbf{x}(\mathbf{s})|}{|\mathbf{t} - \mathbf{s}|^\alpha} - \frac{|\mathbf{y}(\mathbf{t}) - \mathbf{y}(\mathbf{s})|}{|\mathbf{t} - \mathbf{s}|^\alpha}\right|\right\},$$

where $0 < \alpha \leq 1$, with an obvious corresponding form for scalars in $\mathbf{H}_{0,\alpha}(\Omega)$.

In Hölder spaces $\mathbf{H}_{k,\alpha}(\Omega)^3$ of vector functions with appropriate derivatives up to order $k$ Hölder continuous, a metric can be defined in terms of the above, and similarly for $\mathbf{H}_{k,\alpha}(\Omega)$. For example, considering $\mathbf{H}_{1,\alpha}(\Omega)$ defined such that $x(\mathbf{t}) \in \mathbf{H}_{1,\alpha}(\Omega)$ if $\nabla x \in \mathbf{H}_{0,\alpha}(\Omega)^3$, then the metric in $\mathbf{H}_{1,\alpha}(\Omega)$ is defined for any $\mathbf{t}, \mathbf{s} \in \Omega$, $\mathbf{t} \neq \mathbf{s}$ as [22, pp. 40, 214]

$$\begin{aligned} d^{1,\alpha}(x, y) =& d_\infty(x, y) + d^{0,\alpha}(\nabla x, \nabla y) \\ =& \sup_{\mathbf{t}\in\Omega}\{|x(\mathbf{t}) - y(\mathbf{t})|\} + \sup_{\mathbf{t}\in\Omega}\{|\nabla_\mathbf{t} x(\mathbf{t}) - \nabla_\mathbf{t} y(\mathbf{t})|\} \\ &+ \sup_{\mathbf{t},\mathbf{s}\in\Omega}\left\{\left|\frac{|\nabla_\mathbf{t} x(\mathbf{t}) - \nabla_\mathbf{s} x(\mathbf{s})|}{|\mathbf{t} - \mathbf{s}|^\alpha} - \frac{|\nabla_\mathbf{t} y(\mathbf{t}) - \nabla_\mathbf{s} y(\mathbf{s})|}{|\mathbf{t} - \mathbf{s}|^\alpha}\right|\right\}. \end{aligned}$$

9. For scalar functions of $n$ variables in the Sobolev space $\mathbf{W}_p^k(\Omega)$ the usual metric is

$$d_p^k(x, y) = \left(\sum_{|\alpha| \leq k} \int_\Omega |D^\alpha x(\mathbf{t}) - D^\alpha y(\mathbf{t})|^p \, d\Omega\right)^{1/p}.$$

For vector functions in the Sobolev space $\mathbf{W}_p^k(\Omega)^m$ a metric can be defined as

$$d_p^k(\mathbf{x}, \mathbf{y}) = \left(\sum_{\alpha \leq k} \int_\Omega |L^\alpha(\nabla)\mathbf{x}(\mathbf{t}) - L^\alpha(\nabla)\mathbf{y}(\mathbf{t})|^p \, d\Omega\right)^{1/p}$$

for $L^\alpha$ appropriately defined. For example, in the space $\mathbf{W}_2^1(\Omega)^3 = \mathbf{H}(\text{curl}, \Omega)$ considered in the previous section, $k = 1$, $L^1 = \nabla\times$, and $L^0 = 1$, such that

$$\begin{aligned} &d_{\mathbf{H}(\text{curl},\Omega)}(\mathbf{x}, \mathbf{y}) \\ &= \left(\sum_{\alpha=0}^{1} \int_\Omega |L^\alpha(\nabla)\mathbf{x}(\mathbf{t}) - L^\alpha(\nabla)\mathbf{y}(\mathbf{t})|^2 \, d\Omega\right)^{1/2} \end{aligned}$$

$$= \left( \int_\Omega |\mathbf{x}(\mathbf{t}) - \mathbf{y}(\mathbf{t})|^2 \, d\Omega + \int_\Omega |\nabla \times \mathbf{x}(\mathbf{t}) - \nabla \times \mathbf{y}(\mathbf{t})|^2 \, d\Omega \right)^{1/2}$$

$$= \left( (d_2(\mathbf{x}, \mathbf{y}))^2 + (d_2(\nabla \times \mathbf{x}, \nabla \times \mathbf{y}))^2 \right)^{1/2}$$

with $d_2$ being the $\mathbf{L}^2(\Omega)^m$ metric. Similarly,

$$d_{\mathbf{H}(\mathrm{div},\Omega)}(\mathbf{x}, \mathbf{y}) = \left( (d_2(\mathbf{x}, \mathbf{y}))^2 + (d_2(\nabla \cdot \mathbf{x}, \nabla \cdot \mathbf{y}))^2 \right)^{1/2},$$

$$d_{\mathbf{H}(\mathrm{curl},\mathrm{div},\Omega)}(\mathbf{x}, \mathbf{y})$$

$$= \left( (d_2(\mathbf{x}, \mathbf{y}))^2 + (d_2(\nabla \times \mathbf{x}, \nabla \times \mathbf{y}))^2 + (d_2(\nabla \cdot \mathbf{x}, \nabla \cdot \mathbf{y}))^2 \right)^{1/2}.$$

A metric space is then properly defined by specifying both the set and the metric, $(S, d)$, although $S$ alone will usually denote the space if a certain metric is understood. The sets $\mathbf{C}^n$, $\mathbf{l}^p$, $\mathbf{l}^2(\mu)$, $\mathbf{C}(\Omega)^m$, $\mathbf{M}(\Omega)^m$, $\mathbf{BC}(\Omega)^m$, $\mathbf{C}^k(\Omega)^m$, $\mathbf{L}^p(\Omega)^m$, $\mathbf{H}_{k,\alpha}(\Omega)^m$, and $\mathbf{W}_p^k(\Omega)^m$ are then all metric spaces under the metrics mentioned above. Proofs that the above metrics satisfy the axioms of a metric are left as exercises.

Some of the following inequalities are useful for proving the triangle inequality $(d)$ for a given metric and are indeed generally useful in many areas of analysis. The first two are generally known as *Minkowski's inequalities*, whereas the second two are known as *Hölder's inequalities*. We have [21]

$$\left( \sum_{i=1}^n |x_i \pm y_i|^p \right)^{1/p} \leq \left( \sum_{i=1}^n |x_i|^p \right)^{1/p} + \left( \sum_{i=1}^n |y_i|^p \right)^{1/p},$$

$$\left( \int |x(t) \pm y(t)|^p \, dt \right)^{1/p} \leq \left( \int |x(t)|^p \, dt \right)^{1/p} + \left( \int |y(t)|^p \, dt \right)^{1/p}$$

for $1 \leq p < \infty$, and

$$\sum_{i=1}^n |x_i y_i| \leq \left( \sum_{i=1}^n |x_i|^p \right)^{1/p} \left( \sum_{i=1}^n |y_i|^q \right)^{1/q},$$

$$\int |x(t) y(t)| \, dt \leq \left( \int |x(t)|^p \, dt \right)^{1/p} \left( \int |y(t)|^q \, dt \right)^{1/q}$$

for $1 < p, q < \infty$, and $p^{-1} + q^{-1} = 1$. The upper summation limit may be infinite, and for the infinite-dimensional cases it is assumed that the sums and integrals converge, i.e., $\sum_{i=1}^\infty |x_i|^p < \infty$, $\sum_{i=1}^\infty |y_i|^p < \infty$, $\int |x(t)|^p \, dt < \infty$, and $\int |y(t)|^p \, dt < \infty$. For $p = q = 2$ the latter two are also known as *Cauchy–Schwarz inequalities*.

The Euclidean metric $d_2$ for both vectors and functions is the most widely used metric in engineering applications. The metric $d_2$ provides an average distance between two elements in a space. It is especially useful for defining a reasonable measure of the error between an element and its approximation.

### 2.2.1 Continuity and Convergence

The definition of a metric also allows one to generalize the concepts of continuity and convergence, which are important in the work to follow.

#### Continuity and Uniform Continuity

Consider the usual definition of continuity for real-valued functions (mappings) on a real interval. The function $f$ defined on a set $S$ is *continuous* at $t_0 \in S$ if, given any $\epsilon > 0$, there is a $\delta > 0$ such that $|f(t) - f(t_0)| < \epsilon$ whenever $|t - t_0| < \delta$. Note that $\delta = \delta(\epsilon, t_0)$. The function is said to be continuous (on a domain) if it is continuous at each point in its domain. If $\delta$ depends only on $\epsilon$ but not the point $t_0$, then the function is said to be *uniformly continuous* (on its domain). The following definition [23, p. 61] generalizes this concept for arbitrary elements of a metric space.

**Definition 2.2.** *Let $f : S_i \to S_j$ be a mapping of the metric space $(S_i, d_i)$ into the metric space $(S_j, d_j)$. The mapping $f$ is said to be continuous at the point $t_0 \in S_i$ if for every $\epsilon > 0$ there exists a number $\delta(t_0, \varepsilon) > 0$ such that*

$$d_j(f(t), f(t_0)) < \epsilon$$

*whenever $d_i(t, t_0) < \delta$. A mapping that is continuous at every point in its domain is said to be continuous.*

Note that one may say that a mapping is continuous, denoting a global quality, or continuous at a point, implying a local property. *Uniform continuity* is a stronger type of global continuity.

**Definition 2.3.** *Let $f : S_i \to S_j$ be a mapping of the metric space $(S_i, d_i)$ into the metric space $(S_j, d_j)$. The mapping $f$ is said to be uniformly continuous if for each $\epsilon > 0$ there exists a number $\delta(\epsilon) > 0$ such that for any two points $t_0$ and $t$ in $S_i$, $d_j(f(t), f(t_0)) < \epsilon$ whenever $d_i(t, t_0) < \delta$.*

It is a standard proof in advanced calculus that a function continuous on a closed bounded interval $[a, b]$ is uniformly continuous on $[a, b]$. This also holds for closed bounded regions in $\mathbf{R}^n$.

To develop some intuition concerning the difference between continuity and uniform continuity, consider the function $f(t) = \sin(1/t)$ for $t \in (0, 1)$. The function is continuous at every point on $(0, 1)$ and so $f$ is continuous on $(0, 1)$. It can be appreciated that the function is not uniformly continuous on $(0,1)$ by considering two points $t_1 = 1/n\pi$ and $t_2 = 1/(n+1/2)\pi$. It is easy to see that

$$d(f(t_1), f(t_2)) = \left|\sin\left(\frac{1}{t_1}\right) - \sin\left(\frac{1}{t_2}\right)\right| = 1,$$

whereas

$$d(t_1, t_2) = |t_1 - t_2| = \frac{1}{n(2n+1)\pi},$$

which can be made arbitrarily small by taking $n$ large.

In general, if a function is fairly "well behaved" on its domain, it will be uniformly continuous.

### Piecewise Continuity

Another global notion of continuity that occurs often in applied problems is that of *piecewise continuity.*

**Definition 2.4.** *A mapping is piecewise continuous on a domain if it is continuous at all but a finite number of points in the domain and if at each point of discontinuity right and left limits exist and the function's value is either the right or left limit of the function at that point.*

In applied electromagnetics unknown quantities are often approximated in numerical solutions by piecewise continuous functions. Also, piecewise continuity is often assumed as a minimal requirement for field quantities to satisfy. For example, normal electric-field components may be continuous on either side of a dielectric boundary but discontinuous across the boundary, resulting in a piecewise continuous electric field.

We next discuss Hölder continuity, which, as described in Section 1.3.4, is very useful in many electromagnetics applications.

### Hölder and Lipschitz Continuity

**Definition 2.5.** *A function $f$ defined on a bounded interval $(a,b)$ is said to be $\alpha$-Hölder continuous on $(a,b)$ if there is a constant $k > 0$ such that*

$$|f(t) - f(s)| \le k\,|t-s|^{\alpha}$$

*for any $t, s \in (a,b)$. For $\alpha = 1$ the function is said to be Lipschitz continuous.*

*A function $f(\mathbf{t})$ of $n$ variables $(\mathbf{t} = (t_1, t_2, \ldots, t_n) \in \Omega)$ is said to be $\alpha_1 \cdots \alpha_n$-Hölder continuous if for any two points $\mathbf{t}, \mathbf{s} \in \Omega$ there are constants $k_i > 0$ such that*

$$|f(\mathbf{t}) - f(\mathbf{s})| \le \sum_{i=1}^{n} k_i\,|t_i - s_i|^{\alpha_i},$$

*where* $0 < \alpha_i \leq 1$. *If* $f(\mathbf{t})$ *is* $\alpha_1 \cdots \alpha_n$*-Hölder continuous then it is* $\alpha_j$*-Hölder continuous uniformly with respect to each variable* $t_j$*:*

$$|f(t_1, \ldots, t_j, \ldots, t_n) - f(t_1, \ldots, s_j, \ldots, t_n)| \leq k_j |t_j - s_j|^{\alpha_j},$$

*where* $j = 1, \ldots, n$.

*For* $m$*-tuple vector functions one also considers* $\mathbf{f}(\mathbf{t}) \in \mathbf{H}_{0,\alpha}(\Omega)^m$*, defined by the condition*

$$|\mathbf{f}(\mathbf{t}) - \mathbf{f}(\mathbf{s})| \leq k |\mathbf{t} - \mathbf{s}|^{\alpha},$$

*where* $0 < \alpha \leq 1$.

Note that $\alpha = 1$ (Lipschitz) defines the strongest continuity statement, and for any $0 < \alpha \leq 1$ Hölder continuous functions are uniformly continuous. Moreover, $\alpha_p$-Hölder continuity is stronger than $\alpha_r$-Hölder continuity for $\alpha_p > \alpha_r$, and $\alpha > 1$ implies $f(t)$ is a constant,[15] the most continuous of functions.

It can be seen that Hölder continuity in $\mathbf{H}_{0,\alpha}(\Omega)$ is stronger than continuity in $\mathbf{C}(\Omega)$, yet weaker than differentiability in $\mathbf{C}^1(\Omega)$. However, Hölder continuity of continuously differentiable functions in $\mathbf{H}_{1,\alpha}(\Omega)$ is stronger than differentiability in $\mathbf{C}^1(\Omega)$. In general, $\mathbf{H}_{k,\alpha}(\Omega) \subset \mathbf{C}^k(\Omega)$.

Hölder continuous functions arise in electromagnetic applications involving weakly singular improper integrals, as discussed in Chapter 1, and singular integral equations [4], [5], and play a significant role in problems of applied electromagnetics leading to Fredholm integral equations with logarithmic-type and Cauchy-type singular kernels. They also are extensively utilized in the analysis of fields and currents near surfaces containing edges, where fields and currents are Hölder continuous with a discontinuous derivative.

## Convergence

The notion of a metric also allows one to generalize the concept of convergence using sequences. A sequence of real elements $\{x_1, x_2, \ldots\}$ in $S$ is said to *converge* to an element $x \in S$ if, given any real $\epsilon > 0$, there exists an integer $N(\epsilon)$ such that $|x - x_n| < \epsilon$ for all $n \geq N$. This states that given $n$ large enough, $x_n$ is arbitrarily close to $x$. If a sequence is not convergent, it is *divergent*. The following definition generalizes this concept of convergence for sequences of generally complex, vector-valued multivariable functions using metrics.

[15] To see this, let $\alpha = 1 + \varepsilon$, $\varepsilon > 0$. Then $|f(t) - f(s)| \leq k|t - s|^{\alpha}$ implies

$$\frac{|f(t) - f(s)|}{|t - s|} \leq k|t - s|^{\varepsilon}.$$

As $t \to s$ the left side becomes $|f'|$ and the right side approaches 0. Therefore, in this limit we have $|f'| \leq 0$, which implies $f' = 0$, and so $f$ is a constant.

**Definition 2.6.** *A sequence of elements $x_1, x_2, \ldots$ in a metric space $(S, d)$ is said to be convergent if there is an element $x \in S$ such that for every $\epsilon > 0$ there is an integer $N(\epsilon)$ such that*

$$d(x_n, x) < \epsilon$$

*whenever $n > N$.*

This is sometimes denoted as $\lim_{n\to\infty} d(x_n, x) = 0$ or simply $d(x_n, x) \to 0$. The element $x$ is said to be the limit of the sequence, written as $\lim_{n\to\infty} x_n = x$ or $x_n \to x$. Some properties of sequences follow [24]:

- The limit of a convergent sequence is unique, i.e., if $\lim_{n\to\infty} x_n = x$ and $\lim_{n\to\infty} x_n = y$, then $x = y$.
- A convergent sequence is bounded; however, a bounded sequence may not be convergent.
- Any subsequence of a convergent sequence is convergent and converges to the same limit as the sequence.

### Cauchy Convergence

An extremely important convergence criteria is *Cauchy* convergence.

**Definition 2.7.** *A sequence $\{x_n\}$ in a metric space $(S, d)$ is said to be a Cauchy sequence (meaning the sequence converges in the Cauchy sense) if for each $\epsilon > 0$ there exists an integer $N(\epsilon)$ such that*

$$d(x_n, x_m) < \epsilon$$

*for all integers $n, m > N$.*

This is also written as $\lim_{n,m\to\infty} d(x_n, x_m) \to 0$, and it is equivalent to uniform convergence in $k$ of $d(x_{m+k}, x_m)$ to 0 as $m$ tends to infinity. Cauchy sequences are sometimes called *fundamental* sequences. Note that $n$ and $m$ are unrelated and independent. For Cauchy convergence it is not enough for consecutive terms to become close together (although if the distance between consecutive terms goes to zero "fast enough" a sequence can be shown to be Cauchy). It can be shown that any convergent sequence is also a Cauchy sequence.

For sequences of numbers or constant vectors, the above definitions of convergence usually suffice. For sequences of functions, there are many other notions to consider.

### Pointwise and Uniform Convergence

**Definition 2.8.** *Let a sequence of functions $\{f_n(t)\}$ be defined on a set $S$. The sequence is said to converge pointwise to $f(t)$ on a set $S_1 \subset S$ if for each $t \in S_1$ and for each $\epsilon > 0$ there exists $N(\epsilon, t)$ such that*

$$|f(t) - f_n(t)| < \epsilon$$

*for all $n > N$ (for each $t \in S_1$, $f_n(t) \to f(t)$).*

Therefore, pointwise convergence is simply examining convergence of a sequence at given points, amounting to convergence of a sequence of numbers (the function values) associated with those points.

Pointwise convergence is in some sense not a very strong condition, because $N(\epsilon, t_1)$ and $N(\epsilon, t_2)$ required for convergence may differ greatly even for $t_1$ near $t_2$. A stronger type of convergence is *uniform convergence* as defined below.

**Definition 2.9.** *Let a sequence of functions $\{f_n(t)\}$ be defined on a set $S$. The sequence is said to converge uniformly to $f(t)$ on a set $S_1 \subset S$ if for each $\epsilon > 0$ there exists $N(\epsilon)$ such that*

$$|f(t) - f_n(t)| < \epsilon$$

*for all $n > N$ and all $t \in S_1$.*

The principal difference between pointwise convergence and uniform convergence is that for the latter, $N$ does not depend on $t$. Therefore, for uniform convergence and given some $\varepsilon$, all the graphs of $f_n(t)$ for $n > N$ lie within a strip of width $\epsilon$ of the limit graph $f(t)$.

Some important properties relating to uniform convergence are the following:

- Uniform convergence implies pointwise convergence.
- If $\{f_n(t)\}$ converges uniformly to $f(t)$ on a *bounded* region $\Omega$, then

$$\lim_{n\to\infty} \int_\Omega f_n(t)\, d\Omega = \int_\Omega f(t)\, d\Omega.$$

  That is, the limit operation and the integration operation may be interchanged. This is not generally true for pointwise convergence or for uniform convergence on an unbounded interval.

A reformulation of the definition for uniform convergence can be obtained using the sup-metric for functions.

**Definition 2.10.** *Let a sequence of functions $\{f_n(t)\}$ be defined on a set $S$. The sequence is said to converge uniformly to $f(t)$ on a bounded region $\Omega$ if for each $\epsilon > 0$ there exists $N(\epsilon)$ such that*

$$d_\infty(f(t), f_n(t)) = \sup_{t \in \Omega} |f_n(t) - f(t)| < \epsilon$$

*for all $n > N$.*

Therefore, uniform convergence implies convergence in the sup-metric, and vice versa.

### Mean and Mean-Square Convergence

In the definitions of pointwise and uniform convergence we considered the criteria $d(f(t), f_n(t)) < \epsilon$ for all $n > N$. Both definitions involved considering the metric for some values of $t$, with the difference being that for pointwise convergence we fix $t$ and $\epsilon$ and then find $N(t, \epsilon)$, while for uniform convergence we fix $\epsilon$ and find $N(\epsilon)$ valid for all $t$ of interest. Two other important types of convergence are often used which consider the integral of the difference of the functions over the domain of interest. This idea brings about the notion of two functions being similar in an average sense.

**Definition 2.11.** *Let a sequence of functions $\{f_n(t)\}$ be defined on a set $S$. The sequence is said to converge in the mean sense to $f(t)$ on a bounded region $\Omega$ if for each $\epsilon > 0$ there exists $N(\epsilon)$ such that*

$$d_1(f(t), f_n(t)) = \int_\Omega |f(t) - f_n(t)| \, d\Omega < \epsilon$$

*for all $n > N$.*

**Definition 2.12.** *Let a sequence of functions $\{f_n(t)\}$ be defined on a set $S$. The sequence is said to converge in the mean-square sense to $f(t)$ on a bounded region $\Omega$ if for each $\epsilon > 0$ there exists $N(\epsilon)$ such that*

$$d_2(f(t), f_n(t)) = \left( \int_\Omega |f(t) - f_n(t)|^2 \, d\Omega \right)^{1/2} < \epsilon$$

*for all $n > N$.*

We can also use the notation $d_2(f, f_n) \to 0$, and similarly for $d_1$, $d_\infty$.

Note that Cauchy convergence for functions can be in the mean or mean-square sense as well; for each $\epsilon > 0$ there exists an integer $N(\epsilon)$ such that $d_{1,2}(f_n, f_m) < \epsilon$ for all integers $n, m > N$ (however, Cauchy convergence in this metric does not imply uniform convergence). For sequences of continuous functions, uniform convergence implies pointwise, mean, and mean-square convergence.

There is an important theorem relating convergence and continuity:

**Theorem 2.1.** *Consider the mapping $f : (S_i, d_i) \to (S_j, d_j)$. The function is continuous if and only if*

$$f(\lim_{n\to\infty} x_n) = \lim_{n\to\infty} f(x_n)$$

*for every convergent sequence $\{x_n\} \subset (S_i, d_i)$.*

Therefore, for convergent sequences, continuous functions and limits may be interchanged. Another way to state this is that the mapping is continuous at $x \in S_i$ if and only if $x_n \to x$ implies $f(x_n) \to f(x)$. Note that a metric is itself a continuous mapping $d(x, y) : S \to \mathbf{R}$, i.e., $x_n \to x$ and $y_n \to y$ imply $d(x_n, y_n) \to d(x, y)$.

In summary, for sequences of numbers or fixed vectors, we have the notions of convergence and Cauchy convergence. Convergence is the stronger of the two, in that convergence implies Cauchy convergence, but the converse is not necessarily true. For sequences of functions we have pointwise convergence, uniform (Cauchy) convergence, mean convergence, and mean-square convergence, with uniform convergence the strongest type of convergence.

### 2.2.2 Denseness, Closure, and Compactness

Here we introduce several definitions that will be used later.

**Definition 2.13.** *Consider two subsets $X, Y$ of a metric space $S$ such that $X \subset Y$. The set $X$ is said to be dense in $Y$ if, for each $y \in Y$ and each $\epsilon > 0$, there exists an element $x \in X$ such that $d(x, y) < \epsilon$.*

This states that every element in $Y$ can be approximated (in the sense of the metric) arbitrarily closely by elements of the set $X$. Another, even more transparent statement is the following: $X$ dense in $Y$ means for every $y \in Y$ there is a sequence $\{x_n\} \subset X$ such that $x_n \to y$. In some sense the set $X$ (which is contained in $Y$) is "close" to $Y$, although $Y$ may possess an infinite number of elements that are not in $X$. Trivially, a set is always dense in itself.

#### Examples

1. The set $(a, b)$ is dense in $[a, b]$.

2. The set of all polynomials on $\Omega$, denoted by $P(\Omega)$, is dense in the set of real-valued continuous functions $\mathbf{C}(\Omega)$ under the metrics $d_\infty$ and $d_2$. This implies that any real-valued continuous function can be approximated arbitrarily closely by polynomials. This is also known as the *Weierstrass approximation theorem* [23], which can be extended componentwise to vector functions $\mathbf{x}(\mathbf{t})$.

3. The set of continuous functions $\mathbf{C}(\Omega)$ is dense in $\mathbf{L}^2(\Omega)$ under the $d_2$ metric. Therefore, any $\mathbf{L}^2(\Omega)$ function can be approximated arbitrarily closely in the $d_2$ sense by continuous functions. Note that since $P(\Omega)$ is dense in $\mathbf{C}(\Omega)$, and $\mathbf{C}(\Omega)$ is dense in $\mathbf{L}^2(\Omega)$, then $P(\Omega)$ is dense in $\mathbf{L}^2(\Omega)$. This example can be generalized for any $1 \leq p < \infty$ such that $\mathbf{C}(\Omega)$ is dense in $\mathbf{L}^p(\Omega)$ under the $d_p$ metric ($\mathbf{C}(\Omega)$ is not dense in $\mathbf{L}^\infty(\Omega)$).

4. The set of the $k$-times continuously differentiable functions $\mathbf{C}^k(\Omega)$ is dense in $\mathbf{C}(\Omega)$. It automatically follows from the previous example that $\mathbf{C}^k(\Omega)$ is dense in $\mathbf{L}^p(\Omega)$ under the $d_p$ metric, and it is obvious that $\mathbf{C}^k(\Omega)$ is not dense in $\mathbf{L}^\infty(\Omega)$.

5. The set of all sequences with a finite number of nonzero terms is dense in $\mathbf{l}^p$ for all $1 \leq p < \infty$.

**Definition 2.14.** *A point $x$ is a limiting point of a set $S$ if any neighborhood of $x$ contains at least one point $y \in S$ such that $y \neq x$.*

**Definition 2.15.** *The closure of a set $S$ is the set $\widetilde{S}$ consisting of the limits of all sequences (all limiting points) that can be constructed from $S$.*

The closure of $S$ is the set of all points of $S$ together with the limits of all convergent sequences in $S$. The closure of a geometrical set is usually simple: If $S = (a, b)$, then $\widetilde{S} = [a, b]$. If $\Omega$ is an open region bounded by surface $\Gamma$, then $\widetilde{\Omega} = \Omega \cup \Gamma$. Closure can also be used to define the concept of denseness; given two subsets $X, Y$ of a metric space $S$ such that $X \subset Y$, the set $X$ is said to be *everywhere dense* in $Y$ if $\widetilde{X} = Y$. This means that there is a neighborhood of a point $y \in Y$ that contains points from the set $X$. The set $X$ is said to be *nowhere dense* in $Y$ if $Y \setminus \widetilde{X}$ is everywhere dense in $Y$.

**Definition 2.16.** *A set $X$ is said to be countable if its elements can be put in one-to-one correspondence with the integer numbers, i.e., $X = \{x_1, x_2, \ldots\}$.*

**Definition 2.17.** *A space $Y$ is called separable if a countable set $X \subset Y$ is everywhere dense in this space.*

If $Y$ is a separable space, then for each $y \in Y$ and each $\epsilon > 0$ there is an element $x_n$ of the countable set $X$ such that $d(x_n, y) < \epsilon$.

**Examples**

1. $\mathbf{R}^n$ is separable. Indeed, the countable set of ordered $n$-tuples of real numbers $\mathbf{x} = (x_1, x_2, \ldots, x_n)$ is everywhere dense in $\mathbf{R}^n$.

2. $\mathbf{C}(\Omega)$ is separable under the infinity metric.

3. If $Q$ is a subset of $\mathbf{l}^p$ such that it contains all sequences that have only a finite number of nonzero elements, then the countable set $X$

$$X = \{(x_1, x_2, \ldots, x_n, 0, 0, \ldots), x_i \in Q,\ i = 1, \ldots, n, n \in N\}$$

is everywhere dense in $\mathbf{l}^p$, and it follows that $\mathbf{l}^p$ is separable in the usual metric $d_p$ for any $1 \le p < \infty$. The space $\mathbf{l}^\infty$ is not separable.

4. $\mathbf{L}^p(\Omega)$ is separable in the usual metric $d_p$. This can be shown using a continuous mapping $\mathbf{l}^p \to \mathbf{L}^p$ that preserves separability. The space $\mathbf{L}^\infty(\Omega)$ is not separable.

5. One can show that $\mathbf{c}_0, \mathbf{c}, \mathbf{C}^k(\Omega), \mathbf{W}_p^k(\Omega)$ are separable spaces and that $\mathbf{H}_{0,\alpha}(\Omega)$ is a nonseparable space.

Next we introduce the concepts of closed sets, bounded sets, and compact sets.

**Definition 2.18.** *A set $S$ is closed if $\widetilde{S} = S$.*

Any set containing a finite number of elements is necessarily closed.

**Definition 2.19.** *A set $S$ is bounded if there exists a number $K$ such that $d(x, y) < K$ for all $x, y \in S$.*

**Definition 2.20.** *A set $S$ is (sequentially) compact if each sequence of elements in $S$ (not necessarily convergent) has a subsequence that converges to an element of $S$.*

The set $\mathbf{R}$ is not compact since, for example, the sequence $1, 2, \ldots, n, \ldots$ has no convergent subsequence in $\mathbf{R}$. The set $(0, 1] \subset \mathbf{R}$ is also not compact, since, for example, the sequence $\{1, 1/2, 1/3, 1/4, \ldots\}$ does not have a convergent subsequence in $(0, 1]$. However, it can be shown that the closed set $[0, 1] \subset \mathbf{R}$ is compact. Note that a compact set is necessarily closed and bounded, and in $\mathbf{C}^n$ every bounded, closed set is compact. A compact set with a metric is called a *compact metric space,* and it can be shown that every compact metric space is separable.

The following forms the definition of compactness in some spaces.

- $X \subset \mathbf{C}(\Omega)^m$: $X$ is a compact set if and only if functions $\mathbf{x}(\mathbf{t}) \in X$ are uniformly bounded[16] and equicontinuous[17] (Arzela criterion).

---

[16] There is a constant $M$ such that $|\mathbf{x}(\mathbf{t})| \le M$ for all $\mathbf{x}(\mathbf{t}) \in X$ and $\mathbf{t} \in \Omega$.

[17] For any $\varepsilon > 0$ there is $\delta = \delta(\varepsilon)$ such that $|\mathbf{x}(\mathbf{t}_1) - \mathbf{x}(\mathbf{t}_2)| < \varepsilon$ for any $\mathbf{t}_1, \mathbf{t}_2 \in \Omega$ and any function $\mathbf{x}(\mathbf{t}) \in X$, whenever $|\mathbf{t}_1 - \mathbf{t}_2| < \delta$.

- $X \subset \mathbf{L}^p(\Omega)^m$ $(p \geq 1)$: $X$ is a compact set if and only if $X$ is bounded in $\mathbf{L}^p(\Omega)^m$, and for any $\varepsilon > 0$ there is $\delta > 0$ such that

$$\int_{\Omega} |\mathbf{x}(\mathbf{t}+\tau) - \mathbf{x}(\mathbf{t})|^p \, d\Omega < \varepsilon$$

for $|\tau| < \delta$ (Riesz criterion).

- $X \subset \mathbf{l}^p$ $(p \geq 1)$: $X$ is a compact set if and only if $X$ is bounded and for any $\varepsilon > 0$ there is a number $N = N(\varepsilon)$ such that

$$\sum_{i=N}^{\infty} |x_i|^p < \varepsilon$$

for all $\mathbf{x} = \{x_1, x_2, \ldots, x_n, \ldots\} \subset X$.

The ideas of a separable space, a set dense in another set, and a bounded set should be relatively straightforward from the above definitions. The concept of compactness for a set is somewhat abstract. It is introduced here since we are interested in compact operators throughout the text, and the definition of a compact operator and that of a compact set are similar.

Compact sets are in a sense very well behaved. This nice behavior is related to completeness, which is introduced in the next section. In a compact space, even "poorly" behaved sequences (i.e., divergent sequences) have at least one "well behaved" (i.e., convergent) subsequence. Divergent sequences in noncompact spaces may not have any convergent subsequences.

### 2.2.3 Completeness

An extremely important property of metric spaces is that of *completeness.*

**Definition 2.21.** *A metric space $(S, d)$ is said to be complete if each Cauchy sequence in $(S, d)$ is a convergent sequence in $(S, d)$.*

Because every convergent sequence in a metric space is a Cauchy sequence, in a complete metric space a sequence converges if and only if it is a Cauchy sequence. This is important in practice, since it is much easier to prove that a sequence is a Cauchy sequence than to prove that a sequence converges.

The importance of the "in" part of the definition is best shown by an example. Consider the space $S = \{x \in \mathbf{R} : 0 < x \leq 1\}$ with the metric $d(x,y) = |x - y|$, and the sequence $\{1/n\} = \{1, 1/2, 1/3, \ldots\} \subset S$. The sequence seems to converge to 0, but 0 is not in the space $(S, d)$; therefore, it does not converge in the sense of the space we are considering. It is easy to show that $\{1/n\}$ is Cauchy,[18] therefore, $(S,d)$ as considered above is incomplete. In this case the space can be completed if it is enlarged to include the origin. It can be shown that $S = \{x \in \mathbf{R} : 0 \leq x \leq 1\}$ with the metric $d(x,y) = |x-y|$ is complete, that is, every Cauchy sequence in $(S,d)$ converges.

Another common example is the set of rational numbers. It is easy to find a Cauchy sequence in this set that does not converge to an element in the set (but that converges to an irrational number). If the set is enlarged to contain the irrational numbers as well as the rational numbers, the set can be shown to be complete.

### Examples

We state without proof that the following are complete spaces:

1. Any compact metric space.

2. The space of complex numbers $\mathbf{C}$ with the usual metric $d(x,y) = |x-y|$.

3. The $\mathbf{l}^p$ sequence space with the usual metric $d_p(x,y) = \left(\sum_{i=1}^{\infty} |x_i - y_i|^p\right)^{1/p}$, $1 \leq p < \infty$.

4. The spaces of sequences $\mathbf{c}_0$, $\mathbf{c}$, and $\mathbf{l}^\infty$ with the sup-metric $d_\infty(x,y) = \sup_{i \in (1,\infty)} |x_i - y_i|$.

5. The space of sequences $\mathbf{l}^2(\mu)$ with the usual metric $d_2^\mu(x,y) = (\sum_{i=1}^{\infty} |x_i - y_i|^2 \, i^\mu)^{1/2}$, $\mu \geq 0$.

6. The spaces $\mathbf{M}(\Omega)^m$, $\mathbf{C}(\Omega)^m$, and $\mathbf{BC}(\Omega)^m$ with the sup-metric $d_\infty(\mathbf{x},\mathbf{y}) = \sup_{\mathbf{t}\in\Omega} |\mathbf{x}(\mathbf{t}) - \mathbf{y}(\mathbf{t})|$.

7. The space of continuous differentiable functions $\mathbf{C}^k(\Omega)$ with the metric $d_\infty^k(x,y) = \sum_{|\alpha| \leq k} \sup_{\mathbf{t}\in\Omega} \{|D^\alpha x(\mathbf{t}) - D^\alpha y(\mathbf{t})|, \ |\alpha| \leq k\}$.

   Furthermore, we have the following:

8. The space $(\mathbf{C}(\Omega)^m, d_2)$ is incomplete, where $d_2$ is the Euclidean metric. The incompleteness of this space is due to the fact that a discontinuous function can be approximated in the mean-square ($d_2$) sense as the limit of continuous functions. To see this, consider the one-dimensional case $\mathbf{C}\,(0,2)$ and the sequence $\{f_n\}$ shown in Figure 2.5,

$$f_n(t) = \begin{cases} 0, & t \leq 1 - \frac{1}{n}, \\ n\,(t-1)+1, & 1 - \frac{1}{n} < t < 1, \\ 1, & t \geq 1. \end{cases}$$

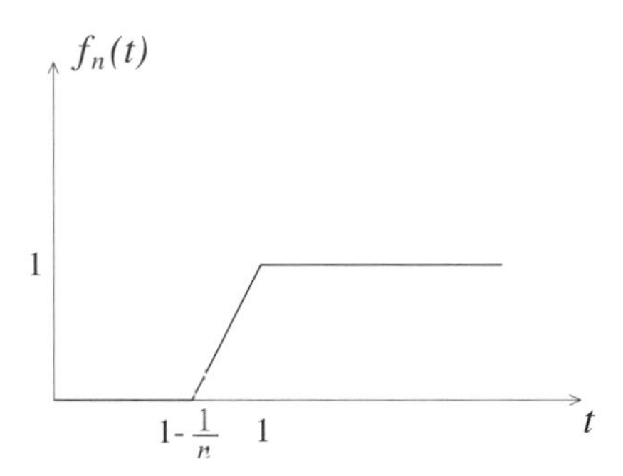


Figure 2.5: A sequence that is Cauchy yet does not converge in $\mathbf{C}(0,2)$, demonstrating that $\mathbf{C}(0,2)$ is not complete.

The sequence is Cauchy (in the $d_2$ metric, i.e., $d_2(f_n, f_m) \to 0$[19]) and continuous, but the sequence does not converge. Again, we emphasize that by converge we implicitly mean to an element in the space under consideration. Indeed, the *limit* of the sequence is the step function,

$$f(t) = \begin{cases} 1, & t > 1, \\ 0, & 0 < t \leq 1 \end{cases}$$

which is discontinuous and therefore not in $\mathbf{C}(0,2)$. Therefore, even though $d_2(f_n, f) \to 0$, $f \notin \mathbf{C}(0,2)$, and so the Cauchy sequence does not converge and hence the space $(\mathbf{C}(a,b), d_2)$ is incomplete (the sequence is not Cauchy under the $d_\infty$ metric, so that there is no contradiction with Example 6).

If the space of continuous functions is enlarged to contain all limit functions, it will be complete. That is the genesis of the important space $\mathbf{L}^p(\Omega)$.

---

[18]Let $N = 2/\epsilon$. For $n, m > N$, $1/n, 1/m < \epsilon/2$. Therefore, $|1/n - 1/m| \leq 1/n + 1/m < \epsilon$ for all $n, m > N$.

[19]For $\mathbf{C}(0,2)$ with $m > n$,

$$\begin{aligned} d_2(f_n, f_m) &= \int_0^2 |f_n(t) - f_m(t)|^2 \, dt \\ &= \int_{1-\frac{1}{n}}^{1-\frac{1}{m}} [(n(t-1)+1) - 0]^2 \, dt \\ &\quad + \int_{1-\frac{1}{m}}^{1} [(n(t-1)+1) - (m(t-1)+1)]^2 \, dt \\ &= \frac{1}{3m^2} \frac{m^2 - 2nm + n^2}{n} \to 0 \text{ as } m, n \to \infty. \end{aligned}$$

9. The space $\mathbf{L}^p(\Omega)^m$ $(1 \le p < \infty)$ with the usual metric[20]
   $d_p(\mathbf{x}, \mathbf{y}) = \left(\int_\Omega |\mathbf{x}(\mathbf{t}) - \mathbf{y}(\mathbf{t})|^p d\Omega\right)^{1/p}$ is complete.

10. The space $\mathbf{L}^\infty(\Omega)^m$ with the sup-metric
    $d_\infty(\mathbf{x}, \mathbf{y}) = \sup_{\mathbf{t}\in\Omega} |\mathbf{x}(\mathbf{t}) - \mathbf{y}(\mathbf{t})|$ is complete.

11. The Hölder space $\mathbf{H}_{0,\alpha}(\Omega)^m$ $(0 < \alpha \le 1)$ with the metric

$$d^{0,\alpha}(\mathbf{x}, \mathbf{y}) = \sup_{\mathbf{t}\in\Omega}\{|\mathbf{x}(\mathbf{t}) - \mathbf{y}(\mathbf{t})|\} + \sup_{\mathbf{t},\mathbf{s}\in\Omega}\left\{\left|\frac{|\mathbf{x}(\mathbf{t}) - \mathbf{x}(\mathbf{s})|}{|\mathbf{t} - \mathbf{s}|^\alpha} - \frac{|\mathbf{y}(\mathbf{t}) - \mathbf{y}(\mathbf{s})|}{|\mathbf{t} - \mathbf{s}|^\alpha}\right|\right\}$$

    is complete.

12. The Sobolev space $\mathbf{W}_p^k(\Omega)$ $(1 \le p < \infty)$ with the usual metric $d_p^k(x, y) = \left(\sum_{|\alpha|\le k} \int_\Omega |D^\alpha x(\mathbf{t}) - D^\alpha y(\mathbf{t})|^p d\Omega\right)^{1/p}$ for $k \ge 0$ ($k$ an integer) is complete. In the vector case $\mathbf{W}_p^k(\Omega)^m$ it can be shown that $\mathbf{H}(\mathrm{div}, \Omega)$, $\mathbf{H}(\mathrm{curl}, \Omega)$, and $\mathbf{H}(\mathrm{curl}, \mathrm{div}, \Omega)$ are complete under the metrics described in Section 2.2.

## 2.3 Linear Space

Although many further topological concepts could be introduced for metric spaces, we will instead move on to linear spaces and subsequently work within that framework.

**Definition 2.22.** *A linear space (or vector space) $S$ is a set of elements for which there are defined operations of addition and scalar multiplication*[21] *satisfying the following axioms for all $x, y, z \in S$ and $\alpha, \beta \in \mathbf{R}$ or $\mathbf{C}$:*

a. $x + y = y + x$ *(commutativity).*

b. $(x + y) + z = x + (y + z)$ *(associativity).*

c. *$S$ contains an element $0$ such that $x + 0 = x$.*

d. *For every $x$ there exists an element $y$, called the negative of $x$, such that $x + y = 0$.*

e. $\alpha(x + y) = \alpha x + \alpha y$ *(distributivity).*

f. $(\alpha + \beta)x = \alpha x + \beta x$ *(distributivity).*

g. $1 \cdot x = x$.

h. $\alpha(\beta x) = (\alpha\beta)x$.

[20] Theoretically if the integral is taken in the Riemann sense the space is incomplete, but complete when taken in the Lebesgue sense. In most applications we can nevertheless use the ordinary Riemann integral.

[21] A linear space equipped with a suitably defined operation of multiplication between elements in the space is called an *algebra*.

The space is said to be a *real linear space* if $\alpha, \beta \in \mathbf{R}$, and a *complex linear space* if $\alpha, \beta \in \mathbf{C}$. In static electromagnetic problems one usually works with a real linear space, while in dynamic time-harmonic or transform-domain problems it is usually necessary to work in a complex linear space.

Elements of the space are often called vectors, which may refer to traditional vectors as in $\mathbf{C}^n$ or functions (e.g., $\mathbf{L}^p(\Omega)^m$). The sets $\mathbf{C}^k(\Omega)^m$ and $\mathbf{L}^p(\Omega)^m$, among many others, are still called vector spaces or linear spaces, although some authors use the name *function space* to highlight the fact that the elements of the space are functions rather than classical vectors. In what follows we will sometimes prefer the term linear space because of its generality.

**Examples Related to Linear Spaces**

1. The set $\mathbf{C}^n$ ($\mathbf{R}^n$) is a linear space under the usual rules for vector addition and scalar multiplication.

   (a) If $A = (a_1, a_2) \in \mathbf{R}^2$ and $B = (b_1, b_2) \in \mathbf{R}^2$, then $A + B = \{(a_1 + b_1, a_2 + b_2)\} \in \mathbf{R}^2$.

   (b) $A + B = B + A$.

   (c) If $z_1, z_2 \in \mathbf{C}$, then $z_1 + z_2 \in \mathbf{C}$.

   (d) If $z_1, z_2 \in \mathbf{C}$ and $\alpha, \beta \in \mathbf{C}$, then $\alpha z_1 + \beta z_2 \in \mathbf{C}$.

   (e) If $z_1, z_2 \in \mathbf{C}$ and $\alpha, \beta \in \mathbf{C}$, then $(z_1, z_2, \alpha z_1 + \beta z_2) \in \mathbf{C}^3$.

2. The sets $\mathbf{C}(\Omega)^m$, $\mathbf{M}(\Omega)^m$, $\mathbf{BC}(\Omega)^m$, $\mathbf{C}^k(\Omega)^m$, $\mathbf{L}^p(\Omega)^m$, $\mathbf{H}_{k,\alpha}(\Omega)^m$, and $\mathbf{W}_p^k(\Omega)^m$ are linear spaces under the usual method of adding functions and multiplying scalars and functions. The sets of all real-valued (complex-valued) sequences $\mathbf{c}_0$, $\mathbf{c}$, $\mathbf{l}^\infty$, $\mathbf{l}^p$, $\mathbf{l}^2(\mu)$ are linear spaces under the usual method of adding and multiplying real (complex) numbers.

3. More generally, complex $n$-tuple vector functions $\mathbf{f} = (\mathbf{f}_1(\mathbf{t}), \mathbf{f}_2(\mathbf{t}), \ldots, \mathbf{f}_n(\mathbf{t}))$ on $\Omega$ form a linear space, without the need to impose additional conditions such as continuity or square-integrability. Such a space is too general to be of much utility in applied problems. As such, additional constraints are necessary to produce sharp results for the theory to follow. The space $\mathbf{L}^2(\Omega)^m$ in particular will occupy a role of central importance throughout this text.

## 2.3.1 Linear Subspace

Further building upon our set structure, we introduce the concept of a *linear subspace.*

**Definition 2.23.** *A subset $M \subset S$ of a linear space $S$ is called a linear subspace of $S$ if $\alpha x + \beta y \in M$ for $x, y \in M$ and $\alpha, \beta \in \mathbf{R}$ for real spaces or $\alpha, \beta \in \mathbf{C}$ for complex spaces.*

A linear subspace is itself a linear space, and, somewhat trivially, a linear space $S$ is a linear subspace of itself. Subspaces play a role of importance in the study of projections to follow.

**Definition 2.24.** *A linear subspace $M$ is said to be closed*[22] *(in $S$) if, whenever a sequence of vectors in $M$ converges to a limit, the limit of the sequence belongs to $M$.*

Since completeness is an extremely important property, it is of interest to determine when subsets of a complete space will be complete.

**Theorem 2.2.** *If $(S, d)$ is a complete metric space and $(M, d)$ a subspace of $(S, d)$, then $(M, d)$ is complete if and only if $M$ is a closed set in $(S, d)$.*

### Examples of Linear Subspaces

1. Any plane through the origin comprises a linear subspace of $\mathbf{R}^3$.
2. In $\mathbf{C}(\Omega)$ the subset

   $$K = \{x \in \mathbf{C}(\Omega) : \int_\Omega x(\mathbf{t})\, d\Omega = 1\}$$

   is not a linear subspace, but the subset

   $$K = \{x \in \mathbf{C}(\Omega) : \int_\Omega x(\mathbf{t})\, d\Omega = 0\}$$

   is a linear subspace of $\mathbf{C}(\Omega)$.
3. $\mathbf{C}(\Omega)$ is a linear subspace of $\mathbf{L}^p(\Omega)$ since a sum of two continuous functions is itself continuous. The space $\mathbf{C}(\Omega)$ does not form a closed linear subspace, since a sequence of continuous functions may converge to a discontinuous function (depending on the metric).
4. $\mathbf{C}^k(\Omega)$ is a linear subspace of $\mathbf{C}(\Omega)$.
5. For $f \in \mathbf{C}^2(a, b)$, the set of all real-valued solutions of the equation

   $$\frac{d^2 f}{dt^2} - f = 0$$

   on $a \leq t \leq b$ is a (two-dimensional) linear subspace of $\mathbf{C}^2(a, b)$.

---

[22] A variety of terminology is used in the literature; the most common alternative is to say that $M$ satisfying Definition 2.23 is a *linear manifold*, and then defining a linear subspace as being a closed linear manifold.

6. The set of solutions to a homogeneous system of $n$ linear equations in $n$ unknowns forms a linear subspace (of dimension less than $n$) of $\mathbf{C}^n$.

In dealing with the important topic of projections in later sections, it is useful to have the concept of a *sum* of spaces and a *direct sum* of spaces.

**Definition 2.25.** *Let $M_1$ and $M_2$ be linear spaces. The sum of $M_1$ and $M_2$, denoted by $M_1 + M_2$, is the set made up of all elements $x = x_1 + x_2$, where $x_1 \in M_1$, $x_2 \in M_2$. The direct sum of $M_1$ and $M_2$, denoted by $M_1 \oplus M_2$, is the set of ordered pairs $(x_1, x_2)$ where $x_1 \in M_1$, $x_2 \in M_2$. If every element in $M$ has a unique representation of the form $x + y$, where $x \in M_1$ and $y \in M_2$,*[23] *then $M$ is (isomorphic to) the direct sum of $M_1$ and $M_2$, written as*[24] $M = M_1 \oplus M_2$.

As an example, $\mathbf{R}^3 = M_1 \oplus M_2$, where $M_1$ is the $xy$-plane and $M_2$ is the $z$-axis. Another decomposition is $\mathbf{R}^3 = M_1 \oplus M_2 \oplus M_3$, where $M_1, M_2, M_3$ are the $x$-, $y$-, and $z$- axes, respectively.

### 2.3.2 Dimension of a Linear Space, Basis

So far we have the idea of sets of elements (a space), the distance between elements in the set (a metric), a special set of elements for which addition and scalar multiplication are defined (a linear space), and some important subsets (linear subspace). Next we need to generate a measure of the size of the space itself, called the dimension. Before defining dimension, we need the concept of *linear independence.*

**Definition 2.26.** *A set of elements $x_1, x_2, \ldots, x_n \in S$ is said to be linearly dependent if there exist constants $\alpha_1, \alpha_2, \ldots, \alpha_n$, not all zero, such that*

$$\alpha_1 x_1 + \cdots + \alpha_n x_n = 0.$$

*If no such constants can be found, the set $x_1, x_2, \ldots, x_n$ is said to be linearly independent.*

In $\mathbf{R}^2$, two vectors are linearly dependent if they are proportional to one another, i.e., $x_1 = -(\alpha_2/\alpha_1)\, x_2$. For infinite sets of elements, independence indicates that every finite subset of elements is independent.

We are now ready to define dimension.

[23] This will occur if and only if $M_1$ and $M_2$ are disjoint.

[24] The equality $M = M_1 \oplus M_2$ is in the sense that $M$ and $M_1 \oplus M_2$ are isomorphic (see Section 3.1.3) since $M_1 \oplus M_2$ consists of ordered pairs whereas $M$ does not. The mapping $(x_1, x_2) \to x_1 + x_2$ provides the desired isomorphism from $M_1 \oplus M_2$ to $M$. If $M_1$ and $M_2$ are not disjoint, then there is a mapping from $M_1 \oplus M_2$ to $M$, although this mapping is not an isomorphism.

**Definition 2.27.** *A linear space $S$ is said to be $n$-dimensional if $S$ contains $n$ linearly independent elements, while every set of $n+1$ elements in $S$ is dependent. The space is infinite-dimensional if $S$ contains $n$ independent elements for each positive integer $n = 1, 2, \ldots$.*

The (vector) spaces $\mathbf{R}^n$ and $\mathbf{C}^n$ are $n$-dimensional. The (sequence) spaces $\mathbf{c}_0$, $\mathbf{c}$, $\mathbf{l}^\infty$, $\mathbf{l}^p$, $\mathbf{l}^2(\mu)$ are infinite-dimensional, as are the (function) spaces $\mathbf{C}(\Omega)^m$, $\mathbf{M}(\Omega)^m$, $\mathbf{BC}(\Omega)^m$, $\mathbf{C}^k(\Omega)^m$, $\mathbf{L}^p(\Omega)^m$, $\mathbf{H}_{k,\alpha}(\Omega)^m$, and $\mathbf{W}_p^k(\Omega)^m$.

Next, the concept of a *basis* is introduced. For finite-dimensional spaces the definition is straightforward.

**Definition 2.28.** *A basis of an $n$-dimensional space $S$ is any set of $n$ linearly independent elements in $S$.*

The concept of a basis is extremely important. It allows one to *uniquely* expand any element in a space in terms of the basis elements of the space. To see this, consider an $n$-dimensional space $S$ with $x_1, x_2, \ldots, x_n$ as a basis. The set $\{x_1, x_2, \ldots, x_n\}$ is therefore linearly independent, and the set $\{x, x_1, x_2, \ldots, x_n\}$ is necessarily dependent. As such, there exist constants $\alpha_0, \alpha_1, \ldots, \alpha_n$ not all zero such that

$$\alpha_0 x + \alpha_1 x_1 + \cdots + \alpha_n x_n = 0.$$

Without loss of generality, assume that the coefficient $\alpha_0$ is not zero. Then

$$x = (\alpha_1 x_1 + \cdots + \alpha_n x_n)/\alpha_0.$$

The expansion must be unique. To prove uniqueness, consider

$$x = (\beta_1 x_1 + \cdots + \beta_n x_n) = (\gamma_1 x_1 + \cdots + \gamma_n x_n),$$

i.e., two different expansions of the element $x$ with respect to the same basis. We get

$$[(\beta_1 - \gamma_1)x_1 + \cdots + (\beta_n - \gamma_n)x_n] = 0,$$

and since the set $\{x_1, x_2, \ldots, x_n\}$ is linearly independent, the coefficients must be zero, i.e., $\beta_i = \gamma_i$.

For infinite-dimensional spaces the situation is a bit more tricky—not every infinite set of linearly independent elements is a basis (e.g., start with an infinite number of elements that form a basis, and remove one element. An infinite set of linearly independent elements remains, although it is no longer a basis). A basis for an infinite-dimensional space would, though, necessarily consist of an infinite number of linearly independent elements. To facilitate a definition of basis for the infinite-dimensional case, we need to introduce the concept of *span*.

**Definition 2.29.** *Let $S_1$ be a subset of a linear space $S$. The set of all finite linear combinations of elements in $S_1$ is called the span of $S_1$, denoted by span $S_1$.*

For example, in two-dimensional space let $S_1 = \{\mathbf{x}, \mathbf{y}\}$, where $\mathbf{x} = (\alpha, 0)$ and $\mathbf{y} = (0, \beta)$. Then span $S_1 = \mathbf{R}^2$, the $xy$-plane. The set $S_1$ may contain an infinite number of elements; the span of $S_1$ then consists of all *finite* combinations of the (infinite set of) elements of $S_1$. The set span $S_1$ is said to be *spanned* by $S_1$, or $S_1$ is said to be the spanning set for span $S_1$. The set span $S_1$ is a linear subspace of $S$. If $S_1$ possesses $n$ elements, its span is a linear subspace of dimension less than or equal to $n$.

Now, for infinite-dimensional spaces we can define a basis in the following way, which also holds for finite-dimensional spaces.

**Definition 2.30.** *A linearly independent set $B$ that spans a linear space $S$ (span $B = S$, i.e., $B$ is a spanning set for $S$) is said to be a (Hamel) basis for $S$.*

As an example, the vectors

$$\mathbf{e}_1 = (1, 0, 0), \ \mathbf{e}_2 = (0, 1, 0), \ \mathbf{e}_3 = (0, 0, 1)$$

form a basis in $\mathbf{R}^3$ (span$(\mathbf{e}_1, \mathbf{e}_2, \mathbf{e}_3) = \mathbf{R}^3$ and the vectors are obviously linearly independent). As another example, consider the differential equation

$$x'' - \alpha x = 0,$$

where $\alpha \neq 0$. The set of all solutions form a two-dimensional linear space, the basis for which is $\{e^{\alpha t}, e^{-\alpha t}\}$, so that every solution can be written as $x(t) = c_1 e^{\alpha t} + c_2 e^{-\alpha t}$.

In finite-dimensional spaces the concept of a Hamel basis is similar to the idea of a coordinate system, and the use of a Hamel basis is sufficient for the representation of elements in finite-dimensional spaces. A Hamel basis for a finite-dimensional space has a finite spanning set, while an infinite-dimensional space must have an infinite spanning set. For reasons associated with span being made of *finite* linear combinations, a Hamel basis is not too useful for applications in infinite-dimensional spaces. To have something really useful, one needs another concept of basis, which will be introduced later in the context of an inner product space.

## 2.4 Normed Linear Space

The metric introduced the concept of the distance between two elements of a space, and dimension corresponds to the size of the space. The concept of a *norm* allows one to define the concept of the size of an element in a space.

**Definition 2.31.** *A set $S$ is a normed linear space (real or complex) on which there is defined a real-valued function $||x||$, called the norm of $x$, with the properties*

a. $||x|| \geq 0$ *and* $||x|| = 0$ *if and only if* $x = 0$,

*b.* $||\alpha x|| = |\alpha| \cdot ||x||$ *(homogeneity),*

*c.* $||x + y|| \leq ||x|| + ||y||$ *(triangle inequality), for all* $x, y \in S$.

A normed linear space is called a real normed space if $\alpha \in \mathbf{R}$ and a complex normed space if $\alpha \in \mathbf{C}$. The triangle inequality ($c$) is used to obtain the *second triangle inequality:*

$$||x - y|| \geq ||x|| - ||y||.$$

This inequality implies the continuity of a norm as shown below.

Similar to the case of metrics, we can define $p$-norms, which can be recovered from the $p$-metric as $\|\mathbf{x}\|_p = d_p(\mathbf{x}, \mathbf{0})$. Note, though, that not every metric space is a normed space (there exist metrics $d$ such that $d(\mathbf{x}, \mathbf{0})$ is not a norm). However, every normed linear space is a metric space with the metric $d(\mathbf{x}, \mathbf{y}) = ||\mathbf{x} - \mathbf{y}||$, denoted as the *natural metric.*

All of the metrics detailed in the examples in Section 2.2 lead to common norms for the spaces considered. For example, from pp. 84–87 we have the following norms.

## Examples

1. For $\mathbf{C}$, the common norm is $\|x\| = d(x, 0) = |x|$ (absolute value).

2. For $\mathbf{C}^n$, $\mathbf{x} = (x_1, x_2, \ldots, x_n) \in \mathbf{C}^n$, common norms are

$$\|\mathbf{x}\|_1 = d_1(\mathbf{x}, \mathbf{0}) = \sum_{i=1}^{n} |x_i| \qquad (\textit{one-norm}),$$

$$\|\mathbf{x}\|_2 = d_2(\mathbf{x}, \mathbf{0}) = \left( \sum_{i=1}^{n} |x_i|^2 \right)^{1/2} \qquad (\textit{two-norm}),$$

$$\|\mathbf{x}\|_\infty = d_\infty(\mathbf{x}, \mathbf{0}) = \max_i |x_i| \qquad (\textit{infinity- or max-norm}),$$

which are specific cases of the $p$-norm

$$\|\mathbf{x}\|_p = d_p(\mathbf{x}, \mathbf{0}) = \left( \sum_{i=1}^{n} |x_i|^p \right)^{1/p}.$$

For vectors the two-norm gives the usual "magnitude" of the vector.

The infinity-norm $\|\mathbf{x}\|_\infty$ is also defined for spaces $\mathbf{c}_0$, $\mathbf{c}$, and $\mathbf{l}^\infty$, and it can be shown that for any $p \geq 1$ and any $\mathbf{x} \in \mathbf{l}^p$, the inequality $\|\mathbf{x}\|_\infty \leq \|\mathbf{x}\|_p$ is true. If a sequence of elements $\{x_n\}$ is convergent to $x$ in the norm of $\mathbf{l}^p$, then $\{x_n\}$ is pointwise convergent to $x$. However, pointwise convergence does not imply convergence in the norm of $\mathbf{l}^p$.

The usual norm for sequences of vectors of $m$ variables in $\mathbf{l}^p_m$ is defined by

$$\|\mathbf{x}\|_p = d_p(\mathbf{x}, \mathbf{0}) = \left( \sum_{i=1}^{\infty} |\mathbf{x}_i|^p \right)^{1/p},$$

where $\mathbf{x}_i = \{x^i_1, x^i_2, \ldots, x^i_m\}$.

3. For the sequence spaces $\mathbf{l}^2(\mu)$ $(\mu \geq 0)$ the norm is

$$||\mathbf{x}||^\mu_2 = d^\mu_2(\mathbf{x}, \mathbf{0}) = \left( \sum_{i=1}^{\infty} |x_i|^2 i^\mu \right)^{1/2},$$

which, for $\mu = 0$, is equivalent to the $\|\mathbf{x}\|_2$ norm. The norm in $\widetilde{\mathbf{l}^2}$ space is defined by $||\mathbf{x}||^1_2$.

4. For all (real or complex) functions in $\mathbf{M}(\Omega)^m$, $\mathbf{C}(\Omega)^m$, and $\mathbf{BC}(\Omega)^m$, the *sup-norm* is

$$||\mathbf{x}||_\infty = d_\infty(\mathbf{x}, \mathbf{0}) = \sup_{\mathbf{t}\in\Omega} |\mathbf{x}(\mathbf{t})|.$$

A sequence of functions $\{\mathbf{x}_n(\mathbf{t})\}$ norm-convergent to $\mathbf{x}(\mathbf{t})$ in the above spaces is uniformly convergent to $\mathbf{x}(\mathbf{t})$.

5. In $\mathbf{L}^p(\Omega)^m$ the $p$-norm is

$$\|\mathbf{x}\|_p = d_p(\mathbf{x}, \mathbf{0}) = \left( \int_\Omega |\mathbf{x}(\mathbf{t})|^p d\Omega \right)^{1/p},$$

where $1 \leq p < \infty$, with $\|\mathbf{x}\|_\infty = \sup_{\mathbf{t}} |\mathbf{x}(\mathbf{t})|$, which can also be used for $\mathbf{C}(\Omega)^m$.

6. For sequences of $m$-tuple vector-valued functions of $n$ variables in $\mathbf{L}^p(\Omega)^m$, the $p$-norm is

$$\|\mathbf{x}\|_p = d_p(\mathbf{x}, \mathbf{0}) = \left( \sum_{i=1}^{\infty} \int_\Omega |\mathbf{x}_i(\mathbf{t})|^p d\Omega \right)^{1/p},$$

where $\mathbf{x}_i(\mathbf{t}) = (x^i_1(\mathbf{t}), x^i_2(\mathbf{t}), \ldots, x^i_m(\mathbf{t}))$.

7. In $\mathbf{C}^k(\Omega)$ $(0 \leq k < \infty)$ the sup-norm is defined by

$$\|x\|_\infty^k = d_\infty^k(x, 0) = \sum_{|\alpha| \leq k} \sup_{\mathbf{t} \in \Omega} \{|D^\alpha x(\mathbf{t})|\},$$

for all $\mathbf{t} \in \Omega$, where $|\alpha| \leq k$. It can be shown that a sequence of functions $\{D^\alpha x_n(\mathbf{t}),\ |\alpha| = k\}$ in $\mathbf{C}^k(\Omega)$ is uniformly convergent to $D^\alpha x(\mathbf{t}),\ |\alpha| = k$.

8. In Hölder spaces $\mathbf{H}_{0,\alpha}(\Omega)^m$ a norm is defined for any $\mathbf{t}, \mathbf{s} \in \Omega, \mathbf{t} \neq \mathbf{s}$, such that

$$||\mathbf{x}||_{0,\alpha} = d^{0,\alpha}(\mathbf{x}, \mathbf{0}) = \sup_{\mathbf{t} \in \Omega} |\mathbf{x}(\mathbf{t})| + \sup_{\mathbf{t},\mathbf{s} \in \Omega} \left\{ \frac{|\mathbf{x}(\mathbf{t}) - \mathbf{x}(\mathbf{s})|}{|\mathbf{t} - \mathbf{s}|^\alpha} \right\},$$

where $0 < \alpha \leq 1$.

9. The norm in the Sobolev space of scalar $n$-variable functions of order $k$ ($k$ an integer, $k \geq 0$), $\mathbf{W}_p^k(\Omega)$, can be defined as

$$\begin{aligned} ||x||_p^k = d_p^k(x, 0) &= \left( \sum_{|\alpha| \leq k} \int_\Omega |D^\alpha x(\mathbf{t})|^p d\Omega \right)^{1/p} \\ &= \left( \sum_{|\alpha| \leq k} ||D^\alpha x(\mathbf{t})||_p^p \right)^{1/p}, \end{aligned}$$

where $||\cdot||_p^p$ is the $\mathbf{L}^p$ norm raised to the $p$th power. In the Sobolev space $\mathbf{W}_p^k(\Omega)$ a sequence of functions $\{x_n(\mathbf{t})\}$ converges to $x(\mathbf{t})$ if and only if the sequence $\{D^\alpha x_n(\mathbf{t})\}$ converges to $D^\alpha x(\mathbf{t})$ in $\mathbf{L}_p(\Omega)$ as $n \to \infty$ for all $|\alpha| \leq k$.

   For $m$-tuple vector functions a norm in $\mathbf{W}_p^k(\Omega)^m$ can be found by replacing $D^\alpha$ with an appropriate $L^\alpha(\nabla)$. For instance, in $\mathbf{W}_2^1(\Omega)^3$

$$\begin{aligned} ||\mathbf{x}||_{\mathbf{H}(\mathrm{curl},\Omega)} &= d_{\mathbf{H}(\mathrm{curl},\Omega)}(\mathbf{x}, \mathbf{0}) = \left( ||\mathbf{x}||_2^2 + ||\nabla \times \mathbf{x}||_2^2 \right)^{1/2}, \\ ||\mathbf{x}||_{\mathbf{H}(\mathrm{div},\Omega)} &= d_{\mathbf{H}(\mathrm{div},\Omega)}(\mathbf{x}, \mathbf{0}) = \left( ||\mathbf{x}||_2^2 + ||\nabla \cdot \mathbf{x}||_2^2 \right)^{1/2}, \\ ||\mathbf{x}||_{\mathbf{H}(\mathrm{curl},\mathrm{div},\Omega)} &= d_{\mathbf{H}(\mathrm{curl},\mathrm{div},\Omega)}(\mathbf{x}, \mathbf{0}) \\ &= \left( ||\mathbf{x}||_2^2 + ||\nabla \times \mathbf{x}||_2^2 + ||\nabla \cdot \mathbf{x}||_2^2 \right)^{1/2}. \end{aligned}$$

The sets $\mathbf{C}^n, \mathbf{l}^p, \mathbf{l}^2(\mu), \mathbf{C}^k(\Omega)^m$, $\mathbf{L}^p(\Omega)^m$, $\mathbf{H}_{0,\alpha}(\Omega)^m$, and $\mathbf{W}_p^k(\Omega)^m$ are all normed spaces under the norms listed above.

Minkowski's inequalities stated previously can be written in a simple form using norms as

$$\|x \pm y\|_p \leq \|x\|_p + \|y\|_p$$

for $1 \leq p \leq \infty$, and Hölder's inequality can be written as

$$\|xy\|_1 \leq \|x\|_p \|y\|_q$$

for $1 \leq p \leq \infty$ and $p^{-1} + q^{-1} = 1$.

Since a metric can be obtained from a norm, in a normed space the concept of convergence and completeness can be given in terms of a norm rather than a metric by simply replacing $d(\mathbf{x}, \mathbf{y})$ with $\|\mathbf{x} - \mathbf{y}\|$. A simpler, although less precise, notation will be used for convergence in the remainder of the text. The notation $x_n \to x$ will imply that the sequence $x_n \in S$ converges to $x \in S$ in the sense of the particular norm associated with the space, meaning $\|x_n - x\| \to 0$ as $n \to \infty$. Therefore, for a normed space $(S, \|\cdot\|)$,

- A sequence of elements $\{x_n\}$ is convergent to $x$ if $\|x_n - x\| \to 0$ ($n \to \infty$ implied).
- A sequence of elements $\{x_n\}$ (functions, numbers, etc.) is said to be a fundamental or Cauchy sequence if $\|x_n - x_m\| \to 0$ as $n, m \to \infty$.
- A sequence of functions $\{f_n(t)\}$ is said to converge uniformly to $f(t)$ if $\|f_n(t) - f(t)\|_\infty \to 0$ as $n \to \infty$ (pointwise convergence cannot be stated using norms).
- A sequence of functions $\{f_n(t)\}$ is said to converge in the mean sense to $f(t)$ if $\|f_n(t) - f(t)\|_1 \to 0$ as $n \to \infty$.
- A sequence of functions $\{f_n(t)\}$ is said to converge in the mean-square sense to $f(t)$ if $\|f_n(t) - f(t)\|_2 \to 0$ as $n \to \infty$.

Some properties of norms based on the second triangle inequality are the following [25]:

- If $||x_n - x|| \to 0$, then $||x_n|| \to ||x||$ as $n \to \infty$ (continuity of a norm).
- If $||x_n - x|| \to 0$ and $||x_n - y_n|| \to 0$, then $||y_n - x|| \to 0$ as $n \to \infty$.
- If $||x_n - x|| \to 0$ and $||y_n - y|| \to 0$, then $||x_n - y_n|| \to ||x - y||$ as $n \to \infty$.

### 2.4.1 Completeness, Banach Space

Similar to metric spaces, a normed space $S$ is said to be complete if every Cauchy sequence in $S$ converges (in the sense of the norm) to an element of $S$. A complete normed space is called a *Banach space*.

One can also view a complete normed space as the completion of some incomplete normed space [25].

**Theorem 2.3.** *For every linear normed space $S$ there exists a Banach space $S_c$ such that $S \subset S_c$, with $S$ everywhere dense in $S_c$, and for all $x \in S$, $\|x\|_S = \|x\|_{S_c}$.*

$S_c$ is said to be the *completion* of $S$. As an example [25], consider a function $x(\mathbf{t}) \in \mathbf{C}^k(\widetilde{\Omega})$ and the norm

$$||x||_p^k = \left( \sum_{|\alpha| \leq k} ||D^\alpha x||_p^p \right)^{1/p}$$

for $k \geq 0$. The linear normed space $\mathbf{C}^k(\widetilde{\Omega})$ with the norm $||x||_p^k$ is incomplete (in the sup-norm it is complete), but the completion of $\mathbf{C}^k(\widetilde{\Omega})$ in the norm $||x||_p^k$ is complete and defined as the Sobolev space $\mathbf{W}_p^k(\Omega)$.

## 2.5 Inner Product Space

### 2.5.1 Inner Product

Adding further useful structure to the collections of sets discussed so far, we introduce the concept of an *inner product* and an *inner product space.* An inner product defines the concept of two elements being orthogonal to each other. Many practical engineering problems are set in an inner product space. Sometimes inner product spaces are called pre-Hilbert spaces, and in $\mathbf{R}^n$ the inner product is often called a *dot product.*

**Definition 2.32.** *Consider a linear space $S$. An inner product on $S$ is a mapping $S \times S \to \mathbf{C}$ that associates to each ordered pair $x, y \in S$ a scalar denoted by $\langle x, y \rangle$ with the properties*

*a.* $\langle x, x \rangle \geq 0$ *and* $\langle x, x \rangle = 0$ *if and only if* $x = 0$,

*b.* $\langle x, y \rangle = \overline{\langle y, x \rangle}$ *(Hermitian property),*

*c.* $\langle \alpha x, y \rangle = \alpha \langle x, y \rangle$,

*d.* $\langle x + y, z \rangle = \langle x, z \rangle + \langle y, z \rangle$,

*where $z \in S$ and $\alpha$ is an arbitrary scalar.*

The overbar denotes complex conjugate, where for real inner product spaces the complex conjugate operation is simply omitted. The above definition can be used to show the four useful results

(i) $\langle x, \alpha y \rangle = \overline{\alpha} \langle x, y \rangle$,

(ii) $\langle x, x \rangle \in \mathbf{R}$,

(iii) $\langle x, 0 \rangle = 0$,

(iv) if $\langle x, z \rangle = \langle y, z \rangle$ for all $z \in S$, then $x = y$.

With the definition of an inner product on a linear space, one may define the concept of an inner product space.

**Definition 2.33.** *An inner product space is a linear space $S$ with an inner product defined on $S$.*

A finite-dimensional real (complex) inner product space is also known as a *Euclidean* (*unitary*) space.

It can be shown that the function $\langle x, x\rangle^{1/2}$ satisfies the properties of a norm, and so

$$\|x\| = \langle x, x\rangle^{1/2}$$

is said to be the *norm induced by the inner product.* Therefore, every inner product space is also a normed space; however, the converse is not necessarily true. In the succeeding *whenever a norm is used in an inner product space it is the norm induced by the inner product.*

### Examples of Inner Product Spaces

1. $\mathbf{C}^n$ with $\langle \mathbf{x}, \mathbf{y}\rangle = \sum_{i=1}^{n} x_i\overline{y_i} = \mathbf{x}\cdot\overline{\mathbf{y}}$.
2. $\mathbf{l}^2$ with $\langle \mathbf{x}, \mathbf{y}\rangle = \sum_{i=1}^{\infty} x_i\overline{y_i}$.
3. $\mathbf{l}^2(\mu)$ $(\mu \geq 0)$ with $\langle \mathbf{x}, \mathbf{y}\rangle_\mu = \sum_{i=1}^{\infty} x_i\overline{y_i}i^\mu$.
4. $\mathbf{L}^2(a, b)$ and $\mathbf{C}(a, b)$ with $\langle x, y\rangle = \int_a^b x(t)\overline{y(t)}\,dt$.
5. $\mathbf{L}^2(\Omega)^m$ and $\mathbf{C}(\Omega)^m$ with $\langle \mathbf{x}, \mathbf{y}\rangle = \int_\Omega \mathbf{x}(\mathbf{t})\cdot\overline{\mathbf{y}(\mathbf{t})}\,d\Omega$.
6. $\mathbf{C}^k(\Omega)$ and $\mathbf{W}_2^k(\Omega)$ $(k \geq 0,\ k$ an integer) for scalar functions of $n$ variables with

$$\langle x, y\rangle_k = \int_\Omega \sum_{|\alpha| \leq k} D^\alpha x(\mathbf{t})\,\overline{D^\alpha y(\mathbf{t})}\,d\Omega$$

and for $m$-tuple vector functions in $\mathbf{C}^k(\Omega)^m$ and $\mathbf{W}_2^k(\Omega)^m$ with an appropriate $L^\alpha(\nabla)$ replacing $D^\alpha$ and using a dot product. In particular, for $\mathbf{W}_2^1(\Omega)^3$

(a) $\mathbf{H}(\mathrm{curl}, \Omega)$ with

$$\langle \mathbf{x}, \mathbf{y}\rangle = \int_\Omega \left[\mathbf{x}(\mathbf{t})\cdot\overline{\mathbf{y}(\mathbf{t})} + (\nabla \times \mathbf{x}(\mathbf{t}))\cdot\left(\nabla \times \overline{\mathbf{y}(\mathbf{t})}\right)\right] d\Omega,$$

(b) $\mathbf{H}(\mathrm{div}, \Omega)$ with

$$\langle \mathbf{x}, \mathbf{y}\rangle = \int_\Omega \left[\mathbf{x}(\mathbf{t})\cdot\overline{\mathbf{y}(\mathbf{t})} + (\nabla \cdot \mathbf{x}(\mathbf{t}))\left(\nabla \cdot \overline{\mathbf{y}(\mathbf{t})}\right)\right] d\Omega,$$

(c) $\mathbf{H}(\mathrm{curl}, \mathrm{div}, \Omega)$ with

$$\langle \mathbf{x}, \mathbf{y} \rangle = \int_\Omega \Big[\mathbf{x}(\mathbf{t})\cdot\overline{\mathbf{y}(\mathbf{t})} + (\nabla \times \mathbf{x}(\mathbf{t}))\cdot\left(\nabla \times \overline{\mathbf{y}(\mathbf{t})}\right)$$
$$+ (\nabla \cdot \mathbf{x}(\mathbf{t}))\left(\nabla \cdot \overline{\mathbf{y}(\mathbf{t})}\right)\Big] d\Omega.$$

For the above inner product spaces the norm induced by the inner product is the two-norm, e.g., $\|\mathbf{x}\| = \langle \mathbf{x}, \mathbf{x} \rangle^{1/2} = (\int_\Omega |\mathbf{x}(\mathbf{t})|^2 \, d\Omega)^{1/2} = \|\mathbf{x}\|_2$ for $\mathbf{L}^2(\Omega)^m$ and $\mathbf{C}(\Omega)^m$. The inner products in the last examples are not the only possible inner products for those sets, only perhaps the most useful ones. These will be called the usual inner products. Note that $\mathbf{l}^p$, $\mathbf{L}^p(\Omega)^m$, and $\mathbf{W}_p^k(\Omega)^m$ are not inner product spaces unless $p = 2$. This is easily seen from the following theorem, known as the *parallelogram law.*

**Theorem 2.4.** *If $S$ is an inner product space, then*

$$\|x + y\|^2 + \|x - y\|^2 = 2\|x\|^2 + 2\|y\|^2$$

*for all $x, y \in S$.*

The proof follows from

$$\begin{aligned} \|x + y\|^2 + \|x - y\|^2 &= \langle x + y, x + y \rangle + \langle x - y, x - y \rangle \\ &= 2\langle x, x \rangle + 2\langle y, y \rangle \\ &= 2\|x\|^2 + 2\|y\|^2 . \end{aligned}$$

We will also state without proof that in a normed space if the norm satisfies the parallelogram law, then there exists a unique inner product that generates the norm, making the space an inner product space as well.

A very useful inequality called the *Cauchy–Schwarz–Bunjakowsky inequality* (CSB) can be written in the form

$$|\langle x, y \rangle| \le \sqrt{\langle x, x \rangle}\sqrt{\langle y, y \rangle} = \|x\| \cdot \|y\|$$

with equality if and only if $x$ and $y$ are dependent.

An important property of inner products is continuity. Let $S$ be an inner product space. Given an element $y \in S$ and a sequence $\{x_n\} \subset S$ converging to an element $x \in S$, i.e., $\|x_n - x\| \to 0$, then $\langle x_n, y \rangle \to \langle x, y \rangle$ (see Definition 2.36). This can be written as

$$\lim_{n\to\infty} \langle x_n, y \rangle = \left\langle \lim_{n\to\infty} x_n, y \right\rangle,$$

emphasizing the fact that the limit and the inner product can be interchanged. This also implies that convergent series and inner products may be interchanged. Note also that this implies $\|x_n\| \to \|x\|$, which means that the limit operation and the norm may also be interchanged.

The proof of continuity of the inner product follows from Schwarz's inequality. Indeed, write $\langle x_n, y \rangle \to \langle x, y \rangle$ as $\langle x_n, y \rangle - \langle x, y \rangle \to 0$, or $\langle x_n - x, y \rangle \to 0$. By Schwarz's inequality $|\langle x_n - x, y \rangle| \le \|x_n - x\| \cdot \|y\|$, but $\|x_n - x\| \to 0$ so that $\langle x_n - x, y \rangle \to 0$.

### 2.5.2 Pseudo Inner Product, Reaction

In functional analysis the concept of an inner product is extremely important. This is also true in electromagnetic applications, but another, somewhat related concept is also very useful, known as a *pseudo inner product* [26, p. 42].

**Definition 2.34.** *Consider a linear space $S$. A pseudo inner product on $S$ is a mapping $S \times S \to \mathbf{C}$ that associates to each ordered pair $x, y \in S$ a scalar denoted by $\langle x, y \rangle_p$ with the properties*

*a.* $\langle x, y \rangle_p = \langle y, x \rangle_p$,

*b.* $\langle \alpha x, y \rangle_p = \alpha \langle x, y \rangle_p$,

*c.* $\langle x + y, z \rangle_p = \langle x, z \rangle_p + \langle y, z \rangle_p$,

*where $z \in S$ and $\alpha$ is an arbitrary scalar.*

Sometimes the pseudo inner product is called a *symmetric product.*

Compared to the inner product, there is no use of the complex conjugate for the pseudo inner product. In addition, we no longer require $\langle x, x \rangle_p > 0$; in fact, for complex-valued $x$, $\langle x, x \rangle_p$ is not even real-valued. Thus the pseudo inner product cannot generate a norm. The above definition can be used to show $\langle x, \alpha y \rangle_p = \alpha \langle x, y \rangle_p$.

For scalar functions $x(t), y(t) \in \mathbf{L}^2(a, b)$, a typical pseudo inner product is

$$\langle x, y \rangle_p = \int_a^b x(t) y(t)\, dt$$

with its vector analog in $\mathbf{L}^2(\Omega)^m$ being

$$\langle \mathbf{x}, \mathbf{y} \rangle_p = \int_\Omega \mathbf{x}(\mathbf{t}) \cdot \mathbf{y}(\mathbf{t})\, d\Omega. \tag{2.12}$$

Although the pseudo inner product does not generate a Hilbert (or Banach) space, because of its relationship with the concept of reaction it will be very useful later in the text. It is fortunate that while a pseudo inner product space lacks many important properties of an inner product space, it does generate enough mathematical structure to be very useful in electromagnetic applications. Also, for time-domain fields (which are assumed real), the pseudo inner product is an actual inner product (if we add the axiom $\langle x, x \rangle_p \geq 0$ and $\langle x, x \rangle_p = 0$ if and only if $x = 0$ to Definition 2.34), which is also the case in electro- and magnetostatics.

### 2.5.3 Strong and Weak Convergence

Noting that every inner product space is a normed space, and that a norm provides a measure of convergence, one can discuss convergence in inner product spaces. If particular, two types of convergence are defined, *strong*[25] and *weak* convergence.

**Definition 2.35.** *(Strong convergence) A sequence of elements $x_1, x_2, \ldots$ in an inner product space $S$ is said to be strongly convergent to an element $x \in S$ (denoted $x_n \to x$) if*

$$\|x_n - x\| \to 0.$$

Strong convergence is of course in the sense of a particular norm, which in turn can be defined by choice of the inner product. For instance, in $\mathbf{L}^2(\Omega)^m$ with the usual inner product, $\mathbf{x}_n \to \mathbf{x}$ means

$$\left( \int_\Omega |\mathbf{x}_n(\mathbf{t}) - \mathbf{x}(\mathbf{t})|^2 d\Omega \right)^{1/2} \to 0,$$

i.e., mean-square convergence.

**Definition 2.36.** *(Weak convergence) A sequence of elements $x_1, x_2, \ldots$ in an inner product space $S$ is said to be weakly convergent to an element $x \in S$ (denoted by $x_n \xrightarrow{w} x$) if*

$$\langle x_n, y \rangle \to \langle x, y \rangle$$

*for all $y \in S$.*

As an example, in $H = \mathbf{L}^2(\Omega)^m$ with the usual inner product, $\mathbf{x}_n \xrightarrow{w} \mathbf{x}$ means

$$\int_\Omega \mathbf{x}_n(\mathbf{t}) \cdot \overline{\mathbf{y}(\mathbf{t})}\, d\Omega \to \int_\Omega \mathbf{x}(\mathbf{t}) \cdot \overline{\mathbf{y}(\mathbf{t})}\, d\Omega$$

for all $\mathbf{y} \in H$, or $\int_\Omega (\mathbf{x}_n(\mathbf{t}) - \mathbf{x}(\mathbf{t})) \cdot \overline{\mathbf{y}(\mathbf{t})}\, d\Omega \to 0$.

It is possible to relate strong and weak convergence, as stated in the following theorem.

**Theorem 2.5.** *A strongly convergent sequence is weakly convergent to the same limit (i.e., $x_n \to x$ implies $x_n \xrightarrow{w} x$).*

The proof of the theorem follows from continuity of the inner product. The converse is not generally true, but it can be shown that if $x_n \xrightarrow{w} x$ and $\|x_n\| \to \|x\|$, then $x_n \to x$.

---

[25] We previously called strong convergence simply convergence.

### 2.5.4 Completeness, Hilbert Space

One of the most useful classes of spaces from a practical standpoint are complete inner product spaces, known as *Hilbert spaces*.

**Definition 2.37.** *An inner product space complete in the norm* $||x|| = \langle x, x\rangle^{1/2}$ *is called a Hilbert space.*

We can also understand a Hilbert space as the completion of an inner product space. For instance, the completion of $\mathbf{C}(\Omega)^m$ using the inner product $\langle \mathbf{x}, \mathbf{y}\rangle = \int_\Omega \mathbf{x}(\mathbf{t})\cdot\overline{\mathbf{y}(\mathbf{t})}\, d\Omega$ and associated norm is the Hilbert space of Lebesgue-measurable functions $\mathbf{L}^2(\Omega)^m$ (see Example 8 in Section 2.2.3). In a similar manner, the space $\mathbf{C}^k(\widetilde{\Omega})$ of $k$-times continuously differentiable complex-valued functions defined on a closed domain $\widetilde{\Omega}$ with the inner product $\langle x, y\rangle_k = \sum_{|\alpha|\leq k} \int_\Omega D^\alpha x(\mathbf{t})\, \overline{D^\alpha y(\mathbf{t})}\, d\Omega$ and associated norm is incomplete, the completion of which is the Sobolev space $\mathbf{W}_2^k(\Omega)$.

To say a Hilbert space $H$ is complete means that every Cauchy sequence in $H$ converges to an element of $H$, i.e., for every $\{x_n\} \subset H$ such that $\|x_n - x_m\| \to 0$, there exists $x \in H$ such that $\|x_n - x\| \to 0$. Since an inner product defines a norm, every Hilbert space is a Banach space.

It is an easy exercise to see that if a closed linear subspace $M$ is contained in a Hilbert space $H$, i.e., $M \subset H$, then $M$ itself is a Hilbert space (see Theorem 2.2).

#### Examples of Hilbert Spaces

1. $\mathbf{C}^n$ with $\langle \mathbf{x}, \mathbf{y}\rangle = \sum_{i=1}^n x_i\overline{y_i}$.

2. $\mathbf{l}^2$ with $\langle \mathbf{x}, \mathbf{y}\rangle = \sum_{i=1}^\infty x_i\overline{y_i}$.

3. $\mathbf{l}^2(\mu)$ $(\mu \geq 0)$ with $\langle \mathbf{x}, \mathbf{y}\rangle_\mu = \sum_{i=1}^\infty x_i\overline{y_i}i^\mu$.

4. $\mathbf{L}^2(a, b)$ with $\langle x, y\rangle = \int_a^b x(t)\overline{y(t)}\, dt$.

5. $\mathbf{L}^2(\Omega)^m$ with $\langle \mathbf{x}, \mathbf{y}\rangle = \int_\Omega \mathbf{x}(\mathbf{t})\cdot\overline{\mathbf{y}(\mathbf{t})}\, d\Omega$.

6. $\mathbf{H}^k(\Omega)$ $(\equiv \mathbf{W}_2^k(\Omega))$ for scalar functions of $n$ variables, $k$ an integer $\geq 0$, with

$$\langle x, y\rangle_k = \sum_{|\alpha|\leq k} \int_\Omega D^\alpha x(\mathbf{t})\, \overline{D^\alpha y(\mathbf{t})}\, d\Omega$$

   and for $m$-tuple vector functions in $\mathbf{H}^k(\Omega)^m$ $(\equiv \mathbf{W}_2^k(\Omega)^m)$ with an appropriate $L^\alpha(\nabla)$ replacing $D^\alpha$. In particular, $\mathbf{H}(\text{curl}, \Omega)$, $\mathbf{H}(\text{div}, \Omega)$, and $\mathbf{H}(\text{curl}, \text{div}, \Omega)$ are Hilbert spaces with the inner products shown on p. 111.

The inner product $\langle \cdot, \cdot \rangle$ in a Hilbert space is a continuous function with respect to convergence in norm: If $x_n \to x$ and $y_n \to y$, then $\lim_{n\to\infty} \langle x_n, y_n \rangle = \langle x, y \rangle$. Indeed, using Schwarz's inequality, $|\langle x_n, y_n \rangle - \langle x, y \rangle| \leq ||x_n - x|| \cdot ||y_n|| + ||x|| \cdot ||y_n - y||$. However, $||x_n - x|| \to 0$ and $||y_n - y|| \to 0$ as $n \to \infty$.

In addition to Hilbert spaces, we will often be interested in dense linear subspaces of Hilbert spaces, especially $\mathbf{C}(\Omega)$. Also, note that any finite-dimensional inner product space is a Hilbert space.

### 2.5.5 Orthogonality

In two and three dimensions the concept of vectors being perpendicular to one another is familiar. A definition that encompasses that notion, and generalizes it to inner product spaces, is the idea of *orthogonality.*

**Definition 2.38.** *Two elements of an inner product space are said to be orthogonal if* $\langle x, y \rangle = 0$.

Sometimes if two elements are orthogonal the notation $x \perp y$ is used. Broadening this idea to a set of elements, we say that a set of elements $x_m$ is an *orthogonal set* if $\langle x_i, x_j \rangle = 0$ whenever $i \neq j$. If, in addition, $\langle x_i, x_i \rangle = 1$ ($\|x_i\| = 1$), the set is said to be *orthonormal.* In this case the Kronecker delta-function[26] is often used, so that a set is orthonormal if $\langle x_i, x_j \rangle = \delta_{ij}$. Obviously the set $\{e_1,\ e_2,\ e_3\}$ is an orthonormal set. The set of vectors $e_1 = (1, 0, 0, \dots)$, $e_2 = (0, 1, 0, \dots)$, $e_3 = (0, 0, 1, \dots), \dots$ is an example of an orthonormal set in the space $\mathbf{l}^2$.

An equality that is sometimes useful is the Pythagorean formula.

**Theorem 2.6.** *For any pair of orthogonal elements* $x, y$,

$$\|x + y\|^2 = \|x\|^2 + \|y\|^2 .$$

This can easily be proved using the parallelogram law.

In addition to elements of a space being orthogonal, we can discuss the notion of subsets of spaces being orthogonal.

**Definition 2.39.** *Let* $M$ *be a linear subspace in a Hilbert space* $H$. *The orthogonal complement of* $M$, *denoted as* $M^{\perp}$, *is defined to be the set of all* $x \in H$ *such that* $\langle x, y \rangle = 0$ *for all* $y \in M$.

The orthogonal complement of $M$ is therefore the set of all elements in $H$ that are orthogonal to every element of $M$. Note that $M^{\perp}$ is a

[26] The Kronecker delta-function is defined as

$$\delta_{ij} \equiv \begin{cases} 1 & \text{if } i = j, \\ 0 & \text{if } i \neq j. \end{cases}$$

closed linear subspace of $H$, regardless of whether or not $M$ itself is closed. Therefore, by Theorem 2.2, $M^{\perp}$ is itself a Hilbert space whether or not $M$ is a Hilbert space.

To show $M^{\perp}$ is closed we use continuity of the inner product. Let $\{x_n\}$ be a sequence in $M^{\perp}$ converging to an element $x \in H$. We can show that in fact $x \in M^{\perp}$ (and so $M^{\perp}$ is closed) since $\langle x_n, y \rangle = 0$ for all $y \in M$ and $0 = \langle x_n, y \rangle \to \langle x, y \rangle$ so that $x \in M^{\perp}$. Note that we don't require completeness for the definition of an orthogonal complement; we simply work in a Hilbert space since that is the most important space for applications.

As a simple example, let $H = \mathbf{R}^3$ and define $M = \{\alpha \mathbf{e}_1, \beta \mathbf{e}_2\}$, i.e., the set of all vectors in the "$xy$"-plane. Then $M^{\perp} = \{\gamma \mathbf{e}_3\}$, i.e., the set of all vectors directed along the "$z$"-axis. As an example from a function space, let $H = \mathbf{L}^2(-a, a)$ and define the closed linear subspace $M$ as the set of all even functions (those for which $f(t) = f(-t)$ for all $t$). Then $M^{\perp}$ is the set of all odd functions (those for which $f(t) = -f(-t)$ for all $t$).

Additional useful properties follow, some of which require completeness.

**Theorem 2.7.** *Let $M$ be a linear subspace in $H$. Then*

*a.* $M^{\perp\perp} = \widetilde{M}$, where $\widetilde{M}$ is the closure of $M$,

*b.* if $M$ is closed, then $M^{\perp\perp} = M$,

*c.* $M^{\perp} = \{0\}$ if and only if $M$ is dense in $H$,

*d.* $\{0\}^{\perp} = H$ and $H^{\perp} = \{0\}$,

*e.* if $M$ is closed and $M^{\perp} = \{0\}$, then $M = H$.

From ($c$) we see that the concept of orthogonal complement helps to provide an understanding of what it means for one set to be dense in another set. Previously we said that $X \subset Y$ dense in $Y$ meant that every element in $Y$ could be approximated arbitrarily closely by elements in the subset $X$. Here we see that if a subspace $M$ is dense in $H$, then only the zero element is orthogonal to every element in $M$.

Concerning sums of closed subspaces, we have the following.

**Theorem 2.8.** *Let $M$ and $N$ be closed linear subspaces in $H$. If $M \perp N$, then $M \oplus N$ is a closed linear subspace in $H$.*

This can be proved using the Pythagorean theorem and the completeness of $M$ and $N$.

### 2.5.6 Expansions and Projections

Armed with the concepts of inner products, norms, and orthogonality, we can now investigate an important topic in applied analysis and, in fact, one of the main topics of this book. In general, it concerns when one element of a space can be represented, or at least approximated, by a set of elements within the space. A familiar example in a function space is the representation of $x(t) \in \mathbf{L}^2(-\pi, \pi)$ by a Fourier series,

$$x(t) = a_0 + \sum_{n=1}^{\infty} a_n \cos(nt) + b_n \sin(nt),$$

where the *expansion functions* of course belong to the space, i.e.,

$$1, \cos(nt), \sin(nt) \in \mathbf{L}^2(-\pi, \pi)$$

and $a_0, a_n, b_n$ are known as *expansion coefficients.* Another simple example is the representation of the vector $\mathbf{x} = (x_1, x_2, x_3) \in \mathbf{R}^3$ by the three "unit" vectors $\{\mathbf{e}_1, \mathbf{e}_2, \mathbf{e}_3\} \subset \mathbf{R}^3$,

$$\mathbf{x} = c_1\mathbf{e}_1 + c_2\mathbf{e}_2 + c_3\mathbf{e}_3.$$

In both cases equality between the element and the expansion is achieved, at least in the norm sense. Alternatively, consider the approximations

$$\widetilde{x}(t) \equiv \alpha_0 + \sum_{n=1}^{N} \alpha_n \cos(nt) + \beta_n \sin(nt) \simeq x(t) \tag{2.13}$$

for functions, where $N < \infty$, and

$$\widetilde{\mathbf{x}} \equiv \gamma_1\mathbf{e}_1 + \gamma_2\mathbf{e}_2 \simeq \mathbf{x} \tag{2.14}$$

for vectors in $\mathbf{C}^3$. In both cases we are approximating the element $x$ using an insufficient (incomplete) set of elements. The quality of the given approximation depends on the element to be approximated. For example, if $x(t)$ in (2.13) is similar to a harmonic function with frequency $\omega \simeq n \leq N$, then $\widetilde{x}(t)$ may be a very good approximation to $x(t)$. By "good" we mean $\|x - \widetilde{x}\| = \epsilon$, where $\epsilon \geq 0$ is a very "small" number, $\epsilon \ll 1$. Similarly, in (2.14) if the component $x_3$ of the vector $\mathbf{x}$ is very small, then the vector $\widetilde{\mathbf{x}}$ may be a very good approximation to the vector $\mathbf{x}$, i.e., $\|\mathbf{x} - \widetilde{\mathbf{x}}\| = \epsilon$.

An important issue in either an exact representation or an approximation is developing a method for determining the expansion coefficients that lead to the best possible approximation, or to an exact representation if possible. The idea is to determine the expansion coefficients to minimize the error $\|x - \widetilde{x}\|$ as follows.

Consider $N+1$ elements of $H$: the element $x \in H$, and a set of mutually orthonormal elements $y_m \in H$, $m = 1, \ldots, N$. Combine the set of elements $y_m$ as $\widetilde{x} = \sum_{m=1}^{N} \alpha_m y_m$. The sum $\sum_{m=1}^{N} \alpha_m y_m$ generates a closed linear subspace $M \subset H$ consisting of different elements associated with different sequences of coefficients $\{\alpha_m\}$. Define the difference of $x$ and $\widetilde{x}$ as an error, $e = x - \widetilde{x}$, and consider the norm (size) of the error, $\|e\| = \|x - \widetilde{x}\|$. Squaring the norm of the error leads to

$$\begin{aligned} \|x - \widetilde{x}\|^2 &= \langle x - \widetilde{x}, x - \widetilde{x}\rangle = \langle x, x\rangle + \langle \widetilde{x}, \widetilde{x}\rangle - \langle \widetilde{x}, x\rangle - \langle x, \widetilde{x}\rangle \\ &= \|x\|^2 + \sum_{m=1}^{N} |\alpha_m|^2 - \sum_{m=1}^{N} \alpha_m \overline{\langle x, y_m\rangle} - \sum_{m=1}^{N} \overline{\alpha_m} \langle x, y_m\rangle . \end{aligned}$$

The right side can be rewritten by completing the square, leading to

$$\|x - \widetilde{x}\|^2 = \|x\|^2 + \sum_{m=1}^{N} |\alpha_m - \langle x, y_m\rangle|^2 - \sum_{m=1}^{N} |\langle x, y_m\rangle|^2 .$$

All the terms in the sum are nonnegative, leading to the conclusion that the choice of coefficients that minimize the norm of the error is

$$\alpha_m = \langle x, y_m\rangle .$$

These coefficients are called *generalized Fourier coefficients,* and the sum

$$\widetilde{x} = \sum_{m=1}^{N} \langle x, y_m\rangle\, y_m$$

is called the *expansion* of $x$ with respect to the set $y_m$. It represents the best possible approximation to $x$ by $y_m \in M \subset H$. It is easy to show that the error is orthogonal to the approximation,[27] i.e.,

$$\langle e, \widetilde{x}\rangle = 0.$$

Orthogonality between the error and the expansion agrees intuitively with the following geometric picture from three-dimensional space: Assume an arbitrary vector $\mathbf{x} \in \mathbf{R}^3$ is given by $\mathbf{x} = c_1\mathbf{e}_1 + c_2\mathbf{e}_2 + c_3\mathbf{e}_3$ for certain given values $c_1, c_2, c_3$. Writing $\widetilde{\mathbf{x}} = \gamma_1\mathbf{e}_1 + \gamma_2\mathbf{e}_2$, where $\gamma_1, \gamma_2$ are the Fourier coefficients $\gamma_1 = \langle \mathbf{x}, \mathbf{e}_1\rangle = c_1$, $\gamma_2 = \langle \mathbf{x}, \mathbf{e}_2\rangle = c_2$, leads to an error $\mathbf{e} = \mathbf{x} - \widetilde{\mathbf{x}} = c_3\mathbf{e}_3$ and, of course, to $\langle \mathbf{e}, \widetilde{\mathbf{x}}\rangle = \langle c_3\mathbf{e}_3, c_1\mathbf{e}_1 + c_2\mathbf{e}_2\rangle = 0$.

At this point it is not clear how good our approximation is; all we know is that

$$\|x - \widetilde{x}\|^2 = \|x\|^2 - \sum_{m=1}^{N} |\langle x, y_m\rangle|^2 ,$$

which immediately leads us to see that

$$\sum_{m=1}^{N} |\langle x, y_m\rangle|^2 \leq \|x\|^2 \tag{2.15}$$

[27]Indeed, $\langle e, \widetilde{x}\rangle = \langle x - \widetilde{x}, \widetilde{x}\rangle = \left\langle x - \sum_{i=1}^{N} \alpha_i y_i, \sum_{j=1}^{N} \alpha_j y_j \right\rangle = \sum_{j=1}^{N} \alpha_j \langle x, y_j\rangle - \sum_{i=1}^{N}\sum_{j=1}^{N} \alpha_i \alpha_j \langle y_i, y_j\rangle = \sum_{j=1}^{N} \left(\alpha_j^2 - \alpha_j^2\right) = 0.$

since $\|x - \widetilde{x}\|^2 \geq 0$. The above inequality is known as *Bessel's inequality* (which also holds if $N = \infty$). It is easy to see that if equality is achieved (in which case

$$\sum_{m=1}^{N} |\langle x, y_m \rangle|^2 = \|x\|^2 \tag{2.16}$$

is known as *Parseval's equality* for the finite or infinite sum), then

$$\|x - \widetilde{x}\|^2 = 0$$

and $\widetilde{x}$ is equal to $x$ (as always, in the sense of the norm). For any $x \in H$, Bessel's inequality implies

$$\lim_{m \to \infty} \langle x, y_m \rangle = 0$$

since $\sum_{m=1}^{\infty} \beta_m < \infty$ implies $\lim_{m \to \infty} \beta_m = 0$. Therefore, if $y_m$ is an orthonormal set, then $y_m \xrightarrow{w} 0$.

An important theorem in analysis is the *projection theorem*, which is related to the concept of best approximation in the following way. Best approximation considered approximately representing an element of a Hilbert space $H$ by elements of a linear subspace $M$ in $H$ (generated by the sum $\sum_{m=1}^{N} \alpha_m y_m$ as discussed above). The projection theorem allows an element of $H$ to be *exactly* represented by an element of a closed linear subspace and an element of its orthogonal complement.

**Theorem 2.9.** *(Projection theorem) Let $M$ be a closed linear subspace in $H$. Any element of $H$ can be written in a unique way as the sum of an element in $M$ and an element in $M^{\perp}$ (i.e., any $z \in H$ can be written as $z = x_0 + y_0$, where $x_0 \in M$ and $y_0 \in M^{\perp}$). Moreover, $H = M \oplus M^{\perp}$ and $\|z\|^2 = \|x_0\|^2 + \|y_0\|^2$.*

Importantly, the element $x_0 \in M$ is the unique element in $M$ that is *closest* to $z$. By closest we mean $\|z - x_0\| \leq \|z - x\|$ for all $x \in M$. The element $x_0$ is called the projection of $z$ onto $M$. For any $x \in M$, $||x_0||^2 + ||x - x_0||^2 = ||x||^2$ and $||x_0|| \leq ||x||$.

A related theorem is useful for the spectral problems to be considered in Chapter 4.

**Theorem 2.10.** *(Orthogonal structure theorem) Let $M = M_1 \oplus M_2 \oplus \cdots \subset H$, where the $M_i$ are all mutually orthogonal ($M_i \perp M_j$ for $i \neq j$), closed linear subspaces of $H$. Then each $x \in M$ can be written uniquely as $x = \sum_n x_n$, where $x_n \in M_n$, and $\|x\|^2 = \sum_n \|x_n\|^2$.*

### 2.5.7 Basis of a Hilbert Space

While the projection theorem allows us to exactly represent any element of $H$ by decomposing the space, we can determine an exact representation of any element in $H$ in terms of other elements of $H$ called basis elements. With the concept of orthogonality established we are ready to introduce the most important notion for a basis, which relates in some way to a "perfect" expansion. Working generally in the context of an infinite-dimensional Hilbert space, first recall that the algebraic concept of a Hamel basis $B = \{x_1, x_2, \dots\}$ shows that any $x \in H$ can be approximated to any desired accuracy by a finite linear combination of (Hamel) basis elements. Specifically, for each $x \in H$ and every $\epsilon > 0$ there exist $N(\epsilon)$ and scalars $\alpha_1, \alpha_2, \dots, \alpha_n$ (depending on $\epsilon$ as well) such that

$$\left\| x - \sum_{n=1}^{N} \alpha_n x_n \right\| < \epsilon.$$

To remove the dependence of $\alpha_i$ on $\epsilon$ we introduce another notion of basis. For convenience we work with orthonormal sets, noting that any orthogonal set is easily changed into an orthonormal set.

**Definition 2.40.** *An orthonormal set of elements $B = \{x_n\}$ is said to be a (Schauder) basis for a Hilbert space $H$ if each element $y \in H$ can be written in a unique way as*

$$y = \sum_{n=1}^{\infty} \langle y, x_n \rangle \, x_n.$$

The coefficients $\alpha_n = \langle y, x_n \rangle$ are called *generalized Fourier coefficients.* This is the usual notion for a basis in an infinite-dimensional Hilbert space and agrees with the algebraic notion of a basis for finite-dimensional spaces. For (norm) convergence of the series $\sum_{n=1}^{\infty} \alpha_n x_n$ in $H$ we have the following.

**Theorem 2.11.** *(Riesz–Fischer theorem) Let $\{x_n\}$ be a basis for a Hilbert space $H$. The series*

$$\sum_{n=1}^{\infty} \alpha_n x_n$$

*converges in norm if and only if $\{\alpha_n\} \in \mathbf{l}^2$, i.e.,*

$$\sum_{n=1}^{\infty} |\alpha_n|^2 < \infty.$$

Recall that $\mathbf{l}^1 \subset \widetilde{\mathbf{l}^2} \subset \mathbf{l}^2$.

Theorem 2.9 (projection theorem) provided the idea of uniquely representing an element of $z \in H$ as an element $x_0$ of a closed linear subspace $M \subset H$, and an element $y_0 \in M^{\perp}$. With the concept of a basis established, the next theorem provides the form of the elements $x_0$, $y_0$.

**Theorem 2.12.** *Let $M$ be a closed linear subspace in $H$ with orthonormal basis $\{x_n\}$, and let $\{y_n\}$ be an orthonormal basis for $M^{\perp}$. Any element $z \in H$ can be written in a unique way as the sum of an element $x_0 \in M$ and an element $y_0 \in M^{\perp}$ (by the projection theorem), with*

$$x_0 = \sum_n \langle z, x_n \rangle x_n,$$

$$y_0 = \sum_n \langle z, y_n \rangle y_n.$$

Some other useful concepts relating to bases for Hilbert spaces follow.

**Theorem 2.13.** *Let $B$ be an orthonormal set in a Hilbert space $H$. The elements of $B$ are linearly independent.*

**Proof.** Let $\{x_n\}$ be a finite set from $B$ and consider $\alpha_1 x_1 + \alpha_2 x_2 + \cdots + \alpha_n x_n = 0$. Then $0 = \langle 0, x_1 \rangle = \langle \alpha_1 x_1 + \alpha_2 x_2 + \ldots + \alpha_n x_n, x_1 \rangle = \alpha_1 \langle x_1, x_1 \rangle + \alpha_2 \langle x_2, x_1 \rangle + \cdots + \alpha_n \langle x_n, x_1 \rangle = \alpha_1$. Repeating for all $n$ shows that $\alpha_1, \alpha_2, \ldots, \alpha_n = 0$ and so the set $\{x_1, x_2, \ldots, x_n\}$ is linearly independent. ■

**Definition 2.41.** *An orthonormal set $B$ in a Hilbert space $H$ is called maximal if and only if $x \perp B$ implies $x = 0$.*

Another way to say this is that a set $B = \{x_n\}$ in a Hilbert space $H$ is maximal if there is no element $y \in H$ such that $\{y, x_n\}$ is an orthonormal set (the set $\{y, x_n\}$ is not independent). Note that any set of $n$ linearly independent (orthonormal) elements in an $n$-dimensional space is maximal.

**Theorem 2.14.** *A maximal orthonormal set $B$ in $H$ is an orthonormal basis for $H$.*

When $B = \{x_n\}$ is an orthonormal basis (a maximal set) for $H$, $y = \sum_{n=1}^{\infty} \langle y, x_n \rangle x_n$ for all $y \in H$ such that Parseval's equality $\sum_{n=1}^{\infty} |\langle y, x_n \rangle|^2 = \|y\|^2$ holds. The scalar product in this representation is defined as $\langle x, y \rangle = \sum_{n=1}^{\infty} \langle x, x_n \rangle \overline{\langle y, x_n \rangle}$ for any $x, y \in H$.

The above concepts are summarized in the next theorem. It is stated for separable Hilbert spaces, although an analogous theorem can be stated for nonseparable spaces.

**Theorem 2.15.** *Let $H$ be a Hilbert space and $\{x_n\}$ an orthonormal basis for $H$. Then, for every $y \in H$,*

$$y = \sum_n \langle y, x_n \rangle x_n,$$

$$\|y\|^2 = \sum_n |\langle y, x_n \rangle|^2 .$$

*If $y \perp \{x_n\}$, then $y = 0$.*

While equality is, in general, in the norm sense, other notions of convergence (uniform or pointwise) may also hold.

Multidimensional bases are generalized in the following way.

**Theorem 2.16.** *If $\{x_1(t), x_2(t), \dots\}$ is an orthonormal basis for $\mathbf{L}^2(a, b)$, then $\{X_{ij}(t, s)\} = \{x_i(t)x_j(s)\}$, $1 \leq i, j$, is an orthonormal basis for $\mathbf{L}^2((a, b) \times (a, b))$.*

More generally, if $\{x_1, x_2, \dots\}$ is an orthonormal basis for $\mathbf{L}^2(\Omega)$ and $\{y_1, y_2, \dots\}$ is an orthonormal basis for $\mathbf{L}^2(\Lambda)$, then $\{x_i y_j\}$, $1 \leq i, j$, is an orthonormal basis for $\mathbf{L}^2(\Omega \times \Lambda)$.

The proof of the theorem is interesting in that it uses the fact that the only element orthogonal to a basis is the zero element.

**Proof.** Given that the set $\{x_i(t)\}$ is orthonormal in $\mathbf{L}^2(a, b)$, i.e., $\langle x_i, x_j \rangle = \int_a^b x_i(t)\overline{x_j(t)}\, dt = \delta_{ij}$, it is easy to see the set $\{x_i(t)x_j(s)\}$ is orthonormal. Indeed, the inner product is

$$\begin{aligned}
\langle x_i x_j, x_n x_m \rangle &= \int_a^b \int_a^b x_i(t)x_j(s)\overline{x_n(t)x_m(s)}\, ds\, dt \\
&= \int_a^b x_j(s)\overline{x_m(s)}\, ds \int_a^b x_i(t)\overline{x_n(t)}\, dt = \delta_{jm}\delta_{in}.
\end{aligned}$$

To show that the set $\{x_i(t)x_j(s)\}$ is an orthonormal basis, we let $f \in \mathbf{L}^2((a, b) \times (a, b))$ and show that vanishing of the Fourier coefficients of the function $f$ implies the function is zero, i.e., if $\langle f, x_i x_j \rangle = 0$, then $f = 0$. To proceed,

$$\begin{aligned}
0 = \langle f, x_i x_j \rangle &= \int_a^b \int_a^b f(t, s)\overline{x_n(t)x_m(s)}\, ds\, dt \\
&= \int_a^b ds\, \overline{x_m(s)} \int_a^b f(t, s)\overline{x_n}(t)\, dt.
\end{aligned}$$

Since the "inner" integral (as a function of $s$) is orthogonal to every basis element $x_m$, the inner integral vanishes for almost every $s$. Because $\int_a^b f(t, s)\overline{x_n}(t)\, dt = 0$ for each $s$, then $f$ is orthogonal to the basis elements $x_n$, implying $f = 0$ a.e. ∎

**Examples of Bases:**[28]

1. In $\mathbf{L}^2(-\pi,\pi)$ the set $\left\{\frac{1}{\sqrt{2\pi}}e^{int}\right\}$, $n=0,\pm1,\pm2,\ldots$ is a basis.

2. In $\mathbf{L}^2(-\pi,\pi)$ the set $\left\{\sqrt{\frac{1}{2\pi}}, \sqrt{\frac{1}{\pi}}\cos(nt), \sqrt{\frac{1}{\pi}}\sin(nt)\right\}$, $n=1,2,\ldots$ is a basis.

3. In $\mathbf{L}^2(0,\pi)$ the set $\left\{\sqrt{\frac{2}{\pi}}\sin(nt)\right\}$, $n=1,2,\ldots$ is a basis.

4. In $\mathbf{L}^2(0,\pi)$ the set $\left\{\sqrt{\frac{1}{\pi}}, \sqrt{\frac{2}{\pi}}\cos(nt)\right\}$, $n=1,2,\ldots$ is a basis.

5. In $\mathbf{L}^2(0,1)$ the set $\left\{e^{i2\pi nt}\right\}$, $n=0,\pm1,\pm2,\ldots$ is a basis.

6. In $\mathbf{l}^2$ the infinite set $e_1=(1,0,0,\ldots)$, $e_2=(0,1,0,\ldots)$, $\ldots$ is a basis (called the standard basis).

7. In $\mathbf{C}^n$ the set $\mathbf{e}_1=(1,0,0,\ldots,0)$, $\mathbf{e}_2=(0,1,0,\ldots,0)$, $\mathbf{e}_n=(0,0,0,\ldots,1)$ is a basis (called the standard basis).

8. In $\mathbf{L}^2\left((-\pi,\pi)\times(-\pi,\pi)\right)$ the set $\left\{\frac{1}{2\pi}e^{i(nt+ms)}\right\}$, $n,m=0,\pm1,\pm2,\ldots$ is an orthonormal basis, and the set $\left\{\frac{2}{\pi}\sin(nt)\sin(mt)\right\}$, $n,m=1,2,\ldots$ is a basis for $\mathbf{L}^2\left((0,\pi)\times(0,\pi)\right)$.

Many examples of eigenfunction bases are considered in Chapters 4 and 5 and in Part II of the text.

[28] For vector-valued functions we can expand each scalar component in the appropriate scalar set.

# Bibliography

[1] Debnath, L. and Mikusiński, P. (1999). *Introduction to Hilbert Spaces with Applications*, San Diego: Academic Press.

[2] Cessenat, M. (1996). *Mathematical Methods in Electromagnetism. Linear Theory and Applications*, Singapore: World Scientific.

[3] Zhang, W.X. (1991). *Engineering Electromagnetism: Functional Methods*, New York: Ellis Horwood.

[4] Muskhelishvili, N.I. (1953). *Singular Integral Equations: Boundary Problems of Function Theory and Their Application to Mathematical Physics*, Groningen: P. Noordhoff.

[5] Gakhov, F.D. (1966). *Boundary Value Problems*, Oxford: Pergamon Press.

[6] Hsiao, G.C. and Kleinman, R.E. (1997). Mathematical foundations for error estimation in numerical solutions of integral equations in electromagnetics, *IEEE Trans. Antennas Propagat.*, Vol. 45, no. 3, pp. 316–328, March.

[7] Prudnikov, A.P., Brychkov, Yu.A., and Marichev, O.I. (1986). *Integrals and Series,* Vol. 1, New York: Gordon and Breach Science Publishers.

[8] Abramowitz, M. and Stegun, I.A. (1972). *Handbook of Mathematical Functions*, New York: Dover.

[9] Mittra, R. and Lee, S.W. (1971). *Analytical Techniques in the Theory of Guided Waves,* New York: The Macmillan Company.

[10] Veliev, E.I. and Shestopalov, V.P. (1980). Excitation of a circular array of cylinders with longitudinal slits, *Radiophysics and Quantum Electronics (Engl. Transl.).* Vol. 23, no. 2, pp. 144–151.

[11] Koshparenok, V.N., Melezhik, P.N., Poyedinchuk, A.Y., and Shestopalov, V.P. (1983). Rigorous solution to the 2D problem of

diffraction by a finite number of curved screens, *USSR J. Comput. Maths. Mathem. Physics (Engl. Transl.)*. Vol. 23, no. 1, pp. 140–151.

[12] Kirilenko, A.A. and Yashina, N.P. (1980). Rigorous mathematical analysis and electrodynamic characteristics of the diaphragm in a circular waveguide, *Radiophysics and Quantum Electronics (Engl. Transl.)*. Vol. 23, no. 11, pp. 897–903.

[13] Bliznyuk, N.B. and Nosich, A.I. (1998). Limitations and validity of the cavity model in disk patch antenna simulations, *Proc. Intl. Symp. Antennas, JINA-98,* Nice, France.

[14] Veliev, E.I., Veremey, V.V., and Matsushima, A. (1993). Electromagnetic wave diffraction by conducting flat strips, *Trans. IEE Japan,* Vol. 113-A, no. 3, pp. 139–146.

[15] Tuchkin, Yu. A. (1985). Wave scattering by unclosed cylindrical screen of arbitrary profile with the Dirichlet boundary condition, *Soviet Physics Doklady (Engl. Transl.)*. Vol. 30, pp. 1027–1030.

[16] Tuchkin, Yu. A. (1987). Wave scattering by unclosed cylindrical screen of arbitrary profile with the Neumann boundary condition, *Soviet Physics Doklady (Engl. Transl.),* Vol. 32, pp. 213–216.

[17] Kirilenko, A.A., Rud', L.A., and Tkachenko, V.I. (1996). Semi-inversion method for an accurate analysis of rectangular waveguide H-plane angular discontinunities, *Radio Science,* Vol. 31, no. 5, pp. 1271–1280.

[18] Nosich, A.I., Okuno, Y., and Shiraishi, T. (1996). Scattering and absorption of E- and H-polarized plane waves by a circularly curved resistive strip, *Radio Science,* Vol. 31, no. 6, pp. 1733–1742.

[19] Chumachenko, V.P. (1989). Substantiation of one method of solving two-dimensional problems concerning the diffraction of electromagnetic waves by polygon structures. Uniqueness theorem, *Soviet J. Commun. Technol. Electronics (Engl. Transl.),* Vol. 34, no. 15, pp. 140–143.

[20] Nosich, A.I. (1993). Green's function—dual series approach in wave scattering from combined resonant scatterers, in Hashimoto, M., Idemen, M., and Tretyakov, O.A. (editors), *Analytical and Numerical Methods in Electromagnetic Wave Scattering,* Tokyo, Science House, pp. 419–469.

[21] Cloud, M.J. and Drachman, B.C. (1998). *Inequalities with Applications to Engineering,* New York: Springer-Verlag.

[22] Colton, D. and Kress, R. (1998). *Inverse Acoustic and Electromagnetic Scattering Theory,* New York: Springer-Verlag.

[23] Naylor, A.W. and Sell, G.R. (1982). *Linear Operator Theory in Engineering and Science*, 2nd ed., New York: Springer-Verlag.

[24] Buck, R.C. (1978). *Advanced Calculus*, 3rd ed., New York: McGraw-Hill.

[25] Berezansky, Y.M., Sheftel, Z.G., and Us, G.F. (1996). *Functional Analysis,* Vol. 1, Basel: Birkhäuser Verlag.

[26] Mrozowski, M. (1997). *Guided Electromagnetic Waves, Properties and Analysis*, Somerset, England: Research Studies Press.

[27] Friedman, B. (1956). *Principles and Techniques of Applied Mathematics*, New York: Dover.

[28] Akhiezer, N.I. and Glazman, I.M. (1961, 1963). *Theory of Linear Operators in Hilbert Space*, Vol. I and II, New York: Dover.

[29] Shilov, G.E. (1974). *Elementary Functional Analysis*, New York: Dover.

[30] Friedman, A. (1982). *Foundations of Modern Analysis*, New York: Dover.

[31] Krall, A.M. (1973). *Linear Methods in Applied Analysis*, Reading, MA: Addison-Wesley.

[32] Stakgold, I. (1999). *Green's Functions and Boundary Value Problems*, 2nd ed., New York: John Wiley and Sons.

[33] Dudley, D.G. (1994). *Mathematical Foundations for Electromagnetic Theory*, New York: IEEE Press.

[34] Reddy, B.D. (1998). *Introductory Functional Analysis*, New York: Springer-Verlag.

[35] Reddy, B.D. (1986). *Functional Analysis and Boundary-Value Problems: An Introductory Treatment,* Essex, England: Longman Scientific & Technical.

[36] Gohberg, I. and Goldberg, S. (1980). *Basic Operator Theory,* Boston: Birkhäuser.

[37] Chae, S.B. (1980). *Lebesgue Integration,* New York: Marcel Dekker.

[38] Boccara, N. (1990). *Functional Analysis. An Introduction for Physicists,* San Diego: Academic Press.

[39] Kantorovich, L.V. and Akilov, G.P. (1964). *Functional Analysis in Normed Spaces*, Oxford: Pergamon Press.

[40] Goffman, C. and Pedrick, G. (1983). *First Course in Functional Analysis*, 2nd ed., New York: Chelsea Publ. Co.

[41] Kohman, B. and Trench, W.F. (1971). *Elementary Multivariable Calculus*, New York: Academic Press.

[42] Curtiss, J.H. (1978). *Introduction to Functions of a Complex Variable*, New York: Marcel Dekker.

[43] Courant, R. (1974). *Introduction to Calculus and Analysis*, Vol. II, New York: Wiley.

[44] Smirnov, Yu.G. (1991). On the Fredholm problem for the diffraction by a plane, but perfectly conducting screen, *Sov. Physics Dokl.,* Vol. 36, pp. 512–513.

[45] Smirnov, Yu.G. (1991). Fredholmness of systems of pseudo differential equations in the problem of diffraction on a bounded screen, *Differential Eqs.*, Vol. 28, pp. 130–136.

[46] Smirnov, Yu.G. (1992). Solvability of integrodifferential equations in the problem of diffraction by a perfectly conducting plane screen, *Sov. J. Comm. Tech. Elec.*, Vol. 37, pp. 118–121.

[47] Smirnov, Yu.G. (1994). The solvability of vector integro-differential equations for the problem of the diffraction of an electromagnetic field by screens of arbitrary shape, *Comp. Maths Math. Phys.*, Vol. 34, no. 10, pp. 1265–1276.

# 3

# Introductory Linear Operator Theory

In this chapter we apply concepts of functional analysis, especially those concepts related to Hilbert and Banach spaces, to introduce basic operator theory relevant to applied electromagnetics. We begin with the definition of a linear operator and provide examples of common operators that arise in physical problems. We next define linear functionals as a special class of linear operators. Linear functionals occur quite often in electromagnetics and are very useful in theoretical investigations and in formulating problems to be solved numerically. In addition, the concept of a linear functional, in conjunction with the Riesz representation theorem, gives an appropriate motivation for introducing the important concept of an adjoint operator. Next, the class of self-adjoint operators is discussed, as well as the broader category of normal operators. We will see later that self-adjoint operators, and especially compact self-adjoint operators, have very nice mathematical properties that can be usefully exploited. Definite operators are then discussed, which themselves are contained within the class of self-adjoint or symmetric operators, and lead to a new notion of convergence. Compact operators are introduced, both at the function and sequence (infinite matrix) levels, and examples from applied mathematics and electromagnetics are provided. The important topic of the operator inverse is then covered, which naturally arises in the consideration of solving operator equations and, primarily for differential operators, is related to the concept of a Green's function. Projection operators are briefly discussed, followed by a statement of the Fredholm alternative, which details solvability conditions for certain operator equations.

# 3.1 Linear Operators

## 3.1.1 General Concepts

We will define a *linear operator* as a linear mapping between linear spaces.

**Definition 3.1.** *A mapping $A$ from $X$ to $Y$, denoted as $A : X \to Y$, is called a linear operator (linear mapping, linear transformation) if for all $x$ and $y$ in the domain of $A$ (defined below) and $\alpha, \beta \in \mathbf{C}$,*

$$A(\alpha x + \beta y) = \alpha Ax + \beta Ay,$$

*where $Ax, Ay \in Y$.*

Note that a linear operator maps elements of $X$ to elements of $Y$, and often $X$ is equal to $Y$. As an example of a familiar operator, the mapping $A : \mathbf{C}^n \to \mathbf{C}^m$, given by an $m \times n$ complex-valued matrix maps $n$-component vectors in $\mathbf{C}^n$ into $m$-component vectors in $\mathbf{C}^m$.

The elements on which the operator acts, and the elements it produces, are given special names.

**Definition 3.2.** *The domain of a linear operator $A : X \to Y$, denoted as $D_A$, is simply the set of elements for which the mapping $A$ is defined.*

It is convenient in the following to restrict our linear spaces $X$ and $Y$ to be Hilbert spaces, denoted as $H_1$ and $H_2$, respectively, although in a few instances we will consider Banach-space formulations.

For operators acting *on* a Hilbert space $H_1$, $D_A = H_1$. Often we will be interested in operators acting *in* a Hilbert space $H_1$, specifically on a closed linear subspace $M$ of a Hilbert space. In this case $D_A = M \subset H_1$, $D_A \neq H_1$. For example, the differential operator $A : \mathbf{L}^2(a,b) \to \mathbf{L}^2(a,b)$ defined by $(Ax)(t) = dx(t)\,/\,dt = x'(t)$ cannot have as its domain all of $\mathbf{L}^2(a,b)$, since many functions in this space are not differentiable, or even continuous.

Defining a domain is an important part of defining an operator; an operator is completely defined by specifying *both* its domain and its action on elements of the domain. Furthermore, often properties of an operator are very sensitive to the choice of the domain. This is especially true for differential operators, where boundary conditions are part of the domain specification.

The most important operators acting *in* Hilbert spaces are *densely defined.*

**Definition 3.3.** *An operator $A : H_1 \to H_2$ acting in $H_1$ is said to be densely defined if $D_A$ is an everywhere dense subset of $H_1$, i.e., $\widetilde{D}_A = H_1$.*

Recall from Definition 2.13 that this means that every element in $H_1$ can be approximated arbitrarily closely by elements of the set $D_A$, and any neighborhood of an arbitrary point of $H_1$ contains points from the set $D_A$. Thus, in this case while $D_A$ is not the whole space $H_1$ ($H_1$ may, in fact, contain infinitely many elements not in $D_A$), $D_A$ is "close" to $H_1$ in this sense.

Unless otherwise noted, when $D_A \neq H_1$ we will work with operators on densely defined subspaces of $H_1$.

**Definition 3.4.** *The range (or image space) of a linear operator $A : H_1 \to H_2$, denoted as $R_A$, is the set of elements $y \in H_2$ resulting from $Ax$ for all possible $x \in D_A$.*

The dimension of the range of $A$ is called the *rank* of $A$, and for $Ax = y$, $y$ is called the *image* of $x$. Note that the range of $A$ need not be all of $H_2$. A mapping $A : H_1 \to H_2$ for which $R_A = H_2$ is said to be *onto* (surjective), whereas if $R_A \subset H_2$ is a proper subspace of $H_2$ then the mapping is *into*. Therefore, $A : H_1 \to H_2$ may be *in* or *on* $H_1$, and *into* or *onto* $H_2$.

**Definition 3.5.** *The null space (or kernel) of a linear operator $A : H_1 \to H_2$, denoted as $N_A$, is the set of all elements $x \in D_A$ for which $Ax = 0$.*

As an example, the common vector identity $\nabla \times \nabla \psi = \mathbf{0}$ indicates that the range of the gradient operator must be in the null space of the curl operator, i.e., $R_\nabla \subseteq N_{\nabla \times}$. The dimension of the null space is called the *nullity*. It is easy to see that the range and null space of an operator are linear subspaces of Hilbert spaces.

An important class of operators are those having *finite rank*.[1] This is a different concept from that of an operator acting on a finite-dimensional domain. It can be shown that for $A : H_1 \to H_2$ a linear operator and $H_1$ finite-dimensional, then $R_A$ is a finite-dimensional subspace of $H_2$ (and $\dim(H_1) = \text{nullity}\,(A) + \text{rank}\,(A)$). It then follows that every operator acting on a finite-dimensional domain is a finite-rank operator, yet the converse is not necessarily true. In the following we reserve the name "finite-dimensional operator" for the special subset of finite-rank operators that act on finite-dimensional spaces.

## Boundedness and Continuity of Operators

A very important classification of operators pertains to their *boundedness*.

**Definition 3.6.** *A linear operator $A : H_1 \to H_2$ is bounded on its domain if, for all $x \in H_1$, there exists a number $k > 0$ such that*[2]

$$\|Ax\|_2 \leq k \|x\|_1 .$$

[1] Also known as *finite-dimensional operators*.

[2] As always when working in Hilbert spaces, we exclusively use the norm induced by the inner product, with the subscript (often omitted) indicating the associated space ($H_1$ or $H_2$).

We use $x \in H_1$ rather than $x \in D_A$ since it is shown in Section 3.4.1 that for a bounded operator $A$, $D_A = H_1$, at least by extension.

Note that bounded operators map bounded sets in $D_A$ into bounded sets in $R_A$. This is a somewhat different notion than the calculus idea of a bounded function, where a function is said to be bounded if the range of the function is a bounded set.

If an operator is not bounded, it is said to be *unbounded*. To show that an operator $A$ is unbounded it is sufficient to find a sequence of elements $\mathbf{x} = \{x_1, x_2, \ldots\}$ with $x_n \in D_A$ such that $||x_n||_1 \leq M$ for some $M$ and $||Ax_n||_2 \to \infty$ as $n \to \infty$ (see, e.g., Example 8 in this section). The most important class of unbounded operators relevant to electromagnetics is differential operators.

The definition of boundedness of an operator brings about the concept of an operator norm (as an abstraction of size).

**Definition 3.7.** *The norm of an operator $A : H_1 \to H_2$, denoted as $\|A\|$, is the smallest number $k$ that satisfies $\|Ax\|_2 \leq k \|x\|_1$ for all $x \in H_1$. This can be stated as*

$$\|A\| = \sup_{\|x\|_1 \neq 0} \frac{\|Ax\|_2}{\|x\|_1} = \sup_{\|x\|_1 \leq 1} \|Ax\|_2 = \sup_{\|x\|_1 = 1} \|Ax\|_2 .$$

It is convenient to note that $\|Ax\|_2 \leq \|A\| \|x\|_1$. In the following we will generally omit the subscript on the norm.

A fundamental property that divides operators into two classes is continuity, or lack thereof.

**Definition 3.8.** *A linear operator $A : H_1 \to H_2$ is said to be continuous at $x_0 \in H_1$ if for every $\epsilon > 0$ there exists a number $\delta > 0$ such that $\|Ax - Ax_0\| < \varepsilon$ whenever $\|x - x_0\| < \delta$.*

The next theorem states an extremely important fact concerning continuous operators.

**Theorem 3.1.** *If $A : H_1 \to H_2$ is a continuous linear operator at $x_0 \in H_1$, and if $x_n \to x_0$, then $Ax_n \to Ax_0$.*

This theorem states that if $\lim_{n\to\infty} x_n = x_0$, then $\lim_{n\to\infty} Ax_n = A \lim_{n\to\infty} x_n$, indicating that the operator and the limit operation may be interchanged.

**Proof.** Assume $x_n \to x$ (i.e., $\|x_n - x\| \to 0$) and show $\|Ax_n - Ax\| \to 0$ (i.e., $Ax_n \to Ax$). Indeed, $\|Ax_n - Ax\| = \|A(x_n - x)\| \leq \|A\| \; \|x_n - x\|$. Because $\|x_n - x\| \to 0$, then $\|Ax_n - Ax\| \to 0$, where we assume $\|A\| < \infty$. ■

A very important relationship exists between the continuity and boundedness of a linear operator.

**Theorem 3.2.** *A linear operator $A : H_1 \to H_2$ is continuous if and only if it is bounded.*

**Proof.** Assume $A$ is bounded and $x_0$ an arbitrary point in $H_1$. Then $\|Ax - Ax_0\| = \|A(x - x_0)\| \leq k \|x - x_0\|$. Thus if $\|x - x_0\| < \delta$, setting $\delta = \varepsilon/k$ gives $\|A(x - x_0)\| < \varepsilon$. Then $A$ is continuous at an arbitrary point and so is continuous on its domain.

Now assume $A$ is continuous, and in particular take $x_0 = 0$. Then for $\varepsilon = 1$ there exists a $\delta > 0$ such that $\|Ax\| < 1$ whenever $\|x\| < \delta$. For $x \neq 0$ let $z = \beta x$ with $\beta = \delta / \|x\|$. Then $\|z\| = \|\beta x\| = \|(\delta\, x) / \|x\|\| = \delta$. Therefore, $1 > \|Az\| = \|A(\beta x)\| = |\beta| \|Ax\|$ and $\|Ax\| < 1/\beta = (1/\delta) \|x\|$ and so $A$ is bounded. Of course, for $x = 0$, $\|A0\| = 0 < \|x\|$. ∎

In summary, continuity at 0 implies boundedness, which implies continuity at an arbitrary point $x$.

**Theorem 3.3.** *If a linear operator $A : H_1 \to H_2$ is continuous at one point, it is continuous on its domain.*

If a linear operator is continuous, it is uniformly continuous, since $\delta$ in the above proof is independent of $x_0$.

For linear operators (but not most ordinary functions) boundedness implies continuity, and conversely. Unbounded operators are then related to some sort of discontinuity of the operator. For unbounded operators, $x_n \to x_0$ does not imply $Ax_n \to Ax_0$.

It can be shown that the space of all bounded operators $A : X_1 \to X_2$, where $X_{1,2}$ are Banach or Hilbert spaces, is itself a Banach space [1, p. 43]. This is important since later we need the concept of convergence of a series of operators (see, e.g., Section 4.4), and in a complete space absolute convergence, $\sum_n \|A^n\|$, implies convergence of $\sum_n A^n$.

**Theorem 3.4.** *If $A : H_1 \to H_2$, and $B : H_2 \to H_3$ are bounded linear operators, then the composition $BA : H_1 \to H_3$ is bounded, and $\|BA\| \leq \|B\| \|A\|$.*

**Proof.** $\|BAx\| \leq \|B\| \|Ax\| \leq \|B\| \|A\| \|x\|$, proving boundedness of $BA$. The inequality follows from $\|BAx\| \leq \|B\| \|A\| \|x\|$ and the definition of $\|BA\|$. ∎

Finally, there is an interesting statement about continuity of operators on finite-dimensional spaces.

**Theorem 3.5.** *Consider a linear operator $A : H_1 \to H_2$. If $H_1$ is finite-dimensional, then $A$ is continuous (and so bounded).*

**Proof.** Let $\{x_n\}_{n=1}^N$ be a basis for $H_1$, and take any $0 \neq x \in H_1$ as $x = \sum_{n=1}^N \alpha_n x_n$. Then $\|Ax\| = \left\|\sum_{n=1}^N \alpha_n Ax_n\right\| \leq \sum_{n=1}^N |\alpha_n| \|Ax_n\|$. Let

$K = \max_n \{\|Ax_n\|\}$. Then $\|Ax\| \leq K \sum_{n=1}^{N} |\alpha_n|$. It can be shown (proof is omitted here; it is at least reasonable since the sum is finite) that an $M$ exists such that $\sum_{n=1}^{N} |\alpha_n| \leq M \|x\|$ and so $\|Ax\| \leq KM \|x\|$, which shows that $A$ is bounded and hence continuous. ■

### 3.1.2 Examples Related to Linear Operators

1. Ordinary real-valued functions $x(t)$ can be thought of as operators $x : \mathbf{R} \to \mathbf{R}$. However, only linear functions are themselves linear operators. For example, the function $x(t) = t$ for $t \in \mathbf{R}$ defines a linear operator, yet $x(t) = \sin(t)$ (or $t + 1, t^2, e^t$, etc.) does not.

2. The common "real-part" and "imaginary-part" operators $A_r, A_i : \mathbf{C} \to \mathbf{R}$ defined by

$$\begin{aligned} A_r x &\equiv \mathrm{Re}(x), \\ A_i x &\equiv \mathrm{Im}(x), \\ D_{A_r} &= D_{A_i} \equiv \mathbf{C} \end{aligned}$$

are easily seen to be nonlinear operators. The complex conjugation operator $A_{cc} : \mathbf{C} \to \mathbf{C}$ defined by

$$\begin{aligned} A_{cc} x &\equiv \overline{x}, \\ D_{A_{cc}} &\equiv \mathbf{C}, \end{aligned}$$

is also nonlinear.

3. The identity operator $I : H \to H$ defined by

$$\begin{aligned} Ix &\equiv x, \\ D_I &\equiv H, \end{aligned}$$

is a bounded linear operator with $\|I\| = 1$, $R_I = H$, and $N_I = \{0\}$.

4. The null operator $0 : H \to H$ defined by

$$\begin{aligned} 0x &\equiv 0, \\ D_0 &\equiv H, \end{aligned}$$

is a bounded linear operator with $\|0\| = 0$, $R_0 = \{0\}$, and $N_0 = H$.

5. A bounded linear operator $A : H \to H$ defined by

$$\begin{aligned} Ax &= y, \\ D_A &\equiv H, \end{aligned}$$

where $\{x_1, x_2, \dots\}$ is an orthonormal basis for $H_1$, and $\{y_1, y_2, \dots\}$ is an orthonormal basis for $H_2$. If $\langle x_i, y_j \rangle = \delta_{ij}$, the two basis sets are called *bi-orthonormal*.

6. As a finite-dimensional version of a matrix operator, the operator $A : \mathbf{C}^n \to \mathbf{C}^m$ (bounded since $\mathbf{C}^n$ is finite-dimensional) is uniquely represented, for $\{x_1, x_2, \dots, x_n\}$ a basis of $\mathbf{C}^n$ and $\{y_1, y_2, \dots, y_m\}$ a basis of $\mathbf{C}^m$, by the $m \times n$ matrix $[a_{ij}]$ where $a_{ij} = \langle Ax_j, y_i \rangle$. Conversely, every $m \times n$ matrix $[a_{ij}]$ where $a_{ij} = \langle Ax_j, y_i \rangle$ uniquely defines a bounded linear operator $A : \mathbf{C}^n \to \mathbf{C}^m$ with respect to the given bases. Exactly the same result is obtained for a more general mapping $A : H_1 \to H_2$, where $H_1, H_2$ are any two finite-dimensional Hilbert spaces.

   For operators on finite-dimensional spaces some explicit equations for the norm may be developed in terms of the matrix representation of the operator. For instance, considering the operator $A : \mathbf{C}^n \to \mathbf{C}^m$ and using the norm $\|\mathbf{x}\|_1 = \sum_{j=1}^n |x_j|$ lead to the matrix norm[5]

$$\|A\|_1 = \max_{1 \le j \le n} \sum_{i=1}^m |a_{ij}|, \tag{3.1}$$

   which obviously corresponds to summing the absolute values of elements of each column and taking the largest such sum. This is called the *column-sum norm.* Alternatively, the $\infty$-norm $\|\mathbf{x}\|_\infty = \max_{1 \le j \le n} |x_j|$ leads to

$$\|A\|_\infty = \max_{1 \le i \le m} \sum_{j=1}^n |a_{ij}|,$$

   which corresponds to summing the absolute values of elements of each row and taking the largest such sum (*row-sum norm*). Obviously the relationship $\|A\|_1 = \|A^T\|_\infty$ holds. The two-norm $\|\mathbf{x}\|_2 = (\sum_{j=1}^n |x_j|^2)^{1/2}$ leads to $\|A\|_2 = \sqrt{\lambda_{\max}}$, where $\lambda_{\max}$ is the largest eigenvalue of $A^*A$, which is sometimes called the *spectral norm* of $A$.

   The above are known as *subordinate matrix norms*, since they arise directly from an associated vector norm. Other matrix norms, which come directly from Definition 2.31, rather than the operator norm Definition 3.7, are sometimes found to be useful. Two common

[5] To see this, consider

$$\|A\mathbf{x}\| = \sum_{i=1}^m \left| \sum_{j=1}^n a_{ij} x_j \right| \le \sum_{j=1}^n \left( \sum_{i=1}^m |a_{ij}| \right) |x_j| \le \max_{1 \le j \le n} \sum_{i=1}^m |a_{ij}| \, \|\mathbf{x}\|.$$

The maximum over all $\mathbf{x}$ with $\|\mathbf{x}\| = 1$ leads to the desired formula.

norms from this latter class are the *Schur norm*, defined as

$$\|A\|_S = \left( \sum_{i=1}^{m} \sum_{j=1}^{n} |a_{ij}|^2 \right)^{1/2},$$

and the *maximum norm* $\|A\|_M = \max_{i,j} |a_{ij}|$. The Schur norm satisfies $\|AB\| \leq \|A\| \, \|B\|$, while the maximum norm does not. Both are easy to compute but are generally less useful than the subordinate norms. Some discussion of these concepts, as well as the role of the *condition number*[6] in numerical matrix analysis, can be found in [2, pp. 176–183].

7. Consider the operators $A, B : \mathbf{C}^n \to \mathbf{C}^n$ associated with matrices $[a_{ij}]_{i,j=1}^n$ and $[b_{ij}]_{i,j=1}^n$, respectively. The composition (product) $C = AB$ of operators $A$ and $B$ is associated with the product of corresponding matrices:

$$(C\mathbf{x})_{i=1}^n = (AB\mathbf{x})_{i=1}^n = \sum_{j=1}^{n} \left( \sum_{k=1}^{n} a_{ik} b_{kj} \right) x_j = \sum_{j=1}^{n} c_{ij} x_j,$$

where $c_{ij} = \sum_{k=1}^{n} a_{ik} b_{kj}$. If $A$ and $B$ are bounded in $\mathbf{C}^n$, then by Theorem 3.4 the product $C$ is bounded in $\mathbf{C}^n$. Indeed,

$$\begin{aligned} \|C\mathbf{x}\| &= \sum_{i=1}^{n} \left| \sum_{j=1}^{n} \left( \sum_{k=1}^{n} a_{ik} b_{kj} \right) x_j \right| \\ &\leq \max_{1 \leq k \leq n} \sum_{i=1}^{n} |a_{ik}| \max_{1 \leq j \leq n} \sum_{k=1}^{n} |b_{kj}| \, \|\mathbf{x}\|. \end{aligned}$$

Similarly, the product $C = AB : \mathbf{l}^2 \to \mathbf{l}^2$ of bounded in $\mathbf{l}^2$ operators $A$ and $B$ corresponding to the infinite matrices $[a_{ik}]$ and $[b_{kj}]$, assuming $\sum_{i=1}^{\infty} \sum_{k=1}^{\infty} |a_{ik}|^2 < \infty$ and $\sum_{k=1}^{\infty} \sum_{j=1}^{\infty} |b_{kj}|^2 < \infty$, is bounded in $\mathbf{l}^2$. This property of bounded operators can be used to simplify the proof of boundedness and compactness of the product operator obtained in some electromagnetic problems, where each matrix element is obtained in terms of infinite series (Galerkin-type projection techniques often result in such internal summations).

8. The differential operator $A : \mathbf{L}^2(a, b) \to \mathbf{L}^2(a, b)$ defined by

$$(Ax)(t) \equiv \frac{dx(t)}{dt} = x'(t),$$
$$D_A \equiv \left\{ x : x, x' \in \mathbf{L}^2(a, b) \right\}$$

[6] The condition number of a matrix $A$ is defined as $\|A\| \left\| A^{-1} \right\|$ using a subordinate matrix norm.

is unbounded. To see this, consider the sequence of functions $x_n(t) = \cos nt,\ n = 1, 2, \ldots$, defined on the interval $[-\pi, \pi]$. Then

$$||x_n|| = \left( \int_{-\pi}^{\pi} (\cos nt)^2 dt \right)^{1/2} = \sqrt{\pi}$$

and

$$||Ax_n|| = \left( \int_{-\pi}^{\pi} (n \sin nt)^2 dt \right)^{1/2} = n\sqrt{\pi},$$

resulting in $||Ax_n|| = n||x_n||$. Clearly, $||Ax_n|| \to \infty$ as $n \to \infty$. Moreover, all differential operators are usually unbounded, although distributional spaces can be constructed where they are, in fact, bounded.

As mentioned previously, this operator cannot have as its domain all of $\mathbf{L}^2(a, b)$ since many functions in this space are not differentiable, or even continuous. Note that $N_A$ is the set of constant functions.

9. The multiplication operator $A : \mathbf{C}(\Omega) \to \mathbf{C}(\Omega)$ ($\mathbf{C}(\Omega)$ is a Banach space in the max-norm, $\Omega \subset \mathbf{R}^n$) defined by

$$(Ax)\,(\mathbf{t}) \equiv f(\mathbf{t})x(\mathbf{t})$$

is a bounded linear operator assuming $k = \max_{\mathbf{t} \in \Omega} |f(\mathbf{t})| < \infty$ with $||A|| \le k$. Boundedness follows from the inequality

$$|f(\mathbf{t})x(\mathbf{t})| \le ||f||\,||x||,$$

where $||f|| = \max_{\mathbf{t} \in \Omega} |f(\mathbf{t})|$ and $||x|| = \max_{\mathbf{t} \in \Omega} |x(\mathbf{t})|$. Also, the multiplication operator $A$ is bounded in the Hilbert space $\mathbf{L}^2(\Omega)$ for $f$ continuous,

$$||Ax||^2 = \int_{\Omega} |f(\mathbf{t})x(\mathbf{t})|^2 d\Omega \le k^2 ||x||^2,$$

where $k = \max_{\mathbf{t} \in \Omega} |f(\mathbf{t})| < \infty$.

10. The Banach-space integral operator $A : \mathbf{C}(\Omega) \to \mathbf{C}(\Omega)$ defined by

$$(Ax)(\mathbf{t}) \equiv \int_{\Omega} k(\mathbf{t}, \mathbf{s}) x(\mathbf{s})\, d\Omega,$$

where $k(\mathbf{t}, \mathbf{s}) \in \mathbf{C}(\Omega \times \Omega)$ and $\Omega \subset \mathbf{R}^n$, with the max-norm

$$||k|| = \max_{\mathbf{t} \in \Omega} \int_{\Omega} |k(\mathbf{t}, \mathbf{s})|\, d\Omega < \infty,$$

is bounded with

$$||A|| \le \max_{\mathbf{t} \in \Omega} \int_{\Omega} |k(\mathbf{t}, \mathbf{s})|\, d\Omega.$$

Indeed, using the inequality of the previous example,

$$|(Ax)(\mathbf{t})| \le \int_\Omega |k(\mathbf{t},\mathbf{s})||x(\mathbf{s})|\,d\Omega \le \max_{\mathbf{t}\in\Omega} \int_\Omega |k(\mathbf{t},\mathbf{s})|\,d\Omega\ ||x||$$

and, in fact, it can be shown the equality holds.

Similarly, it can be shown that the weakly singular integral operator

$$(Ax)(\mathbf{t}) \equiv \int_\Omega \frac{k(\mathbf{t},\mathbf{s})}{|\mathbf{t}-\mathbf{s}|^\alpha} x(\mathbf{s})\,d\Omega$$

is bounded in $\mathbf{C}(\Omega)$ for $\alpha < n$ (for $\alpha = 0$ this reduces to the previous case).

11. The Hilbert-space integral operator $A : \mathbf{L}^2(\Omega) \to \mathbf{L}^2(\Omega)$ defined by

$$(Ax)(\mathbf{t}) \equiv \int_\Omega k(\mathbf{t},\mathbf{s})x(\mathbf{s})\,d\Omega$$

with kernel $k(\mathbf{t},\mathbf{s}) \in \mathbf{L}^2(\Omega \times \Omega)$ and the two-norm

$$||k||^2 = \int_\Omega \int_\Omega |k(\mathbf{t},\mathbf{s})|^2\,d\Omega_s\ d\Omega_t < \infty,$$

where $\Omega \subset \mathbf{R}^n$, is bounded with[7]

$$\|A\|^2 \le \int_\Omega \int_\Omega |k(\mathbf{t},\mathbf{s})|^2\,d\Omega_s\ d\Omega_t.$$

The above follows from the inequality

$$\begin{aligned}\|Ax\|^2 = \int_\Omega |(Ax)(\mathbf{t})|^2 d\Omega_t &= \int_\Omega \left| \int_\Omega k(\mathbf{t},\mathbf{s})x(\mathbf{s})\,d\Omega_s \right|^2 d\Omega_t \\ &\le \int_\Omega \int_\Omega |k(t,s)|^2\,d\Omega_s\ d\Omega_t\ ||x||^2.\end{aligned}$$

Such operators are known as *Hilbert–Schmidt operators*, and the kernel $k(\mathbf{t},\mathbf{s})$ is said to be of *Hilbert–Schmidt type.*

More generally, a bounded operator $A : H \to H$ is called a Hilbert–Schmidt operator if, given $\{x_n\}$ an orthonormal basis of $H$,

$$\sum_n \|Ax_n\|^2 < \infty.$$

[7] A sufficient condition of boundedness of the integral operator in $\mathbf{L}^2(\Omega)$ is $\int_\Omega \int_\Omega |k(\mathbf{t},\mathbf{s})|^2 d\Omega_s\,d\Omega_t < \infty$. Also, the integral operator is bounded in $\mathbf{L}^2(\Omega)$ if there is a number $M < \infty$ such that $\int_\Omega |k(\mathbf{t},\mathbf{s})|\,d\Omega_s \le M$ and $\int_\Omega |k(\mathbf{t},\mathbf{s})|\,d\Omega_t \le M$. This condition of boundedness is also sufficient.

12. The integration (*Volterra*) operator $A : \mathbf{L}^2(a,b) \to \mathbf{L}^2(a,b)$ defined by

$$(Ax)(t) \equiv \int_a^t x(s)\, ds$$

is a linear bounded operator for $-\infty < a < b < \infty$. This is just a special case of the last example ($\Omega = (a,b)$) with $k(t,s) = H(t-s)$, the Heaviside function, where $H(t-s) = 1$ for $t > s$, and 0 for $t < s$.

13. Let $k(t,s)$ be bounded, i.e., $k(t,s) \in \mathbf{L}^\infty((a,b) \times (a,b))$. The weakly singular integral operator

$$(Ax)(t) \equiv \int_a^b \frac{k(t,s)}{|t-s|^\alpha} x(s)\, ds$$

is bounded in $\mathbf{L}^p(a,b)$ for $1 < p < \infty$ and $0 \leq \alpha < 1$ [3, p. 57]. Moreover, with $\alpha = 1$ the operator, taken in the sense of the Cauchy principal value, is bounded in $\mathbf{L}^p(a,b)$ for $1 < p < \infty$. As a special case, the Hilbert transform operator defined as

$$(Ax)(t) \equiv \frac{1}{\pi} \int_{-\infty}^{\infty} \frac{x(s)}{(t-s)}\, ds$$

is bounded in $\mathbf{L}^p(-\infty,\infty)$ for $1 < p < \infty$.

Generalizing, the weakly singular integral operator defined by

$$(Af)(\mathbf{s}) \equiv \int_\Omega \frac{k(\mathbf{t},\mathbf{s})}{|\mathbf{t}-\mathbf{s}|^\alpha} x(\mathbf{s})\, d\Omega$$

with $0 \leq \alpha < n$ and where $k(\mathbf{t},\mathbf{s})$ is bounded on $\Omega \times \Omega$, is bounded in $\mathbf{L}^2(\Omega)$ for $\Omega \subset \mathbf{R}^n$ [4, p. 162]. This also holds for integrals of the form

$$(Af)(\mathbf{s}) \equiv \int_\Gamma \frac{k(\mathbf{t},\mathbf{s})}{|\mathbf{t}-\mathbf{s}|^\alpha} x(\mathbf{s})\, d\Gamma,$$

where $\Gamma$ is a smooth $n$-dimensional surface in an $(n+1)$-dimensional space. Therefore, in a finite region of three-dimensional space, the usual free-space Green's function kernel

$$k(\mathbf{r},\mathbf{r}') = (e^{-ik|\mathbf{r}-\mathbf{r}'|})/(4\pi|\mathbf{r}-\mathbf{r}'|)$$

generates a bounded operator in $\mathbf{L}^2(\Omega)$ and $\mathbf{L}^2(\Gamma)$.

14. The singular Cauchy integral operator defined by

$$(Ax)(t) \equiv \frac{1}{\pi i} \int_\Gamma \frac{x(s)}{s-t}\, ds$$

is bounded on the space $\mathbf{H}_{0,\alpha}(\Gamma)$ for every $\alpha$ such that $0 < \alpha < 1$, where $\Gamma$ is a smooth line [3], [5].

15. Consider the integral operators $A_1$ and $A_2$ acting in the Banach space $\mathbf{C}(a,b)$ with kernels $k_1(t,\tau)$ and $k_2(\tau,s)$, respectively [6]. The integral product operator $A = A_1A_2$ for any function $x \in \mathbf{C}(a,b)$ is defined as

$$\begin{aligned}(Ax)(t) = (A_1A_2x)(t) &= \int_a^b k_1(t,\tau)\ (A_2x)(\tau)\,d\tau \\ &= \int_a^b k_1(t,\tau)\left(\int_a^b k_2(\tau,s)x(s)\,ds\right)d\tau \\ &= \int_a^b \left(\int_a^b k_1(t,\tau)\ k_2(\tau,s)\,d\tau\right)x(s)\,ds \\ &= \int_a^b k^{(2)}(t,s)x(s)\,ds.\end{aligned}$$

Thus, $A = A_1A_2$ is an integral operator with the kernel $k^{(2)}(t,s) = \int_a^b k_1(t,\tau)\ k_2(\tau,s)\,d\tau$. If the integral operators $A_1$ and $A_2$ are bounded in $\mathbf{C}(a,b)$ (i.e., $||A_1||$, $||A_2|| < \infty$), then the integral product operator $A = A_1A_2$ is bounded in $\mathbf{C}(a,b)$ (this simply follows from $||Ax|| \leq ||A_1||\ ||A_2||\ ||x||$). Also, it can be shown if the integral operators $A_1$ and $A_2$ are bounded in the space $\mathbf{L}^2(a,b)$, then the integral product operator $A = A_1A_2$ is bounded in $\mathbf{L}^2(a,b)$.

The integral product operator has been utilized in the method of overlapping regions (see Section 9.5) and in the Schwarz's iterative method where the resulting kernel is determined by subsequent iterations. Indeed, if we set $k_1 = k_2 = k$ such that $k^{(2)}(t,s) = \int_a^b k(t,\tau)\ k(\tau,s)\,d\tau$, then

$$k^{(n)}(t,s) = \int_a^b k(t,\tau)\ k^{(n-1)}(\tau,s)\,d\tau, \qquad n > 2.$$

### 3.1.3 Isomorphisms

Since we have the concept that an operator takes elements of one space to those of another space, it is worthwhile to consider some relationships between those spaces. The following definitions are useful.

- Two linear spaces $X$ and $Y$ are said to be *isomorphic* if there is a one-to-one, onto linear map $A : X \to Y$.
- Two normed spaces $X$ and $Y$ are said to be (topologically) isomorphic if there is a continuous one-to-one, onto linear map $A : X \to Y$ such that $A^{-1}$ is also continuous. Furthermore, the spaces are said to be *isometrically isomorphic* or *unitarily equivalent* if $A$ is unitary, i.e.,

$A$ preserves norms ($\|Ax\| = \|x\|$ for all $x \in X$). It can be shown that two Hilbert spaces are isomorphic if and only if they are isometrically isomorphic.

An operator $A : X \to Y$ that connects two isomorphic spaces is called an *isomorphism*. If two spaces are isomorphic to one another, then they are essentially the same space, at least at an abstract level.

As a simple example, the normed spaces $\mathbf{C}$ and $\mathbf{R}^2$ are easily seen to be isometrically isomorphic, with $A : \mathbf{C} \to \mathbf{R}^2$ the mapping that takes $z = x + iy \in \mathbf{C}$ into the ordered pair $(\operatorname{Re} z, \operatorname{Im} z) = (x, y) \in \mathbf{R}^2$. Indeed, $\|Az\| = \sqrt{x^2 + y^2} = |z| = \|z\|$.

One reason for introducing the concept of spaces being isomorphic to one another is to see the connection between function spaces and sequence spaces. For example, consider the space $\mathbf{L}^2(-\pi, \pi)$. Every $f \in \mathbf{L}^2(-\pi, \pi)$ can be written as a Fourier series

$$f = \sum_{n=-\infty}^{\infty} a_n \frac{e^{inx}}{\sqrt{2\pi}},$$

where $a_n = \left\langle f, e^{inx}/\sqrt{2\pi} \right\rangle = (1/\sqrt{2\pi}) \int_{-\pi}^{\pi} f(x) e^{-inx} dx$, and it can be easily shown that $\{a_n\} \in \mathbf{l}^2$. Let the operator $A : \mathbf{L}^2(-\pi, \pi) \to \mathbf{l}^2$ be the mapping that takes $f \in \mathbf{L}^2(-\pi, \pi)$ to the sequence $\{a_n\}_{n=-\infty}^{\infty}$ as

$$(Af)(x) = \frac{1}{\sqrt{2\pi}} \int_{-\pi}^{\pi} f(x) e^{-inx} dx = \{a_n\}$$

with the inverse mapping $A^{-1} : \mathbf{l}^2 \to \mathbf{L}^2(-\pi, \pi)$ taking $\{a_n\}$ to $f$ as

$$A^{-1}(\{a_n\}) = \sum_{n=-\infty}^{\infty} a_n \frac{e^{inx}}{\sqrt{2\pi}} = f(x).$$

It is clear that $A$ is linear, one-to-one (if $a_n = \left\langle f, e^{inx}/\sqrt{2\pi} \right\rangle = 0$ for all $n$, then $f = 0$ since $\left\{ e^{inx}/\sqrt{2\pi} \right\}$ is complete in $\mathbf{L}^2(-\pi, \pi)$), and onto (every element of $\mathbf{l}^2$ corresponds to an infinite series, representing a function in $\mathbf{L}^2$; see Section 2.5.7). Therefore, $\mathbf{L}^2(-\pi, \pi)$ and $\mathbf{l}^2(-\infty, \infty)$ are isomorphic with the Fourier series operator providing the required isomorphism. Since both spaces are Hilbert spaces, they are isometrically isomorphic.[8] Furthermore, any separable Hilbert space of functions is isomorphic to $\mathbf{l}^2$ [7, p. 47].

[8] $A$ is unitary since

$$\|Af\|^2 = \langle Af, Af \rangle = \sum_n |a_n|^2 = \sum_n \left| \left\langle f, \frac{e^{inx}}{\sqrt{2\pi}} \right\rangle \right|^2 = \|f\|^2,$$

where the last equality is obtained from Parseval's equality (see Section 2.5.6) and it is clear that $A^{-1}$ exists and is continuous.

# 3.2 Linear Functionals

## 3.2.1 General Concepts

A linear operator maps elements of one linear space into another linear space, or perhaps a space into itself. A linear functional maps elements of a linear space into $\mathbf{C}$, the set of complex numbers. We concentrate on mappings of Hilbert spaces.

**Definition 3.9.** *A mapping $l$ from $H$ to $\mathbf{C}$ (denoted as $l : H \to \mathbf{C}$ or $l(H, \mathbf{C})$) is called a linear functional if, for all $x, y \in D_l$, there correspond numbers $l(x), l(y) \in \mathbf{C}$ such that*

$$l(\alpha x + \beta y) = \alpha l(x) + \beta l(y),$$

*where $\alpha, \beta \in \mathbf{C}$.*

Under this definition ordinary *linear* functions $x(t)$ that map $t \in \mathbf{C}$ into $\mathbf{C}$ are linear functionals. More often the term functional is used for "functions of functions," e.g., $l : H \to \mathbf{C}$ where perhaps $H = \mathbf{L}^2(a, b)$. As an example, for $x \in \mathbf{L}^2(a, b)$, $\int_a^b x(t)\,dt$ is a linear functional $l : \mathbf{L}^2(a, b) \to \mathbf{C}$.

**Definition 3.10.** *A linear functional $l$ is called continuous if the mapping $l : H \to \mathbf{C}$ is continuous on $H$.*

If a linear functional $l$ is continuous at one point, then it is continuous everywhere on its domain.

**Definition 3.11.** *A linear functional $l : H \to \mathbf{C}$ is bounded on its domain if there exists a number $k \in \mathbf{R}$ such that for all elements $x \in D_l$,*

$$\|l(x)\| = |l(x)| \leq k \|x\|.$$

As with the more general operators previously discussed, the definition of boundedness of a functional brings about the concept of the norm of a functional.

**Definition 3.12.** *The norm of a linear functional $l : H \to \mathbf{C}$, denoted as $\|l\|$, is the smallest number $k$ that satisfies*

$$|l(x)| \leq k \|x\|$$

*for all $x \in D_l$.*

In parallel with operators, continuity, boundedness, domain, range, and null space of linear functionals are defined in obvious ways from the corresponding definitions for linear operators. In addition, as with linear operators, continuity and boundedness of a linear functional go hand in hand. In this way a linear functional $l : H \to \mathbf{C}$ can be thought of as a special case of a linear operator $A : H_1 \to H_2$. It can be shown that every linear functional $l : \mathbf{C}^n \to \mathbf{C}$ is bounded (see Theorem 3.5). The space of all bounded linear functionals $l : H \to \mathbf{C}$ is known as the *dual space* of $H$.

**Riesz Representation Theorem**

An important theoretical use of linear functionals is in defining *adjoint* operators. Before proceeding, note that for $y$ a fixed element of an inner product space, it is easy to see that $l(x) = \langle x, y \rangle$ is a bounded linear functional (although $l(x) = \langle y, x \rangle$ for $y$ a fixed element is not linear). Indeed, linearity and continuity follow directly from the definition of the inner product. Therefore, for each element of an inner product space there is a natural functional given by an inner product. In Hilbert spaces the converse is also true.

**Theorem 3.6.** *(Riesz representation theorem) Let $l$ be a bounded linear functional on a Hilbert space $H$. There is a unique element $y \in H$ such that $l(x) = \langle x, y \rangle$ for all $x \in H$. Moreover, $\|l\| = \|y\|$.*

The Riesz representation theorem guarantees that every bounded linear functional on $\mathbf{L}^2(a,b)$ has the form

$$l(x) = \int_a^b x(t)\overline{y(t)}\, dt$$

with the norm $||l|| = ||y|| = (\int_a^b |y(t)|^2 dt)^{1/2}$ for some $y(t) \in \mathbf{L}^2(a,b)$ and all $x(t) \in \mathbf{L}^2(a,b)$. Similarly, every bounded linear functional on $\mathbf{l}^2$ has the form

$$l(\mathbf{x}) = \sum_{i=1}^{\infty} x_i \overline{y_i}$$

with the norm $||l|| = ||\mathbf{y}|| = (\sum_{i=1}^{\infty} |y_i|^2)^{1/2}$ for some $\mathbf{y} = \{y_1, y_2, \dots\} \in \mathbf{l}^2$ and all $\mathbf{x} = \{x_1, x_2, \dots\} \in \mathbf{l}^2$, and every bounded linear functional on $\mathbf{C}^n$ has the form

$$l(\mathbf{x}) = \sum_{i=1}^{n} x_i \overline{y_i}$$

with the norm $||l|| = ||\mathbf{y}|| = \left(\sum_{i=1}^{n} |y_i|^2\right)^{1/2}$ for some $\mathbf{y} = \{y_1, y_2, \dots, y_n\} \in \mathbf{C}^n$ and all $\mathbf{x} = \{x_1, x_2, \dots, x_n\} \in \mathbf{C}^n$.

### 3.2.2 Examples Related to Linear Functionals

1. Consider a linear operator $A : \mathbf{L}^2(\Omega) \to \mathbf{L}^2(\Omega)$. A bounded linear functional on $\mathbf{L}^2(\Omega)$ is

$$l(x) = \langle Ax, y \rangle = \int_\Omega (Ax)(\mathbf{t})\overline{y(\mathbf{t})}\, d\Omega$$

for some $y(\mathbf{t}) \in \mathbf{L}^2(\Omega)$ and all $x(\mathbf{t})$ in the domain $D_A = \{x : x(\mathbf{t}), (Ax)(\mathbf{t}) \in \mathbf{L}^2(\Omega)\}$. If the adjoint operator $A^* : \mathbf{L}^2(\Omega) \to \mathbf{L}^2(\Omega)$

exists (see Section 3.3), then

$$l(x) = \langle Ax, y \rangle = \langle x, A^* y \rangle = \int_\Omega x(\mathbf{t}) \overline{(A^* y)(\mathbf{t})}\, d\Omega,$$

where $y(\mathbf{t})$ is in the domain $D_{A^*} = \{y : y(\mathbf{t}), (A^* y)(\mathbf{t}) \in \mathbf{L}^2(\Omega)\}$.

2. The linear functional $\phi_c : \mathbf{C}(a,b) \to \mathbf{C}$ defined by $\phi_c(f) \equiv f(c)$, where $c \in [a,b]$, can be written in terms of a delta-function (in the distributional sense; note $\delta \notin \mathbf{C}(\Omega), \mathbf{L}^2(\Omega))$ as

$$\phi_c(f) = \langle f, \delta \rangle = \int_a^b f(x)\delta(x-c)\, dx.$$

3. As an example in electromagnetics, scalar components of the magnetic vector potential $A_\alpha$ at some *fixed* point in space $\mathbf{r}$ caused by an electric current in free space are linear functionals of the current density $J_\alpha$,

$$A_\alpha(J_\alpha) = \left\langle \mu J_\alpha, \frac{e^{jk|\mathbf{r}-\mathbf{r}'|}}{4\pi|\mathbf{r}-\mathbf{r}'|} \right\rangle = \int_\Omega \mu\, J_\alpha(\mathbf{r}') \frac{e^{-jk|\mathbf{r}-\mathbf{r}'|}}{4\pi|\mathbf{r}-\mathbf{r}'|}\, d\Omega',$$

as are various electric- and magnetic field components. Alternatively, for a fixed $J_\alpha$, $A_\alpha$ can be thought of as a nonlinear functional of the permittivity function $\varepsilon$ via the wavenumber $k$.

4. The voltage $V_{ab}$ between two fixed points $a, b$ is a linear functional of the electric field $\mathbf{E}$ via

$$V_{ab}(\mathbf{E}) = -\int_b^a \mathbf{E}(\mathbf{r}) \cdot d\mathbf{l}.$$

As further examples, the total charge $Q$ in a volume $\Omega$ is a linear functional of the charge density $\rho$ via $Q(\rho) = \int_\Omega \rho(\mathbf{r})\, d\Omega$, and electrostatic power $P$ is a linear functional (of $\mathbf{E}$ or $\mathbf{J}$) via $P = \int_\Omega \mathbf{E} \cdot \mathbf{J}\, d\Omega$. Numerous other examples can obviously be found.

5. Consider the operator $A : \mathbf{l}^2 \to \mathbf{l}^2$ associated with the infinite matrix $[a_{ij}]$ and defined by $(A\mathbf{x})_{i=1}^\infty = \sum_{j=1}^\infty a_{ij} x_j$. Then a linear functional $l(\mathbf{x})$ is defined on $\mathbf{l}^2$ as

$$l(\mathbf{x}) \equiv \langle A\mathbf{x}, \mathbf{y} \rangle = \sum_{i=1}^\infty \sum_{j=1}^\infty a_{ij} x_j \overline{y_i}$$

for some $\mathbf{y} \in \mathbf{l}^2$ and all $\mathbf{x}$ in the domain $D_A = \{\mathbf{x} : \mathbf{x}, A\mathbf{x} \in \mathbf{l}^2\}$. If the operator $A$ is bounded assuming $\sum_{i=1}^\infty \sum_{j=1}^\infty |a_{ij}|^2 < \infty$ (Hilbert–Schmidt operator), then the functional $l(\mathbf{x})$ is bounded on $\mathbf{l}^2$.

6. Consider the vector differential operator $A : \mathbf{L}^2(\Omega)^3 \to \mathbf{L}^2(\Omega)^3$ defined by

$$(A\mathbf{x})(\mathbf{t}) \equiv \nabla \times \nabla \times \mathbf{x}(\mathbf{t}),$$
$$D_A \equiv \left\{\mathbf{x} : \mathbf{x}(\mathbf{t}), \nabla \times \nabla \times \mathbf{x}(\mathbf{t}) \in \mathbf{L}^2(\Omega)^3\right\}.$$

A linear functional $l(\mathbf{x})$ is defined by

$$l(\mathbf{x}) \equiv \langle A\mathbf{x}, \mathbf{y}\rangle = \int_\Omega (\nabla \times \nabla \times \mathbf{x}) \cdot \overline{\mathbf{y}}\, d\Omega$$

for some 3-tuple $\mathbf{y}(\mathbf{t}) \in \mathbf{L}^2(\Omega)^3$ and all 3-tuples $\mathbf{x}(\mathbf{t}) \in D_A$.

## 3.3 Adjoint Operators

### 3.3.1 General Concepts: Bounded Operators

The Riesz representation theorem is useful for defining in a natural way the important concept of an operator *adjoint* to a given operator. Consider first a bounded linear operator $A : H_1 \to H_2$. For every $y \in H_2$, it can be seen that $l(x) = \langle Ax, y\rangle_2$ is a bounded linear functional $l : H_2 \to \mathbf{C}$ for all $x \in H_1$. Therefore, by the Riesz representation theorem there exists a unique $y^* \in H_1$ such that for all $x \in H_1$,

$$\langle Ax, y\rangle_2 = \langle x, y^*\rangle_1 .$$

Because $y^*$ depends on $y$, we introduce a new operator $A^* : H_2 \to H_1$, called the *adjoint* of $A$, defined by $A^*y = y^*$, leading to

$$\langle Ax, y\rangle_2 = \langle x, A^*y\rangle_1 .$$

Usually we will drop the subscripts on the inner products.

Therefore, if $A$ is a bounded linear operator, by the Riesz representation theorem a unique adjoint exits. It can be shown that[9] $\|A\| = \|A^*\|$, and so the adjoint of a bounded linear operator is a bounded. Some convenient properties relating to adjoints of bounded linear operators are the following.

- $\|A^*A\| = \|AA^*\| = \|A\|^2$
- $(A+B)^* = A^* + B^*$
- $(\alpha A)^* = \overline{\alpha} A^*$

---

[9] This simply follows from

$$||A|| = \sup_{||x||=||y||=1} |\langle Ax, y\rangle| = \sup_{||x||=||y||=1} |\langle x, A^*y\rangle| = ||A^*||.$$

- $(CA)^* = A^*C^*$, where $C : H_2 \to H_3$
- $(A^*)^* = A$

where in the above $A, B : H_1 \to H_2$. Relationships between the null space and range of an operator and those of its adjoint are given by the next theorem.

**Theorem 3.7.** *Let $A : H_1 \to H_2$ be a bounded linear operator. Then*

*a.* $N_A = (R_{A^*})^\perp$

*b.* $N_{A^*} = (R_A)^\perp$

*c.* $\widetilde{R_A} = (N_{A^*})^\perp$

*d.* $\widetilde{R_{A^*}} = (N_A)^\perp$

*where $\widetilde{R_A}$ is the closure of $R_A$.*

From Theorem 2.7 we see that since the null space is always closed, if the range is closed then the last two properties are repetitive with the first two.

### 3.3.2 Examples of Operator Adjoints: Bounded Operators

1. The identity and zero operators have simple adjoints; $I^* = I$, $0^* = 0$.

2. It was shown before that a bounded operator $A : C^n \to C^n$, with $\{x_1, x_2, \ldots x_n\}$ an orthonormal basis for $\mathbf{C}^n$, is represented by the matrix $[a_{ij}]$, where $a_{ij} = \langle Ax_j, x_i \rangle$. From the definition $\langle Ax_j, x_i \rangle = \langle x_j, A^* x_i \rangle$ one can see that the adjoint operator $A^* : C^n \to C^n$ is represented in the basis $\{x_n\}$ by the matrix $[\overline{a_{ji}}]$, the conjugate transpose[10] of the matrix $[a_{ij}]$.

   The same result is obtained for the operator $A : \mathbf{l}^2 \to \mathbf{l}^2$ defined by $(A\mathbf{x})_{i=1}^\infty = \sum_{j=1}^\infty a_{ij} x_j$, corresponding to the infinite matrix $[a_{ij}]$ with $\sum_{i=1}^\infty \sum_{j=1}^\infty |a_{ij}|^2 < \infty$ (Hilbert–Schmidt operator).

3. The multiplication operator $A : \mathbf{L}^2(\Omega) \to \mathbf{L}^2(\Omega)$ defined by

$$(Ax)(\mathbf{t}) \equiv f(\mathbf{t}) x(\mathbf{t})$$

   is a bounded linear operator assuming $k = \max_{\mathbf{t} \in \Omega} |f(\mathbf{t})| < \infty$ where $\Omega \subset \mathbf{R}^n$. Then

$$\langle Ax, y \rangle = \int_\Omega f(\mathbf{t}) x(\mathbf{t}) \overline{y(\mathbf{t})} \, d\Omega = \langle x, \overline{f} \, y \rangle = \langle x, A^* y \rangle ,$$

[10]Note that this is true only when $\{x_n\}$ is an orthonormal basis for $\mathbf{C}^n$. When $\{x_n\}$ is merely orthogonal, then the matrix representation of $A^*$ is given by a similarity transformation (see Section 4.3.1) of the conjugate transpose matrix $[\overline{a_{ji}}]$.

where

$$(A^* y)(\mathbf{t}) = \overline{f(\mathbf{t})} y(\mathbf{t}).$$

Therefore, for the operator "multiplication by $f(\mathbf{t})$," the adjoint is "multiplication by $\overline{f(\mathbf{t})}$."

4. The integral operator $A : \mathbf{L}^2(\Omega)^m \to \mathbf{L}^2(\Omega)^m$ defined by

$$(A\mathbf{x})(\mathbf{t}) \equiv \int_\Omega \underline{\mathbf{k}}(\mathbf{t}, \mathbf{s}) \cdot \mathbf{x}(\mathbf{s}) \, d\Omega,$$

where $\Omega \subset \mathbf{R}^n$ with $\underline{\mathbf{k}}(\mathbf{t}, \mathbf{s}) \in \mathbf{L}^2 (\Omega \times \Omega)^{m \times m}$ (in the sense that each component of the $(m \times m)$-tuple dyadic kernel $\underline{\mathbf{k}}(\mathbf{t}, \mathbf{s})$ belongs to $\mathbf{L}^2 (\Omega \times \Omega)$), is bounded. The adjoint is found from

$$\begin{aligned} \langle A\mathbf{x}, \mathbf{y} \rangle &= \int_\Omega \int_\Omega \underline{\mathbf{k}}(\mathbf{t}, \mathbf{s}) \cdot \mathbf{x}(\mathbf{s}) \, d\Omega_s \cdot \overline{\mathbf{y}(\mathbf{t})} \, d\Omega_t \\ &= \int_\Omega \mathbf{x}(\mathbf{s}) \cdot \overline{\int_\Omega \overline{\underline{\mathbf{k}}^\top (\mathbf{t}, \mathbf{s})} \cdot \ \mathbf{y}(\mathbf{t}) \, d\Omega_t} \, d\Omega_s = \langle \mathbf{x}, A^* \mathbf{y} \rangle \end{aligned}$$

such that the adjoint operator is

$$(A^* \mathbf{y})(\mathbf{t}) = \int_\Omega \overline{\underline{\mathbf{k}}^\top (\mathbf{s}, \mathbf{t})} \cdot \ \mathbf{y}(\mathbf{s}) \, d\Omega$$

with the adjoint kernel $\underline{\mathbf{k}}^* (\mathbf{t}, \mathbf{s}) = \overline{\underline{\mathbf{k}}^\top (\mathbf{s}, \mathbf{t})}$, the conjugate of the transpose of the original kernel with the variables interchanged. For example, the dyadic Green's function for static fields

$$\underline{\mathbf{G}}(\mathbf{r}, \mathbf{r}') = \underline{\mathbf{I}} \, \frac{1}{4\pi R} = \underline{\mathbf{k}}(\mathbf{r}, \mathbf{r}')$$

(with $R = |\mathbf{r} - \mathbf{r}'|$) provides a self-adjoint (see Section 3.4) bounded kernel in $\mathbf{R}^n$, $n \geq 2$.

Similarly, the dyadic Green's function

$$\underline{\mathbf{G}}(\mathbf{r}, \mathbf{r}') = \underline{\mathbf{I}} \, \frac{e^{-jkR}}{4\pi R} = \underline{\mathbf{k}}(\mathbf{r}, \mathbf{r}')$$

has adjoint

$$\underline{\mathbf{k}}^* (\mathbf{r}, \mathbf{r}') = \underline{\mathbf{I}} \, \frac{e^{+jkR}}{4\pi R}.$$

In the scalar case of $A : \mathbf{L}^2(\Omega) \to \mathbf{L}^2(\Omega)$ defined by

$$(Ax)(\mathbf{t}) \equiv \int_\Omega k(\mathbf{t}, \mathbf{s}) x(\mathbf{s}) \, d\Omega,$$

the adjoint operator is

$$(A^*y)(\mathbf{t}) = \int_\Omega \overline{k(\mathbf{s},\mathbf{t})}y(\mathbf{s})\, d\Omega = \int_\Omega k^*(\mathbf{t},\mathbf{s})y(\mathbf{s})\, d\Omega$$

with the adjoint kernel $k^*(\mathbf{t},\mathbf{s}) = \overline{k(\mathbf{s},\mathbf{t})}$, the conjugate of the original kernel with the variables interchanged.

5. The integration operator $A : \mathbf{L}^2(a,b) \to \mathbf{L}^2(a,b)$ defined by

$$(Ax)(t) \equiv \int_a^t x(s)\, ds$$

is a linear bounded operator. Its adjoint is found as a special case of the previous example ($\Omega = (a,b)$) with $k(t,s) = H(t-s)$. Therefore, the adjoint operator is $k^*(t,s) = H(s-t)$, leading to

$$(A^*y)(t) = \int_t^b y(s)\, ds.$$

### 3.3.3 General Concepts: Unbounded Operators

Before considering the adjoint of the differentiation operator, we need to investigate the concept of an adjoint for unbounded operators. Since the Riesz representation theorem no longer holds, we are no longer guaranteed that for an element $y \in H_2$ an element $y^* \in H_1$ exists such that $\langle Ax, y\rangle = \langle x, y^*\rangle$ for all $x \in H_1$. Often, though, such an element exists, and we obtain the adjoint $A^*$ as before; $A^*y = y^*$ such that $\langle Ax, y\rangle = \langle x, A^*y\rangle$. In general, such an adjoint may not be unique. If the operator is densely defined, which is the case of primary interest here, the adjoint operator is unique.[11]

Actually, in some cases we need to be more careful. If $A$ is bounded, then $\langle Ax, y\rangle = \langle x, A^*y\rangle$ actually *defines* the adjoint. If $A$ is unbounded, then this relationship merely describes the "action" of the operator but does not always lead to the correct identification of the domain $D_{A^*}$. For an unbounded, densely defined operator, a rigorous definition of the adjoint is the following.

**Definition 3.13.** *Let $A : H \to H$ be a densely defined operator. Then*

$$D_{A^*} \equiv \{z \in H : \exists y \in H \text{ such that } \langle Ax, z\rangle = \langle x, y\rangle\,,\ \forall x \in D_A\}\,,$$
$$A^*z \equiv y.$$

[11]Indeed, if $\langle Ax, y\rangle = \langle x, y^*\rangle$ were true for two different $y^*$, say $y_1^*$ and $y_2^*$, then $\left\langle x, y_1^* - y_2^*\right\rangle = 0$ for all $x \in D_A$. Because $D_A$ is dense in $H_1$, only the zero element is orthogonal to every element in $D_A$ (by Theorem 2.7(c)); hence $y_1^* = y_2^*$.

However, in most practical cases the simpler procedure described below leads to the correct domain of $A^*$.

In summary, all bounded linear operators have a unique adjoint, and for all unbounded, densely defined linear operators that possess an adjoint, the adjoint is unique. All unbounded differential operators considered in this text possess adjoints.

### 3.3.4 Examples of Operator Adjoints: Unbounded Operators

1. Consider the differentiation operator $A : \mathbf{L}^2(a,b) \to \mathbf{L}^2(a,b)$ defined by

$$(Ax)(t) \equiv \frac{d}{dt}x(t) = x'(t),$$
$$D_A \equiv \left\{x : x(t), x'(t) \in \mathbf{L}^2(a,b),\ x(a) = 0\right\}.$$

The adjoint operator can be found by integration by parts:

$$\begin{aligned}\langle Ax, y\rangle &= \int_a^b x'(t)\overline{y(t)}\,dt \\ &= x(b)\overline{y(b)} - x(a)\overline{y(a)} - \int_a^b x(t)\overline{y'(t)}\,dt \\ &= \langle x, A^*y\rangle .\end{aligned}$$

Therefore, the adjoint differentiation operator is

$$(A^*y)(t) = -\frac{d}{dt}y(t),$$
$$D_{A^*} = \left\{y : y(t), y'(t) \in \mathbf{L}^2(a,b),\ y(b) = 0\right\};$$

hence $D_{A^*} \neq D_A$ and $A^*x = -Ax$ for all $x \in D_A \cap D_{A^*}$. Note that we actually get a conjugate-adjoint boundary condition on $\overline{y}$, which is easily converted into an adjoint boundary condition on $y$.

2. In the previous example, if we define

$$D_A \equiv \left\{x : x(t), x'(t) \in \mathbf{L}^2(a,b),\ x(a) = x(b) = 0\right\},$$

then $A^*x = -Ax$ for all suitable $x$, but no boundary conditions result to restrict $D_{A^*}$, i.e.,

$$D_{A^*} = \left\{y : y(t), y'(t) \in \mathbf{L}^2(a,b)\right\} \supset D_A.$$

Furthermore, if $D_A$ is such that no boundary conditions are imposed on $x$, then boundary conditions $y(a) = y(b) = 0$ will be induced on the adjoint.

3. Consider the differentiation operator $A : \mathbf{L}^2(a,b) \to \mathbf{L}^2(a,b)$ defined by

$$(Ax)(t) \equiv \frac{d^2}{dt^2}x(t) = x''(t),$$
$$D_A \equiv \left\{x : x(t),\ x'(t),\ x''(t) \in \mathbf{L}^2(a,b),\ x(a) = x(b) = 0\right\}.$$

This can be thought of as the one-dimensional Laplacian operator. The adjoint operator can be found by integration by parts twice (equivalently Green's second theorem)

$$\begin{aligned}\langle Ax, y\rangle &= \int_a^b \frac{d^2}{dt^2}x(t)\overline{y(t)}\,dt \qquad (3.2)\\ &= \left(\frac{d}{dt}x(t)\right)\overline{y(t)}\Big|_a^b - x(t)\frac{d}{dt}\overline{y(t)}\Big|_a^b + \int_a^b x(t)\frac{d^2}{dt^2}\overline{y(t)}\,dt\\ &= \langle x, A^*y\rangle.\end{aligned}$$

Therefore,

$$(A^*y)(t) = \frac{d^2}{dt^2}y(t) = (Ay)(t),$$
$$D_{A^*} = \left\{y : y(t), y'(t), y''(t) \in \mathbf{L}^2(a,b),\ y(a) = y(b) = 0\right\} = D_A$$

such that $A^* = A$.

4. Similar to the last example, the Laplacian operator $A : \mathbf{L}^2(\Omega) \to \mathbf{L}^2(\Omega)$ defined by

$$(Ax)(\mathbf{t}) \equiv \nabla^2 x(\mathbf{t}),$$
$$D_A \equiv \left\{x : x(\mathbf{t}), \nabla^2 x(\mathbf{t}) \in \mathbf{L}^2(\Omega),\ x(\mathbf{t})|_\Gamma = 0\right\},$$

where $\Gamma$ is the smooth boundary of $\Omega \subset \mathbf{R}^n$, is such that $A^* = A$. Indeed, from Green's second theorem

$$\begin{aligned}\langle Ax, y\rangle &= \int_\Omega \overline{y}\nabla^2 x\,d\Omega\\ &= \oint_\Gamma \left(\overline{y}\frac{\partial x}{\partial n} - x\frac{\partial \overline{y}}{\partial n}\right) dS + \int_\Omega x\nabla^2\overline{y}\,d\Omega\\ &= \langle x, A^*y\rangle.\end{aligned}$$

5. Consider the vector differential operator $A : \mathbf{L}^2(\Omega)^3 \to \mathbf{L}^2(\Omega)^3$ defined by

$$(A\mathbf{x})(\mathbf{t}) \equiv \nabla \times \nabla \times \mathbf{x}(\mathbf{t}),$$
$$D_A \equiv \left\{\mathbf{x} : \mathbf{x}(\mathbf{t}), \nabla \times \nabla \times \mathbf{x}(\mathbf{t}) \in \mathbf{L}^2(\Omega)^3,\ \mathbf{n} \times \mathbf{x}(\mathbf{t})|_\Gamma = \mathbf{0}\right\}.$$

The adjoint operator can be found using the second vector Green's theorem

$$\begin{aligned}\langle A\mathbf{x}, \mathbf{y}\rangle &= \int_{\Omega} \overline{\mathbf{y}} \cdot (\nabla \times \nabla \times \mathbf{x}) \; d\Omega \\ &= \oint_{\Gamma} ((\mathbf{n} \times \mathbf{x}) \cdot (\nabla \times \overline{\mathbf{y}}) - (\mathbf{n} \times \overline{\mathbf{y}}) \cdot (\nabla \times \mathbf{x})) \; dS \\ &\quad + \int_{\Omega} \mathbf{x} \cdot (\nabla \times \nabla \times \overline{\mathbf{y}}) \; d\Omega \\ &= \langle \mathbf{x}, A^* \mathbf{y}\rangle ,\end{aligned}$$

where

$$\begin{aligned}(A^*\mathbf{y})(\mathbf{t}) &= \nabla \times \nabla \times \mathbf{y}(\mathbf{t}) = (A\mathbf{y})(\mathbf{t}), \\ D_{A^*} &= \left\{\mathbf{y} : \mathbf{y}(\mathbf{t}), \nabla \times \nabla \times \mathbf{y}(\mathbf{t}) \in \mathbf{L}^2(\Omega)^3, \, \mathbf{n} \times \mathbf{y}(\mathbf{t})|_{\Gamma} = \mathbf{0}\right\} \\ &= D_A.\end{aligned}$$

Thus, $A^* = A$. This example can be generalized for a $(3 \times 3)$-tuple dyadic function (rank 2 tensor) $\underline{\mathbf{y}}(\mathbf{t})$ using the second vector-dyadic Green's theorem as follows,

$$\begin{aligned}\left\langle A\mathbf{x}, \underline{\mathbf{y}}\right\rangle &= \int_{\Omega} (\nabla \times \nabla \times \mathbf{x}) \cdot \overline{\underline{\mathbf{y}}} \; d\Omega \\ &= \oint_{\Gamma} ((\mathbf{n} \times \mathbf{x}) \cdot \left(\nabla \times \overline{\underline{\mathbf{y}}}\right) - (\nabla \times \mathbf{x}) \cdot \left(\mathbf{n} \times \overline{\underline{\mathbf{y}}}\right)) \; dS \\ &\quad + \int_{\Omega} \mathbf{x} \cdot \left(\nabla \times \nabla \times \overline{\underline{\mathbf{y}}}\right) \; d\Omega \\ &= \left\langle \mathbf{x}, A^* \underline{\mathbf{y}}\right\rangle ,\end{aligned}$$

where again we find $A^* = A$. We understand $\underline{\mathbf{y}}(\mathbf{t}), \nabla \times \nabla \times \underline{\mathbf{y}}(\mathbf{t}) \in \mathbf{L}^2(\Omega)^{3\times 3}$ in the sense that components of the dyadic function and components of the differential operator belong to $\mathbf{L}^2(\Omega)$, and that $\mathbf{L}^2(\Omega)^{3\times 3}$ represents a direct sum of spaces $\mathbf{L}^2(\Omega)$.

6. Consider the vector differential operator (curl) $A : \mathbf{L}^2(\Omega)^3 \to \mathbf{L}^2(\Omega)^3$ defined by

$$\begin{aligned}(A\mathbf{x})(\mathbf{t}) &\equiv \nabla \times \mathbf{x}(\mathbf{t}), \\ D_A &\equiv \left\{\mathbf{x} : \mathbf{x}(\mathbf{t}), \nabla \times \mathbf{x}(\mathbf{t}) \in \mathbf{L}^2(\Omega)^3, \, \mathbf{n} \times \mathbf{x}(\mathbf{t})|_{\Gamma} = \mathbf{0}\right\},\end{aligned}$$

where $D_A$ is a subspace of $H(\text{curl}, \Omega)$. The adjoint operator can be found using the vector identity $\nabla \cdot (\mathbf{x} \times \mathbf{y}) = \mathbf{y} \cdot \nabla \times \mathbf{x} - \mathbf{x} \cdot \nabla \times \mathbf{y}$ and the divergence theorem as

$$\begin{aligned}(A^*\mathbf{y})(\mathbf{t}) &= \nabla \times \mathbf{y}(\mathbf{t}), \\ D_{A^*} &= \left\{\mathbf{y} : \mathbf{y}(\mathbf{t}), \nabla \times \mathbf{y}(\mathbf{t}) \in \mathbf{L}^2(\Omega)^3\right\}.\end{aligned}$$

In a similar manner, the adjoint of the differential operator (divergence) $A : \mathbf{L}^2(\Omega)^3 \to \mathbf{L}^2(\Omega)$ defined by

$$\begin{aligned}(A\mathbf{x})(\mathbf{t}) &\equiv \nabla \cdot \mathbf{x}(\mathbf{t}),\\ D_A &\equiv \left\{\mathbf{x} : \mathbf{x}(\mathbf{t}) \in \mathbf{L}^2(\Omega)^3, \nabla \cdot \mathbf{x}(\mathbf{t}) \in \mathbf{L}^2(\Omega), \mathbf{n} \cdot \mathbf{x}(\mathbf{t})|_\Gamma = 0\right\},\end{aligned}$$

where $D_A$ is a subspace of $H(\text{div}, \Omega)$, is found from the vector identity $\nabla \cdot (\psi \mathbf{x}) = \mathbf{x} \cdot \nabla\psi + \psi \nabla \cdot \mathbf{x}$ and the divergence theorem as

$$\begin{aligned}(A^*\psi)(\mathbf{t}) &= -\nabla\psi(\mathbf{t}),\\ D_{A^*} &= \left\{\psi : \psi(\mathbf{t}), \nabla\psi(\mathbf{t}) \in \mathbf{L}^2(\Omega)\right\}.\end{aligned}$$

Therefore, the formal adjoint[12] of curl is curl, the formal adjoint of divergence is negative gradient, and the formal adjoint of gradient is negative divergence.

## General Procedure for Determining Operator Adjoints

It is worthwhile to summarize the procedure for obtaining the adjoint of a differential operator. Let $L$ be a general differential operator (scalar, vector, or dyadic) with domain $D_L$; the specification of $D_L$ may include boundary conditions $B(x) = \eta$, $x \in D_L$. One applies an appropriate Green's second theorem, or related vector identity corresponding to integration by parts, to the pair $x, \overline{y}$, where $y$ is conjugated to account for the conjugate in the inner product. The domain of the adjoint is determined by requiring that the "integrated terms" in the Green's theorem (i.e., the boundary terms) vanish when the homogeneous boundary condition $B(x) = 0$ is enforced (regularity of the functions in $D_{L^*}$ is assumed). This homogeneous condition is applied even if the given boundary conditions are inhomogeneous, i.e., if $\eta \neq 0$ [8, pp. 180, 185], [9, pp. 72–73]. This leads to conjugate-adjoint boundary conditions $B^*_{ca}(\overline{y}) = 0$, which are then converted to adjoint boundary conditions $B^*$ on $y$, $B^*(y) = \overline{B^*_{ca}(\overline{y})} = 0$ (adjoint boundary conditions are always homogeneous). The adjoint operator is then easily recognized from the remaining terms. In a real space the conjugate is omitted.

To generalize this for any linear operator $L$ consider the expression

$$\int_\Omega (Lx)\, \overline{y}\, d\Omega = \int_\Omega x\, \overline{(L^*y)}\, d\Omega + \int_\Gamma J(x, \overline{y})\, d\Gamma, \tag{3.3}$$

which is known as a *generalized Green's theorem* [10, p. 870] (for $n = 1$, $\Omega$ is a line segment and the boundary term is replaced with a term

[12] The formal adjoint is the adjoint without considering the associated domains, i.e., concentrating on the formal action of the operator (see Section 3.4).

involving the two endpoints of the line). The term involving $J$ is known as the *bilinear concomitant* or the *conjunct*, and the operator $L^*$ is called the *formal adjoint* of $L$. The domain of $L^*$ is determined by requiring $\int_\Gamma J(x,\overline{y})\,d\Gamma = 0$ for $x \in D_L$ (again, if $B(x) = \eta \neq 0$, we enforce $B(x) = 0$ in the conjunct) leading to conjugate-adjoint boundary conditions on $\overline{y}$, which are then converted to adjoint conditions on $y \in D_{L^*}$. With the conjunct so removed we have $\langle Lx, y\rangle = \langle x, L^*y\rangle$ as expected. Note that the conjunct is simply the "integrated term" in an $n$-dimensional integration-by-parts procedure.

If $L^*$ is a real operator,[13] as is often the case, then the generalized Green's theorem becomes

$$\int_\Omega (Lx)\,\overline{y}\,d\Omega = \int_\Omega x\,(L^*\overline{y})\,d\Omega + \int_\Gamma J(x,\overline{y})\,d\Gamma. \tag{3.4}$$

This form can represent any of the Green's second theorems listed in Appendix A.4. For example, if $L = \nabla^2 = L^*$ (formally) and $J(x,\overline{y}) = \overline{y}\,dx/dn - x\,d\overline{y}/dn$ where $d/dn$ is the outward normal derivative on $\Gamma$, we then have the standard Green's second theorem for scalars (substitute $h = \overline{y}$ to get the usual form without conjugation). In one dimension, $L = d/dt = -L^*$ and $J(x,\overline{y}) = x\overline{y}$ lead to the usual integration by parts formula. In the vector case (see, e.g., [11, Ch. III] for related material) (3.4) becomes

$$\int_\Omega (\mathbf{Lx})\cdot\overline{\mathbf{y}}\,d\Omega = \int_\Omega \mathbf{x}\cdot(\mathbf{L}^*\overline{\mathbf{y}})\,d\Omega + \int_\Gamma \mathbf{J}(\mathbf{x},\overline{\mathbf{y}})\cdot d\Gamma. \tag{3.5}$$

For $\mathbf{L} = \nabla\times\nabla\times = \mathbf{L}^*$ and $\mathbf{J}(\mathbf{x},\overline{\mathbf{y}}) = \mathbf{x}\times\nabla\times\overline{\mathbf{y}} + (\nabla\times\mathbf{x})\times\overline{\mathbf{y}}$, one has the usual vector second Green's theorem.

## 3.4 Self-Adjoint, Symmetric, Normal, and Unitary Operators

### 3.4.1 General Concepts

In discussing self-adjoint operators we will naturally restrict our attention to linear operators $A : H \to H$. We include operators acting on $H$ and on domains dense in $H$.

#### Self-Adjoint Operators

**Definition 3.14.** *An operator $A$ is called self-adjoint if $A = A^*$.*

Note by this we mean $D_A = D_{A^*}$ *and* $A^*x = Ax$ for all $x \in D_A = D_{A^*}$. That is, the operators $A$ and $A^*$ act on the same set of elements of $H$, and

[13] An operator $L$ is called *real* if $\overline{Lx} = L\overline{x}$.

the action of $A$ and $A^*$ on those elements is identical. Such operators are also called *Hermitian*, a name usually reserved for the finite-dimensional (matrix) case.

Obviously, for self-adjoint operators $\langle Ax, y\rangle = \langle x, Ay\rangle$ for all $x, y \in D_A = D_{A^*}$. Moveover, self-adjoint operators belong to a larger class of *symmetric* operators. In the following we focus on densely defined operators, since for this class of operator if an adjoint exists it is unique.

### Symmetric Operators

**Definition 3.15.** *A densely defined operator $A$ is called symmetric (or formally self-adjoint) if $Ax = A^*x$ for all $x \in D_A \subseteq D_{A^*}$.*

This is equivalent to $\langle Ax, y\rangle = \langle x, Ay\rangle$ for all $x, y \in D_A$. It turns out that for bounded operators the difference between symmetry and self-adjointness is not important; however, this is not the case for unbounded operators. To see this we need to introduce the concept of an *operator extension*.

**Definition 3.16.** *A linear operator $B$ is said to be the extension of the linear operator $A$ if $D_A \subseteq D_B$ and $Ax = Bx$ for all $x \in D_A$.*

It can be shown that if a bounded operator is not defined *on* a Hilbert space, but is defined on a linear subspace *in* a Hilbert space, it can always be extended to the whole space (see, e.g., [12, pp. 325–326]). For most purposes then we may take $D_A = H$ for bounded operators $A : H \to H$, in which case the operator always acts *on* the Hilbert space and $D_{A^*} = H$. Therefore, we conclude a bounded operator $A$ is self-adjoint (at least under an extension) if it is symmetric.

It is clear that every self-adjoint operator is symmetric and that every bounded symmetric operator is self-adjoint. It is possible for an unbounded, symmetric operator to be nonself-adjoint, even if it is densely defined, as the examples in Section 3.3.4 illustrate.

Since in general unbounded operators cannot be extended to the whole space, even if they are densely defined, then for unbounded operators symmetry does not imply self-adjointness. With self-adjointness being a much stronger condition than mere symmetry, it is fortunate that many differential operators of interest in electromagnetics are not only symmetric but are in fact self-adjoint (dynamic problems in lossless unbounded regions being a notable exception, since radiation conditions don't lead to a self-adjoint operator).

Some basic properties relating to self-adjoint operators are given below.

**Theorem 3.8.** *If $A : H \to H$ is a bounded linear operator, then $A^*A$, $AA^*$, and $A + A^*$ are self-adjoint.*

**Proof.** $(A^*A)^* = A^*A^{**} = A^*A$, $(AA^*)^* = A^{**}A^* = AA^*$, and $(A+A^*)^* = A^* + A^{**} = A^* + A = A + A^*$. ■

**Theorem 3.9.** *Let $A : H \to H$ be a bounded linear operator. Then there exist unique self-adjoint operators $S$ and $T$ such that $A = S + iT$ and $A^* = S - iT$.*

In particular,

$$\begin{aligned} S &= (A + A^*)/2, \\ T &= (A - A^*)/2i, \end{aligned}$$

a property used in the method of characteristics [13]. To prove that $S$ and $T$ are self-adjoint operators, we use the properties of adjoint operators:

$$\begin{aligned} S^* &= \left(\frac{A + A^*}{2}\right)^* = \frac{A^* + A^{**}}{2} = \frac{A + A^*}{2} = S, \\ T^* &= \left(\frac{A - A^*}{2i}\right)^* = \frac{A^* - A^{**}}{-2i} = \frac{A - A^*}{2i} = T. \end{aligned}$$

Note that if $A$ is self-adjoint, then $S = A$ and $T = 0$.

**Theorem 3.10.** *Let $A, B : H \to H$ be bounded self-adjoint operators. The product operator $AB : H \to H$ is self-adjoint if and only if the operators commute, i.e., $AB = BA$.*

The fact that $\langle Ax, x\rangle \in \mathbf{R}$ for a self-adjoint linear operator leads to a special characterization of its norm (it is easy to see that for $H$ a complex Hilbert space, if $\langle Ax, x\rangle \in \mathbf{R}$ then $A$ is symmetric, and conversely if $A$ is symmetric then $\langle Ax, x\rangle \in \mathbf{R}$).

**Theorem 3.11.** *Let $A : H \to H$ be a bounded linear self-adjoint operator. Then*

$$\|A\| = \sup_{\|x\|=1} |\langle Ax, x\rangle| .$$

Finally, we introduce the idea of a *normal* operator and a *unitary* operator.

### Normal Operators

**Definition 3.17.** *A bounded linear operator $A : H \to H$ is said to be normal if*

$$AA^* = A^*A.$$

We introduce

$$[A, B] \equiv AB - BA,$$

called the *commutator* of operators $A$ and $B$ [6].[14] Then the operator $A$ is normal if $[A, A^*] = AA^* - A^*A = 0$.

**Theorem 3.12.** *A bounded linear operator $A : H \to H$ is normal if and only if $[S, T] = 0$ ($S$, $T$ defined above).*

**Proof.** First assume $A$ is normal. Then

$$[S, T] = \left[\frac{A + A^*}{2}, \frac{A - A^*}{2i}\right] = \frac{A^*A - AA^*}{2i} = 0.$$

Now assume $[S, T] = 0$. Then $[A, A^*] = 0$ by the equation above, hence $A$ is normal. ■

**Theorem 3.13.** *A bounded linear operator $A : H \to H$ is normal if and only if $\|Ax\| = \|A^*x\|$ for all $x \in H$.*

**Proof.** First assume $A$ is normal. Then for all $x \in H$, $\langle AA^*x, x\rangle = \langle A^*x, A^*x\rangle$. Also, $\langle AA^*x, x\rangle = \langle A^*Ax, x\rangle = \langle Ax, Ax\rangle$. Therefore, $\|A^*x\| = \|Ax\|$. Now assume $\|A^*x\| = \|Ax\|$ for all $x \in H$ and show $A$ is normal. Indeed, from the above one immediately gets $\langle (AA^* - A^*A)\, x, x\rangle = 0$ for all $x \in H$. Since it can be shown that if $\langle Bx, x\rangle = 0$ for all $x$ in a complex Hilbert space $H$, then $B = 0$, one obtains $AA^* = A^*A$ as desired. ■

Note that if $A : H \to H$ is self-adjoint then it is normal, although the converse is not generally true.

### Unitary Operators

**Definition 3.18.** *Let $A : H \to H$ be a bounded linear operator. The operator $A$ is said to be an unitary operator if and only if*

$$AA^* = A^*A = I,$$

*i.e., $A^* = A^{-1}$.*

Obviously, unitary operators are normal. It is easy to see that if $A$ is unitary, then $\|A\| = 1$. Furthermore,

$$\|Ax\|^2 = \langle Ax, Ax\rangle = \langle x, A^*Ax\rangle = \langle x, Ix\rangle = \|x\|^2,$$

which, using polarization, is shown to also give

$$\langle Ax, Ay\rangle = \langle x, y\rangle$$

[14]Operators $A$ and $B$ for which $[A, B] = 0$ are called *commuting operators.*

for all $x, y \in H$. That is, unitary operators preserve size, distances, and angles (more generally unitary operators preserve topological structure). The range $R_A$ of the unitary operator $A : H \to H$ coincides with that of $H$ in a finite-dimensional space; however, this is not necessarily true in an infinite-dimensional Hilbert space.

**Examples**

1. The operator $M : \mathbf{L}^2(a,b) \to \mathbf{L}^2(a,b)$, which multiplies $f \in \mathbf{L}^2(a,b)$ by $e^{i\omega t}$, where $\omega \in \mathbf{R}$, is a unitary operator.[15] In essence, this multiplication "rotates" the function $f$, since if $f = |f|\, e^{i\theta_f}$, then $e^{i\omega t} f = |f|\, e^{i(\theta_f + \omega t)}$.

2. The Fourier transformation operator $F : \mathbf{L}^2(-\infty,\infty) \to \mathbf{L}^2(-\infty,\infty)$ defined by
$$F(f)(k) \equiv \frac{1}{\sqrt{2\pi}} \int_{-\infty}^{\infty} f(x)\, e^{-ikx} dx$$
where $-\infty < k < \infty$, is a one-to-one, onto, unitary operator.

3. The operator $A : \mathbf{C}^2 \to \mathbf{C}^2$, which rotates $\mathbf{x} = (x_1, x_2) \in \mathbf{C}^2$ by the angle $\theta$ to produce $\mathbf{y} = (y_1, y_2) \in \mathbf{C}^2$, is a unitary operator with matrix representation (in the standard basis)
$$\begin{bmatrix} \cos\theta & \sin\theta \\ -\sin\theta & \cos\theta \end{bmatrix}.$$

Figure 3.1 depicts the interrelationships among some of the operators discussed here. Self-adjoint operators may be further classified as to their definiteness, as discussed in the next section.

### 3.4.2 Examples Relating to Self-Adjointness

1. The identity $(Ix = x)$ and zero $(0x = 0)$ operators are self-adjoint.

2. The bounded operator $A : H \to H$, represented by the matrix $[a_{ij}]$, where $a_{ij} = \langle Ax_j, x_i \rangle$ and $\{x_1, x_2, \ldots\}$ is an orthonormal basis for $H$, is self-adjoint if and only if $[a_{ij}] = [\overline{a_{ji}}]$, i.e., the matrix representation of the operator is Hermitian.

3. The multiplication operator $A : \mathbf{L}^2(\Omega) \to \mathbf{L}^2(\Omega)$ defined by
$$(Ax)(\mathbf{t}) \equiv f(\mathbf{t}) x(\mathbf{t})$$
is self-adjoint if and only if $f(\mathbf{t}) = \overline{f(\mathbf{t})}$, i.e., $f(\mathbf{t})$ is real-valued.

---

[15] Indeed, $\left\| e^{i\omega t} f \right\|^2 = \int \left| e^{i\omega t} f \right|^2 dt = \int |f|^2 \, dt = \|f\|^2$.

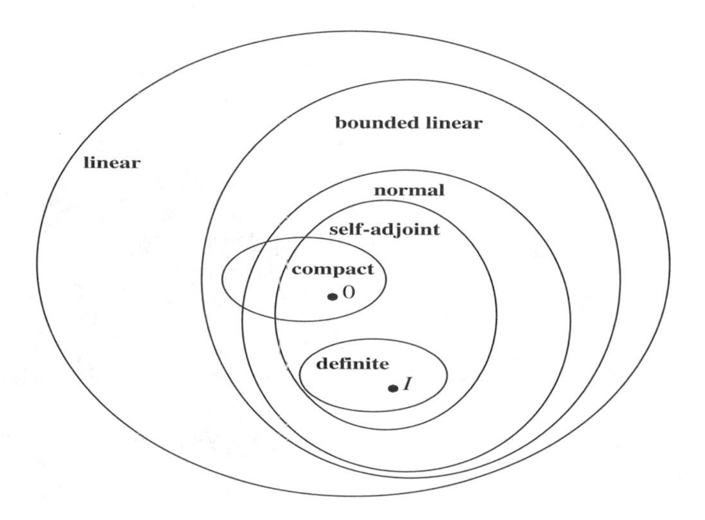


Figure 3.1: Depiction of the interrelationships among various classes of linear operators on an infinite-dimensional space. The zero and identity operators are represented by 0 and $I$, respectively.

4. The bounded integral operator $A : \mathbf{L}^2(\Omega)^m \to \mathbf{L}^2(\Omega)^m$ defined by

$$(A\mathbf{x})(\mathbf{t}) \equiv \int_\Omega \underline{\mathbf{k}}(\mathbf{t},\mathbf{s}) \cdot \mathbf{x}(\mathbf{s})\, d\Omega,$$

where $\underline{\mathbf{k}}(\mathbf{t},\mathbf{s}) \in \mathbf{L}^2(\Omega \times \Omega)^{m\times m}$, is self-adjoint if and only if $\underline{\mathbf{k}}(\mathbf{t},\mathbf{s}) = \underline{\mathbf{k}}^\top(\mathbf{s},\mathbf{t})$.

As a special case, the bounded integral operator $A : \mathbf{L}^2(\Omega) \to \mathbf{L}^2(\Omega)$ defined by

$$(Ax)(\mathbf{t}) \equiv \int_\Omega k(\mathbf{t},\mathbf{s}) x(\mathbf{s})\, d\Omega,$$

where $k(\mathbf{t},\mathbf{s}) \in \mathbf{L}^2(\Omega \times \Omega)$ (Hilbert–Schmidt kernel), is self-adjoint if and only if $k(\mathbf{t},\mathbf{s}) = \overline{k(\mathbf{s},\mathbf{t})}$. Every bounded self-adjoint operator in $\mathbf{L}^2(\Omega)$ can be represented as an integral operator, with perhaps the kernel $k(\mathbf{t},\mathbf{s})$ considered in the sense of a generalized function.

5. The integration operator $A : \mathbf{L}^2(a,b) \to \mathbf{L}^2(a,b)$ defined by $(Ax)(t) \equiv \int_a^t x(s)\, ds$ is a linear bounded operator, but it is not self-adjoint.

6. The differentiation operator

$$(Ax)(t) \equiv \frac{d}{dt}x(t) = x'(t),$$
$$D_A \equiv \left\{x : x(t), x'(t) \in \mathbf{L}^2(a,b),\ x(a) = 0\right\}$$

is not self-adjoint since

$$\left(A^* x\right)(t) = -\frac{d}{dt}x(t),$$
$$D_{A^*} = \left\{x : x(t), x'(t) \in \mathbf{L}^2(a,b),\ x(b) = 0\right\}.$$

7. The differentiation operator $A : \mathbf{L}^2(a,b) \to \mathbf{L}^2(a,b)$ defined by

$$(Ax)(t) \equiv \frac{d^2}{dt^2}x(t) = x''(t),$$
$$D_A \equiv \left\{x : x(t), x'(t),\ x''(t) \in \mathbf{L}^2(a,b),\ x(a) = x(b) = 0\right\}$$

(one-dimensional Laplacian) is self-adjoint as shown in the previous section.

8. The Laplacian operator $A : \mathbf{L}^2(\Omega) \to \mathbf{L}^2(\Omega)$ defined by

$$(Ax)(\mathbf{t}) \equiv \nabla^2 x(\mathbf{t}),$$
$$D_A \equiv \left\{x : x(\mathbf{t}), \nabla^2 x(\mathbf{t}) \in \mathbf{L}^2(\Omega),\ x(\mathbf{t})|_\Gamma = 0\right\}$$

is self-adjoint as shown in the previous section.

9. The vector differential operator $A : \mathbf{L}^2(\Omega)^3 \to \mathbf{L}^2(\Omega)^3$ defined by

$$(A\mathbf{x})(\mathbf{t}) \equiv \nabla \times \nabla \times \mathbf{x}(\mathbf{t}),$$
$$D_A \equiv \left\{\mathbf{x} : \mathbf{x}(\mathbf{t}), \nabla \times \nabla \times \mathbf{x}(\mathbf{t}) \in \mathbf{L}^2(\Omega)^3,\ \mathbf{n} \times \mathbf{x}(\mathbf{t})|_\Gamma = \mathbf{0}\right\}$$

is self-adjoint as shown in the previous section. The generalization to a dyadic function gives a self-adjoint operator as well.

10. From the previous section we see that curl, divergence, and gradient are not self-adjoint, although curl is formally self-adjoint.

11. Electrodynamic problems involving unbounded lossless regions often lead to nonself-adjoint operators. For instance, let the Helmholtz operator $A$ be defined by

$$Ax \equiv \left(\nabla^2 + k^2\right)x,$$
$$D_A \equiv \left\{x : x \in \mathbf{C}^2\left(\mathbf{R}^3\right),\ \lim_{r\to\infty} r\left[\frac{\partial x}{\partial r} + ikx\right] = 0\right\},$$

i.e., functions in the domain of $A$ satisfy the outgoing radiation condition

$$\lim_{r\to\infty} r\left[\frac{\partial x}{\partial r} + ikx\right] = 0, \tag{3.6}$$

where we assume $k \in \mathbf{R}$. It is easy to see that the operator $A$ is not self-adjoint. Indeed, from Green's second theorem,

$$\begin{aligned}\langle Ax, y\rangle &= \int_{\mathbf{R}^3} \overline{y}\left(\nabla^2 x + k^2 x\right) d\Omega \\ &= \int_{\mathbf{R}^3} x\left(\nabla^2 \overline{y} + k^2 \overline{y}\right) d\Omega + \oint_{\Gamma_\infty} \left(\overline{y}\frac{\partial x}{\partial n} - x\frac{\partial \overline{y}}{\partial n}\right) d\Gamma = \langle x, Ay\rangle\end{aligned}$$

with the last equality being valid if the boundary integral (conjunct) vanishes. With $x$ satisfying the outgoing radiation condition (3.6), the boundary integral over $\Gamma_\infty$ will vanish if $y$ satisfies the incoming radiation condition

$$\lim_{r\to\infty} r\left[\frac{\partial y}{\partial r} - iky\right] = 0. \tag{3.7}$$

Therefore, the operator $A$ is symmetric (formally self-adjoint) but not self-adjoint, with functions in the domain of the adjoint obeying an incoming radiation condition which is conjugate to (3.6).

### 3.4.3 Pseudo Adjoints, Pseudo Symmetry, and Reciprocity

#### Pseudo Adjoints

In many electromagnetics problems the most natural form resembling an inner product is the pseudo inner product $\langle x, y\rangle_p$, described by (2.34). Recall that unless the underlying space is real, this is not an actual inner product, and it does not generate a normed space or associated completeness properties. Nevertheless, we can introduce a *pseudo-adjoint operator* $A^{p*} : X_2 \to X_1$ satisfying

$$\langle Ax, y\rangle_{p,2} = \langle x, A^{p*}y\rangle_{p,1} \tag{3.8}$$

for the operator $A : X_1 \to X_2$ where $X_{1,2}$ are pseudo inner product spaces (see Section 2.5.2). If $X_{1,2}$ are real spaces, then the pseudo inner product is the same as the inner product, and $A^{p*} = A^*$.

As with inner product spaces, we can consider an operator $A : X_1 \to X_1$ to be *formally pseudo self-adjoint* or *pseudo symmetric* if

$$\langle Ax, y\rangle_p = \langle x, Ay\rangle_p \tag{3.9}$$

for all $x, y \in D_A$, and *pseudo self-adjoint* if, in addition, $D_{A^{p*}} = D_A$.

As an important electromagnetic example, consider the integral operator (see Section 1.3.4) defined by

$$(Ax)(\mathbf{r}) \equiv \int_\Omega k(\mathbf{r}, \mathbf{r}')x(\mathbf{r}')\, d\Omega'.$$

To determine the pseudo-adjoint operator, consider

$$\begin{aligned}\langle Ax, y\rangle_p &= \int_\Omega \int_\Omega k(\mathbf{r}, \mathbf{r}')x(\mathbf{r}')\, d\Omega' y(\mathbf{r})\, d\Omega \\ &= \int_\Omega x(\mathbf{r}') \int_\Omega k(\mathbf{r}, \mathbf{r}')\; y(\mathbf{r})\, d\Omega\; d\Omega' = \langle x, A^{p*}y\rangle_p\end{aligned}$$

(we assume that the order of integration may be interchanged, justified by properties of the kernel). Therefore, the pseudo-adjoint operator is found to be

$$\left(A^{p*}y\right)(\mathbf{r}) = \int_\Omega k(\mathbf{r}', \mathbf{r})y(\mathbf{r}')\, d\Omega' = \int_\Omega k^{p*}(\mathbf{r}, \mathbf{r}')y(\mathbf{r}')\, d\Omega',$$

and the pseudo-adjoint kernel is

$$k^{p*}(\mathbf{r}, \mathbf{r}') = k(\mathbf{r}', \mathbf{r})$$

(recall from the examples in Section 3.3.2 that if we were using an actual inner product we would obtain the adjoint kernel as $k^*(\mathbf{r}, \mathbf{r}') = \overline{k(\mathbf{r}', \mathbf{r})}$).

For example, consider the scalar, three-dimensional free-space Green's function (1.41),

$$g(\mathbf{r}, \mathbf{r}') = \frac{e^{-jk|\mathbf{r}-\mathbf{r}'|}}{4\pi|\mathbf{r}-\mathbf{r}'|} = g(\mathbf{r}', \mathbf{r}) = g^{p*}(\mathbf{r}', \mathbf{r}).$$

For $\mathbf{r}$, $\mathbf{r}' \in \Omega \subset \mathbf{R}^3$ we see that integral operators of the form

$$(Af)(\mathbf{r}) = \int_\Omega g(\mathbf{r}, \mathbf{r}')\; f(\mathbf{r}')\; d\Omega'$$

are pseudo self-adjoint, with bounded (weakly singular and so compact) kernel $g(\mathbf{r}, \mathbf{r}')$. The Green's function kernel represents an outgoing spherical wave, as does its pseudo adjoint, while the adjoint kernel

$$\overline{g(\mathbf{r}, \mathbf{r}')} = \frac{e^{+jk|\mathbf{r}-\mathbf{r}'|}}{4\pi|\mathbf{r}-\mathbf{r}'|}$$

represents an incoming spherical wave, where we assume $k \in \mathbf{R}$.

Referring to Example 4 in Section 3.3.2, and using reciprocity of the free-space dyadic Green's function[16] $\underline{\mathbf{G}}(\mathbf{r}, \mathbf{r}')$ (weakly singular and so compact for the vector potential (1.72), and singular for the electric field (1.79)),

[16]Various free-space dyadic Green's functions are of interest (e.g., for vector potentials and for electric and magnetic fields). In all cases we have $\underline{\mathbf{G}}^\top(\mathbf{r}, \mathbf{r}') = \underline{\mathbf{G}}(\mathbf{r}', \mathbf{r})$. This also holds for any inhomogeneous reciprocal medium where $\underline{\mathbf{G}}$ directly relates the field quantity and the source, e.g.,

$$\mathbf{E}(\mathbf{r}) = \int_\Omega \underline{\mathbf{G}}\left(\mathbf{r}, \mathbf{r}'\right) \cdot \mathbf{J}\left(\mathbf{r}'\right) d\Omega'.$$

we see that the dyadic Green's function $\underline{\mathbf{G}}(\mathbf{r},\mathbf{r}')$ is a pseudo self-adjoint kernel of the pseudo self-adjoint operator

$$(A\mathbf{f})(\mathbf{r}) = \int_{\Omega} \underline{\mathbf{G}}(\mathbf{r},\mathbf{r}') \cdot \mathbf{f}(\mathbf{r}')\, d\Omega'.$$

While the concept of a pseudo inner product is very convenient in electromagnetic problems, for the complex linear spaces of primary interest it does not generally lead to the wide array of powerful analysis tools that apply to inner product spaces. In particular, the pseudo inner product does not lead to Hilbert spaces. This is also seen for the matrix representation of operators on finite-dimensional spaces.[17] However, pseudo self-adjointness is sufficient to prove several important results (e.g., the stationary nature of secondary functionals, etc.), and on real spaces the pseudo inner product is a true inner product.

### Pseudo Symmetry and Reciprocity

For a reciprocal medium the Lorentz reciprocity theorem (1.20) provides a form that is very useful for a variety of applications and exhibits pseudo symmetry (symmetry under a pseudo inner product) of field operators. For instance, consider in the vicinity of the origin an inhomogeneous, anisotropic medium characterized by symmetric dyadics, $\underline{\varepsilon} = \underline{\varepsilon}^{\top}, \underline{\mu} = \underline{\mu}^{\top}$ (reciprocal medium). We assume the space is unbounded, and as $r \to \infty$ the medium becomes free space. Let $\mathbf{E}_1, \mathbf{H}_1$ and $\mathbf{E}_2, \mathbf{H}_2$ be the fields caused by sources $\mathbf{J}_{e1}, \mathbf{J}_{m1}$ and $\mathbf{J}_{e2}, \mathbf{J}_{m2}$, respectively. Then, (1.20) becomes

$$\langle \mathbf{E}_2 \cdot \mathbf{J}_{e1} \rangle_p - \langle \mathbf{H}_2 \cdot \mathbf{J}_{m1} \rangle_p = \langle \mathbf{E}_1 \cdot \mathbf{J}_{e2} \rangle_p - \langle \mathbf{H}_1 \cdot \mathbf{J}_{m2} \rangle_p . \tag{3.10}$$

We can think of $\mathbf{E}$ ($\mathbf{H}$) in terms of a linear operator $\mathbf{L}_{\mathbf{E}}$ ($\mathbf{L}_{\mathbf{H}}$) that produces the electric (magnetic) field from current sources[18] $\mathbf{J}_e, \mathbf{J}_m$. Let

$$\Psi_i = \begin{bmatrix} \mathbf{J}_{ei} \\ \mathbf{J}_{mi} \end{bmatrix}, \qquad F_i = \begin{bmatrix} \mathbf{E}_i \\ -\mathbf{H}_i \end{bmatrix},$$

and consider the operator[19]

$$\mathbf{L} = \begin{bmatrix} L_{\mathbf{E}} \\ -L_{\mathbf{H}} \end{bmatrix} = \begin{bmatrix} L_{e,e} & L_{e,m} \\ -L_{m,e} & -L_{m,m} \end{bmatrix}$$

[17] Indeed, Hermitian matrices (the matrix analog of self-adjointness) have very special properties (such as real eigenvalues, basis of eigenvectors, etc.), whereas complex-valued matrices symmetric about the main diagonal, termed *complex-symmetric matrices* (the matrix analog of pseudo self-adjointness), do not possess especially useful properties.

[18] For instance, (1.86) provides a form for $L_{\mathbf{E}(\mathbf{H})}$ in the special case of isotropic, homogeneous media.

[19] Using this notation, application of the operators follows the usual rules of matrix multiplication.

such that $\mathbf{L}\mathbf{\Psi}_i = \mathbf{F}_i$, or

$$\begin{bmatrix} L_{e,e} & L_{e,m} \\ -L_{m,e} & -L_{m,m} \end{bmatrix} \begin{bmatrix} \mathbf{J}_{ei} \\ \mathbf{J}_{mi} \end{bmatrix} = \begin{bmatrix} \mathbf{E}_i \\ -\mathbf{H}_i \end{bmatrix}.$$

Then from (3.10) we have

$$\langle \mathbf{L}\mathbf{\Psi}_2, \mathbf{\Psi}_1 \rangle_p = \langle \mathbf{\Psi}_2, \mathbf{L}\mathbf{\Psi}_1 \rangle_p$$

such that the operator is pseudo-symmetric. In this case it is obvious that symmetry under a pseudo inner product is simply a statement of reciprocity. As mentioned previously, for static fields the pseudo inner product is a legitimate inner product since we deal with real-valued quantities. In this case the operator $\mathbf{L}$ is symmetric (self-adjoint if the domains are appropriately chosen).

If desired, the two independent Maxwell curl equations from (1.9) can be written in operator form as

$$\mathbf{L}_M \mathbf{F} = \begin{pmatrix} -i\omega\underline{\varepsilon}\cdot & -\nabla\times\underline{\mathbf{I}}\cdot \\ -\nabla\times\underline{\mathbf{I}}\cdot & i\omega\underline{\mu}\cdot \end{pmatrix} \begin{pmatrix} \mathbf{E} \\ -\mathbf{H} \end{pmatrix} = \mathbf{\Psi} = \begin{pmatrix} \mathbf{J}_e \\ \mathbf{J}_m \end{pmatrix},$$

which is useful for formal manipulations. Comparing with $\mathbf{L}\mathbf{\Psi} = \mathbf{F}$, the operator $\mathbf{L}_M$ is obviously the formal inverse to $\mathbf{L}$, i.e., $\mathbf{L}\mathbf{L}_M = \mathbf{L}_M\mathbf{L} = I$; therefore, $\mathbf{L}^{-1} = \mathbf{L}_M$. If reciprocity holds, then $\mathbf{L}_M$ is pseudo-symmetric.

## 3.5 Definiteness and Convergence in Energy

### 3.5.1 General Concepts

Since for self-adjoint or merely symmetric operators $\langle Ax, x \rangle \in \mathbf{R}$, certain further distinctions may be made.

**Definition 3.19.** *A self-adjoint or symmetric operator $A : H \to H$ is said to be nonnegative if*

$$\langle Ax, x \rangle \geq 0$$

*for all $x \in H$, denoted as $A \geq 0$. An operator is said to be (strictly) positive if $\langle Ax, x \rangle > 0$ for all $x \neq 0$ in $H$, denoted as $A > 0$. An operator is said to be positive definite if there is a constant $k > 0$ such that*

$$\langle Ax, x \rangle \geq k \, \|x\|^2$$

*for all $x \in H$.*

*Nonpositive*, *negative*, and *negative definite* operators can be similarly defined, and in general we will call an operator *definite* if $\langle Ax, x \rangle \gtrless 0$ (i.e., a strict inequality exists). It is clear that positive definite operators are also positive, but the converse is not necessarily true. A symmetric or self-adjoint operator that doesn't fit into one of these categories is called *indefinite*. In this case $\langle Ax, x\rangle$ will be positive for some $x \in H$ and negative for other $x \in H$.

An operator $A$ satisfying the definition for positive-definiteness is also said to be *bounded below*. An operator $A$ is said to be *bounded above* if there is a constant $K$ such that $\langle Ax, x\rangle \leq K \|x\|^2$ for all $x \in H$. An operator on a Hilbert space is bounded if and only if it is bounded above and below [12, p. 364]. We also have the following theorem.

**Theorem 3.14.** *Let $A : H \to H$ be a bounded linear operator. Then, the operators $A^*A$ and $AA^*$ are nonnegative. If $Ax = 0$ has only the trivial solution $x = 0$, then $A^*A$ is positive.*

**Proof.** $\langle A^*Ax, x\rangle = \langle Ax, Ax\rangle = \|Ax\|^2 \geq 0$, $\langle AA^*x, x\rangle = \langle A^*x, A^*x\rangle = \|A^*x\|^2 \geq 0$. Clearly, $\|Ax\|^2 > 0$ unless $Ax = 0$, but this would imply $x = 0$. ■

Note that the product of two commuting positive operators is positive, but the product of two positive operators is not necessarily a positive operator. Furthermore, a *square root* of a positive operator $A$ is a self-adjoint operator $B$ such that $B^2 = A$.

If an operator is not symmetric, then generally $\langle Ax, x\rangle \in \mathbf{C}$. In this case there is a concept related to definiteness.

**Definition 3.20.** *An operator $A : H \to H$ is dissipative if, for all $x \in D_A$,*

$$\operatorname{Im}\langle Ax, x\rangle \geq 0. \tag{3.11}$$

If $A$ is bounded, then (3.11) is equivalent to the condition that its imaginary component $(A - A^*)/2i$ (See Theorem 3.9) be nonnegative.

In accordance with the desire to make Hilbert spaces analogous to geometrical spaces, it is worthwhile to note an analogy between operators on a Hilbert space and numbers in the complex plane. Under this analogy the adjoint plays the role of complex conjugation, self-adjoint operators are analogous to real numbers ($A = A^*$), and unitary operators are analogous to complex numbers having unit magnitude ($AA^* = I$). Also, positive (negative) operators are analogous to positive (negative) numbers.

### Definiteness and Fields

Considering electric currents and using the operator $L_{e,e}$, which produces the electric field from the electric current, it is clear that

$$\langle L_{e,e}\mathbf{J}_e, \mathbf{J}_e\rangle = \langle \mathbf{E}, \mathbf{J}_e\rangle = \int_\Omega \mathbf{E}(\mathbf{r})\cdot\overline{\mathbf{J}}_e(\mathbf{r})\, d\Omega$$

has units of *watts* and represents a complex power density. In general $L_{e,e}$ is pseudo-symmetric, and so $\langle L_{e,e}\mathbf{J}_e, \mathbf{J}_e \rangle$ is generally not even real-valued, let alone strictly positive or negative. However, for static fields $L_{e,e}$ is symmetric and $\langle L_{e,e}\mathbf{J}_e, \mathbf{J}_e \rangle$ will be definite under fairly general conditions. For instance, for applied currents $\mathbf{J}_e = \mathbf{J}_e^i$, $\langle L_{e,e}\mathbf{J}_e, \mathbf{J}_e \rangle < 0$ for a source supplying power. For induced currents in a passive medium, $\mathbf{J}_e = \underline{\sigma_e} \cdot \mathbf{E}$ and $\langle L_{e,e}\mathbf{J}_e, \mathbf{J}_e \rangle > 0$, representing dissipated power.

### 3.5.2 Examples Relating to Definiteness

1. The identity operator $I : H \to H$ defined by $Ix \equiv x$ is a bounded, self-adjoint positive operator since $\langle Ix, x \rangle = \langle x, x \rangle > 0$ for all $x \neq 0$ in $H$.

2. The multiplication operator $A : \mathbf{L}^2(a,b) \to \mathbf{L}^2(a,b)$ defined by

$$(Ax)(t) \equiv f(t)x(t),$$

where $f(t)$ is a bounded real-valued nonnegative (positive) function, is a bounded, self-adjoint nonnegative (positive) operator. Indeed,

$$\langle Ax, x \rangle = \int_a^b f(t)x(t)\overline{x(t)}\, dt = \int_a^b f(t)\left|x(t)\right|^2 dt \geq 0$$

if $f(t) \geq 0$ on $[a,b]$.

3. The operator $A : \mathbf{C}^2 \to \mathbf{C}^2$ with matrix representation $\begin{bmatrix} 1 & 2 \\ 2 & 1 \end{bmatrix}$ is indefinite, as can be seen by examining $\langle Ax, x \rangle$ with $x = \begin{bmatrix} 1 & -1 \end{bmatrix}^T$ ($\langle Ax, x \rangle = -2 < 0$) and $x = \begin{bmatrix} 1 & 1 \end{bmatrix}^T$ ($\langle Ax, x \rangle = 6 > 0$).

4. The differentiation operator $A : \mathbf{L}^2(0,b) \to \mathbf{L}^2(0,b)$ (with $0 < b < \infty$) defined by

$$\begin{aligned} (Ax)(t) &\equiv -\frac{d^2}{dt^2}x(t) = -x''(t), \\ D_A &\equiv \left\{x : x(t), x'(t), x''(t) \in \mathbf{L}^2(0,b),\ x(0) = x(b) = 0\right\} \end{aligned}$$

is an unbounded, self-adjoint, positive operator. To see this, consider

$$\begin{aligned} \langle Ax, x \rangle &= -\int_0^b x''(t)\overline{x(t)}\, dt \\ &= -x'(t)\overline{x(t)}\Big|_0^b + \int_0^b x'(t)\overline{x'(t)}\, dt = \int_0^b \left|x'(t)\right|^2 dt \geq 0, \end{aligned}$$

showing that $A$ is at least nonnegative. Because $x(0) = 0$,

$$|x(t)|^2 = \left|\int_0^t x'(\tau)\, d\tau\right|^2 \leq \int_0^t 1^2 d\tau \int_0^t \left|x'(\tau)\right|^2 d\tau \leq t\int_0^b \left|x'(\tau)\right|^2 d\tau.$$

Therefore,

$$\int_0^b |x'(\tau)|^2 \, d\tau = \langle Ax, x \rangle \geq \frac{|x(t)|^2}{t} > 0,$$

and so $\langle Ax, x \rangle > 0$.

5. The negative Laplacian operator $A : \mathbf{L}^2(\Omega) \to \mathbf{L}^2(\Omega)$ defined by

$$\begin{aligned} (Ax)(\mathbf{t}) &\equiv -\nabla^2 x(\mathbf{t}), \\ D_A &\equiv \left\{ x : x(\mathbf{t}), \nabla^2 x(\mathbf{t}) \in \mathbf{L}^2(\Omega), \, x(\mathbf{t})|_S = 0 \right\} \end{aligned}$$

is an unbounded, self-adjoint, nonnegative operator. Indeed, from the first scalar Green's theorem

$$\begin{aligned} \langle Ax, x \rangle &= -\int_\Omega \overline{x} \nabla^2 x \, d\Omega \\ &= -\oint_S \overline{x} \frac{\partial x}{\partial n} dS + \int_\Omega \overline{\nabla x} \cdot \nabla x \, d\Omega = \int_\Omega |\nabla x|^2 \, d\Omega \geq 0. \end{aligned}$$

6. The vector differential operator $A : \mathbf{L}^2(\Omega)^3 \to \mathbf{L}^2(\Omega)^3$ defined by

$$\begin{aligned} (A\mathbf{x})(\mathbf{t}) &\equiv \nabla \times \nabla \times \mathbf{x}(\mathbf{t}), \\ D_A &\equiv \Big\{ \mathbf{x} : \mathbf{x}(\mathbf{t}), \nabla \times \mathbf{x}(\mathbf{t}), \nabla \times \nabla \times \mathbf{x}(\mathbf{t}) \in \mathbf{L}^2(\Omega)^3, \\ &\qquad \mathbf{n} \times \mathbf{x}(\mathbf{t})|_S = \mathbf{0} \} \end{aligned}$$

is an unbounded, self-adjoint, nonnegative operator. To see that the operator $A$ is nonnegative, consider the first vector Green's theorem,

$$\begin{aligned} \langle A\mathbf{x}, \mathbf{x} \rangle &= \int_\Omega \overline{\mathbf{x}} \cdot (\nabla \times \nabla \times \mathbf{x}) \; d\Omega \\ &= -\oint_S (\mathbf{n} \times \overline{\mathbf{x}}) \cdot (\nabla \times \mathbf{x}) \; dS + \int_\Omega \overline{(\nabla \times \mathbf{x})} \cdot (\nabla \times \mathbf{x}) \; d\Omega \\ &= \int_\Omega |\nabla \times \mathbf{x}|^2 \; d\Omega \geq 0. \end{aligned}$$

### 3.5.3 Convergence in Energy

We previously introduced the notion of strong and weak convergence for inner product spaces (see Section 2.5.3). Recall that a sequence of elements $x_1, x_2, \ldots$ in an inner product space $S$ is said to be strongly convergent to an element $x \in S$ (denoted $x_n \to x$) if $\|x_n - x\| \to 0$, and weakly convergent to an element $x \in S$ (denoted as $x_n \overset{w}{\to} x$) if $\langle x_n, y \rangle \to \langle x, y \rangle$ for all $y \in S$. For positive operators (and positive definite operators as a special case) we can introduce another measure of convergence by using a special inner product and norm called the *energy inner product* and *energy norm* [4, p. 91].

**Definition 3.21.** *Consider a positive operator* $A : H \to H$. *The energy inner product is a mapping* $H \times H \to \mathbf{C}$ *that associates, for a given* $A$, *an ordered pair* $x, y \in D_A$ *with a scalar denoted by*

$$\langle x, y \rangle_e \equiv \langle Ax, y \rangle \tag{3.12}$$

*obeying the usual properties of an inner product.*

Since the energy inner product obeys the usual inner product rules, it induces a norm, called the energy norm. Denoting the energy norm as $\|x\|_e$, we obtain

$$\|x\|_e = \sqrt{\langle x, x \rangle_e} = \sqrt{\langle Ax, x \rangle}. \tag{3.13}$$

We define the Hilbert space $H_e$, associated with a particular positive operator $A$, as an energy inner product space complete in the norm $\|x\|_e$. Within this space we have the concept of an orthonormal set,

$$\langle x_n, x_m \rangle_e = \delta_{nm}. \tag{3.14}$$

An orthonormal set $\{x_n\}$ is a basis of $H_e$ if each $y \in H_e$ can be written in a unique way as

$$y = \sum_{n=1}^{\infty} \langle y, x_n \rangle_e x_n.$$

Because $\|x\|_e^2 = \langle Ax, x \rangle \geq k \|x\|^2$ for positive definite operators, it is easy to see that

$$\|x\| \leq \frac{1}{\sqrt{k}} \|x\|_e .$$

Now we are ready to define the concept of *convergence in energy.*

**Definition 3.22.** *A sequence of elements* $x_1, x_2, \ldots$ *in a Hilbert space* $H$ *is said to converge in energy to an element* $x \in H$ *(denoted* $x_n \xrightarrow{e} x$*) if* $\|x_n - x\|_e \to 0$.

We have the following interrelationships among convergence.

**Theorem 3.15.** *Assume* $\|Ax_n\|$ *is bounded. Then*

a. *strong convergence implies convergence in energy,*

b. *convergence in energy implies* $Ax_n \xrightarrow{w} Ax$,

c. *if* $A$ *is positive definite, convergence in energy implies strong convergence.*

**Proof.** (a) $\|x_n - x\|_e^2 = |\langle A(x_n - x), (x_n - x) \rangle| \leq \|A(x_n - x)\| \|x_n - x\| = \|Ax_n - Ax\| \|x_n - x\| \leq (\|Ax_n\| + \|Ax\|) \|x_n - x\|$. Since $\|x_n - x\| \to 0$ by strong convergence, and $\|Ax_n\|$ is bounded, then $\|x_n - x\|_e \to 0$.

(b) $\|x_n - x\|_e^2 \to 0$ from (a) implies $|\langle A(x_n - x), (x_n - x) \rangle| \to 0$ for $x \in H$, which leads to $Ax_n \xrightarrow{w} Ax$ as $x_n \xrightarrow{w} x$.

(c) Because $\|x_n - x\| \leq (1/\sqrt{k}) \|x_n - x\|_e$, (c) is proved. ■

## 3.6 Compact Operators

### 3.6.1 General Concepts

*Compact (completely continuous) operators* comprise an important subset of bounded operators, especially within the theory of integral equations. Compact operators are operators that have a finite-dimensional, or an almost finite-dimensional, range. In infinite-dimensional spaces compact operators have the simplest properties of all infinite-dimensional operators. This is especially true when one considers the spectral theory covered in the next chapter, where it is shown that the basic elements of spectral theory for compact self-adjoint operators are similar to those from the spectral theory of matrix operators. Also note that the theory of linear compact operators is a generalization of Fredholm's theory of integral equations. In particular, compactness is a very important concept in formulating boundary value problems of applied electromagnetics leading to Fredholm integral equations. The solution of operator equations involving compact operators is discussed in a later section.

**Definition 3.23.** *A bounded linear operator $A : H_1 \to H_2$ is compact*[20] *if, for each bounded sequence $\{x_n\} \subset H_1$, there is a subsequence $\{x_{n_i}\}$ such that $\{Ax_{n_i}\}$ converges in $H_2$.*

Note that every compact linear operator is bounded, but the converse is not necessarily true (for example, consider the identity operator). It can be shown that every linear bounded operator with finite-dimensional range[21] is compact. Therefore, necessarily, $A : \mathbf{C}^n \to \mathbf{C}^m$ is compact (e.g., matrix operators are compact).

Information on the range of a compact operator is given in the following.

**Theorem 3.16.** *Let $A : H_1 \to H_2$ be a compact linear operator. Given any $\varepsilon > 0$, there exists a finite-dimensional subspace $M$ of $R_A$ such that*

$$\inf\{\|Ax - m\|\} \leq \varepsilon \|x\|$$

*for $m \in M$, $x \in H_1$.*

Therefore, the finite-dimensional subspace $M$ is very close to (in a sense within $\varepsilon$ of the possibly infinite-dimensional) $R_A$.

The next theorem provides some interesting properties of the important operator $I - A$ for $A$ compact; these properties are useful for considering solutions of second-kind operator equations.

**Theorem 3.17.** *Let $T = I - A$, with $I$ the identity operator on a Hilbert space $H$, and $A : H \to H$ a compact operator. Then*

---

[20] A frequently used alternate definition is that an operator $A$ is compact if it maps any bounded set of $H_1$ into a compact set (i.e., one having compact closure) of $H_2$.

[21] A linear operator with a finite-dimensional range is also called a finite-rank operator.

*a.* $N_T$, $N_{T^*}$ *are finite-dimensional and have the same dimension.*

*b.* $H - R_{T^*} = N_T$, $H - R_T = N_{T^*}$.

*c.* $R_T = H$ *if and only if* $N_T = \{0\}$, *and* $R_{T^*} = H$ *if and only if* $N_{T^*} = \{0\}$.

*d.* $R_T$, $R_{T^*}$ *are closed subspaces of* $H$.

Thus, the operator $I - A$ for $A$ compact has particularly nice geometric properties.

We will state without proof some properties of compact operators.

- Let $A, B : H_1 \to H_2$ be compact linear operators and $C : H_3 \to H_1$, $D : H_2 \to H_3$ be bounded operators. Then $A + B$, $AC$, and $DA$ are compact linear operators.

  In practical applications it is sometimes convenient to represent an operator as a product of operators. Compactness is then shown for the product operator by simply showing that one operator is bounded and the other is compact. Also, the fact that $A+B$ is compact is often used in the theory of integral equations (for example, see [14, p. 346]). In general, any linear combination of compact linear operators is also a compact operator.

- An operator $A : H_1 \to H_2$ is compact if and only if its adjoint $A^* : H_2 \to H_1$ is compact. If $AA^*$ is compact, then $A$ is also compact.

- Let $\{A_n\} : H_1 \to H_2$ be a sequence of compact operators. If

$$\|A_n - A\| \to 0$$

  for some $A : H_1 \to H_2$, then $A$ is compact (this important property is often used to prove compactness).

The above property provides insight into the fact that compact operators either have finite rank or are nearly finite rank, since the limit of a sequence of finite-rank (and hence compact) operators is compact. Furthermore, any compact operator in a Hilbert space can be approximated in norm arbitrarily closely by a sequence of operators of finite rank [15, p. 507]. In fact, on a Hilbert space, and most Banach spaces [16, pp. 183–184], a compact operator can be expressed as the sum of a finite-rank (degenerate) operator and an operator with a small norm.

The next property is also useful for proving compactness [12, p. 358].

- If $A$ is a compact operator in $H$, and $\{x_n\}$ is an orthonormal sequence in $H$, then $\lim_{n\to\infty} Ax_n = 0$.

It is also seen that compact operators are very well behaved when applied in the "forward direction."

- An operator $A$ is compact if and only if $x_n \overset{w}{\rightarrow} x$ implies $Ax_n \rightarrow Ax$.

Recall that this means that if $A$ is compact, weak convergence, i.e., $\langle x_n, y\rangle \rightarrow \langle x, y\rangle$ for all $y \in H$, implies strong convergence, $\|Ax_n - Ax\| \rightarrow 0$. As an example, for a sequence of functions $\{x_n(t)\}$ in $\mathbf{L}^2(a,b)$ with the usual inner product, weak convergence

$$\int_a^b (x_n(t) - x(t))\,\overline{y(t)}\,dt \rightarrow 0$$

for all $y \in H$ implies convergence in the norm[22]

$$\left(\int_a^b |Ax_n(t) - Ax(t)|^2 dt\right)^{1/2} \rightarrow 0.$$

Thus, while bounded linear operators map strongly convergent sequences into strongly convergent sequences (and for operators boundedness is equivalent to continuity), compact linear operators map weakly convergent sequences into strongly convergent sequences. Compact operators can then be thought of as, in a sense, strongly bounded operators, or equivalently, strongly continuous operators—hence the name *completely continuous operators*. However, the inverse of a compact operator is unbounded, as shown in Theorem 3.37, and this fact presents some difficulties in applications.

Finally, we have the following.

- An operator $A$ is compact if for any sequences $x_n$ and $y_n$ weakly converging to $x$ and $y$ ($x_n \overset{w}{\rightarrow} x$, $y_n \overset{w}{\rightarrow} y$) the inner product $\langle Ax, y\rangle$ is a weakly continuous function of $x$ and $y$ such that

$$\lim_{n\rightarrow\infty} \langle Ax_n, y_n\rangle = \langle Ax, y\rangle\,.$$

### 3.6.2 Examples Relating to Compactness

1. The identity operator ($Ix = x$) on an infinite-dimensional space is bounded but not compact ($I$ is compact only on a finite-dimensional space), but the zero operator ($0x = 0$) is compact. The fact that the identity operator $I$ is not compact is relevant to the theory of Fredholm operators represented in the form $I - A$, where $A$ is a compact operator, leading to the existence of the bounded inverse operator $(I - A)^{-1}$.

[22] As a concrete example, on $H = \mathbf{L}^2(0,\pi)$ the sequence $\{x_n\} = \{\sin(nt)\}$ converges weakly to 0, i.e., $\int_0^\pi (\sin(nt) - 0)\,\overline{y(t)}\,dt \rightarrow 0$ for $y \in H$ (try $y = 1$ or $y = t$ as a simple test). The sequence does not converge in norm to 0, since $(\int_0^\pi (\sin^2(nt) - 0)\,dt)^{1/2} = \sqrt{\pi/2}$. However, subsequent to application of the compact operator $(Ax)(t) \equiv \int_0^t x(s)\,ds$ we have $\{Ax_n\} = \{\frac{1-\cos tn}{n}\}$, which converges strongly to 0, i.e., $\left(\int_a^b |-\frac{\cos tn - 1}{n} - 0|^2 dt\right)^{1/2} \rightarrow 0$.

2. The operator $A : \mathbf{l}^2 \to \mathbf{l}^2$ corresponding to the infinite matrix $[a_{ij}]$, where $\sum_{i=1}^{\infty}\sum_{j=1}^{\infty}|a_{ij}|^2 < \infty$, is bounded and compact[23] in $\mathbf{l}^2$. The proof is based on the previously stated criterion that a sequence of finite-rank (compact) operators $A_n$ is convergent in the $\mathbf{l}^2$ norm to the operator $A$ such that $||A_n - A|| \to 0$ [17, p. 86]. As a special case, the operator corresponding to an infinite diagonal matrix $\text{diag}[\lambda_1, \lambda_2, \lambda_3, \ldots]$ for which $\lambda_n \to 0$ as $n \to \infty$ is compact (and self-adjoint if all $\lambda$'s are real-valued).

3. The multiplication operator $A : \mathbf{L}^2(\Omega) \to \mathbf{L}^2(\Omega)$ defined by

$$(Ax)(\mathbf{t}) \equiv f(\mathbf{t})x(\mathbf{t})$$

is a bounded linear operator (assuming $k = \max_{\mathbf{t}\in\Omega}|f(\mathbf{t})| < \infty$) that is not compact (unless $f \equiv 0$) [17, pp. 87–88].

4. The integral operator $A : \mathbf{L}^2(\Omega)^m \to \mathbf{L}^2(\Omega)^m$ defined by

$$(A\mathbf{x})(\mathbf{t}) \equiv \int_{\Omega} \underline{\mathbf{k}}(\mathbf{t},\mathbf{s}) \cdot \mathbf{x}(\mathbf{s})\, d\Omega,$$

where $\Omega \subset \mathbf{R}^n$ with $\underline{\mathbf{k}}(\mathbf{t},\mathbf{s}) \in \mathbf{L}^2(\Omega\times\Omega)^{m\times m}$ (in the sense that each component of the $(m \times m)$-tuple dyadic kernel $\underline{\mathbf{k}}(\mathbf{t},\mathbf{s})$ belongs to $\mathbf{L}^2(\Omega\times\Omega)$), is bounded, and it is compact if and only if each component of the integral operator $(A\mathbf{x})(\mathbf{t})$ is compact in $\mathbf{L}^2(\Omega)$.

As a special case, the integral operator $A : \mathbf{L}^2(\Omega) \to \mathbf{L}^2(\Omega)$ defined by

$$(Ax)(\mathbf{t}) \equiv \int_{\Omega} k(\mathbf{t},\mathbf{s})x(\mathbf{s})\, d\Omega$$

with $k(\mathbf{t},\mathbf{s}) \in \mathbf{L}^2(\Omega\times\Omega)$ is bounded (Hilbert–Schmidt kernel, $||A|| \leq ||k||$) and compact.[24]

A Hilbert–Schmidt kernel $k(\mathbf{t},\mathbf{s})$ may itself be (and often is) unbounded as a function of $\mathbf{t}$ and $\mathbf{s}$, yet it always generates a compact operator (more generally, all Hilbert–Schmidt operators are compact). Also, the conditions for a kernel to be Hilbert–Schmidt are sufficient, but not necessary, for an integral operator to be compact.

This example can be also generalized to show that a bounded in $\mathbf{L}^p(\Omega)$ $(1 < p < \infty)$ integral operator $A : \mathbf{L}^p(\Omega) \to \mathbf{L}^p(\Omega)$ with $||A|| \leq ||k||$ and $||k||^p = \int_{\Omega}\int_{\Omega}|k(\mathbf{t},\mathbf{s})|^p\, d\Omega_s\, d\Omega_t < \infty$ is compact.

[23] The condition $\sum_{i=1}^{\infty}\sum_{j=1}^{\infty}|a_{ij}|^2 < \infty$ for compactness of matrix operators from $\mathbf{l}^2$ into $\mathbf{l}^2$ is often used in applications. However, this condition is a very strong one, and it is not necessary that every compact operator from $\mathbf{l}^2$ into $\mathbf{l}^2$ satisfy this condition.

[24] The proof of compactness is based on the approximation of $||k||$ by a sequence of continuous degenerate kernels $k_n(t,s)$ such that $||k - k_n|| \to 0$ and the fact that finite-rank operators $A_n$ are compact. Using $||A|| \leq ||k||$ for a bounded operator, $||A - A_n|| \leq ||k - k_n|| \to 0$, then $A_n \to A$ and $A$ is compact (for a complete proof, see [16, p. 183] and [4, pp. 155–158]).

5. The integration operator $A : \mathbf{L}^2(a,b) \to \mathbf{L}^2(a,b)$ defined by $(Ax)(t) \equiv \int_a^t x(s)\,ds$ is compact.

6. The weakly singular integral operator defined by

$$(Af)(\mathbf{s}) \equiv \int_\Omega \frac{k(\mathbf{t},\mathbf{s})}{|\mathbf{t}-\mathbf{s}|^m} f(\mathbf{t})\,d\Omega,$$

where $\Omega \subset \mathbf{R}^n, 0 \leq m < n$, and $k(\mathbf{t},\mathbf{s})$ is bounded on $\Omega \times \Omega$, is compact in $\mathbf{L}^2(\Omega)$ [4, p. 160]. This also holds for integrals of the form

$$(Af)(\mathbf{s}) = \int_\Gamma \frac{k(\mathbf{t},\mathbf{s})}{|\mathbf{t}-\mathbf{s}|^m} f(\mathbf{t})\,d\Gamma$$

where $\Gamma$ is a smooth $n$-dimensional surface in an $(n+1)$ dimensional space.

Therefore, the usual free-space Green's function kernel

$$g(\mathbf{t},\mathbf{s}) = \frac{e^{-ik|\mathbf{t}-\mathbf{s}|}}{4\pi|\mathbf{t}-\mathbf{s}|}$$

in a bounded region $\Omega \subset \mathbf{R}^3$ generates a compact (nonself-adjoint) operator

$$(Af)(\mathbf{s}) = \int_\Omega g(\mathbf{t},\mathbf{s}) f(\mathbf{t})\,d\Omega$$

in $\mathbf{L}^2(\Omega)$. Similarly, in two dimensions the Green's function

$$g(\mathbf{t},\mathbf{s}) = \frac{1}{4i} H_0^{(2)}(k|\mathbf{t}-\mathbf{s}|)$$

generates a weakly singular ($H_0^{(2)}(x) \sim \ln(x/2)$ as $x \to 0$) compact integral operator in $\mathbf{L}^2(\Omega)$, $\Omega \subset \mathbf{R}^2$. However, if the kernel has the form of sufficiently many derivatives of $g$, then the resulting operator will not be compact.

Note if $0 < m < n/2$, then the operator with a weak singularity is a Hilbert–Schmidt operator in $\mathbf{L}^2(\Omega)$ such that

$$\int_\Omega \int_\Omega \frac{d\Omega_t\,d\Omega_s}{|\mathbf{t}-\mathbf{s}|^{2m}} < \infty.$$

Also note that the one-dimensional Cauchy singular operator

$$(Af)(s) = \int_\Gamma \frac{f(t)}{|t-s|}\,dt$$

is bounded in $\mathbf{L}^p(\Gamma)$ for $1 < p < \infty$ but not compact in $\mathbf{L}^p(\Gamma)$. However, the operator

$$A(gf) - gA(f) = \int_\Gamma \frac{g(t)-g(s)}{|t-s|} f(t)\,dt$$

where $H$ is an infinite-dimensional space, is uniquely represented by an infinite matrix with respect to a given basis. To see this, let $\{x_1, x_2, \dots\}$ be an orthonormal basis for $H$. Then, for any $x \in H$, $x = \sum_{i=1}^{\infty} \langle x, x_i \rangle x_i$ and by continuity[3] and linearity, $Ax = \sum_{i=1}^{\infty} \langle x, x_i \rangle Ax_i = y$. Taking the inner product of both sides with $x_j$, $j = 1, 2, \dots$, leads to the system of equations

$$\sum_{i=1}^{\infty} \langle x, x_i \rangle \langle Ax_i, x_j \rangle = \langle y, x_j \rangle, \qquad j = 1, 2, \dots,$$

which can be written in matrix form as

$$\begin{bmatrix} \langle Ax_1, x_1 \rangle & \langle Ax_2, x_1 \rangle & \cdots \\ \langle Ax_1, x_2 \rangle & \langle Ax_2, x_2 \rangle & \cdots \\ \vdots & \vdots & \ddots \end{bmatrix} \begin{bmatrix} \langle x, x_1 \rangle \\ \langle x, x_2 \rangle \\ \vdots \end{bmatrix} = \begin{bmatrix} \langle y, x_1 \rangle \\ \langle y, x_2 \rangle \\ \vdots \end{bmatrix}.$$

Therefore, the bounded linear operator $A : H \to H$ defined by $Ax = y$ is represented in the basis $\{x_1, x_2, \dots\}$ by the infinite matrix $[a_{ij}]$, where $a_{ij} = \langle Ax_j, x_i \rangle$. It can be shown that $\sum_{i=1}^{\infty} |a_{ij}|^2 < \infty$ and $\sum_{j=1}^{\infty} |a_{ij}|^2 < \infty$. Conversely, an infinite matrix $[a_{ij}]$, where $\sum_{i=1}^{\infty} \sum_{j=1}^{\infty} |a_{ij}|^2 < \infty$ defines a bounded[4] linear operator $A : \mathbf{l}^2 \to \mathbf{l}^2$, with respect to a certain basis, with

$$\|A\|^2 \leq \sum_{i=1}^{\infty} \sum_{j=1}^{\infty} |a_{ij}|^2 .$$

The number $\left( \sum_{i=1}^{\infty} \sum_{j=1}^{\infty} |a_{ij}|^2 \right)^{1/2}$ is called the *absolute norm* of the operator $A$.

Similarly, a bounded operator $A : H_1 \to H_2$ defined by $Ax = y$, $D_A \equiv H_1$, where $H_1, H_2$ are infinite-dimensional spaces, is uniquely represented with respect to given bases by an infinite matrix $[a_{ij}]$, with $a_{ij} = \langle Ax_j, y_i \rangle$ such that

$$\begin{bmatrix} \langle Ax_1, y_1 \rangle & \langle Ax_2, y_1 \rangle & \cdots \\ \langle Ax_1, y_2 \rangle & \langle Ax_2, y_2 \rangle & \cdots \\ \vdots & \vdots & \ddots \end{bmatrix} \begin{bmatrix} \langle x, x_1 \rangle \\ \langle x, x_2 \rangle \\ \vdots \end{bmatrix} = \begin{bmatrix} \langle y, y_1 \rangle \\ \langle y, y_2 \rangle \\ \vdots \end{bmatrix},$$

[3] While any linear operator can be interchanged with a finite sum, boundedness (continuity) of $A$ is required to allow the free interchange $A \sum_{i=1}^{\infty} \alpha_i x_i = \sum_{i=1}^{\infty} \alpha_i Ax_i$.

[4] The conditions $\sum_{i=1}^{\infty} |a_{ij}|^2 < \infty$ and $\sum_{j=1}^{\infty} |a_{ij}|^2 < \infty$ are not sufficient for $A$ to be bounded; the condition $\sum_{i=1}^{\infty} \sum_{j=1}^{\infty} |a_{ij}|^2 < \infty$ is a sufficient but not necessary condition for $A$ to be bounded. It can also be shown that the conditions $\sum_{i=1}^{\infty} |a_{ij}| < K$ and $\sum_{j=1}^{\infty} |a_{ij}| < K$, where $K$ does not depend on $i$ or $j$, are also sufficient conditions for $A$ to be bounded.

pp. 181–182]).[25] If $\Omega$ is not bounded, then $A : \mathbf{C}(\Omega) \to \mathbf{C}(\Omega)$ is a bounded operator that is not compact. This is important in the analysis of Fourier integrals and convolution-type operators with the kernel $k(\mathbf{t}, \mathbf{s}) = k(\mathbf{t} - \mathbf{s})$.

10. Consider the operator $A : H^{(n)} \to H^{(n)}$ given by a matrix of operators $A_{ij} : H \to H$ such that

$$A = \begin{bmatrix} A_{11} & A_{12} & \cdots & A_{1n} \\ A_{21} & A_{22} & \cdots & A_{2n} \\ \vdots & \vdots & \ddots & \vdots \\ A_{n1} & A_{n2} & \cdots & A_{nn} \end{bmatrix},$$

where $H^{(n)} = H \oplus H \oplus \cdots \oplus H$ ($n$ times) and $H$ may be infinite-dimensional. The operator $A : H^{(n)} \to H^{(n)}$ is compact if and only if each operator $A_{ij} : H \to H$ is compact. Many problems of applied electromagnetics lead to a coupled system of integral equations where an operator can be represented by a matrix of individual integral operators. Usually, if diagonal operators $A_{ii}$ are compact, then off-diagonal operators $A_{ij}$ $(i \neq j)$ will be compact as well.

## 3.7 Continuity and Compactness of Matrix Operators

In this section we present some criteria governing properties of continuity and compactness of matrix operators defined on sequence spaces. In particular, this becomes useful when a problem is reduced from a function space to a matrix level. There are important theorems for matrix operators that can be directly applied to electromagnetic problems to justify existence and uniqueness of solutions by means of properties of matrix operators. In addition, it is often easier to work with matrix operators than with the corresponding operators on the level of a function space.

[25] Basically, to show that the integral operator $A$ is compact in $\mathbf{C}(a, b)$ it is sufficient to prove that the following conditions are satisfied for its kernel $k(t, s)$:

$$\max_{a \le t \le b} \int_a^b |k(t, s)| \, ds < \infty,$$

$$\lim_{\delta \to 0} \int_a^b |k(t + \delta, s) - k(t, s)| ds = 0, \qquad a \le t \le b.$$

### 3.7.1 Matrix Operators in $\mathbf{l}^2$

A sufficient condition for compactness of $A : \mathbf{l}^2 \to \mathbf{l}^2$ corresponding to the infinite matrix $[a_{ij}]$ is discussed in Section 3.6.2, Example 2,

$$\sum_{i=1}^{\infty}\sum_{j=1}^{\infty} |a_{ij}|^2 < \infty, \tag{3.15}$$

which is a strong criterion in the sense that not every compact operator will satisfy this condition. In general, operators that meet the requirement (3.15) are Hilbert–Schmidt- and Fredholm-type operators with finite absolute norm.

For the special case

$$\sum_{i=1}^{\infty}\sum_{j=1}^{\infty} |a_{ij}|^2 < 1, \tag{3.16}$$

there is a theorem related to solvability of matrix equations of the second kind, $(I - A)\,x = y$ [17].

**Theorem 3.18.** *The infinite system of equations*

$$x_i - \sum_{j=1}^{\infty} a_{ij} x_j = y_i,$$

$i = 1, 2, \ldots,$ *where elements of the matrix* $[a_{ij}]$ *are such that (3.16) is satisfied, has a unique solution* $\xi = (\xi_1, \xi_2, \ldots)$ *in* $\mathbf{l}^2$ *for every* $y = (y_1, y_2, \ldots) \in \mathbf{l}^2$. *The truncated system of equations*

$$x_i - \sum_{j=1}^{n} a_{ij} x_j = y_i,$$

$i = 1, 2, \ldots, n,$ *has a unique solution* $(x_1^{(n)}, \ldots, x_n^{(n)})$ *and, furthermore,* $(x_1^{(n)}, \ldots, x_n^{(n)}, 0, 0, \ldots)$ *converges to* $\xi$ *in* $\mathbf{l}^2$ *as* $n \to \infty$.

In problems of applied electromagnetics where the solution is restricted by a finite energy condition, such that the sequence space $\widetilde{\mathbf{l}}^2$ is encountered for the coefficients in an eigenfunction expansion (see Example 2 in Section 2.1.3), a sufficient condition for compactness of a matrix operator $A : \widetilde{\mathbf{l}}^2 \to \widetilde{\mathbf{l}}^2$ can be written as [18], [19]

$$\sum_{i=1}^{\infty}\sum_{j=1}^{\infty} |a_{ij}|^2 \frac{|i|}{|j|} < \infty.$$

Other conditions for continuity and compactness are discussed in [20]. For example, consider a matrix operator $A : \mathbf{l}^2 \to \mathbf{l}^2$ corresponding to an infinite matrix of elements $[a_{ij}]$. Then

$$||Ax||^2 = \sum_{i=1}^{\infty} a_{ij}x_j\bar{a}_{ik}\bar{x}_k \leq \sup_{1\leq k<\infty} \sum_{j=1}^{\infty}\sum_{i=1}^{\infty} |a_{ij}a_{ik}|||x||^2 = c||x||^2,$$

where $c = \sup_{1\leq k<\infty} \sum_{j=1}^{\infty}\sum_{i=1}^{\infty} |a_{ij}a_{ik}|$. The operator $A : \mathbf{l}^2 \to \mathbf{l}^2$ is bounded (continuous) if $||A|| \leq c$, which results in the condition

$$||A|| \leq \left( \sup_{1\leq k<\infty} \sum_{j=1}^{\infty}\sum_{i=1}^{\infty} |a_{ij}a_{ik}| \right)^{1/2} < \infty. \tag{3.17}$$

The matrix operator $A : \mathbf{l}^2 \to \mathbf{l}^2$ is completely continuous (compact) if, in addition to (3.17), we also have

$$||A|| = \lim_{s\to\infty} \sup_{1\leq k<\infty} \sum_{j=1}^{\infty}\sum_{i=s}^{\infty} |a_{ij}a_{ik}| = 0. \tag{3.18}$$

The following lemma gives a criterion for boundedness of $A : \mathbf{l}^2 \to \mathbf{l}^2$ [20].

**Lemma 3.19.** *(Schur) Assume that* $[a_{ij}] = [\overline{a_{ji}}]$ *and* $\sup_i \sum_{j=1}^{\infty} |a_{ij}| \leq c$. *Then the matrix operator* $A : \mathbf{l}^2 \to \mathbf{l}^2$ *is bounded, such that* $||A|| \leq c$.

**Proof.** For a symmetric operator $Ax = A^*x$ it can be shown that

$$|\langle Ax, x\rangle| = \left| \sum_{i=1}^{\infty}\sum_{j=1}^{\infty} a_{ij}x_j\overline{x_i} \right| \leq \sup_{1\leq i<\infty} \sum_{j=1}^{\infty} |a_{ij}|||x||^2 \leq c||x||^2.$$

Therefore, $||A|| \leq c$. ■

The criteria for continuity and compactness of matrix operators in $\mathbf{l}^2$ discussed here have been used in the analysis of various electromagnetic problems leading to an infinite system of linear algebraic equations of the second kind. See, for example, [21]–[29], among others.

In the next section we summarize some useful theorems for continuity and compactness of matrix operators acting in different sequence spaces. Some applications of these theorems are shown in Section 9.5.

### 3.7.2 Gribanov's Theorems

According to [30], we can define a class of continuous matrix operators $S(\mathbf{l}^1 \to \mathbf{l}^2)$ that continuously map the sequence space $\mathbf{l}^1$ into $\mathbf{l}^2$. Within $S(\mathbf{l}^1 \to \mathbf{l}^2)$ we define a class of completely continuous (compact) operators $V(\mathbf{l}^1 \to \mathbf{l}^2)$, such that $V(\mathbf{l}^1 \to \mathbf{l}^2) \subset S(\mathbf{l}^1 \to \mathbf{l}^2)$. For a matrix operator $A$

corresponding to an infinite matrix with elements $[a_{ij}]$, we define a matrix operator $A_{ii}$ as

$$A_{ii} \equiv \begin{pmatrix} a_{11} & a_{12} & \dots & a_{1i} & 0 & 0 & \dots \\ a_{21} & a_{22} & \dots & a_{2i} & 0 & 0 & \dots \\ \dots & \dots & \dots & \dots & \dots & \dots & \dots \\ a_{i1} & a_{i2} & \dots & a_{ii} & 0 & 0 & \dots \\ 0 & 0 & \dots & 0 & 0 & 0 & \dots \\ 0 & 0 & \dots & 0 & 0 & 0 & \dots \\ \dots & \dots & \dots & \dots & \dots & \dots & \dots \end{pmatrix}.$$

**Definition 3.24.** *A matrix operator $A \in S(\mathbf{l}^1 \to \mathbf{l}^2)$ is called $\omega$-continuous if*

$$\lim_{i \to \infty} ||A - A_{ii}|| = 0.$$

*The class of $\omega$-continuous operators is denoted by $V(\omega, \mathbf{l}^1 \to \mathbf{l}^2)$, and $V(\omega, \mathbf{l}^1 \to \mathbf{l}^2) \subseteq V(\mathbf{l}^1 \to \mathbf{l}^2)$.*

Classes of continuous $S(\mathbf{l}^1 \to \mathbf{l}^2)$, $\omega$-continuous $V(\omega, \mathbf{l}^1 \to \mathbf{l}^2)$, and completely continuous $V(\mathbf{l}^1 \to \mathbf{l}^2)$ matrix operators can be generalized to $S(\mathbf{l}^1 \to \mathbf{l}^p)$, $V(\omega, \mathbf{l}^1 \to \mathbf{l}^p)$, and $V(\mathbf{l}^1 \to \mathbf{l}^p)$, $p \geq 1$, respectively, and other sequence spaces used in the theorems to follow. Below we summarize some of the theorems presented in [30] for continuity and $\omega$-continuity of matrix operators acting in certain sequence spaces.[26]

**Theorem 3.20.** *If a matrix operator $A$ is $\omega$-continuous from $\mathbf{l}^1$ into $\mathbf{l}^p$ for $p \geq 1$, i.e., $A \in V(\omega, \mathbf{l}^1 \to \mathbf{l}^p)$, then $A$ is a completely continuous operator from $\mathbf{l}^1$ into $\mathbf{l}^p$, $A \in V(\mathbf{l}^1 \to \mathbf{l}^p)$, and $V(\omega, \mathbf{l}^1 \to \mathbf{l}^p) \subset V(\mathbf{l}^1 \to \mathbf{l}^p)$.*

*(a) A matrix operator $A$ is continuous from $\mathbf{l}^1$ into $\mathbf{l}^p$, $A \in S(\mathbf{l}^1 \to \mathbf{l}^p)$, if and only if*

$$||A|| = \sup_{1 \leq j < \infty} \left( \sum_{i=1}^{\infty} |a_{ij}|^p \right)^{1/p} < \infty.$$

*(b) A matrix operator $A$ is $\omega$-continuous from $\mathbf{l}^1$ into $\mathbf{l}^p$, $A \in V(\omega, \mathbf{l}^1 \to$*

[26] A criterion for compactness of a matrix operator $A$ acting in $\mathbf{l}^2$ has been introduced by Gilbert in [31],

$$\lim_{k \to \infty} \sup_{\sum_{j=1}^{\infty} x_j^2 \leq 1} \sup_{\sum_{i=1}^{\infty} y_i^2 \leq 1} \left| \sum_{i=1}^{\infty} \sum_{j=1}^{\infty} a_{ij} x_j y_i - \sum_{i=1}^{k} \sum_{j=1}^{k} a_{ij} x_j y_i \right| = 0,$$

and was generalized by Gribanov [30] for a class of $\omega$-continuous matrix operators in arbitrary sequence spaces.

$\mathbf{l}^p$), *if and only if* [27]

$$||A|| = \lim_{k\to\infty} \sup_{k\le j<\infty} \sum_{i=1}^{\infty} |a_{ij}|^p = 0.$$

The next theorem concerns properties of matrix operators in the space of bounded sequences $\mathbf{l}^\infty$ [32], [30].

**Theorem 3.21.** *If a matrix operator $A$ is $\omega$-continuous from $\mathbf{l}^\infty$ into $\mathbf{l}^\infty$, $A \in V(\omega, \mathbf{l}^\infty \to \mathbf{l}^\infty)$, then $A$ is a completely continuous operator from $\mathbf{l}^\infty$ into $\mathbf{l}^\infty$, $A \in V(\mathbf{l}^\infty \to \mathbf{l}^\infty)$, and $V(\omega, \mathbf{l}^\infty \to \mathbf{l}^\infty) \subset V(\mathbf{l}^\infty \to \mathbf{l}^\infty)$.*

*(a) A matrix operator $A$ is continuous from $\mathbf{l}^\infty$ into $\mathbf{l}^\infty$, $A \in S(\mathbf{l}^\infty \to \mathbf{l}^\infty)$, if and only if*

$$||A|| = \sup_{1\le i<\infty} \sum_{j=1}^{\infty} |a_{ij}| < \infty.$$

*(b) A matrix operator $A$ is $\omega$-continuous from $\mathbf{l}^\infty$ into $\mathbf{l}^\infty$, $A \in V(\omega, \mathbf{l}^\infty \to \mathbf{l}^\infty)$, if and only if*

$$||A|| = \lim_{k\to\infty} \sup_{k\le i<\infty} \sum_{j=1}^{\infty} |a_{ij}| = 0.$$

The next theorem gives criteria for matrix operators acting from $\mathbf{l}^1$ into $\mathbf{l}^\infty$.

**Theorem 3.22.** *(a) In order for a matrix operator $A$ to be a continuous operator from $\mathbf{l}^1$ into $\mathbf{l}^\infty$, $A \in S(\mathbf{l}^1 \to \mathbf{l}^\infty)$, it is necessary and sufficient that*

$$||A|| = \sup_{1\le i,j<\infty} |a_{ij}| < \infty.$$

---

[27] A particular case for the compactness of $A : \mathbf{l}^1 \to \mathbf{l}^1$ is discussed in [17], such that

$$||A|| = \lim_{k\to\infty} \sup_{k\le j<\infty} \sum_{i=1}^{\infty} |a_{ij}| = 0. \tag{3.19}$$

Note that Gribanov's theorems for continuity and $\omega$-continuity presented in this section, and the above criterion (3.19) for compactness of $A : \mathbf{l}^1 \to \mathbf{l}^1$, represent necessary and sufficient conditions for continuity and compactness of matrix operators acting in certain sequence spaces. Similar to the sufficient condition (strong condition) for compactness of $A : \mathbf{l}^2 \to \mathbf{l}^2$ (3.15), there is a sufficient condition for compactness of $A : \mathbf{l}^1 \to \mathbf{l}^1$,

$$\sum_{i=1}^{\infty} \sum_{j=1}^{\infty} |a_{ij}| < \infty. \tag{3.20}$$

This condition is much stronger than (3.19), and not every compact matrix operator $A$: $\mathbf{l}^1 \to \mathbf{l}^1$ satisfies condition (3.20).

*(b) In order for a matrix operator $A$ to be an $\omega$-continuous operator from $\mathbf{l}^1$ into $\mathbf{l}^\infty$, $A \in V(\omega, \mathbf{l}^1 \to \mathbf{l}^\infty)$, it is necessary and sufficient that*

$$\lim_{k\to\infty} \sup_{1\le j<\infty} \sup_{k\le i<\infty} |a_{ij}| = 0,$$

$$\lim_{k\to\infty} \sup_{1\le i<\infty} \sup_{k\le j<\infty} |a_{ij}| = 0.$$

Finally, there is a theorem for continuity and $\omega$-continuity of matrix operators acting from $\mathbf{l}^p$ into $\mathbf{l}^\infty$, $p > 1$.

**Theorem 3.23.** *If a matrix operator $A$ is $\omega$-continuous from $\mathbf{l}^p$ into $\mathbf{l}^\infty$, $A \in V(\omega, \mathbf{l}^p \to \mathbf{l}^\infty)$, then $A$ is a completely continuous operator from $\mathbf{l}^p$ into $\mathbf{l}^\infty$, $A \in V(\mathbf{l}^p \to \mathbf{l}^\infty)$, and $V(\omega, \mathbf{l}^p \to \mathbf{l}^\infty) \subset V(\mathbf{l}^p \to \mathbf{l}^\infty)$.*

*(a) A matrix operator $A$ is continuous from $\mathbf{l}^p$ into $\mathbf{l}^\infty$, $A \in S(\mathbf{l}^p \to \mathbf{l}^\infty)$, if and only if*

$$||A|| = \sup_{1\le i<\infty} \left( \sum_{j=1}^{\infty} |a_{ij}|^q \right)^{1/q} < \infty,$$

*where $q$ is given by the relationship $1/p + 1/q = 1$.*

*(b) A matrix operator $A$ is an $\omega$-continuous operator from $\mathbf{l}^p$ into $\mathbf{l}^\infty$, $A \in V(\omega, \mathbf{l}^p \to \mathbf{l}^\infty)$, if and only if*

$$||A|| = \lim_{k\to\infty} \sup_{k\le i<\infty} \sum_{j=1}^{\infty} |a_{ij}|^q = 0.$$

The properties of continuity and complete continuity (compactness) of matrix operators based on Gribanov's theorems, in connection with problems of uniqueness and solvability of infinite systems of linear equations, were extensively studied and reported in the Russian and Ukrainian literature; see, for example, [33]–[38].

## 3.8 Closed and Closable Operators

In this brief section we discuss, for completeness, the concept of closed and closable operators.

**Definition 3.25.** *The operator $A : H_1 \to H_2$ acting on a domain $D_A$ is said to be closed if, for $\{x_n\} \subset D_A$, $x_n \to x$ and $y_n = Ax_n \to y$, then $x \in D_A$ and $y = Ax$.*

In some sense, the property of being closed is a weak form of continuity. For instance, let $A$ be continuous and let $x_n \to x$, i.e., $\lim_{n\to\infty} x_n = x$. Then, $Ax = A \lim_{n\to\infty} x_n = \lim_{n\to\infty} Ax_n$, such that we can apply $A$ term by term (we can interchange $A$ and the limit operator), and the resulting sequence will converge. Now, assume that $A$ is merely closed and that $x_n \to x$. If $A$ is applied term by term and the resulting sequence $Ax_n$ converges to something (it need not), then $Ax = A \lim_{n\to\infty} x_n = \lim_{n\to\infty} Ax_n$.

To summarize, we can freely interchange continuous operators and limits. We can interchange closed operators and limits as long as the sequence obtained by the term-by-term action of the operator converges. It can be shown that adjoint operators are always closed, from which it follows that self-adjoint operators are always closed.

Alternatively, an operator $A : H_1 \to H_2$ is closed if its *graph,* defined to be the set of pairs

$$\Gamma(A) = \{(x, Ax) : x \in D_A\},$$

is a closed subspace of $H_1 \oplus H_2$.

If $\Gamma(A)$ is not closed, then it may have a closed extension (the requirement is that $\widetilde{\Gamma(A)}$ be the graph of an operator). In this case the operator is said to be *closable.* It can be shown that symmetric operators are closable. Furthermore, every continuous, i.e., bounded, linear operator is closed, but not every closed linear operator is continuous. For example, derivative operators are often closed, but are not bounded.

Some properties related to the concept of a closed operator are the following.

- A closed operator on a closed domain is bounded.
- A closed operator does not necessarily have a closed domain or range.
- The null space of a closed operator is closed.

## 3.9 Invertible Operators

### 3.9.1 General Concepts

The idea of an *operator inverse* arises naturally from the operator equation $Ax = b$, where one wants to solve for $x$ given $A$ and $b$. Division by an operator is not defined, but if $A$ possesses a continuous inverse $A^{-1}$, then this equation has a unique solution, $x = A^{-1}b$, and this solution is a continuous function of $b$. Although in practice this is not the method by which one would numerically solve an equation of the form $Ax = b$, the concept of an inverse is a very important analytical tool in the theory of linear operators, both on finite-dimensional and infinite-dimensional spaces.

**Definition 3.26.** *An operator $A : H_1 \to H_2$ is said to be invertible if there exists an operator $A^{-1} : H_2 \to H_1$ such that*

$$A^{-1}Ax = x,$$

*for all* $x \in H_1$, *and*

$$AA^{-1}y = y$$

*for all* $y \in H_2$. *The operator* $A^{-1}$ *is called the inverse of* $A$.

Note that $A^{-1}A = I_1$, the identity operator of $H_1$, and $AA^{-1} = I_2$, the identity operator of $H_2$. It is easy to see that if an inverse exists, it is unique.[28]

We can also observe that if $A : H_1 \to H_2$ is invertible, with the inverse given by $A^{-1} : H_2 \to H_1$, then $A^{-1} : H_2 \to H_1$ is invertible, with the inverse being $A : H_1 \to H_2$.

Next we characterize some properties relating to invertible operators.

**Definition 3.27.** *An operator* $A : H_1 \to H_2$ *is said to be one-to-one (injective) if no two distinct elements of* $H_1$ *are mapped to the same element of* $H_2$.

This states that $A : H_1 \to H_2$ is one-to-one if and only if $Ax_1 \neq Ax_2$ whenever $x_1 \neq x_2$ for $x_1, x_2 \in H_1$.

The next theorem presents an important statement concerning one-to-one linear operators.

**Theorem 3.24.** *A linear operator* $A : H_1 \to H_2$ *is one-to-one if and only if* $N_A = \{0\}$.

**Proof.** Assume $A$ is one-to-one and let $x \in N_A$. Then $Ax = 0$, but $A0 = 0$, and since we assumed $A$ is one-to-one, then $x = 0$ and so $N_A = \{0\}$. Now assume $N_A = \{0\}$. We need to show that if two points $x_1, x_2 \in H_1$ map into the same $y \in H_2$, then $x_1 = x_2$. Indeed, $Ax_1 = Ax_2$ implies $A(x_1 - x_2) = 0$, which in turn implies $(x_1 - x_2) \in N_A = \{0\}$. Therefore, $x_1 = x_2$. ■

If $N_A = \{0\}$ then $A$ is said to have a *trivial null space.*

It seems obvious that if an operator is not one-to-one then it cannot be (uniquely) invertible. Indeed, one-to-oneness is a necessary condition for an operator to have an inverse. Necessary and sufficient conditions for $A$ to be invertible are given next.

**Theorem 3.25.** *An operator* $A : H_1 \to H_2$ *is invertible if and only if it is one-to-one and onto* ($R_A = H_2$).

Such an operator, which is one-to-one (injective) and onto (surjective), is said to be *bijective.* From the above it can be seen that if $A : H_1 \to H_2$ is onto, then it is invertible if and only if $Ax = 0$ implies $x = 0$.

[28] Let two inverses of an operator $A$ be denoted by $A_1^{-1}$ and $A_2^{-1}$, and $y$ be any point of $H_2$. Then $A_1^{-1}y = A_1^{-1}I_2y = A_1^{-1}AA_2^{-1}y = I_1A_2^{-1}y = A_2^{-1}y$.

For matrix operators we have the following useful criterion. A matrix operator $A : \mathbf{C}^n \to \mathbf{C}^n$ having entries $a_{ij} \in \mathbf{C}$ is said to be *strictly diagonally dominant* if

$$\sum_{\substack{j=1 \\ j \neq i}}^{n} |a_{ij}| < |a_{ii}|$$

for $i = 1, 2, \ldots, n$. In this case we have the following theorem.

**Theorem 3.26.** *If $A : \mathbf{C}^n \to \mathbf{C}^n$ is strictly diagonally dominant, then $A^{-1} : \mathbf{C}^n \to \mathbf{C}^n$ exists.*

**Proof.** Let $A : \mathbf{C}^n \to \mathbf{C}^n$ be a strictly diagonally dominant matrix, and suppose that the inverse does not exist. Then, there exists $\mathbf{0} \neq \mathbf{x} \in \mathbf{C}^n$ such that $A\mathbf{x} = \mathbf{0}$. We therefore have $\sum_{j=1}^{n} a_{mj}x_j = 0$ for any $1 \leq m \leq n$. Let $x_m = \max_{1 \leq i \leq n} |x_i|$, and write $a_{mm}x_m = -\sum_{\substack{j=1 \\ j \neq m}}^{n} a_{mj}x_j$. Then, $|a_{mm}|\,|x_m| = |\sum_{\substack{j=1 \\ j \neq m}}^{n} a_{mj}x_j| \leq |x_m| \sum_{\substack{j=1 \\ j \neq m}}^{n} |a_{mj}|$, which contradicts the assumption on diagonal dominance. ■

## Basic Properties of Inverses

An important theorem in functional analysis is the following for bounded operators, which is typically cast in terms of Banach spaces.

**Theorem 3.27.** *(Open mapping theorem) A continuous bijection of one Banach space onto another Banach space has a continuous inverse.*

A merely one-to-one mapping (not necessarily onto) would perhaps seem to be invertible, yet the inverse can act on elements of $H_2$ that can not be "reached" from $H_1$. One remedy of this problem is the restriction $D_{A^{-1}} = R_A$. Such an operator is intrinsically one-to-one and onto. In the following, when we state that the operator $A : H_1 \to H_2$ is invertible, we implicitly assume either that $A$ is onto or that we are defining the domain of the inverse in a meaningful way. In this way the convenient conclusion is that *an operator $A$ is invertible if and only if $Ax = 0$ implies $x = 0$.* Note though that unless $D_{A^{-1}} = R_A = H_2$, the open mapping theorem does not necessarily apply.

**Theorem 3.28.** *If $A : H_1 \to H_2$ is an invertible linear operator, then $A^{-1} : H_2 \to H_1$ is a linear operator.*

**Proof.** Let $y_1 = Ax_1$, $y_2 = Ax_2$, where $x_1, x_2 \in H_1$. Then $A^{-1}(y_1 + y_2) = A^{-1}(Ax_1 + Ax_2) = A^{-1}A(x_1 + x_2) = x_1 + x_2 = A^{-1}y_1 + A^{-1}y_2$. Similarly, $A^{-1}(\alpha y_1) = A^{-1}(\alpha Ax_1) = A^{-1}A(\alpha x_1) = \alpha x_1 = \alpha A^{-1}y_1$. ■

**Theorem 3.29.** *If $A : H_1 \to H_2$ is invertible and $\{x_1, x_2, \ldots, x_n\} \subset H_1$ is a linearly independent set, then $\{Ax_1, Ax_2, \ldots, Ax_n\} \subset H_2$ is a linearly independent set.*

**Proof.** Assume $\alpha_1 Ax_1 + \alpha_2 Ax_2 + \cdots + \alpha_n Ax_n = 0$; then $A(\alpha_1 x_1 + \alpha_2 x_2 + \cdots + \alpha_n x_n) = 0$. Because $A$ is invertible, $(\alpha_1 x_1 + \alpha_2 x_2 + \cdots + \alpha_n x_n) = 0$, which implies $\alpha_1 = \alpha_2 = \cdots = \alpha_n = 0$ since $\{x_n\}$ is a linearly independent set. Then we see that $\{Ax_1, Ax_2, \ldots, Ax_n\}$ are linearly independent. ■

**Theorem 3.30.** *If $A : H_1 \to H_2$ and $B : H_2 \to H_3$ are invertible, then $BA : H_1 \to H_3$ is invertible with $(BA)^{-1} = A^{-1}B^{-1}$.*

**Proof.** If $A$ and $B$ are invertible, $(A^{-1}B^{-1})(BA) = A^{-1}(B^{-1}B)A = A^{-1}I_2 A = A^{-1}A = I_1$. Then clearly $BA$ is invertible ($(BA)^{-1}(BA) = I_1$ and $(BA)(BA)^{-1} = I_3$) with $(BA)^{-1} = A^{-1}B^{-1}$. ■

**Theorem 3.31.** *If $A$ is a positive definite invertible operator, $A^{-1}$ is positive definite.*

**Proof.** Consider $Ax = y$ for all $y \in D_{A^{-1}}$ and some $x \in H$. Then $\langle A^{-1}y, y\rangle = \langle A^{-1}Ax, Ax\rangle = \langle x, Ax\rangle \geq 0$ (since $A$ is positive definite). ■

**Theorem 3.32.** *If $A$ is closed and $A^{-1}$ exists, then $A^{-1}$ is closed.*

**Proof.** Consider a sequence $\{y_n\} \subset R_A$ such that $\lim_{n\to\infty} y_n = y$ and $\lim_{n\to\infty} x_n = \lim_{n\to\infty} A^{-1}y_n = x$. To prove that $A^{-1}$ is closed we need to show that $y \in R_A$ and $A^{-1}y = x$. Since $A$ is closed, $\lim_{n\to\infty} y_n = y$, and $\lim_{n\to\infty} x_n = x$, then $x \in D_A$ and $Ax = y$. Therefore, $y \in R_A$ and $A^{-1}y = x$ (the inverse of $A$ exists). ■

**Theorem 3.33.** *If $A$ is closed and $A^{-1}$ exists, then $R_A$ is closed if and only if $A^{-1}$ is bounded.*

**Proof.** First assume that $A^{-1}$ is bounded and prove that $R_A$ is closed. Consider a sequence $\{y_n\} \subset R_A$ such that $\lim_{n\to\infty} y_n = y$. The fact that $A^{-1}$ is bounded shows that $x_n = A^{-1}y_n$ is a Cauchy sequence such that $\lim_{n\to\infty} x_n = x$ with $\{x_n\} \subset D_A$ and $\lim_{n\to\infty} Ax_n = y$. Since $A$ is closed (according to the definition, $x \in D_A$ and $Ax = y$), this leads to the conclusion that $y \in R_A$ and therefore $R_A$ is closed.

Now we assume that $R_A$ is closed and prove that $A^{-1}$ is bounded. The proof is based on the fact that a closed operator on a closed domain is bounded [12, p. 332]. Since $A^{-1}$ is closed (from the above theorem) and $R_A$ is closed, then $A^{-1}$ is bounded. ■

**Theorem 3.34.** *Let $A : H_1 \to H_2$ be a bounded operator that is onto ($R_A = H_2$). If $A$ has a bounded inverse, then $A^*$ is invertible with $(A^*)^{-1} = (A^{-1})^*$.*

**Proof.** Comparing the following inner products for any $x \in H_1$ and $y \in H_2$,

$$\left\langle y, \left(A^{-1}\right)^* A^* x\right\rangle = \left\langle A^{-1}y, A^* x\right\rangle = \left\langle AA^{-1}y, x\right\rangle = \langle y, x\rangle$$

and

$$\left\langle y, A^*(A^{-1})^* x\right\rangle = \left\langle Ay, (A^{-1})^* x\right\rangle = \left\langle A^{-1}Ay, x\right\rangle = \langle y, x\rangle,$$

we conclude that $\left(A^{-1}\right)^* A^* x = A^*(A^{-1})^* x = x$; hence $(A^*)^{-1} = \left(A^{-1}\right)^*$. ■

**Theorem 3.35.** *If $A$ is a bounded self-adjoint operator, and $A^{-1}$ is a bounded operator, then $A^{-1}$ is self-adjoint.*

**Proof.** The proof simply follows from the previous theorem: $\left(A^{-1}\right)^* = (A^*)^{-1} = A^{-1}$. ■

The next theorem is quite important, since it states necessary and sufficient conditions for an operator to have a bounded inverse.

**Theorem 3.36.** *A linear operator $A : H_1 \to H_2$ possesses a bounded inverse defined on $R_A$, $A^{-1} : R_A \to H_1$, if and only if there exists $m > 0$ such that $\|Ax\| \geq m \|x\|$ for all $x \in H_1$. Furthermore, $\left\|A^{-1}y\right\| \leq (1/m) \|y\|$.*

Such operators $A$ are called bounded below operators (sometimes called *bounded away from zero*) (also see Section 3.5 for bounded below and above operators). Note that $A$ itself does not need to be bounded.

**Proof.** Assume $\|Ax\| \geq m \|x\|$ is true and show $A$ is invertible with a bounded inverse. If $Ax = 0$, then $\|0\| = 0 \geq m \|x\|$, so $\|x\| \leq 0$, but from the properties of a norm $\|x\| \geq 0$, so $\|x\| = 0$, which implies $x = 0$. Since the null space is trivial, then $A$ is one-to-one, and since we are defining the domain of the inverse as the range of $A$ (inherently onto), then $A$ is invertible. To show the inverse is bounded, let $x = A^{-1}y$, which leads to $\|Ax\| = \left\|AA^{-1}y\right\| = \|y\| \geq m \|x\| = m \left\|A^{-1}y\right\|$, or $\left\|A^{-1}y\right\| \leq (1/m) \|y\|$, and so $A^{-1}$ is bounded. Now assume $A$ possesses a bounded inverse and show there exists an $m > 0$ such that $\|Ax\| \geq m \|x\|$. Indeed, if $A^{-1}$ exists and is bounded, then $\left\|A^{-1}y\right\| = \|x\| \leq w \|y\| = w \|Ax\|$ or $\|Ax\| \geq (1/w) \|x\|$. ■

Since every linear operator on a finite-dimensional space is bounded, then if $A : H^n \to H^n$ is invertible, where $H^n$ is finite-dimensional, then $A^{-1}$ is bounded.

### Well Posed Problems

Bounded below operators are important in defining what are known as *well posed problems.* The operator equation $Ax = b$ is called well posed if there exists a unique solution $x$ that depends continuously on the "data" $b$. It is obvious what role existence and uniqueness play in a problem being well posed. The requirement that the solution be a continuous function of the data is important because one does not want small errors in the data to result in large errors in the solution.

If a problem is not well posed it is said to be *ill posed.* When $A$ is bounded below, it possesses a continuous (bounded) inverse, such that $x = A^{-1}b$ depends continuously on $b$, and the problem is well posed [55] (see also Section 3.11).

The next theorem shows that compact operators on infinite-dimensional spaces are not bounded below and hence do not have a continuous inverse.

**Theorem 3.37.** *If $A : H \to H$ for $H$ an infinite-dimensional space is compact and invertible, then its inverse is unbounded.*

**Proof.** Let $\{x_n\}$ be an infinite orthonormal sequence in $H$. If $A : H \to H$ is compact, then $\lim_{n\to\infty} Ax_n = 0$ (see Section 3.6) and, therefore, the operator $A : H \to H$ is not bounded below. Based on Theorem 3.36, we conclude that the inverse operator $A^{-1} : H \to H$, where $A$ is compact, is unbounded. ■

Therefore, while in the "forward direction" compact operators have a "nice" effect, in the "inverse direction" just the opposite is true. An analogy is that integration tends to smooth out functions, while the inverse operation, differentiation, tends to make functions less smooth. However, on a finite-dimensional space the inverse of a compact operator is bounded, by Theorem 3.5.

Examples of the above relevant to electromagnetics are first-kind compact integral operators with Green's function kernels, where the inverse is a differential (unbounded) operator associated with a wave equation.

### Resolvent Operators

An operator with very desirable properties in both the forward and inverse directions is the operator having the form $A - \lambda I$, where $A$ is compact and $\lambda \neq 0$ (recall that $I$ is not compact). In preparation for examining this type of operator, we state the following for $A - \lambda I$, with $A$ not necessarily compact.

**Theorem 3.38.** *The operator $A - \lambda I$, with $A : H \to H$ bounded and $|\lambda| > \|A\|$, is invertible with bounded inverse. Furthermore,*

$$(A - \lambda I)^{-1} = -\sum_{n=0}^{\infty} \frac{1}{\lambda^{n+1}} A^n$$

*and*

$$\left\|(A-\lambda I)^{-1}\right\| \leq (|\lambda| - \|A\|)^{-1}.$$

The above is known as the *Neumann expansion* for $(A-\lambda I)^{-1}$. In general, the operator $(A-\lambda I)^{-1}$ is known as the *resolvent operator.* Before proving the above, it is convenient to prove the following.

**Theorem 3.39.** *If $A : H \to H$ is a bounded operator such that $\|A\| < 1$, then $(A-I)^{-1}$ exists and is bounded on $H$, and*

$$(A-I)^{-1} = -\sum_{n=0}^{\infty} A^n,$$

*which is convergent in norm. Furthermore, $\left\|(A-I)^{-1}\right\| \leq (1-\|A\|)^{-1}$.*

**Proof.** The series $\sum_{n=0}^{\infty} \|A\|^n$ converges for $\|A\| < 1$ (geometric series, $\sum_{n=0}^{\infty} \|A\|^n = (1-\|A\|)^{-1}$). Because $\|A^n\| \leq \|A\|^n$, then the series $\sum_{n=0}^{\infty} \|A^n\|$ converges. When a series of the form $\sum_n \|x\|$ converges, the series $\sum_n x$ is called *absolutely convergent*. As discussed on p. 133, the space of all bounded operators on $H$ is itself a complete (Banach) space, and in a complete space absolute convergence implies convergence in norm. Thus, $\sum_{n=0}^{\infty} A^n$ converges in norm. To show that the inverse is given by $-\sum_{n=0}^{\infty} A^n$, assume this is the case and form $(A-I)(A-I)^{-1} = -(A-I)\left(I + A + A^2 + \cdots + A^n\right) = -\left(A^{n+1} - I\right)$. As $n \to \infty$, $A^{n+1} \to 0$ since $\|A\| < 1$, proving the relationship. Finally, we have $\|(A-I)^{-1}\| = \|\sum_{n=0}^{\infty} A^n\| \leq \sum_{n=0}^{\infty} \|A\|^n = (1-\|A\|)^{-1}$. ■

The above theorem also holds for $A : X \to X$, with $X$ being a Banach space. Now, by forming $(A-\lambda I)^{-1} = -(1/\lambda)\left(I - \frac{1}{\lambda}A\right)^{-1}$ we see from the last theorem that the series converges if $\|(1/\lambda)A\| < 1$, or $\|A\| < \lambda$, proving Theorem 3.38.

For $A$ compact we have the following.

**Theorem 3.40.** *For $A : H \to H$ a compact operator and $\lambda \neq 0$, where $\lambda$ is not an eigenvalue of $A$, then the operator $(A-\lambda I)^{-1}$ is bounded.*

One proof of this theorem follows from Theorem 4.13.

In particular, the case $\lambda = 1$ often arises in the theory of integral equations, where one finds that these so-called "identity plus compact" operators have very desirable properties.

### Right and Left Inverses

Sometimes an operator is not invertible in the sense described above, but may possess what is known as a *right inverse* or a *left inverse*. These concepts are not often utilized in applied electromagnetics, but a study of these operators helps to clarify the geometrical structure of invertible linear operators on Hilbert spaces.

**Definition 3.28.** *An operator* $A : H_1 \to H_2$ *is said to be left invertible if an operator* $A_L^{-1} : H_2 \to H_1$ *exists such that*

$$A_L^{-1} A = I_1,$$

*where* $I_1$ *is the identity operator on* $H_1$*. The operator* $A_L^{-1} : H_2 \to H_1$ *is said to be the left inverse of* $A : H_1 \to H_2$*.*

**Theorem 3.41.** *An operator* $A : H_1 \to H_2$ *is left invertible if and only if it is one-to-one.*

**Proof.** First show that $A$ is left invertible if it is one-to-one. Indeed, assuming $A$ is one-to-one, then $A : H_1 \to R_A$ is invertible with $A^{-1} : R_A \to H_1$. Let $A_L^{-1} : H_2 \to H_1$ be any extension of $A^{-1}$. Then $A_L^{-1} A = I_1$. Now assume that $A$ is left invertible and show it is one-to-one. Assume $Ax_1 = Ax_2$. Then $x_1 = A_L^{-1}(Ax_2) = A_L^{-1} A x_2 = x_2$. ■

It can be seen from the proof that the left inverse is not necessarily unique, since the extension is not necessarily unique. If the operator $A$ is onto, then $A_L^{-1}$ is unique.

**Definition 3.29.** *An operator* $A : H_1 \to H_2$ *is said to be right invertible if an operator* $A_R^{-1} : H_2 \to H_1$ *exists such that*

$$A A_R^{-1} = I_2,$$

*where* $I_2$ *is the identity operator on* $H_2$*. The operator* $A_R^{-1} : H_2 \to H_1$ *is said to be the right inverse of* $A : H_1 \to H_2$*.*

**Theorem 3.42.** *An operator* $A : H_1 \to H_2$ *is right invertible if and only if it is onto* ($R_A = H_2$)*.*

**Proof.** It is easy to see that $A$ is right invertible if it is onto. Now assume that $A$ is right invertible and show it is onto. Assuming $y \in H_2$, then $y = A A_R^{-1} y = Ax$, so that $y$ is in the range of $A$. Because $y \in H_1$, $y \in R_A$, and $y$ is arbitrary, then $R_A = H_2$. ■

The right inverse is not necessarily unique, but if the operator is one-to-one then the right inverse is unique.

**Theorem 3.43.** *If an operator* $A : H_1 \to H_2$ *is both right and left invertible, then the operator is invertible with* $A^{-1} = A_L^{-1} = A_R^{-1}$*.*

**Proof.** The proof is trivial since if $A$ has both a right and left inverse it is one-to-one and onto and hence invertible. ■

### 3.9.2 Examples Relating to Inverses

1. The identity operator ($Ix \equiv x$) on an infinite-dimensional space is invertible (bounded, yet not compact) with $I^{-1} = I$. The identity operator on a finite-dimensional space is bounded, compact, and invertible. The zero operator ($0x \equiv 0$) is not invertible.

2. The operator $A : \mathbf{l}^2 \to \mathbf{l}^2$ corresponding to the infinite matrix $[a_{ij}]$, where $\sum_{i=1}^{\infty}\sum_{j=1}^{\infty}|a_{ij}|^2 < \infty$, is bounded and compact. Then, if $A$ is invertible, $A^{-1}$ is unbounded. Note also that by Theorem 3.40, $(I - A)$ is invertible and the inverse operator $(I - A)^{-1}$ is bounded. This result is used in the solution of second-kind infinite systems of linear equations.

3. The operator $A : \mathbf{l}^2 \to \mathbf{l}^2$, defined by $A(\mathbf{x}) \equiv \{x_n/n^{\alpha}\}_{n=1}^{\infty}$, where $\alpha > 1/2$ and $\mathbf{x} = \{x_1, x_2, \dots\}$, is bounded since

$$||A\mathbf{x}|| = \left(\sum_{n=1}^{\infty}\frac{|x_n|^2}{n^{2\alpha}}\right)^{1/2} \leq \left(\sum_{n=1}^{\infty}|x_n|^2\right)^{1/2} = ||\mathbf{x}||.$$

   The operator $A$ is also invertible: $A^{-1}(\mathbf{x}) = \{n^{\alpha}x_n\}_{n=1}^{\infty}$, but its inverse is not bounded. Indeed, considering the sequence $\{x_n\}$ of elements with $||\mathbf{x}|| = 1$, it is clear that $||A^{-1}(\mathbf{x})|| = n^{\alpha} \to \infty$ as $n \to \infty$.

4. The bounded operator $A : \mathbf{C}^n \to \mathbf{C}^n$ is uniquely represented by the $n \times n$ matrix $[a_{ij}]$ where $a_{ij} = \langle Ax_j, x_i\rangle$, with $\{x_1, x_2, \dots, x_n\}$ a basis of $\mathbf{C}^n$. The operator is invertible (one-to-one and onto) if and only if $\det[a_{ij}] \neq 0$.

5. The bounded multiplication operator $A : \mathbf{L}^2(a,b) \to \mathbf{L}^2(a,b)$ defined by

$$(Ax)(t) \equiv f(t)x(t)$$

   with $k = \max_{a\leq t\leq b}|f(t)| < \infty$ has as its inverse the division operator,

$$(A^{-1}x)(t) = \frac{1}{f(t)}x(t)$$

   assuming $f(t) \neq 0$.

6. The differentiation operator $A : \mathbf{L}^2(a,b) \to \mathbf{L}^2(a,b)$ defined by

$$(Ax)(t) \equiv \frac{d}{dt}x(t) = x'(t),$$
$$D_A \equiv \left\{x : x(t), x'(t) \in \mathbf{L}^2(a,b)\right\}$$

   is not one-to-one (and so cannot be invertible), since there exist infinitely many functions, all differing from one another by a constant, that have the same image. That is, $N_A = \{c\} \neq \{0\}$. However, if $a = -\infty$ and $b = \infty$, then constants are not in $D_A$ and the operator is invertible. Also, for $a, b$ finite, if we change the domain of $A$ to be

$$D_A \equiv \left\{x : x(t), x'(t) \in \mathbf{L}^2(a,b),\ x(a) = 0\right\},$$

then $A$ is invertible, $A^{-1}$ being of course the integration operator

$$(A^{-1}x)(t) = \int_a^t x(s)\,ds.$$

### 3.9.3 Green's Functions and Green's Operators

At this point it is worthwhile to draw special attention to operators inverse to differential operators, leading to the concept of a Green's function and a Green's operator. Green's functions for electromagnetics have been discussed in some detail in Chapter 1. They are also extensively discussed for an important class of second-order differential operators in Chapter 5 and in various sections in Part II. The discussion here is intended to establish the general concept of a Green's function as the kernel of an integral operator, the Green's operator, which is inverse to an invertible differential operator. The development is mostly formal and presented at a somewhat abstract level to avoid details that would draw attention from the main concepts.

**Scalar Problems**

#### Green's Operators

Consider a linear $n$th-order differential operator $L : \mathbf{L}^2(a,b) \to \mathbf{L}^2(a,b)$

$$-L \equiv p_n(t)\frac{d^n}{dt^n} + p_{n-1}(t)\frac{d^{n-1}}{dt^{n-1}} + \cdots + p_1(t)\frac{d}{dt} + p_0(t),$$

$$D_L \equiv \left\{ \begin{array}{c} \psi : \frac{d^m\psi}{dt^m} \in \mathbf{L}^2(a,b), \quad m = 0,1,2,\ldots,n, \\ B_1(\psi) = \eta_1, \quad B_2(\psi) = \eta_2 \end{array} \right\}$$

and the differential equation

$$(L - \lambda)\psi = \phi, \tag{3.21}$$

where $\lambda$ is a complex parameter, $\psi = \psi(t) \in D_{L-\lambda} = D_L$, and $\phi = \phi(t) \in R_{L-\lambda}$. In the following we will also use the notation $L_\lambda \equiv L - \lambda$. Boundary (and/or initial) conditions are

$$\begin{aligned}
B_1(\psi) &= \alpha_{11}\psi(a) + \cdots + \alpha_{1n}\psi^{(n-1)}(a) + \beta_{11}\psi(b) + \cdots + \beta_{1n}\psi^{(n-1)}(b) \\
&= \eta_1, \\
&\vdots \\
B_n(\psi) &= \alpha_{n1}\psi(a) + \cdots + \alpha_{nn}\psi^{(n-1)}(a) + \beta_{n1}\psi(b) + \cdots + \beta_{nn}\psi^{(n-1)}(b) \\
&= \eta_n,
\end{aligned} \tag{3.22}$$

with $\alpha_{i,j}, \beta_{i,j} \in \mathbf{R}$. We assume $L - \lambda$ is invertible (the example on p. 189 shows that not all differential operators are invertible). Formally solving (3.21) by applying the inverse operator leads to

$$\psi = (L - \lambda)^{-1} \phi.$$

Linearity of the differential operator implies linearity of $(L - \lambda)^{-1}$ by Theorem 3.28; therefore, one would expect that the solution $\psi$ should be representable as a superposition integral[29]

$$\psi(t) = (L - \lambda)^{-1} \phi(t) = \int_a^b g(t, t', \lambda)\, \phi(t')\, dt'. \tag{3.23}$$

This is also consistent with the fact that we would expect the inverse of a differential operator to be an integral operator. The function $g(t, t', \lambda)$ is called the *Green's function* of the operator $L - \lambda$. Often one considers the case where $\lambda = 0$, such that $g(t, t', 0) = g(t, t')$ is the Green's function.

If we substitute the solution (3.23) into (3.21), we obtain

$$\begin{aligned}
\phi(t) = (L - \lambda)\psi(t) &= (L - \lambda) \int_a^b g(t, t', \lambda)\, \phi(t')\, dt' \\
&= \int_a^b \left[(L - \lambda)\, g(t, t', \lambda)\right] \phi(t')\, dt',
\end{aligned}$$

and we can make the identification

$$(L - \lambda)\, g(t, t', \lambda) = \delta(t - t'). \tag{3.24}$$

This will formally serve as the defining equation for the Green's function $g$ in the sense of distributions. Also note that regardless of the possible inhomogeneity of the boundary conditions associated with $L$, the Green's function is taken to satisfy the homogeneous boundary conditions

$$B_1(g) = \cdots = B_n(g) = 0.$$

[29] If $(a, b) = (-\infty, \infty)$, then the form of this solution is correct. In the presence of boundaries it may be necessary to add additional boundary contributions to the solution. These will occur when the boundary conditions that the Green's function and the quantity $\psi$ obey differ from one another.

The inverse operator is called a *Green's operator*, denoted by $G_\lambda$, such that

$$\left((L-\lambda)^{-1}\phi\right)(t) = (G_\lambda \phi)(t) = \int_a^b g(t,t',\lambda)\,\phi(t')\,dt',$$

which has as its kernel the associated Green's function.

## Adjoint Green's Operators

One can go through the same procedure with the adjoint equation

$$(L-\lambda)^* \psi^* = \phi^* \tag{3.25}$$

(note $(L-\lambda)^* = L^* - \overline{\lambda}$) where the domain of $L_\lambda^* \equiv (L-\lambda)^*$ is obtained from requiring that the bilinear concomitant in the generalized Green's theorem (3.3) vanish, leading to adjoint homogeneous boundary conditions

$$B_1^*(\psi^*) = \cdots = B_n^*(\psi^*) = 0$$

on $\psi^*$. The operator $(L-\lambda)^*$ satisfies

$$\langle (L-\lambda)\xi, \zeta\rangle = \langle \xi, (L-\lambda)^* \zeta\rangle$$

for $\xi \in D_L$, $\zeta \in D_{L^*}$.

The adjoint equation (3.25) leads to an *adjoint Green's operator*

$$\left(((L-\lambda)^*)^{-1}\psi\right)(t) = (G_\lambda^* \psi)(t) = \int_a^b g^*(t,t',\lambda)\,\psi(t')\,dt'$$

in terms of an *adjoint Green's function* $g^*(t,t',\lambda)$ that satisfies

$$(L-\lambda)^* g^*(t,t',\lambda) = \delta(t-t')$$

subject to adjoint homogeneous boundary conditions $B_1^*(g^*) = \cdots = B_n^*(g^*) = 0$. Forming $\langle L_\lambda \psi, \psi^*\rangle = \langle \psi, L_\lambda^* \psi^*\rangle$, it is seen that $\langle G_\lambda \psi, \phi\rangle = \langle \psi, G_\lambda^* \phi\rangle$, which leads to

$$\overline{g^*}(t,t',\lambda) = g(t',t,\lambda). \tag{3.26}$$

As an example, for the symmetric operator $L_\lambda = -d^2/dt^2 - \lambda$, we obtain the scalar free-space Green's functions

$$g(t,t') = \frac{e^{-i\sqrt{\lambda}\,|t-t'|}}{2i\sqrt{\lambda}},$$

$$g^*(t,t') = -\frac{e^{+i\sqrt{\overline{\lambda}}\,|t-t'|}}{2i\sqrt{\overline{\lambda}}}.$$

When $L_\lambda$ is self-adjoint, then $g^* = g$ so that $\overline{g}(t,t',\lambda) = g(t',t,\lambda)$.

is compact in $\mathbf{L}^p(\Gamma)$, where $g(t)$ is a continuous function on $\Gamma$ ($\Gamma$ is a sufficiently smooth (Lyapunov) curve). This is an important result in the theory of one-dimensional singular operators.

7. The operator $A : \mathbf{L}^2(\Omega)^3 \to \mathbf{L}^2(\Omega)^3$ defined by

$$(A\mathbf{J})(\mathbf{r}) \equiv \int_s \mathbf{n} \times \left[\nabla \frac{e^{-ik|\mathbf{r}-\mathbf{r}'|}}{4\pi |\mathbf{r}-\mathbf{r}'|} \times \mathbf{J}(\mathbf{r}')\right] dS',$$

which arises from the magnetic field integral equation (1.118), is compact [14, pp. 346–348].

8. Consider the integral operator $A : \mathbf{L}^2(-a,a) \to \mathbf{L}^2(-a,a)$ defined by

$$(Ax)(t) \equiv \int_{-a}^{a} x(s)k(t,s)w(s)\,ds$$

with the kernel $k(t,s) = \ln|t-s|$ and with the weight function $w(s) = 1/\sqrt{a^2-s^2}$. The operator

$$(Ax)(t) = \int_{-a}^{a} x(s)\frac{\ln|t-s|}{\sqrt{a^2-s^2}}\,ds$$

is compact in $\mathbf{L}^2(-a,a)$ and, moreover, $\ln|t-s|$ is a Hilbert–Schmidt kernel with respect to the weight function $w$ such that

$$\int_{-a}^{a}\int_{-a}^{a} \frac{|\ln|t-s||^2}{\sqrt{a^2-s^2}\sqrt{a^2-t^2}}\,ds\,dt < \infty.$$

9. The integral operator $A : \mathbf{C}(\Omega) \to \mathbf{C}(\Omega)$, defined by

$$(Ax)(\mathbf{t}) \equiv \int_\Omega k(\mathbf{t},\mathbf{s})x(\mathbf{s})\,d\mathbf{s},$$

where $\Omega \subset \mathbf{R}^n$ and $k(\mathbf{t},\mathbf{s}) \in \mathbf{C}(\Omega \times \Omega)$, is bounded with

$$\|A\| = \max_{\mathbf{t}\in\Omega} \int_\Omega |k(\mathbf{t},\mathbf{s})|\,d\Omega$$

assuming $\|k\| = \max_{\mathbf{t}\in\Omega}\int_\Omega |k(\mathbf{t},\mathbf{s})|\,d\mathbf{s} < \infty$. Also, suppose that there are constants $c$ and $m < n$ such that

$$|k(\mathbf{t},\mathbf{s})| \le c|\mathbf{t}-\mathbf{s}|^{-m}$$

for $\mathbf{t},\mathbf{s} \in \Omega$ and $\mathbf{t} \neq \mathbf{s}$. Then, $A$ is compact in $\mathbf{C}(\Omega)$. The proof is based on the Arzela criterion for a compact set (see Section 2.2.2; also [16,

to consider the adjoint problem and utilize some form of generalized Green's theorem, e.g., (3.3).

As an illustration of obtaining an explicit solution, assume $L_\lambda$ is real and formally self-adjoint (a common case in electromagnetics, e.g., $L = d^2/dt^2$). Then the generalized Green's theorem (3.4) is

$$\int_a^b (L_\lambda \psi)\, g\, dt = \int_a^b \psi\, (L_\lambda g)\; dt + J(\psi, g)\big|_a^b . \tag{3.28}$$

In terms of an inner product, (3.28) can be written as

$$\langle L_\lambda \psi, \overline{g} \rangle = \langle \psi, \overline{L_\lambda g} \rangle + J(\psi, g)\big|_a^b .$$

Using (3.21) and (3.24), and the sifting property of the delta-function, one obtains the solution as

$$\psi(t') = \int_a^b \phi(t)\, g(t, t', \lambda)\; dt - J(\psi, g)\big|_a^b . \tag{3.29}$$

It is important to note that any Green's function satisfying (3.24) will work in the above. The judicious choice of boundary conditions (always of a homogeneous form or satisfying some fitness condition) on $g$ do affect the conjunct term, and often a preferable set of conditions is dictated by the goal of simplifying or eliminating the conjunct $J$.

### Green's Operators as Improper Integral Operators

Although the procedure leading to (3.29) usually provides the correct solution in the scalar case, a few comments are required in generalizing the method to other problems of interest in electromagnetics.

The generalized Green's theorem will be valid for suitably defined elements, i.e., if $\psi, g, L_\lambda \psi$, and $L_\lambda g$ are suitably well behaved; in practice this means adequately differentiable. Regarding the discussion leading up to (1.52), one notes that the correct procedure that ensures the elements utilized in the Green's theorem are suitably well behaved is to apply the theorem to the intervals $[a, t' - \varepsilon)$ and $(t + \varepsilon, b]$, thereby omitting the point $t = t'$ from the integration. If $\Omega = [a, b]$ and $\Omega_0 = (t' - \varepsilon, t' + \varepsilon)$, then (3.28) becomes

$$\lim_{\varepsilon \to 0} \int_{\Omega - \Omega_0} (L_\lambda \psi)\, g\, dt$$
$$= \lim_{\varepsilon \to 0} \int_{\Omega - \Omega_0} \psi\, (L_\lambda g)\; dt + \lim_{\varepsilon \to 0} \left\{ J(\psi, g)\big|_a^{t' - \varepsilon} + J(\psi, g)\big|_{t' + \varepsilon}^b \right\} .$$

In this case the first term on the right side vanishes (since $L_\lambda g = 0$ on $\Omega - \Omega_0$) and, replacing $L_\lambda \psi$ with $\phi$, one obtains

$$\lim_{\varepsilon \to 0} \int_{\Omega - \Omega_0} \phi(t)\, g(t, t', \lambda)\; dt = \lim_{\varepsilon \to 0} \left\{ J(\psi, g)\big|_a^{t' - \varepsilon} + J(\psi, g)\big|_{t' + \varepsilon}^b \right\} .$$

By reorganizing the conjunct term as

$$J(\psi, g)|_a^{t'-\varepsilon} + J(\psi, g)|_{t'+\varepsilon}^b = J(\psi, g)|_a^b - J(\psi, g)|_{t'-\varepsilon}^{t'+\varepsilon},$$

we have

$$\lim_{\varepsilon\to 0} J(\psi, g)|_{t'-\varepsilon}^{t'+\varepsilon} = -\lim_{\varepsilon\to 0}\int_{\Omega-\Omega_0} \phi(t)\, g(t, t', \lambda)\, dt + J(\psi, g)|_a^b, \tag{3.30}$$

where the limit on the conjunct term on the right side has been removed since $\psi, g$ are assumed regular at the endpoints. By noting the similarity of (3.30) and (3.29), one expects

$$\lim_{\varepsilon\to 0} J(\psi, g)|_{t'-\varepsilon}^{t'+\varepsilon} = -\psi(t'), \tag{3.31}$$

leading to

$$\psi(t') = \lim_{\varepsilon\to 0}\int_{\Omega-\Omega_0} \phi(t)\, g(t, t', \lambda)\, dt - J(\psi, g)|_a^b. \tag{3.32}$$

If the Green's function is weakly singular, then the solution (3.29) should be interpreted in the sense of (3.32).

The condition (3.31) characterizes the nature of the singularity of $g$ [39, p. 10]. For a given problem one knows the conjunct and the properties of the Green's function, and (3.31) may be explicitly evaluated. For example, with $L = d^2/dt^2 = L^*$, (3.28) is simply Green's second theorem in one dimension, where the conjunct is $J(\psi, g) = g\, d\psi/dx - \psi\, dg/dx$. Explicitly, $(d^2/dt^2)\, g(t, t') = \delta(t - t')$ leads to $dg(t, t')/dt|_{t=t'-\varepsilon}^{t=t'+\varepsilon} = 1$ and continuity of $g$ at $t = t'$, such that (3.31) is satisfied.

As described in Section 1.3.4, for most scalar and vector problems the limiting procedure is not necessary (in the sense that the correct answer nevertheless emerges from the sifting property of the delta-function), although the meaning of any resulting improper integral needs to be properly understood. In the dyadic case the limiting procedure, or some other rigorous method, needs to be used to obtain correct results.

Finally, for a general nonsymmetric problem we use (3.3) applied to $\psi, g^*$,

$$\int_a^b (L_\lambda \psi)\, \overline{g^*}\, dt = \int_a^b \psi\, \overline{(L_\lambda^* g^*)}\, dt + J\left(\psi, \overline{g^*}\right)\Big|_a^b. \tag{3.33}$$

Formally substituting $L_\lambda \psi = \phi$, $L_\lambda^* g^* = \delta$ into (3.33) yields

$$\psi(t') = \int_\Omega \phi(t)\, \overline{g^*}(t, t')\, d\Omega - J\left(\psi, \overline{g^*}\right)\Big|_a^b$$

such that the solution is obtained in terms of the conjugate-adjoint Green's function. In this one-dimensional case, specific examples are shown in Section 5.3.

## Systems of Dyadic Operators

### Dyadic Green's Operators

Because electromagnetic theory is inherently a vector theory, one often needs to consider vector and dyadic differential operators and corresponding dyadic Green's operators with dyadic Green's function kernels.

Toward this end, we next consider a $2 \times 2$ matrix system of dyadic differential operators[31] $\mathbf{L} : H \to H$ and the matrix differential operator equation[32]

$$(\mathbf{L} - \lambda \mathbf{I}) \, \mathbf{\Psi} = \mathbf{\Phi}, \tag{3.34}$$

where

$$\mathbf{L} = \begin{bmatrix} \underline{\mathbf{L}}_{1,1} & \underline{\mathbf{L}}_{1,2} \\ \underline{\mathbf{L}}_{2,1} & \underline{\mathbf{L}}_{2,2} \end{bmatrix}, \qquad \mathbf{I} = \begin{bmatrix} \underline{\mathbf{I}} & \underline{\mathbf{0}} \\ \underline{\mathbf{0}} & \underline{\mathbf{I}} \end{bmatrix}$$

and

$$\mathbf{\Psi} = \begin{bmatrix} \mathbf{\Psi}_1(\mathbf{r}) \\ \mathbf{\Psi}_2(\mathbf{r}) \end{bmatrix} \in D_{\mathbf{L}}, \qquad \mathbf{\Phi} = \begin{bmatrix} \mathbf{\Phi}_1(\mathbf{r}) \\ \mathbf{\Phi}_2(\mathbf{r}) \end{bmatrix} \in R_{\mathbf{L}-\lambda\mathbf{I}}$$

with $\mathbf{\Psi}_i$, $\mathbf{\Phi}_i$ vectors, $i = 1, 2$. Assuming $\mathbf{L}_\lambda \equiv (\mathbf{L} - \lambda \mathbf{I})$ is invertible, and repeating the procedure followed in the one-dimensional case, we identify $\mathbf{L}_\lambda^{-1} = \mathbf{G}_\lambda$ as the inverse (Green's) operator ($\mathbf{L}_\lambda^{-1} \mathbf{L}_\lambda \mathbf{\Psi} = \mathbf{\Psi} = \mathbf{L}_\lambda^{-1} \mathbf{\Phi}$). The solution of (3.34) is obtained as

$$\begin{aligned} \mathbf{\Psi}(\mathbf{r}) &= (\mathbf{G}_\lambda \mathbf{\Phi})(\mathbf{r}) = \int_\Omega \mathbf{g}(\mathbf{r}, \mathbf{r}', \lambda) \, \mathbf{\Phi}(\mathbf{r}') \, d\Omega' \\ &= \int_\Omega \begin{bmatrix} \underline{\mathbf{g}}_{1,1}(\mathbf{r}, \mathbf{r}', \lambda) & \underline{\mathbf{g}}_{1,2}(\mathbf{r}, \mathbf{r}', \lambda) \\ \underline{\mathbf{g}}_{2,1}(\mathbf{r}, \mathbf{r}', \lambda) & \underline{\mathbf{g}}_{2,2}(\mathbf{r}, \mathbf{r}', \lambda) \end{bmatrix} \cdot \begin{bmatrix} \mathbf{\Phi}_1(\mathbf{r}') \\ \mathbf{\Phi}_2(\mathbf{r}') \end{bmatrix} d\Omega', \end{aligned} \tag{3.35}$$

where $\mathbf{G}_\lambda$ is a Green's operator with (matrix) Green's function kernel $\mathbf{g}(\mathbf{r}, \mathbf{r}', \lambda)$, the matrix entries of which are dyadic Green's functions. Explicitly using the scalar product notation, one can also write the above as

$$\begin{aligned} \mathbf{\Psi}_1(\mathbf{r}) &= \int_\Omega \underline{\mathbf{g}}_{1,1}(\mathbf{r}, \mathbf{r}', \lambda) \cdot \mathbf{\Phi}_1(\mathbf{r}') \, d\Omega' + \int_\Omega \underline{\mathbf{g}}_{1,2}(\mathbf{r}, \mathbf{r}', \lambda) \cdot \mathbf{\Phi}_2(\mathbf{r}') \, d\Omega', \\ \mathbf{\Psi}_2(\mathbf{r}) &= \int_\Omega \underline{\mathbf{g}}_{2,1}(\mathbf{r}, \mathbf{r}', \lambda) \cdot \mathbf{\Phi}_1(\mathbf{r}') \, d\Omega' + \int_\Omega \underline{\mathbf{g}}_{2,2}(\mathbf{r}, \mathbf{r}', \lambda) \cdot \mathbf{\Phi}_2(\mathbf{r}') \, d\Omega'. \end{aligned} \tag{3.36}$$

---

[31] The treatment here generally follows [58, Sec. 1.1]; see also [10]. The generalization to an $n \times n$ system is straightforward, as is the reduction to a single dyadic or vector equation.

[32] By this notation we mean

$$\mathbf{\Theta} \mathbf{\Lambda} = \begin{bmatrix} \underline{\mathbf{\Theta}}_{1,1} & \underline{\mathbf{\Theta}}_{1,2} \\ \underline{\mathbf{\Theta}}_{2,1} & \underline{\mathbf{\Theta}}_{2,2} \end{bmatrix} \cdot \begin{bmatrix} \underline{\mathbf{\Lambda}}_1 \\ \underline{\mathbf{\Lambda}}_2 \end{bmatrix} = \begin{bmatrix} \underline{\mathbf{\Theta}}_{1,1} \cdot \underline{\mathbf{\Lambda}}_1 + \underline{\mathbf{\Theta}}_{1,2} \cdot \underline{\mathbf{\Lambda}}_2 \\ \underline{\mathbf{\Theta}}_{2,1} \cdot \underline{\mathbf{\Lambda}}_1 + \underline{\mathbf{\Theta}}_{2,2} \cdot \underline{\mathbf{\Lambda}}_2 \end{bmatrix}$$

for elements of $\mathbf{\Theta}$, $\mathbf{\Lambda}$ being dyadics. For example, in electromagnetics the case $\underline{\mathbf{\Theta}}_{i,j} = \alpha_{i,j} \nabla \times \underline{\mathbf{I}}$ often occurs. In the case of vector or scalar entries, appropriate multiplication rules apply.

As in the scalar case treated previously, if the region of interest $\Omega$ is finite then (3.36) may need to be augmented by boundary terms (the conjunct in a generalized Green's theorem), in this case boundary integrals, representing contributions from induced currents not accounted for by the Green's function. The presence or absence of these contributions will depend on the boundary conditions satisfied by the Green's function and the fields, and an explicit solution is obtained by utilizing an appropriate Green's theorem.

If we substitute the solution (3.36) into (3.34) we have

$$\begin{aligned}\mathbf{\Phi}(\mathbf{r}) = \mathbf{L}_\lambda \mathbf{\Psi}(\mathbf{r}) &= \mathbf{L}_\lambda \int_\Omega \mathbf{g}(\mathbf{r}, \mathbf{r}', \lambda)\, \mathbf{\Phi}(\mathbf{r}')\, d\Omega' \\ &= \int_\Omega [\mathbf{L}_\lambda \mathbf{g}(\mathbf{r}, \mathbf{r}', \lambda)]\, \mathbf{\Phi}(\mathbf{r}')\, d\Omega'\end{aligned} \tag{3.37}$$

and we can make the identification

$$\mathbf{L}_\lambda \mathbf{g}(\mathbf{r}, \mathbf{r}', \lambda) = \mathbf{I}\,\delta(\mathbf{r} - \mathbf{r}'). \tag{3.38}$$

For notational convenience we set $\lambda = 0$ such that the individual dyadic Green's functions are seen to satisfy

$$\begin{aligned}\underline{\mathbf{L}}_{1,1} \cdot \underline{\mathbf{g}}_{1,1} + \underline{\mathbf{L}}_{1,2} \cdot \underline{\mathbf{g}}_{2,1} &= \underline{\mathbf{I}}\,\delta(\mathbf{r} - \mathbf{r}'), \\ \underline{\mathbf{L}}_{1,1} \cdot \underline{\mathbf{g}}_{1,2} + \underline{\mathbf{L}}_{1,2} \cdot \underline{\mathbf{g}}_{2,2} &= \underline{\mathbf{0}}, \\ \underline{\mathbf{L}}_{2,1} \cdot \underline{\mathbf{g}}_{1,1} + \underline{\mathbf{L}}_{2,2} \cdot \underline{\mathbf{g}}_{2,1} &= \underline{\mathbf{0}}, \\ \underline{\mathbf{L}}_{2,1} \cdot \underline{\mathbf{g}}_{1,2} + \underline{\mathbf{L}}_{2,2} \cdot \underline{\mathbf{g}}_{2,2} &= \underline{\mathbf{I}}\,\delta(\mathbf{r} - \mathbf{r}'),\end{aligned} \tag{3.39}$$

which can be decoupled to yield

$$\begin{aligned}\left(\underline{\mathbf{L}}_{1,1} - \underline{\mathbf{L}}_{1,2} \cdot \underline{\mathbf{L}}_{2,2}^{-1} \cdot \underline{\mathbf{L}}_{2,1}\right) \cdot \underline{\mathbf{g}}_{1,1} &= \underline{\mathbf{I}}\,\delta(\mathbf{r} - \mathbf{r}'), \\ \left(\underline{\mathbf{L}}_{2,2} - \underline{\mathbf{L}}_{2,1} \cdot \underline{\mathbf{L}}_{1,1}^{-1} \cdot \underline{\mathbf{L}}_{1,2}\right) \cdot \underline{\mathbf{g}}_{2,2} &= \underline{\mathbf{I}}\,\delta(\mathbf{r} - \mathbf{r}')\end{aligned}$$

assuming $\underline{\mathbf{L}}_{i,i}^{-1}$ are invertible. Note also that the original equations (3.34) can be decoupled as

$$\begin{aligned}\left(\underline{\mathbf{L}}_{1,1} - \underline{\mathbf{L}}_{1,2} \cdot \underline{\mathbf{L}}_{2,2}^{-1} \cdot \underline{\mathbf{L}}_{2,1}\right) \cdot \mathbf{\Psi}_1 &= \mathbf{\Phi}_1 - \underline{\mathbf{L}}_{1,2} \cdot \underline{\mathbf{L}}_{2,2}^{-1} \cdot \mathbf{\Phi}_2, \\ \left(\underline{\mathbf{L}}_{2,2} - \underline{\mathbf{L}}_{2,1} \cdot \underline{\mathbf{L}}_{1,1}^{-1} \cdot \underline{\mathbf{L}}_{1,2}\right) \cdot \mathbf{\Psi}_2 &= \mathbf{\Phi}_2 - \underline{\mathbf{L}}_{2,1} \cdot \underline{\mathbf{L}}_{1,1}^{-1} \cdot \mathbf{\Phi}_1.\end{aligned}$$

It can be seen that one advantage of the Green's function approach is that the source terms in the decoupled equations for the Green's functions are simpler than for the field quantities themselves.

### Adjoint Dyadic Green's Operators

For the adjoint problem,

$$\mathbf{L}_\lambda^* \mathbf{\Psi}^* = \mathbf{\Phi}^* \tag{3.40}$$

and we identify $(\mathbf{L}_\lambda^*)^{-1} = \mathbf{G}_\lambda^*$ as the adjoint Green's operator with matrix kernel having entries $\underline{\mathbf{g}}_{\alpha,\beta}^*(\mathbf{r},\mathbf{r}',\lambda)$, leading to the solution

$$\begin{aligned}\mathbf{\Psi}^*(\mathbf{r}) &= (\mathbf{G}_\lambda^*\mathbf{\Phi}^*)(\mathbf{r}) = \int_\Omega \mathbf{g}^*(\mathbf{r},\mathbf{r}',\lambda)\,\mathbf{\Phi}^*(\mathbf{r}')\,d\Omega' \qquad (3.41)\\ &= \int_\Omega \begin{bmatrix} \underline{\mathbf{g}}_{1,1}^*(\mathbf{r},\mathbf{r}',\lambda) & \underline{\mathbf{g}}_{1,2}^*(\mathbf{r},\mathbf{r}',\lambda) \\ \underline{\mathbf{g}}_{2,1}^*(\mathbf{r},\mathbf{r}',\lambda) & \underline{\mathbf{g}}_{2,2}^*(\mathbf{r},\mathbf{r}',\lambda)\end{bmatrix} \cdot \begin{bmatrix}\mathbf{\Phi}_1^*(\mathbf{r}') \\ \mathbf{\Phi}_2^*(\mathbf{r}')\end{bmatrix} d\Omega',\end{aligned}$$

where $\mathbf{g}^*$ satisfies

$$\mathbf{L}_\lambda^*\mathbf{g}^*(\mathbf{r},\mathbf{r}',\lambda) = \mathbf{I}\delta(\mathbf{r}-\mathbf{r}') \qquad (3.42)$$

subject to adjoint boundary conditions. Also $\langle \mathbf{G}_\lambda\mathbf{\Psi}, \mathbf{\Psi}^*\rangle = \langle \mathbf{\Psi}, \mathbf{G}_\lambda^*\mathbf{\Psi}^*\rangle$, leading to

$$\underline{\mathbf{g}}_{\alpha,\beta}^*(\mathbf{r},\mathbf{r}',\lambda) = \overline{\underline{\mathbf{g}}_{\beta,\alpha}^\top}(\mathbf{r}',\mathbf{r},\lambda). \qquad (3.43)$$

For example, the operator $L_\lambda = -\nabla^2 - \lambda$ leads to the free-space Green's function

$$g(\mathbf{r},\mathbf{r}') = \frac{e^{-i\sqrt{\lambda}|\mathbf{r}-\mathbf{r}'|}}{4\pi|\mathbf{r}-\mathbf{r}'|},$$

which satisfies an outgoing radiation condition (see Section 1.3.4), and the adjoint Green's function

$$g^*(\mathbf{r},\mathbf{r}') = \frac{e^{+i\sqrt{\overline{\lambda}}|\mathbf{r}-\mathbf{r}'|}}{4\pi|\mathbf{r}-\mathbf{r}'|},$$

which satisfies an incoming radiation condition.

The preceding comments in the scalar case concerning the need to consider the adjoint problem hold for systems of vector or dyadic operators as well; if an appropriate Green's second theorem is available, one can express the solution in terms of the Green's function, without consideration of the adjoint problem, assuming $\mathbf{L}_\lambda$ is invertible. If $\mathbf{L}_\lambda$ is real and formally self-adjoint, then the generalized Green's theorem (3.4) is

$$\int_\Omega (\mathbf{L}_\lambda\mathbf{\Psi})\,\mathbf{g}\,d\Omega = \int_\Omega \mathbf{\Psi}\,(\mathbf{L}_\lambda\mathbf{g})\,d\Omega + \int_\Gamma \mathbf{J}(\mathbf{\Psi},\mathbf{g})\cdot d\mathbf{\Gamma}. \qquad (3.44)$$

Using (3.34) and (3.38) one obtains the solution as

$$\mathbf{\Psi}(\mathbf{r}') = \int_\Omega \mathbf{\Phi}(\mathbf{r})\,\mathbf{g}(\mathbf{r},\mathbf{r}',\lambda)\,d\Omega - \int_\Gamma \mathbf{J}(\mathbf{\Psi},\mathbf{g})\cdot d\mathbf{\Gamma}. \qquad (3.45)$$

As discussed in Section 1.3.5, the procedure leading to (3.45) may not result in the correct solution because of the strength of the dyadic Green's function source-point singularity. A proper procedure is to apply (3.44) to $\Omega - \Omega_0$, where $\Omega_0$ contains the point $\mathbf{r} = \mathbf{r}'$. We then have

$$\lim_{\Omega_0 \to 0} \int_{\Omega - \Omega_0} (\mathbf{L}_\lambda \mathbf{\Psi}) \, \mathbf{g} \, d\Omega = \lim_{\Omega_0 \to 0} \int_{\Omega - \Omega_0} \mathbf{\Psi} \, (\mathbf{L}_\lambda \mathbf{g}) \, d\Omega + \lim_{\Omega_0 \to 0} \int_{\Gamma} \mathbf{J}(\mathbf{\Psi}, \mathbf{g}) \cdot d\mathbf{\Gamma} + \lim_{\Omega_0 \to 0} \int_{\Gamma_0} \mathbf{J}(\mathbf{\Psi}, \mathbf{g}) \cdot d\mathbf{\Gamma}, \tag{3.46}$$

where the first term on the right side vanishes since the point $\mathbf{r} = \mathbf{r}'$ is excluded from the integration. We obtain the solution as

$$\lim_{\Omega_0 \to 0} \int_{\Gamma_0} \mathbf{J}(\mathbf{\Psi}, \mathbf{g}) \cdot d\mathbf{\Gamma} = \lim_{\Omega_0 \to 0} \int_{\Omega - \Omega_0} \mathbf{\Phi} \mathbf{g} \, d\Omega - \int_{\Gamma} \mathbf{J}(\mathbf{\Psi}, \mathbf{g}) \cdot d\mathbf{\Gamma}. \tag{3.47}$$

Comparing with (3.45) we identify

$$\lim_{\Omega_0 \to 0} \int_{\Gamma_0} \mathbf{J}(\mathbf{\Psi}, \mathbf{g}) \cdot d\mathbf{\Gamma} = \mathbf{\Psi}(\mathbf{r}').$$

**Examples**

Consider Maxwell's differential equations for a region $\Omega \subset \mathbf{R}^3$ bounded by surface $\Gamma$ (see, e.g., (1.13)),

$$\begin{aligned} \nabla \times \mathbf{E}(\mathbf{r}) &= -i\omega\mu \mathbf{H}(\mathbf{r}) - \mathbf{J}_m(\mathbf{r}), \\ \nabla \times \mathbf{H}(\mathbf{r}) &= \ \ i\omega\varepsilon \mathbf{E}(\mathbf{r}) + \mathbf{J}_e(\mathbf{r}) \end{aligned}$$

subject to $\mathbf{n} \times \mathbf{E}|_\Gamma = \mathbf{0}$. One can consider the above in the form (3.34), where

$$\mathbf{L} = \begin{bmatrix} i\omega\varepsilon \underline{\mathbf{I}} & -\nabla \times \underline{\mathbf{I}} \\ \nabla \times \underline{\mathbf{I}} & i\omega\mu \underline{\mathbf{I}} \end{bmatrix}, \quad \mathbf{\Psi} = \begin{bmatrix} \mathbf{E}(\mathbf{r}) \\ \mathbf{H}(\mathbf{r}) \end{bmatrix}, \quad \mathbf{\Phi} = \begin{bmatrix} -\mathbf{J}_e(\mathbf{r}) \\ -\mathbf{J}_m(\mathbf{r}) \end{bmatrix}$$

and $\lambda = 0$ (note the dyadic identity $(\nabla \times \underline{\mathbf{I}}) \cdot \mathbf{\Psi} = \nabla \times \mathbf{\Psi}$). In this case we have

$$\mathbf{g}(\mathbf{r}, \mathbf{r}') = \begin{bmatrix} \underline{\mathbf{g}}_{e,e}(\mathbf{r}, \mathbf{r}') & \underline{\mathbf{g}}_{e,m}(\mathbf{r}, \mathbf{r}') \\ \underline{\mathbf{g}}_{m,e}(\mathbf{r}, \mathbf{r}') & \underline{\mathbf{g}}_{m,m}(\mathbf{r}, \mathbf{r}') \end{bmatrix},$$

and the formal solution (3.36) is

$$\begin{aligned} \mathbf{E}(\mathbf{r}) &= -\int_\Omega \underline{\mathbf{g}}_{e,e}(\mathbf{r}, \mathbf{r}') \cdot \mathbf{J}_e(\mathbf{r}') \, d\Omega' - \int_\Omega \underline{\mathbf{g}}_{e,m}(\mathbf{r}, \mathbf{r}') \cdot \mathbf{J}_m(\mathbf{r}') \, d\Omega', \\ \mathbf{H}(\mathbf{r}) &= -\int_\Omega \underline{\mathbf{g}}_{m,e}(\mathbf{r}, \mathbf{r}') \cdot \mathbf{J}_e(\mathbf{r}') \, d\Omega' - \int_\Omega \underline{\mathbf{g}}_{m,m}(\mathbf{r}, \mathbf{r}') \cdot \mathbf{J}_m(\mathbf{r}') \, d\Omega', \end{aligned} \tag{3.48}$$

where the Green's dyadics satisfy $\mathbf{Lg}(\mathbf{r},\mathbf{r}') = \mathbf{I}\delta(\mathbf{r}-\mathbf{r}')$, leading to

$$\begin{aligned} -\nabla\times\underline{\mathbf{g}}_{m,e} + i\omega\varepsilon\underline{\mathbf{g}}_{e,e} &= \underline{\mathbf{I}}\,\delta(\mathbf{r}-\mathbf{r}'), \\ -\nabla\times\underline{\mathbf{g}}_{m,m} + i\omega\varepsilon\underline{\mathbf{g}}_{e,m} &= \underline{\mathbf{0}}, \\ \nabla\times\underline{\mathbf{g}}_{e,e} + i\omega\mu\underline{\mathbf{g}}_{m,e} &= \underline{\mathbf{0}}, \\ \nabla\times\underline{\mathbf{g}}_{e,m} + i\omega\mu\underline{\mathbf{g}}_{m,m} &= \underline{\mathbf{I}}\,\delta(\mathbf{r}-\mathbf{r}'), \end{aligned} \tag{3.49}$$

subject to $\mathbf{n}\times\underline{\mathbf{g}}_{e,e}\Big|_\Gamma = \underline{\mathbf{0}}$ and $\mathbf{n}\times\underline{\mathbf{g}}_{e,m}\Big|_\Gamma = \underline{\mathbf{0}}$.

Decoupling the above leads to

$$\begin{aligned} \nabla\times\nabla\times\underline{\mathbf{g}}_{e,e} - \omega^2\mu\varepsilon\underline{\mathbf{g}}_{e,e} &= i\omega\mu\underline{\mathbf{I}}\,\delta(\mathbf{r}-\mathbf{r}'), \\ \nabla\times\nabla\times\underline{\mathbf{g}}_{m,m} - \omega^2\mu\varepsilon\underline{\mathbf{g}}_{m,m} &= i\omega\varepsilon\underline{\mathbf{I}}\,\delta(\mathbf{r}-\mathbf{r}'), \end{aligned} \tag{3.50}$$

where $\mathbf{n}\times\underline{\mathbf{g}}_{e,e}\Big|_\Gamma = \underline{\mathbf{0}}$ and $\mathbf{n}\times\nabla\times\underline{\mathbf{g}}_{m,m}\Big|_\Gamma = \underline{\mathbf{0}}$, with $\underline{\mathbf{g}}_{e,m}$ and $\underline{\mathbf{g}}_{m,e}$ subsequently determined from the solution of (3.50) using the relationships (3.49). Alternatively, wave equations

$$\begin{aligned} \nabla\times\nabla\times\underline{\mathbf{g}}_{e,m} - \omega^2\mu\varepsilon\underline{\mathbf{g}}_{e,m} &= \nabla\times\underline{\mathbf{I}}\,\delta(\mathbf{r}-\mathbf{r}'), \\ \nabla\times\nabla\times\underline{\mathbf{g}}_{m,e} - \omega^2\mu\varepsilon\underline{\mathbf{g}}_{m,e} &= -\nabla\times\underline{\mathbf{I}}\,\delta(\mathbf{r}-\mathbf{r}') \end{aligned} \tag{3.51}$$

subject to $\mathbf{n}\times\underline{\mathbf{g}}_{e,m}\Big|_\Gamma = \underline{\mathbf{0}}$ and $\mathbf{n}\times\nabla\times\underline{\mathbf{g}}_{m,e}\Big|_\Gamma = \underline{\mathbf{0}}$ can be solved, and the remaining quantities determined from (3.49). It can also be seen from the defining equations that the field quantities satisfy

$$\begin{aligned} \nabla\times\nabla\times\mathbf{E}(\mathbf{r}) - k^2\mathbf{E}(\mathbf{r}) &= -i\omega\mu\mathbf{J}_e(\mathbf{r}) - \nabla\times\mathbf{J}_m(\mathbf{r}), \\ \nabla\times\nabla\times\mathbf{H}(\mathbf{r}) - k^2\mathbf{H}(\mathbf{r}) &= -i\omega\varepsilon\mathbf{J}_m(\mathbf{r}) + \nabla\times\mathbf{J}_e(\mathbf{r}) \end{aligned}$$

subject to $\mathbf{n}\times\mathbf{E}|_\Gamma = \mathbf{0}$, $\mathbf{n}\times\nabla\times\mathbf{H}|_\Gamma = \mathbf{0}$.

Without detailing the adjoint problem, it can be seen that the symmetry relation (3.43) reduces to

$$\underline{\mathbf{g}}_{\alpha,\beta}(\mathbf{r},\mathbf{r}') = (-1)^{\alpha+\beta}\,\underline{\mathbf{g}}^{\top}_{\beta,\alpha}(\mathbf{r}',\mathbf{r}) \tag{3.52}$$

since $\nabla\times\nabla\times$ subject to boundary conditions of the form discussed is a real, self-adjoint operator. If radiation conditions, rather than boundary conditions, are specified, the operator is real and symmetric, and the same conclusion holds per the discussion subsequent to (3.27). The sign change is obvious from the form of the defining equations.

Finally, consider the scalar (transmission line) system (see Chapter 7)

$$\begin{aligned} \frac{dv(z)}{dz} &= -i\omega L\, i(z) - i_m(z), \\ \frac{di(z)}{dz} &= -i\omega C\, v(z) + i_e(z), \end{aligned} \tag{3.53}$$

which can be analyzed as a special case of the above. We have (3.34) with $\lambda = 0$,

$$\mathbf{L} = \begin{bmatrix} -i\omega C & -\frac{d}{dz} \\ \frac{d}{dz} & i\omega L \end{bmatrix}, \quad \mathbf{\Psi} = \begin{bmatrix} v(z) \\ i(z) \end{bmatrix}, \quad \mathbf{\Phi} = \begin{bmatrix} -i_e(z) \\ -i_m(z) \end{bmatrix},$$

leading to

$$\mathbf{g}(z, z') = \begin{bmatrix} g_{v,v}(z, z') & g_{v,i}(z, z') \\ g_{i,v}(z, z') & g_{i,i}(z, z') \end{bmatrix}$$

with the solution of (3.36) as

$$\begin{aligned} v(z) &= -\int_a^b g_{v,v}(z, z')\, i_e(z)\, dz' - \int_a^b g_{v,i}(z, z')\, i_m(z)\, dz', \\ i(z) &= -\int_a^b g_{i,v}(z, z')\, i_e(z)\, dz' - \int_a^b g_{i,i}(z, z')\, i_m(z)\, dz'. \end{aligned} \tag{3.54}$$

The Green's functions are seen to satisfy

$$\mathbf{L}\mathbf{g}(z, z') = \mathbf{I}\delta(z - z'),$$

where $\mathbf{I} = \begin{bmatrix} 1 & 0 \\ 0 & 1 \end{bmatrix}$ is the usual identity matrix, leading to

$$\begin{aligned} -\frac{d}{dz} g_{i,v} - i\omega C g_{v,v} &= \delta(z - z'), \\ -\frac{d}{dz} g_{i,i} - i\omega C g_{v,i} &= 0, \\ \frac{d}{dz} g_{v,v} + i\omega L g_{i,v} &= 0, \\ \frac{d}{dz} g_{v,i} + i\omega L g_{i,i} &= \delta(z - z'). \end{aligned}$$

Decoupling the above leads to

$$\begin{aligned} \left(\frac{d^2}{dz^2} + \omega^2 LC\right) g_{v,v}(z, z') &= -i\omega L\delta(z - z'), \\ \left(\frac{d^2}{dz^2} + \omega^2 LC\right) g_{i,i}(z, z') &= -i\omega C\delta(z - z'), \end{aligned}$$

where the voltage and current quantities satisfy

$$\begin{aligned} \left(\frac{d^2}{dz^2} + \omega^2 LC\right) v(z) &= -i\omega L i_e - i_m', \\ \left(\frac{d^2}{dz^2} + \omega^2 LC\right) i(z) &= +i\omega C i_m + i_e'. \end{aligned}$$

Note that

$$g_{\alpha,\beta}(z, z') = (-1)^{\alpha+\beta} g_{\beta,\alpha}(z', z).$$

## 3.10 Projection Operators

We have previously introduced the projection theorem (p. 120), which provides the idea of uniquely representing an element $z \in H$ as $z = x_0 + y_0$, where $x_0 \in M$, with $M$ a closed linear subspace of $H$, and $y_0 \in M^\perp$. After introducing the concept of a basis for an infinite-dimensional system, it was stated that $x_0 = \sum_n \langle z, x_n \rangle x_n$ and $y_0 = \sum_n \langle z, y_n \rangle y_n$, where $\{x_n\}$ is an orthonormal basis for $M$ and $\{y_n\}$ is an orthonormal basis for $M^\perp$. We can now cast this into operator form using *projection operators.*

**Definition 3.30.** *Let $M$ be a closed linear subspace of a Hilbert space $H$, and let $z = x_0 + y_0$, where $z \in H$, $x_0 \in M$, and $y_0 \in M^\perp$. An operator $P$ defined on $H$ is said to be the orthogonal projection onto $M$ if*

$$P(z) = P(x_0 + y_0) = x_0.$$

Projection operators are linear, bounded operators with[33] $\|P\| \le 1$. In addition, $R_P = M$ and $N_P = M^\perp$ (note that $N_p \perp R_p$ and $H = R_P \oplus N_P$). Also, $I - P$ on $H$ is an orthogonal projection onto $M^\perp$, i.e.,

$$(I - P)(z) = z - x_0 = y_0$$

with $R_{I-P} = M^\perp$ and $N_{I-P} = M$.

Some properties of projection operators follow.

- Projection operators are *idempotent.*[34]
- A bounded operator is a projection if and only if it is idempotent and self-adjoint.
- A projection operator is a nonnegative operator.
- The sum $P + Q$ of two projection operators $P$ and $Q$ is a projection if and only if $PQ = 0$.
- The product $PQ$ of two projection operators $P$ and $Q$ is a projection if and only if[35] $[P, Q] = 0$.
- A projection onto $M$, a closed linear subspace of $H$, is compact if and only if $M$ is finite-dimensional.
- If $P$ is a projection onto $M$, a closed linear subspace of $H$, then for all $x \in H$, $\langle Px, x \rangle = ||Px||^2$.
- $N_p$ and $R_p$ are closed linear subspaces of $H$.

From the projection theorem one can see that the orthogonal projection $P$ onto $M$ is unique.

[33] The Pythagorean formula gives $\|Pz\|^2 = \|x_0\|^2 = \|z - y_0\|^2 = \|z\|^2 - \|y_0\|^2 \le \|z\|^2$.

[34] An operator $A$ is idempotent if $A^2 = A$.

[35] Recall that $[P, Q] = PQ - QP$ is the commutator of operators $P$ and $Q$.

### Examples of Projection Operators

Consider an element $z \in H$ and a closed linear subspace $M$ of $H$ with $\{x_n\}$ an orthonormal basis for $M$. The projection operator onto $M$ is defined as $Pz = \sum_n \langle z, x_n \rangle x_n$.

1. The zero operator is a projection operator onto $\{0\}$ ($M = \{0\}$, $M^\perp = H$).

2. The identity operator is a projection operator onto $H$ ($M = H$, $M^\perp = \{0\}$),
$$Iz = z = \sum_n \langle z, x_n \rangle x_n.$$
Obviously, $\|I\| = 1$, and, in fact, $\|P\| = 1$ for every nonzero projection operator since $Px = x$ for all $x \in M$.

3. On $H = \mathbf{R}^3$ with the usual $xyz$-coordinate system, let $M$ be the $xy$-plane with its standard basis $\{\mathbf{e}_1, \mathbf{e}_2\}$. The orthogonal projection of $\gamma = (3\mathbf{e}_1, 2\mathbf{e}_2, 5\mathbf{e}_3)$ onto $M$ is $P\gamma = \sum_{n=1}^{2} \langle \gamma, \mathbf{e}_n \rangle \mathbf{e}_n = (3\mathbf{e}_1, 2\mathbf{e}_2)$.

4. On $H = \mathbf{L}^2(-\pi, \pi)$ let $M^e$ be the closed linear subspace consisting of all even functions in $H$, defined by $x(t) = x(-t)$ for all $t$. Then $(M^e)^\perp = M^o$, the space of all odd functions in $H$ (i.e., $x(t) = -x(-t)$ for all $t$). Then, letting $x(t) \in H$,
$$P^e x = \frac{1}{2}[x(t) + x(-t)]$$
is an orthogonal projection onto $M^e$, and
$$P^o x = (I - P^e)x = \frac{1}{2}[x(t) - x(-t)]$$
is an orthogonal projection onto $M^o$.

5. On $H = \mathbf{L}^2(-\pi, \pi)$ with basis
$$\left\{\sqrt{1/(2\pi)}, \sqrt{1/\pi}\cos(nt), \sqrt{1/\pi}\sin(nt), n = 1, 2, \ldots\right\}$$
let
$$x_0^e = \sqrt{\tfrac{1}{2\pi}}, \qquad x_n^e = \sqrt{\tfrac{1}{\pi}}\cos(nt),$$
$$x_0^o = 0, \qquad x_n^o = \sqrt{\tfrac{1}{\pi}}\sin(nt),$$
$n = 1, 2, \ldots$. Let $M^e$ be the closed linear subspace consisting of all even functions in $H$, for which $\{x_n^e\}$ is a basis. Then $(M^e)^\perp = M^o$, the space of all odd functions in $H$, for which $\{x_n^o\}$ is a basis ($\{x_n^e\} \cup \{x_n^o\}$ is a basis for $H$). Then, letting $x(t) \in H$,
$$P^e x = \sum_{n=0}^{\infty} \langle x, x_n^e \rangle x_n^e$$

is the projection operator onto $M^e$, and

$$P^o x = (I - P^e)\, x = \sum_{n=0}^{\infty} \langle x, x_n^e + x_n^o \rangle\, x_n - \sum_{n=0}^{\infty} \langle x, x_n^e \rangle\, x_n^e$$
$$= \sum_{n=0}^{\infty} \langle x, x_n^o \rangle\, x_n^o$$

is the projection operator onto $M^o$. Since the projection is unique,

$$P^e x = \frac{1}{2}\left[x(t) + x(-t)\right] = \sum_{n=0}^{\infty} \langle x, x_n^e \rangle\, x_n^e,$$

and similarly for $P^o$.

Projection operators are extensively used to show geometrical properties of spectral problems. In fact, advanced concepts in spectral theory are often cast in terms of projections. Therefore, much of the next chapter could be studied using projection operators, although for the purposes of studying spectral problems in applied electromagnetics these concepts will not be pursued much further. The interested reader can consult [23, Ch. 6], [56], [7]. For applications of projection operators in numerical solutions of electromagnetic operator equations, see [57], [43].

## 3.11 Solution of Operator Equations

In this section we consider the solution of linear equations at an operator level. In general, we are interested in operator equations of the form

$$(A - \lambda I)\, x = y, \tag{3.55}$$

where $A : H \to H$ is a linear operator, $x \in D_{A-\lambda I} = D_A \subseteq H$, and $y \in R_{A-\lambda I} \subseteq H$. If $\lambda \neq 0$, we call (3.55) an operator equation of the *second kind*, whereas if $\lambda = 0$, we have an operator equation of the *first kind*. Obviously, if the resolvent operator exists, formally we have the solution $x = (A - \lambda I)^{-1}\, y$, which is continuously dependent on $y$ if the resolvent is bounded. Furthermore, if $\lambda$ is an eigenvalue of $A$, we cannot solve (3.55) for $y \neq 0$, since the homogeneous form defines the eigenvalue problem, $(A - \lambda_n I)\, x_n = 0$.

### 3.11.1 Existence of Solutions

In general, it is difficult to state solvability theorems for (3.55) unless considerable restrictions are placed on $A$. In the following we present a solvability theorem known as the *Fredholm alternative*, which we state for the operator equation $Ax = y$ (this is not necessarily a first-kind equation since $A$ may have the form $A = B - \lambda I$). Operators satisfying the Fredholm alternative are called *Fredholm operators*. The alternative will be stated for operators in Hilbert spaces although a similar alternative holds for operators in Banach spaces.

Of course, for the equation $Ax = y$ to make sense we must have $x \in D_A$ and $y \in R_A$. In particular, the requirement $y \in R_A$ is important if $R_A \neq H$. Note also that if $A$ is positive definite, the solution of $Ax = y$ can be shown to be equivalent to minimization of a quadratic functional [4, Ch. 5], although this topic will not be discussed here.

**Theorem 3.44.** *(Fredholm alternative) Let $A : H \to H$ be a bounded linear Fredholm operator. Then either*

*I. for every $y, g \in H$ the nonhomogeneous equations*

$$\begin{aligned} Ax &= y, \\ A^* f &= g \end{aligned} \tag{3.56}$$

*have unique solutions $x, f$, and the corresponding homogeneous equations*

$$\begin{aligned} Ax &= 0, \\ A^* f &= 0 \end{aligned} \tag{3.57}$$

*have only the trivial solutions $x = 0$ and $f = 0$,* ***or***

*II. the homogeneous equations*

$$\begin{aligned} Ax &= 0, \\ A^* f &= 0 \end{aligned} \tag{3.58}$$

*have the same number of (nontrivial) linearly independent solutions, $x_1, x_2, \ldots, x_n$ and $f_1, f_2, \ldots, f_n$, respectively, $n \geq 1$, and the nonhomogeneous equations*

$$\begin{aligned} Ax &= y, \\ A^* f &= g \end{aligned}$$

*are not solvable for all $y, g$. A solution (nonunique) exists if and only if $y$ and $g$ are such that $y \perp N_{A^*}$ and $g \perp N_A$, respectively.*

The first part of the alternative is easily understood from Theorem 3.25 and its subsequent discussion, because if the homogeneous equations (3.57) have only trivial solutions, then $A^{-1}$ and $(A^*)^{-1}$ exist, and so (3.56) can be uniquely solved as $x = A^{-1}y$ and $f = (A^*)^{-1} g$. The fact that both the original equation and the adjoint equation behave in the same manner, in terms of being solvable or not, can be seen to be reasonable, at least for the special case of the Fredholm operator $I - K$ with $K$ compact (discussed next), from Theorem 3.17.

The second part of the alternative is less obvious, because in the case of nontrivial homogeneous solutions to $Ax = 0$, solvability of $Ax = y$ depends, perhaps unexpectedly, on properties of the adjoint operator $A^*$. The necessity of the condition $y \perp N_{A^*}$ for $Ax = y$ to be solvable in this case is seen from the following. Consider $Ax = y$ and $A^*f = 0$, and form the inner product $\langle Ax, f\rangle = \langle y, f\rangle$. Using $\langle y, f\rangle = \langle Ax, f\rangle = \langle x, A^*f\rangle = \langle x, 0\rangle = 0$ we see that we must have the condition $\langle y, f\rangle = 0$, i.e., $y \perp N_{A^*}$. The sufficiency of this condition for Fredholm operators can also be proved, but will be omitted here. Of course, if $A$ is self-adjoint, the Fredholm alternative is simplified, and no aspect of the adjoint operator need be considered.

For Theorem 3.44 to be useful, we need to know what types of operators satisfy the Fredholm alternative. Three classes of Fredholm operators are presented below.

We first state that the operator $A - \lambda I$, where $\lambda \neq 0$ and $A$ is a compact operator, is a Fredholm operator, such that solvability of the equation $(A - \lambda I)\, x = y$ can be examined using the above alternative.

**Theorem 3.45.** *Let $A : H \to H$ be a compact linear operator and let $A_\lambda \equiv A - \lambda I$, where $\lambda \neq 0$. Then $A_\lambda$ satisfies the Fredholm alternative.*

The proof is shown in [40, p. 451] in a Banach-space setting and in [12, p. 359] in a Hilbert space. An important case is provided by $\lambda = 1$. Alternatively, since $\lambda \neq 0$, one can consider the operator $(1/\lambda)\, A - I$ or $K - I$, where $(1/\lambda)\, A = K$ is compact if $A$ is compact. See also Theorem 4.34.

Theorem 3.45 is most often applied to examine solvability of second-kind integral equations involving compact integral operators. The above theorem can be extended in the sense that $A_\lambda$ satisfies the Fredholm alternative if $(A_\lambda)^n$, $n \geq 1$, is compact [41, p. 138]. Also note that $\dim N_A = \dim N_{A^*} < \infty$.

From the above it is easily seen that for $A$ compact and $\lambda \neq 0$, if $(A - \lambda I)\, x = 0$ has only the trivial solution $x = 0$ (meaning $\lambda$ is not an eigenvalue of $A$), then $(A - \lambda I)\, x = y$ has a unique solution given by $x = (A - \lambda I)^{-1}\, y$. From Theorem 3.40 one can see that the solution depends continuously on $y$, and so the problem is well posed.

The next theorem characterizes when the operator (matrix) equation $Ax = b$ is solvable, where $A : H^n \to H^m$ is bounded and compact (necessarily so since $H^n$ and $H^m$ are finite-dimensional Hilbert spaces) with $x \in H^n$ and $b \in H^m$.

**Theorem 3.46.** *The operator $A : H^n \to H^m$ satisfies the Fredholm alternative.*

Usually this is stated for $A : \mathbf{C}^n \to \mathbf{C}^m$, leading to the matrix system $Ax = b$ where $A$ is an $m \times n$ complex matrix and $b$ is an $m \times 1$ matrix. The most familiar case is of course when $m = n$, such that $Ax = b$ is uniquely solvable if and only if $\det A \neq 0$, implying that $Ax = 0$ has only the trivial solution.

Lastly, in the case of a differential operator we have the following [12, p. 229].

**Theorem 3.47.** *The self-adjoint Sturm–Liouville differential operator L (see (5.1), (5.2), and (5.4)) satisfies the Fredholm alternative.*

Note also that the nonself-adjoint Sturm–Liouville operator with homogeneous boundary conditions, considered in Section 5.3, also satisfies the Fredholm alternative. The solution of operator equations is further discussed in Section 4.4 from the viewpoint of eigenfunction expansions.

### 3.11.2 Convergence of Solutions

While it is not the intent of this text to present a through discussion of numerical solution techniques, it is worthwhile to briefly list a few important facts concerning the numerical solution of operator equations (see, for instance, [14], [42], and [43]).

Recall from Sections 2.5.6 and 2.5.7 that given a Hilbert space $H$ with an orthonormal basis $\{x_n\}$, any $x \in H$ may be written as a norm-convergent series $x = \sum_{n=1}^{\infty} \langle x, x_n \rangle x_n$. This can be viewed as defining an $N$-term approximation $x \simeq x^N$,

$$x^N = \sum_{n=1}^{N} \langle x, x_n \rangle x_n, \tag{3.59}$$

where $x^N$ lies in the $N$-dimensional subspace of $H$ spanned by $\{x_n\}_{n=1}^{N}$. The error in the approximation is defined as $e^N \equiv x - x^N$, and because $\{x_n\}$ is a basis for $H$, $\lim_{N\to\infty} \left\|e^N\right\| = \lim_{N\to\infty} \left\|x - x^N\right\| = 0$. Furthermore, the error is orthogonal to the approximation, i.e.,

$$\left\langle x - x^N, x^N \right\rangle = 0. \tag{3.60}$$

For finite $N$, the coefficient $\langle x, x_n \rangle$ in (3.59) provides the minimum error in the approximation of $x$ (i.e., this coefficient provides the best possible approximation to $x$ in the $N$-dimensional subspace spanned by $\{x_n\}_{n=1}^{N}$).

Now, consider determining the unknown $x$ from the general operator equation

$$Ax = y,$$

where $A : H_1 \to H_2$ with $\{x_n\}$ and $\{y_n\}$ bases for the spaces $H_1$ and $H_2$, respectively. We form the approximations of $x \in H_1$ and $y \in H_2$ as[36]

$$x^N = \sum_{n=1}^{N} \alpha_n x_n, \tag{3.61}$$

$$y^N = \sum_{n=1}^{N} \langle y, y_n \rangle y_n,$$

where $\alpha_n$ are unknown coefficients to be determined, and we are assured that $y^N \to y \in H_2$ in norm. The general idea is to determine the solution of

$$Ax^N = y^N$$

for a given $N$, and consider under what conditions this solution converges to the solution of the original equation (i.e., when does $x^N \to x$).

Since the summation is finite, $Ax^N = \sum_{n=1}^{N} \alpha_n Ax_n$. Noting that $Ax_n \in H_2$, we have $Ax_n = \sum_{m=1}^{\infty} \langle Ax_n, y_m \rangle y_m \simeq (Ax_n)^N = \sum_{m=1}^{N} \langle Ax_n, y_m \rangle y_m$, so that $Ax^N = y^N$ becomes

$$\sum_{n=1}^{N} \alpha_n \sum_{m=1}^{N} \langle Ax_n, y_m \rangle y_m = \sum_{n=1}^{N} \langle y, y_n \rangle y_n,$$

leading to the matrix system

$$\sum_{n=1}^{N} \alpha_n \langle Ax_n, y_m \rangle = \langle y, y_m \rangle \tag{3.62}$$

for $m = 1, 2, \ldots, N$. The solution of (3.62) provides the coefficients $\alpha_n$, leading to determination of $x^N$ as in (3.61).

For a given, finite-value $N$, the coefficients $\alpha_n$ obtained from (3.62) do not necessarily provide the best approximation of $x$ in $H_1$. That is, if we knew the exact solution $x$ by other means, we would find that the coefficients $\alpha_n$ do not generally agree with the coefficients $\langle x, x_n \rangle$ in (3.59). Therefore, the approximation of the solution $x$ is not the best possible approximation in $H_1$; however, we do obtain the best possible approximation of $Ax \in H_2$. Moreover, if we increase $N$, we do not necessarily know that the coefficients $\alpha_n$ converge to $\langle x, x_n \rangle$, although they may do so.

Alternatively, if we take

$$\begin{aligned} x &= \sum_{n=1}^{\infty} \alpha_n x_n, \\ y &= \sum_{n=1}^{\infty} \langle y, y_n \rangle y_n, \end{aligned} \tag{3.63}$$

---

[36] We take the upper summation limit as $N$ in both cases, which will lead to an $N \times N$ matrix system. However, this is not necessary, and one may generate an $N \times M$ matrix system solvable by least-squares or comparable methods.

then if $A$ is bounded (to permit the interchange of $A$ and the summation in (3.63)), we directly obtain the infinite matrix system

$$\sum_{n=1}^{\infty} \alpha_n \langle Ax_n, y_m \rangle = \langle y, y_m \rangle \tag{3.64}$$

for $m = 1, 2, \ldots$. In theory, this infinite system yields the correct coefficients $\alpha_n = \langle x, x_n \rangle$ to make the equality in (3.63) valid. However, in order to solve (3.64) in practice, this system is truncated to size $N \times N$, leading to (3.62); either way we do not have assurance that the coefficients $\alpha_n$ converge to $\langle x, x_n \rangle$.

### Method of Moments

In electromagnetics, the matrix equation (3.62) is usually obtained by a procedure known as the *method of moments* [44]. In this case we form an approximation

$$x \simeq x^N = \sum_{n=1}^{N} \alpha_n x_n,$$

where $\{x_n\} \in D_A$ are linearly independent functions called *expansion functions*. We therefore have $Ax \simeq \sum_{n=1}^{N} \alpha_n Ax_n$, and, rather than expanding $Ax_n$ in the basis of $H_2$ as above, one forms a matrix equation by forcing the *residual* (error) $\sum_{n=1}^{N} \alpha_n Ax_n - y$ to be orthogonal to a set of *weighting* (or *testing*) functions $\{y_n\} \subset R_A$, i.e.,

$$\left\langle \sum_{n=1}^{N} \alpha_n Ax_n - y, y_m \right\rangle = 0$$

for $m = 1, 2, \ldots, N$. This is equivalent to

$$\sum_{n=1}^{N} \alpha_n \langle Ax_n, y_m \rangle = \langle y, y_m \rangle, \tag{3.65}$$

which is the same as (3.62). If the expansion and testing functions are identical, then this is known as *Galerkin's method.* In practice a pseudo inner product is often used in (3.65).

The interpretation of the method-of-moments procedure is somewhat different from the former method, because we don't necessarily require $\{x_n\}$ or $\{y_n\}$ to be bases, or even orthogonal. However, as with the former method, we are finding the best approximation of $Ax \in R_A$ for a given set $\{y_n\}$ and in this sense minimizing the norm of the residual error. Therefore, to achieve a reasonably accurate solution of the operator equation in an engineering sense, we may not necessarily require the expansion functions to be complete in $D_A$, but they should be such that $Ax_n$ is complete in $R_A$ [45].

### Convergence of Solutions

Unfortunately, in applications it is often not possible to show in a rigorous manner convergence of the approximation $x^N \rightarrow x$. However, two important and practical cases can be distinguished which lead to guaranteed convergence. We first consider the special case of a positive definite operator [46], [4].

**Theorem 3.48.** *Consider the operator equation $Ax = y$, where $A : H \rightarrow H$ is a positive definite operator. If $\{x_n\} = \{y_n\}$ forms an orthonormal basis for $H_e$, the Hilbert space associated with the energy norm (3.13), then*

$$x^N = \sum_{n=1}^{N} \alpha_n x_n \rightarrow x$$

*in the norm of $H$, where $\alpha_n$ are determined by the solution of (3.65).*

**Proof.** Since $A$ is positive definite, we rewrite (3.65) in terms of the energy inner product (3.12) as

$$\sum_{n=1}^{N} \alpha_n \langle x_n, x_m \rangle_e = \langle x, x_m \rangle_e \tag{3.66}$$

for $m = 1, 2, \ldots, N$. Because $\{x_n\}$ is an orthonormal set in $H_e$, then $\alpha_n = \langle x, x_m \rangle_e$ and we have $x^N = \sum_{n=1}^{N} \langle x, x_n \rangle_e x_n$. Using completeness of $\{x_n\}$ in $H_e$ leads to[37] $\lim_{N\rightarrow\infty} \|x^N - x\|_e = 0$. By Theorem 3.15, since $A$ is positive definite, convergence in energy leads to strong (norm) convergence in $H$, i.e., $\lim_{N\rightarrow\infty} \|x^N - x\| = 0$. ∎

In electromagnetics, it often occurs that static problems lead to positive definite operators [47], such as the Laplacian operator considered in Chapter 6.

If the operator equation in question is a second-kind Fredholm equation, then convergence can also be established [14, Ch. 5], [42], [43, Ch. 5] (see also Theorem 3.18).

**Theorem 3.49.** *Consider the second-kind operator equation $(\lambda I - K)x = y$, where $K : H \rightarrow H$ is a compact operator and $\lambda$ is not an eigenvalue of $K$. If $\{x_n\}$ and $\{y_n\}$ are orthonormal bases for $H$, then*

$$x^N = \sum_{n=1}^{N} \alpha_n x_n \rightarrow x$$

*in norm, where $\alpha_n$ are determined by the solution of (3.65), with $A = (\lambda I - K)$.*

---

[37] If $A$ is merely positive, then we must stop here and settle for weak convergence (by Theorem 3.15, part b). See also Theorem 3.14, which concerns forming a positive operator from any bounded operator that possesses an adjoint.

In fact, the above statement is more restrictive than it need be, and certain relaxations on the expansion and testing functions may be allowed. Thus, second-kind compact integral equations that admit solutions may often be solved by discretization, where convergence of the solution can be established. This important fact is the impetus of many regularization schemes.

Under the conditions of Theorem 3.49, (3.65) is simply a truncated version of (3.64). To examine convergence at the matrix level, consider the infinite second-kind matrix equation

$$x_i - \sum_{j=1}^{\infty} k_{ij} x_j = y_i, \tag{3.67}$$

$i = 1, 2, \ldots$, where the matrix $[k_{ij}]$ is compact. We write (3.67) as

$$X - KX = Y$$

where $X = [x_1, x_2, \ldots]^\top$, $Y = [y_1, y_2, \ldots]^\top$, and $K = [k_{ij}]_{i,j=1}^{\infty}$. We also consider the "truncated" matrix system

$$X^n - K^n X^n = Y$$

where $K^n$ is filled with zeroes outside of the $nxn$ square. Then,

$$\begin{aligned}
(X - KX) - (X^n - K^n X^n) &= 0, \\
X - X^n &= KX - K^n X^n \\
&= KX - K^n X^n + K^n X - K^n X, \\
(X - X^n) + K^n (X^n - X) &= (K - K^n) X.
\end{aligned}$$

Therefore

$$(X - X^n) = (I - K^n)^{-1} (K - K^n) X,$$

and, upon taking norms,

$$\begin{aligned}
\|(X - X^n)\| &= \left\| (I - K^n)^{-1} (K - K^n) X \right\| \\
&\leq \left\| (I - K^n)^{-1} \right\| \|K - K^n\| \|X\|,
\end{aligned}$$

leading to

$$\frac{\|(X - X^n)\|}{\|X\|} \leq \left\| (I - K^n)^{-1} \right\| \|K - K^n\|$$

which can provide an error estimate for $X^n$. Furthermore, it can be shown that[38] $\left\| (I - K^n)^{-1} - (I - K)^{-1} \right\| \to 0$, and so as $n \to \infty$ the first

[38] If $A : H \to H$ is bounded and invertible, $B : H \to H$ is bounded, and $\|A - B\| < \frac{1}{\|A^{-1}\|}$, then [17, p. 71] $B$ is invertible and

$$\left\| A^{-1} - B^{-1} \right\| \leq \frac{\left\| A^{-1} \right\|^2 \|A - B\|}{1 - \|A^{-1}\| \|A - B\|}.$$

term on the right side becomes $\| (I - K)^{-1} \|$, which is bounded by Theorem 3.40 (this is where compactness enters into the problem). Obviously $\|K - K^n\| \to 0$ and so $X^n \to X$ in norm. The proof can also be done at the function space level if one can approximate the operator $K$ by a sequence of finite-rank operators $K^n$ such that $\|K - K^n\| \to 0$.

---

Replacing $A$ with $(I - K)$ and $B$ with $(I - K^n)$ proves the statement.

# Bibliography

[1] Krall, A.M. (1973). *Linear Methods in Applied Analysis*, Reading, MA: Addison-Wesley.

[2] Stoer, J. and Bulirsch, R. (1980). *Introduction to Numerical Analysis*, New York: Springer-Verlag.

[3] Mikhlin, S.G. and Prössdorf, S. (1986). *Singular Integral Operators*, Berlin: Academie-Verlag.

[4] Mikhlin, S.G. (1970). *Mathematical Physics, An Advanced Course*, Amsterdam: North-Holland.

[5] Muskhelishvili, N.I. (1953). *Singular Integral Equations: Boundary Problems of Function Theory and Their Application to Mathematical Physics*, Groningen: P. Noordhoff.

[6] Berezansky, Y.M., Sheftel, Z.G., and Us, G.F. (1996). *Functional Analysis,* Vol. 1, Basel, Switzerland: Birkhäuser Verlag.

[7] Reed, M. and Simon, B. (1980). *Methods of Mathematical Physics I: Functional Analysis*, San Diego: Academic Press.

[8] Lanczos, C. (1997). *Linear Differential Operators*: New York: Dover.

[9] Dudley, D.G. (1994). *Mathematical Foundations for Electromagnetic Theory,* New York: IEEE Press.

[10] Morse, P.M. and Feshbach, H. (1953). *Methods of Theoretical Physics*, Vol. I, New York: McGraw-Hill.

[11] Naimark, M.A. (1967). *Linear Differential Operators*, Part I, New York: Frederick Ungar.

[12] Stakgold, I. (1999). *Green's Functions and Boundary Value Problems*, 2nd ed., New York: Wiley.

[13] Harrington, R.F. and Mautz, J.R. (1971). Theory of characteristic modes for conducting bodies, *IEEE Trans. Antennas Propagat.*, Vol. AP-19, pp. 622–628.

[14] Jones, D.S. (1994). *Methods in Electromagnetic Wave Propagation*, 2nd ed., New York: IEEE Press.

[15] Byron, F.W. and Fuller, R.W. (1970). *Mathematics of Classical and Quantum Physics*, Vol. II, Reading, MA: Addison-Wesley.

[16] Hutson, V. and Pym, J.S. (1980). *Applications of Functional Analysis and Operator Theory*, London: Academic Press.

[17] Gohberg, I. and Goldberg, S. (1980). *Basic Operator Theory*, Boston: Birkhäuser.

[18] Shestopalov, V.P., Kirilenko, A.A., and Masalov, S.A. (1984). *Convolution-Type Matrix Equations in the Theory of Diffraction,* Kiev: Nauk. Dumka ( in Russian).

[19] Shestopalov, V.P. and Sirenko, Yu.K. (1989). *Dynamic Theory of Gratings,* Kiev: Nauk. Dumka ( in Russian).

[20] Ramm, A.G. (1986). *Scattering by Obstacles,* Dordrecht, Holland: D. Reidel Publishing.

[21] Koshparenok, V.N., Melezhik, P.N., Poedinchuk, A.Y., and Shestopalov, V.P. (1983). Rigorous solution to the 2D problem of diffraction by a finite number of curved screens, *USSR J. Comput. Maths. Mathem. Physics (*Engl. transl.*).* Vol. 23, no. 1, pp. 140–151.

[22] Koshparenok, V.N., Melezhik, P.N., Poedinchuk, A.Y., and Shestopalov, V.P. (1985). Rigorous theory of open two-dimensional resonators with dielectric inclusions, *Izv. Vuzov. Radiofizika,* v. 28, pp. 1311–1321.

[23] Koshparenok, V.N., Melezhik, P.N., Poedinchuk, A.Y., and Shestopalov, V.P. (1986). Method of Riemann-Hilbert problem in the spectral theory of open two-dimensional resonators, *Soviet J. of Communications Technology and Electronics (*Engl. transl.) Vol. 31, no. 6, pp. 4*5–52.*

[24] Sirenko, Yu.K. (1983). Justification of semi-inversion method of matrix operators in the problems of wave diffraction, *Zhurn. Vych. Matem. i Matem. Fiziki,* Vol. 23, pp. 1381–1391.

[25] Tuchkin, Yu. A. (1985). Wave scattering by an open cylindrical screen of arbitrary profile with Dirichlet boundary conditions, *Soviet Physics Doklady,* Vol. 30, pp. 1027–1030.

[26] Tuchkin, Yu. A. (1987). Wave scattering by an open cylindrical screen of arbitrary profile with the Neumann boundary condition, *Soviet Physics Doklady,* Vol. 32, pp. 213–216.

[27] Poedinchuk, A.E., Tuchkin, Ya.A., and Shestopalov, V.P. (1998). The method of the Riemann–Hilbert problem in the theory of diffraction by shells of arbitrary cross section, *Computational Mathematics and Mathematical Physics,* Vol. 38, no. 8, pp. 1260–1273.

[28] Nosich, A.I. (1993). Green's function—dual series approach in wave scattering from combined resonant scatterers, in Hashimoto, M., Idemen, M. and Tretyakov, O.A. (ed.), *Analytical and Numerical Methods in Electromagnetic Wave Scattering,* Tokyo: Science House, pp. 419–469.

[29] Veliev, E.I. and Veremey, V.V. (1993). Numerical-analytical approach for the solution to the wave scattering by polygonal cylinders and flat strip structures, in Hashimoto, M., Idemen, M. and Tretyakov, O.A. (ed.), *Analytical and Numerical Methods in Electromagnetic Wave Scattering,* Tokyo: Science House, pp. 470–519.

[30] Gribanov, Yu. I. (1963). The coordinate spaces and infinite systems of linear equations. III, *Izv. Vuzov.—Mathematics,* Vol. 34, no. 3, pp. 27–39 (in Russian).

[31] Hilbert, D. (1924). *Grundzüge einer allgemeinen Theorie der linearen Integralgleichungen,* Leipzig, pp. 164–174.

[32] Gribanov, Yu. I. (1962). The coordinate spaces and infinite systems of linear equations. I, *Izv. Vuzov.—Mathematics,* Vol. 29, no. 4, pp. 38–48 (in Russian).

[33] Rud', L.A., Sirenko, Yu.K., Shestopalov, V.P., and Yashina, N.P. (1986). Algorithms of the spectral problem solution connected with open waveguide resonators, *Preprint no. 318,* Institute of Radiophysics and Electronics, Ukr. Acad. of Sciences, Kharkov, Ukraine.

[34] Kirilenko, A.A., Rud', L.A., and Tkachenko, V.I. (1996). Semi-inversion method for an accurate analysis of rectangular waveguide H-plane angular discontinunities, *Radio Science,* Vol. 31, no. 5, pp. 1271–1280.

[35] Chumachenko, V.P. (1989). Substantiation of one method of solving two-dimensional problems concerning the diffraction of electromagnetic waves by polygon structures. Uniqueness theorem, *Soviet J. Commun. Technol. Electronics* (Engl. Transl.), Vol. 34, no. 15, pp. 140–143.

[36] Petrusenko, I.V., Yakovlev, A.B., and Gnilenko, A.B. (1994). Method of partial overlapping regions for the analysis of diffraction problems, *IEE Proc.—Microw. Antennas Propag.,* Vol. 141, no. 3, pp. 196–198, June.

[37] Petrusenko, I.V. (1996). New technique for solving of electrodynamics singular integral equations, *Proc. MMET'96,* Kharkov, Ukraine, pp. 125–127.

[38] Petrusenko, I.V. (1998). The method of analysis of the diffraction on step discontinuity, *Proc. MMET'98,* Kharkov, Ukraine, pp. 369–371.

[39] Van Bladel, J. (1985). *Electromagnetic Fields*, Washington: Hemisphere.

[40] Kreyszig, E. (1978). *Introductory Functional Analysis with Applications*, New York: Wiley.

[41] Shilov, G.E. (1974). *Elementary Functional Analysis*, New York: Dover.

[42] Atkinson, K.E. (1976). *A Survey of Numerical Methods for the Solution of Fredholm Integral Equations of the Second Kind*, Philadelphia: SIAM.

[43] Peterson, A.F., Ray, S.L., and Mittra, R. (1998). *Computational Methods for Electromagnetics*, New York: IEEE Press.

[44] Harrington, R.F. (1993). *Field Computation by Moment Methods*, New York: IEEE Press.

[45] Sarkar, T.K., Djordjević, A.R., and Arvas, E. (1985). On the choice of expansion and weighting functions in the numerical solution of operator equations, *IEEE Trans. Antennas Propagat.*, Vol. AP-33, pp. 988–996, Sept.

[46] Mikhlin, S.G. (1965). *The Problem of the Minimum of a Quadratic Functional*, San Francisco: Holden-Day.

[47] Dudley, D.G. (1985). Error minimization and convergence in numerical methods, *Electromagnetics*, Vol. 5, pp. 89–97.

[48] Friedman, B. (1956). *Principles and Techniques of Applied Mathematics*, New York: Dover.

[49] Akhiezer, N.I. and Glazman, I.M. (1961 & 1963). *Theory of Linear Operators in Hilbert Space*, Vols. I and II, New York: Dover.

[50] Naylor, A.W. and Sell, G.R. (1982). *Linear Operator Theory in Engineering and Science*, 2nd ed., New York: Springer-Verlag.

[51] Mrozowski, M. (1997). *Guided Electromagnetic Waves, Properties and Analysis*, Somerset, England: Research Studies Press.

[52] Debnath, L. and Mikusiński, P. (1999). *Introduction to Hilbert Spaces with Applications*, San Diego: Academic Press.

[53] Zabreyko, P.P., Koshelev, A.I., Krasnosel'skii, M.A., Mikhlin, S.G., Rakovshchik, L.S., and Stet'senko, V.Ya. (1975). *Integral Equations—A Reference Text*, Leyden, The Netherlands: Noordhoff International Publishing.

[54] Colton, D. and Kress, R. (1983). *Integral Equation Methods in Scattering Theory*, New York: Wiley.

[55] Lavrent'ev, M.M. and Savel'ev, L.Ya. (1995). *Linear Operators and Ill-Posed Problems*, New York: Plenum Publishing.

[56] DeVito, C.L. (1990). *Functional Analysis and Linear Operator Theory*, New York: Addison-Wesley.

[57] Steele, C.W. (1997). *Numerical Computation of Electric and Magnetic Fields*, 2nd ed., New York: Chapman & Hall.

[58] Felson, L.B. and Marcuvitz, N. (1994). *Radiation and Scattering of Waves*, New York: IEEE Press.

# 4

# Spectral Theory of Linear Operators

In this chapter we examine the spectral properties of operators commonly encountered in electromagnetics. We emphasize the role of eigenvalues and eigenfunctions in spectral theory since these quantities play an important role in many applications, as both mathematical and physical entities. The primary goal of this chapter is to present elements from the spectral theory of operators on infinite-dimensional spaces. Operators on finite-dimensional spaces and their associated matrix representations are also briefly covered, as this material forms an appropriate starting point for discussion.

After considering the finite-dimensional case, we then show that for certain classes of operators acting on infinite-dimensional spaces, simple results may be obtained. In particular, when the operator is compact and self-adjoint, the spectral theory becomes relatively simple. Similar results are obtained if the operator is unbounded, self-adjoint, and positive with compact inverse. Fortunately, these two cases occur quite frequently in physical applications. For example, the first case frequently arises in the study of integral equations pertaining to shielded electrodynamic problems and in the integral equation formulation of electrostatic problems. The second class encompasses various differential operators that commonly occur in electromagnetics applications.

The chapter begins with the presentation of the basic concepts behind spectral elements, including the point, continuous, and residual spectrum. Next, spectral properties of various operators are presented, including discussions of various eigenvalue problems. Spectral representations and expansion theorems are then presented for different classes of operators. After a section on functions of operators, the chapter concludes with a discussion of the solution of operator equations using spectral methods.

## 4.1 General Concepts

Consider a linear operator $A : H \to H$ mapping a Hilbert space $H$ to itself. The *spectral problem* consists, in part, of examining solutions of the equation

$$Ax = \lambda x, \tag{4.1}$$

where $\lambda \in \mathbf{C}$ is known as an *eigenvalue* (or *characteristic value*) and $x \in D_A$, $x \neq 0$, is called an *eigenvector* (or often *eigenfunction* if $H$ is a function space). We refer to this as a *standard eigenvalue problem.* It is convenient for later work to write the above as

$$(A - \lambda I)\, x = 0, \tag{4.2}$$

where $I$ is the identity operator on $H$.

Another, more general formulation is possible, involving two operators $A, B : H \to H$, given as

$$Ax = \lambda Bx \tag{4.3}$$

or

$$(A - \lambda B)\, x = 0 \tag{4.4}$$

and called a *generalized eigenvalue problem.* Of course, (4.3) reduces to (4.1) when operator $B$ is the identity operator on $H$.

Finally, a third formulation commonly occurs in electromagnetic applications, known as a *nonstandard eigenvalue problem.* It is formulated as

$$A(\gamma)x = 0, \tag{4.5}$$

where $\gamma$ is called a nonstandard eigenvalue and $x \neq 0$ is a nonstandard eigenvector. The operator $A(\gamma)$, dependent on the variable $\gamma$, is called an *operator-valued function* (of $\gamma$). The standard and generalized eigenvalue problems can also be viewed as special cases of (4.5), where, for instance, we may consider the nonstandard eigenvalue as a parameter $\gamma$ such that

$$(A(\gamma) - \lambda\,(\gamma)\, I)\, x = 0.$$

Often we seek the value of the nonstandard eigenvalue $\gamma$ such that $\lambda(\gamma) = 0$.

**Example 4.1.**

Consider the microstrip transmission line depicted in Figure 4.1. The strip is infinitely thin and perfectly conducting, and exists over $z \in (-\infty, \infty)$. We seek guided modes of the transmission line in the form

$$\mathbf{E}\,(\mathbf{r},t) = \mathbf{E}\,(x, y)\, e^{i(\omega t - \beta z)},$$
$$\mathbf{H}\,(\mathbf{r},t) = \mathbf{H}\,(x, y)\, e^{i(\omega t - \beta z)},$$

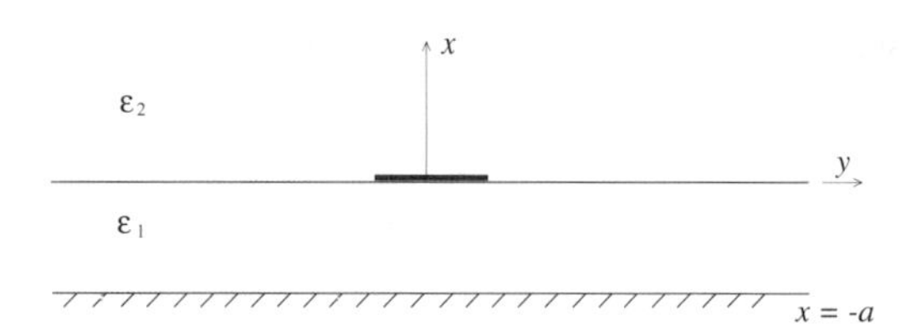


Figure 4.1: Geometry of a microstrip transmission line as considered in Example 4.1. The determination of the guided modes of the microstrip line can be associated with the spectral properties of a self-adjoint positive operator.

where $\omega$ is the radian frequency and $\beta$ is the propagation constant, with $\omega, \beta \in \mathbf{R}$. Letting $\nabla_\beta \times \mathbf{x}$, $\nabla_\beta \cdot \mathbf{x}$, and $\nabla_\beta x$ be the differential operators obtained from curl, divergence, and gradient, respectively, upon the substitution $\partial/\partial z \to -i\beta$, the problem of determining guided microstrip modes can be formed as the spectral problem [1]

$$A(\beta)\,\mathbf{E} = \omega^2(\beta)\,\mathbf{E},$$

where $A : \mathbf{L}^2(\Omega)^3 \to \mathbf{L}^2(\Omega)^3$ is defined as

$$A(\beta)\,\mathbf{E} \equiv \frac{1}{\varepsilon}\left\{\nabla_\beta \times \nabla_\beta \times \mathbf{E} - \varepsilon \nabla_\beta \nabla_\beta \cdot \varepsilon \mathbf{E}\right\},$$

$$D_A \equiv \left\{ \begin{array}{c} \mathbf{E} : \mathbf{E}, \nabla_\beta \times \mathbf{E}, \nabla_\beta \times \nabla_\beta \times \mathbf{E} \in \mathbf{L}^2(\Omega)^3, \\ \nabla_\beta \cdot \varepsilon \mathbf{E} \in \mathbf{H}_{0,1}, \, \mathbf{n} \times \mathbf{E}|_\Gamma = \mathbf{0} \end{array} \right\},$$

with $\Gamma$ being the contour of both the strip and the ground plane. It can be shown that $A(\beta)$ is self-adjoint and positive, so that eigenvalues $\lambda(\beta) = \omega^2(\beta)$ are real-valued and positive[1] (see Section 4.2.6).

Furthermore, $A$ in (4.5) may depend on a large number of independent variables (in electromagnetics these may be frequency, transform-domain wavenumbers, geometrical values, constitutive quantities, etc.), and usually $A$ is nonlinear in these variables. In particular, (4.5) can be written as $A(\tau,\gamma)x = 0$, where one seeks a solution $\gamma(\tau)$. This often occurs in resonance problems, where $\gamma$ represents a natural frequency and $\tau$ is a vector of geometrical and constitutive quantities. The nonstandard formulation also commonly occurs in complicated waveguiding problems, where $\gamma$ would represent a propagation constant. In this chapter we concentrate on the standard eigenvalue problem, although both the generalized and nonstandard problems are also discussed.

[1]In [1] it is also established that the essential spectrum (for our purposes continuous spectrum; see Section 4.1.2) of $A$ is identical to the essential spectrum of the grounded dielectric without the strip and that guided microstrip modes exist.

### 4.1.1 Operators on Finite-Dimensional Spaces

Beginning with the finite-dimensional case, consider (4.1) where $A \in \mathbf{C}^{n\times n}$ is an $n \times n$ complex matrix (representing a finite-dimensional operator with respect to some basis, or the projection of an infinite-dimensional operator onto a finite-dimensional space) and $x \in \mathbf{C}^n = D_A$. The eigenvalues of $A$ are the solutions $\lambda \in \mathbf{C}$ of (4.1), which is equivalent to the set of values $\lambda$ for which $A - \lambda I$ is not invertible. For every eigenvalue $\lambda$, (4.2) has nontrivial solutions $x$, which are the eigenvectors corresponding to that particular eigenvalue. The set of all eigenvalues of $A$ is called the *spectrum* of $A$, denoted as $\sigma(A)$, i.e.,

$$\sigma(A) \equiv \left\{ \lambda : (A - \lambda I)^{-1} \text{ does not exist} \right\}.$$

Because we are interested in nontrivial solutions of (4.2), we get as a necessary condition

$$\det(A - \lambda I) = 0, \tag{4.6}$$

which leads to an $n$th-order polynomial known as the *characteristic polynomial.* By the fundamental theorem of algebra, (4.6) has $n$ roots, which form the spectrum[2] of $A$.

Once a particular eigenvalue has been determined, its corresponding eigenvector(s) can be found from (4.2). Those eigenvectors, together with the zero vector, form what is known as the *eigenspace* of $A$ for that particular eigenvalue (i.e., the null space of the operator $A - \lambda I$ for a given $\lambda$). The case of the union of all eigenspaces spanning $H$ is very important, as is discussed later in the spectral theorems.

The *algebraic multiplicity* of an eigenvalue is the number of times that eigenvalue appears as a root of (4.6) (i.e., its multiplicity as a zero of the characteristic polynomial), and its *geometric multiplicity* is the dimension of the corresponding eigenspace. It can be shown that the geometric multiplicity of an eigenvalue cannot exceed its algebraic multiplicity.

It is convenient to recall that the determinant of a matrix is equal to the product of its eigenvalues, counting algebraic multiplicity, i.e.,

$$\det(A) = \prod_{i=1}^{n} \lambda_i$$

such that if any eigenvalue has the value 0, then $\det(A) = 0$ and $A$ is not invertible.

Regarding the location of the eigenvalues we have the following theorem.

---

[2]It is important to work within complex Hilbert spaces even when dealing with real-valued matrices, because if we make the restriction $\lambda \in \mathbf{R}$ eigenvalues may not exist.

**Theorem 4.1.** *(Gerschgorin circle theorem) Every eigenvalue of the operator $A : \mathbf{C}^n \to \mathbf{C}^n$ lies in one of the complex domains*

$$|\lambda - a_{ii}| \leq \sum_{\substack{j=1 \\ j \neq i}} |a_{ij}|$$

*for $i = 1, 2, \ldots, n$.*

**Proof.** Let $\lambda$ be an eigenvalue that does not lie in one of the specified domains. Then, $|\lambda - a_{ii}| > \sum_{\substack{j=1 \\ j \neq i}} |a_{ij}|$ and so $(A - \lambda I)$ is diagonally dominant. By Theorem 3.26, $(A - \lambda I)$ has an inverse, which contradicts the assumption that $\lambda$ is an eigenvalue. ■

Furthermore, if the elements of the matrix $A$ are continuous functions of a parameter $\gamma$, such that the eigenvalue problem is

$$(A(\gamma) - \lambda(\gamma) I) x = 0,$$

then $\det (A(\gamma))$ is a continuous function of $\gamma$ and the eigenvalues $\lambda(\gamma)$ vary continuously with $\gamma$.

For all other values of $\lambda$, $A - \lambda I$ is invertible such that (4.2) has only the trivial solution $x = 0$. These values of $\lambda$ make up the *resolvent set*, denoted as $\rho(A)$, which is the complement of the spectrum, i.e.,

$$\rho(A) \equiv \mathbf{C} - \sigma(A).$$

### 4.1.2 Operators on Infinite-Dimensional Spaces

It turns out that the spectrum of a linear operator $A : H \to H$, where $H$ is an infinite-dimensional Hilbert space, is much more complicated than for an operator acting on a finite-dimensional space. Considering the operator $A : H \to H$ and motivated by (4.2), it is useful to examine properties of the linear operator $A_\lambda \equiv A - \lambda I$, where $I$ is the identity operator on $H$. We denote the domain of $A_\lambda$ as $D_{A-\lambda I} = D_A$, which will be assumed dense in $H$. The range of $A_\lambda$ will depend on $\lambda$ and may or may not be dense[3] in $H$. As discussed in the previous chapter, the operator inverse to $A_\lambda$ is called the *resolvent operator* (or simply the *resolvent*), denoted as

$$R_\lambda(A) \equiv (A - \lambda I)^{-1}.$$

For a given operator $A$, $R_\lambda(A)$ is an operator-valued function of $\lambda$, analytic on $\rho(A)$, with $R_{\lambda=0}(A)$ being simply the inverse operator $A^{-1}$. If $R_\lambda(A)$ exists for a particular $\lambda$, then that $\lambda$ cannot be an eigenvalue since (4.2) would only have trivial solutions. Therefore, it is not surprising that spectral properties of an operator $A$ can be determined by considering properties of $R_\lambda(A)$ for various values[4] of $\lambda$:

---

[3]Recall from Theorem 2.7 that $R_{A-\lambda I}$ is dense in $H$ if and only if $(R_{A-\lambda I})^\perp = \{0\}$. The dimension of $(R_{A-\lambda I})^\perp$ is called the *deficiency* of $\lambda$ for obvious reasons.

[4]This classification scheme is not unique. Here we follow [2, pp. 125–127] and [37, pp. 194–195], among others.

1. **Values of $\lambda$ for which $R_\lambda(A)$ does not exist** (i.e., $A - \lambda I$ is not invertible and therefore (4.2) has a nontrivial solution).

   The set of these values of $\lambda$ forms the *point spectrum* of $A$, denoted as $\sigma_p(A)$. Such values of $\lambda$ are called eigenvalues of $A$, and for a given eigenvalue $\lambda$ the corresponding nontrivial solutions $x \in H$ are eigenvectors corresponding to that eigenvalue. Eigenvalues of $A$ can be considered to be poles of the resolvent operator $R_\lambda(A)$.

2. **Values of $\lambda$ for which $R_\lambda(A)$ exists** (i.e., $A - \lambda I$ is invertible and (4.2) has only the trivial solution $x = 0$):

   (a) **Closure of the range of $A - \lambda I$ is $H$** (i.e., the range of $A - \lambda I$ is dense in $H$):

      i. if the above condition holds and $R_\lambda(A)$ is bounded, then such a value $\lambda$ is called a *regular value* of $A$. Regular values are not part of the spectrum of $A$, and therefore regular values make up the resolvent set $\rho(A)$.

      ii. if the above condition holds and $R_\lambda(A)$ is unbounded (not continuous), then such a value $\lambda$ is part of the *continuous spectrum* of $A$, denoted as $\sigma_c(A)$.

   (b) **Closure of the range of $A - \lambda I$ is a proper subset**[5] **of $H$.**

      The set of such values of $\lambda$ forms the *residual spectrum* of $A$, denoted as $\sigma_r(A)$.

The total spectrum is then

$$\sigma(A) = \mathbf{C} - \rho(A) = \sigma_r(A) \cup \sigma_p(A) \cup \sigma_c(A).$$

Note that we associate with operators on complex finite-dimensional spaces a point spectrum and a resolvent set, while the additional notions of continuous and residual spectrum only arise in the infinite-dimensional case. The residual spectrum does not usually occur in electromagnetic applications,[6] while the point and continuous spectrum are extremely important in the analysis to follow.

Because the above criteria are fairly easy to apply, except possibly the denseness condition, and since we will not be concerned with the residual spectrum here, then for our purposes it will be sufficient to check the existence of $R_A(\lambda)$—or the lack thereof. This will divide the continuous spectrum and resolvent set from the point spectrum. If $R_A(\lambda)$ exists, then its boundedness can be used to separate the continuous spectrum from the resolvent set.

---

[5] A set $X$ can be a subset of itself. To exclude this case, if $X$ is a subset of $Y$ and $X \neq Y$, we say $X$ is a *proper* subset of $Y$.

[6] For an example relating to the residual spectrum, see the analysis of the shift operator in [2, pp. 126–127].

In all examples presented here the residual spectrum will be empty (the range of $A - \lambda I$ will be dense in $H$); in a few later theorems the lack of a residual spectrum will be shown to be true for several classes of operators important in electromagnetic applications (see Theorems 4.18 and 4.26). It is worthwhile to note that for a linear operator on a complex Hilbert space, the spectrum is never empty.

Finally, since we are primarily interested in the point and continuous spectrum, and since we will later identify elements of the continuous spectrum as having some properties similar to those of eigenvalues, we introduce the concept of the *approximate spectrum* [2, p. 127].

**Definition 4.1.** *Consider a linear operator $A : H \to H$. A value $\lambda$ is in the approximate spectrum of $A$, denoted by $\sigma_a(A)$, if there exists a sequence of elements $x_n \in D_A$ with $\|x_n\| = 1$ such that $\|(A - \lambda I)x_n\| < 1/n$. The elements $x_n$ are called approximate eigenvectors of $A$, and the corresponding values $\lambda \in \sigma_a(A)$ are said to be approximate eigenvalues of $A$.*

**Theorem 4.2.** *The approximate spectrum of a linear operator $A : H \to H$ contains the point and continuous spectrum of $A$, but not any elements of the resolvent set.*

**Proof.** Let $\lambda_n \in \sigma_p(A)$, with corresponding eigenfunction $x_n$, $\|x_n\| = 1$. Because $(A - \lambda_n I)x_n = 0$, the desired inequality is seen to hold for all finite $n$. Now, let $\lambda \in \sigma_c(A)$. Because by definition $(A - \lambda I)^{-1}$ is unbounded, there exists a sequence of functions $y_n$ such that $\|(A - \lambda I)^{-1} y_n\| > n\|y_n\|$. With $x_n = (A - \lambda I)^{-1} y_n$ we have $\|x_n\| > n\|(A - \lambda I)x_n\|$, or $\|(A - \lambda I)x_n\| < (1/n)\|x_n\|$ and so with $\|x_n\| = 1$, $\lambda$ clearly belongs to the approximate spectrum. If $\lambda \in \rho(A)$, then $(A - \lambda I)^{-1}$ is bounded and so $\lambda$ cannot be in the approximate spectrum. ■

Elements of the residual spectrum may or may not be in the approximate spectrum.

### Examples of Point Spectrum, Continuous Spectrum, and Resolvent Set

1. For the operator $A : H \to H$ defined by $Ax \equiv \alpha x$, where $\alpha \in \mathbf{C}$ (i.e., multiplication by a complex number, $A = \alpha I$), all vectors $x \in H$, $x \neq 0$ are eigenvectors with eigenvalues $\lambda = \alpha$. The only point in the spectrum is $\lambda = \alpha$, which has infinite multiplicity assuming $H$ is infinite dimensional. Therefore, the resolvent set is $\rho(A) = \mathbf{C} - \{\alpha\}$. In particular, for $\alpha = 1$ we have the identity operator, and for $\alpha = 0$ we have the zero operator.

2. For the matrix operator $\begin{bmatrix} -5 & 0 \\ 1 & 2 \end{bmatrix}$, eigenvalues and corresponding

eigenvectors are determined from (4.6) and (4.2) as

$$\lambda = 2, \qquad x = \begin{bmatrix} 0 \\ \alpha \end{bmatrix},$$

$$\lambda = -5, \qquad x = \begin{bmatrix} -7\beta \\ \beta \end{bmatrix},$$

where $\alpha, \beta \neq 0$, such that $\rho(A) = \mathbf{C} - \{2, -5\}$.

3. The multiplication operator $A : \mathbf{L}^2(a,b) \to \mathbf{L}^2(a,b)$ defined by

$$(Ax)(t) \equiv f(t)x(t)$$

is a linear bounded operator assuming $\max_t |f(t)| < \infty$.

First, consider the case of $f(t) = c$, where $c \in \mathbf{C}$ is a constant. Then

$$(A - \lambda I)^{-1} x(t) = (c - \lambda)^{-1} x(t)$$

such that the resolvent fails to exist for $\lambda = c$, which is an eigenvalue of $A$ (any nonzero $x \in \mathbf{L}^2(a,b)$ is a corresponding eigenvector). For all other values of $\lambda$ the resolvent exists and is bounded, and therefore the continuous spectrum is empty and the resolvent set is $\rho(A) = \mathbf{C} - \{c\}$.

Next, consider the case where $f(t) = t$. Then, by considering

$$(A - \lambda I)^{-1} x(t) = (t - \lambda)^{-1} x(t),$$

it is clear that the resolvent generally exists but is unbounded if $a \leq \lambda \leq b$. Therefore, every $\lambda \in [a,b]$ is in the continuous spectrum. For all other values of $\lambda$ the resolvent exists and is bounded, and so the point spectrum is empty and every value $\lambda \notin [a,b]$ is in the resolvent set (note that although $(t-\lambda)\,\delta(t-\lambda) = 0$, $\delta(t-\lambda) \notin \mathbf{L}^2$ and so $\delta(t-\lambda)$ is not an eigenvector).

4. The differential operator $A : \mathbf{L}^2(0,\infty) \to \mathbf{L}^2(0,\infty)$ defined by[7]

$$(Ax)(t) \equiv \frac{d}{dt}x(t),$$

$$D_A \equiv \left\{ x : x, \frac{d}{dt}x(t) \in \mathbf{L}^2(0,\infty) \right\}$$

is a linear unbounded operator. Consider the equation

$$(A - \lambda I)\, x = (x' - \lambda x) = -y.$$

For $y = 0$ and $\text{Re}(\lambda) < 0$, $(A - \lambda I)\, e^{\lambda t} = 0$ so that $e^{\lambda t}$ is an eigenfunction of $A$ with eigenvalue $\lambda$ (with $\text{Re}(\lambda) < 0$; for $\text{Re}(\lambda) >$

[7]Strictly speaking, $x$ must be absolutely continuous on $(0, \infty)$.

0, $e^{\lambda t} \notin \mathbf{L}^2(0,\infty)$ and so in this case $e^{\lambda t}$ is not an eigenfunction). Therefore, $\sigma_p(A)$ is that part of the complex $\lambda$-plane with $\mathrm{Re}(\lambda) < 0$ (the left half-plane).[8]

Next, consider the solution of $(A - \lambda I)\,x = (x' - \lambda x) = -y$ for $y \neq 0$. The formal solution is[9]

$$x(t) = \int_t^\infty e^{\lambda(t-\tau)} y(\tau)\, d\tau$$

and so the resolvent operator is formally identified by

$$R_\lambda(A)y = \int_t^\infty e^{\lambda(t-\tau)} y(\tau)\, d\tau.$$

This operator exists and is continuous for all $\lambda$ such that $\mathrm{Re}(\lambda) > 0$, and so this part of the complex $\lambda$-plane (the right half-plane) comprises the resolvent set.

Finally, considering the case $\lambda = 0$ ($x(t) = \int_t^\infty y(\tau)\, d\tau$) it can be seen that the resolvent exists but is not continuous (i.e., not bounded; compare with Example 12, Section 3.1.2), and so $\lambda = 0$ is in the continuous spectrum. This also remains true for $\mathrm{Re}(\lambda) = 0$, so $\sigma_c(A)$ is the imaginary axis of the complex $\lambda$-plane. Because $\sigma(A) = \mathbf{C} = \rho(A) \cup \sigma_r(A) \cup \sigma_p(A) \cup \sigma_c(A)$, we see that the residual spectrum is empty, i.e., $\sigma_r(A) = \emptyset$. Figure 4.2 depicts the sets $\sigma(A)$ and $\rho(A)$ in the complex $\lambda$-plane.

5. Consider the differential operator $A : \mathbf{L}^2(0,a) \to \mathbf{L}^2(0,a)$ defined by

$$(Ax)(t) \equiv -\frac{d^2}{dt^2} x(t),$$
$$D_A \equiv \left\{ x : x, \frac{d^2}{dt^2} x \in \mathbf{L}^2(0,a), x(0) = x(a) = 0 \right\}.$$

Eigenfunctions are found to be $\sin \sqrt{\lambda_n} t$, with corresponding eigenvalues $\lambda_n = (n\pi/a)^2$, $n = 1, 2, \ldots$ (defining a branch of the square root $\sqrt{\lambda_n}$ leads to, for example, only positive integers $n$). The point spectrum, which in this case is discrete, is $\sigma_p(A) = \{\lambda_n\}$, the residual and continuous spectrum are empty, and the resolvent set is $\rho(A) = \mathbf{C} \setminus \{\lambda_n\}$. This is depicted in Figure 4.3.

---

[8] Note that although the point spectrum is often discrete, this example shows that this need not be the case.

[9] Indeed, differentiating $x$ we obtain

$$\frac{d}{dt}x(t) = \int_t^\infty \frac{d}{dt} e^{\lambda(t-\tau)} y(\tau)\, d\tau - y(t) = \lambda \int_t^\infty e^{\lambda(t-\tau)} y(\tau)\, d\tau - y(t) = \lambda x(t) - y(t).$$

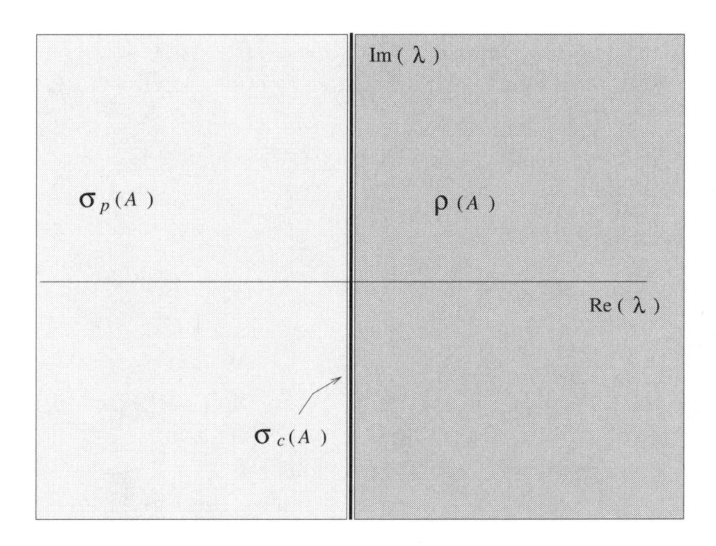


Figure 4.2: Complex $\lambda$-plane depicting the sets $\sigma(A)$ and $\rho(A)$ for the differential operator in Example 4. The point spectrum $\sigma_p(A)$ is the left half-plane, the resolvent set $\rho(A)$ is the right half-plane, and the continuous spectrum $\sigma_c(A)$ is the imaginary axis.

It should be noted that eigenfunctions of differential operators never satisfy nonhomogeneous boundary conditions, since this generally wouldn't allow for the determination of $\lambda$; they satisfy either homogeneous conditions, periodic conditions, or some sort of fitness conditions as described in Chapter 5.

## 4.2 Spectral Properties of Operators

In this section we present some spectral properties important in both theory and applications. We are particularly interested in linear independence, orthogonality, and especially completeness of the eigenvectors of a given operator class in a particular space. Even for general linear operators some strong statements can be made, although the best situations arise when the operators belong to some appropriately restricted class.

### 4.2.1 General Properties

It is very important to know when eigenvectors of an operator form a linearly independent set. The next theorem addresses this issue.

**Theorem 4.3.** *For a linear operator $A : H \to H$, eigenvectors $x_1, x_2, \ldots, x_n$ corresponding to distinct eigenvalues $\lambda_1, \lambda_2, \ldots, \lambda_n$ form a linearly independent set in $H$.*

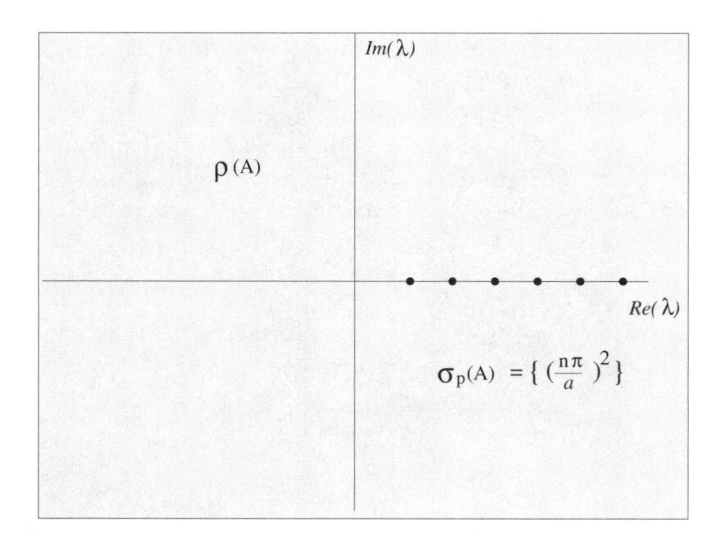


Figure 4.3: Complex $\lambda$-plane depicting the sets $\sigma(A)$ and $\rho(A)$ for the differential operator in Example 5. The point spectrum $\sigma_p(A) = \sigma(A)$ is the discrete set of points $\lambda_n = (n\pi/a)^2$, $n = 1, 2, \ldots$, and the resolvent set $\rho(A)$ is everything else.

**Proof.** Assume that $\{x_1, x_2, \ldots, x_n\}$ forms a linearly dependent set and that $x_m$ is the first element that can be written as a linear combination of the preceding (and thus linearly independent) elements $\{x_1, x_2, \ldots, x_{m-1}\}$ so that $x_m = \sum_{i=1}^{m-1} \alpha_i x_i$. Applying the operator $(A - \lambda_m I)$ to this sum we get

$$(A - \lambda_m I)\, x_m = 0 = (A - \lambda_m I) \sum_{i=1}^{m-1} \alpha_i x_i = \sum_{i=1}^{m-1} \alpha_i (\lambda_i - \lambda_m)\, x_i.$$

Because the elements on the right form a linearly independent set that sums to zero with coefficients $\alpha_i(\lambda_i - \lambda_m)$, then these coefficients must all vanish. We then have $\alpha_i(\lambda_i - \lambda_m) = 0$, but $\lambda_i - \lambda_m \neq 0$ by assumption and so $\alpha_i = 0$, $i = 1, \ldots, m-1$. But then $x_m = 0$, which contradicts the fact that $x_m$ is an eigenvector. Therefore, we have shown that $\{x_1, x_2, \ldots, x_n\}$ forms a linearly independent set. ■

A linearly independent set can be converted to an orthonormal set by the *Gram–Schmidt procedure*, as detailed in Appendix C.

**Theorem 4.4.** *If a linear operator* $A : H \to H$ *has eigenvalues* $\lambda_n$ *and eigenvectors* $x_n$*, i.e.,* $Ax_n = \lambda_n x_n$*, then* $\lambda_n - \alpha$ *and* $x_n$ *are eigenvalues and eigenvectors, respectively, of the operator* $B : H \to H$ *defined as* $B \equiv A - \alpha I$*.*

**Proof.** Let the eigenvalues and eigenvectors of $A$ be $\lambda_n$ and $x_n$, respectively, and let the eigenvalues and eigenvectors of $(A - \alpha I)$ be $\gamma_n$ and $y_n$, respectively. Then $(A - \alpha I)\, y_n = \gamma_n y_n$ and $Ay_n - \alpha y_n = \gamma_n y_n$, such that $Ay_n = (\gamma_n + \alpha) y_n$, which shows that $y_n$ is an eigenvector of $A$ with eigenvalue $\gamma_n + \alpha$. Therefore, $y_n = x_n$ and $\gamma_n + \alpha = \lambda_n$, or $\gamma_n = \lambda_n - \alpha$. ∎

The next theorem shows that parts of the spectrum of an operator are often related to (perhaps different) parts of the spectrum of the adjoint operator.

**Theorem 4.5.** *If $\lambda$ is in the residual spectrum of a linear operator $A : H \to H$ with deficiency $m$, then $\overline{\lambda}$ is an eigenvalue of $A^*$ of multiplicity $m$.*

That is, if $\lambda \in \sigma_r(A)$, then $\overline{\lambda} \in \sigma_p(A^*)$.

**Proof.** For $\lambda \in \sigma_r(A)$, $R_{A-\lambda I}$ is not dense in $H$ (by definition), and so $(R_{A-\lambda I})^{\perp}$ is not empty. By Theorem 3.7, $(R_{A-\lambda I})^{\perp} = N_{(A-\lambda I)^*} = N_{A^* - \overline{\lambda} I}$, and so $N_{A^* - \overline{\lambda} I}$ is not empty; hence $\overline{\lambda} \in \sigma_p(A^*)$. The multiplicity argument is fairly obvious. ∎

More generally, one can show the following [3, p. 228].

**Theorem 4.6.** *For any bounded linear operator $A : H \to H$,*

$$\sigma(A^*) = \overline{\sigma}(A). \tag{4.7}$$

**Proof.** If $\lambda \in \rho(A)$, then $R_A(\lambda)$ is bounded, with the range of the invertible operator $(A - \lambda I)$ dense in $H$. The operator $(A - \lambda I)$ then satisfies the conditions of Theorem 3.34, such that $(A - \lambda I)^* = \left(A^* - \overline{\lambda} I\right)$ has range dense in $H$ and is invertible with bounded inverse. Therefore, $\overline{\lambda} \in \rho(A^*)$, and we have $\sigma(A^*) \subseteq \overline{\sigma}(A)$. Reversing the process by starting with $\lambda \in \rho(A^*)$ leads to $\sigma(A^*) \subseteq \overline{\sigma}(A)$ and hence $\sigma(A^*) = \overline{\sigma}(A)$. ∎

It is important to note that this result applies to the entire spectrum; spectral elements $\sigma_p$, $\sigma_r$, $\sigma_c$ do not necessarily individually satisfy (4.7). In fact, the presence of a nonempty residual spectrum complicates relating eigenvalues of $A$ to those of $A^*$. It turns out, however, that the point spectrum of an operator is often related to the point spectrum of its adjoint, as the following theorem states (see also Theorem 4.24 and [2, p. 90]).

**Theorem 4.7.** *Let $A, A^* : H \to H$ have empty residual spectra, and let $A$ have eigenvalues $\lambda_n$ and corresponding eigenvectors $x_n$ (i.e., $Ax_n = \lambda_n x_n$). Then $\overline{\lambda}_n$ are eigenvalues of the adjoint operator $A^* : H \to H$ corresponding to adjoint eigenvectors $y_n$ (i.e., $A^* y_n = \overline{\lambda}_n y_n$), and*

$$(\lambda_n - \lambda_m) \langle x_n, y_m \rangle = 0. \tag{4.8}$$

**Proof.** Assume that $\lambda_n, x_n$ are eigenvalues and eigenvectors, respectively, of $A : H \to H$ and $y$ an arbitrary function in the domain of $A^*$. Then $(A - \lambda_n I)\, x_n = 0$ and $\langle y, (A - \lambda_n I)\, x_n \rangle = 0$ and, upon noting

$\langle y, Ax \rangle = \langle A^* y, x \rangle$ we have $\left\langle (A - \lambda_n I)^* y, x_n \right\rangle = \left\langle \left(A^* - \overline{\lambda}_n I\right) y, x_n \right\rangle = 0$. Thus, either $\left(A^* - \overline{\lambda}_n I\right) y = 0$, indicating $y$ is an eigenvector of $A^*$ with eigenvalue $\overline{\lambda}_n$, or $\left(A^* - \overline{\lambda}_n I\right) y = f \neq 0$. If the latter is true and $y$ is arbitrary, then $\langle f, x_n \rangle = 0$ for $f$ arbitrary, implying $x_n = 0$. Because this cannot occur, the former statement must be true.[10]

To prove the relationship among the eigenvectors of $A$ and $A^*$, start with $Ax_n = \lambda_n x_n$ and $A^* y_m = \overline{\lambda}_m y_m$ (under suitable ordering of the eigenvalues). Taking inner products we have $\langle Ax_n, y_m \rangle = \langle \lambda_n x_n, y_m \rangle = \lambda_n \langle x_n, y_m \rangle$ and $\langle A^* y_m, x_n \rangle = \left\langle \overline{\lambda}_m y_m, x_n \right\rangle = \overline{\lambda}_m \langle y_m, x_n \rangle$, where the latter becomes $\langle x_n, A^* y_m \rangle = \lambda_m \langle x_n, y_m \rangle$. Forming $0 = \langle Ax_n, y_m \rangle - \langle x_n, A^* y_m \rangle = \lambda_n \langle x_n, y_m \rangle - \lambda_m \langle x_n, y_m \rangle$ leads to $(\lambda_n - \lambda_m) \langle x_n, y_m \rangle = 0$, or $\langle x_n, y_m \rangle = 0$ for $\lambda_n \neq \lambda_m$. ■

Note that (4.8) can be interpreted as

$$N_{A-\lambda_n I} \perp N_{A-\lambda_m I}$$

for $\lambda_n \neq \lambda_m$. In particular, for a self-adjoint operator the set of eigenvectors (if any even exist) corresponding to distinct eigenvalues forms an orthogonal set. For nonself-adjoint operators, the set of eigenvectors (if any even exist) and adjoint eigenvectors corresponding to distinct eigenvalues forms a biorthogonal set.

### Rayleigh Quotient

Assuming $x_1, x_2, \ldots, x_n$ are orthogonal eigenvectors corresponding to different eigenvalues $\lambda_1, \lambda_2, \ldots, \lambda_n$ of $A : H \to H$, one can obtain the eigenvalues using the *Rayleigh quotient* representation,

$$\lambda_n = \frac{\langle Ax_n, x_n \rangle}{\langle x_n, x_n \rangle}. \tag{4.9}$$

Of course, for (4.9) to be directly useful for calculation of the eigenvalue $\lambda_n$ one must know the eigenvector $x_n$. However, if we consider the functional

$$F(x) \equiv \frac{\langle Ax, x \rangle}{\langle x, x \rangle}, \tag{4.10}$$

then it can be shown that for many types of operators (not necessarily bounded) the functional $F$ is stationary at $x_n$, in the sense that first-order errors in approximating the element $x_n$ lead to second-order errors in calculating the eigenvalue $\lambda_n$. That is, if $x = x_n + O(\varepsilon)$, then $F(x) = \lambda_n + O(\varepsilon^2)$.

---

[10] Lack of a residual spectrum is important for the following reason: By definition, $f \in R_{A^* - \overline{\lambda}_n I}$. By excluding the possibility of a residual spectrum the range of $A^* - \overline{\lambda}_n I$ is dense in $H$, assuming $\overline{\lambda}_n$ is not an eigenvalue of $A^*$. The conclusion that $\langle f, x_n \rangle = 0$ for $f$ arbitrary implies $x_n = 0$ is only valid if $f$ is an arbitrary element of $H$, or of a space dense in $H$.

When $L_\lambda$ is real and self-adjoint (a common case in electromagnetics), one can obtain a further, actually more useful, result, namely

$$g(t,t',\lambda) = g(t',t,\lambda). \tag{3.27}$$

To see this, note that if $L_\lambda$ is self-adjoint then $L_\lambda^* g^*(t,t',\lambda) = L_\lambda g^*(t,t',\lambda) = \delta(t-t')$, and $\overline{L_\lambda g^*(t,t',\lambda)} = L_\lambda \overline{g^*}(t,t',\lambda) = \delta(t-t')$, since $L_\lambda$ is real. From the last equality one sees that if the boundary conditions on $g$ and those on $\overline{g^*}$ are identical, then $g(t,t',\lambda) = \overline{g^*}(t,t',\lambda)$, proving (3.27) from (3.26). The statement regarding the boundary conditions can be seen to hold since the boundary conditions on $g$ and those on $g^*$ will be identical by self-adjointness of $L_\lambda$ and, assuming conditions of the form (3.22) on $g^*$ with $\alpha_{i,j}, \beta_{i,j} \in \mathbf{R}$, $0 = B_i(g) = B_i(g^*) = \overline{B_i(g^*)} = B_i\left(\overline{g^*}\right)$.

It is important to note that (3.27) also holds for $L_\lambda$ real and formally self-adjoint (symmetric) if the boundary conditions on $g$ and those on $\overline{g^*}$ are identical. For the operator leading to the free-space Green's function ($L = -\nabla^2$) this is the case, since the radiation condition for $g$ (outgoing) is the conjugate of the radiation condition for $g^*$ (incoming).[30] In electromagnetics one can also determine (3.27) based on reciprocity. If the medium is nonreciprocal, $L_\lambda$ will not be self-adjoint and (3.27) will generally not hold.

It is worthwhile to point out that if $L_\lambda$ is self-adjoint, one does not need to consider the adjoint problem; it is even more important to note that often the adjoint problem does not need to be considered even if $L_\lambda$ is nonself-adjoint. Assuming $L_\lambda$ is invertible, all one needs is an appropriate Green's theorem (which will be available whenever the operator is formally self-adjoint) applied to the elements $\psi, g$ (see Section 1.3.4 and Chapter 5), whereby one obtains an explicit solution. The various Green's theorems are usually stated for $L$ real and formally self-adjoint (e.g. $L = d^2/dt^2$, $\nabla^2$, $\nabla \times \nabla \times$); since this is the usual case in electromagnetics, it is usually not necessary to consider the adjoint problem.

However, if the homogeneous equation $L_\lambda \psi_0 = 0$ admits nonhomogeneous solutions, then $L_\lambda \psi = \phi$ cannot be solved uniquely, and perhaps not at all. To see this, note that if $L_\lambda \psi_0 = 0$ has nontrivial solutions, then by Theorem 3.25 $L_\lambda$ is not invertible. Therefore, the Green's operator will not exist. However, in Section 3.11 we show that under these conditions $L_\lambda \psi = \phi$ will have a (nonunique) solution if and only if $\phi \perp \psi_0^*$, where $\psi_0^*$ are solutions of the adjoint homogenous equation, $L_\lambda^* \psi_0^* = 0$. In this case, what is known as an *extended* or *generalized* Green's operator can be found (see [39, p. 12] for details). Therefore, in these cases consideration of the adjoint problem will be necessary to ascertain solvability of the equations.

In summary, if $L$ is formally self-adjoint (symmetric), then one need not consider the adjoint problem if $L$ is invertible. Of course, if $L$ is self-adjoint, the adjoint and original problems coincide. In more general cases one needs

[30] However, to work in an $\mathbf{L}^2$ space we consider the space to have small loss.

If $F(x)$ is stationary at all $x_n$, then all eigenvalues $\lambda_n$ can be approximated (to first order) using appropriate trial functions for $x_n$. This is a common method in applications. Furthermore, if $A$ is self-adjoint with a discrete spectrum, then various eigenvalues can be obtained from (4.10) by minimizing $F(x)$. For instance, the smallest eigenvalue of $A$ is equal to $\min_{x \in D_A} F(x)$, with the minimum attained when $x = x_n$ [2, p. 209]. Also, if $A$ is bounded, the largest eigenvalue can be obtained as $\max_{x \in D_A} F(x)$. Often one can go further and develop so-called minimax theorems for obtaining all eigenvalues $\lambda_n$ as minimizations of $F(x)$ over specially chosen sets $\{x\}$ within $D_A$. See, e.g., [4, pp. 419–421] for a detailed discussion.

**Example 4.2.**

Consider a source-free cavity occupying a region $\Omega$ with perfectly conducting boundary surface $\Gamma$ and filled with a homogeneous medium characterized by $\mu, \varepsilon$. The electromagnetic field satisfies

$$\begin{aligned} \nabla \times \nabla \times \mathbf{E} &= k^2 \mathbf{E}, \\ \nabla \cdot \mathbf{E} &= 0, \\ \mathbf{n} \times \mathbf{E}|_\Gamma &= \mathbf{0}, \end{aligned}$$

where $k^2 = k_n^2 = \omega_n^2 \mu\varepsilon > 0$ represents real-valued eigenvalues of the operator $\nabla \times \nabla \times$ corresponding to eigenfunctions $\mathbf{E} = \mathbf{E}_n$. Then, the functional

$$\omega_n^2(\mathbf{E}) = \frac{1}{\mu\varepsilon} \frac{\langle \nabla \times \nabla \times \mathbf{E}, \mathbf{E} \rangle}{\langle \mathbf{E}, \mathbf{E} \rangle} = \frac{1}{\mu\varepsilon} \frac{\int_\Omega \nabla \times \nabla \times \mathbf{E} \cdot \mathbf{E} \, d\Omega}{\int_\Omega \mathbf{E} \cdot \mathbf{E} \, d\Omega} \tag{4.11}$$

is identically satisfied for $\mathbf{E} = \mathbf{E}_n$ and stationary [5, p. 332] for all "trial" eigenfunctions $\mathbf{E}_n^t = \mathbf{E}_n + \delta \mathbf{E}_n$ such that $\mathbf{n} \times \mathbf{E}_n^t|_\Gamma = \mathbf{0}$.

### 4.2.2 Bounded Operators

If a linear operator is bounded, then it turns out that its spectrum is bounded as well.[11]

**Theorem 4.8.** *Let $A : H \to H$ be a bounded linear operator. The spectrum $\sigma(A)$ is a compact (therefore, necessarily closed and bounded) subset of the complex plane lying within the closed region $|z| \le \|A\|$.*

It can be easily seen that if $A : H \to H$ is a bounded linear operator and $|z| > \|A\|$, then $z$ is in the resolvent set of $A$. Therefore, for a bounded linear operator the resolvent set is never empty.

A more restrictive theorem relating to the eigenvalues themselves is easier to prove.

[11] See Theorem 4.1 for an explicit formula pertaining to operators on finite-dimensional spaces.

**Theorem 4.9.** *If $\lambda$ is an eigenvalue of a linear bounded operator $A : H \to H$, then $|\lambda| \leq \|A\|$.*

**Proof.** Starting with $Ax = \lambda x$, we obtain $\|Ax\| = \|\lambda x\| = |\lambda| \|x\|$. However, $\|Ax\| \leq \|A\| \|x\|$ and so $|\lambda| \|x\| \leq \|A\| \|x\|$, implying $|\lambda| \leq \|A\|$. ∎

The concept of a *spectral radius* is motivated by the preceding.

**Definition 4.2.** *The spectral radius of an operator, denoted as $r_\sigma(A)$, is the radius of the smallest closed disk, centered at the origin of the complex plane, containing $\sigma(A)$. The spectral radius is given by*

$$r_\sigma(A) = \sup_{\lambda \in \sigma(A)} |\lambda| .$$

The next theorem shows the connection between bounded operators and their eigenvalues.

**Theorem 4.10.** *A linear operator $A : H \to H$ such that $\sigma_p(A) = \sigma(A)$ is continuous (bounded) if and only if the set $\{|\lambda_1|, |\lambda_2|, \dots\}$ is bounded. Furthermore, $\|A\| \geq r_\sigma(A)$.*

**Proof.** Assume $A : H \to H$ is bounded. Then $\{|\lambda_n|\}$ is bounded by Theorem 4.9. Now assume $\{|\lambda_n|\}$ is a bounded set. Because $\|Ax\| = \|\lambda x\| = |\lambda| \|x\|$ and $|\lambda|$ is bounded, then by Definition 3.6, $A$ is bounded. ∎

### 4.2.3 Invertible Operators

An important observation can be made about the spectral properties of invertible operators.

**Theorem 4.11.** *Let $A : H \to H$ be an invertible linear operator with eigenvalues $\lambda$ and corresponding eigenvectors $x$ (recall that if $A$ is invertible, $N_A = \{0\}$ from Theorem 3.24, and so $\lambda \neq 0$). Then, $A^{-1} : H \to H$ has eigenvalues $1/\lambda$ and corresponding eigenvectors $x$.*

**Proof.** It is given that we have $Ax = \lambda x$ with $\lambda, x \neq 0$. Because $A^{-1}$ exists we have $A^{-1}Ax = A^{-1}\lambda x$ such that $x = \lambda A^{-1}x$, or $A^{-1}x = (1/\lambda)x$. ∎

Furthermore, if $A$ and $A^{-1}$ are bounded, then

$$\sigma(A^{-1}) = \sigma^{-1}(A).$$

The following theorem shows a relationship pertaining to similarity transformations, which will be discussed in Section 4.3.1 in conjunction with the theory of matrix representations of operators.

**Theorem 4.12.** *Let $A : H \to H$ be a linear operator and let the linear operator $T : H \to H$ be invertible. The operators $A$ and $TAT^{-1}$ (which are called similar) have the same eigenvalues.*

**Proof.** Let $\lambda$ be an eigenvalue of $A$ with eigenvector $x$. Then we have $TAT^{-1}(Tx) = TAx = T\lambda x = \lambda(Tx)$, showing that $\lambda$ is an eigenvalue of $TAT^{-1}$. If $\lambda$ is an eigenvalue of $TAT^{-1}$ with eigenvector $y$, then $TAT^{-1}y = \lambda y$, and so $AT^{-1}y = \lambda T^{-1}y$, showing that $\lambda$ is an eigenvalue of $A$. ■

### 4.2.4 Compact Operators

Compact operators need not have any eigenvalues (as a common example the integral operator $A : \mathbf{L}^2(0,1) \to \mathbf{L}^2(0,1)$ defined by $(Af)(t) \equiv \int_0^t f(\tau)\,d\tau$ is Hilbert–Schmidt, and hence compact, but does not possess any eigenvalues[12]). However, if eigenvalues do exist, then the following theorems are useful.

**Theorem 4.13.** *Every spectral value $\lambda \neq 0$ of a compact linear operator $A : H \to H$ is an eigenvalue.*

**Proof.** If $R_\lambda(A) = (A - \lambda I)^{-1}$ does not exist, then $\lambda \in \sigma_p(A)$ by definition. Now, let $\lambda \neq 0$ and assume that $R_\lambda(A)$ exists. Then $(A - \lambda I)x = 0$ implies $x = 0$, and so the Fredholm alternative 3.44 shows that $(A - \lambda I)x = y$ is uniquely solvable for all $y \in H$. Therefore, $R_\lambda(A)$ is defined on all $y \in H$ and is bounded by the open mapping theorem (Theorem 3.27), and $\lambda \in \rho(A)$. ■

Therefore, the continuous spectrum and residual spectrum of a compact operator either are empty or are the zero element.

**Theorem 4.14.** *Let $A : H \to H$ be a compact linear operator. Then, the null space of $A - \lambda I$, $\lambda \neq 0$, is finite dimensional.*

**Theorem 4.15.** *Let $A : H \to H$ be a compact linear operator. Then, for $\alpha > 0$ the number of eigenvalues $\lambda$ such that $|\lambda| \geq \alpha$ is finite.*

### 4.2.5 Self-Adjoint Operators

Self-adjoint operators may not have any eigenvalues (recall the multiplication operator example, p. 226), but if they do possess eigenvalues, then the eigenvalues and corresponding eigenvectors have some very desirable properties.

[12]To see this consider $\lambda = 0$ as a possible eigenvalue, leading to $Ax = 0x = 0$. For the operator in question this indicates $x = 0$. For some $\lambda \neq 0$ as a possible eigenvalue, the equation $Ax = \lambda x$ implies $x(0) = 0$ and leads to $x = \lambda x'$, the only solution of which is $x = 0$.

**Theorem 4.16.** *Let $A : H \to H$ be a self-adjoint linear operator. Then*

a. *the eigenvalues of $A$ are real, and*

b. *the eigenvectors corresponding to distinct eigenvalues are orthogonal.*

**Proof.** (a) Let $x$ be an eigenvector corresponding to eigenvalue $\lambda$, satisfying $Ax = \lambda x$. Then $\langle Ax, x\rangle = \langle \lambda x, x\rangle = \lambda \langle x, x\rangle$. Also, since $A$ is self-adjoint, $\langle Ax, x\rangle = \langle x, Ax\rangle = \langle x, \lambda x\rangle = \overline{\lambda} \langle x, x\rangle$. Therefore, since $\langle x, x\rangle > 0$, $\lambda = \overline{\lambda}$, proving[13] $\lambda \in \mathbf{R}$. (b) Let $x_1, x_2$ be eigenvectors corresponding to distinct eigenvalues $\lambda_1$, $\lambda_2$ of the self-adjoint operator $A$, so that we have $Ax_1 = \lambda_1 x_1$ and $Ax_2 = \lambda_2 x_2$. Then $\lambda_1 \langle x_1, x_2\rangle = \langle Ax_1, x_2\rangle = \langle x_1, Ax_2\rangle = \langle x_1, \lambda_2 x_2\rangle = \overline{\lambda_2} \langle x_1, x_2\rangle = \lambda_2 \langle x_1, x_2\rangle$, where the last equality comes about since eigenvalues are real. We have $(\lambda_1 - \lambda_2) \langle x_1, x_2\rangle = 0$, and, since $\lambda_1 \neq \lambda_2$, then $\langle x_1, x_2\rangle = 0$ as desired. ■

Unfortunately, all eigenvectors of a self-adjoint operator are not necessarily mutually orthogonal, only those corresponding to distinct eigenvalues (although within each eigenspace we can orthogonalize the set). If an operator is real and self-adjoint, then, if eigenvectors exist, they either are real-valued or can be made so,[14] although in the latter case orthogonality will generally be lost.

Furthermore, beyond considerations of the eigenvalues themselves, we have the following.

**Theorem 4.17.** *The spectrum $\sigma(A)$ of a bounded self-adjoint linear operator $A : H \to H$ is real and is a subset of the interval $[-\|A\|, \|A\|]$. Furthermore, for $\lambda \in \rho(A)$,*

$$\|R_\lambda (A)\| \leq \frac{1}{|\operatorname{Im} \lambda|}.$$

The proof is fairly simple for the eigenvalues themselves, but more difficult for the entire spectrum.

The above indicates that for a self-adjoint operator, every $\lambda$ with $\operatorname{Im} \lambda \neq 0$ is in the resolvent set. We also have the following theorem regarding the residual spectrum.

**Theorem 4.18.** *The residual spectrum $\sigma_r(A)$ of a self-adjoint linear operator $A : H \to H$ is empty.*

---

[13] Obviously this is also true if $A$ is merely symmetric. Note also that the converse is also true, i.e., that if an eigenvalue $\lambda$ is real-valued, then $A$ is at least symmetric. To see this assume $\lambda \in \mathbf{R}$, and note that $\langle Ax, x\rangle = \langle \lambda x, x\rangle = \lambda \langle x, x\rangle = \overline{\lambda} \langle x, x\rangle = \langle x, \lambda x\rangle = \langle x, Ax\rangle$.

[14] Let $A : H \to H$ be real and self-adjoint, with $x$ being an eigenvector of $A$ with corresponding (real) eigenvalue $\lambda$. Then $(A - \lambda I)(x - \overline{x}) = (A - \lambda I)\overline{x} = \overline{(A - \lambda I)\overline{x}} = (A - \lambda I) x = 0$, and therefore $(x - \overline{x})$ is a (real-valued) eigenvector of $A$.

The proof follows as a special case of Theorem 4.26 presented later. Note that, at most, the spectrum of a self-adjoint linear operator will contain a discrete and a continuous part, and both will be bounded on the real axis if the operator is bounded.

For unbounded linear operators, where the distinction between self-adjointness and symmetry is important, we have similar results for operators that are merely symmetric, namely that the point spectrum and continuous spectrum are real (the residual spectrum, if it exists, need not be real) and that eigenvectors corresponding to distinct eigenvalues are orthogonal.

### 4.2.6 Nonnegative and Positive Operators

The eigenvalues of nonnegative and positive operators (recall that these are special cases of self-adjoint, or at least symmetric, operators) have the expected spectral properties detailed in the next theorem.

**Theorem 4.19.** *All eigenvalues of a nonnegative linear operator are nonnegative. All eigenvalues of a positive linear operator are positive.*

**Proof.** Consider $Ax = \lambda x$, with $A$ nonnegative and $x \neq 0$. Then $0 \leq \langle Ax, x \rangle = \lambda \langle x, x \rangle = \lambda \|x\|^2$, which shows that $\lambda \geq 0$ as required. Repeating for $\langle Ax, x \rangle > 0$ leads to $\lambda > 0$. ■

Similar statements can be made about nonpositive and negative operators and about the entire spectrum $\sigma(A)$. For a bounded dissipative operator ($\operatorname{Im} \langle Ax, x \rangle \geq 0$) we have the following.

**Theorem 4.20.** *The eigenvalues of a bounded dissipative linear operator lie in the half-plane* $\operatorname{Im} \lambda \geq 0$.

**Proof.** Consider $Ax = \lambda x$, with $A$ dissipative and $x \neq 0$. Then $0 \leq \operatorname{Im} \langle Ax, x \rangle = \operatorname{Im} (\lambda \langle x, x \rangle) = \operatorname{Im} (\lambda) \|x\|^2$, which shows that $\operatorname{Im} \lambda \geq 0$ as required. ■

Furthermore, it can be shown that $\sigma(A)$ lies in the half-plane $\operatorname{Im} \lambda \geq 0$.

### 4.2.7 Compact Self-Adjoint Operators

It was shown previously that compact operators and self-adjoint operators may not possess any eigenvalues; however, if eigenvalues do exist, then they have some convenient mathematical properties. When one considers the intersection of these two classes of operators, namely compact self-adjoint operators, the situation changes quite a bit as the following theorems demonstrate. This will also be seen in later sections on spectral expansions.

The first thing to state about compact self-adjoint operators is that they do possess an eigenvalue, at least one.

**Theorem 4.21.** *Let $A : H \to H$ be a compact self-adjoint linear operator on a nontrivial Hilbert space. Then $A$ has an eigenvalue $\lambda$ with $|\lambda| = \|A\|$.*

Therefore, compact self-adjoint operators have at least one eigenvalue. This fact is also true for compact normal operators.

**Theorem 4.22.** *Let $\{\lambda_n\}$ be the set of distinct nonzero eigenvalues of a compact, self-adjoint linear operator. Then, either $\{\lambda_n\}$ is a finite set, or $\lim_{n\to\infty} \lambda_n = 0$.*

**Proof.** Assume $A$ is a compact self-adjoint operator with infinitely many distinct eigenvalues $\lambda_n$, and corresponding eigenvectors $x_n$, where the set $\{x_n\}$ is orthogonal simply by the distinctness of the eigenvalues since $A$ is self-adjoint. Assume further that the eigenvectors are normalized to become an orthonormal set. By the discussion in Section 2.5.6, orthonormal sequences are weakly convergent to 0, and from Section 3.6.1 we have $0 = \lim_{n\to\infty} \|Ax_n\|^2 = \lim_{n\to\infty} \langle Ax_n, Ax_n\rangle = \lim_{n\to\infty} \langle \lambda_n x_n, \lambda_n x_n\rangle = \lim_{n\to\infty} \lambda_n^2 \|x_n\| = \lim_{n\to\infty} \lambda_n^2$. ■

Note that the proof relies on compactness to convert the weakly convergent sequence into a strongly convergent sequence. Theorem 4.22 also follows from Theorems 4.14 and 4.15.

The last theorem indicates a potential problem with compact self-adjoint operators, namely that their eigenvalues $\lambda_n$ tend toward the origin in the complex plane as $n \to \infty$. In later sections, when expansions of operators and operator inverses are discussed, we will see that in this case the fact that the inverse of a compact operator on an infinite-dimensional space is unbounded (as discussed in the previous chapter) is analogous to division by an eigenvalue tending toward the origin.

Similar to Theorem 4.13 we have the following theorem [6, p. 24].

**Theorem 4.23.** *Every spectral value $\lambda \neq 0$ of a compact self-adjoint linear operator $A : H \to H$ is an eigenvalue of finite multiplicity that can only accumulate at $\lambda = 0$. Conversely, a self-adjoint operator having these properties is compact.*

### 4.2.8 Normal and Unitary Operators

Concerning the spectral properties of normal and unitary operators, we have the following.[15]

The first theorem provides a relationship between eigenvalues and eigenvectors of a normal operator and those of its adjoint.

---

[15]Recall that because self-adjoint operators are always normal, the following theorems apply to self-adjoint operators as well.

**Theorem 4.24.** *Let $A : H \to H$ be a bounded normal linear operator. If $\lambda$ is an eigenvalue of $A$ with $x \in H$ its corresponding eigenvector, then $\overline{\lambda}$ is an eigenvalue of $A^*$ and the corresponding eigenvector of $A^*$ is $x$. Furthermore, $N_{A-\lambda I} = N_{A^*-\overline{\lambda} I}$.*

**Proof.** If $A$ is normal, then $A - \lambda I$ is also normal. From Theorem 3.13 we see that $\|(A-\lambda I)x\| = 0$ if and only if $\left\|\left(A^* - \overline{\lambda} I\right)x\right\| = 0$. From Definition 2.31 this indicates that $\left(A^* - \overline{\lambda} I\right)x = 0$. ∎

The next theorem extends the orthogonality of eigenvectors corresponding to distinct eigenvalues in the self-adjoint case (Theorem 4.16) to the case of normal operators.

**Theorem 4.25.** *Let $A : H \to H$ be a normal linear operator. Then $N_{A-\lambda_n I} \perp N_{A-\lambda_m I}$ for $\lambda_n \neq \lambda_m$.*

**Proof.** Let $x_n \in N_{A-\lambda_n I}$ and $x_m \in N_{A-\lambda_m I}$, $x_n, x_m \neq 0$. From $\langle Ax_n, x_m \rangle = \langle x_n, A^* x_m \rangle$, we obtain $\langle \lambda_n x_n, x_m \rangle = \left\langle x_n, \overline{\lambda_m} x_m \right\rangle$. Then $(\lambda_n - \lambda_m)\langle x_n, x_m \rangle = 0$, which implies $\langle x_n, x_m \rangle = 0$ since $\lambda_n - \lambda_m \neq 0$. ∎

The next theorem is worthwhile to note since it makes any consideration of the residual spectrum for normal operators unnecessary.

**Theorem 4.26.** *The residual spectrum of a normal linear operator is empty.*

**Proof.** From Section 4.1.2 (see p. 223) it is enough to show that if $A - \lambda I$ is invertible, then $R_{A-\lambda I}$ is dense in $H$. To show this, assume $A - \lambda I$ is invertible (and so one-to-one) and let $y$ be orthogonal to $R_{A-\lambda I}$. We then have $0 = \langle (A-\lambda I)x, y \rangle = \left\langle x, \left(A^* - \overline{\lambda} I\right)y \right\rangle$ for all $x \in H$. It then follows from Definition 2.41 that $\left(A^* - \overline{\lambda} I\right)y = 0$, and so $y \in N_{A^*-\overline{\lambda} I}$. From Theorem 4.24 it follows that $y \in N_{A-\lambda I}$. Because $A - \lambda I$ is one-to-one, then $N_{A-\lambda I} = \{0\}$, and therefore $y = 0$. We then have $R_{A-\lambda I}^{\perp} = \{0\}$, and by Theorem 2.7(c) $R_{A-\lambda I}$ is dense in $H$. ∎

Furthermore, for typical differential operators the residual spectrum is empty, but this is not a general result.

The next theorem provides a nice geometrical picture of the eigenvalues of unitary operators.

**Theorem 4.27.** *All eigenvalues of a unitary linear operator $A : H \to H$ are complex numbers of modulus 1. Furthermore, $\sigma(A)$ lies on the unit circle $|\lambda| = 1$. For $|\lambda| \neq 1$,*

$$\|R_\lambda(A)\| \leq \frac{1}{|1 - |\lambda||}.$$

**Proof.** We prove only the first part of the theorem. Let $A$ be unitary and let $\lambda$ be an eigenvalue of $A$ with corresponding eigenvector $x \in H$. Then $\langle Ax, Ax \rangle = \langle \lambda x, \lambda x \rangle = |\lambda|^2 \|x\|^2$. Also, $\langle Ax, Ax \rangle = \langle x, A^* Ax \rangle = \langle x, x \rangle = \|x\|^2$, leading to $|\lambda| = 1$. ∎

### 4.2.9 Generalized Eigenvalue Problems

While attention is usually focused on the standard eigenvalue problem, generalized and nonstandard eigenvalue problems, and eigenvalue problems utilizing a pseudo inner product, are also of interest in electromagnetics. In this and the next two sections we briefly present some results concerning more general eigenvalue problems. Much of the material follows that presented in [7, Ch. 3] and [2].

Recall that the generalized eigenvalue problem is given as

$$Ax = \lambda Bx \tag{4.12}$$

or, equivalently,

$$(A - \lambda BI)\, x = 0. \tag{4.13}$$

In this case it is not enough for the operators $A$ and $B$ to be self-adjoint for the spectrum to be real.

**Theorem 4.28.** *Let $A, B : H \to H$ be self-adjoint (or symmetric) linear operators such that $\langle Ax, x\rangle \neq 0$ or $\langle Bx, x\rangle \neq 0$. Then eigenvalues corresponding to $Ax = \lambda Bx$ are real-valued.*

**Proof.** First assume that $\langle Bx, x\rangle \neq 0$. Then, given $Ax = \lambda Bx$ we have $\langle Ax, x\rangle = \lambda \langle Bx, x\rangle$ and $\langle x, Ax\rangle = \overline{\lambda} \langle x, Bx\rangle$. Since $A$ and $B$ are symmetric, then $\lambda \langle Bx, x\rangle = \overline{\lambda} \langle Bx, x\rangle$, or $\left(\lambda - \overline{\lambda}\right) \langle Bx, x\rangle = 0$. Therefore, $\left(\lambda - \overline{\lambda}\right) = 0$, which proves eigenvalues $\lambda$ are real if $\langle Bx, x\rangle \neq 0$. The value of $\langle Ax, x\rangle$ is immaterial. Now assume that $\langle Ax, x\rangle \neq 0$. Then, given $Ax = \lambda Bx$ with $\lambda \neq 0$ we have $(1/\lambda) \langle Ax, x\rangle = \langle Bx, x\rangle$ and $(1/\overline{\lambda}) \langle x, Ax\rangle = \langle x, Bx\rangle$. Since $A$ and $B$ are symmetric, then $(1/\lambda) \langle Ax, x\rangle = (1/\overline{\lambda}) \langle Ax, x\rangle$, or $\left((1/\lambda) - (1/\overline{\lambda})\right) \langle Ax, x\rangle = 0$. Therefore, $\left((1/\lambda) - (1/\overline{\lambda})\right) = 0$, which proves that eigenvalues $\lambda$ are real if $\langle Ax, x\rangle \neq 0$. The value of $\langle Bx, x\rangle$ is immaterial. ∎

Therefore, if $A$ and $B$ are self-adjoint, or at least symmetric, and either $A$ or $B$ is definite (positive, positive definite, negative, or negative definite as defined in Section 3.5), then the eigenvalue problem (4.12) admits only real eigenvalues. For $A$ and $B$ self-adjoint or symmetric, complex eigenvalues can occur only if both $A$ and $B$ are indefinite.

As with the ordinary eigenvalue problem, one can investigate orthogonality of eigenvectors for the generalized eigenvalue problem.

**Theorem 4.29.** *Let $A, B : H \to H$ be self-adjoint (or symmetric) operators. Then eigenvectors corresponding to $Ax = \lambda Bx$ satisfy the orthogonality relationships*

$$\left(\lambda_n - \overline{\lambda}_m\right) \langle Bx_n, x_m\rangle = 0,$$

$$\left(\frac{1}{\lambda_n} - \frac{1}{\overline{\lambda}_m}\right) \langle Ax_n, x_m\rangle = 0$$

*($\lambda_n, \lambda_m \neq 0$ in the last expression).*

**Proof.** We have $\langle Ax_n, x_m \rangle = \langle \lambda_n Bx_n, x_m \rangle = \lambda_n \langle Bx_n, x_m \rangle = \langle x_n, Bx_m \rangle$ and $\langle Ax_m, x_n \rangle = \langle \lambda_m Bx_m, x_n \rangle = \lambda_m \langle Bx_m, x_n \rangle$. Using both expressions we obtain $\overline{\langle Ax_m, x_n \rangle} = \overline{\lambda_m \langle Bx_m, x_n \rangle} = \overline{\lambda}_m \langle x_n, Bx_m \rangle = \lambda_n \langle x_n, Bx_m \rangle$; therefore, $\left(\lambda_n - \overline{\lambda}_m\right) \langle x_n, Bx_m \rangle = 0$ and using self-adjointness of $B$ we have the first result. A similar procedure leads to the second relationship. ■

If one of the operators $A$ or $B$ is definite, then $\lambda_n \in \mathbf{R}$ and the orthogonality relation holds for distinct eigenvalues $\lambda_n \neq \lambda_m$.

**Example 4.3.**

Consider Maxwell's equations applied to the interior of a source-free cavity $\Omega$ bounded by a perfectly conducting, closed surface $\Gamma$,

$$\begin{aligned} \nabla \times \mathbf{E}(\mathbf{r}) &= -i\omega\mu\mathbf{H}(\mathbf{r}), \\ \nabla \times \mathbf{H}(\mathbf{r}) &= \ i\omega\varepsilon\mathbf{E}(\mathbf{r}), \\ \mathbf{n} \times \mathbf{E}|_\Gamma &= \mathbf{0}. \end{aligned} \tag{4.14}$$

This set of equations can be described in operator form as

$$-i\begin{pmatrix} 0 & \nabla\times \\ -\nabla\times & 0 \end{pmatrix}\begin{pmatrix} \mathbf{E} \\ \mathbf{H} \end{pmatrix} = \omega\begin{pmatrix} \varepsilon & 0 \\ 0 & \mu \end{pmatrix}\begin{pmatrix} \mathbf{E} \\ \mathbf{H} \end{pmatrix}$$

or

$$A\begin{pmatrix} \mathbf{E} \\ \mathbf{H} \end{pmatrix} = \lambda B\begin{pmatrix} \mathbf{E} \\ \mathbf{H} \end{pmatrix}$$

with the operators $A, B : \mathbf{L}^2(\Omega)^6 \to \mathbf{L}^2(\Omega)^6$ defined by

$$\begin{aligned} A &\equiv -i\begin{pmatrix} 0 & \nabla\times \\ -\nabla\times & 0 \end{pmatrix}, \qquad B \equiv \begin{pmatrix} \varepsilon & 0 \\ 0 & \mu \end{pmatrix}, \\ D_A &\equiv \left\{\mathbf{x} : \mathbf{x}, \nabla \times \mathbf{x} \in \mathbf{L}^2(\Omega)^6, \mathbf{n} \times \mathbf{x}|_\Gamma = \mathbf{0}\right\}, \\ D_B &\equiv \left\{\mathbf{x} : \mathbf{x} \in \mathbf{L}^2(\Omega)^6\right\}. \end{aligned}$$

The operators $A$ and $B$ can be seen to be self-adjoint, and since $B$ is clearly positive definite for $\varepsilon$, $\mu > 0$, by Theorem 4.28 we see that the resonance frequencies $\lambda_n = \omega_n$ are real. Eigenfunctions $\mathbf{x}_n = \begin{pmatrix} \mathbf{E}_n \\ \mathbf{H}_n \end{pmatrix}$, corresponding to distinct eigenvalues $\omega_n$, are orthogonal in the sense

$$(\omega_n - \omega_m)\langle B\mathbf{x}_n, \mathbf{x}_m \rangle = (\omega_n - \omega_m)\int_\Omega \left(\varepsilon\mathbf{E}_n \cdot \overline{\mathbf{E}}_m + \mu\mathbf{H}_n \cdot \overline{\mathbf{H}}_m\right) d\Omega = 0.$$

If $A$ and/or $B$ is not symmetric, then we have the following.

**Theorem 4.30.** *If $x_n$ are eigenvectors of $Ax = \lambda Bx$, with corresponding eigenvalues $\lambda_n$, then $\overline{\lambda}_n$ are eigenvalues of the adjoint equation $A^*y = \lambda^* B^* y$ corresponding to adjoint eigenvectors $y_n$ (i.e., $\lambda_n^* = \overline{\lambda}_n$), and*

$$(\lambda_n - \lambda_m) \langle Bx_n, y_m \rangle = 0,$$
$$\left( \frac{1}{\lambda_n} - \frac{1}{\lambda_m} \right) \langle Ax_n, y_m \rangle = 0$$

*($\lambda_n, \lambda_m \neq 0$ in the last expression).*

As with Theorem 4.7, we assume that the residual spectrum is empty. The proof follows along the same lines as in the previous cases. Moreover, the result for eigenvalues holds for the entire spectrum,

$$\sigma(A, B) = \overline{\sigma}(A^*, B^*).$$

### 4.2.10 Eigenvalue Problems under a Pseudo Inner Product

We now restate three theorems presented in previous sections, this time using a pseudo inner product. Since a pseudo inner product does not generate a Hilbert space, here we consider linear operators $A, B : X \to X$, where $X$ is a pseudo inner product space (see Section 2.5.2).

For the ordinary eigenvalue problem $Ax = \lambda x$, Theorem 4.7 becomes the following.

**Theorem 4.31.** *If $x_n$ are eigenvectors of the linear operator $A : X \to X$, with corresponding eigenvalues $\lambda_n$ (i.e., $Ax_n = \lambda_n x_n$), then $\lambda_n$ are also eigenvalues of the pseudo adjoint operator $A^{p*}(X)$ corresponding to adjoint eigenvectors $y_n$ (i.e., $A^{p*} y_n = \lambda_n y_n$), and*

$$(\lambda_n - \lambda_m) \langle x_n, y_m \rangle_p = 0.$$

The proof follows along the same lines as the proof of Theorem 4.7, and, in fact, more generally

$$\sigma(A) = \sigma(A^{p*}).$$

For a pseudo self-adjoint problem ($A = A^{p*}$ and, upon suitable ordering, $y_n = x_n$), we have the orthogonality relationship $(\lambda_n - \lambda_m) \langle x_n, x_m \rangle_p = 0$.

For a generalized eigenvalue problem $Ax = \lambda Bx$, Theorem 4.29 becomes the following.

**Theorem 4.32.** *Let $A, B : X \to X$ be pseudo self-adjoint (or pseudo symmetric) operators. Then eigenvectors corresponding to $Ax = \lambda Bx$ satisfy* *the orthogonality relationships*

$$(\lambda_n - \lambda_m) \langle Bx_n, x_m \rangle_p = 0,$$
$$\left(\frac{1}{\lambda_n} - \frac{1}{\lambda_m}\right) \langle Ax_n, x_m \rangle_p = 0$$

*($\lambda_n, \lambda_m \neq 0$ in the last expression).*

For the case of a generalized nonpseudo self-adjoint problem, Theorem 4.30 becomes the next result.

**Theorem 4.33.** *If $x_n$ are eigenvectors of $Ax = \lambda Bx$, with corresponding eigenvalues $\lambda_n$ (i.e., $Ax_n = \lambda_n Bx_n$), then $\lambda_n$ are eigenvalues of the pseudo adjoint equation $A^{p*}y = \lambda^{p*} B^{p*} y$ (i.e., $\lambda_n^{p*} = \lambda_n$) corresponding to pseudo adjoint eigenvectors $y_n$, and*

$$(\lambda_n - \lambda_m) \langle Bx_n, y_m \rangle_p = 0,$$
$$\left(\frac{1}{\lambda_n} - \frac{1}{\lambda_m}\right) \langle Ax_n, y_m \rangle_p = 0$$

*($\lambda_n, \lambda_m \neq 0$ in the last expression).*

Again, the more general relationship

$$\sigma(A, B) = \sigma(A^{p*}, B^{p*})$$

holds, and we assume that the residual spectrum is empty in all cases.

### 4.2.11 Pseudo Self-Adjoint, Nonstandard Eigenvalue Problems

Recall that the nonstandard eigenvalue problem (4.5) is given as

$$A(\gamma)x = 0, \tag{4.15}$$

where $\gamma$ is called a nonstandard eigenvalue and $x$ is a nonstandard eigenvector. Usually $A$ is a nonlinear operator-valued function of $\gamma$.

In the general case, not much can be said concerning the spectral properties of the nonstandard eigenvalues $\gamma$ and eigenfunctions $x$. Furthermore, recalling that pseudo self-adjointness is a far weaker property (for complex spaces) than self-adjointness, it is interesting and very useful to note that if a nonstandard eigenvalue problem is such that $A(\gamma)$ is pseudo self-adjoint, then it is easy to obtain a stationary functional for the nonstandard eigenvalue $\gamma$. In fact, that stationary functional is related to a Galerkin solution of the operator equation. The analysis presented here follows [8], where a more general problem is treated.

We assume that the operator $A(\gamma)$ in the nonstandard problem is pseudo self-adjoint, $\langle Ax, y\rangle_p = \langle y, Ax\rangle_p$, which is physically associated with reciprocity. A variational expression for a nonstandard eigenvalue $\gamma$ is obtained from

$$F(\gamma, x) = \langle x, A(\gamma)x\rangle_p = 0.$$

We form the first variation of $F$ by replacing $\gamma$ and $x$ with $\gamma + \delta\gamma$ and $x + \delta x$, respectively. Therefore,

$$F(\gamma + \delta\gamma, x + \delta x) = \langle x + \delta x, A(\gamma + \delta\gamma)(x + \delta x)\rangle_p = F(\gamma, x) + \delta F. \quad (4.16)$$

Defining $A' \equiv \partial A(\gamma) / \partial\gamma = [A(\gamma + \delta\gamma) - A(\gamma)] / \delta\gamma$ [9, Ch. 4], which approaches the derivative of $A$ as $\delta\gamma \to 0$, and neglecting second-order terms result in

$$\delta F = \delta\gamma \langle x, A'x\rangle_p + \langle x, A\delta x\rangle_p \quad (4.17)$$

where we have used (4.15). Exploiting the pseudo self-adjointness of $A$, $\langle x, A\delta x\rangle_p = \langle Ax, \delta x\rangle_p$, the second term in (4.17) vanishes, resulting in

$$\delta F = \delta\gamma \langle x, A'x\rangle_p. \quad (4.18)$$

Therefore, $\delta\gamma = 0$ is a stationary point of the functional $F$ (unless it happens that $\langle x, A'(\gamma)x\rangle_p = 0$), such that at $\delta\gamma = 0$ the first variation of $F$ vanishes. Thus, first-order errors in $x$ result in second-order errors in $\gamma$.

In a method-of-moments solution of $Ax = 0$,

$$\langle x, A(\gamma)x\rangle_p = \langle A(\gamma)x, x\rangle_p = 0$$

is simply Galerkin's method, which is widely used to solve a variety of integral equations occurring in applied electromagnetics. It is easy to see that if $x$ is approximated by a series of basis functions, $x \sim \widetilde{x} = \sum_{i=1}^{N} a_i x_i$, then $\langle A(\gamma)x, x\rangle_p = 0$ becomes a stationary (nonlinear) functional equation

$$H(\gamma) = \det[Z(\gamma)] = 0,$$

where $Z$ is the matrix with elements $Z_{ij} = \langle A(\gamma)x_j, x_i\rangle_p$. Since for electromagnetic applications in reciprocal media the resulting operators are generally pseudo self-adjoint (see the footnote on p. 162), natural resonance and spectral problems typically exhibit this stationary characteristic. An example is provided in Section 8.3.

We assume that the equation $H(\gamma) = 0$ leads to the first-order root $\gamma = \gamma_0$. A second-order root is obtained by imposing the additional condition $\partial H(\gamma) / \partial\gamma = H'(\gamma) = 0$. The system of equations

$$H(\gamma) = H'(\gamma) = 0$$

(with some additional constraints) has been applied for the numerical determination of various critical points in waveguiding problems in [10] and references therein. The above procedure also holds if $A$ is self-adjoint, $\langle Ax, y\rangle = \langle y, Ax\rangle$, by replacing $\langle \cdot, \cdot\rangle_p$ with $\langle \cdot, \cdot\rangle$.

### 4.2.12 Steinberg's Theorems for Compact Operators

Several theorems relating to the spectral properties of identity-plus-compact Fredholm operators are presented by Steinberg in [11] (see also [12]). In the following theorems $A$ is a compact operator on a Banach space $B$.

**Theorem 4.34.** *(Analytic Fredholm theorem) If $A(\gamma)$ is an analytic family of compact operators*[16] *for $\gamma \in \Omega$, where $\Omega$ is an open, connected subset of the complex plane, then either*

- $(I - A(\gamma))$ *is nowhere invertible in $\Omega$, or else*
- $(I - A(\gamma))^{-1}$ *is meromorphic in $\Omega$.*

If the second result holds, then there exists a discrete set of values $\{\gamma_n\}$ that are the poles of the inverse operator. In this case $(I - A(\gamma))^{-1}$ exists and is analytic at all points $\gamma \neq \gamma_n$, and for $\gamma = \gamma_n$ the equation

$$(I - A(\gamma))\, x = 0$$

has a nontrivial solution in $B$.

Note that the set $\{\gamma_n\}$ may be empty. Therefore, Theorem 4.34 is not an existence theorem for poles. Rather, it guarantees that if the second condition holds, then if singularities are present in $\Omega$, they will be pole singularities.

**Example 4.4.**

Consider scattering from a perfectly conducting, finite-sized object having a sufficiently smooth surface $S$ located in free space. The magnetic field integral equation (1.118)

$$\int_S \left[\underline{\mathbf{G}}_{me}(\mathbf{r}, \mathbf{r}') \cdot \mathbf{J}_e^s(\mathbf{r}')\right] \times \mathbf{n}\; dS' + \frac{1}{2}\mathbf{J}_e^s(\mathbf{r}) = \mathbf{n} \times \mathbf{H}^i(\mathbf{r}), \qquad \mathbf{r} \in S, \quad (4.19)$$

can be used to determine the induced current density $\mathbf{J}_e^s$ caused by an excitation $\mathbf{H}^i$, which can, in turn, be used to determine the scattered field. From (1.79),

$$\begin{aligned}\underline{\mathbf{G}}_{me}(\mathbf{r}, \mathbf{r}') \cdot \mathbf{J}_e^s(\mathbf{r}') &= \left(\nabla g(\mathbf{r}, \mathbf{r}') \times \underline{\mathbf{I}}\right) \cdot \mathbf{J}_e^s(\mathbf{r}') \\ &= \nabla g(\mathbf{r}, \mathbf{r}') \times \mathbf{J}_e^s(\mathbf{r}'),\end{aligned}$$

where $g = \left(e^{-jkR}\right) / (4\pi R)$ is the usual free-space scalar Green's function and $k = \omega\sqrt{\mu_0 \varepsilon_0}$, we obtain the integral equation

$$\mathbf{J}_e^s(\mathbf{r}) - 2\int_S \mathbf{n} \times \left(\nabla g(\mathbf{r}, \mathbf{r}') \times \mathbf{J}_e^s(\mathbf{r}')\right)\, dS' = 2\mathbf{n} \times \mathbf{H}^i(\mathbf{r}), \qquad \mathbf{r} \in S. \quad (4.20)$$

[16] This means that $A(\gamma)$ is analytic and compact for each $\gamma \in \Omega$.

This can be written as

$$(I - A(k))\, \mathbf{J}_e^s(\mathbf{r}) = 2\mathbf{n} \times \mathbf{H}^i(\mathbf{r})$$

for $\mathbf{r} \in S$, where the operator $A(k) : \mathbf{L}^2(\Omega)^3 \to \mathbf{L}^2(\Omega)^3$, defined as

$$(A\mathbf{J})(\mathbf{r}) \equiv 2 \int_s \mathbf{n} \times \left[ \nabla \frac{e^{-ik|\mathbf{r}-\mathbf{r}'|}}{4\pi |\mathbf{r} - \mathbf{r}'|} \times \mathbf{J}(\mathbf{r}') \right] dS'$$

for suitable currents $\mathbf{J}$, is compact as discussed in Section 3.6.2 and clearly analytic in the finite $k$-plane. $(I - A)$ is invertible for some $k$ and therefore, by the above theorem, $(I - A(k))^{-1}$ is a meromorphic operator-valued function of $k$. This establishes the existence of a discrete set of nonstandard eigenvalues $\{k_n\} \in \mathbf{C}$, at which points the inverse does not exist. It is well known that this set is not empty.

Viewed as a spectral problem,

$$(I - A(k))\, \mathbf{x}_n = \lambda_n(k)\, \mathbf{x}_n,$$

the values of $k_n$ are such that $\lambda_n(k_n) = 0$. These points lead to complex-valued resonance frequencies $\omega_n = k_n / \sqrt{\mu_0 \varepsilon_0}$, which are called the *exterior resonances* of the object and which may be determined from

$$\det(I - A(k_n)) = 0.$$

In many cases it is assumed that the poles are simple, which can be proven for $S$ corresponding to a spherical surface. By the Fredholm alternative (Theorem 3.44), because $(I - A(k))^{-1}$ does not exist at $k = k_n$, nontrivial solutions $\mathbf{x}_n$ and $\mathbf{y}_n$ exist such that

$$(I - A(k))\, \mathbf{x}_n = 0,$$
$$(I - A^*(k))\, \mathbf{y}_n = 0.$$

The eigenfunctions and adjoint eigenfunctions $\mathbf{x}_n$ and $\mathbf{y}_n$, respectively, form a bi-orthogonal set (although not necessarily a basis) in $\mathbf{L}^2(\Omega)^3$.

For $k \neq k_n$, the solution of the integral equation is

$$\mathbf{J}_e^s(\mathbf{r}) = 2(I - A(k))^{-1}\, \mathbf{n} \times \mathbf{H}^i(\mathbf{r}),$$

and since poles are the only singularities in the finite $k$-plane, the Mittag–Leffler theorem [13] can be used to expand the inverse operator in terms of the associated spectral parameters (see [14], and, in particular, [15]). This important result, originally obtained from the magnetic field integral equation by a somewhat different method in [16], was used to establish the theoretical basis for the *singularity expansion method* developed by Baum [17]. By the uniqueness theorem the exterior resonances are associated with the object, and not the mathematical formulation.

For some additional applications in electromagnetic scattering, see [18].

The next theorem extends the above result to include the dependence of the inverse on a nonspectral parameter $\tau \in \mathbf{R}$.

**Theorem 4.35.** *Let $A(\gamma, \tau)$ be a family of compact operators analytic in $\gamma \in \Omega$, where $\Omega$ is an open, connected subset of the complex plane, and let $A(\gamma, \tau)$ be jointly continuous in $(\gamma, \tau)$ for each $(\gamma, \tau) \in \Omega \times \mathbf{R}$. Then,*

- *if $(I - A(\gamma, \tau))$ is somewhere invertible for each $\tau$, $(I - A(\gamma, \tau))^{-1}$ is meromorphic in $\gamma$ for each $\tau$,*
- *if $\gamma_0$ is not a pole of $(I - A(\gamma, \tau_0))^{-1}$, $(I - A(\gamma, \tau))^{-1}$ is jointly continuous in $(\gamma, \tau)$ at $(\gamma_0, \tau_0)$, and*
- *the poles $\gamma(\tau)$ of $(I - A(\gamma, \tau))^{-1}$ depend continuously on $\tau$ and can appear and disappear only at the boundary of $\Omega$ (which may extend to infinity).*

In typical electromagnetics applications, $\tau$ represents a geometrical parameter, and $\gamma$ represents frequency in a three-dimensional problem or propagation constant in a two-dimensional problem. The above theorem can be applied to establish (a) the existence of a discrete set (possibly empty) of nonstandard eigenvalues $\gamma_n$ for a given $\tau$, (b) the continuity of the inverse operator with respect to spectral and structural parameters, and (c) the continuity of the spectral values $\gamma_n(\tau)$ as a function of $\tau$. For example, this theorem can be used to show that the exterior resonances $k_n$ described in Example 4.4 smoothly vary as the shape of the scatterer $S$ smoothly varies.

In electromagnetics problems involving lossy media or infinite media, the domain of analyticity $\Omega$ often is constrained by the presence of branch point singularities, as the next example demonstrates.

**Example 4.5.**

As in Example 4.4, consider scattering from a perfectly conducting, finite-sized object having a sufficiently smooth surface $S$, but this time let the object reside within an infinite parallel-plate region with plate separation $a$. As in the previous example, the magnetic field integral equation (1.118)

$$\int_S \left[\underline{\mathbf{G}}_{me}(\mathbf{r}, \mathbf{r}') \cdot \mathbf{J}_e^s(\mathbf{r}')\right] \times \mathbf{n} \, dS' + \frac{1}{2}\mathbf{J}_e^s(\mathbf{r}) = \mathbf{n} \times \mathbf{H}^i(\mathbf{r}), \qquad \mathbf{r} \in S, \quad (4.21)$$

can be used to determine the induced current density $\mathbf{J}_e^s$ caused by an excitation $\mathbf{H}^i$. In this case, the Green's dyadic $\underline{\mathbf{G}}_{me}$ accounts for the infinite parallel plates (see Sections 8.1.2 and 8.3.2) and is given by (8.197) with (8.203) and (8.208),

$$\begin{aligned}\underline{\mathbf{G}}_{me}(\mathbf{r}, \mathbf{r}') &= \nabla \times \underline{\mathbf{G}}_\pi(\mathbf{r}, \mathbf{r}') \\ &= \nabla \times \left(\widehat{\mathbf{x}}\widehat{\mathbf{x}}\, G_\pi^E(\mathbf{r}, \mathbf{r}') + (\widehat{\mathbf{y}}\widehat{\mathbf{y}} + \widehat{\mathbf{z}}\widehat{\mathbf{z}})\, G_\pi^H(\mathbf{r}, \mathbf{r}')\right),\end{aligned}$$

where

$$G_{\pi}^{H}(\mathbf{r},\mathbf{r}') = \frac{1}{2ia}\sum_{n=1}^{\infty}\sin\left(\frac{n\pi}{a}x\right)\sin\left(\frac{n\pi}{a}x'\right)H_0^{(2)}(\psi_n\rho),$$
$$G_{\pi}^{E}(\mathbf{r},\mathbf{r}') = \frac{1}{4ia}\sum_{n=0}^{\infty}\epsilon_n\cos\left(\frac{n\pi}{a}x\right)\cos\left(\frac{n\pi}{a}x'\right)H_0^{(2)}(\psi_n\rho), \tag{4.22}$$

with

$$\psi_n = \sqrt{k^2 - \left(\frac{n\pi}{a}\right)^2}.$$

We obtain the operator equation

$$(I - A(k))\,\mathbf{J}_e^s(\mathbf{r}) = 2\mathbf{n}\times\mathbf{H}^i(\mathbf{r})$$

for $\mathbf{r}\in S$. The operator $A(k) : \mathbf{L}^2(\Omega)^3 \to \mathbf{L}^2(\Omega)^3$, defined as

$$(A\mathbf{J})(\mathbf{r}) \equiv -2\int_s \left[\underline{\mathbf{G}}_{me}(\mathbf{r},\mathbf{r}')\cdot\mathbf{J}(\mathbf{r}')\right]\times\mathbf{n}\,dS'$$

for suitable currents $\mathbf{J}$, is compact but clearly not analytic in the finite $k$-plane, since the Green's dyadic is a function of the parallel-plate waveguide dispersion function $\psi_n$.

Concentric regions $\Omega_n$ may be defined in the complex $k$-plane wherein $A(k)$ is analytic,

$$\Omega_0 \equiv \left\{k : |k| < \frac{\pi}{a}\right\},$$
$$\Omega_1 \equiv \left\{k : \frac{\pi}{a} < |k| < \frac{2\pi}{a}\right\},$$
$$\vdots$$
$$\Omega_n \equiv \left\{k : \frac{n\pi}{a} < |k| < \frac{(n+1)\pi}{a}\right\},$$

such that $(I - A(k))^{-1}$ will be meromorphic in each $\Omega_n$. The poles of the resolvent, $k_n$, may appear or disappear at the values $k = n\pi/a$, which defines the cutoff condition for parallel-plate waveguide modes (at cutoff forward and backward propagating modes intersect, and below cutoff they become evanescent).

## 4.3 Expansions and Representations, Spectral Theorems

In this section we are concerned with simple representations of linear operators. We first discuss the matrix representation of linear operators on finite-dimensional spaces, where it is often of interest to obtain as simple a matrix representation as possible. Conditions under which an operator has a diagonal matrix representation are stated. If a diagonal matrix representation is not possible, then other fairly simple representations are discussed.

We will see that the prime consideration for this topic is the choice of an appropriate basis set. For example, in Section 3.1.2 we found that a bounded linear operator $A : H \to H$ has the matrix representation

$$\begin{bmatrix} \langle Ax_1, x_1 \rangle & \langle Ax_2, x_1 \rangle & \cdots \\ \langle Ax_1, x_2 \rangle & \langle Ax_2, x_2 \rangle & \cdots \\ \vdots & \vdots & \ddots \end{bmatrix}$$

with respect to the basis $\mathbf{x} = \{x_1, x_2, \dots\}$ of $H$ (the matrix has finite (infinite) dimension if $H$ is finite- (infinite-) dimensional). It is clear that if the basis $\mathbf{x}$ is such that $\langle Ax_i, x_j \rangle = 0$ for $i \neq j$, the matrix representation of $A$ will be diagonal. Conditions for this to occur will be discussed, but as a preview if $x_i$ is an eigenvector of $A$ corresponding to the eigenvalue $\lambda_i$, then $\langle Ax_i, x_j \rangle = \langle \lambda_i x_i, x_j \rangle = \lambda_i \langle x_i, x_j \rangle$. If $\langle x_i, x_j \rangle = 0$ for $i \neq j$, the off-diagonal entries of the matrix will be zero. However, there also needs to be enough eigenvectors, so that they form a basis for the space $H$.

This section begins by considering the matrix representation of operators on finite-dimensional spaces with respect to different bases. Operators that have a diagonal representation with respect to some (eigenvector) basis are identified, and similarity transformations from an arbitrary basis to an eigenvector basis are discussed. Operators on infinite-dimensional spaces are considered next. Although one may view the infinite-dimensional case in the same way as the finite-dimensional case—that is, as a question of finding a diagonal matrix representation of the operator—it is more useful to focus on the (essentially equivalent) eigenfunction expansion problem. Operators on infinite-dimensional spaces that possess a complete eigenfunction basis are identified, and associated expansions are discussed.

### 4.3.1 Operators on Finite-Dimensional Spaces

Because the spectral theory of operators on finite-dimensional spaces is much simpler than that for operators on infinite-dimensional spaces (recall that in the finite-dimensional case $\sigma_c$ and $\sigma_r$ do not exist, and $\sigma_p$ is purely discrete), and since electromagnetic problems are often ultimately cast as numerical matrix problems, we begin this section with some elements from the spectral theory of matrices. For generality, we work within finite-dimensional Hilbert spaces $H_1^n, H_2^m$, with the superscript indicating the dimension of the space.

### Equivalent and Similar Matrices

Recall from the examples in Section 3.1.2 that the bounded operator $A : H_1^n \to H_2^m$ is uniquely represented, for $\mathbf{x} = \{\mathbf{x}_1, \mathbf{x}_2, \ldots, \mathbf{x}_n\}$ a basis of $H_1^n$ and $\mathbf{y} = \{\mathbf{y}_1, \mathbf{y}_2, \ldots, \mathbf{y}_m\}$ a basis of $H_2^m$, by the $m \times n$ matrix $[a_{ij}]_{\mathbf{x},\mathbf{y}} \in \mathbf{C}^{m\times n}$, where $a_{ij} = \langle Ax_j, y_i\rangle$ and the subscript indicates the basis pairs.[17] If $\mathbf{x}, \mathbf{y}$ are the standard bases in $H_1^n = \mathbf{C}^n, H_2^m = \mathbf{C}^m$, respectively, then $[a_{ij}]_{\mathbf{x},\mathbf{y}}$ is called the *standard matrix representation* of the linear operator $A$. The matrix representation of a given operator will, of course, be different for different bases, and it is often desirable to achieve a matrix representation that is as simple as possible, perhaps triangular or even diagonal in form.

It can be shown that for a given linear operator $A : H_1^n \to H_2^m$, represented as $[a_{ij}]_{\mathbf{x},\mathbf{y}}$ in the pair of bases $\{x_n\}$ and $\{y_n\}$, and as $[a_{ij}]_{\xi,\zeta}$ in the pair of bases $\{\xi_n\}$ and $\{\zeta_n\}$, the two matrix operator representations are related as

$$[a_{ij}]_{\mathbf{x},\mathbf{y}} = [P]_{\zeta\to\mathbf{y}}\, [a_{ij}]_{\xi,\zeta}\, [Q]_{\xi\to\mathbf{x}}^{-1}. \tag{4.23}$$

In the above, the $m \times m$ matrix $[P]_{\zeta\to\mathbf{y}}$ and the $n \times n$ matrix $[Q]_{\xi\to\mathbf{x}}$ are nonsingular *transition matrices* from the bases $\zeta$ to $\mathbf{y}$ and $\xi$ to $\mathbf{x}$, respectively[19, p. 127].[18] Note also that if $[Q]_{\xi\to\mathbf{x}}$ is a nonsingular transition matrix from the basis $\xi$ to the basis $\mathbf{x}$, then inversion of the matrix reverses the operation, i.e., $[Q]_{\xi\to\mathbf{x}}^{-1}$ is the transition matrix from the basis $\mathbf{x}$ to the basis $\xi$. If the bases are orthonormal, then $[P]$ and $[Q]$ are unitary (therefore, $[P]\,[P]^* = [I]$); by Example 2 in Section 3.3.2 if the bases are orthonormal, then the inverse of $[P]$ or $[Q]$ is easily computed as the conjugate-transpose matrix.

The two matrices $[a_{ij}]_{\mathbf{x},\mathbf{y}}$ and $[a_{ij}]_{\xi,\zeta}$ in (4.23) are said to be *equivalent*, in the sense that the set of all matrix representations of $A$ with respect to different pairs of bases defines an equivalence class of matrices that all represent the "parent" operator $A : H_1^n \to H_2^m$ in $\mathbf{C}^{m\times n}$.

Now, for simplicity, consider a finite-dimensional operator $A : H^n \to H^n$, represented as $[a_{ij}]_{\mathbf{x}}$ in the basis $\mathbf{x}$ ($a_{ij} = \langle A\mathbf{x}_j, \mathbf{x}_i\rangle$ ) and as $[a_{ij}]_\xi$ in the basis $\xi$ ($a_{ij} = \langle A\xi_j, \xi_i\rangle$). The two operator representations (matrices) are related as [19, p. 130]

$$[a_{ij}]_{\mathbf{x}} = [P]\, [a_{ij}]_\xi\, [P]^{-1}, \tag{4.24}$$

where the $n\times n$ matrix $[P] = [P]_{\xi\to\mathbf{x}}$ is a nonsingular transition matrix from the basis $\xi$ to the basis $\mathbf{x}$. When two matrices are related as in (4.24), they are said to be *similar*. Similar matrices have the same rank, determinant, and eigenvalues, and related eigenvectors.

[17] The space of bounded operators $H_1^n \to H_2^m$ and the space $\mathbf{C}^{m\times n}$ are isomorphic, and so one can work with the operator, or a matrix representation of the operator, in an equivalent manner.

[18] Explicitly, the matrix $[P]_{\zeta\to\mathbf{y}}$ is such that the two bases are related as $\zeta_j = \sum_{i=1}^m P_{ij}\, \mathbf{y}_i$, $j = 1, 2, \ldots, m$, and similarly for $[Q]$.

**Spectral Representations (Matrix Form)**

The primary reason for introducing the concept of similar matrices is to facilitate an understanding of when and how one obtains a "simple" matrix representation of a given operator on a finite-dimensional space. The simplest possible matrix has a diagonal form, and the next theorem provides conditions leading to a diagonal matrix representation of an operator $A : H^n \to H^n$ (alternately, one may say that the matrix representation of the operator in some basis is *diagonalizable*, or simply that a given matrix is diagonalizable). A clear way to envision this problem is to recognize that the set of all matrix representations of $A$ with respect to different bases defines an equivalence class of similar matrices. The problem is then to find the simplest matrix in this class, which will undoubtedly most clearly show the salient features of the "parent" operator.

**Theorem 4.36.** *Let $A : H^n \to H^n$ be a linear operator on a finite-dimensional complex Hilbert space. If $A$ has $n$ distinct eigenvalues $\{\lambda_n\}$, then $A$ has a diagonal matrix representation with respect to a basis of eigenvectors*[19] *of $A$.*

Actually, the last theorem provides a sufficient but unnecessary condition. More generally, we do not actually require $n$ distinct eigenvalues; instead we require that the eigenvectors of $A$ form a basis for the space $H^n$.

**Theorem 4.37.** *Let $A : H^n \to H^n$ be a linear operator on a finite-dimensional, complex Hilbert space. Then if $A$ has $n$ linearly independent eigenvectors $\{\mathbf{x}_n\}$(i.e., an eigenbasis of $H^n$), corresponding to the (not necessarily distinct) eigenvalues $\{\lambda_n\}$, $A$ has a diagonal matrix representation with respect to its eigenbasis.*

Such an operator is said to have *simple structure* (or, equivalently, to be represented by a simple matrix). The elements on the main diagonal of the resulting matrix representation are exactly the eigenvalues of the operator.

The next theorem states precisely when an operator on a finite-dimensional space will possess an eigenbasis.

**Theorem 4.38.** *(Complex spectral theorem for finite-dimensional spaces) Let $A : H^n \to H^n$ be a linear operator on a finite-dimensional, complex Hilbert space. Then $H^n$ has an orthonormal basis of eigenvectors $\{\mathbf{x}_n\}$ if and only if $A$ is normal.*

Therefore, if $A : H^n \to H^n$ is normal (or self-adjoint), it has a diagonal matrix representation with respect the eigenbasis $\mathbf{e} = \{\mathbf{x}_n\}$,

$$[A]_{\mathbf{e}} = \operatorname{diag}[\lambda_1, \lambda_2, \ldots, \lambda_n].$$

[19] A basis of a space consisting of eigenvectors of an operator is called an *eigenbasis*.

It also turns out that two self-adjoint operators $A$ and $B$ can be simultaneously diagonalized (by a single basis) if they commute, i.e., if $[A, B] = AB - BA = 0$. The above theorem holds for $H^n$ being a real space if the condition that $A$ be normal is replaced with the condition that $A$ be self-adjoint (symmetric). For a more complete discussion, see [20, p. 133].

It is again useful to recall that all self-adjoint operators are normal, but normal operators may not be self-adjoint. If $A$ is self-adjoint, then the resulting diagonal matrix representation has real entries on the main diagonal.

It can also be shown that if $A : H^n \to H^n$ admits a diagonal matrix representation, then any other nondiagonal matrix representation of the same operator (that is, a matrix representation with respect to a basis $\xi$ that is not an eigenbasis) may be diagonalized by (4.24).

**Example 4.6.**

To demonstrate these concepts, consider the self-adjoint operator $A : \mathbf{C}^2 \to \mathbf{C}^2$ defined by

$$A \begin{bmatrix} \alpha_1 \\ \alpha_2 \end{bmatrix} \equiv \begin{bmatrix} \alpha_1 + \alpha_2 \\ \alpha_1 - \alpha_2 \end{bmatrix}$$

and the two orthonormal bases $(\mathbf{y}_1, \mathbf{y}_2)$ and $(\xi_1, \xi_2)$, where

$$\mathbf{y}_1 = \begin{bmatrix} 1 \\ 0 \end{bmatrix}, \qquad \mathbf{y}_2 = \begin{bmatrix} 0 \\ 1 \end{bmatrix}$$

is the standard basis and

$$\xi_1 = \frac{1}{\sqrt{2}} \begin{bmatrix} 1 \\ 1 \end{bmatrix}, \qquad \xi_2 = \frac{1}{\sqrt{2}} \begin{bmatrix} -1 \\ 1 \end{bmatrix}.$$

Then the matrix representations with respect to the two bases are

$$[A]_{\mathbf{y}} = \begin{bmatrix} 1 & 1 \\ 1 & -1 \end{bmatrix}, \qquad [A]_{\xi} = \begin{bmatrix} 1 & -1 \\ -1 & -1 \end{bmatrix}.$$

The transition matrix from the basis $(\xi_1, \xi_2)$ to the basis $(\mathbf{y}_1, \mathbf{y}_2)$ is found to be

$$[P]_{\xi \to \mathbf{y}} = \frac{1}{\sqrt{2}} \begin{bmatrix} 1 & -1 \\ 1 & 1 \end{bmatrix}$$

such that (4.24) is easily seen to be satisfied, i.e., $[A]_{\mathbf{y}} = [P]_{\xi \to \mathbf{y}} [A]_{\xi} [P]_{\xi \to \mathbf{y}}^{-1}$.

Neither basis yields a diagonal matrix representation. Because the operator is self-adjoint, and so necessarily normal, we can find an eigenbasis that will yield a diagonal matrix representation. Alternatively, we can define a similarity transformation from, say, the basis $(\mathbf{y}_1, \mathbf{y}_2)$ to the eigenbasis that will result in a diagonal matrix representation.

The eigenvalues and eigenvectors of the operator $A$ are found to be

$$\lambda_{1,2} = \pm\sqrt{2}, \qquad \mathbf{e}_{1,2} = \begin{bmatrix} 1 \\ \pm\sqrt{2} - 1 \end{bmatrix},$$

which, when normalized to form an orthonormal eigenbasis, become

$$\mathbf{x}_1 = \frac{1}{2\sqrt{1 - 2^{-1/2}}} \begin{bmatrix} 1 \\ \sqrt{2} - 1 \end{bmatrix}, \qquad \mathbf{x}_2 = \frac{1}{2\sqrt{1 + 2^{-1/2}}} \begin{bmatrix} 1 \\ -\sqrt{2} - 1 \end{bmatrix}.$$

Noting that the transition matrix from the eigenbasis $\mathbf{e}$ to the basis $\mathbf{y}$ is

$$[P]_{\mathbf{e}\to\mathbf{y}} = \begin{bmatrix} \frac{1}{2\sqrt{1-2^{-1/2}}} & \frac{1}{2\sqrt{1+2^{-1/2}}} \\ \frac{\sqrt{2}-1}{2\sqrt{1-2^{-1/2}}} & \frac{-\sqrt{2}-1}{2\sqrt{1+2^{-1/2}}} \end{bmatrix} = [P]_{\mathbf{y}\to\mathbf{e}},$$

then, using (4.24), we obtain

$$[A]_{\mathbf{e}} = [P]\,[A]_{\mathbf{y}}\,[P]^{-1} = \begin{bmatrix} \sqrt{2} & 0 \\ 0 & -\sqrt{2} \end{bmatrix},$$

the desired diagonal matrix representation of the operator $A$.

In summary, every finite-dimensional normal operator is similar to a diagonal matrix operator. As such, every finite-dimensional normal operator $A : H^n \to H^n$ can be represented as a multiplication operator, $\left(PAP^{-1}\mathbf{x}\right)_i = \lambda_i x_i$. This fact also holds for normal operators on infinite-dimensional spaces.

### Jordan Canonical Form

It is seen that Theorem 4.38 characterizes explicitly when an operator on a finite-dimensional, complex Hilbert space will have a diagonal matrix representation. If an operator is not normal, a diagonal matrix representation will not be possible. In this case it is of interest to investigate what other fairly simple forms the matrix representation of such a linear operator may assume, which brings up the concept of *generalized eigenvectors* and the *Jordan canonical form.* For completeness this topic is included here, although the treatment is very brief. This material does, however, relate in an analogous way to the topic of generalized eigenvectors in infinite-dimensional space problems.

Consider a linear operator on a finite-dimensional space $A : H \to H$ with an eigenvalue $\lambda_n$. Recall that $\mathbf{0} \neq \mathbf{x}_n \in H$ is an eigenvector of $A$ corresponding to an eigenvalue $\lambda_n$ if $(A - \lambda_n I)\,\mathbf{x}_n = 0$. To generalize this concept, an element $\mathbf{x}_{n,m-1} \in H$ is called a *generalized eigenvector* (or *root vector*) of rank $m$ of the operator $A$ corresponding to the eigenvalue $\lambda_n$ if

$$(A - \lambda_n I)^m \,\mathbf{x}_{n,m-1} = 0,$$
$$(A - \lambda_n I)^{m-1} \,\mathbf{x}_{n,m-1} \neq 0,$$

where $m$ is a positive integer.[20] The rank $m$ cannot be higher than the algebraic multiplicity of the corresponding eigenvalue $\lambda_n$. Note that every (ordinary) eigenvector of $A$ is a generalized eigenvector of rank 1 ($\mathbf{x}_{n,0} \equiv \mathbf{x}_n$). We will call generalized eigenvectors having rank $m > 1$ *associated eigenvectors*, in order to distinguish them from ordinary eigenvectors.[21] The system of generalized eigenvectors is called the *root system* of $A$ and is composed of the union of the ordinary and associated eigenvectors.

The next theorem highlights the importance of generalized eigenvectors.

**Theorem 4.39.** *Consider the linear operator $A : H^n \to H^n$, where $H^n$ is a complex, finite-dimensional Hilbert space. Then the maximal set of all linearly independent generalized eigenvectors of $A$ forms a basis (generalized eigenbasis) of $H^n$.*

Therefore, in $H^n$ this basis, also called the *root basis*, exists for any operator $A : H^n \to H^n$. For infinite-dimensional space this is not necessarily the case.

Furthermore, let us call a basis of $H^n$ a *Jordan basis* for $A$ if, with respect to this basis, the matrix representation of $A$ has a block-diagonal form

$$\begin{bmatrix} M_1 & & 0 \\ & \ddots & \\ 0 & & M_n \end{bmatrix}, \tag{4.25}$$

where the matrices $M_i$ are lower-triangular, e.g.,

$$M_i = \begin{bmatrix} \lambda_i & 0 & 0 & 0 \\ 1 & \lambda_i & 0 & 0 \\ 0 & 1 & \lambda_i & 0 \\ 0 & 0 & 1 & \lambda_i \end{bmatrix}.$$

The dimension of the matrix $M_i$ is equal to the algebraic multiplicity of $\lambda_i$, and, in the case of simple eigenvalues, the matrix representation of $A$ is diagonal.

For any complex-valued matrix operator $A : H^n \to H^n$ there is a basis of generalized eigenvectors that is a Jordan basis. Therefore, any $A : H^n \to H^n$ with $H^n$ complex admits a matrix representation of the form (4.25), which is fairly simple. Matrix representations of $A$ in some other basis can be brought into the Jordan form by a similarity transformation, representing a change of basis to the Jordan basis. Unfortunately, while the

[20] Note that we define an operator raised to a power as $A^n x = \underbrace{A A A \cdots A}_{n \text{ times}} x$, where for this to make sense $R_A \subseteq D_A$ (see Section 4.4).

[21] It can be shown that self-adjoint operators only possess generalized eigenvectors of rank 1, i.e., that they only possess ordinary eigenvectors (see p. 259). However, this does not imply that the dimension of the eigenspace associated with an eigenvalue is necessarily unity.

Jordan basis is made up of generalized eigenvectors, its actual construction is somewhat complicated and will not be discussed here (see, for instance, [19, Ch. 6]).

### Spectral Representations

In preparation for examining eigenfunction expansions in the next section, we consider the *spectral representation* of an operator acting on a finite-dimensional space. Assume $A : H^n \to H^n$ admits a diagonal representation $[A]_{\mathbf{e}}$ with respect to an eigenbasis $\mathbf{e} = \{\mathbf{x}_1, \mathbf{x}_2, \ldots, \mathbf{x}_n\}$, and write

$$[A]_{\mathbf{e}} = \sum_{i=1}^{n} \lambda_i \left[\widetilde{P}_i\right],$$

where $[\widetilde{P}_i]$ is the matrix for which all entries are zero except the $i$th element on the main diagonal ($\sum_{i=1}^{n} [\widetilde{P}_i] = I_{n\times n}$). Then, the representation of $A$ in an arbitrary basis $\mathbf{x}$ is

$$[A]_{\mathbf{x}} = [P] \left( \sum_{i=1}^{n} \lambda_i \left[\widetilde{P}_i\right] \right) [P]^{-1} = \sum_{i=1}^{n} \lambda_i \left[P_i\right],$$

where $[P_i] \equiv [P] [\widetilde{P}_i] [P]^{-1}$, with $[P] = [P]_{\mathbf{e}\to\mathbf{x}}$ being the nonsingular transition matrix from the basis $\mathbf{e}$ to the basis $\mathbf{x}$. In operator form we have

$$A = \sum_{i=1}^{n} \lambda_i P_i. \tag{4.26}$$

Note the simple geometric meaning presented by the representation (4.26): The space $H^n$ decomposes into $n$ orthogonal subspaces $H_1, H_2, \ldots, H_n$, each $H_i$ being associated[22] with the *ith* eigenvector $\mathbf{x}_i$. For any $\mathbf{x} \in H^n$,

$$P_i \mathbf{x} = \alpha_i \mathbf{x}_i \in H_i.$$

Therefore, the $P_i : H^n \to H_i$ are seen to be projection operators from the space $H^n$ to the subspace $H_i$ (note that[23] $\sum_{i=1}^{n} P_i = I_{n\times n}$). This representation for $A$ is called a *weighted sum of projections* and is depicted in Figures 4.4 and 4.5.

The above decomposition for normal or self-adjoint operators on a finite-dimensional space $H^n$ can be extended to normal and self-adjoint operators on infinite-dimensional spaces, most transparently using Riemann–Stieltjes integrals. The procedure is, nevertheless, quite complicated and so will be omitted here, although if we add the condition that $A$ is compact we are back to a simple situation, as shown in the next section.

---

[22] The null space for each $\lambda_i$ may not be one-dimensional, but the space spanned by $\mathbf{x}_i$, i.e., $H_i$, is one-dimensional. Therefore, more than one $H_i$ may be associated with a single eigenvalue.

[23] $\sum_{i=1}^{n} P_i = I_{n\times n}$ is called a *resolution of the identity*.

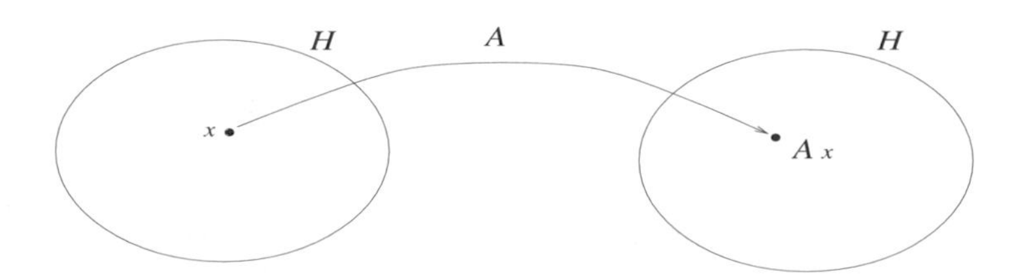


Figure 4.4: Depiction of the mapping $A : H \to H$.

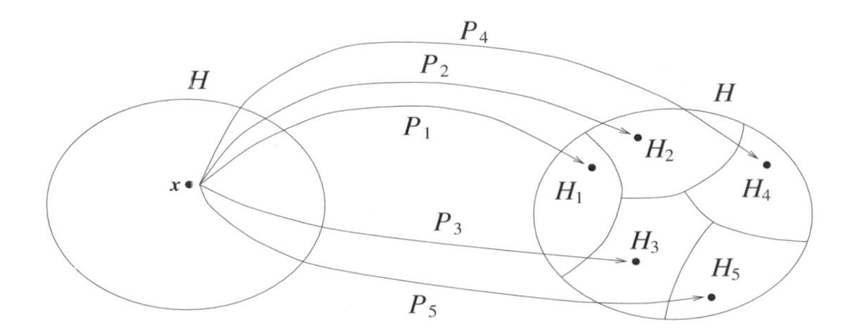


Figure 4.5: Decomposition of a simple operator $A : H \to H$ as $A = \sum_{i=1}^{n} \lambda_i P_i$, where $H$ is finite-dimensional. The space $H$ is decomposed into $n$ orthogonal subspaces, $H = H_1 \oplus H_2 \oplus \cdots \oplus H_n$ ($n = 5$ in this figure), where $H_i = N_{A-\lambda_i I}$ and $P_i\mathbf{x} = \langle \mathbf{x}, \mathbf{x}_i \rangle \mathbf{x}_i \in H_i$.

**Example 4.7.**

As an illustration of a spectral representation, consider the operator in the example on p. 251, where

$$[A]_{\mathbf{y}} = \sum_{i=1}^{2} \lambda_i [P_i] = \sqrt{2}\,[P_1] - \sqrt{2}\,[P_2]$$

and

$$[P_1] = [P]^{-1} \begin{bmatrix} 1 & 0 \\ 0 & 0 \end{bmatrix} [P],$$

$$[P_2] = [P]^{-1} \begin{bmatrix} 0 & 0 \\ 0 & 1 \end{bmatrix} [P].$$

This form is particularly useful for constructing functions of operators and matrices, as shown in Section 4.4.

A form related to (4.26) is easily developed in the finite-dimensional case and also extends to the infinite-dimensional case for compact normal operators. Assume that $A : H^n \to H^n$ is a simple operator, i.e., that it possesses an eigenbasis. Then, any $\mathbf{x} \in H^n$ can be written as

$$\mathbf{x} = \sum_{i=1}^{n} \langle \mathbf{x}, \mathbf{x}_i \rangle \mathbf{x}_i,$$

where $\mathbf{x}_i$ is the eigenvector corresponding to the eigenvalue $\lambda_i$, i.e., $A\mathbf{x}_i = \lambda_i \mathbf{x}_i$. With $A\mathbf{x} = \sum_{i=1}^{n} \langle \mathbf{x}, \mathbf{x}_i \rangle A\mathbf{x}_i = \sum_{i=1}^{n} \lambda_i \langle \mathbf{x}, \mathbf{x}_i \rangle \mathbf{x}_i$, one may write the operator $A$ as

$$A = \sum_{i=1}^{n} \lambda_i \langle \cdot, \mathbf{x}_i \rangle \mathbf{x}_i. \tag{4.27}$$

If we identify $P_i : H^n \to H_i$ as the (projection) operator that takes $\mathbf{x} \in H^n$ into the subspace spanned by $\mathbf{x}_i$, i.e., $P_i\mathbf{x} = \langle \mathbf{x}, \mathbf{x}_i \rangle \mathbf{x}_i$, then the spectral representations (4.26) and (4.27) are seen to be equivalent. However, (4.26) is usually preferred for theoretical constructions. The identity operator is given by

$$I_{n \times n} = \sum_{i=1}^{n} \langle \cdot, \mathbf{x}_i \rangle \mathbf{x}_i,$$

where it is sometimes convenient to write

$$I_{n \times n} = \sum_{i=1}^{n} \mathbf{x}_i \overline{\mathbf{x}}_i$$

such that $I_{n \times n}\mathbf{x} = \mathbf{x} = \sum_{i=1}^{n} \mathbf{x}_i \overline{\mathbf{x}}_i \cdot \mathbf{x}$.

Finally, it is important to make a connection between the eigenvalues and eigenvectors of a linear operator on a finite-dimensional complex space $A : H^n \to H^n$, and the eigenvalues and eigenvectors of its matrix representation $[a_{ij}]_{\mathbf{y}}$ with respect to some basis $\mathbf{y}$. It can be shown [19, pp. 152–153] that

- the spectrum of an operator and the spectrum of its matrix representation coincide, and
- an element $\mathbf{x} \in H^n$ is an eigenvector of $A$ if and only if the representation of $\mathbf{x}$ with respect to a basis of $H^n$ is an eigenvector of the matrix representation of $A$ with respect to the same basis.

However, for operators acting on infinite-dimensional spaces projected onto a finite-dimensional subspace, eigenvalues of the actual operator may or may not be well approximated by eigenvalues of the resulting matrix, depending on the chosen finite subspace. For example, an operator on an infinite-dimensional space may have no eigenvalues, yet an $n \times n$ matrix, which in some approximate way is supposed to represent the operator, always has $n$ eigenvalues.

### 4.3.2 Operators on Infinite-Dimensional Spaces

#### Spectral Expansions

We have already seen that the spectrum of an operator acting on an infinite-dimensional space $H$ may be much more complicated than that of an operator acting on a finite-dimensional space. For the finite-dimensional case, one usually is interested in a simple matrix representation of a given operator. To determine the simplest matrix representation possible one needs to know the eigenvalues and eigenvectors of the operator. Although for the infinite-dimensional case one may investigate matrix representations, it is more often desirable to consider the possibility of representing an element of $H$ in terms of eigenvectors of $A$. The following theorems provide sufficient conditions for when such representations are possible.

Since operators that are merely compact, or merely self-adjoint, need not possess any eigenvalues, in the general case one needs to concentrate on compact self-adjoint operators to obtain sufficiently strong results. In the case of typical boundary value problems, self-adjointness alone is often sufficient to get similarly strong results because the inverse operator is often compact and self-adjoint.

The first theorem concerns the representation of an element $x \in H$ in terms of the eigenvectors corresponding to nonzero eigenvalues of an operator $A : H \to H$ [21, pp. 188–191].

**Theorem 4.40.** *(Hilbert–Schmidt theorem) Let $A : H \to H$ be a compact, self-adjoint linear operator acting on an infinite-dimensional Hilbert space $H$. Then there exists an orthonormal system of eigenvectors $\{u_n\}$ corresponding to nonzero eigenvalues $\{\lambda_n\}$ such that every $x \in H$ can be uniquely represented as*

$$x = x_0 + \sum_{n=1}^{\infty} \langle x, u_n \rangle \, u_n,$$

*where $x_0$ satisfies $Ax_0 = 0$. Furthermore, if $\{\lambda_n\}$ is an infinite set of distinct eigenvalues, then $\lim_{n\to\infty} \lambda_n = 0$.*

Note that the expansion coefficients are the generalized Fourier coefficients $\alpha_n = \langle x, u_n \rangle$. If we denote by $S$ the space spanned by $\{u_n\}$ ($S$ is then a closed linear subspace of $H$), by Theorem 2.9 (projection) we see that the described representation is simply a decomposition of $x$ into an element $\sum_{n=1}^{\infty} \alpha_n u_n \in S$ and an element $x_0 \in S^{\perp}$.

Although the set $\{u_n\}$ does not form a basis for $H$ unless $N_A = \{0\}$,[24] it is clear that it does form a basis for the range of $A$, such that for any $x \in H$,

$$Ax = \sum_{n} \lambda_n \langle x, u_n \rangle \, u_n.$$

[24]Note that $\{u_n\}$ will also be a basis for $H$ if $R_A$ is dense in $H$.

Such a set $\{u_n\}$ is called a *basic* system of eigenvectors.

Next we consider the complete set of eigenvectors (including those corresponding to eigenvalues $\lambda_n = 0$), which are found to form a basis for $H$.

**Theorem 4.41.** *(Spectral theorem for compact self-adjoint operators) Let $A : H \to H$ be a compact, self-adjoint (or more generally, normal) linear operator acting on an infinite-dimensional Hilbert space $H$. Then there exists an orthonormal basis for $H$ of eigenvectors $\{x_n\}$ with corresponding eigenvalues $\{\lambda_n\}$. For every $x \in H$,*

$$x = \sum_n \langle x, x_n \rangle x_n$$

*and*

$$Ax = \sum_n \lambda_n \langle x, x_n \rangle x_n.$$

To gain insight into the above, let $\{\nu_n\}$ be a basis for the Hilbert space $N_A$ consisting of eigenvectors associated with eigenvalues $\lambda_n = 0$, and note that $\{x_n\} = \{u_n\} \cup \{v_n\}$ is an orthonormal basis for $H$. Then $x_0 = \sum_n \langle x, \nu_n \rangle \nu_n$ where $Ax_0 = 0$ [43, p. 101]. From Theorem 3.17, $N_{A-\lambda_n I}$ is finite-dimensional for $\lambda_n \neq 0$, although $N_A$ will be infinite-dimensional. It can then be observed that the space $H$ is decomposed as $H = N_A \bigoplus_n N_{A-\lambda_n I}$.

Although Theorem 4.41 is fundamentally important, it is often convenient to use the notation of Theorem 4.40 and explicitly separate out the term $x_0$, such that

$$\begin{aligned} x &= x_0 + \sum_{n=1}^{\infty} \langle x, u_n \rangle u_n, \\ Ax &= \sum_{n=1}^{\infty} \lambda_n \langle x, u_n \rangle u_n, \end{aligned} \tag{4.28}$$

where $Ax_0 = 0$. In this form we are concerned only with the basic system of eigenvectors $u_n$, corresponding to nonzero eigenvalues. As usual, all equalities are understood in the norm sense.

It is worthwhile to note that the converse of the spectral theorem is also true [23, p. 117].

**Theorem 4.42.** *Let $\{x_n\}$ be an orthonormal set in $H$ and $\{\lambda_n\}$ a sequence of real numbers that is either finite or that converges to zero. The linear operator defined by $Ax \equiv \sum_n \lambda_n \langle x, x_n \rangle x_n$ is compact and self-adjoint.*

Because compact operators $A : H \to H$, where $H$ is infinite-dimensional, are necessarily bounded, they can always be represented by an infinite matrix. From the properties of compact operators, and those of self-adjoint operators, we see that the matrix representation of a compact, self-adjoint operator $A : H \to H$ with respect to the eigenbasis $\{x_n\}$ is a diagonal matrix with real-valued eigenvalues on the main diagonal, tending towards zero as $n \to \infty$. Furthermore, it can be shown that two commuting compact, self-adjoint operators $A, B : H \to H$ possess a common eigenbasis (see Theorem 3.12).

Finally, if an operator $A : H \to H$ is compact but not self-adjoint, we have the following expansion theorem.

**Theorem 4.43.** *Let $A : H \to H$ be a compact operator. Then, there exist orthonormal sets $\{x_n\}$ and $\{y_n\}$ and nonnegative real numbers $\{s_n\}$ with $s_n \to 0$ such that*

$$Ax = \sum_n s_n \langle x, x_n \rangle y_n.$$

It turns out that $\{x_n\}$ are eigenvectors or $A^*A$, $\{y_n\}$ are eigenvectors of $AA^*$, and $\{s_n^2\}$ are eigenvalues of $A^*A$ (and also of $AA^*$). The $s_n$ are called *singular values* of the operator $A$, and the series $Ax$ is known as a *singular-value decomposition* of $A$. The foundation of this representation is that while $A$ is compact and not necessarily normal, $A^*A$ and $AA^*$ are compact and self-adjoint by Theorem 3.8 and the discussion in Section 3.6.1.

### Jordan Chain

Unfortunately, while for operators acting on finite-dimensional spaces one can always obtain a basis of generalized eigenvectors (Jordan basis) by Theorem 4.39, the same is not true for the infinite-dimensional case. However, if generalized eigenvectors exist, then they are defined as in the finite-dimensional case, i.e., an element $0 \neq x_{n,m-1} \in H$ is a generalized eigenvector (root vector) of rank $m$ of the operator $A : H \to H$ corresponding to an eigenvalue $\lambda_n$ if

$$\begin{aligned} (A - \lambda_n I)^m \, x_{n,m-1} &= 0, \\ (A - \lambda_n I)^{m-1} \, x_{n,m-1} &\neq 0, \end{aligned} \tag{4.29}$$

where $m$ is a positive integer. Every (ordinary) eigenvector of $A$ is a generalized eigenvector of rank 1 ($x_{n,0} \equiv x_n$), and we again call the generalized eigenvectors having rank $m > 1$ associated eigenvectors to distinguish them from ordinary eigenvectors. The root system of $A$ is defined as the union of the ordinary and associated eigenvectors. This system may or may not form a basis for $H$, depending on the operator[25].

In practice, to determine the associated eigenvectors, one starts with an eigenvector $x_n$ satisfying $(A - \lambda_n I)\, x_n = 0$. If the equation

$$(A - \lambda_n I)\, x_{n,1} = x_n$$

[25] An example of a differential operator that leads to a root basis is provided in Section 5.3.2.

has a solution $x_{n,1}$, then $x_{n,1}$ is a generalized eigenvector of rank 2; more specifically, an associated eigenvector associated with the eigenvalue $\lambda_n$ and eigenvector $x_n$. Continuing, if

$$(A - \lambda_n I)\, x_{n,2} = x_{n,1}$$

is solvable, then $x_{n,2}$ is another associated eigenvector (rank 3), associated with the eigenvalue $\lambda_n$ and eigenvector $x_n$. In general, we consider

$$(A - \lambda_n I)\, x_{n,k} = x_{n,k-1}$$

such that the chain $\{x_n, x_{n,1}, x_{n,2}, \ldots x_{n,j}\}$ consisting of the ordinary and associated eigenvectors is called a Jordan chain of length $j+1$.

It is worthwhile to note that if generalized eigenvectors of a self-adjoint operator exist, then they have rank 1 (i.e., no associated eigenvectors exist).[26]

### 4.3.3 Spectral Expansions Associated with Boundary Value Problems

Theorems in the previous section detailed sufficient conditions for eigenfunctions of an operator to form a basis in $H$. Fortunately, many operators that do not satisfy the previous theorems also lead to an eigenbasis. Important examples are unbounded operators corresponding to self-adjoint boundary value problems. We make the following observation.

*The eigenfunctions of a self-adjoint boundary value problem on a Hilbert space $H$ form an orthonormal basis of $H$.*

The term "boundary value problem" in common usage typically means a differential operator together with some specified boundary conditions. Since the boundary conditions define (in part) the domain of the operator, they are part of the definition of the operator, such that by "self-adjoint boundary value problem" we actually refer to a self-adjoint differential operator, which we will denote by $L$. The above classification of self-adjoint boundary value problems relies on the fact that such operators have a compact self-adjoint inverse (integral) operator $L^{-1}$ on $H$, the Green's operator, with an associated orthonormal eigenbasis by Theorem 4.41. By Theorem 4.11 the eigenfunctions of the inverse operator $L^{-1}$ are also eigenfunctions of $L$ (zero eigenvalues are not an issue), and therefore $L$ possesses an eigenbasis for $H$.

---

[26] Assume $A : H \to H$ is self-adjoint, with $x$ an associated eigenvector of rank 2 corresponding to eigenvalue $\lambda$. Then, $0 = \left\langle x, (A - \lambda I)^2 x \right\rangle = \langle (A - \lambda I)\, x, (A - \lambda I)\, x \rangle = \|(A - \lambda I)\, x\|^2$, which implies, from Definition 2.31, that $(A - \lambda I)\, x = 0$ which is a contradiction, and so no generalized eigenvectors of rank 2 exist. Therefore, no generalized eigenvectors of rank higher than 1 exist.

The above conclusion holds for problems on a bounded region of space, $\Omega \subset \mathbf{R}^n$. If $\Omega$ is unbounded the inverse will not, in general, be compact, and Theorem 4.41 is not applicable. However, if the differential operator is self-adjoint and certain fitness conditions are satisfied at infinity, then the eigenfunctions (proper and improper) form an eigenbasis (possibly discrete and continuous) for $H$. This is discussed in more detail in later sections (see, e.g., Sections 5.2 and 8.1).

In particular, we have the following [4, pp. 396–397].

**Theorem 4.44.** *The pth-order differential operator* $L{:}\,\mathbf{L}^2_w(a,b) \to \mathbf{L}^2_w(a,b)$ *defined by*

$$-L \equiv a_p \frac{d^p}{dt^p} + \cdots + a_1 \frac{d}{dt} + a_0,$$

$$D_L \equiv \left\{ x{:}\, \frac{d^j x}{dt^j} \in \mathbf{L}^2_w(a,b) \text{ for } j = 1,2,\ldots,p,\ B_a(x) = B_b(x) = 0 \right\},$$

*where* $a, b$ *are finite,* $p$ *is even,* $a_n \in \mathbf{C}^p(a,b)$ *are real-valued coefficients with* $a_p \neq 0$*, and where* $B_{a,b}$ *are given as*

$$B_a(x) = \alpha_1 x(a) + \alpha_2 x'(a) + \cdots + a_p x^{(p-1)}(a) = 0,$$
$$B_b(x) = \beta_1 x(b) + \beta_2 x'(b) + \cdots + \beta_p x^{(p-1)}(b) = 0,$$

*with* $\alpha = (\alpha_1, \ldots, \alpha_p)$ *and* $\beta = (\beta_1, \ldots, \beta_p)$ *independent vectors in* $\mathbf{R}^p$*, is self-adjoint (and so forms a self-adjoint boundary value problem) on* $H = \mathbf{L}^2_w(a,b)$.

Therefore, the eigenvalue problem

$$Lx_n = \lambda_n x_n$$

with $L$ defined as in Theorem 4.44 leads to an orthonormal eigenbasis $\{x_n\}$ of $H$ and to the expansion $x = \sum_n \langle x, x_n \rangle x_n$ for any $x(t) \in H = \mathbf{L}^2_w(a,b)$. In the above, $H$ is a weighted $\mathbf{L}^2$ space, where the weight $w$ (possibly unity) depends on the coefficients and is chosen so that the operator is self-adjoint. The inner product for the space is

$$\langle f, g \rangle = \int_a^b f(\xi) g(\xi) w(\xi)\, d\xi.$$

Second-order operators of this type are extensively studied in Chapter 5, where self-adjointness is shown and a sketch of the completeness proof is presented.

For scalar partial differential operators we have the following theorem [4, pp. 542–543] (see Example 8 in Section 3.4.2).

**Theorem 4.45.** *The negative Laplacian operator* $L : \mathbf{L}^2(\Omega) \to \mathbf{L}^2(\Omega)$ *defined by*

$$(Lx)(\mathbf{t}) \equiv -\nabla^2 x(\mathbf{t}),$$
$$D_L \equiv \left\{x : x(\mathbf{t}), -\nabla^2 x(\mathbf{t}) \in \mathbf{L}^2(\Omega),\ x(\mathbf{t})|_\Gamma = 0\right\},$$

*where* $\Gamma$ *is the sufficiently smooth boundary of the region* $\Omega \subset \mathbf{R}^3$*, forms a self-adjoint boundary value problem on* $\mathbf{L}^2(\Omega)$*.*

As such, the orthonormal eigenbasis defined by

$$-\nabla^2 x_n = \lambda_n x_n$$

leads to the expansion $x = \sum_n \langle x, x_n\rangle x_n$ for any $x(\mathbf{t}) \in H = \mathbf{L}^2(\Omega)$. This topic is discussed further in Section 6.3 for electrostatic problems, where self-adjointness is shown and the completeness proof outlined. Additionally, these eigenfunctions are important in separation-of-variables solutions for many partial differential equations, as shown in Section 5.4.

Furthermore, for the vector differential operator $\nabla \times \nabla \times$ we have a similar theorem (see Example 9, Section 3.4.2).

**Theorem 4.46.** *The vector differential operator* $L : \mathbf{L}^2(\Omega)^3 \to \mathbf{L}^2(\Omega)^3$ *defined by*

$$(L\mathbf{x})(\mathbf{t}) \equiv \nabla \times \nabla \times \mathbf{x}(\mathbf{t}),$$
$$D_L \equiv \left\{\mathbf{x} : \mathbf{x}(\mathbf{t}), \nabla \times \nabla \times \mathbf{x}(\mathbf{t}) \in \mathbf{L}^2(\Omega)^3,\ \mathbf{n} \times \mathbf{x}(\mathbf{t})|_\Gamma = \mathbf{0}\right\},$$

*where* $\Gamma$ *is the sufficiently smooth boundary of the region* $\Omega \subset \mathbf{R}^3$*, forms a self-adjoint boundary value problem on* $\mathbf{L}^2(\Omega)^3$*.*

Therefore, the orthonormal eigenbasis defined by

$$\nabla \times \nabla \times \mathbf{x}_n = \lambda_n \mathbf{x}_n$$

with appropriate vector boundary conditions leads to the expansion $\mathbf{x} = \sum_n \langle \mathbf{x}, \mathbf{x}_n\rangle \mathbf{x}_n$ for any $\mathbf{x}(\mathbf{t}) \in H = \mathbf{L}^2(\Omega)^3$. The same result holds for the vector eigenvalue problem

$$\nabla^2 \mathbf{x}_n = \lambda_n \mathbf{x}_n.$$

The situation is quite different if $\Omega$ is an unbounded region of space.

Because we deal with vector eigenfunctions, boundary conditions generally need to be specified for both tangential and normal field components to uniquely identify the desired eigenfunctions. For example, vector Dirichlet conditions are $\mathbf{x}_n|_\Gamma = \mathbf{0}$, although the boundary conditions $\mathbf{n} \times \mathbf{x}_n|_\Gamma = \mathbf{0}$ and $\nabla \cdot \mathbf{x}_n|_\Gamma = 0$ are generally more useful in electromagnetics applications (see Chapters 9 and 10). Boundary conditions do not need to be imposed on all three scalar components of $\mathbf{x}_n$ if the eigenfunction is required to satisfy homogeneous Maxwell's equations. In such cases it is only necessary to impose tangential conditions.

The above theorems concerning self-adjoint differential operators are extremely important because they justify many eigenfunction expansion methods used in electromagnetic applications.

### 4.3.4 Spectral Expansions Associated with Integral Operators

We now consider some spectral expansions arising from integral operators involving the free-space scalar Green's function. For all of the resulting eigenfunctions, completeness in certain spaces is well known and can be established using nonspectral techniques. Therefore, as with the boundary value problems just considered, identification of the (integral) operator as being compact and self-adjoint is not required, although some of the operators fall into this category. Many of the same expansions are obtained from a differential operator (Sturm–Liouville) approach in Section 5.5.

#### Chebyshev Polynomials of the First Kind

Consider the weighted Hilbert space $\mathbf{L}_w^2(-a,a)$ with weight $w = \left(a^2 - x^2\right)^{-\frac{1}{2}}$ and inner product

$$\langle f, g \rangle = \int_{-a}^{a} f(x) g(x) \frac{dx}{\sqrt{a^2 - x^2}}.$$

The compact, self-adjoint integral operator $A : \mathbf{L}_w^2(-a,a) \to \mathbf{L}_w^2(-a,a)$ defined by[27]

$$(Af)(x) \equiv \int_{-a}^{a} f(x') \frac{\ln(|x - x'|)}{\sqrt{a^2 - x'^2}}\, dx'$$

arises, for example, in the study of static and quasi-static electromagnetic diffraction by a strip or by a slot in a perfectly conducting infinite screen. For two-dimensional dynamic problems these types of logarithmic kernels represent the source-point singularity associated with the two-dimensional principal Green's function (Hankel function). The eigenvalue problem

$$Af = \lambda f$$

leads to *first-kind Chebyshev polynomials* $T_n(x/a)$ as eigenfunctions, with corresponding eigenvalues $\lambda_0 = -\pi \ln(2/a)$ for $n = 0$, and $\lambda_n = -\pi/n$ for

[27] $\ln(|x - \xi|)$ is the natural restriction of the two-dimensional, static free-space Green's function $\ln(|\mathbf{r} - \xi'|)$ to a planar surface.

$n \neq 0$ (note $N_A = \{0\}$ if $a \neq 2$), i.e.,

$$\int_{-a}^{a} \frac{\ln|x-x'|}{\sqrt{a^2-x'^2}} T_n(x'/a)\,dx' = \begin{cases} -\pi\ln(2/a)T_0(x/a), & n=0, \\ -\frac{\pi}{n}T_n(x/a), & n>0. \end{cases} \tag{4.30}$$

The eigenfunctions are normalized as

$$\frac{\varepsilon_n}{\pi}\langle T_n, T_m\rangle = \frac{\varepsilon_n}{\pi}\int_{-a}^{a} T_n(x/a)T_m(x/a)\frac{1}{\sqrt{a^2-x^2}}\,dx = \delta_{nm}, \tag{4.31}$$

where $\varepsilon_0 = 1$ and $\varepsilon_n = 2$ for $n \neq 0$, and the set of Chebyshev polynomials $\left\{\sqrt{\varepsilon_n/\pi}T_n(x/a)\right\}$, $n = 0, 1, 2, \ldots$, forms a basis in $\mathbf{L}^2_w(-a,a)$. Therefore, any function $f(x) \in \mathbf{L}^2_w(-a,a)$ can be expanded as

$$f = \frac{1}{\pi}\sum_{n=0}^{\infty} \varepsilon_n \langle f, T_n\rangle T_n$$

and

$$Af = \frac{1}{\pi}\sum_{n=0}^{\infty} \varepsilon_n\lambda_n \langle f, T_n\rangle T_n.$$

For example, in $\mathbf{L}^2_w(-a,a)$ the equation

$$Ag = f$$

can be solved by writing $f = (1/\pi)\sum_{n=0}^{\infty}\varepsilon_n\langle f, T_n\rangle T_n$, leading to

$$g = \frac{1}{\pi}\sum_{n=0}^{\infty}\frac{\varepsilon_n}{\lambda_n}\langle f, T_n\rangle T_n. \tag{4.32}$$

Convergence in the form $\lim_{N\to\infty}\|g - g_N\| = 0$, where

$$g_N \equiv (1/\pi)\sum_{n=0}^{N}(\varepsilon_n/\lambda_n)\langle f, T_n\rangle T_n,$$

depends on the condition $\sum_n (1/\lambda_n^2)\left|\langle f, T_n\rangle\right|^2 < \infty$ by the Riesz–Fischer theorem (Theorem 2.11). Since $\lambda_n \to 0$ as $n \to \infty$, which is a general property of compact operators, convergence problems may occur depending on the properties of $f$. This type of problem is typical of first-kind operator equations involving compact operators.

### Chebyshev Polynomials of the Second Kind

Consider the weighted Hilbert space $\mathbf{L}^2_w(-a,a)$ with weight $w = \left(a^2-x^2\right)^{\frac{1}{2}}$ and inner product

$$\langle f, g\rangle = \int_{-a}^{a} f(x)g(x)\sqrt{a^2-x^2}\,dx.$$

As with the previous case, the operator $A : \mathbf{L}^2_w(-a,a) \to \mathbf{L}^2_w(-a,a)$ defined by

$$(Af)(x) \equiv \frac{d^2}{dx^2} \int_{-a}^{a} f(x') \ln(|x-x'|) \sqrt{a^2 - x'^2}\, dx'$$

arises in the study of strip and slot diffraction. The eigenvalue problem

$$Af = \lambda f$$

leads to *second-kind Chebyshev polynomials* $U_n(x/a)$ as eigenfunctions, with corresponding eigenvalues $\lambda_n = \pi(n+1)$, i.e.,

$$\frac{d^2}{dx^2} \int_{-a}^{a} U_n(x'/a) \ln|x-x'| \sqrt{a^2 - x'^2}\, dx' = \pi(n+1) U_n(x/a).$$

The eigenfunctions are normalized as

$$\frac{2}{\pi a^2} \langle U_n, U_m \rangle = \frac{2}{\pi a^2} \int_{-a}^{a} U_n(x/a) U_m(x/a) \sqrt{a^2 - x^2}\, dx = \delta_{nm},$$

and the set of Chebyshev polynomials $\left\{ \sqrt{2/(\pi a^2)} U_n(x/a) \right\}$, $n = 0, 1, 2, \ldots$, forms a basis in $\mathbf{L}^2_w(-a,a)$. Therefore, any function $f(x) \in \mathbf{L}^2_w(-a,a)$ can be expanded as

$$f = \frac{2}{\pi a^2} \sum_{n=0}^{\infty} \langle f, U_n \rangle U_n$$

and

$$Af = \frac{2}{\pi a^2} \sum_{n=0}^{\infty} \lambda_n \langle f, U_n \rangle U_n.$$

### Laguerre Polynomials

The weighted Hilbert space $\mathbf{L}^2_w(0,\infty)$ with weight $e^{-x}/\sqrt{x}$ and inner product

$$\langle f, g \rangle = \int_{0}^{\infty} f(x) g(x) \frac{1}{\sqrt{x}} e^{-x}\, dx$$

along with the integral operator $A : \mathbf{L}^2_w(0,\infty) \to \mathbf{L}^2_w(0,\infty)$ defined by

$$(Af)(x) \equiv \int_{0}^{\infty} f(x') K_0(|x-x'|) \frac{e^{-x'}}{\sqrt{x'}}\, dx'$$

is encountered in the study of E-polarized wave diffraction by a half-space [24]. The kernel

$$K_0(x) = -\frac{i\pi}{2} H_0^{(2)}(-ix) = \frac{i\pi}{2} H_0^{(1)}(ix)$$

is the modified Bessel function [25], sometimes called the *MacDonald function.* In this case we have the generalized eigenvalue problem

$$Af = \lambda e^{-x} f,$$

which leads to $L_n^{-1/2}(2x)$ as eigenfunctions with $\lambda_n = (\pi/\sqrt{2})\gamma_n^E$ as the corresponding eigenvalues, i.e.,

$$\int_0^\infty L_n^{-1/2}(2x') K_0(|x-x'|) \frac{e^{-x'}}{\sqrt{x'}} dx' = \frac{\pi}{\sqrt{2}} \gamma_n^E e^{-x} L_n^{-1/2}(2x).$$

The functions $L_n^v$ are *associated Laguerre polynomials* ($v > -1$, and for $v = 0$, $L_n^0 = L_n$ are called *Laguerre polynomials*), and

$$\gamma_n^E = \Gamma(n + 1/2) / \Gamma(n+1),$$

where $\Gamma$ is the gamma-function. The eigenfunctions are normalized as

$$\frac{1}{\gamma_n^E} \left\langle L_n^{-1/2}, L_m^{-1/2} \right\rangle = \frac{1}{\gamma_n^E} \int_0^\infty L_n^{-1/2}(x) L_m^{-1/2}(x) \frac{e^{-x}}{\sqrt{x}} dx = \delta_{nm},$$

and the set $\left\{ \sqrt{1/\gamma_n^E} L_n^{-1/2}(x) \right\}$, $n = 0, 1, 2, \ldots$, forms a basis in $\mathbf{L}_w^2(0, \infty)$. Therefore, any function $f(x) \in \mathbf{L}_w^2(0, \infty)$ can be expanded as

$$f = \frac{1}{\gamma_n^E} \sum_{n=0}^\infty \left\langle f, L_n^{-1/2} \right\rangle L_n^{-1/2}.$$

In a similar manner, the operator $A : \mathbf{L}_w^2(0,\infty) \to \mathbf{L}_w^2(0,\infty)$ defined by

$$(Af)(x) \equiv \left( \frac{d^2}{dx^2} - 1 \right) \int_0^\infty f(x') K_0(|x-x'|) \sqrt{x'} e^{-x'} dx'$$

arises in the study of H-polarized wave diffraction by a half-space [24], where the weight is $\sqrt{x} e^{-x}$ with the inner product

$$\langle f, g \rangle = \int_0^\infty f(x) g(x) \sqrt{x} e^{-x} dx.$$

The generalized eigenvalue problem

$$Af = \lambda e^{-x} f$$

leads to $L_n^{1/2}(2x)$ as eigenfunctions with $\lambda_n = -\pi\sqrt{2}\gamma_n^H$ as the corresponding eigenvalues, i.e.,

$$\left( \frac{d^2}{dx^2} - 1 \right) \int_0^\infty L_n^{1/2}(2x') K_0(|x-x'|) \sqrt{x'} e^{-x'} dx' = -\pi\sqrt{2}\gamma_n^H e^{-x} L_n^{1/2}(2x)$$

where $\gamma_n^H = \Gamma(n+3/2)/\Gamma(n+1)$. The eigenfunctions are normalized as

$$\frac{1}{\gamma_n^H}\left\langle L_n^{1/2}, L_m^{1/2}\right\rangle = \frac{1}{\gamma_n^H}\int_0^\infty L_n^{1/2}(x)\, L_m^{1/2}(x)\sqrt{x}e^{-x}dx = \delta_{nm}$$

and the set $\left\{\sqrt{1/\gamma_n^H}L_n^{1/2}(x)\right\}$, $n = 0, 1, 2, \ldots$, forms a basis in $\mathbf{L}^2_w(0,\infty)$. Therefore, any function $f(x) \in \mathbf{L}^2_w(0,\infty)$ can be expanded as

$$f = \frac{1}{\gamma_n^H}\sum_{n=0}^{\infty}\left\langle f, L_n^{1/2}\right\rangle L_n^{1/2}.$$

Expansions involving Chebyshev and Laguerre polynomials are further considered in Section 5.5.

### Circular Exponential Functions

Consider the Hilbert space $\mathbf{L}^2(0,2\pi)$ with inner product

$$\langle f, g\rangle = \int_0^{2\pi} f(\phi)\overline{g(\phi)}\, d\phi.$$

The integral operator $A : \mathbf{L}^2(0,2\pi) \to \mathbf{L}^2(0,2\pi)$ defined by[28]

$$(Af)(\phi) \equiv \int_0^{2\pi} f(\phi')H_0^{(1,2)}\left(2ka\sin\frac{|\phi-\phi'|}{2}\right)d\phi', \tag{4.33}$$

where $H_0^{(1,2)}$ is the zeroeth-order Hankel function (see Section 5.4.1), arises in the study of scattering by circular cross-section cylinders. The eigenvalue problem

$$Af = \lambda f$$

leads to complex exponentials $e^{in\phi}$ as eigenfunctions, $n = 0, \pm1, \pm2, \ldots$, with corresponding eigenvalues $\lambda_n = 2\pi J_n(ka)\, H_n^{(1,2)}(ka)$, i.e.,

$$\int_0^{2\pi} e^{in\phi'}H_0^{(1,2)}\left(2ka\sin\frac{|\phi-\phi'|}{2}\right)d\phi' = 2\pi J_n(ka)\, H_n^{(1,2)}(ka)\, e^{in\phi}.$$

The eigenfunctions are normalized as

$$\frac{1}{2\pi}\left\langle e^{in\phi}, e^{im\phi}\right\rangle = \frac{1}{2\pi}\int_0^{2\pi} e^{in\phi}e^{-im\phi}d\phi = \delta_{nm},$$

[28] $H_0^{(1,2)}(2ka\sin(|\varphi-\xi|/2))$ is the natural restriction of the two-dimensional, dynamic Green's function $H_0^{(1,2)}(k|\mathbf{r}-\xi|)$ to the surface of a circular cylinder having radius $a$. We assume $k \in \mathbf{R}$.

and form the well known basis $\{(1/\sqrt{2\pi})e^{in\phi}\}$, $n = 0, \pm 1, \pm 2, \ldots$, in $\mathbf{L}^2(0, 2\pi)$. Any function $f(\phi) \in \mathbf{L}^2(0, 2\pi)$ can be expanded as

$$f = \frac{1}{2\pi} \sum_{n=-\infty}^{\infty} \langle f, e^{in\phi} \rangle \, e^{in\phi}$$

and

$$Af = \frac{1}{2\pi} \sum_{n=-\infty}^{\infty} \lambda_n \langle f, e^{in\phi} \rangle \, e^{in\phi}.$$

**Spherical Harmonics**

Consider the Hilbert space $\mathbf{L}^2(\Gamma)$, where $\Gamma$ is a spherical surface having radius $a$, with inner product

$$\langle f, g \rangle = \int_0^{\pi} \int_0^{2\pi} f(\theta, \phi) \overline{g(\theta, \phi)} a^2 \sin\theta \, d\theta \, d\phi.$$

The integral operator $A : \mathbf{L}^2(\Gamma) \to \mathbf{L}^2(\Gamma)$ defined by

$$(Af)(\theta, \phi) \equiv \int_0^{\pi} \int_0^{2\pi} \frac{e^{-ik|\mathbf{r}-\mathbf{r}'|}}{4\pi |\mathbf{r} - \mathbf{r}'|} f(\theta', \phi') a^2 \sin\theta' \, d\theta' \, d\phi' \tag{4.34}$$

for $\mathbf{r}|_{\Gamma}$ arises in the study of scattering by spheres, where we assume $k \in \mathbf{R}$. The eigenvalue problem

$$Af = \lambda f$$

leads to spherical harmonics $Y_{n,m}(\theta, \phi)$ (see Section 5.4.2 for more details on spherical harmonics) as eigenfunctions,[29] with corresponding eigenvalues

[29] From the expansion (5.172) we have

$$\begin{aligned}
&\int_0^{\pi} \int_0^{2\pi} \frac{e^{-ik|\mathbf{r}-\mathbf{r}'|}}{4\pi |\mathbf{r} - \mathbf{r}'|} Y_{p,q}(\theta', \phi') a^2 \sin\theta' d\theta' d\phi' \\
&= \int_0^{\pi} \int_0^{2\pi} (-ik) \sum_{n=0}^{\infty} j_n(ka)\, h_n^{(2)}(ka) \sum_{m=-n}^{n} Y_{n,m}(\theta, \phi) \overline{Y}_{n,m}(\theta', \phi') \\
&\qquad \cdot Y_{p,q}(\theta', \phi') a^2 \sin\theta' d\theta' d\phi' \\
&= \left(-ika^2\right) \sum_{n=0}^{\infty} j_n(ka)\, h_n^{(2)}(ka) \sum_{m=-n}^{n} Y_{n,m}(\theta, \phi) \int_0^{\pi} \int_0^{2\pi} \overline{Y}_{n,m}(\theta', \phi') \\
&\qquad \cdot Y_{p,q}(\theta', \phi') \sin\theta' d\theta' d\phi' \\
&= \left(-ika^2\right) \sum_{n=0}^{\infty} j_n(ka)\, h_n^{(2)}(ka) \sum_{m=-n}^{n} Y_{n,m}(\theta, \phi)\, \delta_{np} \delta_{mq} \\
&= \left(-ika^2\right) j_p(ka)\, h_p^{(2)}(ka)\, Y_{p,q}(\theta, \phi),
\end{aligned}$$

where the orthonormality relation (5.164) has been used.

$\lambda_n = -ikj_n(ka)\, h_n^{(2)}(ka)$, i.e.,

$$\int_0^{\pi}\int_0^{2\pi} \left.\frac{e^{-ik|\mathbf{r}-\mathbf{r}'|}}{4\pi|\mathbf{r}-\mathbf{r}'|}\right|_{\mathbf{r}\in\Gamma} Y_{n,m}(\theta',\phi')\sin\theta'\,d\theta'\,d\phi' = (-ik)\, j_n(ka)\, h_n^{(2)}(ka)\, Y_{n,m}(\theta,\phi),$$

where $m = \{\dots,-2,-1,0,1,2,\dots\}$ and $n \geq m$. The eigenfunctions are normalized as

$$\begin{aligned}\langle Y_{n,m}, Y_{n',m'}\rangle &= \int_0^{2\pi}\int_0^{\pi} Y_{n,m}(\theta,\phi)\,\overline{Y}_{n',m'}(\theta,\phi)\,a^2\sin\theta\,d\theta\,d\phi \\ &= a^2\delta_{nn'}\delta_{mm'},\end{aligned} \tag{4.35}$$

forming the well known basis $\{Y_{n,m}(\theta,\phi)\}$ in $\mathbf{L}^2(\Gamma)$.

Any function $f(\theta,\phi)$ square-integrable on a sphere of radius $a$ can be expanded as

$$f(\theta,\phi) = \sum_{n=0}^{\infty}\sum_{m=-n}^{n} \langle f, Y_{n,m}\rangle\, Y_{n,m}(\theta,\phi)$$

and

$$Af = \sum_{n=0}^{\infty}\sum_{m=-n}^{n} \lambda_n \langle f, Y_{n,m}\rangle\, Y_{n,m}(\theta,\phi).$$

### 4.3.5 Generalized Eigenvectors and the Root System

Considering operators on infinite-dimensional Hilbert spaces, Theorem 4.41 provides sufficient conditions under which operators admit simple eigenfunction representations (i.e., the existence of an eigenbasis is assured). In Sections 4.3.3 and 4.3.4, differential and integral operators are considered that also lead to an eigenbasis, even though they do not necessarily satisfy the conditions of Theorem 4.41.

If the eigenfunctions do not form a basis for the space in question, it may happen that the set of generalized eigenfunctions (i.e., the root system) forms a basis in $H$. Further, if the root system is a basis, then one can find a set orthogonal to the root system which is also a basis, and a generalized expansion in terms of the resulting bi-orthogonal set will be valid. This is a rather difficult subject (see, e.g., [26] and [27]), and here we mention only a few results, with particular attention to the compact operator $A : \mathbf{L}^2(\Gamma) \to \mathbf{L}^2(\Gamma)$ defined by

$$(Af)(\mathbf{r}) \equiv \int_\Gamma \frac{e^{-ik|\mathbf{r}-\mathbf{r}'|}}{4\pi|\mathbf{r}-\mathbf{r}'|} f(\mathbf{r}')\,d\Gamma', \tag{4.36}$$

where $\Gamma$ is a smooth closed surface in $\mathbf{R}^3$ and $k > 0$. This topic is also treated briefly for the case of a second-order differential operator in Section 5.3.2.

The following theorem is proved in [26] (see also [28]).

**Theorem 4.47.** *Let $Q : H \to H$ be a nonnegative compact operator, and let $D : H \to H$ be a dissipative nuclear*[30] *operator. Then the system of root vectors of the operator $B = Q + D$ is complete in $H$.*

In particular, this theorem applies to the operator adjoint to (4.36), where

$$(Qf)(\mathbf{r}) = \int_\Gamma \frac{1}{4\pi |\mathbf{r}-\mathbf{r}'|} f(\mathbf{r}')\, d\Gamma',$$

$$(Df)(\mathbf{r}) = \int_\Gamma \frac{\left(e^{+ik|\mathbf{r}-\mathbf{r}'|} - 1\right)}{4\pi |\mathbf{r}-\mathbf{r}'|} f(\mathbf{r}')\, d\Gamma',$$

so that the root vectors of the operator adjoint to (4.36) form a basis in $\mathbf{L}^2(\Gamma)$.

Furthermore, when a compact operator $A : H \to H$ is normal, the root system coincides with the system of ordinary eigenfunctions [28] (see also [29]), that is, all generalized eigenfunctions are of rank 1 (this result is obviously in harmony with Theorem 4.41). While the operator in (4.36) is clearly not self-adjoint (unless $k = 0$ or $ik \in \mathbf{R}$), it is normal under certain circumstances. Indeed, because for $A$ to be normal we must have $AA^* - A^*A = 0$, this condition applied to (4.36) yields

$$\begin{aligned}
&(AA^*f)(\mathbf{r}) - (A^*Af)(\mathbf{r}) \\
&\quad = \int_\Gamma \int_\Gamma \frac{\left(e^{-ik(|\mathbf{r}-\mathbf{r}''|-|\mathbf{r}''-\mathbf{r}'|)} - e^{+ik(|\mathbf{r}-\mathbf{r}''|-|\mathbf{r}''-\mathbf{r}'|)}\right)}{(4\pi)^2 |\mathbf{r}-\mathbf{r}''||\mathbf{r}''-\mathbf{r}'|} f(\mathbf{r}')\, d\Gamma' d\Gamma'' \\
&\quad = \int_\Gamma \int_\Gamma \frac{-\sin k\left(|\mathbf{r}-\mathbf{r}''| - |\mathbf{r}''-\mathbf{r}'|\right)}{2i(4\pi)^2 |\mathbf{r}-\mathbf{r}''||\mathbf{r}''-\mathbf{r}'|} f(\mathbf{r}')\, d\Gamma'\, d\Gamma'' = 0.
\end{aligned}$$

Therefore, we obtain

$$\int_\Gamma \frac{-\sin k\left(|\mathbf{r}-\mathbf{r}''| - |\mathbf{r}'-\mathbf{r}'|\right)}{2i(4\pi)^2 |\mathbf{r}-\mathbf{r}''||\mathbf{r}''-\mathbf{r}'|} f(\mathbf{r}')\, d\Gamma' = 0 \tag{4.37}$$

for $\mathbf{r}, \mathbf{r}'' \in \Gamma$, which is a condition on the surface $\Gamma$. This condition is satisfied if $\Gamma$ is a spherical surface, but not, for instance, if $\Gamma$ is an ellipsoid.

We also have the following theorem [30], which establishes conditions under which all generalized eigenfunctions of a compact operator $A$ corresponding to an eigenvalue $\lambda$ have rank 1.

**Theorem 4.48.** *The eigenspace and the root space of a compact operator $A$, corresponding to an eigenvalue $\lambda$, coincide if and only if*

*a. $\lambda$ is a simple pole of the resolvent $(A - \lambda I)^{-1}$, or*

[30] An operator $D$ is called *nuclear* if $\sum s_n < \infty$, where $s_n$ are the eigenvalues of $(D^*D)^{1/2}$.

*b.* $(A - \lambda I)^2 f = 0$ *implies* $(A - \lambda I) f = 0$, *or*

*c. the operator* $(A - \lambda I)$ *does not have zeros in the subspace* $R_{A-\lambda I}$.

It should be noted that it is also possible to show that eigenfunctions of an operator form an eigenbasis using other methods. For instance, it is shown in [31, p. 774] that sufficient conditions for an operator to possess a complete set of eigenfunctions in an $\mathbf{L}^2$ space are that the operator is positive definite, and hence self-adjoint, and that the eigenvalue equation $Lu = \lambda u$ corresponds to a variational principle. Determining completeness of the eigenfunctions for nonself-adjoint operators is, in general, much more difficult, although some further results can be stated for dissipative operators [26, Ch. 6]. For our purposes we will call an operator *simple* if it possesses an orthonormal eigenbasis of ordinary eigenfunctions.

### 4.3.6 Spectral Representations

For a compact, self-adjoint operator $A : H \to H$, or, in general, whenever the ordinary eigenfunctions of $A : H \to H$ form an orthonormal basis for $H$, one may develop a spectral expansion of the operator in terms of eigenfunctions. From Theorem 4.41 with $x = \sum_n \langle x, x_n \rangle x_n$, we identify the form analogous to (4.27),

$$A = \sum_{n=1}^{\infty} \lambda_n \langle \cdot, x_n \rangle x_n. \tag{4.38}$$

Note that $I = \sum_{n=1}^{\infty} \langle \cdot, x_n \rangle x_n$ and that on a function space where $x = x(t)$ one may write, in a distributional sense,[31]

$$\delta(t - t') = \sum_{n=1}^{\infty} x_n(t) \overline{x}_n(t'). \tag{4.39}$$

---

[31]

$$\begin{aligned} x(t) &= \int \delta\left(t - t'\right) x\left(t'\right) \, dt' = \int \sum_{n=1}^{\infty} x_n(t) \overline{x}_n\left(t'\right) x\left(t'\right) \, dt' \\ &= \sum_{n=1}^{\infty} x_n(t) \int \overline{x}_n\left(t'\right) x\left(t'\right) \, dt' = \sum_{n=1}^{\infty} x_n \langle x, x_n \rangle . \end{aligned}$$

In the event of a function space of vectors, $\underline{\mathbf{I}}\delta(\mathbf{t} - \mathbf{t}') = \sum_{n=1}^{\infty} \mathbf{x}_n(\mathbf{t}) \overline{\mathbf{x}}_n(\mathbf{t}')$, and

$$\begin{aligned} \mathbf{x}(\mathbf{t}) &= \int \underline{\mathbf{I}}\delta\left(\mathbf{t} - \mathbf{t}'\right) \cdot \mathbf{x}\left(\mathbf{t}'\right) \, d\mathbf{t}' = \int \sum_{n=1}^{\infty} \mathbf{x}_n(\mathbf{t}) \overline{\mathbf{x}}_n\left(\mathbf{t}'\right) \cdot \mathbf{x}\left(\mathbf{t}'\right) \, d\mathbf{t}' \\ &= \sum_{n=1}^{\infty} \mathbf{x}_n(\mathbf{t}) \int \overline{\mathbf{x}}_n\left(\mathbf{t}'\right) \cdot \mathbf{x}\left(\mathbf{t}'\right) \, d\mathbf{t}' = \sum_{n=1}^{\infty} \mathbf{x}_n \langle \mathbf{x}, \mathbf{x}_n \rangle . \end{aligned}$$

If $P_n : H \to H_n$ is the projection operator that takes $x \in H$ into the one-dimensional subspace spanned by the $n$th eigenfunction $x_n$ (i.e., $P_n x = \langle x, x_n \rangle x_n$, and $P_n P_m = \delta_{nm}$), then the spectral representation is seen to have the same weighted sum of projections form as (4.26),

$$A = \sum_{n=1}^{\infty} \lambda_n P_n \tag{4.40}$$

with

$$x = \sum_{n=1}^{\infty} P_n x, \qquad Ax = \sum_{n=1}^{\infty} \lambda_n P_n x, \qquad I = \sum_{n=1}^{\infty} P_n.$$

If we let the indices $n_j$ be those values of $n$ that correspond to the eigenfunctions $\nu_n = \nu_{n_j}$ (corresponding to eigenvalues $\lambda_{n_j} = 0$), then $\sum_{n_j=1}^{\infty} P_{n_j} x = P_0 x = x_0$, and we have

$$x = P_0 x + \sum_{\substack{n=1 \\ n \neq n_j}}^{\infty} P_n x,$$

in agreement with (4.28).

It can further be shown that if $A$ is bounded, or unbounded but densely defined,[32]

$$A^* = \sum_{n=1}^{\infty} \overline{\lambda}_n P_n$$

or, equivalently,

$$A^* = \sum_{n=1}^{\infty} \overline{\lambda}_n \langle \cdot, x_n \rangle x_n.$$

This representation for the adjoint assumes $A$ possesses an orthonormal basis of eigenvectors (e.g., the formula is useful for compact normal operators) and does not hold for arbitrary operators. Functions of operators in terms of projections, including inversion formulas, are discussed in Section 4.4.

To summarize, compact, self-adjoint operators $A : H \to H$ possess an orthonormal eigenbasis for $H$. The same can be said for compact normal operators, self-adjoint boundary value problems, as well as some noncompact integral operators. Moreover, the stated conditions are sufficient but not necessary.

[32] To see this, let $x = \sum_n \langle x, x_n \rangle x_n$ and $y = \sum_n \langle y, x_n \rangle x_n$, such that $\langle x, y \rangle = \sum_n \langle x, x_n \rangle \overline{\langle y, x_n \rangle}$. With $\langle Ax, y \rangle = \sum_n \lambda_n \langle x, x_n \rangle \overline{\langle y, x_n \rangle}$, we look for $z \in H$ such that $\langle Ax, y \rangle = \langle x, z \rangle$. If such a $z = \sum_n \langle z, x_n \rangle x_n$ exists, then $\overline{\langle z, x_n \rangle} = \lambda_n \overline{\langle y, x_n \rangle}$, or $\langle z, x_n \rangle = \overline{\lambda}_n \langle y, x_n \rangle$. Therefore, if $z = A^* y$, then $A^* y = \sum_n \overline{\lambda}_n \langle y, x_n \rangle x_n$, i.e., $A^* = \sum_{n=1}^{\infty} \overline{\lambda}_n P_n$.

## 4.4 Functions of Operators

It is often necessary to consider functions of operators, such as determining the exponential of an operator $A$. In this section we consider functions of an operator via three methods: spectral representations, series expansions, and the Dunford integral representation.

### 4.4.1 Functions of Operators via Spectral Representations

We assume that the operator $A : H \to H$ is bounded and simple,[33] with $H$ generally infinite-dimensional. Therefore, for every $x \in H$ we have $x = \sum_n \langle x, x_n \rangle x_n$, $Ax = \sum_n \lambda_n \langle x, x_n \rangle x_n$, and $Ix = \sum_n \langle x, x_n \rangle x_n$.[34] In what follows we also use the form (4.40), such that $x = \sum_n P_n x$, $A = \sum_n \lambda_n P_n$, and $I = \sum_n P_i$.

Beginning with operators raised to an integer power, we have $A^2 x \equiv AAx = \sum_n \lambda_n^2 \langle x, x_n \rangle x_n$, and one can see that, generally,

$$A^m x = \sum_n \lambda_n^m \langle x, x_n \rangle x_n \tag{4.41}$$

for any power $m < \infty$. Therefore, for any polynomial $p(t) = \sum_{m=0}^{M} a_m t^m$ (we assume $x = x(t) \in H$),

$$p(A) x = \sum_n p(\lambda_n) \langle x, x_n \rangle x_n \tag{4.42}$$

or

$$p(A) = \sum_n p(\lambda_n) P_n.$$

Generalizing[35] to bounded continuous functions $f(t)$, we define the operator-valued function $f(A)$ by

$$f(A) x = \sum_n f(\lambda_n) \langle x, x_n \rangle x_n \tag{4.43}$$

or

$$f(A) = \sum_n f(\lambda_n) P_n. \tag{4.44}$$

---

[33] If the operator is not simple, its set of eigenfunctions will not necessarily be orthogonal or complete in $H$. In light of Theorem 4.7, it can be seen that in this case an expansion of the form $x = \sum_n \langle x, y_n \rangle x_n$ may be useful, where $y_n$ are eigenfunctions of the adjoint operator.

[34] As usual, when index limits are not specified, we assume a finite set $n = 1, \ldots, N$ for operators on finite-dimensional spaces, and an infinite set $n = 1, 2, \ldots$ for operators on infinite-dimensional spaces.

[35] We can achieve this generalization since any real-valued continuous function can be written as the limit of a sequence of polynomials $\{p_n(t)\}$ via the Weierstrass theorem (see Example 2, Section 2.2.2).

As special cases, if $f(t) = 1$ then $f(A) = I$, and if $f(t) = t$ then $f(A) = A$.

Note also that because we have $A = \sum_n \lambda_n P_n$, then (4.44) implies that if $A$ is a diagonal matrix, $A = \text{diag}[a_{11}, a_{22}, \ldots]$, then $f(A)$ is the diagonal matrix[36]

$$f(A) = \text{diag}\left[f(a_{11}), f(a_{22}), \ldots\right]. \tag{4.45}$$

The spectral representations (4.43) and (4.44) allow one to define such quantities as $\sin(A)$, $\log(A)$, etc., assuming $f(\lambda_n)$ is meaningful. For example, by Theorem 4.19 we see that eigenvalues of a nonnegative operator are nonnegative. Therefore, for a compact, self-adjoint, nonnegative operator $A$, the square root of $A$,

$$A^{1/2} = \sum_n \lambda_n^{1/2} P_n,$$

is well defined. In general, ambiguities can be avoided if $f$ is analytic on $\sigma(A)$.

As another example, the exponential of the operator $A : H \to H$, applied to $x \in H$, is

$$e^{iA} x = \sum_n e^{i\lambda_n} \langle x, x_n \rangle x_n,$$

where $\lambda_n, x_n$ are the eigenvalues and eigenfunctions, respectively, of the operator $A$. Alternatively,

$$e^{iA} = \sum_n e^{i\lambda_n} P_n. \tag{4.46}$$

**Example 4.8.**

Consider a system of $n$ first-order, constant-coefficient differential equations

$$\mathbf{x}'(t) = A\mathbf{x}(t),$$

where $A \in \mathbf{C}^{n\times n}$ is a matrix of constants, $\mathbf{x}$ is an $n$-dimensional vector of functions, and $\mathbf{x}'$ is the vector of derivatives taken individually on elements of $\mathbf{x}$. Initial conditions are given as the vector of constants $\mathbf{x}(0) = \mathbf{x}_0 = [x_0^{(1)}, x_0^{(2)}, \ldots, x_0^{(n)}]^\top \in \mathbf{C}^n$. For the case of one equation ($n = 1$), $A$ is a $1 \times 1$ matrix (a scalar, call it $\alpha$) and the general solution is obviously

$$x(t) = e^{At} x_0 = e^{\alpha t} x_0,$$

[36]Moreover, for any pair of operators $A, B$ related by a similarity transformation, i.e., $A = PBP^{-1}$, then $A^m = \underbrace{\left(PBP^{-1}\right)\left(PBP^{-1}\right)\cdots\left(PBP^{-1}\right)}_{n \text{ times}} = \left(PB^m P^{-1}\right)$ and we see (directly for polynomials and by extension (Weierstrass) to at least continuous functions) that for $f$ defined on $\sigma(A) = \sigma(B)$,

$$f(A) = Pf(B)P^{-1}.$$

where $x_0 = x_0^{(1)}$ is a scalar. In the case of $n$ equations, the general solution is

$$\mathbf{x}(t) = e^{At}\mathbf{x}_0 = \sum_{m=1}^{n} e^{\lambda_m t} \langle \mathbf{x}_0, \mathbf{x}_m \rangle \mathbf{x}_m,$$

where $\lambda_m$ and $\mathbf{x}_m$ are eigenvalues and eigenfunctions of the matrix $A$. If $A$ is diagonal (i.e., if the equations are not coupled), the solution is particularly simple,[37]

$$[x_1(t), x_2(t), \ldots, x_n(t)]^\top = \left[e^{\lambda_1 t} x_0^{(1)}, e^{\lambda_2 t} x_0^{(2)}, \ldots, e^{\lambda_n t} x_0^{(n)}\right]^\top.$$

If $A$ is not diagonal but is at least diagonalizable, then it is easiest to diagonalize $A$ as $D = PAP^{-1}$ such that $\mathbf{z}'(t) = D\mathbf{z}(t)$, solve this new system of equations, and find $\mathbf{x}$ as $\mathbf{x}(t) = P^{-1}\mathbf{z}(t)$.

A particularly important function is inversion, leading to the inverse operator. Letting

$$f(t) = (t - \lambda)^{-1},$$

then

$$f(A) = R_\lambda(A) = \sum_n (\lambda_n - \lambda)^{-1} P_n \tag{4.47}$$

for the resolvent, where obviously if $\lambda = \lambda_n$ we must have $P_n = 0$ for $f(A)$ to possibly be meaningful. For $\lambda = 0$ (note $\lambda_n \neq 0$), $f(t) = t^{-1}$, leading to the inverse operator $f(A) = A^{-1}$,

$$A^{-1} = \sum_n \lambda_n^{-1} P_n \tag{4.48}$$

or

$$A^{-1}x = \sum_n \lambda_n^{-1} \langle x, x_n \rangle x_n. \tag{4.49}$$

It can be shown that for a bounded operator $A$, the spectrum of a function of the operator is equal to the function of the spectrum,

$$\sigma(f(A)) = f(\sigma(A)),$$

which is known as the *spectral mapping theorem*. For example, if $\sigma(A) = \sigma_p(A) = \{\lambda_1, \lambda_2, \ldots\}$, then $\sigma(f(A)) = \{f(\lambda_1), f(\lambda_2), \ldots\}$.

Also, from Theorem 4.10 note that for $A$ bounded, $\|A\| \geq r_\sigma(A) \equiv \sup_{\lambda \in \sigma(A)} |\lambda|$. For $A$ normal it can be shown that equality is achieved, i.e., $\|A\| = r_\sigma(A)$, and further,

$$\|f(A)\| = r_\sigma(f(A)) = \sup_{\lambda \in \sigma(A)} |f(\lambda)|.$$

[37] Note that for a diagonal matrix the eigenvalues are the main diagonal entries and the eigenvectors are the standard vectors, i.e., $\mathbf{x}_1 = [1, 0, \ldots]^\top$, $\mathbf{x}_2 = [0, 1, 0, \ldots]^\top$, ...

The above representations for functions of an operator $f(A)$ are valid assuming the operator can be written as a weighted discrete sum of projections, i.e., that the operator is simple. For the operators considered here, each projection is associated with the span of an eigenfunction. For example, compact normal operators belong to this class. It can be shown that normal operators (not necessarily compact) can be represented as a weighted continuous sum of projections, and formulas similar to a continuous version of (4.44) (i.e., in terms of integrals) can be developed for both bounded and unbounded operators. Furthermore, rather than considering continuous functions, one may develop a theory for generally measurable functions [12, Chs. VII and VIII]. For operators that are not normal, weighted projection methods do not generally apply, and some other method of constructing functions of operators must be used. One such method is to use series expansions.

### 4.4.2 Functions of Operators via Series Expansions

A nonspectral method of constructing functions of operators is via series expansions. For instance, consider a function given by a power series,

$$f(t) = \sum_{n=0}^{\infty} a_n t^n$$

with radius of convergence $R$. Then, if $A$ is bounded with $\|A\| < R$, the series $\sum_{n=0}^{\infty} a_n A^n$ converges[38] and we set

$$f(A) = \sum_{n=0}^{\infty} a_n A^n, \tag{4.50}$$

where $A^0 = I$. For example,

$$e^{i\omega t} = \sum_{n=0}^{\infty} (i\omega)^n t^n / n!,$$

and so[39]

$$e^{i\omega A} = \sum_{n=0}^{\infty} (i\omega)^n A^n / n!.$$

It is interesting to note that $\left(e^{i\omega A}\right)^n = e^{i\omega n A}$ and, in particular, that $e^{i\omega A}$ is invertible with inverse $e^{-i\omega A}$ even if $A$ itself is not invertible (this

[38] Since $\|A^n\| \leq \|A\|^n$ and $\|A\| < R$, then $\sum_{n=0}^{\infty} |a_n| \, \|A^n\|$ is convergent because we are within the radius of convergence of the original power series. When $\sum_{n=0}^{\infty} |a_n| \, \|A^n\|$ is convergent, $\sum_{n=0}^{\infty} a_n A^n$ is called absolutely convergent, which, in a Hilbert (more generally Banach) space implies convergence.

[39] Recalling that power series for trigonometric functions have infinite radii of convergence, these series of operators will converge for any bounded operator.

also shows that $e^{i\omega A}$ is unitary). Exponentials of operators often arise in electromagnetics via propagator matrix approaches used in problems associated with a multilayered medium [32, Ch. 2], and in multiconductor transmission-line problems (see Section 7.4).

### 4.4.3 Functions of Operators via the Dunford Integral Representation

Finally, another method of constructing functions of operators is via a generalization of the Cauchy integral formula. First, consider the case of scalar functions. Let $f(\lambda)$ be analytic in a simply connected region $\Omega$ of the complex $\lambda$-plane, let $\Gamma$ be an arbitrary closed piecewise smooth curve within $\Omega$, and let $z$ be a point interior to $\Gamma$. Then,

$$f(z) = \frac{1}{2\pi i} \int_\Gamma f(\lambda)\,(\lambda - z)^{-1}\, d\lambda \tag{4.51}$$

is known as *Cauchy's integral representation* of the function $f$ [33, p. 37], where the integral is taken in the usual counterclockwise direction.

Now let $A : H \to H$ be bounded and $f(z)$ analytic within a region $\Omega$ of the complex plane, with $\sigma(A) \subset \Omega$. Then

$$f(A) = \frac{1}{2\pi i} \int_\Gamma f(\lambda)\,(\lambda I - A)^{-1}\, d\lambda \tag{4.52}$$

is known as the *Dunford integral representation* [34] of $f(A)$, which is seen to be a generalization of the Cauchy integral representation for ordinary functions to operator-valued functions. In particular,

$$I = \frac{1}{2\pi i} \int_\Gamma (\lambda I - A)^{-1}\, d\lambda$$

and

$$A = \frac{1}{2\pi i} \int_\Gamma \lambda\,(\lambda I - A)^{-1}\, d\lambda.$$

Like the Cauchy integral representation for scalar functions, the Dunford integral representation is primarily useful for theoretical developments rather than computation.

The various definitions of $f(A)$, i.e., (4.44), (4.50), and (4.52), may have different domains of validity. For instance, in order for (4.44) to hold, $A$ must admit a representation as a weighted sum of projections, while for (4.50) to hold, $A$ must be appropriately bounded, such that the series of operators converges. The various definitions will agree whenever they are mutually applicable for a certain operator [22, p. 10]. For instance, let $f$, analytic on $\sigma(A)$ where $A$ is a bounded operator, be given by the power series

$$f(\lambda) = \sum_{n=0}^{\infty} a_n \lambda^n$$

with radius of convergence $R$. Let $\Gamma$ be a circular (radius $r$), positively oriented contour in the $\lambda$-plane such that $\sigma(A)$ is contained within $\Gamma$ (i.e., $r_\sigma(A) < r$). Then

$$f(A) = \frac{1}{2\pi i}\int_\Gamma f(\lambda)(\lambda I - A)^{-1}\,d\lambda$$
$$= \frac{1}{2\pi i}\int_\Gamma \left(\sum_{n=0}^{\infty} a_n\lambda^n\right)(\lambda I - A)^{-1}\,d\lambda$$

is meaningful. Using

$$(\lambda I - A)^{-1} = \sum_{n=0}^{\infty}\frac{1}{\lambda^{n+1}}A^n,$$

which converges by Theorem 3.38 assuming $|\lambda| > \|A\|$, we see that if $r$ satisfies $r_\sigma(A) \le \|A\| < r < R$, then

$$f(A) = \frac{1}{2\pi i}\int_\Gamma \left(\sum_{n=0}^{\infty} a_n\lambda^n\right)\left(\sum_{k=0}^{\infty}\frac{1}{\lambda^{k+1}}A^k\right)d\lambda$$
$$= \sum_{n=0}^{\infty} a_n \sum_{k=0}^{\infty} A^k \frac{1}{2\pi i}\int_\Gamma \lambda^{n-k-1}\,d\lambda,$$

where the interchange of operators is permissible because of the convergence properties of the series. From the residue theorem [33, Sec. 22] we see that if $k \le n-1$ or $k > n$, the integral vanishes, whereas for $k = n$ the integral is $2\pi i$. Therefore,

$$f(A) = \sum_{n=0}^{\infty} a_n A^n,$$

and in this case $f(A)$ defined by the Dunford integral representation (4.52) is in agreement with the series expansion form (4.50).

## 4.5 Spectral Methods in the Solution of Operator Equations

Spectral representations are especially useful for solving inhomogeneous operator equations. In the following analysis the operator $A : H \to H$ is assumed simple (e.g., compact and self-adjoint, in which case $A - \lambda I$ is a Fredholm operator), with an orthonormal eigenbasis $\{x_n\}$ for $H$. The set $\{u_n\} \subseteq \{x_n\}$ consists of eigenfunctions of $A$ corresponding to nonzero eigenvalues, forming a basis of $R_A$, and the set $\{\nu_n\} \subset \{x_n\}$ consists of eigenfunctions corresponding to zero eigenvalues ($A\nu_n = 0$), forming a basis of $N_A$, and $\{x_n\} = \{u_n\} \cup \{\nu_n\}$.

### 4.5.1 First- and Second-Kind Operator Equations

**First-Kind Operator Equation, $N_A = \{0\}$**

Consider a first-kind operator equation of the form

$$Ax = y$$

with $N_A = \{0\}$ (i.e., $\lambda = 0$ is not an eigenvalue of $A$, and so the set $\{u_n\}$ is an orthonormal basis for $H$) and expand $x, y \in H$ as

$$x = \sum_n \langle x, u_n \rangle u_n,$$
$$y = \sum_n \langle y, u_n \rangle u_n,$$

both series being convergent in norm because $\{u_n\}$ is a basis for $H$. Since $Ax = \sum_n \lambda_n \langle x, u_n \rangle u_n = y = \sum_n \langle y, u_n \rangle u_n$, we solve for the coefficients of $x$ as $\langle x, u_n \rangle = (1/\lambda_n) \langle y, u_n \rangle$, leading to the unique solution

$$x = \sum_n \frac{1}{\lambda_n} \langle y, u_n \rangle u_n,$$

which can also be obtained by application of the inverse operator (4.49). By the Riesz–Fischer theorem (Theorem 2.11) $\sum_n (1/\lambda_n) \langle y, u_n \rangle u_n$ will converge if and only if $\sum_n |(1/\lambda_n) \langle y, u_n \rangle|^2 < \infty$, i.e., if $(1/\lambda_n) \langle y, u_n \rangle \in \mathbf{l}^2$.

The solution will not necessarily be continuously dependent on $y$ (i.e., $A^{-1}$ may be unbounded). To see this, change $y$ by a small amount, say $\varepsilon u_k$, where $u_k$ is some eigenfunction of $A$ and $\varepsilon$ is a small parameter (the form $\varepsilon u_k$ is chosen for convenience). Then, with $\widetilde{y} = y + \varepsilon u_k$, the solution of $A\widetilde{x} = \widetilde{y}$ is

$$\widetilde{x} = \sum_n \frac{1}{\lambda_n} \langle \widetilde{y}, u_n \rangle u_n = \sum_n \frac{1}{\lambda_n} \langle y, u_n \rangle u_n + \frac{\varepsilon}{\lambda_k}.$$

Therefore, $\|x - \widetilde{x}\| = |\varepsilon/\lambda_k|$. If $\lambda_k$ is very small, the difference $\|x - \widetilde{x}\|$ can become very large, even for a small perturbation $\varepsilon$ [4, p. 406]. Such is the case when $A$ is compact (and so $A^{-1}$ is unbounded), because $\lambda_k \to 0$ as $k \to \infty$.

For the compact case, or generally where $\lambda_n \to 0$, it is interesting to note that in the numerical solution of first-kind operator equations by discretization methods (e.g., the method of moments), often a reasonably accurate solution can be obtained by a somewhat coarse discretization. As the discretization becomes finer, rather than converging, the solution typically diverges, reflecting the more accurate modeling of the unbounded inverse. Furthermore, often eigenvalues associated with more singular kernels tend toward the origin less quickly compared with eigenvalues associated with smoother kernels. Therefore, the solution of first-kind equations having more singular kernels is sometimes easier from an approximate numerical standpoint, compared with solving first-kind integral equations having smoother kernels.

### First-Kind Operator Equation, $N_A \neq \{0\}$

If $\lambda = 0$ is an eigenvalue of $A$, then it is convenient to write

$$x = x_0 + \sum_n \langle x, u_n \rangle u_n,$$

where $Ax_0 = 0$ and $x_0 = \sum_n \langle x_0, \nu_n \rangle \nu_n$. Since $Ax = \sum_n \lambda_n \langle x, u_n \rangle u_n = y$ and $\{u_n\} \perp \{\nu_n\}$, clearly for $y$ to be in the range of $A$ it must be orthogonal to $\{\nu_n\}$ (i.e., $y \perp N_A$; note that if $N_A = \{0\}$, as in the case considered above, then $y \perp N_A$ trivially). Assuming this is the case, we write $y = \sum_n \langle y, u_n \rangle u_n$. Exploiting orthonormality of the set $\{u_n\}$ leads to $\langle x, u_n \rangle = (1/\lambda_n) \langle y, u_n \rangle$, such that the solution is given by

$$x = x_0 + \sum_n \frac{1}{\lambda_n} \langle y, u_n \rangle u_n,$$

which is, however, nonunique since $x_0$ is an arbitrary element of $N_A$. As before, $\sum_n (1/\lambda_n) \langle y, u_n \rangle u_n$ will converge if and only if $\sum_n |(1/\lambda_n) \langle y, u_n \rangle|^2 < \infty$.

To summarize, the operator equation $Ax = y$, where $A$ is a simple operator, is solvable if $y \perp N_A$. The solution,

$$x = x_0 + \sum_n \frac{1}{\lambda_n} \langle y, u_n \rangle u_n,$$

where $Ax_0 = 0$, converges if $\sum_n |(1/\lambda_n) \langle y, u_n \rangle|^2 < \infty$ and is unique if $N_A = \{0\}$, that is, if $Ax_0 = 0$ implies $x_0 = 0$.

If the operator is not simple, the set of eigenfunctions will not necessarily be mutually orthogonal or complete. In this case, one may expand $x \in H$ in the bi-orthogonal set $\{x_n, y_n\}$ as $x = \sum_n \langle x, y_n \rangle x_n$, with the sets $\{x_n\}$ and $\{y_n\}$ being the eigenfunctions of $A$ and the adjoint operator $A^*$, respectively (of course, completeness of this set and validity of the expansion need to be established). The solution is obtained as

$$x = x_0 + \sum_n \frac{1}{\lambda_n} \langle y, y_n \rangle x_n,$$

and we require $y \perp N_{A^*}$.

### Second-Kind Operator Equation, $\lambda$ Is Not an Eigenvalue of $A$

Consider the second-kind operator equation

$$(A - \lambda I)\, x = y.$$

The case $\lambda = 0$ was considered above, so we will assume $\lambda \neq 0$.

We first consider the case where $\lambda$ is not an eigenvalue of $A$ (i.e., $\lambda \notin \sigma_p(A)$). Therefore, $(A - \lambda I)\, x = 0$ has only the trivial solution, and $N_{A-\lambda I} = \{0\}$.[40] In this case we can make the expansions

$$\begin{aligned} Ax &= \sum_n \lambda_n \langle x, u_n \rangle\, u_n, \\ x &= x_0 + \sum_n \langle x, u_n \rangle\, u_n, \\ y &= y_0 + \sum_n \langle y, u_n \rangle\, u_n, \end{aligned} \tag{4.53}$$

where $x_0$, $y_0 \in N_A$. Exploiting orthonormality of the sets $\{u_n\}$ and $\{\nu_n\}$ leads to $(\lambda_n - \lambda)\langle x, u_n \rangle = \langle y, u_n \rangle$ and $\lambda x_0 = -y_0$, respectively. The solution is therefore

$$x = -\frac{y_0}{\lambda} + \sum_n \frac{1}{(\lambda_n - \lambda)} \langle y, u_n \rangle\, u_n. \tag{4.54}$$

If $N_A = \{0\}$, then the $y_0$ term is omitted and the solution is unique. Otherwise $y_0$ is retained and the solution is not unique. By the Riesz–Fischer theorem (Theorem 2.11) we require $\sum_n |(1/(\lambda_n - \lambda))\langle y, u_n \rangle|^2 < \infty$ for the series to converge. Note that trivially $y \perp N_{A-\lambda I}$.

### Second-Kind Operator Equation, $\lambda$ Is an Eigenvalue of $A$

Now consider the case when $\lambda$ is an eigenvalue of $A$, and let $\lambda = \lambda_m$. For simplicity, assume unit multiplicity of $\lambda_m$. Therefore, $(A - \lambda I)\, x = (A - \lambda_m I)\, x = 0$ has nontrivial solutions in $N_{A-\lambda_m I}$. In this case we again make the expansions (4.53). Exploiting orthonormality of the sets $\{u_n\}$ and $\{\nu_n\}$ leads to $(\lambda_m - \lambda_n)\langle x, u_n \rangle = \langle y, u_n \rangle$ and $\lambda x_0 = -y_0$, respectively. Then,

- if $\langle y, u_m \rangle \neq 0$, we cannot solve for $\langle x, u_m \rangle$, and the equation has no solution,
- if $\langle y, u_m \rangle = 0$, then $\langle x, u_m \rangle$ is arbitrary; call it $\alpha$. The (nonunique) solution is

$$x = -\frac{y_0}{\lambda_m} + \sum_{n \neq m} \frac{1}{(\lambda_n - \lambda_m)} \langle y, u_n \rangle\, u_n + \alpha u_m.$$

[40] Note that this does not imply $N_A = \{0\}$.

By the Riesz–Fischer theorem we need $\sum_{n\neq m} |(1/(\lambda_n - \lambda_m)) \langle y, u_n\rangle|^2 < \infty$ for the series to converge. Note that we require $y \perp N_{A-\lambda_m I}$.

In both of the above cases we require $y \perp N_{A-\lambda I}$ for $y$ to be in the range of the operator, such that a solution will exist. This is in agreement with the Fredholm alternative 3.44 (generally we require $y \perp N_{(A-\lambda I)^*}$, but the adjoint is not needed since we are assuming that the operator is simple[41]).

### 4.5.2 Spectral Methods and Green's Functions

In Section 3.9.3 we showed that a Green's function is the kernel of an integral operator, the Green's operator, which is inverse to a differential operator. In this section we continue the discussion of Green's functions, emphasizing spectral theory. The manipulations are merely formal and intended to highlight salient points of the theory. Explicit formulations are considered in Chapters 1 and 5 and in Part II.

Referring to (3.34), consider the differential operator equation

$$(\mathbf{L}-\lambda I)\,\mathbf{x} = \mathbf{y}, \tag{4.55}$$

where $\mathbf{L}$ is a vector differential operator on a Hilbert space $H$ (for example, $\mathbf{L} = \nabla \times \nabla \times$ or $\nabla^2$), where $\mathbf{x} = \mathbf{x}(\mathbf{r}) \in D_{\mathbf{L}}$ and $\mathbf{y} = \mathbf{y}(\mathbf{r}) \in R_{\mathbf{L}-\lambda I}$. We assume that $\mathbf{x}$ as a vector will generally not be oriented in the same direction as the source vector $\mathbf{y}$, and so we anticipate a Green's operator $\mathbf{G}_\lambda$ (inverse operator to $(\mathbf{L}-\lambda I)$) with a dyadic Green's function kernel $\underline{\mathbf{g}}$, satisfying

$$(\mathbf{L}-\lambda I)\,\underline{\mathbf{g}}(\mathbf{r},\mathbf{r}',\lambda) = \underline{\mathbf{I}}\,\delta(\mathbf{r}-\mathbf{r}') \tag{4.56}$$

such that

$$\mathbf{x}(\mathbf{r}) = (\mathbf{G}_\lambda \mathbf{y})(\mathbf{r}) = \int_\Omega \underline{\mathbf{g}}(\mathbf{r},\mathbf{r}',\lambda)\cdot \mathbf{y}(\mathbf{r}')\, d\Omega'. \tag{4.57}$$

In this case, where we assume that $(\mathbf{L}-\lambda I)$ is invertible, the Green's operator is exactly the resolvent of $\mathbf{L}$, i.e., $\mathbf{G}_\lambda = (\mathbf{L}-\lambda I)^{-1} = R_\lambda(\mathbf{L})$.

To determine a spectral representation, assume that (4.55) forms a self-adjoint boundary value problem, such that the vector eigenfunctions defined by

$$\mathbf{L}\mathbf{x}_n = \lambda_n \mathbf{x}_n$$

form an orthonormal eigenbasis in $H$ subject to some appropriate boundary conditions. The method leading to (4.54) results in the solution of (4.55) as

$$\mathbf{x} = \sum_n \frac{\langle \mathbf{y}, \mathbf{x}_n\rangle}{(\lambda_n - \lambda)} \mathbf{x}_n = \mathbf{G}_\lambda \mathbf{y}, \tag{4.58}$$

[41] See Example 5.14 for an example when the operator is not simple, and associated eigenfunctions are implicated. In such cases the expansion (4.53) needs to be modified to include the associated eigenfunctions.

where for simplicity we assume that $N_{\mathbf{L}} = \{\mathbf{0}\}$ and that $\lambda = 0$ is not an eigenvalue of $\mathbf{L}$. The summation is generally multidimensional. A similar form for the Green's function is also available. Indeed, comparing (4.57) and (4.58) we identify the Green's function as

$$\underline{\mathbf{g}}(\mathbf{r}, \mathbf{r}', \lambda) = \sum_n \frac{\mathbf{x}_n(\mathbf{r})\overline{\mathbf{x}}_n(\mathbf{r}')}{\lambda_n - \lambda}, \tag{4.59}$$

which is known as the *bilinear series* form for $\underline{\mathbf{g}}$.

Using the notation developed in Section 4.4, we can express the Green's operator as a weighted sum of projections

$$\mathbf{G}_\lambda = \sum_n (\lambda_n - \lambda)^{-1} \mathbf{P}_n, \tag{4.60}$$

where $\mathbf{P}_n\mathbf{x} = \langle \mathbf{x}, \mathbf{x}_n \rangle \mathbf{x}_n$ projects $\mathbf{x}$ onto the one-dimensional space spanned by the eigenfunction $\mathbf{x}_n$. This also naturally leads to the idea that the Green's operator is simply one element (the inverse function) in a family of possible functions of the operator $(\mathbf{L} - \lambda I)$.

In higher dimensions the eigenfunction expansion of the Green's function is generally not convergent when the source point $\mathbf{r}'$ coincides with the field point $\mathbf{r}$. In these situations the Green's function is to be considered as an operator within the theory of distributions, such that when substituted as the kernel of an integral expression (like (1.49) or (1.73)) the resulting terms can be integrated term by term, leading to a convergent series.

The representation (4.59) provides the Green's function $\underline{\mathbf{g}}$ given the eigenfunctions and eigenvalues $\mathbf{x}_n$ and $\lambda_n$. Alternatively, given the Green's function one may determine the eigenfunctions and eigenvalues. Assume the series (4.59) represents a *meromorphic function*[42] of $\lambda$ such that the Green's function has simple poles at $\lambda = \lambda_n$ in the complex $\lambda$-plane. Since $\mathbf{L}$ is self-adjoint, these will be on the real $\lambda$-axis. Taking the integral of (4.59) over a closed, counter-clockwise oriented contour $C$ in the complex $\lambda$-plane that encloses all of the poles, conveniently chosen as a circle having

[42] A meromorphic function $f(\lambda)$ is one in which the only singularities of $f$ in the finite $\lambda$-plane are poles. Pole singularities may be absent as well, such that a meromorphic function is a function whose singularities in the finite $\lambda$-plane are "no worse" than poles (i.e., no branch points or essential singularities may exist in the finite $\lambda$-plane).

It is useful to note that the derivative of a meromorphic function $f$ is a meromorphic function $f'$ and that $f$ and $f'$ share the same poles.

radius tending toward infinity in the $\lambda$-plane, leads to[43]

$$\oint_C \underline{\mathbf{g}}(\mathbf{r}, \mathbf{r}', \lambda)\, d\lambda = \sum_n \mathbf{x}_n(\mathbf{r})\overline{\mathbf{x}}_n(\mathbf{r}') \oint_C \frac{d\lambda}{\lambda_n - \lambda} = -2\pi i \sum_n \mathbf{x}_n(\mathbf{r})\overline{\mathbf{x}}_n(\mathbf{r}'), \tag{4.61}$$

where the series (4.59) is viewed as a distribution and integrated term by term, and the last equality is via the residue theorem [33, Sec. 22.1]. The expression (4.61) allows one to obtain the eigenfunctions by considering the poles of the Green's function and its associated residues.

To obtain a further useful expression, consider the expansion of the delta-function $\underline{\mathbf{I}}\,\delta(x-x')$ in eigenfunctions of $\mathbf{L}$ (see the footnote on p. 271),

$$\underline{\mathbf{I}}\,\delta(\mathbf{r}-\mathbf{r}') = \sum_n \mathbf{x}_n(\mathbf{r})\overline{\mathbf{x}}_n(\mathbf{r}'), \tag{4.62}$$

which is called the *spectral representation of the delta-function.* Relations of type (4.62) are also known as *completeness relations*, since taking the (posterior) dot product of (4.62) with $\mathbf{x}(\mathbf{r}')$ and integrating over space lead to the expansion $\mathbf{x}(\mathbf{r}) = \sum_n \langle \mathbf{x}, \mathbf{x}_n \rangle\, \mathbf{x}_n(\mathbf{r})$.

Combining (4.61) and (4.62) we have[44]

$$\frac{1}{2\pi i} \oint_C \underline{\mathbf{g}}(\mathbf{r}, \mathbf{r}', \lambda)\, d\lambda = -\underline{\mathbf{I}}\,\delta(\mathbf{r}-\mathbf{r}'), \tag{4.63}$$

an expression of principal interest as evidenced by its usefulness in later chapters. Throughout the text the integral in relations of the form (4.63) will be taken in a counterclockwise direction.

For unbounded region problems $\underline{\mathbf{g}}$ may not be meromorphic, and often $\underline{\mathbf{g}}$ admits branch-point singularities in the $\lambda$-plane. In this case the discrete set of eigenfunctions is not complete[45] in $H$, and this set must be augmented by a continuum (representing the continuous spectrum of $\mathbf{L}$),[46] leading to

$$\underline{\mathbf{g}}(\mathbf{r}, \mathbf{r}', \lambda) = \sum_n \frac{\mathbf{x}_n(\mathbf{r})\overline{\mathbf{x}}_n(\mathbf{r}')}{\lambda_n - \lambda} + \int_\upsilon \frac{\mathbf{x}(\mathbf{r}, \upsilon)\overline{\mathbf{x}}(\mathbf{r}', \upsilon)}{\lambda(\upsilon) - \lambda}\, d\upsilon, \tag{4.64}$$

[43]The fact that the poles are simple is important, because the residue theorem is utilized in obtaining the second equality in (4.61). For more general operators, poles having multiplicity greater than 1 may be present. In this case, the corresponding residue contributions need to be included and lead to associated eigenfunctions. This is discussed in more detail in Section 5.3.2.

[44]This is a very general expression, such that the equality holds whenever the root system of the operator forms a basis for the space in question, assuming that the Green's function has the proper behavior at infinity. See also Section 5.3.2.

[45]Note that this does not imply that $H$ lacks a discrete orthonormal basis. In fact, every separable $H$ possesses such a basis.

[46]See Section 5.2 for a detailed discussion of the continuous spectrum for a scalar one-dimensional problem.

where $\int_{\upsilon}(\cdot)\,d\upsilon$ represents a multidimensional integration, and again we assume that the poles $\lambda_n$ are simple. Elements of the continuum $\mathbf{x}(\mathbf{r},\upsilon)$ are called improper eigenfunctions (since they do not belong to $H$), with improper eigenvalues $\lambda(\upsilon)$, and satisfy

$$\mathbf{L}\mathbf{x}(\mathbf{r},\upsilon) = \lambda(\upsilon)\,\mathbf{x}(\mathbf{r},\upsilon).$$

In this case the analog of (4.61) and (4.63) is

$$\begin{aligned}\frac{1}{2\pi i}\oint_C \underline{\underline{\mathbf{g}}}(\mathbf{r},\mathbf{r}',\lambda)\,d\lambda &= -\underline{\underline{\mathbf{I}}}\,\delta(\mathbf{r}-\mathbf{r}') \\ &= -\sum_n \mathbf{x}_n(\mathbf{r})\overline{\mathbf{x}}_n(\mathbf{r}') - \int_{\upsilon} \mathbf{x}(\mathbf{r},\upsilon)\overline{\mathbf{x}}(\mathbf{r}',\upsilon)\,d\upsilon.\end{aligned} \tag{4.65}$$

The generalization to the nonself-adjoint case is considered in Section 5.3.2, which includes some comments on the validity of the two equalities in the above equation.

### 4.5.3 Convergence of Nonstandard Eigenvalues in Projection Techniques

It addition to the usefulness of the stationary form presented in Section 4.2.11, it is also possible under certain conditions to ascertain the convergence of spectral parameters obtained by the projection of infinite-dimensional operators onto finite-dimensional spaces. Consider the spectral problem

$$A(\gamma)\,x = 0, \tag{4.66}$$

where $A : B \to B$ is a bounded operator with $B$ an infinite-dimensional Banach space. We assume that $A$ is approximated by the operator $A_n : B_n \to B_n$, such that $A_n \to A$ as $n \to \infty$, where $B_n$ is a finite-dimensional Banach space. In practice, we need to solve

$$A_n(\gamma_n)\,x_n = 0, \tag{4.67}$$

which is the discretized version of (4.66), and we would like to know whether $\gamma_n \to \gamma$ as $n \to \infty$. Indeed, under the conditions that $A$ and $A_n$ are Fredholm operator-valued functions, analytic on an open connected region $\Omega$ of the complex $\gamma$-plane, then [35]

- for every $\gamma \in \sigma(A)$, there exists $\gamma_n \in \sigma(A_n)$ such that $\gamma_n \to \gamma$ as $n \to \infty$.

  More importantly,

- for every $\gamma_n \in \sigma(A_n)$ such that $\gamma_n \to \gamma \in \Omega$, $\gamma \in \sigma(A)$.

Therefore, under the stated conditions the (approximate) nonstandard eigenvalue $\gamma_n$, determined by solving a finite-dimensional version of (4.66), converges to an actual nonstandard eigenvalue $\gamma$ of the original operator. For an application of this result to a dielectric waveguide problem, see [36].

# Bibliography


[1] Bonnet-Ben Dhia, A. and Ramdani, K. (2000). Mathematical analysis of conducting and superconducting transmission lines, *SIAM J. Appl. Math.*, Vol. 60, no. 6, pp. 2087–2113.

[2] Friedman, B. (1956). *Principles and Techniques of Applied Mathematics*, New York: Dover.

[3] Krall, A.M. (1973). *Linear Methods in Applied Analysis*, Reading, MA: Addison-Wesley.

[4] Stakgold, I. (1999). *Green's Functions and Boundary Value Problems*, 2nd ed., New York: John Wiley and Sons.

[5] Harrington, R.F. (1961). *Time-Harmonic Electromagnetic Fields*, New York: McGraw-Hill.

[6] Naimark, M.A. (1968). *Linear Differential Operators*, Part II, New York: Frederick Ungar.

[7] Mrozowski, M. (1997). *Guided Electromagnetic Waves, Properties and Analysis*, Somerset, England: Research Studies Press.

[8] Lindell, I.V. (1982). Variational methods for nonstandard eigenvalue problems in waveguide and resonator analysis, *IEEE Trans. Microwave Theory and Tech.*, Vol. MTT-30, pp. 1194–1204, Aug.

[9] Jones, D.S. (1994). *Methods in Electromagnetic Wave Propagation*, 2nd ed., New York: IEEE Press.

[10] Hanson, G.W. and Yakovlev, A.B. (1999). Investigation of mode interaction on planar dielectric waveguides with loss and gain, *Radio Science*, Vol. 34, no. 6, pp. 1349–1359, Nov.–Dec.

[11] Steinberg, S. (1968). Meromorphic families of compact operators, *Arch. Rat. Mech. Anal.*, Vol. 31, pp. 372–379.

[12] Reed, M. and Simon, B. (1980). *Methods of Mathematical Physics I: Functional Analysis*, San Diego: Academic Press.

[13] Goursat, E. (1959). *Functions of a Complex Variable*, New York: Dover.

[14] (1981). Special issue of *Electromagnetics*, Vol. 1.

[15] Marin, L. (1981). Major results and unresolved issues in singularity expansion method, *Electromagnetics*, Vol. 1, pp. 361–373.

[16] Marin, L. and Latham, R.W. (1972). Analytical properties of the field scattered by a perfectly conducting, finite body, *Interaction Note 92*, Phillips Laboratory Note Series, Kirtland AFB, Jan.

[17] Baum, C.E. (1976). Emerging technology for transient and broad-band analysis and synthesis of antennas and scatterers, *Proc. IEEE*, Vol. 64, Nov.

[18] Cho, S.K. (1990). *Electromagnetic Scattering*, New York: Springer-Verlag.

[19] Lancaster, P. and M. Tismenetsky (1985). *The Theory of Matrices*, New York: Academic Press.

[20] Axler, S. (1997). *Linear Algebra Done Right*, New York: Springer-Verlag.

[21] Debnath, L. and Mikusiński, P. (1999). *Introduction to Hilbert Spaces with Applications*, San Diego: Academic Press.

[22] Dautray, R. and Lions, J.L. (1990). *Mathematical Analysis and Numerical Methods for Science and Technology*, Vol. 3, New York: Springer-Verlag.

[23] Gohberg, I. and Goldberg, S. (1980). *Basic Operator Theory*, Boston: Birkhäuser.

[24] Veliev, E.I. (1999). Plane wave diffraction by a half-plane: A new analytical approach, *J. Electromagnetic Waves and Applications*, Vol. 13, pp. 1439–1453.

[25] Abramowitz, M. and Stegun, I. (1965). *Handbook of Mathematical Functions*, New York: Dover.

[26] Gohberg, I.C. and Krein, M.G. (1969). *Introduction to the Theory of Linear Nonselfadjoint Operators*, Providence, RI: American Mathematical Society.

[27] Ramm, A.G. (1980). Theoretical and practical aspects of singularity and eigenmode expansion methods, *IEEE Trans. Antennas Propagat.*, Vol. AP-28, pp. 897–901, Nov.

[28] Ramm, A.G. (1973). Eigenfunction expansion of a discrete spectrum in diffraction problems, *Radio Eng. Electr. Phys.*, Vol. 18, pp. 364–369.

[29] Dolph, C. L. (1981). On some mathematical aspects of SEM, EEM, and scattering, *Electromagnetics*, Vol. 1, pp. 375–383.

[30] Ramm, A.G. (1972). A remark on the theory of integral equations, *Diff. Eq.*, Vol. 8, pp. 1177–1180.

[31] Morse, P.M. and Feshbach, H. (1953). *Methods of Theoretical Physics*, Vol. I, New York: McGraw-Hill.

[32] Chew, W.C. (1990). *Waves and Fields in Inhomogeneous Media*, New York: IEEE Press.

[33] Dennery, P. and Krzywicki, A. (1995). *Mathematics for Physicists*, New York: Dover.

[34] Dunford, N. and Schwartz (1958). *Linear Operators*, Part I, New York: Wiley.

[35] Vainikko, G.M,. and Karma, O.O. (1974). The convergence rate of approximate methods in the eigenvalue problem when the parameter appears non-linearly, *Zh. Vychisl. Mat. Fiz.*, Vol. 14, no. 6, pp. 1393–1408.

[36] Karchevskii, E.M. (1999). Analysis of the eigenmode spectra of dielectric waveguides, *Comput. Math. and Math. Phy.*, Vol. 39, no. 9, pp. 1493–1498.

[37] Friedman, A. (1982). *Foundations of Modern Analysis*, New York: Dover.

[38] Akhiezer, N.I. and Glazman, I.M. (1961 & 1963). *Theory of Linear Operators in Hilbert Space*, Vols. I and II, New York: Dover.

[39] Shilov, G.E. (1974). *Elementary Functional Analysis*, New York: Dover.

[40] Naylor, A.W. and Sell, G.R. (1982). *Linear Operator Theory in Engineering and Science*, 2nd ed., New York: Springer-Verlag.

[41] Dudley, D.G. (1994). *Mathematical Foundations for Electromagnetic Theory*, New York: IEEE Press.

[42] Tai, C.T. (1994). *Dyadic Green Functions in Electromagnetic Theory*, 2nd ed., New York: IEEE Press.

[43] Young, N. (1988). *An Introduction to Hilbert Space*, Cambridge University Press: Cambridge.

[44] Hansen, W.W. (1935). A new type of expansion in radiation problems, *Phys. Rev.*, Vol. 47, pp. 139–143.

[45] Peterson, A.F., Ray, S.L., and Mittra, R. (1998). *Computational Methods for Electromagnetics*, New York: IEEE Press.

[46] Naimark, M.A. (1967). *Linear Differential Operators*, Part I, New York: Frederick Ungar.

# 5

# Sturm–Liouville Operators

Sturm–Liouville equations arise in many applications of electromagnetics, including in the formulation of waveguiding problems using scalar potentials, and using scalar components of vector fields and potentials. Sturm–Liouville equations are also encountered in separation-of-variables solutions to Laplace and Helmholtz equations, making a connection with certain special functions and classical polynomials. Spectral theory of the Sturm–Liouville operator is well developed and useful and is intertwined with the spectral theory of compact, self-adjoint operators through the inverse and resolvent operators. Accordingly, eigenfunctions of the regular Sturm–Liouville operator are found to be complete in certain weighted spaces of Lebesgue square-integrable functions, and generalizations to accommodate a continuous spectrum in the singular case are possible.

In this chapter basic Sturm–Liouville theory is presented, in both the regular and singular cases, and many examples relevant to electromagnetic problems are included (see also [1]). While the emphasis is on self-adjoint formulations, some material is provided for nonself-adjoint problems, including spectral theory and the possibility of associated completeness relations.

The chapter begins with a description of the regular Sturm–Liouville problem, including spectral properties and expansion theorems. Singular problems are described next, and several examples leading to integral transforms are presented. Nonself-adjoint problems are then discussed, both from the aspect of Green's functions and from spectral theory. Sufficient conditions are given for a nonself-adjoint Sturm–Liouville operator to admit a basis of ordinary eigenfunctions. Special functions associated with singular Sturm–Liouville problems are then described, and the chapter concludes with the identification of some classical orthogonal polynomials as eigenfunctions of singular Sturm–Liouville operators.

## 5.1 Regular Sturm–Liouville Problems

The *regular Sturm–Liouville problem* is formulated for a scalar function $u$ satisfying the second-order ordinary differential equation

$$-\frac{1}{w(x)}\frac{d}{dx}\left[p(x)\frac{du}{dx}\right]+(q(x)-\lambda)\,u=f(x) \tag{5.1}$$

on the finite interval $a \leq x \leq b$ and the boundary conditions[1]

$$\begin{aligned} B_a(u) &= \alpha_1 u(a)+\alpha_2 u'(a)=0, \\ B_b(u) &= \beta_1 u(b)+\beta_2 u'(b)=0. \end{aligned} \tag{5.2}$$

We will restrict the functions in (5.1) such that $p, p', q$, and $w$ are continuous (or piecewise continuous) and real-valued on $[a,b]$, $p(x), w(x) > 0$ on $[a,b]$, and $\alpha_{1,2}, \beta_{1,2} \in \mathbf{R}$ such that $\alpha_1^2+\alpha_2^2>0$ and $\beta_1^2+\beta_2^2>0$. The forcing function $f(x)$ is at least piecewise continuous on $[a,b]$, and generally complex-valued. In this model $u$ is the unknown function and $\lambda \in \mathbf{C}$ is a constant parameter, known for the forced problem and unspecified for the corresponding eigenvalue (spectral) problem.

For reasons discussed below we will work in the space $\mathbf{L}^2_w(a,b)$, defined as the set of all weighted Lebesgue square-integrable functions $u(x)$ on $[a,b]$ such that

$$\int_a^b |u(x)|^2 w(x)\,dx < \infty. \tag{5.3}$$

The Sturm–Liouville operator $L : \mathbf{L}^2_w(a,b) \to \mathbf{L}^2_w(a,b)$ is defined as[2]

$$L \equiv -\frac{1}{w(x)}\frac{d}{dx}\left[p(x)\frac{d}{dx}\right]+q(x), \tag{5.4}$$

$$D_L \equiv \left\{u : u, u'' \in \mathbf{C}^2\,(a,b) \subset \mathbf{L}^2_w(a,b), B_a\,(u) = B_b\,(u) = 0\right\}, \tag{5.5}$$

leading to the concise form for (5.1) as

$$(L-\lambda)\,u=f. \tag{5.6}$$

Because $\mathbf{C}^\infty[a,b]$ is a subset of $D_L$, and $\mathbf{C}^\infty[a,b]$ is dense in $\mathbf{L}^2_w(a,b)$, then the Sturm–Liouville operator is densely defined.

It is also worthwhile to note that (5.1) is equivalent to the general second-order differential equation with variable coefficients

$$a_0(x)\frac{d^2u}{dx^2}+a_1(x)\frac{du}{dx}+a_2(x)u-\lambda u=f\,(x)\,, \tag{5.7}$$

---

[1] These boundary conditions are a special case, called homogeneous and unmixed, of a more general class of possible boundary conditions. See Section 5.3 for more general results.

[2] Such is generally the case if the forcing function $f$ is continuous (which includes the case $f \equiv 0$, representing the homogeneous problem). If $f$ is only piecewise continuous, then we may only require functions in the domain to have piecewise continuous, rather than continuous, second derivatives.

where $a_0(x) = -p(x)/w(x)$, $a_1(x) = (-1/w(x))\, dp/dx$, and $a_2(x) = q(x)$ are continuous or piecewise continuous. It is easily established that $L$ is a real operator, which is important in later developments.

We choose the weighted Hilbert space $\mathbf{L}^2_w(a,b)$ as described above since the operator (5.4) is not formally self-adjoint in the usual inner product for $\mathbf{L}^2(a,b)$, but is self-adjoint, as shown below, in the Hilbert space $\mathbf{L}^2_w(a,b)$ under the weighted inner product

$$\langle u, v\rangle = \int_a^b u(x)\overline{v(x)}w(x)\, dx, \tag{5.8}$$

which, because $w(x) > 0$ for all $x \in [a,b]$, induces the norm

$$\|u\| = \langle u, u\rangle^{1/2} = \sqrt{\int_a^b |u(x)|^2\, w(x)\, dx}. \tag{5.9}$$

As a reminder, equality of elements in this space is in the mean-square sense, i.e., $f = g$ means $\|f - g\| = 0$, and not (necessarily) that $f(x)$ and $g(x)$ are pointwise equivalent for all $x \in [a,b]$.

The adjoint of the operator $L$ can be obtained using integration by parts twice (equivalently Green's second theorem),

$$\begin{aligned}\langle Lu, v\rangle &= \int_a^b \left\{-\frac{1}{w(x)}\frac{d}{dx}\left[p(x)\frac{du}{dx}\right] + q(x)u(x)\right\}\overline{v}(x)w(x)\, dx\\ &= \int_a^b u(x)\left\{-\frac{1}{w(x)}\frac{d}{dx}\left[p(x)\frac{d\overline{v}}{dx}\right] + q(x)\overline{v}(x)\right\}w(x)\, dx\\ &\quad - \left\{p(x)\left[\overline{v}(x)\frac{du}{dx} - u(x)\frac{d\overline{v}}{dx}\right]\right\}\Bigg|_a^b\\ &= \langle u, L^*v\rangle + J(u,v)\big|_a^b,\end{aligned} \tag{5.10}$$

where, as discussed previously, $J(u,v)$ is called the conjunct and $L^*$ is the formal adjoint to $L$. Recalling the procedure for determining the adjoint (see Section 3.3.4), we choose $D_{L^*} = \{v : v \in \mathbf{C}^2[a,b] \subset \mathbf{L}^2_w(a,b),\ B^*_a(v) = B^*_b(v) = 0\}$ where the adjoint boundary conditions $B^*_a(v), B^*_b(v)$ are chosen such that the conjunct vanishes when enforcing (5.2). From the form of (5.2) and of the conjunct, it is easy to see that if we choose the conjugate-adjoint boundary conditions as

$$\begin{aligned}B^*_{ca}(\overline{v}) &= \alpha_1\overline{v}(a) + \alpha_2\overline{v}'(a) = 0,\\ B^*_{ca}(\overline{v}) &= \beta_1\overline{v}(b) + \beta_2\overline{v}'(b) = 0,\end{aligned} \tag{5.11}$$

then $J(u,v)\big|_a^b = 0$. Because $\alpha_{1,2}, \beta_{1,2} \in \mathbf{R}$, conjugating the above conditions leads to the adjoint boundary conditions

$$\begin{aligned}B^*_a(v) &= \alpha_1 v(a) + \alpha_2 v'(a) = 0,\\ B^*_b(v) &= \beta_1 v(b) + \beta_2 v'(b) = 0,\end{aligned} \tag{5.12}$$

and therefore $B_{a,b}(v) = B^*_{a,b}(v)$, such that with this choice $D_{L^*} = D_L$ and the operator $L$ is self-adjoint. The case of more general boundary conditions is presented in Section 5.3 and in [2], [1, Ch. 2].

Actually, for this operator to be self-adjoint we should relax the second-derivative continuity requirement for functions in the domain of $L$. The proper (larger) domain to consider is the set of functions in $\mathbf{L}^2_w$ with absolutely continuous[3] first derivatives and second derivatives (not necessarily continuous) in $\mathbf{L}^2_w$ satisfying the boundary conditions,

$$D_L \equiv \left\{u : u' \text{ a.c.}, u'' \in \mathbf{L}^2_w(a,b), B_a(u) = B_b(u) = 0\right\}.$$

This distinction is somewhat unimportant in most applications.

Moreover, $L$ is a strictly positive operator under certain conditions. Forming $\langle Lu, u\rangle$ and integrating by parts once we have

$$\begin{aligned}\langle Lu, u\rangle &= \int_a^b \left\{-\frac{1}{w(x)}\frac{d}{dx}\left[p(x)\frac{du}{dx}\right] + q(x)u(x)\right\}\overline{u}(x)w(x)\,dx \\ &= \int_a^b \left\{p\left|u'\right|^2 + q\left|u\right|^2 w\right\}dx - p(x)u'(x)\overline{u}(x)\big|_a^b .\end{aligned}$$

Because $p, w > 0$, then if $q > 0$ any boundary conditions that cause the last term to vanish result in $\langle Lu, u\rangle > 0$ for $u \neq 0$; in those cases $L$ is strictly positive. In fact, under these conditions $L$ can be shown to be positive definite [3, pp. 113–114], although this fact is much more difficult to prove.

The above analysis concerns what is known as the *regular Sturm–Liouville problem.* In contradistinction, a *singular Sturm–Liouville problem* is similar to the above, but is defined on either an infinite or semi-infinite interval, is such that some conditions on the coefficients are violated, or some combination thereof. In these first several sections we consider the regular case.

### 5.1.1 Spectral Properties of the Regular Sturm–Liouville Operator

An understanding of the Sturm–Liouville problem necessitates consideration of the spectral properties of the Sturm–Liouville operator (5.4). Recalling from Section 4.2.5 that a self-adjoint operator may not even possess eigenvalues, the spectral properties of the self-adjoint Sturm–Liouville operator are quite remarkable and useful (this is also briefly discussed in Section 4.3.3).

Considering the self-adjoint spectral problem

$$(L - \lambda_n)\,u_n = 0,$$

[3] Although the definition of absolute continuity is not needed here, it is worthwhile to note that absolute continuity is stronger than continuity (Lipschitz functions are absolutely continuous) and that every continuously differentiable function is absolutely continuous.

where $L$ is the Sturm–Liouville operator (5.4) with domain (5.5), we are first led to the following conclusions from Section 4.2.5:

- Eigenvalues $\lambda_n$ of the Sturm–Liouville operator $L$ are real (it can also be shown they are simple[4]).
- Eigenfunctions $u_n$ of $L$ corresponding to different eigenvalues are orthogonal.

Because the eigenfunctions can be made orthonormal we will assume $\langle u_n, u_m \rangle = \delta_{nm}$. Orthogonality is, of course, with respect to the weighted inner product (5.8). It is also easy to demonstrate these properties by direct methods for the operator in question.

Most importantly, the Sturm–Liouville operator has some further spectral properties as implied by the next theorem. We will assume that 0 is not an eigenvalue of $L$, i.e., that $Lu = 0$ subject to $B_a(u) = B_b(u) = 0$ has only the trivial solution[5] $u = 0$. Then, by Theorem 3.24, $L$ is one-to-one and has an inverse defined on its range. Since $L$ is a differential operator, it is not surprising that $L^{-1}$ is an integral (Green's) operator (see Section 3.9.3), as the next theorem asserts.

**Theorem 5.1.** *Let $L$ be the Sturm–Liouville operator (5.4) subject to boundary conditions (5.2), and assume 0 is not an eigenvalue of $L$. Then there exists a continuous, real-valued function $g(x, x') = g(x', x)$ on $a \leq x, x' \leq b$ such that*

$$\left(L^{-1}v\right)(x) = (Gv)(x) = \int_a^b g(x, x')v(x')w(x')\, dx' = \langle g, \overline{v} \rangle$$

*for all $v \in R_L$.*

The proof of this theorem, based on variation of parameters, is omitted here but can be found in [4, pp. 500–501] and (essentially) in [5, pp. 260–262], and in similar texts. Theorem 5.1 could also apply to the resolvent operator

$$\left((L-\lambda)^{-1} v\right)(x) = (G_\lambda v)(x) = \int_a^b g(x, x', \lambda)v(x')w(x')\, dx',$$

where $g(x, x', \lambda)$, the resolvent kernel, is a meromorphic function of $\lambda$, which satisfies

$$(L-\lambda)\, g(x, x', \lambda) = \frac{\delta(x - x')}{w(x)}$$

[4] An eigenvalue is *simple* if any two eigenfunctions corresponding to $\lambda$ are linearly dependent, i.e., the dimension of the eigenspace corresponding to $\lambda$ is unity.

[5] Without this condition the solvability of $Lu = f$ is greatly constrained; see Section 3.11.

($(L-\lambda)$ is assumed invertible, i.e., $\lambda \notin \sigma_p(L)$). In fact, most of the succeeding theory and examples will involve the operator $(L-\lambda)^{-1}$ rather than $L^{-1}$.

In the course of the proof of Theorem 5.1 it becomes clear that $g(x,x')$ is a Hilbert–Schmidt kernel continuous on $[a,b] \times [a,b]$. Therefore,

- $L^{-1}$ is compact and self-adjoint by an example in Section 3.6.2;
- by Theorem 4.41 there exist real eigenvalues $\gamma_n$ and corresponding eigenvectors $u_n$ of $L^{-1}$ (i.e., $L^{-1}u_n = \gamma_n u_n$) such that $\{u_n\}$ forms a basis in $\mathbf{L}^2_w(a,b)$ (or $\{u_n w^{1/2}\}$ forms a basis for $\mathbf{L}^2(a,b)$);
- by Theorem 4.11, the set $\{u_n\}$ consists of eigenvectors of $L$, with corresponding eigenvalues $\lambda_n = (1/\gamma_n)$ $(Lu_n = \lambda_n u_n)$;
- by the Hilbert–Schmidt theorem 4.40, $\lim_{n\to\infty} \gamma_n = 0$, and therefore $\lim_{n\to\infty} \lambda_n = \infty$.

Therefore, the eigenvalues of the Sturm–Liouville operator are real and such that there is a smallest eigenvalue but no largest eigenvalue. Furthermore, for every $u \in \mathbf{L}^2_w(a,b)$ we have

$$u = \sum_{n=1}^{\infty} \langle u, u_n \rangle u_n,$$

where $u_n \in \mathbf{C}^2[a,b]$, and convergence in the norm sense is guaranteed. In addition, for every $u \in D_L$ we have

$$Lu = \sum_{n=1}^{\infty} \lambda_n \langle u, u_n \rangle u_n,$$

where $\sum_{n=1}^{\infty} \lambda_n^2 |\langle u, u_n \rangle|^2 < \infty$ (therefore, by the Riesz–Fischer theorem the expansion for $Lu$ converges).

It can also be shown that if the eigenvalues are ordered as $\lambda_1 < \lambda_2 < \cdots < \lambda_n < \lambda_{n+1} < \cdots$, then the eigenfunction $u_n$ has exactly $n-1$ zeros on $(a,b)$ (not counting the endpoints). This last point is useful in problems involving natural resonances, where the eigenfunctions represent a natural current or electric-field distribution. Of course, if $\lambda = 0$ is an eigenvalue of $L$, then $L^{-1}$ does not exist. Solutions of the Sturm–Liouville equation may still exist, subject to satisfaction of certain solvability conditions as described in Section 3.11.

An important result of Theorem 5.1 and the accompanying discussion is that eigenfunctions of the Sturm–Liouville operator are complete in $\mathbf{L}^2_w(a,b)$, justifying eigenfunction expansion methods involving regular

Sturm–Liouville operators.[6] This statement remains true even if $L$ has 0 as an eigenvalue. Moreover, since the eigenvalues are simple, each null space of an eigenfunction has dimension 1, and by Theorem 2.10 (orthogonal structure) the eigenfunctions decompose the space into a direct sum of one-dimensional eigenspaces.

**Example 5.1.**

Consider the eigenvalue problem

$$(L-\lambda)\,u=0$$

with $L=-d^2/\,dx^2$ ($p=w=1$, and $q=0$ in (5.4)), subject to[7]

$$u(-\pi)=u(\pi), \qquad u'(-\pi)=u'(\pi).$$

It is easy to see that eigenvalues are $\lambda_n=n^2$ with corresponding orthonormal eigenfunctions $(1/\sqrt{2\pi})\,e^{\pm inx}$; obviously these eigenfunctions form a basis for $L^2(-\pi,\pi)$, leading to the familiar Fourier series expansion.

### 5.1.2 Solution of the Regular Sturm–Liouville Problem

The solution of an equation of the form

$$(L-\lambda)\,u=f$$

via the Green's function method is considered in Section 3.9.3. With $B_a(u)=B_b(u)=0$, the Green's function $g(x,x',\lambda)$ is defined as that "function" satisfying

$$(L-\lambda)\,g(x,x',\lambda)=\frac{\delta(x-x')}{w(x)} \tag{5.13}$$

such that $B_a(g)=B_b(g)=0$. Then (5.10) leads to

$$\langle (L-\lambda)\,u,\overline{g}\rangle=\langle u,(L-\lambda)^*\,\overline{g}\rangle+\left.J(u,\overline{g})\right|_a^b,$$

[6] Another important application is conversion of the differential eigenvalue equation $Lu=\lambda u$ into an equivalent integral equation with compact kernel,

$$u(x)=\lambda\int_a^b g(x,x')u(x')w(x')\,dx',$$

although this is not discussed further here.

[7] These *periodic boundary conditions* do not have the form (5.2), yet the above-developed theory holds in this case as well.

where the conjunct can be easily shown to vanish. Noting that $(L-\lambda)^* = \left(L-\overline{\lambda}\right)$ and[8]

$$\left(L-\overline{\lambda}\right)\overline{g}(x,x',\lambda) = \frac{\delta(x-x')}{w(x)},$$

the sifting property of the delta-function formally leads to

$$u(x') = \int_a^b f(x)g(x,x',\lambda)w(x)\,dx.$$

Because[9] $g(x,x',\lambda) = g(x',x,\lambda)$, we have

$$u(x) = \int_a^b f(x')g(x,x',\lambda)w(x')\,dx' \tag{5.14}$$

as the desired solution. Of course, we have assumed that $\lambda$ is not an eigenvalue of $L$; otherwise the resolvent does not exist.

Several methods are commonly used to obtain the explicit form of the Green's function satisfying (5.13) along with $B_a(g) = B_b(g) = 0$. One is the eigenfunction expansion method, which is detailed in the next section. Another method, which we call the *direct method*, is shown below, and a third method, called the *method of scattering superposition*, follows.

### Green's Function via the Direct Method

To determine the Green's function via the direct method, consider the Green's function

$$(L-\lambda)\,g(x,x',\lambda) = \frac{\delta(x-x')}{w(x)}$$

[8]This follows from

$$\begin{aligned}(L-\lambda)\,g(x,x',\lambda) &= \frac{\delta(x-x')}{w(x)} = \overline{(L-\lambda)\,g(x,x',\lambda)}\\ &= \overline{Lg(x,x',\lambda) - \lambda g(x,x',\lambda)} = \overline{Lg(x,x',\lambda)} - \overline{\lambda g(x,x',\lambda)}\\ &= L\overline{g} - \overline{\lambda}\overline{g} = \left(L-\overline{\lambda}\right)\overline{g}(x,x',\lambda)\end{aligned}$$

since $L$ is a real operator, $\overline{Lg} = L\overline{g}$.

[9]In electromagnetics this usually follows from reciprocity. For the problem considered here it can be shown directly by considering

$$\begin{aligned}(L-\lambda)\,g(x,x_1') &= \frac{\delta(x-x_1')}{w(x)},\\ \left(L-\overline{\lambda}\right)\overline{g}(x,x_2') &= \frac{\delta(x-x_2')}{w(x)}\end{aligned}$$

and forming

$$\left\langle (L-\lambda)\,g(x,x_1'),\overline{g}(x,x_2')\right\rangle = \left\langle g(x,x_1'),\left(L-\overline{\lambda}\right)\overline{g}(x,x_2')\right\rangle.$$

We then have $g(x_1',x_2') = g(x_2',x_1')$, which is the desired relationship.

subject to $B_a(g) = B_b(g) = 0$, and the homogeneous equation

$$(L - \lambda)\, g(x, x', \lambda) = 0 \tag{5.15}$$

valid for $x \neq x'$. We divide the interval $[a, b]$ into two $x'$-dependent segments $[a, x')$ and $(x', b]$, and obtain a solution of (5.15) on each segment. Since the operator $L$ is a second-order differential operator, each solution will contain two unknown coefficients for a total of four coefficients to be determined. Enforcement of $B_a(g) = 0$ for the solution on the first segment, and $B_b(g) = 0$ for the solution on the second segment, provides two conditions. The other two conditions required to resolve the four unknowns come from an examination of (5.13).

Although we require the domain of $L$ to consist of twice-differentiable functions with continuous second derivatives as described in Section 5.1, for (5.13) this condition must obviously be relaxed. Without considering the details of casting the problem within the framework of distribution theory, we will simply require the Green's function to be differentiable, and hence continuous, on $[a, b]$. This provides a third condition, known as the *continuity condition*, which is stated as

$$g(x, x', \lambda)\big|_{x=x'-\varepsilon}^{x=x'+\varepsilon} = 0,$$

where $\varepsilon$ is a small positive number.

The fourth condition arises from the observation that the second derivative of the Green's function must provide the singularity $\delta(x - x')$, leading to the conclusion that the first derivative of the Green's function must be a step (Heaviside) function. To explicitly obtain this condition we multiply (5.13) by $w(x)$ and integrate over the region $(x' - \varepsilon, x' + \varepsilon)$, obtaining

$$-\int_{x'-\varepsilon}^{x'+\varepsilon} \frac{d}{dx}\left[p(x)\frac{dg(x, x', \lambda)}{dx}\right] dx + \int_{x'-\varepsilon}^{x'+\varepsilon} (q(x) - \lambda)\, g(x, x', \lambda) w(x)\, dx = 1. \tag{5.16}$$

Because $q, g$, and $w$ are continuous over the domain of integration, the second integral on the left side vanishes in the limit $\varepsilon \to 0$. Performing the integration in the first integral and taking the limit $\varepsilon \to 0$ lead to the fourth condition, known as the *jump condition*,

$$\left.\frac{dg(x, x', \lambda)}{dx}\right|_{x=x'-\varepsilon}^{x=x'+\varepsilon} = -\frac{1}{p(x')}.$$

In summary, to determine a Green's function using the direct method, one solves (5.15) for $x < x'$ and for $x > x'$ and applies the four conditions

$$\begin{aligned} B_a(g) &= 0, \\ B_b(g) &= 0, \\ g(x,x',\lambda)|_{x=x'-\varepsilon}^{x=x'+\varepsilon} &= 0, \\ \left.\frac{dg(x,x',\lambda)}{dx}\right|_{x=x'-\varepsilon}^{x=x'+\varepsilon} + \frac{1}{p(x')} &= 0 \end{aligned} \tag{5.17}$$

to determine the four unknown coefficients.

## Green's Function via Scattering Superposition

Another procedure for obtaining the Green's function is related to the usual method of solving differential equations using the concept of a homogeneous and a particular solution [6, Ch. 2]. In electromagnetics, where the solution of the differential equation has a certain physical meaning, a variant of this method is called *scattering superposition.* In this method, we write the solution of (5.13) as $g = g^0 + g^s$, where $g^0$ is a particular solution of (5.13), sometimes called the principal Green's function (which does not have any unknown coefficients) and $g^s$ is the homogeneous solution of (5.13) on $[a,b]$ containing two undetermined coefficients. Note that, because of the singular nature of the forcing function, the particular solution $g^0$ will satisfy the jump and continuity conditions, whereas the homogeneous solution $g^s$ possesses continuous second derivatives on $[a,b]$. Usually the particular solution is chosen to be the response caused by an impulsive source in an infinite region, and the undetermined coefficients in $g^s$ are chosen such that the sum $g = g^0 + g^s$ will obey the boundary conditions $B_a(g) = B_b(g) = 0$. This method is useful since the "singular term" $g^0$ can be determined once for a certain class of operator, where $g^s$ represents a general solution of the homogeneous equation.

The form of the Green's functions arising from each method will be different, yet equivalent, with perhaps one form more convenient than another for a particular application. We illustrate the above concepts with the following example.

### Example 5.2.

Consider the simple differential equation

$$\left(-\frac{d^2}{dx^2} + \gamma^2\right) u(x) = f(x) \tag{5.18}$$

on $x \in [0,a]$, where $\gamma \in \mathbf{C}$, $\operatorname{Re}\gamma > 0$, and $u(0) = u(a) = 0$. The form (5.18) arises in a variety of electromagnetic applications. The Sturm–Liouville operator (5.4) is simply $L = -d^2/\,dx^2$ with $w(x) = p(x) = 1$ and $q(x) = 0$, such that (5.18) can be written as $(L-\lambda)\,u = f$ with $\lambda = -\gamma^2$. The Green's function satisfies $(L-\lambda)\,g(x,x',\lambda) = \delta(x-x')$,

$$\left(-\frac{d^2}{dx^2}+\gamma^2\right) g(x,x',\gamma) = \delta(x-x'), \tag{5.19}$$

such that $g(0,x',\gamma) = g(a,x',\gamma) = 0$. Following the direct method as described above, the Green's function is found by initially excluding the point $x = x'$, leading to

$$\left(-\frac{d^2}{dx^2}+\gamma^2\right) g(x,x',\gamma) = 0 \tag{5.20}$$

with solutions

$$g(x,x',\gamma) = \begin{cases} A\,(x')\sinh\gamma x + B\,(x')\cosh\gamma x, & x > x', \\ C\,(x')\sinh\gamma x + D\,(x')\cosh\gamma x, & x < x'. \end{cases}$$

Applying the boundary condition $g(0,x',\gamma) = 0$ leads to $D = 0$, and the condition $g(a,x',\gamma) = 0$ leads to $A\sinh\gamma a + B\cosh\gamma a = 0$. The "continuity" condition $g(x,x',\gamma)|_{x=x'-\varepsilon}^{x=x'+\varepsilon} = 0$ and the "jump" condition $(dg(x,x',\gamma)/dx)|_{x=x'-\varepsilon}^{x=x'+\varepsilon} = -1$ lead to two more equations. The remaining unknowns $A, B, C$ can then be easily found, resulting in

$$g(x,x',\gamma) = \frac{\sinh\gamma(a-x_>)\,\sinh\gamma x_<}{\gamma\sinh\gamma a}. \tag{5.21}$$

In (5.21) and throughout the text, $x_>$ and $x_<$ indicate the greater and lesser of the pair $(x,x')$, respectively. For example, if $x > x'$, then $x_> = x$ and $x_< = x'$.

From (5.14) we obtain the solution of (5.18) as

$$u(x) = \int_0^a f(x')g(x,x',\gamma)\,dx' \tag{5.22}$$

for $x \in [0,a]$ and $u(x) \in \mathbf{L}^2(0,a)$.

To illustrate the scattering superposition method, we first determine the principal Green's function $g^0$. We solve

$$\left(-\frac{d^2}{dx^2}+\gamma^2\right) g^0(x,x',\gamma) = \delta(x-x') \tag{5.23}$$

using the direct method,

$$g^0(x,x',\gamma) = \begin{cases} A\,(x')\,e^{-\gamma x} + B\,(x')\,e^{\gamma x}, & x \geq x', \\ C\,(x')\,e^{-\gamma x} + D\,(x')\,e^{\gamma x}, & x \leq x', \end{cases}$$

subject to

$$g^0(x, x', \gamma) \in \mathbf{L}^2(-\infty, \infty),$$
$$g^0(x, x', \gamma)\Big|_{x=x'-\varepsilon}^{x=x'+\varepsilon} = 0,$$
$$\frac{dg^0(x, x', \gamma)}{dx}\Bigg|_{x=x'-\varepsilon}^{x=x'+\varepsilon} + 1 = 0.$$

The condition $g^0(x, x', \gamma) \in \mathbf{L}^2(-\infty, \infty)$ arises from consideration of the singular problem and is discussed in Section 5.2. The first constraint leads to $B = C = 0$ because $\operatorname{Re}\gamma > 0$. The remaining coefficients are determined from the jump and continuity conditions as $A = e^{\gamma x'}/(2\gamma)$ and $D = e^{-\gamma x'}/(2\gamma)$, such that

$$g^0(x, x', \gamma) = \begin{cases} \frac{e^{-\gamma(x-x')}}{2\gamma}, & x > x', \\ \frac{e^{\gamma(x-x')}}{2\gamma}, & x < x', \end{cases}$$

or

$$g^0(x, x', \gamma) = \frac{e^{-\gamma|x-x'|}}{2\gamma}. \tag{5.24}$$

Next we determine the homogeneous solution of

$$\left(-\frac{d^2}{dx^2} + \gamma^2\right) g^s(x, x', \gamma) = 0 \tag{5.25}$$

on $[0, a]$ as

$$g^s(x, x', \gamma) = E \sinh \gamma x + F \cosh \gamma x.$$

The conditions $g(0, x', \gamma) = g(a, x', \gamma) = 0$ are applied to

$$g^0(x, x', \gamma) + g^s(x, x', \gamma) = e^{-\gamma|x-x'|}/(2\gamma) + E \sinh \gamma x + F \cosh \gamma x$$

to yield

$$F = -e^{-\gamma x'}/(2\gamma)$$

and

$$E = (1/(2\gamma \sinh \gamma a)) \left[e^{-\gamma x'} \cosh \gamma a - e^{-\gamma(a-x')}\right],$$

such that the Green's function is obtained as

$$g(x, x', \gamma)$$
$$= \frac{e^{-\gamma|x-x'|}}{2\gamma} + \frac{e^{-\gamma x'} \cosh \gamma a - e^{-\gamma(a-x')}}{2\gamma \sinh \gamma a} \sinh \gamma x - \frac{e^{-\gamma x'}}{2\gamma} \cosh \gamma x. \tag{5.26}$$

This last form is more complicated than (5.21), yet, conveniently, the "singular" term[10] associated with $g^0$ has been separated from the well-behaved term $g^s$.

### 5.1.3 Eigenfunction Expansion Solution of the Regular Sturm–Liouville Problem

As noted above, the eigenfunctions of the regular Sturm–Liouville operator are complete in $\mathbf{L}^2_w(a,b)$; therefore,

- for every $u \in \mathbf{L}^2_w(a,b)$ we have the eigenfunction (generalized Fourier) expansion $u = \sum_{n=1}^{\infty} \langle u, u_n \rangle u_n$,
- for every $u \in D_L$ we have $Lu = \sum_{n=1}^{\infty} \lambda_n \langle u, u_n \rangle u_n$, with $\sum_{n=1}^{\infty} \lambda_n^2 \left| \langle u, u_n \rangle \right|^2 < \infty$.

The eigenfunctions and eigenvalues are defined by $Lu_n = \lambda_n u_n$, where we assume that the eigenfunctions have been cast into orthonormal form, i.e., $\langle u_n, u_m \rangle = \delta_{nm}$.

The solution of $(L - \lambda) u = f$ with $B_a(u) = B_b(u) = 0$ by eigenfunction expansion proceeds as follows. The eigenfunctions are chosen such that $B_a(u_n) = B_b(u_n) = 0$. From the expansion of $u$ and $Lu$ we have

$$(L - \lambda) u = \sum_{n=1}^{\infty} \lambda_n \langle u, u_n \rangle u_n - \lambda \sum_{n=1}^{\infty} \langle u, u_n \rangle u_n = f.$$

Taking the inner product of both sides with $u_m$ and exploiting orthonormality $\langle u_n, u_m \rangle = \delta_{nm}$ lead to[11]
$\langle u, u_m \rangle = \langle f, u_m \rangle / (\lambda_m - \lambda)$, resulting in

$$u(x) = \sum_{n=1}^{\infty} \frac{\langle f, u_n \rangle}{\lambda_n - \lambda} u_n(x). \tag{5.27}$$

This is a generalized Fourier expansion solution with respect to the eigenfunctions of $L$. Of course, the expansion does not exist if $\lambda = \lambda_n$, unless $\langle f, u_n \rangle = 0$.

---

[10] In this one-dimensional example the principal Green's function is actually well behaved at $x = x'$, although second derivatives acting on $g$ induce distributions. In two- and three-dimensional problems the principal Green's function itself exhibits a singularity at $\mathbf{r} = \mathbf{r}'$, as shown in Chapter 1.

[11] Here the interchange of the summation and inner product is valid since the series is convergent (recall that the inner product is continuous; Section 2.5). Another method, which avoids this interchange, is to account for the fact that $L$ is self-adjoint and $\lambda_n \in \mathbf{R}$, such that

$$\begin{aligned} \langle f, u_n \rangle &= \langle (L - \lambda) u, u_n \rangle = \left\langle u, \left(L - \overline{\lambda}\right) u_n \right\rangle \\ &= \left\langle u, \left(\lambda_n - \overline{\lambda}\right) u_n \right\rangle = (\lambda_n - \lambda) \langle u, u_n \rangle . \end{aligned}$$

A similar form for the Green's function is also available. If we take the Green's function to be defined as

$$(L-\lambda)\, g(x,x',\lambda) = \frac{\delta(x-x')}{w(x)}$$

with $B_a(g) = B_b(g) = 0$, then simply substituting $\delta(x-x')/w(x)$ for $f$ in (5.27) leads to[12]

$$g(x,x',\lambda) = \sum_{n=1}^{\infty} \frac{u_n(x)\overline{u}_n(x')}{\lambda_n - \lambda}, \tag{5.28}$$

which is the bilinear series form for $g$.

**Example 5.3.**

Returning to the previous example of the differential equation

$$\left(-\frac{d^2}{dx^2} + \gamma^2\right) u(x) = f(x) \tag{5.29}$$

with $\operatorname{Re}\gamma > 0$ and $u(0) = u(a) = 0$, we cast this as a Sturm–Liouville problem $(L-\lambda)\,u = f$ with $L = -d^2/dx^2$ and $\lambda = -\gamma^2$. Eigenfunctions are found from

$$(L-\lambda_n)\, u_n = -\left(\frac{d^2}{dx^2} + \lambda_n\right) u_n(x) = 0$$

subject to $u_n(0) = u_n(a) = 0$. The eigenfunctions are easily seen to be $u_n(x) = A \sin\sqrt{\lambda_n}x + B\cos\sqrt{\lambda_n}x$. Application of the boundary conditions leads to $B = 0$ and $\sqrt{\lambda_n}a = n\pi$, such that $u_n(x) = A\sin(n\pi x/a)$, with corresponding eigenvalues

$$\lambda_n = \left(\frac{n\pi}{a}\right)^2.$$

When normalized[13] the eigenfunctions become

$$u_n(x) = \sqrt{\frac{2}{a}} \sin\sqrt{\lambda_n}x = \sqrt{\frac{2}{a}} \sin\frac{n\pi}{a}x.$$

---

[12]The delta-function does not belong to the class of admissible functions $f \in \mathbf{L}^2_w(a,b)$ as described at the beginning of the chapter, yet distributionally this substitution is permissible and leads to the correct result.

[13]The normalization is with respect to the inner product,

$$\langle u_n, u_m\rangle = A^2 \int_0^a \sin\frac{n\pi}{a}x \sin\frac{m\pi}{a}x\, dx = \delta_{nm}$$

leading to $A = \sqrt{2/a}$.

Note that the eigenvalues are consistent with the expected property that they be real, have a smallest value, and $\lim_{n\to\infty}\lambda_n = \infty$. The solution of (5.29) is

$$u(x) = \sum_{n=1}^{\infty} \frac{\langle f, u_n \rangle}{\lambda_n - \lambda} u_n(x) = \sum_{n=1}^{\infty} \frac{\frac{2}{a}\int_0^a f(x) \sin\frac{n\pi}{a}x\, dx}{\left(\frac{n\pi}{a}\right)^2 + \gamma^2} \sin\frac{n\pi}{a}x, \tag{5.30}$$

and the expression (5.28) leads to

$$g(x, x', \lambda) = \sum_{n=1}^{\infty} \frac{u_n(x)\overline{u}_n(x')}{\lambda_n - \lambda} = \sum_{n=1}^{\infty} \frac{\sqrt{\frac{2}{a}} \sin\frac{n\pi}{a}x \sqrt{\frac{2}{a}} \sin\frac{n\pi}{a}x'}{\left(\frac{n\pi}{a}\right)^2 + \gamma^2}. \tag{5.31}$$

The substitution of (5.31) into (5.22) leads to (5.30) if the summation and integration operators are interchanged, which is a valid operation because (5.31) is uniformly convergent to an ordinary function for all $x$ and $x'$. In this one-dimensional example, the various forms of the Green's function are "well-behaved" classical functions for all $x$ and $x'$.

Because the Green's function must be unique, from (5.21) we have

$$g(x, x', \lambda) = \sum_{n=1}^{\infty} \frac{\sqrt{\frac{2}{a}} \sin\frac{n\pi}{a}x \sqrt{\frac{2}{a}} \sin\frac{n\pi}{a}x'}{\left(\frac{n\pi}{a}\right)^2 + \gamma^2} = \frac{\sinh\gamma(a - x_>) \sinh\gamma x_<}{\gamma \sinh\gamma a}, \tag{5.32}$$

which must also agree with the scattering superposition solution (5.26).

### 5.1.4 Completeness Relations for the Regular Sturm–Liouville Problem

The representation (5.28) provides the Green's function $g$ given the eigenfunctions and eigenvalues $u_n, \lambda_n$. Alternatively, given the Green's function one may determine the eigenfunctions and eigenvalues. It can be shown that the series (5.28) represents a meromorphic function of $\lambda$ [7, pp. 271–274], such that the Green's function has simple poles at $\lambda = \lambda_n$ in the complex $\lambda$-plane. As $L$ is self-adjoint, these will be on the real $\lambda$-axis. Taking the integral of (5.28) over a closed, counterclockwise-oriented contour $C$ in the complex $\lambda$-plane that encloses all of the poles, chosen as a circle having radius tending toward infinity, leads to

$$\begin{aligned} \oint_C g(x, x', \lambda)\, d\lambda &= \sum_{n=1}^{\infty} \overline{u}_n(x') u_n(x) \oint_C \frac{d\lambda}{\lambda_n - \lambda} \\ &= -2\pi i \sum_{n=1}^{\infty} u_n(x)\overline{u}_n(x'), \end{aligned} \tag{5.33}$$

where the series (5.28) is generally viewed as a distribution and integrated term by term, and the last equality is via the residue theorem [9, Sec. 22.1]. The expression (5.33) allows one to obtain the eigenfunctions by considering the poles of $g$ and associated residues.

To obtain a further useful expression, consider the expansion of the delta-function $\delta(x-x')$ in eigenfunctions of $L$. Since the delta-function is not in $\mathbf{L}^2_w(a,b)$, this is not permissible in the classical sense but is nevertheless acceptable in the distributional sense and leads to the correct result. Proceeding, we have $\delta(x-x')=\sum_{n=1}^{\infty}\alpha_n(x')\,u_n(x)$. Taking the inner product of both sides with $u_m(x)$ and exploiting the sifting property of the delta-function on the left side and orthonormality of the eigenfunctions on the right side, lead to $\alpha_n(x')=w(x')\overline{u_n}(x')$, and therefore[14]

$$\frac{\delta(x-x')}{w(x')}=\sum_{n=1}^{\infty}u_n(x)\overline{u}_n(x'), \tag{5.34}$$

which is the spectral representation of the delta-function. Recall that relations of the type (5.34) are known as completeness relations, because multiplying (5.34) by $u(x')w(x')$ and integrating lead to the expansion $u(x)=\sum_{n=1}^{\infty}\langle u,u_n\rangle\,u_n(x)$.

Combining (5.33) and (5.34), we have[15]

$$\frac{1}{2\pi i}\oint_C g(x,x',\lambda)\,d\lambda=-\frac{\delta(x-x')}{w(x')}=-\sum_{n=1}^{\infty}u_n(x)\overline{u}_n(x'). \tag{5.35}$$

**Example 5.4.**

Returning to the previous example of the differential equation

$$\left(-\frac{d^2}{dz^2}+\gamma^2\right)u(x)=f(x) \tag{5.36}$$

with $\operatorname{Re}\gamma>0$ and $u(0)=u(a)=0$, eigenfunctions are found from

$$-\left(\frac{d^2}{dx^2}+\lambda_n\right)u_n(x)=0$$

---

[14]This representation of the delta-function is with respect to a particular space, in the sense that for $f\in\mathbf{L}^2_w[a,b]$ the relation

$$f(x)=\int_a^b f(x')\delta(x-x')\,dx'$$

with $\delta(x-x')$ given by (5.34) is valid.

If one substitutes (5.34) into the above and interchanges the summation and integral operators, then the expansion $f=\sum_{n=1}^{\infty}\langle f,u_n\rangle\,u_n$ results.

[15]See Section 5.3.2 for a more general development and associated discussion.

subject to $u_n(0) = u_n(a) = 0$ as $u_n(x) = \sqrt{2/a}\sin(n\pi x/a)$, with corresponding eigenvalues $\lambda_n = (n\pi/a)^2$. We see, most directly from (5.34), but also from (5.35) using either form of $g$ in (5.32), that

$$\delta(x - x') = \frac{2}{a}\sum_{n=1}^{\infty} \sin\frac{n\pi}{a}x \,\sin\frac{n\pi}{a}x', \tag{5.37}$$

where the equality is in the distributional sense.

**Example 5.5.**

For the same differential equation as in Example 5.4, but with boundary conditions $u'(0) = u'(a) = 0$, the eigenfunctions are found to be

$$u_n(x) = \sqrt{\frac{\varepsilon_n}{a}}\cos\frac{n\pi}{a}x$$

with corresponding eigenvalues $\lambda_n = (n\pi/a)^2$, $n = 0, 1, 2, \ldots$, where $\varepsilon_n$ is *Neumann's number*, defined as $\varepsilon_n \equiv \begin{cases} 1, & n = 0 \\ 2, & n \neq 0 \end{cases}$. The eigenfunction expansion of the Green's function is then

$$g(x, x', \lambda) = \sum_{n=0}^{\infty} \frac{u_n(x)\overline{u}_n(x')}{\lambda_n - \lambda} = \sum_{n=0}^{\infty} \frac{\sqrt{\frac{\varepsilon_n}{a}}\cos\frac{n\pi}{a}x\,\sqrt{\frac{\varepsilon_n}{a}}\cos\frac{n\pi}{a}x'}{\left(\frac{n\pi}{a}\right)^2 + \gamma^2}, \tag{5.38}$$

and the spectral representation of the delta-function is

$$\delta(x - x') = \sum_{n=0}^{\infty} \frac{\varepsilon_n}{a}\cos\frac{n\pi}{a}x \,\cos\frac{n\pi}{a}x'. \tag{5.39}$$

Both (5.37) and (5.39) are equivalent in the sense that they pertain to the same space, $\mathbf{L}^2(0, a)$.

## 5.2 Singular Sturm–Liouville Problems

In the previous sections the regular Sturm–Liouville problem was investigated. To examine the singular Sturm–Liouville problem we consider the same second-order ordinary differential equation as before,

$$-\frac{1}{w(x)}\frac{d}{dx}\left[p(x)\frac{du}{dx}\right] + (q(x) - \lambda)\,u = f(x) \tag{5.40}$$

on $a < x < b$. In the regular problem we restrict the functions in (5.40) such that $p, p', q, w$ are continuous and real-valued on the finite interval $[a, b]$, and $p(x), w(x) > 0$ on $[a, b]$. In the singular problem the interval

$(a,b)$ may be infinite or semi-infinite, or certain conditions on $p, w$ may not be satisfied. We again work in the space $\mathbf{L}_w^2(a,b)$, defined as the set of Lebesgue-integrable functions $u(x)$ such that

$$\int_a^b |u(x)|^2 w(x)\, dx < \infty$$

with weighted inner product

$$\langle u, v \rangle = \int_a^b u(x)\overline{v(x)}w(x)\, dx \tag{5.41}$$

and norm[16]

$$\|u\| = \langle u, u \rangle^{1/2} = \sqrt{\int_a^b |u(x)|^2\, w(x)\, dx}.$$

As before, the Sturm–Liouville operator $L : \mathbf{L}_w^2(a,b) \to \mathbf{L}_w^2(a,b)$ is defined as

$$L \equiv -\frac{1}{w(x)}\frac{d}{dx}\left[p(x)\frac{d}{dx}\right] + q(x) \tag{5.42}$$

such that (5.40) is written as

$$(L - \lambda)\, u = f. \tag{5.43}$$

The most common classes of singular Sturm–Liouville problems relevant to electromagnetics are such that (i) the interval is infinite or semi-infinite, and/or (ii) $p(x) = 0$ at one or both endpoints. When an endpoint is $\pm\infty$, or if an endpoint is finite but $p(x) = 0$ at the endpoint, then that endpoint is said to be a *singular point.* Furthermore, if $p(x) = 0$ at an endpoint, then $w(x)$ may be unbounded there as well.

In a general sense, the singular Sturm–Liouville problem lacks many of the nice properties of the regular Sturm–Liouville problem, especially regarding the possibility of certain discrete eigenfunction expansions. It may occur that eigenfunctions exist and are complete in the space of interest, in which case the methods of the previous section hold. In other cases, eigenfunctions may fail to exist at all, and yet functions of interest may be expanded in the continuous spectrum of improper eigenfunctions, as in a Fourier transform. It is also possible for the spectrum to have both discrete and continuous components, as shown later in Sections 8.1.6 and 8.1.7.

While the general theory of singular Sturm–Liouville problems is beyond the scope of this book (see, e.g., [8]), in this and the following sections some singular Sturm–Liouville problems associated with electromagnetic applications are discussed. The relevant spectral properties are investigated on a case-by-case basis, and some general conclusions are stated.

[16] For this to be a valid norm we assume $w(x) > 0$ for all $x \in (a,b)$.

Note that on an infinite domain the inverse of the Sturm–Liouville operator is generally not compact (Theorem 5.1 and its consequences concern bounded domains), which is consistent with the possible nonexistence of proper eigenfunctions for infinite-domain problems.

### Transition to a Singular Problem

We begin the discussion of the singular Sturm–Liouville problem by illustrating a transition from a regular to a singular problem. Consider the regular Sturm–Liouville problem (5.18), repeated below as

$$\left(-\frac{d^2}{dx^2}+\gamma^2\right)u(x)=f(x) \tag{5.44}$$

on $[0,a]$, with $a<\infty$, $\operatorname{Re}\gamma>0$, and the boundary conditions $u(0)=u(a)=0$. The normalized eigenfunctions are $u_n(x)=\sqrt{2/a}\sin(n\pi x/a)$ with corresponding eigenvalues $\lambda_n=(n\pi/a)^2$, leading to the eigenfunction expansion of the Green's function (5.31).

As $a$ becomes larger, the spacing between elements of the sequence $\lambda_n=(n\pi/a)^2$ decreases. As $a\to\infty$, the eigenvalues $\lambda_n$ merge into a continuous set $\lambda\in[0,\infty)$. We will see that in this limit eigenfunctions cease to exist and $[0,\infty)$ forms a continuous spectrum.

To see the transformation to a singular problem, let $\zeta_n=n\pi/a$ and consider the interval $[0,a)$. The values of $\zeta_n$ partition the interval $[0,a)$ into $N=a^2/\pi$ uniform subintervals of width $\Delta\zeta_n=\zeta_{n+1}-\zeta_n=\pi/a$. Therefore, as $a\to\infty$, $\zeta_n\to\zeta$, and from (5.31) we have

$$\lim_{a\to\infty}\frac{2}{\pi}\sum_{n=1}^{\infty}\frac{\sin\zeta_n x\,\sin\zeta_n x'}{\zeta_n^2+\gamma^2}\Delta\zeta_n\to\frac{2}{\pi}\int_0^\infty\frac{\sin\zeta x\,\sin\zeta x'}{\zeta^2+\gamma^2}d\zeta. \tag{5.45}$$

The integral in (5.45) can be evaluated using complex analysis. Consider the case $x>x'$, where we manipulate the above as

$$\begin{aligned}g(x,x',\gamma)&=\frac{2}{\pi}\int_0^\infty\frac{\sin\zeta x\,\sin\zeta x'}{\zeta^2+\gamma^2}\,d\zeta=\frac{1}{\pi}\int_{-\infty}^\infty\frac{\sin\zeta x\,\sin\zeta x'}{\zeta^2+\gamma^2}\,d\zeta\\&=\frac{1}{\pi}\int_{-\infty}^\infty\frac{(\sin\zeta x+i\cos\zeta z)\,\sin\zeta x'}{\zeta^2+\gamma^2}\,d\zeta\\&=\frac{i}{\pi}\int_{-\infty}^\infty\frac{e^{-i\zeta x}\,\sin\zeta x'}{\zeta^2+\gamma^2}\,d\zeta.\end{aligned} \tag{5.46}$$

The real-line integration contour over $(-\infty,\infty)$ in (5.46) is closed by a clockwise-oriented semicircle $C_\infty^-$ in the lower-half complex $\zeta$-plane as shown in Figure 5.1. If we write $\zeta^2+\gamma^2=(\zeta+i\gamma)(\zeta-i\gamma)$, the residue

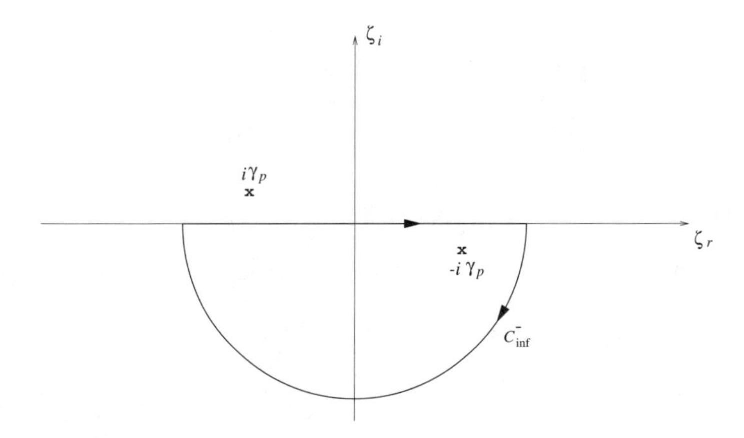


Figure 5.1: Complex $\zeta$-plane integration contour for evaluation of (5.46). Semicircle $C^-_{\text{inf}}$ has infinite radius and "**x**" denotes the position of a pole at $\pm i\gamma_p$.

theorem [9, Sec. 22.1][17] leads to

$$g(x,x',\gamma) = -2\pi i \, \frac{i}{\pi} \, \frac{e^{-i\zeta_p x} \, \sin \zeta_p x'}{(\zeta_p - i\gamma)} \bigg|_{\zeta_p = -i\gamma} - \frac{i}{\pi} \int_{C^-_\infty} \frac{e^{-i\zeta x} \, \sin \zeta x'}{\zeta^2 + \gamma^2} \, d\zeta. \tag{5.47}$$

The integral over the semicircle vanishes by Jordan's lemma [9, Sec. 22],[18] leading to

$$g(x,x',\gamma) = -2\pi i \, \frac{i}{\pi} \, \frac{e^{-i\zeta_p x} \, \sin \zeta_p x'}{(\zeta_p - i\gamma)} \bigg|_{\zeta_p = -i\gamma} = \frac{1}{\gamma} e^{-\gamma x} \sinh \gamma x' \tag{5.48}$$

for $x > x'$. An analogous procedure for $x < x'$ leads to the final expression as

$$g(x,x',\gamma) = \frac{1}{\gamma} e^{-\gamma x_>} \sinh \gamma x_< \tag{5.49}$$

with the associated solution

$$u(x) = \int_0^\infty f(x') g(x,x',\gamma) \, dx' \tag{5.50}$$

[17] Note that in the case $\operatorname{Re}\gamma = 0$ the poles lie on the real $\zeta$-axis such that the integral is undefined. In this situation the proper definition of the integral can be obtained as the limit $\operatorname{Re}\gamma \to 0$ through positive values, physically related to limitingly small material loss. This will place the poles in the 2nd and 4th quadrants rather than on the real axis.

[18] **Jordan's Lemma**: Let $\Gamma_R$ be a semicircle in the upper-half (lower-half) complex plane, and let $f(z)$ be a function that tends uniformly to zero with respect to $\arg(z)$ as $|z| \to \infty$ when $0 \le \arg(z) \le \pi$ ($\pi \le \arg(z) \le 2\pi$). Then, for $\alpha > 0$ ($\alpha < 0$),

$$\lim_{R\to\infty} \int_{\Gamma_R} e^{i\alpha z} f(z) \, dz = 0.$$

for $x \in [0, \infty)$, where $u(x) \in \mathbf{L}^2(0, \infty)$. This solution is similar to (5.22) with $a$ replaced with $\infty$, although in this case the Green's function (5.49) is quite different from the Green's function for finite $a$, (5.21). In fact, the differences between these Green's functions are more significant than may appear at first glance. In the case of finite $a$, (5.21) is a meromorphic function in the $\gamma$-plane, with pole singularities given by $\sinh \gamma a = 0$. In the case $a \to \infty$, (5.49) does not have any pole singularities yet has a branch point at $\gamma = 0$ as described later. The difference in the type of singularity encountered in the Green's function is quite important, such that for finite $a$ the spectrum is purely discrete (as is always the case for the regular Sturm–Liouville problem), whereas in the singular case the spectrum is purely continuous, as shown in the following.

An even simpler method to obtain (5.49) is to examine the limit $a \to \infty$ in (5.21). Indeed, for $x < x'$,

$$\begin{aligned} g(x, x', \gamma) &= \lim_{a \to \infty} \frac{\sinh \gamma(a - x') \sinh \gamma x}{\gamma \sinh \gamma a} \\ &= \lim_{a \to \infty} \frac{1}{\gamma} \frac{e^{\gamma(a-x')} - e^{-\gamma(a-x')}}{e^{\gamma a} - e^{-\gamma a}} \sinh \gamma x \\ &= \frac{1}{\gamma} e^{-\gamma x'} \sinh \gamma x \end{aligned}$$

as expected, again assuming $\operatorname{Re} \gamma > 0$. Note that if $\operatorname{Re} \gamma = 0$ the above limiting procedure is not valid, corresponding to the real-axis integral form (5.46) encountering poles along the path of integration. This leads to an unbounded resolvent, consistent with the fact that these values of $\gamma = i\beta$, $\beta \in (-\infty, \infty)$, are part of the continuous spectrum $\lambda = -\gamma^2 \in [0, \infty)$. This difficulty can again be resolved by taking $\operatorname{Re} \gamma \to 0$ through positive values.

While for this operator on an unbounded interval eigenvalues do not exist and the spectrum is purely continuous, it is important to note that this is specific to the particular operator in question, and not a general result. For instance, $\mathbf{L}^2(0, \infty)$ is separable, and so by Definition 2.17 this space must have a countable basis. Furthermore, we might expect this basis to consist of eigenfunctions associated with some operator. Indeed, the Laguerre polynomials provide such an eigenbasis for $\mathbf{L}^2_w(0, \infty)$ with weight $e^{-x}$ (see Section 5.5). *Hermite functions* play the same role (as a discrete eigenbasis) for $\mathbf{L}^2(-\infty, \infty)$.

### 5.2.1 Classification of Singular Points

The limiting procedure detailed above is convenient and straightforward, but it is often desirable to approach the singular Sturm–Liouville problem using more direct methods. The next theorem is used to divide singular Sturm–Liouville problems into two classes [7, p. 297].

**Theorem 5.2.** *(Weyl's theorem) Consider the homogeneous problem*

$$(L - \lambda)\, u = 0,$$

*where $L$ is the Sturm–Liouville operator (5.4), one of the endpoints is regular, and the other endpoint is singular. Then,*

1. *if for some particular value of $\lambda$ every solution of $(L - \lambda)\, u = 0$ is in $\mathbf{L}^2_w(a, b)$, then for any other value of $\lambda$ every solution $u$ is in $\mathbf{L}^2_w(a, b)$;*

2. *for every $\lambda$ such that $\operatorname{Im} \lambda \neq 0$, there exists at least one solution of $(L - \lambda)\, u = 0$ in $\mathbf{L}^2_w(a, b)$.*

Weyl's theorem divides the singular Sturm–Liouville problem into two mutually exclusive classes:

a. *limit-circle case*: all solutions $u$ are in $\mathbf{L}^2_w(a, b)$ for all $\lambda$,

b. *limit-point case:*

   i. if $\operatorname{Im} \lambda \neq 0$, there exists exactly one solution $u \in \mathbf{L}^2_w(a, b)$;

   ii. if $\operatorname{Im} \lambda = 0$, there is either one solution or no solution in $\mathbf{L}^2_w(a, b)$.

The motivation for the terms *limit-point* and *limit-circle* are clear from the proof of Weyl's theorem [7, pp. 297–301]. If both endpoints are singular, an intermediate (regular) point $l$, such that $a < l < b$, is introduced, dividing the problem into two parts.

Because the limit-point and limit-circle cases are mutually exclusive, one may determine the appropriate class by examining the solution of $(L - \lambda)\, u = 0$ for a single value of $\lambda$. If all solutions are in $\mathbf{L}^2_w(a, b)$, the limit-circle case applies; otherwise the limit-point case applies. In this section we concentrate on the limit-point case, and a later section discusses the limit-circle case.

**Example 5.6.**

To illustrate these concepts for the limit-point case, consider again the differential equation

$$\left(-\frac{d^2}{dx^2} + \gamma^2\right) u(x) = 0 \tag{5.51}$$

on the interval $0 < x < \infty$, resulting in a singular Sturm–Liouville problem $(L - \lambda)\, u = 0$ with $L = -d^2/dx^2$ and $\lambda = -\gamma^2$. We discuss solutions treating $\lambda$ as a parameter in order to classify the singular point at infinity (the endpoint at 0 is regular). For (5.51) we simply check the case $\lambda = 0$, leading to the solution $u = Ax + B$, neither term of which is in $\mathbf{L}^2(0, \infty)$. We therefore have the limit-point case at $\infty$.

To go further, we can check the conditions associated with the limit-point case. For arbitrary $\lambda \neq 0$, the solution is $u = Ae^{i\sqrt{\lambda}x} + Be^{-i\sqrt{\lambda}x}$. Part (ii) of the Weyl classification states that if $\operatorname{Im}\lambda = 0$, there is either one solution or no solution in $\mathbf{L}^2(0,\infty)$. This is easily seen to hold, since if $\lambda \in \mathbf{R}$ with $\lambda < 0$, one solution ($Be^{-i\sqrt{\lambda}x} = Be^{-\alpha x}$, where $\sqrt{\lambda} = -i\alpha$ taking the negative square root) is in $\mathbf{L}^2(0,\infty)$, whereas if $\lambda \geq 0$, no solution is in $\mathbf{L}^2(0,\infty)$. Part (i) of the classification states that if $\operatorname{Im}\lambda \neq 0$, there exists exactly one solution $u \in \mathbf{L}^2(0,\infty)$. This is also easy to confirm by setting $\sqrt{\lambda} = -i\gamma = -i\alpha + \beta$. Then $u = Ae^{i(-i\alpha+\beta)x} + Be^{-i(-i\alpha+\beta)x}$, and only one solution is in $\mathbf{L}^2(0,\infty)$; the second term ($Be^{-\alpha x}e^{-i\beta x}$) if $\alpha > 0$, or the first term if $\alpha < 0$.

In the above classification we have not introduced any boundary conditions in the problem. Of course, to yield a unique solution to the second-order differential equation we must impose appropriate conditions. At a regular point a boundary condition of the form (5.2) must be given. For the limit-point case, though, it can be shown that no boundary condition is necessary at the limit point. Requiring that the solution be in $\mathbf{L}^2_w$ is enough to generate the unique $\mathbf{L}^2_w$ solution (which is usually the one of interest). We call this an *inclusion condition* or a *fitness condition*. Alternatively, one may, if desired, invoke a limiting condition such as

$$\lim_{x\to\infty} u(x) = 0,$$

which will pick out the unique $\mathbf{L}^2_w$ solution [1, pp. 81–82]. Under these conditions the problem is self-adjoint, and we consider such cases as forming a self-adjoint boundary value problem, leading to an eigenbasis (in a general sense, consisting of possibly both discrete (proper) and continuous (improper) eigenfunctions) for the space $H$. The limit-circle case must be treated differently, since all solutions are in $\mathbf{L}^2_w(a,b)$ for all $\lambda$.

**Example 5.7.**

Continuing with the analysis of (5.51), by Weyl's theorem for $\operatorname{Im}\lambda \neq 0$ there exists exactly one solution $u \in \mathbf{L}^2(0,\infty)$. Considering the associated nonhomogeneous problem on $[0,\infty)$

$$\left(-\frac{d^2}{dx^2} + \gamma^2\right) u(x) = f(x), \tag{5.52}$$

where $\operatorname{Re}\gamma > 0$, the Green's function satisfies

$$\left(-\frac{d^2}{dx^2} + \gamma^2\right) g(x,x',\gamma) = \delta(x - x')$$

such that $g(0,x',\gamma) = 0$. For the condition at the limit point either we simply choose the solution by requiring that it be a member of $\mathbf{L}^2(0,\infty)$, or we impose the limiting condition $\lim_{x\to\infty} g(x,x',\gamma) = 0$.

If we initially exclude the point $x = x'$, the Green's function can be found as the solution to

$$\left(-\frac{d^2}{dx^2} + \gamma^2\right) g(x, x', \gamma) = 0,$$

yielding

$$g(x, x', \gamma) = \begin{cases} A\,(x')\, e^{-\gamma x} + B\,(x')\, e^{\gamma x}, & x > x', \\ C\,(x') \sinh \gamma x + D\,(x') \cosh \gamma x, & x < x'. \end{cases}$$

Applying the boundary condition $g(0, x', \gamma) = 0$ leads to $D = 0$. Requiring the solution to be in $\mathbf{L}^2(0, \infty)$ (or alternately that $\lim_{x\to\infty} g(x, x', \gamma) = 0$) leads to $B = 0$. The "jump" and "continuity" conditions lead to two equations for determining the remaining coefficients, resulting in $A = (\sinh \gamma x')/\gamma$ and $C = e^{-\gamma x'}/\gamma$ such that the Green's function is obtained as

$$g(x, x', \gamma) = \frac{1}{\gamma} e^{-\gamma x_>} \sinh \gamma x_<, \tag{5.53}$$

in agreement with (5.49).

If the boundary condition is $\left.\frac{dg(x,x',\gamma)}{dx}\right|_{x=0} = 0$, then we obtain

$$g(x, x', \gamma) = \frac{1}{\gamma} e^{-\gamma x_>} \cosh \gamma x_< \tag{5.54}$$

as the Green's function.

### 5.2.2 Identification of the Continuous Spectrum and Improper Eigenfunctions

In (5.53) and (5.54) one notes that the term $\gamma$ appears, whereas the defining differential equation for the Green's function involves the term $\gamma^2$. Following the treatment in [10, pp. 217–219] and [1, Sec. 3.4], this leads to further insight into the spectral properties of the singular Sturm–Liouville problem. This is illustrated with the following example.

**Example 5.8.**

Consider the limit-point problem

$$(L - \lambda)\, u = f \tag{5.55}$$

on $[0, \infty)$, with $L = -d^2/dx^2$ and subject to $u(0) = 0$. If $\lambda = -\gamma^2$ as in Example 5.7, we have $\gamma = i\sqrt{\lambda}$ with $\operatorname{Im}\sqrt{\lambda} < 0$,[19] such that (5.53) becomes

$$g(x, x', \lambda) = \frac{1}{\sqrt{\lambda}} e^{-i\sqrt{\lambda} x_>} \sin \sqrt{\lambda} x_<. \tag{5.56}$$

[19] We choose this root because it leads to the usual form of traveling waves in later chapters, assuming an $e^{i\omega t}$ time dependence or, more generally, the transform pair (1.7) and (1.8). The choice $\gamma = -i\sqrt{\lambda}$ with $\operatorname{Im}\sqrt{\lambda} > 0$ is equally valid.

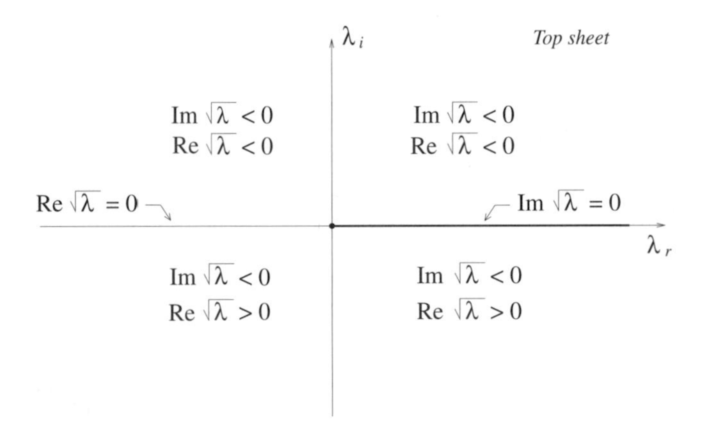


Figure 5.2: Complex $\lambda$-plane indicating the branch point at $\lambda = 0$ (denoted by "•") and the associated branch cut along the positive real axis, which separates Im $\sqrt{\lambda} < 0$ and Im $\sqrt{\lambda} > 0$. Top sheet shown.

The condition Im $\sqrt{\lambda} < 0$ defines a branch cut in the complex $\lambda$-plane as follows. If $\lambda = |\lambda| e^{i\theta}$ with the restriction $2\pi < \theta < 4\pi$, then $\lambda^{1/2} = \sqrt{|\lambda|} e^{i\frac{\theta}{2}}$ with $\pi < \theta/2 < 2\pi$, where $\sqrt{|\lambda|}$ is the positive square root of the real, positive number $|\lambda|$. The condition on $\theta/2$ clearly results in Im $\sqrt{\lambda} < 0$, the desired condition. The restriction $2\pi < \theta < 4\pi$ defines a Riemann sheet [9, Sec. 24] in the complex $\lambda$-plane, which we call the "top" sheet because it is the one of interest. Points defined by $\lambda = |\lambda| e^{i\theta}$ with values of polar angle such that $0 < \theta < 2\pi$ lie on another (the "bottom") Riemann sheet. This motivates the concept of a branch cut along the positive real axis to separate the two Riemann sheets and to keep the function $\sqrt{\lambda}$ single-valued on any given sheet, depicted in Figures 5.2 and 5.3. The range $n\pi < \theta < (n + 2)\pi$ is on the top sheet for $n = 2, 6, 10, \ldots$, and on the bottom sheet for $n = 0, 4, 8, \ldots$.

As noted previously, (5.56) does not have any pole singularities, only a branch point at $\lambda = 0$ and its associated branch cut. It is also clear that the discrete eigenvalue spectrum $\lambda_n = (n\pi/a)^2$, which lies at discrete points along the positive real axis for the case of $a$ finite, has, in the limit $a \to \infty$, coalesced into a continuous spectrum consisting of the positive real axis. The fact that the spectral components, whether discrete or continuous, lie along the real axis is consistent with Theorem 4.16 because the problem is self-adjoint. Furthermore, the residual spectrum is empty, as seen by Theorem 4.18.

The fact that we have a specified branch cut $[0, \infty)$ in (5.56) needs some explanation. Usually, given a multivalued function such as (5.56), the specification of a branch cut is arbitrary from a mathematical standpoint. Enforcement of any legitimate branch cut will result in the function's being

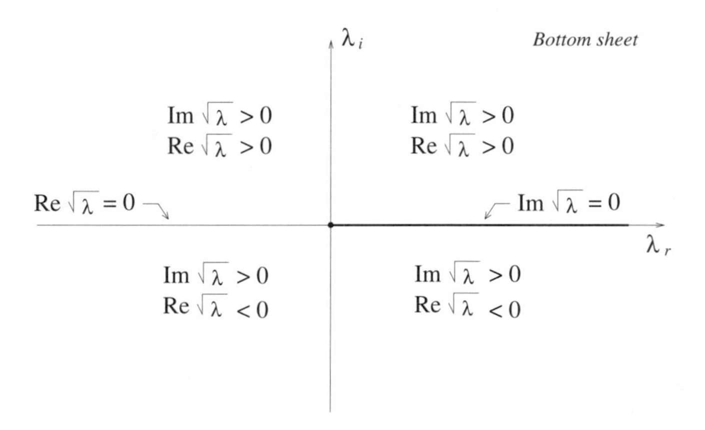


Figure 5.3: Complex $\lambda$-plane indicating the branch point at $\lambda = 0$ (denoted by "•") and the associated branch cut along the positive real axis, which separates Im $\sqrt{\lambda} < 0$ and Im $\sqrt{\lambda} > 0$. Bottom sheet shown.

single-valued on a given Riemann sheet, and the specification of a particular branch cut is usually dictated by physical considerations. In this problem, while any branch cut emanating from $\lambda = 0$ and extending to infinity will enforce single-valued behavior of (5.56), only the branch cut $[0, \infty)$ leads to an $\mathbf{L}^2(0, \infty)$ solution everywhere on a single Riemann sheet. To see this, consider Im $\sqrt{\lambda} < 0$,where $\sqrt{\lambda} = -i\alpha + \beta$, with $\alpha > 0$ and $\beta$ either positive or negative. For $x > x'$ in (5.56), the term containing $e^{-i\sqrt{\lambda}x} = e^{-\alpha x}e^{-i\beta x}$ is in $\mathbf{L}^2(0, \infty)$ everywhere on the top Riemann sheet, regardless of the sign of $\beta$. To see what happens with another choice of branch cut, consider the condition Re $\sqrt{\lambda} > 0$, which defines a branch cut in the complex $\lambda$-plane along the negative real axis.[20] For $x > x'$ with $\sqrt{\lambda} = -i\alpha + \beta$, where $\beta > 0$ and $\alpha$ may be positive or negative, the term containing $e^{-i\sqrt{\lambda}x} = e^{-\alpha x}e^{-i\beta x}$ may or may not be in $\mathbf{L}^2(0, \infty)$ depending on the sign of $\alpha$. Thus, while the Green's function is single-valued, it does not represent the desired solution in $\mathbf{L}^2(0, \infty)$ everywhere on a single Riemann sheet. Alternately, if (5.56) is to represent the limit of (5.21), then one also arrives at $[0, \infty)$ as the proper branch cut for this problem.

[20]Let $\lambda = |\lambda|\, e^{i\theta}$ with the restriction $-\pi < \theta < \pi$, such that $\lambda^{1/2} = \sqrt{|\lambda|}e^{i\frac{\theta}{2}}$ with $-\pi/2 < \theta/2 < \pi/2$. The condition on $\frac{\theta}{2}$ clearly results in Re $\sqrt{\lambda} > 0$, the desired condition. The restriction $-\pi < \theta < \pi$ defines a Riemann sheet such that points defined by $\lambda = |\lambda|\, e^{i\theta}$ with values of polar angle $-\pi < \theta < \pi$ lie on one sheet, whereas points with values of polar angle such that $\pi < \theta < 3\pi$ lie on another sheet. The branch cut is then seen to lie along the negative real axis to separate the two Riemann sheets and keep the function $\sqrt{\lambda}$ single-valued on any given sheet.

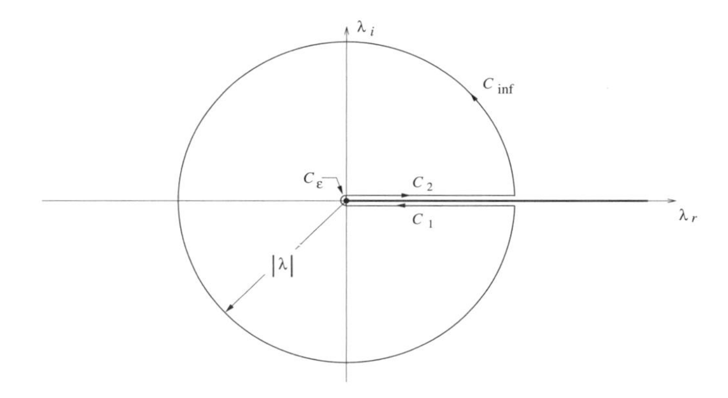


Figure 5.4: Closed integration contour in the $\lambda$-plane appropriate for application of Cauchy's theorem, showing the branch point at $\lambda = 0$ (denoted by "•") and the associated branch cut along the positive real axis. Circle $C_{\text{inf}}$ has infinite radius.

If we consider the integration path in Figure 5.4, Cauchy's theorem [9, Sec. 11] leads to

$$\oint_{C_\infty + C_1 + C_2 + C_\varepsilon} g(x, x', \lambda)\, d\lambda = 0 \tag{5.57}$$

assuming that the integration occurs on the proper sheet such that $g$ vanishes as $|\lambda| \to \infty$. From (5.35) we have

$$\int_{C_\infty} g(x, x', \lambda)\, d\lambda = -2\pi i\, \delta(x - x'), \tag{5.58}$$

which continues to hold in the presence of the branch cut.

It is easy to show that the contribution from $C_\varepsilon$ vanishes as $\varepsilon \to 0$. The contributions from $C_1$ and $C_2$ are obtained as follows. As $\varepsilon \to 0$, $C_1$ becomes a horizontal path just below the real $\lambda$-axis, whereas $C_2$ becomes a horizontal path just above the real $\lambda$-axis. In polar form, $\lambda = re^{i(4\pi - \delta)} = re^{i4\pi^-}$ on $C_1$ and $\lambda = re^{i(2\pi+\delta)} = re^{i2\pi^+}$ on $C_2$, where $\delta$ is a very small positive quantity tending toward zero, $r = |\lambda|$, and the $4\pi$ factor is due to the fact that to get below the real positive axis on the top sheet (so as not to violate the branch cut) one must advance in phase $4\pi$ radians. This leads to $\sqrt{\lambda} = \sqrt{r}e^{i2\pi^-} = \sqrt{r}$ on $C_1$ and $\sqrt{\lambda} = \sqrt{r}e^{i\pi^+} = -\sqrt{r}$ on $C_2$, where, as usual, $\sqrt{r}$ indicates the positive root of $r$. In this way the function $g$ is discontinuous across the branch cut such that oppositely directed integrations along, and infinitely close to, the cut (i.e., $\lim_{\varepsilon \to 0} \oint_{C_1 + C_2} (\cdot)\, d\lambda$) do not cancel, which would be the case if the integrand were continuous[21]

[21] It is important to note that if $g$ were "even" in $\sqrt{\lambda}$, i.e., $g|_{\sqrt{\lambda}=\sqrt{r}} = g|_{\sqrt{\lambda}=-\sqrt{r}}$, then although $\sqrt{\lambda}$ would induce a branch point and associated cut, the Green's function

across the cut. Therefore, for $x > x'$,

$$\begin{aligned}
\lim_{\varepsilon \to 0} \int_{C_1+C_2} & g(x, x', \lambda)\, d\lambda \\
&= \int_{\infty}^{0} \frac{1}{\sqrt{r}} e^{-i\sqrt{r}x} \sin \sqrt{r}x'\, dr + \int_{0}^{\infty} \frac{1}{\sqrt{r}} e^{i\sqrt{r}x} \sin \sqrt{r}x'\, dr \\
&= \int_{0}^{\infty} \frac{1}{\sqrt{r}} e^{i\sqrt{r}x} \sin \sqrt{r}x'\, dr - \int_{0}^{\infty} \frac{1}{\sqrt{r}} e^{-i\sqrt{r}x} \sin \sqrt{r}x'\, dr \\
&= \int_{0}^{\infty} \frac{1}{\sqrt{r}} \sin \sqrt{r}x' \left( e^{i\sqrt{r}x} - e^{-i\sqrt{r}x} \right) dr \\
&= 2i \int_{0}^{\infty} \frac{1}{\sqrt{r}} \sin \sqrt{r}x' \sin \sqrt{r}x\, dr.
\end{aligned} \tag{5.59}$$

The same result is obtained for $x < x'$. Combining with (5.58) we have

$$-2\pi i\, \delta(x - x') + 2i \int_{0}^{\infty} \frac{1}{\sqrt{r}} \sin \sqrt{r}x' \sin \sqrt{r}x\, dr = 0.$$

The substitution $\upsilon = \sqrt{r}$, $d\upsilon = dr\,/(2\sqrt{r})$ leads to

$$\delta(x - x') = \frac{2}{\pi} \int_{0}^{\infty} \sin \upsilon x' \sin \upsilon x\, d\upsilon, \tag{5.60}$$

which is the spectral representation of the delta-function on $\mathbf{L}^2(0, \infty)$ with respect to the continuous spectrum $\upsilon \in [0, \infty)$.[22]

### Identification of Improper Eigenfunctions

It is convenient, although slightly misleading, to consider the integral (5.60) as a continuous (rather than a discrete) superposition of eigenfunctions. Even though

$$-\frac{d^2}{dx^2} \sin \upsilon x = \upsilon^2 \sin \upsilon x$$

such that $\sin \upsilon x$ appears to be an eigenfunction with the corresponding eigenvalue $\upsilon^2$, it is easy to see that $\sin \upsilon x \notin \mathbf{L}^2(0, \infty)$ and so $\sin \upsilon x$ cannot be an eigenfunction for this operator. Actually, for the operator $L = -d^2/dx^2$ on $[0, \infty)$ with $u(0) = 0$, eigenfunctions do not exist, but rather every point $\lambda \in [0, \infty)$ is in the continuous spectrum. We can, though, consider the concept of *improper eigenfunctions* [10, p. 234] such that

would be continuous across the cut such that $\lim_{\varepsilon \to 0} \oint_{C_1+C_2} g(x, x', \lambda)\, d\lambda = 0$. In this case the branch point is said to be *removable*. For example, such would be the case for the regular Sturm–Liouville problem on $[0, a]$ leading to (5.21). In the present case the branch point is nonremovable due to the term $e^{-i\sqrt{\lambda}x}$.

[22] A direct proof that $[0, \infty)$ forms the continuous spectrum is outlined in [86, p. 473], and another method is detailed in [10, pp. 233–236].

$u(x, \upsilon) = \sin \upsilon x$ is an improper eigenfunction with improper eigenvalue $\lambda(\upsilon) = \upsilon^2$. Improper eigenvalues belong to the approximate spectrum, as described in Theorem 4.2. These elements cannot be normalized in the usual way, since, for instance, $\int_0^\infty |\sin \upsilon x|\, dx$ does not exist. However, (5.60) leads to, with a change of variables, the normalization

$$\delta(\upsilon - \nu) = \frac{2}{\pi} \int_0^\infty \sin \upsilon x \sin \nu x \, dx.$$

Although the improper eigenfunctions do not belong to $\mathbf{L}^2(0, \infty)$, they are bounded on $[0, \infty)$.

If we compare (5.60) with (5.37) we see that the term $\sin \upsilon x$ with $\upsilon$ a continuous parameter forms a completeness relation in $\mathbf{L}^2(0, \infty)$, in an analogous way that $\sin(n\pi x/a)$ with $n$ discrete forms a completeness relation for $\mathbf{L}^2(0, a)$ with $a < \infty$. Recall from Section 5.1.3 that for every $u \in \mathbf{L}^2_w(a, b)$ on the finite interval $[a, b]$ we have the eigenfunction expansion $u = \sum_{n=1}^\infty \langle u, u_n \rangle u_n$, interpreted as an expansion of $u(x)$ in terms of the discrete (countably infinite) set $u_n(x)$. One can obtain this expansion by taking the inner product of (5.34) with $u(x)$. As detailed in Section 5.1.3, this leads to the solution of $(L - \lambda) u = f$ as $u(x) = \sum_{n=1}^\infty (1/(\lambda_n - \lambda)) \langle f, u_n \rangle u_n$ and the Green's function $g(x, x', \lambda) = \sum_{n=1}^\infty (1/(\lambda_n - \lambda)) u_n(x) \overline{u}_n(x')$, assuming $\lambda \neq \lambda_n$.

In a similar way, for $u \in \mathbf{L}^2(0, \infty)$, (5.60) has the form of a completeness relation

$$\frac{\delta(x - x')}{w(x')} = \int_\upsilon u(x, \upsilon) \overline{u}(x', \upsilon) \, d\upsilon, \tag{5.61}$$

where $u(x, \upsilon)$ are improper eigenfunctions that satisfy

$$Lu(x, \upsilon) = \lambda(\upsilon) u(x, \upsilon)$$

and $\lambda(\upsilon)$ are improper eigenvalues (often $\lambda(\upsilon) = \upsilon^2$). Taking the inner product

$$\langle f, g \rangle = \int_0^\infty f(x) \overline{g}(x) w(x) \, dx$$

of (5.61) with $u$ leads to the continuous expansion

$$u(x) = \int_\upsilon \langle u, u(x, \upsilon) \rangle \, u(x, \upsilon) \, d\upsilon. \tag{5.62}$$

If[23]

$$Lu = \int_0^\infty \lambda(\upsilon) \langle u, u(x, \upsilon) \rangle \, u(x, \upsilon) \, d\upsilon,$$

[23] Either we directly assume this form of the expansion, analogous to the expansion in Theorem 4.41, or we obtain this form from the expansion of $u$ as in (5.62). In the latter case, to avoid the interchange of the differential operator $L = -d^2/dx^2$ and the spectral

then for $(L-\lambda)\,u=f$ the solution, analogous to (5.27), is

$$u(x)=\int_{\upsilon}\frac{\langle f,u(x,\upsilon)\rangle}{\lambda(\upsilon)-\lambda}u(x,\upsilon)\,d\upsilon, \tag{5.63}$$

where we utilized the orthogonality relation

$$\langle u(x,\upsilon),u(x,p)\rangle=\delta(\upsilon-p). \tag{5.64}$$

The Green's function form analogous to (5.28) is

$$g(x,x',\lambda)=\int_{\upsilon}\frac{u(x,\upsilon)\overline{u}(x,\upsilon)}{\lambda(\upsilon)-\lambda}\,d\upsilon. \tag{5.65}$$

In this case it is easily seen that the appropriate form of the completeness relation, similar to (5.35), is

$$\frac{1}{2\pi i}\oint_{C}g(x,x',\lambda)\,d\lambda=-\frac{\delta(x-x')}{w(x)}=-\int_{\upsilon}u(x,\upsilon)\overline{u}(x',\upsilon)\,d\upsilon. \tag{5.66}$$

In summary,

- improper eigenfunctions $u(x,\upsilon)$ of the operator $L$ satisfy

$$Lu(x,\upsilon)=\lambda(\upsilon)u(x,\upsilon),$$

where $\lambda(\upsilon)$ are improper eigenvalues, which are part of the approximate spectrum $\sigma_a(L)$,

---

integral, we multiply $(L-\lambda)\,u=f$ by $\overline{u}(x,\upsilon)$ and integrate over $[0,\infty)$ to obtain

$$-\int_0^{\infty}\frac{d^2u(x)}{dx^2}\overline{u}(x,\upsilon)\,dx-\lambda\int_0^{\infty}u(x)\overline{u}(x,\upsilon)\,dx=\int_0^{\infty}f(x)\overline{u}(x,\upsilon)\,dx$$
$$=\langle f,u(x,\upsilon)\rangle\,.$$

The first term is converted by a scalar Green's theorem (equivalently, integration-by-parts twice) as

$$-\int_0^{\infty}\frac{\partial^2u(x)}{\partial x^2}\overline{u}(x,\upsilon)\,dx=\left(u(x)\frac{\partial\overline{u}(x,\upsilon)}{\partial x}-\overline{u}(x,\upsilon)\frac{\partial u(x)}{\partial x}\right)\Big|_0^{\infty}-\int_0^{\infty}u(x)\frac{\partial^2\overline{u}(x,\upsilon)}{\partial x^2}dx.$$

The requirement $\lim_{x\to\infty}u(x)=0$ removes the first term on the right side; however, the improper eigenfunction $u(x,\upsilon)$ *does not* vanish as $x\to\infty$. As discussed in [1, p. 116] in electromagnetics usually where $\lim_{x\to\infty}u(x)=0$ then $\lim_{x\to\infty}du(x)/dx=0$, and so we have

$$-\int_0^{\infty}\frac{\partial^2u(x)}{\partial x^2}\overline{u}(x,\upsilon)\,dx=-\int_0^{\infty}u(x)\frac{\partial^2\overline{u}(x,\upsilon)}{\partial x^2}\,dx$$
$$=\lambda(\upsilon)\int_0^{\infty}u(x)\overline{u}(x,\upsilon)\,dx$$

leading to $\langle u,u\,(x,\upsilon)\rangle=(1/(\lambda(\upsilon)-\lambda))\,\langle f,u(x,\upsilon)\rangle$ as desired. Alternatively, one may be able to ascertain that the derivative operator can be simply passed through the integration operator.

- improper eigenfunctions are bounded at infinity but are not square-integrable on infinite or semi-infinite domains,
- improper eigenfunctions are normalized according to

$$\langle u(x,\upsilon), u(x,p)\rangle = \delta(\upsilon - p),$$

which forms the completeness relation (5.61),

- in practice, often we assume that the left-side equality in (5.66) holds, and obtain the completeness relation (i.e., the right-side equality in (5.66)) by integration of the Green's function (as was done in the steps leading to (5.60)). Some comments on the validity of the left-side equality in (5.66) are provided in Section 5.3.2, and
- we require the fitness condition $g \to 0$ on the proper Riemann sheet as $x \to \infty$, which, in turn, requires $\operatorname{Im}\sqrt{\lambda} < 0$ (inducing the branch cut and resulting continuous spectrum). In many physical problems $\lambda = k^2 = \omega^2\mu\varepsilon$, in which case $\operatorname{Im}\sqrt{\lambda} < 0$ is physically associated with material loss. Therefore, there is often is a connection between having (perhaps small) loss and fulfillment of the fitness condition at infinity, which, in turn, is often necessary to obtain a self-adjoint operator. In such cases, even though Theorem 4.41 is not applicable because the inverse operator is not compact, the eigenfunctions (proper and improper) form an eigenbasis for the space in question.[24]

### 5.2.3 Continuous Expansions and Associated Integral Transforms

It turns out that the continuous spectrum of some simple second-order differential operators can be associated with well known integral transforms, a fact that is illustrated in the following examples.

**Example 5.9.**

For the problem considered in Example 5.8,

$$\left(-\frac{d^2}{dx^2} - \lambda\right) u = f$$

on $[0,\infty)$ subject to $u\,(0) = 0$, the improper eigenfunctions and improper eigenvalues were identified as $u(x,\upsilon) = \sqrt{2/\pi}\sin \upsilon x$ and $\lambda(\upsilon) = \upsilon^2$, respectively. The expansion (5.62) becomes

$$u(x) = \sqrt{\frac{2}{\pi}}\int_0^\infty a(\upsilon)\sin \upsilon x\, d\upsilon$$

[24] This is not necessarily true for more general boundary conditions than those considered here (the boundary conditions considered in this section are (5.2) at finite boundaries and fitness conditions at infinity).

with $a(v) = \langle u, u(x,v)\rangle = \sqrt{2/\pi}\int_0^\infty u(x)\sin vx\, dx$. This is nothing more than the *Fourier sine transform*,

$$u(x) = \frac{2}{\pi}\int_0^\infty U(v)\sin vx\, dv,$$
$$U(v) = \int_0^\infty u(x)\sin vx\, dx, \tag{5.67}$$

where the factor $\sqrt{2/\pi}$ in the forward and inverse transform has been shifted to a factor $2/\pi$ in the inverse transform in accordance with typical usage. As indicated by the completeness relation (5.60), any $u(x) \in \mathbf{L}^2(0,\infty)$ can be expanded in the (continuous set) $\sin vx$ as a Fourier sine transform.

Substituting into (5.63) leads to

$$u(x) = \frac{2}{\pi}\int_0^\infty \frac{\langle f, \sin vx\rangle}{v^2+\gamma^2}\sin vx\, dv, \tag{5.68}$$

appropriate for $\lambda = -\gamma^2 \notin [0,\infty)$, i.e., $\lambda$ not part of the continuous spectrum.

The Green's function is, from (5.65),

$$g(x,x',\gamma) = \frac{2}{\pi}\int_0^\infty \frac{\sin vx \sin vx'}{v^2+\gamma^2}\, dv, \tag{5.69}$$

which is equivalent to (5.53), i.e.,

$$g(x,x',\gamma) = \frac{2}{\pi}\int_0^\infty \frac{\sin vx \sin vx'}{v^2+\gamma^2}\, dv = \frac{1}{\gamma}e^{-\gamma x_>}\sinh \gamma x_<. \tag{5.70}$$

Thus, the closed-form Green's function (5.53) and the continuous representation (5.69) for $a \to \infty$ play the same role as (5.21) and the discrete representation (5.31) for finite $a$.

**Example 5.10.**

A problem similar to that considered in Examples 5.8 and 5.9 is the operator $L = -d^2/dx^2$ subject to $u'(0) = 0$ on $[0,\infty)$; an analogous study leads to the Green's function

$$g(x,x',\lambda) = \frac{1}{i\sqrt{\lambda}}e^{-i\sqrt{\lambda}x_>}\cos\sqrt{\lambda}x_< \tag{5.71}$$

and the improper eigenfunctions $u(x,v) = \sqrt{2/\pi}\cos vx$ and improper eigenvalues $\lambda(v) = v^2$, such that (5.61) becomes

$$\frac{2}{\pi}\int_0^\infty \cos vx' \cos vx\, dv = \delta(x-x'). \tag{5.72}$$

The expansion (5.62) is

$$u(x) = \sqrt{\frac{2}{\pi}} \int_0^\infty a(v) \cos vx \, dv$$

with $a(v) = \langle u, u(x,v) \rangle = \sqrt{2/\pi} \int_0^\infty u(x) \cos vx \, dx$, which is nothing more than the *Fourier cosine transform*

$$\begin{aligned} u(x) &= \frac{2}{\pi} \int_0^\infty U(v) \cos vx \, dv, \\ U(v) &= \int_0^\infty u(x) \cos vx \, dx. \end{aligned} \tag{5.73}$$

The solution of $(L - \lambda)\, u = f$ and the Green's function are the same as (5.68) and (5.69), respectively, with cos replacing sin.

**Example 5.11.**

As another example of a singular Sturm–Liouville problem that arises in electromagnetics, consider

$$\left( -\frac{d^2}{dx^2} + \gamma^2 \right) u(x) = f(x) \tag{5.74}$$

on $(-\infty, \infty)$, with $\operatorname{Re} \gamma > 0$. The problem is in the limit-point case at $x = \pm\infty$. For the associated nonhomogeneous problem, the Green's function satisfies

$$\left( -\frac{d^2}{dx^2} + \gamma^2 \right) g(x, x', \gamma) = \delta(x - x'). \tag{5.75}$$

We pick an interior point $-\infty < x' < \infty$ and, since $\operatorname{Re} \gamma > 0$ (which, for $\lambda = -\gamma^2$ with $\gamma = \alpha + i\beta$, $\beta \neq 0$, provides the condition $\operatorname{Im} \lambda \neq 0$), there is one solution of the homogeneous form of (5.74) in $\mathbf{L}^2(-\infty, x')$ and one solution in $\mathbf{L}^2(x', \infty)$. For the condition at the limit points either we simply choose the solution by requiring that it be square-integrable on the appropriate interval, or we impose the limiting conditions

$$\lim_{x \to \infty} g(x, x', \gamma) = \lim_{x \to -\infty} g(x, x', \gamma) = 0.$$

If we initially exclude the point $x = x'$, the Green's function can be found as the solution to

$$\left( -\frac{d^2}{dx^2} + \gamma^2 \right) g(x, x', \gamma) = 0,$$

yielding

$$g(x, x', \gamma) = \begin{cases} A(x')\, e^{-\gamma x} + B(x')\, e^{\gamma x}, & x > x', \\ C(x')\, e^{-\gamma x} + D(x')\, e^{\gamma x}, & x < x'. \end{cases}$$

Applying the limiting conditions

$$\lim_{x\to\infty} g(x,x',\gamma) = \lim_{x\to-\infty} g(x,x',\gamma) = 0$$

leads to $B = C = 0$. The "jump" and "continuity" conditions lead to two equations for determining the remaining coefficients, resulting in $A = e^{\gamma x'}/(2\gamma)$ and $D = e^{-\gamma x'}/(2\gamma)$, such that the Green's function is

$$g(x,x',\gamma) = \begin{cases} \frac{1}{2\gamma}e^{-\gamma(x-x')}, & x \geq x', \\ \frac{1}{2\gamma}e^{\gamma(x-x')}, & x \leq x', \end{cases} \tag{5.76}$$

or

$$g(x,x',\gamma) = \frac{1}{2\gamma}e^{-\gamma|x-x'|}. \tag{5.77}$$

Referring to the discussion appearing after (5.56), if we write (5.75) as $\left(-d^2/dx^2 - \lambda\right) g(x,x',\lambda) = \delta(x-x')$, where $\lambda = -\gamma^2$, then

$$g(x,x',\lambda) = \frac{1}{2i\sqrt{\lambda}}e^{-i\sqrt{\lambda}|x-x'|} \tag{5.78}$$

with $\operatorname{Im}\sqrt{\lambda} < 0$ (consistent with $\gamma = i\sqrt{\lambda} = \alpha + i\beta$, $\alpha > 0$), indicating a branch cut in the $\lambda$-plane along the positive real axis. If we consider the integration path in Figure 5.4, Cauchy's theorem [87, Sec. 11] leads to

$$\oint_{C_\infty + C_1 + C_2 + C_\varepsilon} g(x,x',\lambda)\, d\lambda = 0. \tag{5.79}$$

From (5.66), noting that the spectrum is continuous ($\lambda = 0$ is a branch point and not a pole), we have

$$\int_{C_\infty} g(x,x',\lambda)\, d\lambda = -2\pi i\, \delta(x-x'). \tag{5.80}$$

The contribution from $C_\varepsilon$ vanishes as $\varepsilon \to 0$, and the contributions from $C_1$ and $C_2$ are obtained as in the previous case (see (5.59)). As $\varepsilon \to 0$ on $C_1$, $\lambda = re^{i4\pi^-}$, leading to $\sqrt{\lambda} = \sqrt{r}e^{i2\pi^-} = \sqrt{r}$, and on $C_2$, $\lambda = re^{i2\pi^+}$, $\sqrt{\lambda} = \sqrt{r}e^{i\pi^+} = -\sqrt{r}$, where $r = |\lambda|$. Therefore, for $x > x'$

$$\begin{aligned} \lim_{\varepsilon\to 0}\int_{C_1+C_2} g(x,x',\lambda)\, d\lambda &= \int_\infty^0 \frac{1}{2i\sqrt{r}}e^{-i\sqrt{r}(x-x')}dr + \int_0^\infty \frac{-1}{2i\sqrt{r}}e^{i\sqrt{r}(x-x')}dr \\ &= \int_0^\infty \frac{-1}{2i\sqrt{r}}e^{i\sqrt{r}(x-x')}dr + \int_0^\infty \frac{-1}{2i\sqrt{r}}e^{-i\sqrt{r}(x-x')}dr. \end{aligned}$$

With the substitution $\upsilon = \sqrt{r}$, $d\upsilon = dr/(2\sqrt{r})$, we obtain

$$\begin{aligned}\lim_{\varepsilon\to 0}\int_{C_1+C_2} g(x,x',\lambda)\,d\lambda &= \int_0^\infty ie^{i\upsilon(x-x')}d\upsilon + \int_0^\infty ie^{-i\upsilon(x-x')}d\upsilon \\ &= \int_0^\infty ie^{i\upsilon(x-x')}d\upsilon - \int_0^{-\infty} ie^{i\upsilon(x-x')}d\upsilon \\ &= \int_{-\infty}^\infty ie^{i\upsilon(x-x')}d\upsilon.\end{aligned}$$

Combining with (5.80) we have

$$-2\pi i\,\delta(x-x') + i\int_{-\infty}^\infty e^{i\upsilon(x-x')}d\upsilon = 0,$$

leading to

$$\delta(x-x') = \frac{1}{2\pi}\int_{-\infty}^\infty e^{i\upsilon\,(x-x')}d\upsilon, \tag{5.81}$$

which is the spectral representation of the delta-function with respect to the continuous spectrum $[0,\infty)$. Although this result was obtained for $x > x'$, upon interchanging $x$ and $x'$ and noting that $\delta(x-x') = \delta(x'-x)$, it is easy to see that (5.81) holds for $x < x'$ as well.

The continuous spectrum $[0,\infty)$ is represented by the branch cut integral as before. The integration over $(-\infty,0]$ arises from the nature of the complex exponential; $\upsilon \in (-\infty,0]$ is not part of the continuous spectrum of the operator. The terms $(1/\sqrt{2\pi})e^{i\upsilon x}$ are improper eigenfunctions of the operator $-d^2/dx^2$ with improper eigenvalues $\lambda = \upsilon^2$. Furthermore, for $\upsilon \in (-\infty,\infty)$ the improper eigenvalue ranges over the continuous spectrum, i.e., $\lambda = \upsilon^2 \in [0,\infty)$. The integration over the entire range $(-\infty,\infty)$ does have physical significance, though, as it represents in one dimension a continuous sum of plane-waves.

Recall that for $\mathbf{L}^2(0,\infty)$ the spectral representation of the delta-function (5.60), $(2/\pi)\int_0^\infty \sin\upsilon x' \sin\upsilon x\,d\upsilon = \delta(x-x')$, lead to the Fourier sine transform pair (5.67). Analogously, taking the inner product[25] of (5.81) with $u(x)$ leads to

$$\begin{aligned}u(x) &= \frac{1}{2\pi}\int_{-\infty}^\infty U(\upsilon)e^{i\upsilon x}\,d\upsilon, \\ U(\upsilon) &= \int_{-\infty}^\infty u(x)e^{-i\upsilon x}\,dx,\end{aligned} \tag{5.82}$$

the well known exponential Fourier transform pair, where $U(\upsilon) = \langle u, e^{i\upsilon x}\rangle$. The first relation in (5.82) expresses the expansion of $u(x)$ in the continuous

[25] The appropriate inner product is

$$\langle f,g\rangle = \int_{-\infty}^\infty f(x)\overline{g}(x)\,dx.$$

set of improper eigenfunctions $e^{ivx}$ of the operator $-d^2/dx^2$, with improper eigenvalues $v^2$.

The solution of (5.74) can be written as

$$u(x) = \int_{-\infty}^{\infty} f(x')g(x,x',\gamma)\,dx' \tag{5.83}$$

for $x \in (-\infty, \infty)$, where $u(x) \in \mathbf{L}^2(-\infty, \infty)$, or in the continuous expansion form

$$u(x) = \frac{1}{2\pi}\int_{-\infty}^{\infty} \frac{\langle f, e^{ivx}\rangle}{v^2+\gamma^2} e^{ivx}\,dv, \tag{5.84}$$

which is the analogous version of (5.68) on $(-\infty, \infty)$.

The Green's function is (5.76), or, in spectral form from (5.65),

$$g(x,x',\gamma) = \frac{1}{2\pi}\int_{-\infty}^{\infty} \frac{e^{ivx}e^{-ivx'}}{v^2+\gamma^2}\,dv \tag{5.85}$$

such that

$$g(x,x',\gamma) = \frac{1}{2\pi}\int_{-\infty}^{\infty} \frac{e^{ivx}e^{-ivx'}}{v^2+\gamma^2}\,dv = \begin{cases} \frac{1}{2\gamma}e^{-\gamma(x-x')}, & x > x', \\ \frac{1}{2\gamma}e^{\gamma(x-x')}, & x > x'. \end{cases} \tag{5.86}$$

In summary, the singular Sturm–Liouville problem in the limit-point case may exhibit a pure point spectrum (see, e.g., Sections 5.4 and 5.5), a pure continuous spectrum (e.g., the examples considered in this section), or both point and continuous spectra (see, e.g., Sections 8.1.6 and 8.1.7). It can be shown that if a pure point spectrum exists, then the eigenfunctions form a complete orthonormal set in $\mathbf{L}^2_w$ [7, p. 304]. Problems with an endpoint in the limit-circle case always result in a pure point spectrum.

### 5.2.4 Simultaneous Occurrence of Discrete and Continuous Spectra

To generalize (5.35) and (5.66) for the case of both discrete and continuous spectra, consider

$$(L - \lambda)\,u = f \tag{5.87}$$

along with the proper and improper eigenvalue problems

$$Lu_n = \lambda_n u_n,$$
$$Lu(v) = \lambda(v)u(v),$$

where $u(v) = u(x, v)$, with the orthogonality relations

$$\begin{aligned} \langle u_n, u_m \rangle &= \delta_{nm}, \\ \langle u(v), u(p) \rangle &= \delta(v-p), \\ \langle u_n, u(v) \rangle &= 0. \end{aligned} \tag{5.88}$$

We assume the expansion $u = \sum_n a_n u_n + \int_v a(v)u(v)\,dv$ which, from (5.88), leads to $a_n = \langle u, u_n \rangle$ and $a(v) = \langle u, u(v) \rangle$. We therefore obtain

$$u(x) = \sum_n \langle u, u_n \rangle\, u_n(x) + \int_v \langle u, u(v) \rangle\, u(x, v)\, dv \tag{5.89}$$

as the general expansion of any function $u \in \mathbf{L}^2_w(a, b)$. We similarly expand $Lu$ as $Lu = \sum_n \lambda_n \langle u, u_n \rangle\, u_n + \int_v \lambda(v) \langle u, u(v) \rangle\, u(v)\, dv$. To solve (5.87) we form

$$(L - \lambda)\, u = \sum_n (\lambda_n - \lambda) \langle u, u_n \rangle\, u_n + \int_v (\lambda(v) - \lambda) \langle u, u(v) \rangle\, u(v)\, dv = f$$

such that taking the inner product with $u_m$ leads to $\langle u, u_m \rangle = (1/(\lambda_m - \lambda)) \langle f, u_m \rangle$, and taking the inner product with $u(p)$ leads to $\langle u, u(p) \rangle = (1/(\lambda(p) - \lambda)) \langle f, u(p) \rangle$, or

$$u(x) = \sum_n \frac{\langle f, u_n \rangle}{\lambda_n - \lambda} u_n(x) + \int_v \frac{\langle f, u(v) \rangle}{\lambda(v) - \lambda} u(x, v)\, dv, \tag{5.90}$$

which replaces (5.27) and (5.63) in the general case.

By the same manipulations preceding (5.28) and (5.65) we obtain

$$g(x, x', \lambda) = \sum_n \frac{u_n(x)\overline{u_n}(x')}{\lambda_n - \lambda} + \int_v \frac{u(x, v)\overline{u}(x, v)}{\lambda(v) - \lambda}\, dv \tag{5.91}$$

as the bilinear series form for $g$, leading to the completeness relation

$$\begin{aligned} \frac{1}{2\pi i} \oint_C g(x, x', \lambda)\, d\lambda &= -\frac{\delta(x - x')}{w(x')} \\ &= -\sum_n u_n(x)\overline{u_n}(x') - \int_v u(x, v)\overline{u}(x', v)\, dv \end{aligned} \tag{5.92}$$

replacing (5.35) and (5.66).

## 5.3 Nonself-Adjoint Sturm–Liouville Problems

### 5.3.1 Green's Function Methods in the Nonself-Adjoint Case

Up to this point we have focused on the self-adjoint Sturm–Liouville problem, either regular or singular. In this section we briefly treat the nonself-adjoint problem, since this case occurs in many important electromagnetics applications. We assume all of the conditions, relations, and definitions (5.1)–(5.9) in Section 5.1, with the exception that we generalize the boundary conditions (5.2) somewhat as

$$\begin{aligned} B_a(u) &= \alpha_1 u(a) + \alpha_2 u'(a) = \eta_a, \\ B_b(u) &= \beta_1 u(b) + \beta_2 u'(b) = \eta_b, \end{aligned} \tag{5.93}$$

where $\alpha_{1,2}, \beta_{1,2}, \eta_{a,b} \in \mathbf{C}$. If $\eta_{a,b} = 0$, then (5.93) are said to be homogeneous; otherwise the boundary conditions are inhomogeneous. For homogeneous boundary conditions with $\alpha_{1,2}, \beta_{1,2} \in \mathbf{R}$, (5.93) reduces to (5.2), resulting in a self-adjoint problem. For the homogeneous case with $\alpha_{1,2}, \beta_{1,2} \in \mathbf{C}$, or for the inhomogeneous case, the general form of (5.93) results in a nonself-adjoint Sturm–Liouville problem.

As was previously discussed, we form the adjoint of the Sturm–Liouville operator $L$ using integration by parts twice as in (5.10), repeated below as

$$\begin{aligned} \langle Lu, v \rangle &= \int_a^b \left\{ -\frac{1}{w(x)} \frac{d}{dx} \left[ p(x) \frac{du}{dx} \right] + q(x) u(x) \right\} \overline{v}(x) w(x) \, dx \\ &= \int_a^b u(x) \left\{ -\frac{1}{w(x)} \frac{d}{dx} \left[ p(x) \frac{d\overline{v}}{dx} \right] + q(x) \overline{v}(x) \right\} w(x) \, dx \\ &\quad - \left\{ p(x) \left[ \overline{v}(x) \frac{du}{dx} - u(x) \frac{d\overline{v}}{dx} \right] \right\} \Big|_a^b \\ &= \langle u, L^* v \rangle + J(u, v)|_a^b \, . \end{aligned}$$

Regardless of boundary conditions, we have $L^* = L$, and so the operator $L$ is formally self-adjoint. Although we require $v \in \mathbf{L}_w^2(a, b)$, since $v$ is not specified we are free to choose the adjoint boundary conditions on $v$, denoted as $B_a^*(v)$ and $B_b^*(v)$, such that the conjunct $J$ vanishes. The procedure, described in Section 3.3.4, is as follows. To determine the correct adjoint boundary conditions, we apply homogeneous boundary conditions $B_a(u) = B_b(u) = 0$ in the conjunct (regardless of the value of $\eta_{a,b}$), and choose homogeneous boundary conditions on $\overline{v}$ such that the resulting conjunct vanishes. It is easy to see that if we choose the conjugate-adjoint boundary conditions as

$$\begin{aligned} \overline{B_a^*}(v) &= \alpha_1 \overline{v}(a) + \alpha_2 \overline{v}'(a) = 0, \\ \overline{B_b^*}(v) &= \beta_1 \overline{v}(b) + \beta_2 \overline{v}'(b) = 0, \end{aligned} \tag{5.94}$$

then $J(u, v)|_a^b = 0$. Since we actually need the adjoint (rather than conjugate-adjoint) boundary conditions for the investigation of the domain of $L^*$, we conjugate the above conditions, leading to

$$\begin{aligned} B_a^*(v) &= \overline{\alpha_1} v(a) + \overline{\alpha_2} v'(a) = 0, \\ B_b^*(v) &= \overline{\beta_1} v(b) + \overline{\beta_2} v'(b) = 0. \end{aligned} \tag{5.95}$$

If $\eta_{a,b} = 0$ and $\alpha_{1,2}, \beta_{1,2} \in \mathbf{R}$, then $B_{a,b}(v) = B^*_{a,b}(v)$ and with this choice $D_{L^*} = D_L$ so that the operator $L$ is self-adjoint. This is the case considered previously. If either $\eta_{a,b} \neq 0$ or $\alpha_{1,2}, \beta_{1,2} \notin \mathbf{R}$, or both, then the operator $L$ is not self-adjoint. It is this more general case that we investigate below (see also Section 3.9.3).

With

$$(L - \lambda)\, u = f \tag{5.96}$$

subject to $B_a(u)$ and $B_b(u)$ as the problem to be solved, we consider the Green's function $g(x, x', \lambda)$, defined as satisfying

$$(L - \lambda)\, g(x, x', \lambda) = \frac{\delta(x - x')}{w(x)} \tag{5.97}$$

subject to $B_a(g) = B_b(g) = 0$. For the nonself-adjoint case we also consider the adjoint Green's function, given by

$$(L - \lambda)^*\, g^*(x, x', \lambda) = \frac{\delta(x - x')}{w(x)} \tag{5.98}$$

subject to the homogeneous adjoint boundary conditions $B^*_a(g^*) = B^*_b(g^*) = 0$, which we take to be the same as those chosen for $v$. Following the integration-by-parts procedure in (5.10), we form

$$\langle (L - \lambda)\, u, g^* \rangle = \langle u, (L - \lambda)^*\, g^* \rangle + J(u, g^*)|_a^b$$

where the conjunct is $J(u, g^*) = -\{p(x)[\overline{g^*}(x, x')\, du/dx - u(x)\, d\overline{g^*}(x, x')/dx]\}$. Using (5.98) and the sifting property of the delta-function we obtain

$$\begin{aligned} u(x') = &\int_a^b f(x)\overline{g^*}(x, x', \lambda) w(x)\, dx \\ &+ \left\{ p(x) \left[ \overline{g^*}(x, x', \lambda) \frac{du}{dx} - u(x) \frac{d\overline{g^*}(x, x', \lambda)}{dx} \right] \right\} \Bigg|_{x=a}^{x=b}, \end{aligned} \tag{5.99}$$

which is in terms of the conjugate-adjoint Green's function (for multidimensional problems one uses a generalized Green's theorem as described in Section 3.9.3). From

$$\langle (L - \lambda)\, g(x, x'), g^*(x, x'') \rangle = \langle g(x, x'), (L - \lambda)^*\, g^*(x, x'') \rangle + J(g, g^*)|_a^b$$

and noting that the conjunct vanishes since $B_{a,b}(g)$ and $\overline{B^*_{a,b}}(g^*)$ are homogeneous, we have

$$\overline{g^*}(x', x'', \lambda) = g(x'', x', \lambda). \tag{5.100}$$

The procedure usually followed is that one determines $g$ and then obtains $\overline{g^*}$ from $g$ using (5.100), which is then substituted into (5.99) to complete the solution. If, rather than actually inserting the specific Green's function into (5.99), one wants to obtain a general formula for the solution, the substitution (5.100) can be made directly into the integral term in (5.99). The conjunct term, however, may require special care, although in most cases the substitution is permissible, such that (5.99) becomes

$$u(x') = \int_a^b f(x)g(x',x,\lambda)w(x)\,dx + \left\{p(x)\left[g(x',x,\lambda)\frac{du}{dx} - u(x)\frac{dg(x',x,\lambda)}{dx}\right]\right\}\Bigg|_{x=a}^{x=b}. \tag{5.101}$$

An example where (5.101) would not be applicable is an initial-value problem, where conditions are given as $u(a) = \alpha$ and $u'(a) = \beta$. If one uses (5.101), then the conjunct term at $x = b$ remains unknown, whereas the form (5.99) would be applicable because the homogeneous adjoint boundary conditions on $g^*$ would eliminate the conjunct term at $x = b$. In many cases of interest in electromagnetics, though, homogeneous conditions on $u$ (Dirichlet) or $u'$ (Neumann) are provided, such that (5.101) can be used directly, wherein it is common to change variables and determine the solution as $u(x)$ rather than $u(x')$.

### Direct Method for Real, Symmetric Operators

Since in electromagnetics one is usually concerned with real, symmetric operators (self-adjoint or not), one can solve an equation of the form (5.96) using a Green's theorem without the concept of an adjoint operator if $(L - \lambda)$ is invertible (see Section 3.9.3). As an example, for

$$\left(-\frac{d^2}{dx^2} - \lambda\right)u = f$$

we define the Green's function as $\left(-d^2/dx^2 - \lambda\right)g(x,x') = \delta(x - x')$ and use the one-dimensional form of Green's theorem,

$$\int_a^b \left[\psi_1\frac{d^2\psi_2}{dx^2} - \psi_2\frac{d^2\psi_1}{dx^2}\right]dx = \left(\psi_1\frac{d\psi_2}{dx} - \psi_2\frac{d\psi_1}{dx}\right)\Bigg|_a^b. \tag{5.102}$$

With $\psi_1 = u(x)$ and $\psi_2 = g(x,x',\lambda)$ we have

$$u(x') = \int_a^b f(x)g(x,x',\lambda)w(x)\,dx + \left[g(x,x',\lambda)\frac{du}{dx} - u(x)\frac{dg(x,x',\lambda)}{dx}\right]\Bigg|_{x=a}^{x=b}, \tag{5.103}$$

which is the one-dimensional version of (1.49).

Furthermore, when Green's theorem is applicable, symmetry of the Green's function can be obtained in a direct manner. Indeed, defining

Green's functions as

$$\left(-\frac{d^2}{dx^2}-\lambda\right)g(x,x_1')=\delta(x-x_1'),$$
$$\left(-\frac{d^2}{dx^2}-\lambda\right)g(x,x_2')=\delta(x-x_2'),$$

subject to homogeneous boundary conditions, and substituting into Green's theorem result in

$$\int_a^b\left[g(x,x_1')\frac{d^2g(x,x_2')}{dx^2}-g(x,x_2')\frac{d^2g(x,x_1')}{dx^2}\right]dx$$
$$=\left(g(x,x_1')\frac{dg(x,x_2')}{dx}-g(x,x_2')\frac{dg(x,x_1')}{dx}\right)\Bigg|_a^b$$

or

$$g(x_2',x_1')=g(x_1',x_2'),$$

which is the desired symmetry relationship. Thus, whenever a Green's theorem like (5.102) (or a multidimensional version) is utilized, symmetry of the Green's function is obtained, and (5.101) and (5.103) are equivalent. The procedure using the adjoint approach is more general, though, reducing to the Green's theorem result in most cases (especially those of interest in electromagnetics), but also being capable of handling several other problems, such as those of the initial-value type.

### 5.3.2 Spectral Methods in the Nonself-Adjoint Case

In this section we briefly treat the spectral analysis of the Sturm–Liouville operator, considering the problem

$$(L-\lambda)\,u=f, \tag{5.104}$$

where $L$ has the Sturm–Liouville form but is not necessarily self-adjoint (in fact, we can let $w,p,q$ be complex-valued in (5.4)), acting on an interval $(a,b)$ that is not necessarily bounded. In what follows one must keep in mind that a general nonself-adjoint Sturm–Liouville operator will *not* necessarily admit the spectral representations described below, although such representations will hold in a variety of nonself-adjoint cases. Nonspectral Green's function methods are, however, generally applicable in the nonself-adjoint case.

We consider the eigenvalue and the adjoint eigenvalue problems

$$Lu_n=\lambda_n u_n,$$
$$L^*u_n^*=\lambda_n^*u_n^*,$$

where, from Theorem 4.7, we see that $\lambda_n^* = \overline{\lambda}_n$ with the bi-orthogonality relationship

$$\langle u_n, u_m^* \rangle = \delta_{nm}. \tag{5.105}$$

For generality we also assume improper eigenfunctions may exist, satisfying

$$Lu(v) = \lambda(v)u(v),$$
$$L^*u^*(v) = \lambda^*(v)u^*(v),$$

where $\lambda^*(v) = \overline{\lambda}(v)$ with the orthogonality relations

$$\langle u(v), u^*(p) \rangle = \delta(v - p),$$
$$\langle u_n, u^*(v) \rangle = \langle u(v), u_n^* \rangle = 0.$$

We make the expansion $u = \sum_n a_n u_n + \int_v a(v)u(v)\, dv$, which, from (5.105) and (5.106), leads to $a_n = \langle u, u_n^* \rangle$ and $a(v) = \langle u, u^*(v) \rangle$. Therefore,

$$u = \sum_n \langle u, u_n^* \rangle\, u_n + \int_v \langle u, u^*(v) \rangle\, u(v)\, dv,$$

which replaces (5.89) in this general case. To solve (5.104) we form

$$(L - \lambda)\, u = \sum_n (\lambda_n - \lambda) \langle u, u_n^* \rangle\, u_n + \int_v (\lambda(v) - \lambda) \langle u, u^*(v) \rangle\, u(v)\, dv = f.$$

Taking the inner product with $u_m^*$ leads to $\langle u, u_m^* \rangle = (1/(\lambda_m - \lambda)) \langle f, u_m^* \rangle$; similarly, taking the inner product with $u^*(p)$ leads to $\langle u, u^*(p) \rangle = (1/(\lambda(p) - \lambda))\langle f, u^*(p) \rangle$, or

$$u(x) = \sum_n \frac{\langle f, u_n^* \rangle}{\lambda_n - \lambda} u_n(x) + \int_v \frac{\langle f, u^*(v) \rangle}{\lambda(v) - \lambda} u(x, v)\, dv, \tag{5.106}$$

which replaces (5.90) for the case of a nonself-adjoint Sturm–Liouville operator.

By the same manipulations preceding (5.28), for the nonself-adjoint case we obtain

$$g(x, x', \lambda) = \sum_n \frac{u_n(x)\overline{u_n^*}(x')}{\lambda_n - \lambda} + \int_v \frac{u(x, v)\overline{u^*}(x, v)}{\lambda(v) - \lambda}\, dv \tag{5.107}$$

as the generalized bilinear series form for $g$.

If the validity of (5.106) can be established, a completeness relation can be found for the nonself-adjoint Sturm–Liouville problem. Rather than (5.92), the completeness relation has the form

$$\frac{1}{2\pi i} \oint_C g(x, x', \lambda)\, d\lambda = -\frac{\delta(x - x')}{w(x')}, \tag{5.108}$$

$$-\frac{\delta(x - x')}{w(x')} = -\sum_n u_n(x)\overline{u_n^*}(x') - \int_v u(x, v)\overline{u^*}(x', v)\, dv. \tag{5.109}$$

In general, a nonself-adjoint operator will not admit such a completeness relation, although one can make the following observations:

- (5.108) will hold when the root system[26] of the operator forms a basis for the underlying Hilbert space ($H = \mathbf{L}^2_w(a,b)$ here), assuming the Green's function satisfies Jordan's lemma. In particular, poles of the Green's function need not be simple.
- The equality in (5.109) will hold when the root system of the operator coincides with the system of ordinary eigenfunctions (i.e., all generalized eigenfunctions are of rank 1, such that there are no associated eigenfunctions).
- If the root system is complete yet does not coincide with the system of ordinary eigenfunctions, then, for instance, the Green's function may have poles of multiplicity higher than 1. In this case, generalized eigenfunctions of rank higher than 1 may be implicated [2, pp. 40–41]; in such cases the expansion (5.109) must be augmented with associated eigenfunctions (see Example 5.14).

In the case of a nonself-adjoint operator, completeness of the root system is difficult to ascertain (see [11]), although we do have, for example, Theorem 4.47. However, for nonself-adjoint boundary value problems, a fairly general theorem may be stated [2, pp. 89–90].

**Theorem 5.3.** *Let $L$ be the Sturm–Liouville operator (5.4) (or the differential operator in (5.7)) acting on the finite interval $[a,b]$ and having continuous complex-valued coefficients $w(x), p(x)$, and $q(x)$. Assume boundary conditions*

$$\begin{aligned} B_a(u) &= a_1 u'(a) + b_1 u'(b) + a_0 u(a) + b_0 u(b) = 0, \\ B_b(u) &= c_1 u'(a) + d_1 u'(b) + c_0 u(a) + d_0 u(b) = 0, \end{aligned} \tag{5.110}$$

*where $a_i, b_i \in \mathbf{C}$ for $i = 0, 1$, and suppose that any of the three conditions*

$$\begin{aligned} &i.\ \ a_1 d_1 - b_1 c_1 \neq 0, \\ &ii.\ \ a_1 d_1 - b_1 c_1 = 0, \qquad |a_1| + |b_1| > 0, \\ &\qquad 2(a_1 c_0 + b_1 d_0) \neq \pm (b_1 c_0 + a_1 d_0) \neq 0, \\ &iii.\ \ a_1 = b_1 = c_1 = d_1 = 0, \qquad a_0 d_0 - b_0 c_0 \neq 0 \end{aligned} \tag{5.111}$$

*hold. If the adjoint operator $L^*$ exists, and if all eigenvalues of $L$ have multiplicity 1, any function $f \in \mathbf{L}^2_w(a,b)$ can be expanded in the norm-convergent series*

$$f(x) = \sum_{n=1}^{\infty} \langle f, u_n^* \rangle\, u_n(x),$$

[26] Regarded in a general sense to include improper eigenvalues associated with the continuous spectrum.

*where $u_n$ and $u_n^*$ are eigenfunctions corresponding to the eigenvalues $\lambda_n$ and $\lambda_n^* = \overline{\lambda}_n$ of the operators $L$ and $L^*$, respectively. Similarly, the Green's function may be expanded as*

$$g(x, x', \lambda) = \sum_{n=1}^{\infty} \frac{u_n(x)\overline{u_n^*}(x')}{\lambda_n - \lambda}.$$

This theorem is extremely important, as it establishes the validity of eigenfunction expansions for most scalar differential operators encountered in electromagnetics.[27] For example, the above theorem establishes the existence of an ordinary eigenbasis for a nonself-adjoint Sturm–Liouville operator with any of the familiar boundary conditions

$$\begin{aligned} &i. \quad u'(a) = u'(b) = 0, \\ &ii. \quad u'(a) = u(b) = 0, \\ &iii. \quad u(a) = u(b) = 0. \end{aligned}$$

Actually, the coefficients in (5.7) need not be continuous; piecewise continuity is sufficient. This is fortunate, since in many electromagnetics applications the coefficients are associated with material parameters, which are often piecewise continuous. In practice, the stipulation of unit multiplicity of the eigenvalues is the real limitation. In typical time-harmonic electromagnetics applications, points of higher multiplicity will occur for a set of discrete frequencies (associated with nontrivial modal degeneracies), at which points the expansion may fail.

This theorem provides sufficient, but not necessary, conditions for an ordinary eigenfunction expansion associated with the Sturm–Liouville operator $L$ and adjoint operator $L^*$ to exist in the nonself-adjoint case. The theorem is applicable, for instance, in proving that the eigenfunctions of a parallel-plate waveguide filled with an inhomogeneous, lossy medium form an eigenbasis in $\mathbf{L}^2[a, b]$, except at certain parameter values where eigenvalue degeneracies exist (see Section 8.1, and, in particular, Section 8.1.8).

**Example 5.12.**

Consider the nonself-adjoint eigenvalue problem on $\mathbf{L}^2(0, 1)$

$$\begin{aligned} u'' + \lambda u &= 0, \\ u(0) &= u(1), \\ u'(0) &= -u'(1). \end{aligned}$$

Note that the conditions (5.111) are not satisfied, and, in fact, an eigenbasis does not exist. Indeed, any number $\lambda$ is an eigenvalue, with corresponding

[27] The theorem is also applicable to unbounded region (singular) problems taken in a limiting sense.

eigenfunction $u_n(x) = A\cos\sqrt{\lambda}x + B\sin\sqrt{\lambda}x$. Obviously, the eigenfunctions are not complete in $\mathbf{L}^2(0,1)$.

The next example illustrates a nonself-adjoint problem that satisfies the conditions of Theorem 5.3 and hence admits a representation in terms of ordinary and adjoint eigenfunctions.

**Example 5.13.**

Consider the nonself-adjoint eigenvalue problem on $\mathbf{L}^2(0,1)$

$$\begin{aligned} u'' + \lambda u &= 0, \\ u'(0) &= \alpha u(0), \\ u'(1) &= \beta u(1), \end{aligned}$$

where $\alpha, \beta \in \mathbf{C}$. The adjoint problem is

$$\begin{aligned} u^{*\prime\prime} + \overline{\lambda} u^* &= 0, \\ u^{*\prime}(0) &= \overline{\alpha} u^*(0), \\ u^{*\prime}(1) &= \overline{\beta} u^*(1). \end{aligned}$$

For the special case $\alpha = \beta \neq 0$, the eigenvalues are given by roots of

$$\sin\sqrt{\lambda_n}\left(\sqrt{\lambda_n} + \frac{\alpha^2}{\sqrt{\lambda_n}}\right) = 0,$$

leading to two types of eigenvalues. First, from $\sin\sqrt{\lambda_n} = 0$ we obtain $\lambda_n = (n\pi)^2$, $n = 1, 2, \ldots$ ($n = 0$ cannot occur unless $\alpha = 0$, and $n < 0$ cannot occur by picking an appropriate branch of the square root). Second, the condition $\left(\sqrt{\lambda_n} + \alpha^2/\sqrt{\lambda_n}\right) = 0$ results in the single eigenvalue $\lambda_\alpha = -\alpha^2$. Orthonormal eigenvectors are then

$$\begin{aligned} u_n(x) &= \left\{\sqrt{\frac{2n^2\pi^2}{n^2\pi^2 + \alpha^2}}\left(\cos n\pi x + \frac{\alpha}{n\pi}\sin n\pi x\right) = \overline{u_n^*}(x), \quad n = 1, 2, \ldots, \right. \\ u_\alpha(x) &= \sqrt{\frac{2\alpha}{e^{2\alpha} - 1}} e^{\alpha x} = \overline{u_\alpha^*}(x), \end{aligned}$$

forming a bi-orthogonal set in $\mathbf{L}^2(0,1)$, i.e.,

$$\langle u_n, u_m^* \rangle = \int_0^1 u_n(x)\,\overline{u_m^*}(x)\,dx = \delta_{nm},$$

which also holds if $n$ or $m$ is replaced with $\alpha$.

The conditions of Theorem 5.3 are satisfied, and therefore any function $f \in \mathbf{L}^2(0,1)$ can be expanded in the eigenbasis associated with the bi-orthonormal set $\{u_\alpha, u_n, u^*_\alpha, u^*_n\}$, with the completeness relation on $\mathbf{L}^2(0,1)$ being

$$\delta(x-x') = \sqrt{\frac{2\alpha}{e^{2\alpha}-1}} e^{\alpha x} \sqrt{\frac{2\alpha}{e^{2\alpha}-1}} e^{\alpha x'} + \sum_{n=1}^{\infty} \left\{ \sqrt{\frac{2n^2\pi^2}{n^2\pi^2+\alpha^2}} \left(\cos n\pi x + \frac{\alpha}{n\pi}\sin n\pi x\right) \sqrt{\frac{2n^2\pi^2}{n^2\pi^2+\alpha^2}} \left(\cos n\pi x' + \frac{\alpha}{n\pi}\sin n\pi x'\right) \right\}.$$

From the material in Section 8.1.6, it can be seen that this problem models the vertical part of a parallel-plate impedance waveguide, with impedance planes located at $x = 0, 1$. In this case the eigenfunction $e^{\alpha x}$ represents a surface wave.

The next example illustrates a case where the conditions of Theorem 5.3 are not satisfied, yet the root system of the operator is a basis[28] for $\mathbf{L}^2(0,1)$.

**Example 5.14.**

On $\mathbf{L}^2(0,1)$ consider the nonself–adjoint eigenvalue problem [12]

$$\begin{aligned} u'' + \lambda u &= 0, \\ u(0) &= 0, \\ u'(0) &= u'(1). \end{aligned} \tag{5.112}$$

Eigenvalues are found to be $\lambda_n = (2\pi n)^2$, $n = 0, 1, 2, \ldots$, with ordinary eigenfunctions

$$\begin{aligned} u_0 &= x, & n &= 0, \\ u_n &= \sin 2\pi n x, & n &= 1, 2, \ldots, \end{aligned}$$

[28] In the series of papers [21]–[22], the second-derivative operator

$$D_L \equiv \left\{u : u \in \mathbf{C}^1[0,1], u' \text{ absolutely continuous}, u'' \in \mathbf{L}^2(0,1), B_a(u) = B_b(u) = 0\right\}$$
$$Lu \equiv -u''$$

is considered, where $B_a, B_b$ are defined by (5.110). All such differential operators belong to 13 possible cases, depending on the specified boundary conditions $B_{a,b}$. It is shown that for eight cases the root system is complete in $\mathbf{L}^2(0,1)$. Five of these cases correspond to simple eigenvalues (these cases include the typical Dirichlet and Neumann boundary conditions), and the other three admit higher-multiplicity eigenvalues, leading to associated eigenfunctions. Example 5.14 falls into this latter category. See also [23].

i.e., $u_n$ satisfy

$$\left(-\frac{d^2}{dx^2}-(2\pi n)^2\right)u_n=0,$$
$$u_n(0)=0,\quad u_n'(0)=u_n'(1).$$

Furthermore, for $n>0$ associated eigenfunctions exist (rank= 2) that are found to be

$$u_{n,1}=\frac{x\cos 2n\pi x}{4\pi n},$$

which satisfy

$$\left(-\frac{d^2}{dx^2}-(2\pi n)^2\right)^2 u_{n,1}=0,$$
$$\left(-\frac{d^2}{dx^2}-(2\pi n)^2\right)u_{n,1}=u_n=\sin 2n\pi x\neq 0,$$
$$u_{n,1}(0)=0,\qquad u_{n,1}'(0)=u_{n,1}'(1).$$

In this case the conditions of Theorem 5.3 do not hold and, in fact, an expansion in ordinary eigenfunctions is not possible, although the root system $\{u_0,u_n,u_{n,1}\}$ is a basis for $\mathbf{L}^2(0,1)$. From Theorem 4.7, a bi-orthogonal set is formed from the generalized eigenfunctions of the adjoint problem[29]

$$u_n^{*\prime\prime}+\lambda u_n^*=0,$$
$$u_n^{*\prime}(1)=0,$$
$$u_n^*(1)=u_n^*(0),$$

leading to

$$u_0^*=2,$$
$$u_n^*=16\pi n\cos 2n\pi x,$$
$$u_{n,1}^*=4(1-x)\sin 2n\pi x.$$

Bi-orthonormality relations are[30]

$$\langle u_0,u_0^*\rangle=\langle x,2\rangle=1,$$
$$\langle u_n,u_{m,1}^*\rangle=\langle \sin 2\pi nx,4(1-x)\sin 2m\pi x\rangle=\delta_{nm},$$
$$\langle u_{n,1},u_m^*\rangle=\left\langle\frac{x\cos 2n\pi x}{4\pi n},16\pi m\cos 2m\pi x\right\rangle=\delta_{nm},$$

---

[29] More generally, the bi-orthogonal set comes from the properties of minimal systems (see, e.g., [20, p. 225]).

[30] When associated eigenfunctions are present, orthonormality relations among generalized eigenvectors become more complicated than when only ordinary eigenfunctions are present. Some details can be found in [10, p. 131].

with all other possible combinations resulting in $\langle \cdot, \cdot \rangle = 0$, where $\langle g, h \rangle = \int_0^1 g(x)\, h(x)\, dx$. Any $f \in \mathbf{L}^2(0,1)$ can be expanded in a norm-convergent series of root functions as

$$\begin{aligned} f(x) &= \langle f, u_0^* \rangle u_0 + \sum_{n=1}^{\infty} \langle f, u_{n,1}^* \rangle u_n + \langle f, u_n^* \rangle u_{n,1} \\ &= \langle f, 2 \rangle x + \sum_{n=1}^{\infty} \langle f, 4(1-x) \sin 2n\pi \rangle \sin 2\pi n x \\ &\qquad + \langle f, 16\pi n \cos 2n\pi x \rangle \frac{x \cos 2n\pi x}{4\pi n}. \end{aligned} \tag{5.113}$$

Alternatively, the completeness relation (5.108) leads to the root system starting from the Green's function. In this case the Green's function for (5.112), which satisfies $-\left(d^2/dx^2 + \lambda\right) g(x, x', \lambda) = \delta(x - x')$ subject to the given boundary conditions, is found to be

$$g(x, x', \lambda) = \begin{cases} \frac{\cos\left(\sqrt{\lambda}(1-x')\right) \sin\left(\sqrt{\lambda} x\right)}{\sqrt{\lambda}\left(\cos\sqrt{\lambda} - 1\right)}, & x < x', \\ \frac{\sin\left(\sqrt{\lambda} x'\right) \cos\left(\sqrt{\lambda} x\right)}{\sqrt{\lambda}} + \frac{\sin\sqrt{\lambda} \sin\left(\sqrt{\lambda} x'\right) + \cos\left(\sqrt{\lambda} x'\right)}{\sqrt{\lambda}\left(\cos\sqrt{\lambda} - 1\right)} \sin\left(\sqrt{\lambda} x\right), & x > x'. \end{cases}$$

Considering, for example, the case $x < x'$, it can be shown that[31]

$$\oint_{\Gamma_\lambda} \frac{1}{\sqrt{\lambda}} \frac{\cos\left(\sqrt{\lambda}(1-x')\right) \sin\left(\sqrt{\lambda} x\right)}{\left(\cos\sqrt{\lambda} - 1\right)} d\lambda = \oint_{\Gamma_\gamma} \frac{\cos\left(\gamma(1-x')\right) \sin(\gamma x)}{(\cos\gamma - 1)} d\gamma, \tag{5.114}$$

where $\Gamma_\lambda$ and $\Gamma_\gamma$ are circular contours of infinite radius in the $\lambda$- and $\gamma$-planes, respectively, taken in the usual counterclockwise direction, and $\gamma^2 = \lambda$. Using the identity[32]

$$\frac{1}{\cos\gamma - 1} = -2 \sum_{n=-\infty}^{\infty} \frac{1}{(\gamma - 2n\pi)^2},$$

---

[31] To obtain this result we use

$$\oint_\Gamma f(\lambda)\, d\lambda = \int_{-\pi}^{\pi} f\left(\lambda = re^{i\theta}\right) ire^{i\theta} d\theta,$$

which is valid if $\Gamma$ is a circular contour in the $\lambda$-plane, taken in the counterclockwise direction. Performing the change of variables $\gamma^2 = \lambda$ with $\gamma = r_\gamma e^{i\phi}$ leads to the result (5.114).

[32] One can also define $z = e^{i\sqrt{\lambda}}$ such that

$$\frac{1}{\cos\gamma - 1} = \frac{1}{\frac{1}{2}(z + \overline{z}) - 1} = \frac{1}{\frac{1}{2}\left(z + \frac{1}{z}\right) - 1} = \frac{2z}{(z-1)^2}$$

and proceed with analysis in the $z$-plane, with second-order poles located at $z = 1$.

we see that the Green's function has an infinite number of second-order poles at $\gamma = 2n\pi$. Using the formula for the residue of a function $f$ at an $m$th-order pole $\gamma = \gamma_0$,

$$\mathrm{Res}_m(f) = \frac{1}{(m-1)!}\frac{d^{m-1}}{d\gamma^{m-1}}(\gamma-\gamma_0)^m f(\gamma)\bigg|_{\gamma=\gamma_0}, \tag{5.115}$$

we obtain

$$\begin{aligned}\delta(x-x') &= \frac{-1}{2\pi i}\oint_{\Gamma_\gamma}\frac{\cos(\gamma(1-x'))\sin(\gamma x)}{(\cos\gamma - 1)}\,d\gamma \\ &= u_0u_0^* + \sum_{n=1}^{\infty} u_{n,1}u_n^* + u_nu_{n,1}^* \\ &= 2x + \sum_{n=1}^{\infty}\left(\frac{x\cos 2n\pi x}{4\pi n}\right)16\pi n\cos 2n\pi x' \\ &\qquad + \sin(2\pi nx)\left(4(1-x')\sin 2n\pi x'\right),\end{aligned}$$

consistent with (5.113). Note that the product of eigenfunctions has the form $u_{n,1}u_n^* + u_nu_{n,1}^*$, which is a general result when second-rank generalized eigenfunctions exist [2, p. 41]. See [10, pp. 225–227] for a related example.

Finally, to complete this example we consider the solution of the operator equation

$$(A - \lambda I) f = q, \tag{5.116}$$

where $A : \mathbf{L}^2(0,1) \to \mathbf{L}^2(0,1)$ is the second derivative operator $A = -d^2/dx^2$, and where $\lambda \neq \lambda_n = (2n\pi)^2$. Making the expansions

$$\begin{aligned} f &= \langle f, u_0^*\rangle u_0 + \sum_{n=1}^{\infty}\langle f, u_{n,1}^*\rangle u_n + \langle f, u_n^*\rangle u_{n,1}, \\ g &= \langle g, u_0^*\rangle u_0 + \sum_{n=1}^{\infty}\langle g, u_{n,1}^*\rangle u_n + \langle g, u_n^*\rangle u_{n,1},\end{aligned}$$

and noting

$$Af = \sum_{n=1}^{\infty}\langle f, u_{n,1}^*\rangle \lambda_n u_n + \langle f, u_n^*\rangle(\lambda_n u_{n,1} + u_n),$$

then, upon exploiting bi-orthonormality, the solution of (5.116) is found to be

$$f = \frac{\langle q, u_0^* \rangle}{-\lambda} u_0 + \sum_{n=1}^{\infty} \left( \frac{\langle q, u_{n,1}^* \rangle}{\lambda_n - \lambda} - \frac{\langle q, u_n^* \rangle}{(\lambda_n - \lambda)^2} \right) u_n + \frac{\langle q, u_n^* \rangle}{\lambda_n - \lambda} u_{n,1}$$

$$= \frac{\langle q, 2 \rangle}{-\lambda} x + \sum_{n=1}^{\infty} \left( \frac{\langle q, 4(1-x) \sin 2n\pi x \rangle}{\lambda_n - \lambda} - \frac{\langle q, 16n\pi \cos 2n\pi x \rangle}{(\lambda_n - \lambda)^2} \right) \sin 2n\pi x$$
$$+ \frac{\langle q, 16n\pi \cos 2n\pi x \rangle}{\lambda_n - \lambda} \frac{x \cos 2n\pi x}{4n\pi},$$

which can be compared with (4.54). If we set $q = \delta(x - x')$, then we obtain the spectral representation of the Green's function as

$$g(x, x', \lambda) = \frac{2x}{-\lambda} + \sum_{n=1}^{\infty} \left( \frac{4(1-x') \sin 2n\pi x'}{\lambda_n - \lambda} - \frac{16n\pi \cos 2n\pi x'}{(\lambda_n - \lambda)^2} \right) \sin 2n\pi x$$
$$+ \frac{16n\pi \cos 2n\pi x'}{\lambda_n - \lambda} \frac{x \cos 2n\pi x}{4n\pi},$$

which can be compared with (4.59).

## 5.4 Special Functions Associated with Singular Sturm–Liouville Problems

In this section we discuss some problem formulations that lead to singular Sturm–Liouville equations and to classical special functions. In electromagnetics these equations often arise from separation-of-variables solutions to scalar Laplace or Helmholtz equations in cylindrical and spherical coordinates. Accordingly, we will proceed to demonstrate the separation-of-variables technique and examine the resulting Sturm–Liouville equations. The resulting special functions and orthogonal polynomials are often useful in electromagnetics. Further details concerning the properties of these functions can be found in any text on mathematical physics. For applications to electromagnetic problems see, e.g., [13, Ch. 3].

We consider the homogeneous equation

$$\left(\nabla^2 + k^2\right) u = 0 \tag{5.117}$$

in a domain $\Omega$ with smooth boundary surface $\Gamma$. We assume boundary conditions of the form

$$\alpha u + \beta \frac{\partial u}{\partial n} = \eta \tag{5.118}$$

on $\Gamma$, where $\alpha, \beta, \eta \in \mathbf{C}$. Of particular interest are the following three cases.

1. $k^2 = 0$:

In this case we have Laplace's equation, where generally $\Omega$ is a finite region.[33]

2. $k^2$ is a free (spectral) parameter and $\eta = 0$:

   In this case we have an eigenvalue problem for eigenvalue $k^2$. Assuming that $\Omega$ is finite, we have ordinary (proper) eigenvalues, such as the modes in a closed cavity.

3. $k^2$ is a specified nonzero number:

   In this third case we have a Helmholtz problem corresponding to a specified frequency and medium ($k^2 = \omega^2 \mu \varepsilon$).

The above may be called the "interior" problem, since we are concerned with solutions inside a domain bounded by surface $\Gamma$, where perhaps $\Gamma$ recedes to infinity. In contrast, the "exterior" or "scattering" problem would correspond to the case where the boundary surface $\Gamma$ is finite, enclosing a finite region, and we are interested in the solution outside $\Gamma$, although this topic is not discussed here.

### 5.4.1 Cylindrical Coordinate Problems

Consider the homogeneous scalar Helmholtz equation in cylindrical coordinates

$$\frac{1}{\rho}\frac{\partial}{\partial \rho}\left(\rho \frac{\partial u}{\partial \rho}\right) + \frac{1}{\rho^2}\frac{\partial^2 u}{\partial \phi^2} + \frac{\partial^2 u}{\partial z^2} + k^2 u = 0 \tag{5.119}$$

on a domain $\Omega$ with boundary surface $\Gamma$. We simultaneously consider the case where $k^2$ may be given or a free (eigenvalue) parameter, and we assume boundary conditions of the form (5.118).

We seek a separation-of-variables solution for (5.119) of the form

$$u = R(\rho)\Phi(\phi)Z(z), \tag{5.120}$$

which, by substituting into (5.119), leads to three second-order ordinary differential equations of the Sturm–Liouville type for the separated functions $R, \Phi$, and $Z$, where we assume that the boundary condition is also separable.

To separate (5.119) we substitute (5.120) into (5.119) and divide by $u$, leading to

$$\frac{1}{\rho R}\frac{d}{d\rho}\left(\rho \frac{dR}{d\rho}\right) + \frac{1}{\rho^2 \Phi}\frac{d^2 \Phi}{d\phi^2} + \frac{1}{Z}\frac{d^2 Z}{d z^2} + k^2 = 0, \tag{5.121}$$

where we see that the third term is independent of $\rho$ and $\phi$. This term must also be independent of $z$ if the other terms, which are clearly $z$-independent,

[33] Note that if the corresponding solution $u \neq 0$ exists, it is not, however, an eigenfunction of $-\nabla^2$ corresponding to a zero eigenvalue, unless $\eta = 0$.

are to combine with the third term and sum to zero. Therefore, the third term is independent of $z$ and must be equal to a constant, i.e.,

$$\frac{1}{Z}\frac{d^2 Z}{d z^2} = -k_z^2, \tag{5.122}$$

where $k_z$ is a constant. Substituting this result into (5.121) and multiplying by $\rho^2$ lead to

$$\frac{\rho}{R}\frac{d}{d\rho}\left(\rho\frac{dR}{d\rho}\right) + \frac{1}{\Phi}\frac{d^2\Phi}{d\phi^2} + \left(k^2 - k_z^2\right)\rho^2 = 0, \tag{5.123}$$

where we note that the second term is independent of $\rho$ and $z$ and may, at most, be dependent on $\phi$. Since the other terms are independent of $\phi$ and must combine with the second term and sum to zero, then the second term must be independent of $\phi$ and therefore must be equal to a constant, i.e.,

$$\frac{1}{\Phi}\frac{d^2\Phi}{d\phi^2} = -k_\phi^2, \tag{5.124}$$

where $k_\phi$ is a constant. With (5.124) then (5.123) becomes

$$\frac{\rho}{R}\frac{d}{d\rho}\left(\rho\frac{dR}{d\rho}\right) - k_\phi^2 + \left(k^2 - k_z^2\right)\rho^2 = 0, \tag{5.125}$$

which is clearly dependent only on $\rho$. The three equations (5.122), (5.124), and (5.125) can be solved individually for $Z$, $\Phi$, and $R$, respectively, subject to separated boundary conditions. For convenience we denote

$$k_\rho^2 \equiv k^2 - k_z^2 \tag{5.126}$$

such that the relevant equations to be solved are

$$\left(\rho\frac{d}{d\rho}\left(\rho\frac{d}{d\rho}\right) + \left(k_\rho^2\rho^2 - k_\phi^2\right)\right) R(\rho) = 0, \tag{5.127}$$

$$\left(\frac{d^2}{d z^2} + k_z^2\right) Z(z) = 0, \tag{5.128}$$

$$\left(\frac{d^2}{d\phi^2} + k_\phi^2\right) \Phi(\phi) = 0. \tag{5.129}$$

Of course, with each equation are associated separated boundary conditions, which will be considered as needed.

### Harmonic Equations for Cylindrical Structures

The harmonic equation (5.128) has the form of a homogeneous Sturm–Liouville problem

$$(L - \lambda)\, Z = 0,$$

where $L = -d^2/dz^2$ and $\lambda = k_z^2$. If $k_z \neq 0$, the solution is obviously

$$Z(z) = Ae^{ik_z z} + Be^{-ik_z z} \tag{5.130}$$

or an equivalent form involving $\sin k_z z$ and $\cos k_z z$, with $A$ and $B$ determined by the specified boundary conditions. If $B_a(Z), B_b(Z) = 0$ for a finite interval $z \in [a, b]$, with $B_{a,b}$ corresponding to the form (5.2), then the harmonic equation (5.128) forms a regular homogeneous Sturm–Liouville eigenvalue problem and a corresponding basis of eigenfunctions. If the interval is infinite or semi-infinite we have a singular Sturm–Liouville problem in the limit-point case, as discussed in Section 5.2, leading to a continuum of improper eigenfunctions.

The harmonic equation (5.129) is obviously very similar to (5.128), forming a Sturm–Liouville problem with $L = -d^2/d\phi^2$ and $\lambda = k_\phi^2$, the solution of which is

$$\Phi(\phi) = Ae^{ik_\phi \phi} + Be^{-ik_\phi \phi} \tag{5.131}$$

if $k_\phi \neq 0$; otherwise,

$$\Phi(\phi) = A\phi + B.$$

One difference compared to (5.130) is that in the above equation the range of the variable $\phi$ is essentially finite; although $\phi$ can take on any value, a physical cylindrical structure can be covered by any $2\pi$ range of $\phi$. Often we have a condition like $\Phi(\phi_0) = \Phi(\phi_0 + 2\pi)$, such that $k_\phi = m$, $m$ an integer. This type of condition merely indicates that the field quantities should not be changed by a rotation of $2\pi$ radians.

Moreover, if $\Phi$ satisfies periodic boundary conditions of the form

$$\Phi(0) = \Phi(2\pi), \qquad \Phi'(0) = \Phi'(2\pi),$$

then we obtain eigenvalues $\lambda_m = m^2$ with corresponding orthonormal eigenfunctions $(1/\sqrt{2\pi})e^{\pm imx}$, which obviously form a basis for $L^2(0, 2\pi)$. Other homogeneous conditions of the form (5.2) lead to other eigenbases. Alternatively, both (5.130) and (5.131) may be subject to nonhomogeneous conditions.

### Radial (Bessel's) Equation for Cylindrical Structures

The radial equation (5.127) is known as *Bessel's equation.* For simplicity we set $k_\phi = \nu$ ($\nu$ is not necessarily integer-valued) such that (5.127) is known as Bessel's equation of *order* $\nu$ and *parameter* $k_\rho^2$, rewritten as

$$-\frac{1}{\rho}\frac{d}{d\rho}\left(\rho\frac{dR}{d\rho}\right) + \left(\frac{\nu^2}{\rho^2} - k_\rho^2\right) R = 0, \tag{5.132}$$

which has the form of a Sturm–Liouville equation $(L - \lambda) R = 0$, with

$$L \equiv -\frac{1}{\rho}\frac{d}{d\rho}\left(\rho\frac{d}{d\rho}\right) + \frac{\nu^2}{\rho^2}$$

and $\lambda = k_\rho^2$ (recall that $k_\rho$ is coupled to $k_z$ by (5.126)).

### Singular Points and Solutions of Bessel's Equation

In the following we consider several separate situations that arise frequently in practice, where we consider $a$ to be a finite, positive number. To classify the type of singular endpoint encountered we investigate the case $\lambda(=k_\rho^2) = 0$ and use Theorem 5.2 (Weyl).

1. Bessel's equation of order 0 and parameter $\lambda$ on $[0, a]$:

   We see that the endpoint $\rho = 0$ is a singular point since, for instance, $w = p = 0$ at the endpoint (see (5.4)), violating a condition described in Section 5.1. For $\lambda(=k_\rho^2) = 0$ we have the two independent solutions $R_1(\rho) = 1$ and $R_2(\rho) = \ln \rho$. Therefore, $R_{1,2} \in \mathbf{L}^2_\rho(0, a)$ and we have the limit-circle case at $\rho = 0$.

2. Bessel's equation of order $\nu$ $(0 < \nu < \infty)$ and parameter $\lambda$ on $[0, a]$:

   For $\lambda(=k_\rho^2) = 0$ we have the two independent solutions $R_1(\rho) = \rho^\nu$ and $R_2(\rho) = \rho^{-\nu}$. Therefore, $R_1 \in \mathbf{L}^2_\rho(0, a)$ for any $\nu$, $R_2 \in \mathbf{L}^2_\rho(0, a)$ for $0 < \nu < 1$, and $R_2 \notin \mathbf{L}^2_\rho(0, a)$ for $\nu \geq 1$. Accordingly, if $\nu \geq 1$ we have the limit-point case at $\rho = 0$; otherwise, at $\rho = 0$ we have the limit-circle case.

3. Bessel's equation of order 0 and parameter $\lambda$ on $[a, \infty)$:

   Because $R_1(\rho) = 1$ does not belong to $\mathbf{L}^2_\rho(a, \infty)$ (nor does $R_2(\rho) = \ln \rho$), then the singular point $\rho = \infty$ is in the limit-point case.

4. Bessel's equation of order $\nu$ $(0 < \nu < \infty)$ and parameter $\lambda$ on $[a, \infty)$:

   For $\lambda\,(=k_\rho^2) = 0$ the solution $R_1(\rho) = \rho^\nu$ clearly does not belong to $\mathbf{L}^2_\rho(a, \infty)$, and so we have the limit-point case at $\infty$.

The two independent solutions of (5.132) for $\lambda \neq 0$ and $\nu$ not an integer are

$$J_\nu(\sqrt{\lambda}\rho), \; J_{-\nu}(\sqrt{\lambda}\rho).$$

The function $J_\nu$ is known as *Bessel's function of the first kind of order* $\nu$. For $\lambda \neq 0$ and $\nu$ an integer, the two independent solutions are [9, Sec. 20.3]

$$J_\nu(\sqrt{\lambda}\rho), \; Y_\nu(\sqrt{\lambda}\rho),$$

where $Y_\nu$ is known as *Bessel's function of the second kind*[34] *of order* $\nu$. Because $Y_\nu$ is a linear combination of $J_\nu$ and $J_{-\nu}$, $J_\nu$ and $Y_\nu$ also form a fundamental pair of solutions for noninteger $\nu$. Finally, the independent functions defined by

$$H_\nu^{(1,2)}(\sqrt{\lambda}\rho) = J_\nu(\sqrt{\lambda}\rho) \pm iY_\nu(\sqrt{\lambda}\rho)$$

[34] This function is also known as the *Neumann function.*

are known as *Hankel functions of the first* (+) *and second* (−) *kinds*, of order $\nu$.

Since Bessel functions often arise from cylindrical problems on $\rho \in [0, a]$ or $\rho \in [0, \infty)$, it is worthwhile to briefly consider their asymptotic forms for large and small arguments [14]. Letting $x = \sqrt{\lambda}\rho$, we have[35]

$$J_0(x) \underset{x \to 0}{\to} 1,$$

$$J_\nu(x) \underset{x \to 0}{\to} \frac{1}{\Gamma(\nu+1)}\left(\frac{x}{2}\right)^\nu, \qquad \nu \neq -1, -2, -3, \ldots,$$

$$Y_\nu(x) \underset{x \to 0}{\to} \begin{cases} -\frac{\Gamma(\nu)}{\pi}\left(\frac{2}{x}\right)^\nu & \mathrm{Re}(\nu) > 0, \\ \frac{2}{\pi}\left[\ln\left(\frac{x}{2}\right) + \gamma + \cdots\right], & \nu = 0, \end{cases}$$

where $\gamma$ is Euler's constant ($\gamma \simeq 0.5772$) and

$$J_\nu(x) \underset{|x| \to \infty}{\to} \sqrt{\frac{2}{\pi x}} \cos\left(x - \frac{\pi}{4} - \frac{\nu\pi}{2}\right), \qquad |\arg x| < \pi,$$

$$Y_\nu(x) \underset{|x| \to \infty}{\to} \sqrt{\frac{2}{\pi x}} \sin\left(x - \frac{\pi}{4} - \frac{\nu\pi}{2}\right), \qquad |\arg x| < \pi,$$

$$H_\nu^{(1,2)}(x) \underset{|x| \to \infty}{\to} \sqrt{\frac{2}{\pi x}} e^{\pm i(x - \pi/4 - \nu\pi/2)}, \qquad \begin{array}{c} -\pi < \arg x < 2\pi, \\ -2\pi < \arg x < \pi. \end{array}$$

It is seen then that $J_\nu(\sqrt{\lambda}\rho)$ is regular at $\rho = 0$, whereas $Y_\nu(\sqrt{\lambda}\rho)$ is singular at the origin. Furthermore, $H_\nu^{(1,2)}(\sqrt{\lambda}\rho)$ has the form of an incoming (1) or outgoing (2) cylindrical wave for large $\rho$, assuming an $e^{i\omega t}$ formulation.

To summarize the possible solutions of (5.132) with $\lambda\,(= k_\rho^2)$, we take

$$R(\rho) = AJ_\nu\left(\sqrt{\lambda}\rho\right) + BY_\nu\left(\sqrt{\lambda}\rho\right) \tag{5.133}$$

as the general solution if we want to represent oscillatory solutions, and

$$R(\rho) = AH_\nu^{(1)}\left(\sqrt{\lambda}\rho\right) + BH_\nu^{(2)}\left(\sqrt{\lambda}\rho\right) \tag{5.134}$$

as the general solution if we want to represent traveling wave solutions. If $\lambda = \nu = 0$, we have

$$R(\rho) = A + B\ln\rho \tag{5.135}$$

and if $\lambda = 0$, $\nu \neq 0$,

$$R(\rho) = A\rho^\nu + B\rho^{-\nu}. \tag{5.136}$$

Typical boundary conditions would include physically motivated conditions such as requiring the solution to be finite at $\rho = 0$, leading to $B = 0$

[35] In the following formulas $x$ and $\nu$ may be complex, and $\nu$ is arbitrary. Note that for $\nu$ being an integer, $\Gamma(\nu+1) = \nu!$.

in (5.133). We also may be interested in the range $\rho \in [a, b]$, in which case the second-kind Bessel function may be retained. Another condition, frequently encountered is that the solution should represent an outward-traveling wave, such that $A = 0$ in (5.134).

### Eigenfunctions of Bessel's Equation for a Finite Region

We first consider (5.132) as an eigenvalue problem on $\rho \in [0, a]$ where $a$ is finite, such that it is natural to consider (5.133) as a possible eigenfunction of (5.132). We see that the point $\rho = 0$ is in the limit-circle or limit-point case depending on the value of $\nu$. If it is in the limit-point case ($\nu \geq 1$), we simply require the solution to be in $\mathbf{L}^2_\rho(0, a)$ as the condition at $\rho = 0$, such that $B = 0$. If it is in the limit-circle case ($0 < \nu < 1$), then (in general) the proper procedure to obtain a boundary condition is somewhat complicated (see, e.g., [15, pp. 467–472]). However, based on physical reasons we expect that the eigenfunction should be finite, again leading to $B = 0$. One could also argue that, considering an asymptotic approximation of (5.132) for small $\rho$, one might require $\frac{d}{d\rho}\left(\rho \frac{dR}{d\rho}\right)$ to be finite, which leads to the same conclusion.

At the regular endpoint $\rho = a$ we may require the eigenfunction to vanish at $\rho = a$ (Dirichlet condition), leading to the Dirichlet eigenfunction

$$R_n(\rho) = c_n J_\nu\left(\sqrt{\lambda_n}\rho\right), \tag{5.137}$$

where the eigenvalue $\lambda_n$ is determined from

$$J_\nu\left(\sqrt{\lambda_n} a\right) = 0. \tag{5.138}$$

Alternatively, we may require that the derivative of the eigenfunction vanish at $\rho = a$ (Neumann condition), leading to the eigenfunction

$$R_n(\rho) = c'_n J_\nu\left(\sqrt{\lambda'_n}\rho\right), \tag{5.139}$$

where $\lambda'_n$ satisfies

$$\frac{d}{d\rho} J_\nu\left(\sqrt{\lambda'_n}\rho\right)\Big|_{\rho=a} = 0.$$

For fixed $\nu \geq -1$ the eigenfunctions (5.137) and (5.139) are orthogonal,[36] i.e.,

$$\left\langle J_\nu\left(\sqrt{\lambda_n}\rho\right), J_\nu\left(\sqrt{\lambda_m}\rho\right)\right\rangle = 0, \qquad n \neq m,$$

[36] If $\nu \geq -1$, all zeros of $J_\nu(x)$ are real, and if $\nu \geq 0$, all zeros of $J'_\nu(x)$ are real. If $\nu$ is real, $J_\nu(x)$ and $J'_\nu(x)$ have an infinite number of real zeros, all simple except possibly $z = 0$ [14].

and form a complete orthogonal basis for $\mathbf{L}^2_\rho(0,a)$. For $\nu < -1$, eigenvalues may be complex and the resulting Sturm–Liouville operator is not self-adjoint; therefore, completeness of the eigenset for $\nu < -1$ is generally not established.

The normalization constant for the Dirichlet case is

$$c_n^2 = \int_0^a J_\nu^2\left(\sqrt{\lambda_n}\rho\right)\rho d\rho = \frac{a^2}{2} J_{\nu+1}^2\left(\sqrt{\lambda_n}a\right)$$

and for the Neumann case,

$$c_n'^2 = \int_0^a J_\nu^2\left(\sqrt{\lambda_n'}\rho\right)\rho d\rho = \frac{a^2}{2}\left(1 - \frac{\nu^2}{\sqrt{\lambda_n'}a}\right) J_\nu^2\left(\sqrt{\lambda_n'}a\right).$$

Both sets $\left\{(1/c_n)J_\nu(\sqrt{\lambda_n}\rho)\right\}$ and $\left\{(1/c_n')J_\nu(\sqrt{\lambda_n'}\rho)\right\}$ form orthonormal bases for $\mathbf{L}^2_\rho(0,a)$. In the Dirichlet case we obtain the familiar *Fourier–Bessel expansion* or *Hankel expansion*[37]

$$f(\rho) = \sum_{n=1}^\infty \frac{1}{c_n^2}\left\langle f, J_\nu\left(\sqrt{\lambda_n}\rho\right)\right\rangle J_\nu\left(\sqrt{\lambda_n}\rho\right) \tag{5.140}$$

for any $f \in \mathbf{L}^2_\rho(0,a)$. Since the normalized Bessel functions form an orthonormal eigenbasis for any fixed $\nu \geq -1$, the expansion (5.140) really represents a family of expansions for various values of $\nu$. The completeness relation is

$$\frac{\delta(\rho-\rho')}{\rho'} = \sum_{n=1}^\infty \frac{1}{c_n^2} J_\nu\left(\sqrt{\lambda_n}\rho\right) J_\nu\left(\sqrt{\lambda_n}\rho'\right)$$

on $\mathbf{L}^2_\rho(0,a)$. Analogous results occur for the Neumann eigenfunctions.

**Example 5.15.**

As an example, the equation

$$(L-\lambda)\,u(\rho) = f(\rho)$$

on $[0,a]$, with

$$L \equiv -\frac{1}{\rho}\frac{d}{d\rho}\left(\rho\frac{d}{d\rho}\right) + \frac{\nu^2}{\rho^2}$$

and Dirichlet boundary conditions on $u$, has the solution (see (5.27))

$$u(\rho) = \sum_{n=1}^\infty \frac{1}{c_n^2}\frac{\left\langle f, J_\nu\left(\sqrt{\lambda_n}\rho\right)\right\rangle}{\lambda_n - \lambda} J_\nu\left(\sqrt{\lambda_n}\rho\right) \tag{5.141}$$

[37]The correct inner product is $\langle f, g\rangle = \int_0^a f(\rho)\,g(\rho)\,\rho d\rho$ since we assume $\nu \geq -1$ such that the eigenvalues will be real, and so, consequently, $J_\nu$ is real. For $J_\nu'$ we require $\nu \geq 0$.

with $\lambda_n$ determined implicitly by (5.138). The Green's function for this problem is defined by

$$(L-\lambda)\, g(\rho,\rho',\lambda) = \frac{\delta(\rho-\rho')}{\rho} \tag{5.142}$$

with $g\,(0) = g\,(a) = 0$, leading to the eigenfunction expansion

$$g(\rho,\rho',\lambda) = \sum_{n=1}^{\infty} \frac{1}{c_n^2} \frac{J_\nu\left(\sqrt{\lambda_n}\rho\right) J_\nu\left(\sqrt{\lambda_n}\rho'\right)}{\lambda_n - \lambda}. \tag{5.143}$$

Note, however, that the eigenvalues are not available in closed form.

### Eigenfunctions of Bessel's Equation for a Semi-Infinite Region

For $\rho \in [0,\infty)$ the point at infinity is in the limit-point case. Eigenfunctions will no longer exist, yet improper eigenfunctions exist, as considered in Section 5.2, leading to a continuous expansion. The method described in Section 5.2 can be used to obtain a completeness relation for the continuum,[38]

$$\frac{\delta(\rho-\rho')}{\rho'} = \int_0^\infty J_\nu\,(\upsilon\rho)\, J_\nu\,(\upsilon\rho')\, \upsilon d\upsilon$$

on $\mathbf{L}^2_\rho\,(0,\infty)$ for $\nu > -1/2$. The Fourier–Bessel expansion in this case is therefore

$$f\,(\rho) = \int_0^\infty \langle f, J_\nu\,(\upsilon\rho)\rangle\, J_\nu\,(\upsilon\rho)\, \upsilon d\upsilon$$

for any $f \in \mathbf{L}^2_\rho\,(0,\infty)$. The inner product is the same as in the finite case, except that the integration is over $(0,\infty)$.

**Example 5.16.**

Consider the equation

$$(L-\lambda)\, u\,(\rho) = f\,(\rho)$$

[38]Alternatively, as in the limiting process leading to the result in (5.45), observe that as $a \to \infty$ the eigenvalues are obtained from $J_\nu\left(\sqrt{\lambda_n}a\right) = 0 \to \sqrt{2/(\pi\sqrt{\lambda_n}a)}\cos\left(\sqrt{\lambda_n}a - \pi/4 - \nu\pi/2\right) = 0$, leading to $\sqrt{\lambda_n}a = (n+3/4)\,\pi + \nu\pi/2$. Inserting these values into the normalization constant $c_n^2 = (a^2/2)J_{\nu+1}\left(\sqrt{\lambda_n}a\right)$ and using the large-argument form for the Bessel function lead to $c_n^2 \to (a/\pi)\frac{1}{\upsilon_n}$ where $\upsilon_n = \sqrt{\lambda_n}$. As $a \to \infty$, the eigenvalues coalesce into a continuous spectrum, $\upsilon_n \to \upsilon$, and $\Delta\upsilon_n = \upsilon_{n+1} - \upsilon_n \simeq \pi/a \to d\upsilon$, so that

$$\frac{\delta(\rho-\rho')}{\rho} = \sum_{n=1}^{\infty} \upsilon_n J_\nu\,(\upsilon_n\rho)\, J_\nu\left(\upsilon_n\rho'\right)\Delta\upsilon_n \to \int_0^\infty \upsilon J_\nu\,(\upsilon\rho)\, J_\nu\left(\upsilon\rho'\right) d\upsilon.$$

on $[0, \infty)$ with

$$L \equiv -\frac{1}{\rho}\frac{d}{d\rho}\left(\rho\frac{d}{d\rho}\right) + \frac{\nu^2}{\rho^2}$$

subject to the conditions that $u$ must be finite at $\rho = 0$ and that $u \in \mathbf{L}^2_\rho(0, \infty)$. This equation has the continuous expansion solution

$$u(\rho) = \int_0^\infty \frac{\langle f, J_\nu(\upsilon\rho)\rangle}{\lambda(\upsilon) - \lambda} J_\nu(\upsilon\rho)\,\upsilon\,d\upsilon, \tag{5.144}$$

where it is recognized that the improper eigenfunctions are

$$u(\rho, \upsilon) = \sqrt{\upsilon} J_\nu(\upsilon\rho).$$

With $Lu(\rho, \upsilon) = \lambda(\upsilon)u(\rho, \upsilon)$, we identify $\lambda(\upsilon) = \upsilon^2$. The procedure leading to (5.144) is equivalent to applying the *Fourier–Bessel (Hankel) transform*

$$\begin{aligned} u(\rho) &= \int_0^\infty U(\upsilon) J_\nu(\upsilon\rho)\,\upsilon\,d\upsilon, \\ U(\upsilon) &= \int_0^\infty u(\rho) J_\nu(\upsilon\rho)\,\rho\,d\rho \end{aligned} \tag{5.145}$$

to $(L - \lambda)\,u(\rho) = f(\rho)$.

The Green's function defined by (5.142), finite at $\rho = 0$ and in $\mathbf{L}^2_\rho(0, \infty)$, is obviously

$$g(\rho, \rho', \lambda) = \int_0^\infty \frac{J_\nu(\upsilon\rho)\,J_\nu(\upsilon\rho')}{\upsilon^2 - \lambda}\upsilon\,d\upsilon. \tag{5.146}$$

### General Solution of the Cylindrical Coordinate Problem

Having examined characteristics of the eigenfunctions and solutions of the separated equations (5.127)–(5.129), we now return to the product solution (5.120). We obtain, from (5.130), (5.131), and (5.133)–(5.136), the solution of (5.119) as

$$u(\rho, \phi, \theta) = R(\rho)\Phi(\phi)Z(z). \tag{5.147}$$

If (5.119) is an eigenvalue problem with homogeneous boundary conditions (and possibly certain fitness conditions), then (5.147) is an eigenfunction with corresponding eigenvalue $k^2 = k_\rho^2 + k_z^2$ (for an example, see p. 591).

If (5.119) represents a general boundary value problem, then, in order to match the specified boundary conditions, we often take the solution as a linear combination of product terms. This solution has the form of a discrete sum over the angular variable index $m$ and as either a discrete or continuous sum over $k_z$ or $k_\rho$ (but not both since $k_z$ and $k_\rho$ are connected via (5.126)), i.e.,

$$u(\rho,\phi,\theta)=\sum_m\left(\begin{array}{c}\sum_{k_z}\\ \int_{k_z}\end{array}\right)u_{m,k_z,k_\rho}$$

or

$$u(\rho,\phi,\theta)=\sum_m\left(\begin{array}{c}\sum_{k_\rho}\\ \int_{k_\rho}\end{array}\right)u_{m,k_z,k_\rho}.$$

Terms in the sum are a mixture of eigenfunctions (in those variables corresponding to coordinates having homogeneous or periodic conditions) and general solutions of the corresponding separated equations.

### 5.4.2 Spherical Coordinate Problems

The homogeneous scalar Helmholtz equation (5.117) in spherical coordinates is

$$\frac{1}{r^2}\frac{\partial}{\partial r}\left(r^2\frac{\partial u}{\partial r}\right)+\frac{1}{r^2\sin\theta}\frac{\partial}{\partial\theta}\left(\sin\theta\frac{\partial u}{\partial\theta}\right)+\frac{1}{r^2\sin^2\theta}\frac{\partial^2 u}{\partial\phi^2}+k^2u=0, \quad (5.148)$$

where again we assume a domain $\Omega$ with boundary surface $\Gamma$ and the existence of appropriately separable boundary conditions.

We seek a separation-of-variables solution for (5.148) of the form

$$u=R(r)\Phi(\phi)\Theta(\theta), \quad (5.149)$$

which, by substituting into (5.148), leads to three second-order ordinary Sturm–Liouville differential equations for the separated functions $R$, $\Theta$, and $\Phi$.

With (5.149) substituted into (5.148), dividing by $u$ and multiplying by $r^2\sin^2\theta$ lead to

$$\frac{\sin^2\theta}{R}\frac{\partial}{\partial r}\left(r^2\frac{\partial R}{\partial r}\right)+\frac{\sin\theta}{\Theta}\frac{\partial}{\partial\theta}\left(\sin\theta\frac{\partial\Theta}{\partial\theta}\right)+\frac{1}{\Phi}\frac{\partial^2\Phi}{\partial\phi^2}+k^2r^2\sin^2\theta=0, \quad (5.150)$$

where we see that the third term is independent of $r$ and $\theta$. It must also be independent of $\phi$ if the other terms, which are clearly $\phi$-independent, are to combine with the third term and sum to zero. Therefore, the third term is independent of $\phi$ and must be equal to a constant, i.e.,

$$\frac{1}{\Phi}\frac{d^2\Phi}{d\phi^2}=-k_\phi^2, \quad (5.151)$$

where $k_\phi$ is a constant. Substituting into (5.150) and dividing by $\sin^2\theta$ lead to

$$\frac{1}{R}\frac{\partial}{\partial r}\left(r^2\frac{\partial R}{\partial r}\right)+\frac{1}{\Theta\sin\theta}\frac{\partial}{\partial\theta}\left(\sin\theta\frac{\partial\Theta}{\partial\theta}\right)-\frac{k_\phi^2}{\sin^2\theta}+k^2r^2=0, \quad (5.152)$$

where we note that the second and third terms are independent of $r$ and $\phi$ and may, at most, be dependent on $\theta$. Since the first term is independent of $\theta$ and must combine with the second and third terms to sum to zero, then the second and third terms, taken together, must be independent of $\theta$ and be equal to a constant. Therefore,

$$\frac{1}{\Theta \sin\theta}\frac{d}{d\theta}\left(\sin\theta\frac{\partial\,\Theta}{\partial\,\theta}\right) - \frac{k_\phi^2}{\sin^2\theta} = -k_\theta^2, \tag{5.153}$$

where $k_\theta$ is a constant. With (5.153), (5.152) becomes

$$\frac{1}{R}\frac{d}{dr}\left(r^2\frac{d\,R}{d\,r}\right) - k_\theta^2 + k^2 r^2 = 0, \tag{5.154}$$

which is clearly dependent only on $r$. The three equations (5.151), (5.153), and (5.154) can be solved individually for $\Phi$, $\Theta$, and $R$, respectively, subject to separated boundary conditions. In summary, we have

$$\frac{d}{d\,r}\left(r^2\frac{d\,R}{d\,r}\right) + \left(k^2 r^2 - k_\theta^2\right) R = 0, \tag{5.155}$$

$$\frac{d}{d\theta}\left(\sin\theta\frac{d\,\Theta}{d\,\theta}\right) + \left(k_\theta^2 \sin\theta - \frac{k_\phi^2}{\sin\theta}\right)\Theta = 0, \tag{5.156}$$

$$\frac{d^2\,\Phi}{d\,\phi^2} + k_\phi^2\Phi = 0. \tag{5.157}$$

### Harmonic Equation for Spherical Structures

The harmonic equation (5.157) has the form of a homogeneous Sturm–Liouville problem

$$(L - \lambda)\,\Phi = 0,$$

where $L = -d^2/d\phi^2$ and $\lambda = k_\phi^2$. The solution is obviously

$$\Phi(\phi) = Ae^{ik_\phi\phi} + Be^{-ik_\phi\phi} \tag{5.158}$$

if $k_\phi \neq 0$; otherwise,

$$\Phi\left(\varphi\right) = A\phi + B.$$

As discussed regarding the azimuthal equation for cylindrical problems, if $\Phi(\phi_0) = \Phi(\phi_0 + 2\pi)$, we find that $k_\phi = m$, where $m$ is an integer.

### Angular (Legendre) Equation for Spherical Structures

The angular equation in $\theta$, (5.156), involves the Sturm–Liouville operator (5.4), with $p = \sin\theta$. If $\theta$ is defined from 0 to $\pi$, then $p = 0$ at the endpoints, resulting in a singular equation. With the change of variables

$x = \cos\theta$, such that the interval $\theta \in [0, \pi]$ is transformed to $x \in [-1, 1]$, (5.156) becomes

$$\frac{d}{dx}\left(\left(1 - x^2\right)\frac{d\Theta}{dx}\right) + \left(k_\theta^2 - \frac{k_\phi^2}{1 - x^2}\right)\Theta = 0, \tag{5.159}$$

which is singular at the endpoints $x = \pm 1$. It turns out that (5.159) has solutions that are bounded over $[-1, 1]$, in particular, at $x = \pm 1$, if and only if $\lambda = k_\theta^2 = n(n+1)$ where $n \geq 0$ is an integer. Therefore, in the following we exclusively assume $k_\theta^2 = n(n+1) = \lambda_n$ and consider two cases.[39]

## Legendre's Equation

First, for $k_\phi^2 = 0$ (no azimuthal dependence), (5.159) becomes

$$\frac{d}{dx}\left(\left(1 - x^2\right)\frac{d\Theta}{dx}\right) + n(n+1)\Theta = 0, \tag{5.160}$$

which is known as *Legendre's equation*, having the Sturm–Liouville form (5.4) with $w = 1$, $p = 1 - x^2$, $q = 0$, and $\lambda_n = n(n+1)$. The singular endpoints $x = \pm 1$ are in the limit-circle case, which can be seen from checking the solutions at $\lambda = 0$. The two independent solutions of (5.160) are $P_n(x)$ and $Q_n(x)$, known as *Legendre's polynomials of the first and second kind*, respectively [9, Sec. 31.3] (however, $Q_n$ are not actually polynomials), such that a general solution of (5.156) with $k_\phi^2 = 0$ is

$$\Theta(\theta) = AP_n(\cos\theta) + BQ_n(\cos\theta). \tag{5.161}$$

The first several polynomials $P_n$ are

$$P_0(x) = 1 \qquad P_1(x) = x, \qquad P_2(x) = \frac{1}{2}\left(3x^2 - 1\right),$$
$$P_3(x) = \frac{1}{2}\left(5x^2 - 3x\right),$$

which can be determined from the Rodrigues' formula,

$$P_n(x) = \frac{1}{2^n n!}\frac{d^n}{dx^n}\left(x^2 - 1\right)^n.$$

The recurrence formula

$$(n+1)P_{n+1} = (2n+1)xP_n - nP_{n-1}$$

is also useful in this regard. Note that $|P_n(\cos\theta)| \leq 1$ on $[0, \pi]$.

[39] In some problems the endpoints $\theta = 0, \pi$ are not implicated, such that noninteger values of $n$ may be considered.

The second-kind Legendre polynomials are unbounded at $x = \pm 1$; for instance, $Q_0(x) = (1/2)\log[(1+x)/(1-x)]$. As with the Bessel function $Y_\nu$ considered in the previous section, the condition that an eigenfunction (or field value) be finite for $x \in [-1, 1]$ would remove the second-kind function from consideration,[40] such that the eigenfunctions of (5.160) are the $P_n$ functions with eigenvalues $\lambda_n = n(n+1)$, which form an orthogonal basis for $\mathbf{L}^2(-1, 1)$. It can be shown that

$$\int_{-1}^{1} P_n(x) P_m(x)\, dx = \frac{2}{2n+1}\delta_{nm}$$

such that if $c_n^2 = 2/(2n+1)$, then $\{(1/c_n)P_n(x)\}$ is an orthonormal eigenbasis for $\mathbf{L}^2(-1, 1)$. This leads to the *Fourier–Legendre* expansion formula[41]

$$f(x) = \sum_{n=0}^{\infty} \frac{1}{c_n^2} \langle f, P_n \rangle P_n(x),$$

for any $f \in \mathbf{L}^2(-1, 1)$, and the completeness relation

$$\delta(x - x') = \sum_{n=0}^{\infty} \frac{1}{c_n^2} P_n(x) P_n(x').$$

Legendre polynomials will be encountered in problems with azimuthal symmetry, i.e., $k_\phi = m = 0$, such that there is no field variation in the $\phi$-coordinate.

### Associated Legendre's Equation

For $k_\phi^2 = m^2 \neq 0$,[42] (5.159) is known as the *associated Legendre's equation*, with solutions being *associated Legendre functions* $P_n^m(x)$ and $Q_n^m(x)$, which obviously reduce to the Legendre polynomials when $m = 0$. Therefore, a general solution of (5.156) with $k_\phi^2 = m^2$ is

$$\Theta(\theta) = AP_n^m(\cos\theta) + BQ_n^m(\cos\theta). \tag{5.162}$$

As with the Legendre polynomials, for $P_n^m$ to be bounded we need $k_\theta^2 = n(n+1)$, where $n \geq 0$ is an integer. Furthermore, $Q_n^m$ are not bounded at $x = \pm 1$, and so the appropriate eigenfunctions are $P_n^m(x)$. These eigenfunctions can be obtained from the functions $P_n$ as

$$P_n^m(x) = (-1)^m \left(1 - x^2\right)^{m/2} \frac{d^m}{dx^m} P_n(x)$$

---

[40] The second-kind functions may be needed if the physical problem does not involve the polar axis.

[41] The appropriate inner product is $\langle f, g \rangle = \int_{-1}^{1} f(x)\, g(x)\, dx$ or, in the spherical variable $\theta$, $\langle f, g \rangle = \int_0^\pi f(\cos\theta)\, g(\cos\theta) \sin\theta d\theta$.

[42] Noninteger values of $m$ are sometimes necessary, but this case is not considered here.

such that $P_n^m$ will exist only if $n \geq m$. The eigenvalues associated with $P_n^m$ are the same as for $P_n$, i.e., $n(n+1)$ is an eigenvalue corresponding to an eigenfunction $P_n^m$ if $n \geq m$; otherwise eigenvalues and eigenfunctions do not exist.

For every fixed $m$ the eigenfunctions $P_n^m$ form an orthogonal basis of $\mathbf{L}^2(-1,1)$, and

$$\int_{-1}^{1} P_n^m(x) P_{n'}^m(x)\,dx = \frac{2}{2n+1}\frac{(n+m)!}{(n-m)!}\delta_{nn'}.$$

Therefore, by letting $c_{n,m}^2 = (2/(2n+1))[(n+m)!/(n-m)!]$, the set $\{(1/c_{n,m})P_n^m(x)\}$ is an orthonormal eigenbasis for $L^2(-1,1)$, leading to a family of expansions

$$f(x) = \sum_{n=0}^{\infty} \frac{1}{c_{n,m}^2} \langle f, P_n^m \rangle P_n^m(x)$$

for any $f \in \mathbf{L}^2(-1,1)$ and fixed $m$, and the completeness relation is

$$\delta(x-x') = \sum_{n=0}^{\infty} \frac{1}{c_{n,m}^2} P_n^m(x) P_n^m(x').$$

A useful recurrence formula is

$$(m-n-1)P_{n+1}^m = -(2n+1)xP_n^m + (n+m)P_{n-1}^m.$$

### Spherical Harmonics

Because the eigenfunction set $\{(1/\sqrt{2\pi})e^{\pm im\phi}\}$ forms a basis for $\mathbf{L}^2(0,2\pi)$ in the $\phi$-coordinate, and the set $\{(1/c_{n,m})P_n^m(\cos\theta)\}$ forms a basis for $\mathbf{L}^2(0,\pi)$ in the $\theta$-coordinate for each $m$, then by Theorem 2.16 the product

$$Y_{n,m}(\theta,\phi) \equiv \frac{1}{c_{n,m}\sqrt{2\pi}} e^{im\phi} P_n^m(\cos\theta) \tag{5.163}$$

for $m = \{\cdots, -2, -1, 0, 1, 2, \dots\}$ and $n \geq m$ forms an orthogonal basis for the unit sphere.[43] The functions $Y_{n,m}(\theta,\phi)$ are called *spherical harmonics* and have the orthonormal property

$$\langle Y_{n,m}, Y_{n',m'} \rangle = \int_0^{2\pi}\int_0^{\pi} Y_{n,m}(\theta,\phi)\overline{Y}_{n',m'}(\theta,\phi)\sin\theta\,d\theta\,d\phi = \delta_{nn'}\delta_{mm'}. \tag{5.164}$$

[43] For negative indices,

$$P_n^{-m}(x) = (-1)^m [(n-m)!/(n+m)!] P_n^m(x),$$

such that $Y_{n,-m}(\theta,\phi) = (-1)^m \overline{Y}_{n,m}(\theta,\phi)$.

Alternately, one can view the spherical harmonics as arising from an eigenvalue problem

$$-\left(\frac{1}{\sin\theta}\frac{\partial}{\partial\theta}\left(\sin\theta\frac{\partial}{\partial\theta}\right)+\frac{1}{\sin^2\theta}\frac{\partial^2}{\partial\phi^2}\right)Y_{n,m}(\theta,\phi)=n(n+1)Y_{n,m}(\theta,\phi).$$

Any function $f(\theta,\phi)$ square-integrable on the unit sphere can be expanded as

$$f(\theta,\phi)=\sum_{n=0}^{\infty}\sum_{m=-n}^{n}\langle f,Y_{n,m}\rangle Y_{n,m}(\theta,\phi)$$

with the completeness relationship

$$\delta(\phi-\phi')\,\delta(\cos\theta-\cos\theta')=\sum_{n=0}^{\infty}\sum_{m=-n}^{n}Y_{n,m}(\theta,\phi)\overline{Y}_{n,m}(\theta',\phi').$$

### Radial (Bessel's) Equation for Spherical Structures

Considering now the radial equation (5.155) with $k_\theta^2=n(n+1)$, we obtain

$$\frac{d}{dr}\left(r^2\frac{dR}{dr}\right)+\left[k^2r^2-n(n+1)\right]R=0, \qquad (5.165)$$

which involves the Sturm–Liouville operator (5.4) with $p=w=r^2$, $q=n(n+1)/r^2$, and $\lambda=k^2$. The solution of (5.165) yields the spherical Bessel functions

$$j_n(kr),\quad y_n(kr),\quad h_n^{(1)}(kr),\quad h_n^{(2)}(kr),$$

which are related to the ordinary (cylindrical) Bessel functions as[44]

$$b_n(kr)=\sqrt{\frac{\pi}{2kr}}B_{n+1/2}(kr), \qquad (5.166)$$

where $b$ and $B$ represent the spherical and ordinary (cylindrical) Bessel functions of the same kind, respectively. Thus, the general solution of (5.165) is

$$R(r)=Aj_n(kr)+By_n(kr) \qquad (5.167)$$

or

$$R(r)=Ah_n^{(1)}(kr)+Bh_n^{(2)}(kr), \qquad (5.168)$$

where for $k=0$ the solution is

$$R(r)=Ar^n+Br^{-n-1}. \qquad (5.169)$$

[44]With a change of variables, (5.165) can be converted to (5.132) with noninteger order.

From (5.166) one can see that the qualitative properties of the spherical Bessel functions follow those of the cylindrical Bessel functions. In particular, $j_n$ and $y_n$ represent standing waves, and $h_n^{(1,2)}$ represent inward-(1) and outward-(2) traveling spherical waves. Explicitly, we have

$$j_0(kr) = \frac{\sin kr}{kr} \qquad y_0(kr) = -\frac{\cos kr}{kr},$$
$$h_0^{(1)} = \frac{e^{ikr}}{ikr} \qquad h_0^{(2)} = -\frac{e^{-ikr}}{ikr},$$

and asymptotically,[45] for $x \ll 1, n$,

$$j_0(x) \underset{x\to 0}{\to} 1,$$
$$j_n(x) \underset{x\to 0}{\to} \frac{x^n}{(2n+1)!!},$$
$$y_n(x) \underset{x\to 0}{\to} -\frac{(2n-1)!!}{x^{n+1}},$$

and for $x \gg n(n+1)/2$,

$$j_n(x) \underset{x\to\infty}{\to} \frac{1}{x}\sin\left(x - \frac{n\pi}{2}\right),$$
$$y_n(x) \underset{x\to\infty}{\to} -\frac{1}{x}\cos\left(x - \frac{n\pi}{2}\right),$$
$$h_n^{(1)}(x) \underset{x\to\infty}{\to} -i\frac{e^{i(x-n\pi/2)}}{x},$$
$$h_n^{(2)}(x) \underset{x\to\infty}{\to} i\frac{e^{-i(x-n\pi/2)}}{x},$$

where $(2n+1)!! \equiv (2n+1)(2n-1)(2n-3)\cdots(5)(3)(1)$. At the origin $j_n$ is bounded and $y_n$ is singular, as in the cylindrical case, and a useful recurrence formula is

$$b_{n+1}(x) = \frac{2n+1}{x} b_n(x) - b_{n-1}(x).$$

Many of the properties and applications of spherical Bessel functions follow from the cylindrical case. For instance, if we consider (5.165) as an eigenvalue problem on $r \in [0, a]$, with perhaps $a \to \infty$, we arrive at the same conclusions as those obtained for the cylindrical Bessel functions.

**Example 5.17.**

As an example, on $[0, a]$ with $a$ finite and the Dirichlet condition $R(a) = 0$, we find the eigenfunctions of

$$L = -\frac{1}{r^2}\frac{d}{dr}\left[r^2 \frac{d}{dr}\right] + \frac{n(n+1)}{r^2}$$

[45] In these expressions $x$ may be complex, but we assume $n$ is a nonnegative integer.

to be

$$R_p(r) = A_n j_n\left(\sqrt{\lambda_p}\rho\right) \tag{5.170}$$

with the Dirichlet eigenvalue $\lambda_p$ (because $m$ and $n$ are used above we index the eigenvalues by $p$) determined from[46]

$$j_n\left(\sqrt{\lambda_p}a\right) = 0. \tag{5.171}$$

From the properties of the cylindrical Bessel functions we see that for fixed eigenvalue index $p$ the eigenfunctions are orthogonal,

$$\left\langle j_n\left(\sqrt{\lambda_p}\rho\right), j_{n'}\left(\sqrt{\lambda_p}\rho\right)\right\rangle = 0$$

if $n \neq n'$, and form a complete orthogonal basis of $\mathbf{L}^2_{r^2}(0,a)$.[47] If we denote the normalization constant as

$$c_p^2 = \int_0^a j_n^2\left(\sqrt{\lambda_p}r\right) r^2 dr = \frac{a^3}{2} j_{n+1}^2\left(\sqrt{\lambda_p}a\right),$$

the set $\left\{(1/c_p)\, j_n\left(\sqrt{\lambda_p}r\right)\right\}$ forms an orthonormal basis for $\mathbf{L}^2_{r^2}(0,a)$, and we obtain the spherical *Fourier–Bessel* expansion[48]

$$f(r) = \sum_{p=1}^{\infty} \frac{1}{c_p^2}\left\langle f, j_n\left(\sqrt{\lambda_p}r\right)\right\rangle j_n\left(\sqrt{\lambda_p}r\right)$$

for any $f \in \mathbf{L}^2_{r^2}(0,a)$. We therefore have the completeness relation

$$\frac{\delta(r-r')}{r'^2} = \sum_{p=1}^{\infty} \frac{1}{c_p^2} j_n\left(\sqrt{\lambda_p}r\right) j_n\left(\sqrt{\lambda_p}r'\right),$$

and, for $r \in [0,\infty)$,

$$\frac{\delta(r-r')}{r'^2} = \frac{2}{\pi}\int_0^{\infty} j_n(vr)\, j_n(vr')\, v^2\, dv.$$

By Theorem 2.16, the product of an eigenbasis in the radial coordinate and an eigenbasis for the unit sphere leads to a three-dimensional basis for $\mathbf{L}^2(0,2\pi) \times \mathbf{L}^2(0,\pi) \times \mathbf{L}^2_{r^2}(0,a)$ in spherical coordinates. For example, with $a = \infty$, a useful expansion for the static Green's function is [13, p. 102]

$$\frac{1}{4\pi|\mathbf{r}-\mathbf{r}'|} = \sum_{n=0}^{\infty}\sum_{m=-n}^{n} \frac{1}{2n+1}\frac{r_<^n}{r_>^{n+1}} Y_{n,m}(\theta,\phi)\, \overline{Y}_{n,m}(\theta',\phi')$$

---

[46]The zeros of $j_n(z)$ follow from the zeros of $J_{n+1/2}(z)$.

[47]Due to (5.166) and the requirement $\nu \geq -1$ for zeros of $J_\nu(z)$ to be real, we see that for $n > -1$ the zeros of $j_n(z)$ will be real.

[48]The correct inner product is $\langle f, g\rangle = \int_0^a f(r)\, g(r)\, r^2 dr$.

and for the dynamic Green's function [13, p. 742],

$$\frac{e^{-ik|\mathbf{r}-\mathbf{r}'|}}{4\pi|\mathbf{r}-\mathbf{r}'|} = (-ik)\sum_{n=0}^{\infty} j_n(kr_<)\, h_n^{(2)}(kr_>) \sum_{m=-n}^{n} Y_{n,m}(\theta,\phi)\,\overline{Y}_{n,m}(\theta',\phi'), \tag{5.172}$$

where $r_<$ ($r_>$) is the smaller (larger) of $|\mathbf{r}|$ and $|\mathbf{r}'|$.

Returning to the solution of (5.148), from (5.158), (5.162), and (5.167)–(5.169) we obtain the solution as

$$u(\rho,\phi,\theta) = R(\rho)\Theta(\theta)\,\Phi(\phi). \tag{5.173}$$

If (5.148) represents an eigenvalue problem, (5.173) is an eigenfunction with eigenvalue $k^2$. In the event that (5.148) represents a boundary value problem, we often take the solution to be a linear combination of product terms in order to match the given boundary conditions. The solution is then given as discrete sums over the angular variable indices $m$ and $n$, i.e., $u(r,\theta,\phi) = \sum_n \sum_m u_{m,n}$. Terms in the sum are again a mixture of eigenfunctions (in those variables corresponding to coordinates having homogeneous or periodic conditions) and general solutions of the corresponding separated equations.

## 5.5 Classical Orthogonal Polynomials and Associated Bases

In addition to the Legendre and Bessel operators considered in the previous sections, many other singular Sturm–Liouville operators lead to important functions. Several (Laguerre, Hermite, and Chebyshev) are considered briefly in this section; more details can be found in mathematical physics texts (see, e.g., [9]).

It should also be noted that, rather than considering these polynomials from the standpoint of Sturm–Liouville theory, one can also obtain these functions by applying the Gram–Schmidt orthogonalization procedure to the sequence $\{1, x, x^2, x^3, \ldots\}$ using various inner products. In fact, completeness is also frequently established outside Sturm–Liouville theory.

In the following we consider the eigenvalue problem

$$(L-\lambda_n)\,u_n = 0$$

for several important Sturm–Liouville operators $L$. The Laguerre and Chebyshev polynomials are also considered as eigenfunctions of integral operators in Section 4.3.4.

### 5.5.1 Laguerre Polynomials

The singular *Laguerre* eigenvalue equation is

$$\left[x^{-v}e^{x}\frac{d}{dx}\left(x^{v+1}e^{-x}\frac{d}{dx}\right)+\lambda_n\right]u_n=0$$

on $[0,\infty)$ (limit-circle case at 0, limit-point case at $\infty$), leading to *associated Laguerre polynomials* $(v>-1)$

$$u_n(x)=L_n^v(x)=\frac{1}{n!}x^{-v}e^{x}\frac{d^n}{dx^n}\left(x^{v+n}e^{-x}\right)$$

with $\lambda_n=n$ as eigenvalues. Usually the polynomials $L_n^0=L_n$ are simply called *Laguerre polynomials*. The normalized associated Laguerre polynomials $\{(1/c_n)\,L_n^v(x)\}$ with[49] $c_n^2=\Gamma(n+v+1)/n!$ provide an orthonormal basis of $\mathbf{L}_w^2(0,\infty)$ with weight $x^v e^{-x}$. Completeness is expressed as

$$\frac{\delta\left(x-x'\right)}{x'^{v}e^{-x'}}=\sum_{n=0}^{\infty}\frac{1}{c_n^2}L_n^v(x)L_n^v(x')$$

on $\mathbf{L}_w^2(0,\infty)$.

Note that the Laguerre polynomials satisfy the recurrence equation

$$(n+1)\,L_{n+1}^v(x)=(2n+v+1-x)\,L_n^v(x)-(n+v)\,L_{n-1}^v(x)$$

and that the first few Laguerre polynomials are

$$\begin{aligned}
&L_0^v(x)=1,\qquad L_1^v(x)=v+1-x,\\
&L_2^v(x)=\left(3v+2-4x+v^2-2vx+x^2\right)/2,\\
&L_3^v(x)=\left(11v+6-18x+6v^2-15vx+9x^2+v^3-3v^2x+3vx^2-x^3\right)/6.
\end{aligned}$$

### 5.5.2 Hermite Polynomials

The *Hermite* eigenvalue equation is

$$\left[e^{x^2}\frac{d}{dx}\left(e^{-x^2}\frac{d}{dx}\right)+\lambda_n\right]u_n=0$$

on $(-\infty,\infty)$ (limit-point case at $\pm\infty$), leading to *Hermite polynomials*

$$u_n(x)=H_n(x)=(-1)^n\,e^{x^2}\frac{d^n}{dx^n}\left(e^{-x^2}\right)$$

[49] The *gamma-function* for $n$ an integer is

$$\Gamma(n)=\begin{cases}(n-1)!, & n>0,\\ \infty, & n\le 0.\end{cases}$$

with $\lambda_n = 2n$ as eigenvalues. The set of normalized Hermite polynomials $\{(1/c_n)H_n(x)\}$ with $c_n^2 = \sqrt{\pi}2^n n!$ provide an orthonormal basis of $\mathbf{L}_w^2(-\infty,\infty)$ with weight $e^{-x^2}$. Completeness is expressed as

$$\frac{\delta(x-x')}{e^{-x'^2}} = \sum_{n=0}^{\infty} \frac{1}{c_n^2} H_n(x) H_n(x')$$

on $\mathbf{L}_w^2(-\infty,\infty)$.

The Hermite polynomials satisfy the recurrence equation

$$H_{n+1}(x) = 2xH_n(x) - 2nH_{n-1}(x)$$

with the first few Hermite polynomials being

$$\begin{aligned} H_0(x) &= 1, & H_1(x) &= 2x, \\ H_2(x) &= 4x^2 - 2, & H_3(x) &= 8x^3 - 12x. \end{aligned}$$

Note also that, since

$$\begin{aligned} F\left\{e^{-x^2/2}\right\} &= e^{-\xi^2/2}, \\ F\left\{x^n f(x)\right\} &= (i)^n F\left\{f(x)\right\}^{(n)}, \end{aligned}$$

where $F$ is the Fourier transform (1.7), then

$$\phi_n(x) \equiv H_n(x) e^{-\frac{x^2}{2}}$$

is an eigenfunction of the Fourier transform operator with eigenvalue $(-i)^n$, i.e., $F\{\phi_n\} = (-i)^n \phi_n$, or

$$\int_{-\infty}^{\infty} H_n(\xi) e^{-\frac{\xi^2}{2}} e^{-i\xi x} d\xi = (-i)^n H_n(x) e^{-\frac{x^2}{2}}.$$

Obviously in both the Laguerre and Hermite cases the Sturm–Liouville equation is singular, since the interval of interest is infinite or semi-infinite. The next two Sturm–Liouville equations are singular on finite intervals.

### 5.5.3 Chebyshev Polynomials

#### Chebyshev Polynomials of the First Kind

The *Chebyshev* (or *Tschebycheff*) eigenvalue equation of the first kind is

$$\left[\left(a - x^2\right)^{1/2} \frac{d}{dx}\left(\left(a - x^2\right)^{1/2} \frac{d}{dx}\right) + \lambda_n\right] u_n = 0$$

on $[-a, a]$, leading to first-kind Chebyshev polynomials

$$u_n(x) = T_n(x/a) = \cos\left(n \cos^{-1} x/a\right)$$

for $n = 0, 1, 2, \ldots$, with $\lambda_n = n^2$ as eigenvalues. Often we set $x/a = \cos\theta$ such that $T_n(x/a) = \cos(n\theta)$. The normalized Chebyshev polynomials $\left\{\sqrt{\varepsilon_n/\pi}T_n(x/a)\right\}$, $n = 0, 1, 2, \ldots$, provide an orthonormal[50] basis of $\mathbf{L}^2_w(-a, a)$ with weight $\left(a - x^2\right)^{-1/2}$. Completeness is expressed as

$$\frac{\delta(x - x')}{\left(a - x'^2\right)^{-1/2}} = \frac{2}{\pi}\sum_{n=0}^{\infty}\frac{\varepsilon_n}{2}T_n(x/a)T_n(x'/a)$$

on $\mathbf{L}^2_w(-a, a)$, where $\varepsilon_n$ is Neumann's number, $\varepsilon_n = \left\{\begin{array}{ll} 1, & n = 0 \\ 2, & n \neq 0 \end{array}\right.$.

The first-kind Chebyshev polynomials satisfy the recurrence equation

$$T_{n+1}(x) = 2xT_n(x) - T_{n-1}(x)$$

for $n = 1, 2, \ldots$, with the first few Chebyshev polynomials given by

$$\begin{aligned} &T_0(x) = 1, && T_1(x) = x, \\ &T_2(x) = 2x^2 - 1 && T_3(x) = 4x^3 - 3x. \end{aligned}$$

Note that $|T_n(x)| \leq 1$ for $x \in [-1, 1]$.

### Chebyshev Polynomials of the Second Kind

The Chebyshev eigenvalue equation of the second kind is

$$\left[\left(a - x^2\right)^{-1/2}\frac{d}{dx}\left(\left(a - x^2\right)^{3/2}\frac{d}{dx}\right) + \lambda_n\right]u_n = 0$$

on $[-a, a]$, leading to *second-kind Chebyshev polynomials*

$$u_n(x) = U_n(x/a) = \frac{\sin\left[(n+1)\cos^{-1}x/a\right]}{\sin\left(\cos^{-1}x/a\right)}$$

for $n = 0, 1, 2, \ldots$, with $\lambda_n = n(n+2)$ as eigenvalues $\left(\sin\left(\cos^{-1}x\right) = \sqrt{1 - x^2}\right)$. The Chebyshev polynomials are normalized as $\sqrt{2/\pi}U_n(x/a)$, $n = 0, 1, 2, \ldots$, and provide an orthonormal[51] basis of $\mathbf{L}^2_w(-a, a)$ with

[50] The normalization is

$$\frac{\varepsilon_n}{\pi}\langle T_n, T_m\rangle = \frac{\varepsilon_n}{\pi}\int_{-a}^{a}T_n(x/a)T_m(x/a)\frac{1}{\sqrt{a^2 - x^2}}dx = \delta_{nm}.$$

[51] The normalization is

$$\frac{2}{\pi a^2}\langle U_n, U_m\rangle = \frac{2}{\pi a^2}\int_{-a}^{a}U_n(x/a)U_m(x/a)\sqrt{a^2 - x^2}dx = \delta_{nm}.$$

weight $\left(a-x^2\right)^{1/2}$. Completeness is expressed as

$$\frac{\delta\left(x-x'\right)}{\left(a-x'^2\right)^{1/2}}=\frac{2}{\pi a^2}\sum_{n=0}^{\infty}U_n(x/a)U_n(x'/a)$$

on $\mathbf{L}^2_w\left(-a,a\right)$.

The second-kind Chebyshev polynomials satisfy the recurrence equation

$$U_{n+1}(x)=2xU_n(x)-U_{n-1}(x)$$

for $n=1,2,\ldots$, with the first several polynomials given by

$$U_0(x)=1, \qquad U_1(x)=2x$$
$$U_2(x)=4x^2-1, \qquad U_3(x)=8x^3-4x.$$

Note that $|U_n(x)|\leq n+1$ for $x\in[-1,1]$. Both kinds of Chebyshev functions are useful in electromagnetic problems involving strips and slots.

# Bibliography

[1] Dudley, D.G. (1994). *Mathematical Foundations for Electromagnetic Theory*, New York: IEEE Press.

[2] Naimark, M.A. (1967). *Linear Differential Operators*, Part I, New York: Frederick Ungar.

[3] Mikhlin, S.G. (1970). *Mathematical Physics, An Advanced Course*, Amsterdam: North-Holland.

[4] Naylor, A.W. and Sell, G.R. (1982). *Linear Operator Theory in Engineering and Science*, 2nd ed., New York: Springer-Verlag.

[5] Debnath, L. and Mikusiński, P. (1999). *Introduction to Hilbert Spaces with Applications*, San Diego: Academic Press.

[6] Kreyszig, E. (1993). *Advanced Engineering Mathematics*, 7th ed., New York: Wiley.

[7] Stakgold, I. (1967). *Boundary Value Problems of Mathematical Physics, Vol.* 1, New York: Macmillan.

[8] Coddington, E.A. and Levinson, N. (1984). *Theory of Ordinary Differential Equations*, Malabar: Robert E. Krieger.

[9] Dennery, P. and Krzywicki, A. (1995). *Mathematics for Physicists*, New York: Dover.

[10] Friedman, B. (1956). *Principles and Techniques of Applied Mathematics*, New York: Dover.

[11] Gohberg, I.C. and Krein, M.G. (1969). *Introduction to the Theory of Linear Nonselfadjoint Operators*, Providence, RI: American Mathematical Society.

[12] Ramm, A.G. (1982). Mathematical foundations of the singularity and eigenmode expansion method *J. Math. Anal. Appl.*, Vol. 86, pp. 562–591 (see also *Mathematics Notes,* Note 68, Phillips Laboratory Note Series, Kirtland AFB, Dec. 1980).

[13] Jackson, J.D. (1975). *Classical Electrodynamics*, 2nd ed., New York: John Wiley.

[14] Abramowitz, M. and Stegun, I. (1965). *Handbook of Mathematical Functions*, New York: Dover.

[15] Stakgold, I. (1999). *Green's Functions and Boundary Value Problems*, 2nd ed., New York: Wiley.

[16] Krall, A.M. (1973). *Linear Methods in Applied Analysis*, Reading, MA: Addison-Wesley.

[17] Mrozowski, M. (1997). *Guided Electromagnetic Waves, Properties and Analysis*, Somerset, England: Research Studies Press.

[18] Gohberg, I. and Goldberg, S. (1980). *Basic Operator Theory*, Boston: Birkhäuser.

[19] Morse, P.M. and Feshbach, H. (1953). *Methods of Theoretical Physics*, Vol. I, New York: McGraw-Hill.

[20] Jones, D.S. (1994). *Methods in Electromagnetic Wave Propagation*, 2nd ed., New York: IEEE Press.

[21] Lang, P. and Locker, J. (1989). Spectral theory of two-point differential operators determined by -$D^2$. I. Spectral Properties, *J. Math. Anal. Appl.*, Vol. 141, pp. 538–558, Aug.

[22] Lang, P. and Locker, J. (1990). Spectral theory of two-point differential operators determined by -$D^2$. I. Analysis of Cases, *J. Math. Anal. Appl.*, Vol. 146, pp. 148–191, Feb.

[23] Locker, J. (2000). *Spectral Theory of Non-Self-Adjoint Two-Point Differential Operators*, Providence, RI: American Mathematical Society.

# Part II
# Applications in Electromagnetics

In Part I of the text we discussed the mathematical properties of various classes of operators, along with relevant concepts from functional analysis. In many cases, as these topics were discussed the connection with applied electromagnetics was highlighted. In this part of the book we examine some specific problems in electromagnetics and in each case provide further associations between the electromagnetic formulations and the mathematical characterization of various operators and function spaces.

In many of the problems we consider formulations leading to self-adjoint operators. However, it should be noted that in physical problems, self-adjointness may depend on assumptions or idealizations concerning the state of certain model parameters (i.e., the absence or presence of material loss, etc.). Furthermore, in many situations even idealized problems lead to nonself-adjoint operators. In particular, dynamic problems in an unbounded, lossless space are usually pseudo self-adjoint. Such properties mainly impact spectral formulations.

The first chapter concerns electrostatic potential problems, which were one of the first classes of problems in electromagnetics to be studied using operator-theoretic tools. The formulation of potential problems is often self-adjoint, and basic questions of existence and uniqueness can be settled using Fredholm theory. Spectral properties of the Laplacian operator are fundamental to the study of potentials, and so this topic is treated in some detail. Simple separation-of-variables techniques are discussed, as well as integral formulations leading to first- and second-kind integral equations with compact kernels.

Chapter 7 concerns transmission-line problems considered from the aspect of Sturm–Liouville theory. Although this boundary value approach is not the usual method of transmission-line analysis for engineering purposes, it is nevertheless useful. In particular, this approach is valuable for transmission lines driven by distributed sources and for the analysis of transmission-line radiation and susceptibility. It also provides a one-dimensional example of an electromagnetic waveguide, the analysis of which is fruitfully investigated using the spectral theory of operators and, in many cases, Sturm–Liouville theory. The chapter concludes with a brief treatment of multiconductor transmission lines.

Source-driven waveguiding problems for planar media are discussed in Chapter 8. A general formulation is provided for two-dimensional, multilayered planar waveguides, and specific examples of parallel-plate, impedance-plane, and grounded dielectric structures are described in detail. For these examples the natural waveguide modes of the structure are examined, appropriate Green's functions are derived, and completeness relations and the associated spectral field expansions are detailed. Next, some applications to scattering from dielectric cylinders and conducting strips embedded in a planarly layered medium are described. Three-dimensional sources in planar media are then considered, and dyadic Green's functions for multilayered media are developed via Hertzian potentials.

In Chapter 9 cylindrical waveguiding problems are discussed. We start with some general material on dyadic Green's functions and then formulate integral equations for various scattering problems within a cylindrical waveguide environment. Explicit representations for dyadic Green's functions are provided in the form of partial eigenfunction expansions for the special case of a rectangular waveguide, although the method of analysis is general. Then, more complicated waveguide scattering problems are considered, followed by derivation of the associated Green's dyadics.

Electromagnetic cavities are considered in Chapter 10. Spectral properties of the governing differential operators are discussed, along with vector wavefunctions and spectral expansions. Integral equation techniques are presented, and the reduction of the vector spectral problem to a scalar problem using vector potentials is included.

In most of the application chapters the function space $\mathbf{L}^2$ is used. Often the same problem can be cast in Hölder or Sobolev spaces, although the subsequent analysis is more complicated. It should also be noted that in several of the chapters integral equations are formed. Since a variety of excellent books address the numerical solution of integral equations that arise in electromagnetics (see, e.g., [1] and [2]), here we include only a few representative solutions for integral equations that may be solved exactly. These invariably correspond to canonical geometries, such as infinite flat strips and circular cross-section cylinders.

# 6

# Poisson's and Laplace's Boundary Value Problems: Potential Theory

In this chapter we study the interaction of electrostatic potentials with material media. A typical problem would be to determine the electrostatic field or potential in the vicinity of a conductor or dielectric in the presence of sources. These types of problems are very important for several reasons. First, many important electrical quantities, such as capacitance and resistance, can be determined using the methods developed in this chapter. Also, electrostatic principles are directly utilized in a variety of important technologies, such as photocopying, electrostatic cleaning and purification, and electrostatic protection. Furthermore, slowly varying electrodynamic fields can often be approximated as being electrostatic or quasi-static, greatly simplifying analysis methods.

The chapter begins with a general formulation of problems involving the electrostatic scalar potential. Next, operator and spectral properties of the negative Laplacian are considered. Then, solution techniques for self-adjoint problems are discussed, followed by techniques to treat nonself-adjoint problems. The chapter concludes with a discussion of integral equation techniques for potentials.

## 6.1 Problem Formulation

We will concentrate on solving boundary value problems for the electrostatic scalar potential[1] in an $m$-dimensional open region $\Omega$ ($m \geq 2$) containing isotropic real-valued permittivity, with an $(m-1)$ dimensional boundary $\Gamma$, where $\widetilde{\Omega} = \Omega \cup \Gamma$.

Under time-static conditions ($\partial/\partial t \to 0$), electric and magnetic fields become decoupled and Maxwell's equations (1.1) can be written as

$$\begin{aligned} \nabla \cdot \mathbf{D}(\mathbf{r}) &= \rho_e(\mathbf{r}), \\ \nabla \times \mathbf{E}(\mathbf{r}) &= -\mathbf{J}_m(\mathbf{r}), \\ \nabla \cdot \mathbf{J}_m(\mathbf{r}) &= 0 \end{aligned} \tag{6.1}$$

for electric fields and

$$\begin{aligned} \nabla \cdot \mathbf{B}(\mathbf{r}) &= \rho_m(\mathbf{r}), \\ \nabla \times \mathbf{H}(\mathbf{r}) &= \mathbf{J}_e(\mathbf{r}), \\ \nabla \cdot \mathbf{J}_e(\mathbf{r}) &= 0 \end{aligned} \tag{6.2}$$

for magnetic fields, although usually we set $\rho_m = 0$, $\mathbf{J}_m = \mathbf{0}$. In the absence of magnetic sources the static electric field is caused by static electric charge via Gauss' law, and Faraday's law $\nabla \times \mathbf{E}(\mathbf{r}) = \mathbf{0}$ expresses the conservative nature of the electrostatic field. For magnetostatics, the static magnetic field is caused by steady electric current, and $\nabla \cdot \mathbf{B}(\mathbf{r}) = 0$ expresses the lack of magnetic charge. The solution of the equation containing the source term most directly determines the field, although potentials are often utilized as an intermediate step.

#### Poisson's and Laplace's Equations

Following the method outlined in Section 1.3.2, we note that $\nabla \times \mathbf{E}(\mathbf{r}) = \mathbf{0}$ implies

$$\mathbf{E}(\mathbf{r}) = -\nabla \phi(\mathbf{r}),$$

where $\phi(\mathbf{r})$ is the electrostatic scalar potential. Substituting into Gauss's law leads to

$$-\nabla \cdot \varepsilon(\mathbf{r}) \nabla \phi(\mathbf{r}) = \rho_e(\mathbf{r}), \tag{6.3}$$

which is known as Poisson's equation. If permittivity is constant within $\Omega$, which is assumed here, we have

$$-\nabla^2 \phi(\mathbf{r}) = \frac{\rho_e(\mathbf{r})}{\varepsilon}. \tag{6.4}$$

---

[1] Recall that scalar potentials are also useful for the analysis of static magnetic fields, steady currents, and quasi-static fields, as well as in dynamic problems.

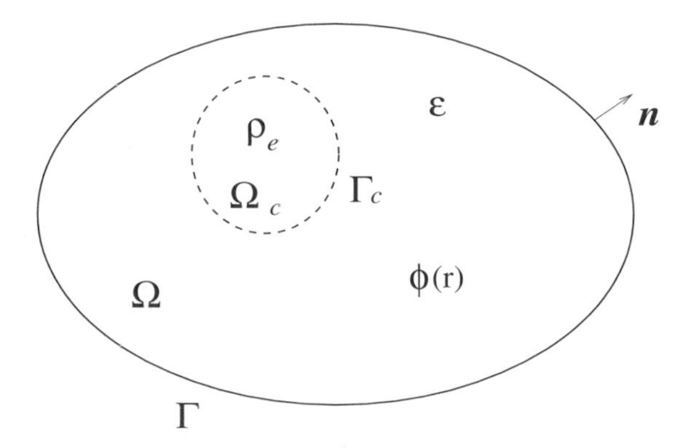


Figure 6.1: Geometry of the interior problem: bounded region $\Omega \subset \mathbf{R}^m$ containing charge density $\rho_e$ in $\Omega_c \subseteq \Omega$, and characterized by permittivity $\varepsilon$, bounded by surface $\Gamma$.

Note that $\rho_e$ is defined in $\Omega$, not $\widetilde{\Omega}$. If the charge density $\rho_e$ is absent, then the governing differential equation is

$$-\nabla \cdot \varepsilon(\mathbf{r})\nabla\phi(\mathbf{r}) = 0, \tag{6.5}$$

which is known as Laplace's equation, and for constant permittivity we have

$$-\nabla^2\phi(\mathbf{r}) = 0. \tag{6.6}$$

We consider two problems, the *interior problem* (Figure 6.1) and the *exterior problem* (Figure 6.2). For the interior problem we have a bounded region $\Omega$ of material characterized by $\varepsilon$, surrounded by a sufficiently smooth boundary[2] $\Gamma$, with $\mathbf{n}$ an outward normal unit vector. Charge density $\rho_e$ may exist in some domain $\Omega_c \subseteq \Omega$, and on the boundary the potential or its normal component will be assumed to be a known, continuous function. We are interested in the potential $\phi$ interior to $\Omega$, or on $\Gamma$, and the media characteristics outside $\Omega$ are immaterial.[3]

For the exterior problem we have an infinite region $\Omega$ of material characterized by $\varepsilon$, with sufficiently smooth finite boundary $\Gamma$, where $\mathbf{n}$ is an outward normal unit vector. Charge density $\rho_e$, if present, must exist in

[2]We do not require $\Omega$ to be simply connected. For example, in two dimensions $\Omega$ could be the annular domain between two concentric circles, as long as both circles form the boundary $\Gamma$.

[3]The media characteristics outside $\Omega$ are immaterial in the sense of the boundary value problem we will consider. Physically, the region exterior to $\Omega$ may influence the boundary values of $\phi$ or $\partial\phi/\partial n$, which need to be specified in this formulation. Later in this chapter an integral equation formulation for a two-region problem will be discussed, where boundary values are unknowns to be determined and which are strongly influenced by the surrounding media.

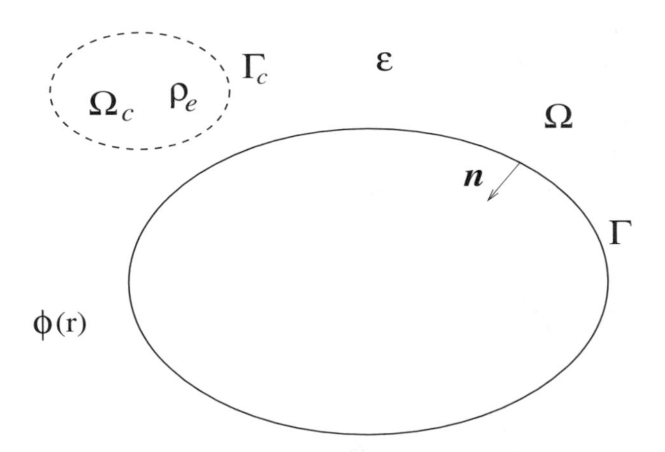


Figure 6.2: Geometry of the exterior problem: unbounded region $\Omega$ containing charge density $\rho_e$ in $\Omega_c \subset \mathbf{R}^m$, and characterized by permittivity $\varepsilon$.

some finite domain $\Omega_c \subset \Omega$, and on the boundary the potential or its normal component will be assumed to be a known, continuous function. We are interested in the potential $\phi$ interior to $\Omega$ or on $\Gamma$, and the media characteristics exterior to $\Omega$ (e.g., in Figure 6.2 depicted as the region surrounded by $\Gamma$ into which $\mathbf{n}$ is directed) are immaterial.

### Boundary and Fitness Conditions

We need to specify boundary conditions for the governing differential equations (6.4) and (6.6) to have unique solutions. For the internal problem it turns out that either specification of the potential

$$\phi(\mathbf{r})|_{\mathbf{r}\in\Gamma} = \eta_1 \in \mathbf{C}(\Gamma)$$

on the boundary (Dirichlet boundary condition), the normal derivative of potential

$$\left.\frac{\partial}{\partial n}\phi(\mathbf{r})\right|_{\mathbf{r}\in\Gamma} = \eta_2 \in \mathbf{C}(\Gamma)$$

on the boundary (Neumann boundary condition), or a mixture of Dirichlet conditions on some part of the boundary and Neumann conditions on the remainder of the boundary is sufficient to yield a unique solution[3, p. 42].[4] In general, we write the boundary condition as

$$B(\phi) = \left.\left(\alpha_1\phi(\mathbf{r}) + \alpha_2\frac{\partial\phi(\mathbf{r})}{\partial n}\right)\right|_{\mathbf{r}\in\Gamma} = \eta(\mathbf{r}) \tag{6.7}$$

[4]For the interior Neumann problem the solution is determined to within an arbitrary constant. In this case uniqueness refers to the fact that the solution is obtained in the form $\phi(\mathbf{r}) + C$, where $C$ is arbitrary. Obviously, in any case the electric field associated with $\phi$ will be unique.

with $\alpha_{1,2} \in \mathbf{C}$, $\eta \in \mathbf{C}(\Gamma)$, which is similar to the boundary condition considered for the one-dimensional Sturm–Liouville problem in Section 5.3. If $\eta = 0$, we call the boundary condition homogeneous; otherwise the boundary condition is said to be inhomogeneous.

For the exterior problem $\Omega$ is an unbounded region. In order to obtain a unique solution to the Poisson or Laplace problem, we need to specify the behavior of $\phi$ at infinity.[5] The appropriate fitness condition to impose that leads to a unique solution of the exterior Dirichlet problem in $m$-dimensional space ($m \geq 2$) is

$$|\phi| \leq \frac{C_1}{r^{m-2}},$$

i.e., $r^{m-2}|\phi|$ must be bounded as $r \to \infty$, where $r$ is the $m$-dimensional radial distance from the origin. For the exterior Neumann problem in $m \geq 3$ dimensions the same condition must be enforced, and the solution is in fact unique [4, pp. 270–272]. In two dimensions the solution to the exterior Neumann problem is unique up to an arbitrary constant.

Since $\rho_e$ is a bounded source having support in the bounded domain $\Omega_c$, where perhaps $\Omega_c = \Omega$ for the interior problem, outside $\Omega_c$ the potential $\phi$ must satisfy Laplace's equation. A function with continuous second derivatives that satisfies Laplace's equation is called a *harmonic function.* For the exterior problem we impose the extra condition that $|\phi| \leq C_1/r^{m-2}$ as $r \to \infty$ in order to classify $\phi$ as harmonic.

In the following we will also need to know the behavior of $\partial\phi/\partial n$ as $r \to \infty$. It can be shown that as $r \to \infty$ first partial derivatives of a harmonic function $\phi$ are bounded by the inequality[6] [4, pp. 269–270]

$$\left|\frac{\partial \phi}{\partial n}\right| \leq \frac{C_2}{r^{m-1}}$$

which can even be tightened for $m = 2$ to be $|\partial\phi/\partial n| \leq C_2/r^2$. Therefore, for the exterior problem we impose the condition $r^{m-2}|\phi| \leq C_1$, and the condition on boundedness of the derivative follows. A potential $\phi$ satisfying the conditions $r^{m-2}|\phi| \leq C_1$ and $r^{m-1}|\partial\phi/\partial n| \leq C_2$, with $C_{1,2} < \infty$, is called *regular at infinity* (see [5, p. 217] for the case $m = 3$).

When considering boundary conditions it should be kept in mind that everywhere on a conductor $\phi$ must be constant (i.e., in this case $\Gamma$ is an equipotential surface) and that $\partial\phi(\mathbf{r})/\partial n$ on the boundary is proportional to boundary charge by (1.5). Therefore, Dirichlet conditions amount to specifying the potential on the boundary, and Neumann conditions amount

---

[5] To see why this is necessary, consider a spherical surface $\Gamma$ having unity radius and the boundary condition $\phi|_\Gamma = 1$. Both $\phi = 1/r$ and $\phi = 1$ satisfy Laplace's equation in $\Omega$ and the given boundary condition on $\Gamma$ yet have very different behavior at infinity, with only one of the solutions being physically realistic.

[6] The corresponding formula for the $\alpha$th generalized derivative is $|D^\alpha \phi| \leq C_\alpha/r^{m-2+\alpha}$.

to specifying the charge on the boundary. Note that arbitrary specification of both $\partial\phi(\mathbf{r})/\partial n$ and $\phi(\mathbf{r})$ everywhere on the boundary overspecifies the problem and does not lead to a solution of Poisson's equation.[7]

Apart from the uniqueness of a solution, the fact that a solution exists at all is obviously of importance. Assuming a sufficiently smooth boundary and boundary condition, the existence of a solution to the interior and exterior Dirichlet and Neumann problems can be shown under fairly general circumstances (see, e.g., [6, pp. 216–222]). Existence of the interior Neumann problem is subject to the significant constraint [4, p. 258]

$$\int_\Omega \frac{\rho_e}{\varepsilon}\, d\Omega + \int_\Gamma \frac{\partial \phi}{\partial n}\, d\Gamma = 0, \tag{6.8}$$

where we again use the outward normal. This constraint can be determined by inserting the pair $\{\phi, 1\}$ into Green's first theorem and utilizing Poisson's equation, or by integrating Poisson's equation over $\Omega$ and applying the divergence theorem. Therefore, for the interior Neumann problem $\partial\phi/\partial n$ cannot be arbitrarily specified, but must be consistent with (6.8).[8] For the Dirichlet problem $\phi$ may be arbitrarily specified on the boundary subject to adequate continuity (in a mathematical sense, of course; physical conditions dictate the actual imposed boundary value for $\phi$).

Various problems associated with Poisson's, and especially Laplace's, equation are considered in detail in many texts on classical electromagnetics, e.g., [3, Chs. 2, 3]. Here we consider a small subset of such problems emphasizing an operator-theoretic approach.

## 6.2 Operator Properties of the Negative Laplacian

In this section we investigate the operator properties of the negative Laplacian, $L = -\nabla^2 : \mathbf{L}^2(\Omega) \to \mathbf{L}^2(\Omega)$, as appropriate for consideration of Poisson's or Laplace's equation. For generality we consider the governing differential equation to be

$$\left(-\nabla^2 - \lambda\right)\phi(\mathbf{r}) = f(\mathbf{r}) \tag{6.9}$$

with $\lambda \in \mathbf{C}$, perhaps $\lambda = 0$, and $f$ specified, subject to the boundary condition

$$B(\phi) = \left(\alpha_1\phi(\mathbf{r}) + \alpha_2\frac{\partial\phi(\mathbf{r})}{\partial n}\right)\Bigg|_{\mathbf{r}\in\Gamma} = \eta(\mathbf{r}). \tag{6.10}$$

[7] The one-dimensional analog of such an overspecified problem would be the differential equation $-d^2y/dz^2 = f$ on $[a,b]$, with $y(a), y(b), y'(a), y'(b)$ given.

[8] Note that in one-dimension Neumann conditions present similar problems. For example, considering $-d^2y/dx^2 = f$ on $[a,b]$ with $y'(a)$ and $y'(b)$ specified, upon integrating we obtain the constraint $y'(b) - y'(a) = -\int_a^b f(x)\,dx$.

For the exterior problem we also require $|\phi| \leq C/r^{m-2}$ as $r \to \infty$. Typical conditions on the forcing function would be to assume that $f$ is at least piecewise continuous in the region of interest, although later we will sometimes require the stronger condition of Hölder continuity.

The associated eigenvalue (spectral) problem is

$$\left(-\nabla^2 - \lambda_n\right) \phi_n(\mathbf{r}) = 0, \tag{6.11}$$

where $\lambda_n$ and $\phi_n$ are unspecified spectral parameters, eigenvalues and eigenfunctions, respectively, subject to the homogeneous boundary condition

$$B(\phi_n) = \left(\alpha_1 \phi_n(\mathbf{r}) + \alpha_2 \frac{\partial \phi_n(\mathbf{r})}{\partial n}\right)\bigg|_{\mathbf{r}\in\Gamma} = 0. \tag{6.12}$$

The eigenvalue problem is considered in this section for the interior problem, since generally for the exterior problem eigenfunctions do not exist, although improper eigenfunctions and eigenvalues may exist. In $m$ dimensions the subscript notation $\phi_n$ typically indicates an $m$-tuple of (perhaps related) parameters (e.g., for $m = 3$, $\phi_n$ represents $\phi_{nkp}$).

The Green's function problem is

$$\left(-\nabla^2 - \lambda\right) g(\mathbf{r}, \mathbf{r}') = \delta(\mathbf{r} - \mathbf{r}') \tag{6.13}$$

with $\lambda \in \mathbf{C}$ specified, perhaps $\lambda = 0$, subject to the boundary condition

$$B(g) = \left(\alpha_1 g(\mathbf{r}, \mathbf{r}') + \alpha_2 \frac{\partial g(\mathbf{r}, \mathbf{r}')}{\partial n}\right)\bigg|_{\mathbf{r}\in\Gamma} = 0 \tag{6.14}$$

valid for both the interior and exterior problems, where for the exterior problem in addition to (6.14) we require that the Green's function be bounded by $|g| \leq C/r^{m-2}$ as $r \to \infty$. Note that in (6.10), (6.12), and (6.14), $\alpha_1 = 1$, $\alpha_2 = 0$ yields the Dirichlet problem and $\alpha_1 = 0$, $\alpha_2 = 1$ results in the Neumann problem. In the following we will consider either the Dirichlet or the Neumann problem, but not the case where both $\alpha_1$ and $\alpha_2$ are simultaneously nonzero.

For the interior Dirichlet problem the domain of $L$, $D_L$, is that subset of $\mathbf{L}^2(\Omega)$ containing functions $\phi$ continuous in $\widetilde{\Omega}$ and twice differentiable with continuous second derivatives[9] in $\Omega$ that satisfy the boundary condition $B(\phi)$, i.e.,

$$D_L \equiv \left\{\phi : \phi \in \mathbf{C}^2(\Omega) \cap \mathbf{C}(\widetilde{\Omega}) \subset \mathbf{L}^2(\Omega), B\left(\phi\right)\right\}.$$

For the interior Neumann problem we require $\phi(\mathbf{r}) \in \mathbf{C}^2(\Omega) \cap \mathbf{C}^1(\widetilde{\Omega})$. For the exterior Dirichlet problem where charge is contained in a finite

[9] In many cases a solution may be obtained in the more restrictive set $\mathbf{C}^2(\widetilde{\Omega})$, rather than in the indicated intersection. Note that if $\rho_e$ is contained in a bounded domain $\Omega_c \subseteq \Omega$, then $\phi$ will be $\mathbf{C}^2$ everywhere in $\Omega$ except on the boundary of $\Omega_c$ ($-\nabla^2\phi = \rho_e/\varepsilon$ inside $\Omega_c$ and $-\nabla^2\phi = 0$ outside $\Omega_c$), in which case $\phi \in \mathbf{C}^2\left(\Omega \backslash \Gamma_c\right) \cap \mathbf{C}(\widetilde{\Omega})$.

region $\Omega_c$, we require $\phi(\mathbf{r}) \in \mathbf{C}^2(\Omega)$ except on $\Gamma$ and $\Gamma_c$, and belonging to $\mathbf{C}(\widetilde{\Omega})$, where we also add the appropriate fitness condition at infinity.

A word of caution regarding the Neumann problem is in order. The Green's function subject to (6.14) with $\alpha_1 = 0$, $\alpha_2 = 1$ (Neumann Green's function) does not generally exist due to the constraint (6.8), as is discussed in more detail later. Other difficulties are also associated with the Neumann problem, such as the existence of a zero eigenvalue of the Laplacian operator. Mathematically, the Dirichlet problem is easier to handle and is often the one physically dictated, and so here we are primarily interested in the Dirichlet problem, although the Neumann problem is discussed in parallel where convenient.

### Self-Adjointness of the Laplacian Operator

Obviously $L$ is an unbounded, real operator. Using the $\mathbf{L}^2(\Omega)$ inner product

$$\langle \phi, \psi \rangle = \int_\Omega \phi(\mathbf{r}) \overline{\psi}(\mathbf{r})\, d\Omega,$$

which induces the norm

$$\|\phi\| = \langle \phi, \phi \rangle^{1/2} = \sqrt{\int_\Omega |\phi(\mathbf{r})|^2\, d\Omega}, \tag{6.15}$$

we easily see that $L$ is self-adjoint under the same restrictions on $\alpha, \eta$ in (6.10) as were necessary in Section 5.1, i.e., $\alpha \in \mathbf{R}$ and $\eta = 0$. To see this we use Green's second theorem in $m$ dimensions, leading to

$$\langle L\phi, \psi \rangle = -\int_\Omega \overline{\psi}(\mathbf{r}) \nabla^2 \phi(\mathbf{r})\, dV = -\int_\Omega \phi \nabla^2 \overline{\psi}\, d\Omega + \oint_\Gamma \left( \phi \frac{\partial \overline{\psi}}{\partial n} - \overline{\psi} \frac{\partial \phi}{\partial n} \right) d\Gamma \tag{6.16}$$

$$= \langle \phi, L\psi \rangle + J(\phi, \psi),$$

where $J$ represents the boundary integral (conjunct). We first notice that the operator is formally self-adjoint, as described in Section 3.4.2. To see when the operator is self-adjoint we need to choose adjoint boundary conditions on $\psi$, denoted as $B^*(\psi)$, to make the boundary integral vanish. Following the procedure detailed in Section 3.3.4, to determine the correct adjoint boundary conditions we apply homogeneous boundary conditions $B(\phi) = 0$ in the boundary integral term (regardless of the value of $\eta$) and choose homogeneous boundary conditions on $\overline{\psi}$ such that the resulting boundary integral vanishes. It is easy to see that it is sufficient to choose the conjugate adjoint boundary conditions as

$$\overline{B^*_{ca}}(\psi) = \alpha_1 \overline{\psi}(\mathbf{r}) + \alpha_2 \frac{\partial \overline{\psi}(\mathbf{r})}{\partial n} \bigg|_{\mathbf{r} \in \Gamma} = 0,$$

in which case $J(\phi, \psi) = 0$. Conjugating the above conditions leads to

$$B^*(\psi) = \overline{\alpha_1}\psi(\mathbf{r}) + \overline{\alpha_2}\frac{\partial\psi(\mathbf{r})}{\partial n}\bigg|_{\mathbf{r}\in\Gamma} = 0.$$

If $\eta = 0$ and $\alpha_{1,2} \in \mathbf{R}$, then $B(\psi) = B^*(\psi)$ and with this choice $D_{L^*} = D_L$, so that the operator $L$ is self-adjoint. If either $\eta \neq 0$ or $\alpha_{1,2} \in \mathbf{C}$, or both, then the operator $L$ is not self-adjoint. Since often $\alpha_{1,2} \in \mathbf{R}$ but $\eta(\mathbf{r}) \neq 0$, the typical Dirichlet ($\alpha_1 = 1$, $\alpha_2 = 0$, $\eta \neq 0$) and Neumann ($\alpha_1 = 0$, $\alpha_2 = 1$, $\eta \neq 0$) problems are not self-adjoint, although homogeneous boundary conditions together with $\rho_e \neq 0$ yield self-adjoint problems of considerable practical interest.

The preceding pertains to the interior problem. For the exterior problem we can add a boundary $\Gamma_\infty$ located at $r = \infty$. Assuming $\phi$ and $\psi$ are regular at infinity, then $\phi\partial\psi/\partial n, \psi\partial\phi/\partial n \sim C/r^{2m-3}$; noting $d\Gamma \sim r^{m-1}$ shows that the boundary integral over $\Gamma_\infty$ vanishes. We are left with the boundary integral over the finite boundary $\Gamma$, leading to the same constraints for self-adjointness as found for the interior problem.

### Definiteness of the Laplacian Operator

Considering the material in Chapters 3 and 4, it is useful to characterize the definiteness of the Laplacian operator. From Green's first theorem we have

$$\langle \phi, L\phi \rangle = -\int_\Omega \phi(\mathbf{r})\nabla^2\phi(\mathbf{r})\, d\Omega = -\oint_\Gamma \phi\nabla\phi \cdot d\mathbf{S} + \int_\Omega \nabla\phi \cdot \nabla\phi\, d\Omega. \quad (6.17)$$

For the special case of homogeneous Dirichlet conditions, the boundary integral vanishes such that

$$\langle \phi, L\phi \rangle = -\int_\Omega \phi(\mathbf{r})\nabla^2\phi(\mathbf{r})\, d\Omega = \int_\Omega |\nabla\phi|^2\, d\Omega > 0, \quad (6.18)$$

and we find for those cases that $L$ is a strictly positive operator. If $\Omega$ is bounded, then $L$ is in fact positive definite [4, pp. 294–295]. For homogeneous Neumann conditions the surface integral also vanishes; however, in this case $\phi = C \neq 0$ can occur such that $\langle \phi, L\phi \rangle = 0$, and so the operator $L$ subject to homogeneous Neumann conditions is merely nonnegative.

In summary, the unbounded, real operator $L = -\nabla^2$ is self-adjoint and strictly positive subject to homogeneous Dirichlet conditions with $\alpha_{1,2} \in \mathbf{R}$, is self-adjoint and nonnegative subject to homogeneous Neumann conditions with $\alpha_{1,2} \in \mathbf{R}$, and is otherwise generally nonself-adjoint. For the interior problem the operator acts on functions $\phi \in \mathbf{C}^2(\Omega) \cap \mathbf{C}(\widetilde{\Omega})$ (or smoother) for the Dirichlet case, and $\phi \in \mathbf{C}^2(\Omega) \cap \mathbf{C}^1(\widetilde{\Omega})$ for the Neumann case, and for the exterior problem $\phi$ must also be regular at infinity.

It is important to note that in the case of nonhomogeneous boundary conditions $B(\phi) = \eta$, one can still work with a self-adjoint Poisson problem. For example, to solve

$$-\nabla^2 \phi = \frac{\rho_e}{\varepsilon},$$
$$B(\phi) = \eta,$$

we may decompose the potential as $\phi = \phi^p + \phi^h$, where $\phi^p$ satisfies

$$\begin{aligned} -\nabla^2 \phi^p &= \frac{\rho_e}{\varepsilon}, \\ B(\phi^p) &= 0, \end{aligned} \tag{6.19}$$

and $\phi^h$ satisfies

$$\begin{aligned} -\nabla^2 \phi^h &= 0, \\ B(\phi^h) &= \eta. \end{aligned} \tag{6.20}$$

Therefore, it is enough to consider Poisson problems with homogeneous conditions $B(\phi) = 0$, which result in self-adjoint positive or nonnegative operators, and Laplace's equation with nonhomogeneous conditions $B(\phi) = \eta$. The solution of Laplace's equation for $\phi^h$ may frequently be obtained using separation of variables, as described in Section 5.4, or integral techniques.

It is also useful to keep in mind that for the interior problem in a simply-connected region, Laplace's equation with homogeneous Dirichlet conditions admits only the trivial solution $\phi = 0$. Furthermore, for the Dirichlet condition $\phi|_\Gamma = \phi_0$, where $\phi_0$ is a constant, then $\phi = \phi_0$ everywhere in the interior region.[10]

## 6.3 Spectral Properties of the Negative Laplacian

We next consider the eigenvalue problem

$$-\nabla^2 \phi_n = \lambda_n \phi_n$$

[10]This can be easily shown from Green's first theorem. Indeed,

$$\int_\Omega \left[\phi \nabla^2 \phi + \nabla \phi \cdot \nabla \phi\right] d\Omega = \oint_\Gamma \phi \nabla \phi \cdot d\mathbf{\Gamma},$$
$$\int_\Omega \left[0 + |\nabla \phi|^2\right] d\Omega = \phi_0 \oint_\Gamma \frac{\partial \phi}{\partial n} d\mathbf{\Gamma}$$
$$= \phi_0 \cdot 0,$$

where $\oint_{\partial\Omega} \partial\phi/\partial n \, d\mathbf{\Gamma} = 0$ by (6.8) and $\Omega$ is simply connected. Therefore, $\nabla \phi(\mathbf{r}) = \mathbf{0}$ for $\mathbf{r} \in \Omega$, and so $\phi = C$, but since $\phi|_\Gamma = \phi_0$ and $\phi$ is continuous, $\phi(\mathbf{r}) = C = \phi_0$ inside $\Omega$. In particular, for homogeneous Dirichlet conditions, $\phi = \phi_0 = 0$. For homogeneous Neumann conditions, $\partial\phi/\partial n = 0$, we again get $\phi$ equal to a constant in $\Omega$, $\phi = C$, although $C$ is not necessarily zero.

on a bounded domain $\Omega$, subject to homogeneous Dirichlet conditions of the form (6.7) with $\alpha_{1,2} \in \mathbf{R}$, such that $L = -\nabla^2 : \mathbf{L}^2(\Omega) \to \mathbf{L}^2(\Omega)$ with $D_L$ described previously is self-adjoint and strictly positive. Since $L$ is self-adjoint, by Theorem 4.2.5 we have $\lambda_n \in \mathbf{R}$, where existence of the eigenvalues is discussed below. Furthermore, eigenfunctions $\phi_n$ of $L$ corresponding to different eigenvalues are orthogonal, and so we'll assume $\langle \phi_n, \phi_m \rangle = \delta_{nm}$. Since $L$ is strictly positive, by Theorem 4.19 we have $\lambda_n > 0$.[11]

With the above established, a statement analogous to Theorem 5.1 for the one-dimensional Sturm–Liouville operator can be formed. Because 0 is not an eigenvalue of $L$, $Lu = 0$ has only the trivial solution $u = 0$.[12] Then, by Theorem 3.24, $L$ is one-to-one and has an inverse defined on its range. As expected, the inverse operator will involve an integration over a Green's function kernel.

**Theorem 6.1.** *Let $L$ be the negative Laplacian operator ($L = -\nabla^2$) acting on a bounded open region $\Omega$ subject to homogeneous Dirichlet conditions. Then there exists a real-valued function $g(\mathbf{r}, \mathbf{r}') = g(\mathbf{r}', \mathbf{r})$ for $\mathbf{r}, \mathbf{r}' \in \Omega$ such that*

$$\left(L^{-1}v\right)(\Omega) = \int_\Omega g(\mathbf{r}, \mathbf{r}')v(\mathbf{r}')\, d\Omega' = \langle g, \overline{v} \rangle$$

*for all $v \in R_L$.*

The proof is omitted here, although properties of the Green's function such as existence, uniqueness, symmetry, and positivity are examined in [7, pp. 130–135]. In this reference it is also established that $L^{-1}$ is a compact operator[13] on $\mathbf{L}^2(\Omega)$ and is, in fact, Hilbert–Schmidt. Therefore,

- by Theorem 4.41 there exist real eigenvalues $\gamma_n$ and corresponding eigenvectors $u_n$ of $L^{-1}$ (i.e., $L^{-1}u_n = \gamma_n u_n$) such that $\{u_n\}$ forms a basis in $\mathbf{L}^2(\Omega)$;

- by Theorem 4.11 $\{u_n\}$ are also the eigenvectors of $L$ (i.e., $u_n = \phi_n$), with corresponding eigenvalues $\lambda_n = 1/\gamma_n$. Because $\lambda_n > 0$, then $\gamma_n > 0$;

---

[11]This is also easily shown directly. Multiplying (6.11) by $\overline{\phi}_n$ and integrating over $\Omega$ we get

$$\lambda_n \|\phi_n\|^2 = -\int_\Omega \overline{\phi}_n \nabla^2 \phi_n d\Omega = \int_\Omega \nabla\phi_n \cdot \nabla\overline{\phi}_n d\Omega = \int_\Omega |\nabla\phi_n|^2\, d\Omega > 0$$

using Green's first theorem. Because $\phi_n \neq 0$, then $\|\phi_n\|^2 > 0$ and $\lambda_n > 0$.

[12]Note that $Lu = 0$ often has nontrivial solutions for nonhomogeneous boundary conditions.

[13]It is easy to see compactness by considering the form of the Green's function. For example, in three dimensions $g(\mathbf{r}, \mathbf{r}') = 1/(4\pi |\mathbf{r} - \mathbf{r}'|) + g^h(\mathbf{r}, \mathbf{r}')$, where $\nabla^2 g^h = 0$. Therefore, $g$ represents a weakly singular kernel that generates a weakly singular integral operator on a bounded region $\Omega$. Then, by an example in Section 3.6.2, $L^{-1}$ is compact on $\mathbf{L}^2(\Omega)$.

- by Theorem 4.40 (Hilbert–Schmidt), $\lim_{n\to\infty} \gamma_n = 0$ and $\lim_{n\to\infty} \lambda_n = \infty$.

From completeness, for every $\phi \in \mathbf{L}^2(\Omega)$ we have

$$\phi = \sum_{n=1}^{\infty} \langle \phi, \phi_n \rangle \phi_n$$

in the norm sense, where $\phi_n \in \mathbf{C}^2(\Omega) \cap \mathbf{C}(\widetilde{\Omega})$. In addition, for every $\phi \in D_L$ we have

$$L\phi = \sum_{n=1}^{\infty} \lambda_n \langle \phi, \phi_n \rangle \phi_n,$$

where $\sum_{n=1}^{\infty} \lambda_n^2 \left| \langle \phi, \phi_n \rangle \right|^2 < \infty$.

For the eigenvalue problem $L\phi_n = \lambda_n \phi_n$ on a bounded domain $\Omega$ subject to homogeneous Neumann conditions (where $L$ is self-adjoint and nonnegative) we have $\lambda_n \in \mathbf{R}$, orthogonality of the eigenfunctions associated with different eigenvalues, and $\lambda_n \geq 0$. Thus, the above comments on the spectral properties of the operator similarly apply for the Neumann problem, although the possibility of $\lambda_n = 0$ complicates the analysis. This was also the case for the Sturm–Liouville problem considered in Chapter 5, where $\lambda_n = 0$ could not, in general, be ruled out.

Taken alone, the domain of $L^{-1}$ could be simply piecewise continuous functions $\rho$ of bounded support, yielding a convergent weakly singular integral representation of a function, which, for the exterior problem, is regular at infinity. However, for its identification as the inverse to the negative Laplacian operator $L$ on its specified domain, we should take the domain of $L^{-1}$ to be functions $\rho$ at least Hölder continuous at all points in $\Omega$, more strongly, with the same $\alpha$ (i.e., $\rho \in \mathbf{H}_{0,\alpha}(\widetilde{\Omega})$), such that the associated potential $\phi = L^{-1}\rho_e$ possesses continuous second partial derivatives in $\Omega$ that satisfy Poisson's equation (i.e., $\phi = L^{-1}\rho_e \in \mathbf{C}^2(\Omega) \cap \mathbf{C}(\widetilde{\Omega})$).

## 6.4 Solution Techniques for Self-Adjoint Problems

### Green's Functions Methods

We consider the solution of

$$(-\nabla^2 - \lambda)\phi(\mathbf{r}) = \frac{\rho_e(\mathbf{r})}{\varepsilon},$$
$$(-\nabla^2 - \lambda)g(\mathbf{r}, \mathbf{r}') = \delta(\mathbf{r} - \mathbf{r}'),$$

subject to $B(\phi) = B(g) = 0$, with $B$ given by (6.7) where $\alpha_{1,2} \in \mathbf{R}$. Substituting these equations into Green's second theorem leads to the solution

(see Section 1.3.4)

$$\phi(\mathbf{r}) = \lim_{\delta \to 0} \int_{\Omega - \Omega_\delta} \frac{\rho_e(\mathbf{r}')}{\varepsilon} g(\mathbf{r}, \mathbf{r}', \lambda)\, d\Omega'. \tag{6.21}$$

If the charge density has support in $\Omega_c \subset \Omega$, then the volume integral extends only over $\Omega_c$, yet (6.21) is valid anywhere in $\Omega$, both inside and outside $\Omega_c$.

Properties of the volume integral (6.21) were discussed in Section 1.3.4. To summarize, consider the source density $\rho_e$ to exist over a bounded region $\Omega_c \subset \Omega$.

- For $\mathbf{r} \notin \Omega_c$ the volume integral represents a proper convergent integral over fixed limits, which can be differentiated arbitrarily often ($\phi \in \mathbf{C}^\infty(\Omega \backslash \widetilde{\Omega}_c)$), with derivatives brought under the integral sign. In this region exterior to the charge density the potential is harmonic. For the exterior problem, since $\rho_e$ is a bounded density, $\phi$ is regular at infinity [5, p. 144].

- For $\mathbf{r} \in \Omega_c$ the improper volume integral associated with the principal Green's function is uniformly convergent to a continuous function $\phi(\mathbf{r})$, and differentiable with first derivatives allowed to be taken under the integral sign, resulting in a function everywhere continuous, i.e., $\phi(\mathbf{r}) \in \mathbf{C}^1(\widetilde{\Omega})$. If the charge density is Hölder continuous ($\rho_e \in \mathbf{H}_{0,\alpha}(\widetilde{\Omega}_c)$), second partial derivatives of the volume integral generally exist as well and are continuous interior to $\Omega_c$ but not on its boundary. These second derivatives satisfy Poisson's equation.

Because for $\mathbf{r} \in \Omega$ (6.21) provides a function at least $\mathbf{C}^1(\widetilde{\Omega})$ that may be differentiated under the integral sign, in obtaining the electric field as $\mathbf{E} = -\nabla\phi$ the gradient operator may be passed inside the integral, resulting in

$$\mathbf{E}(\mathbf{r}) = -\lim_{\delta \to 0} \int_{\Omega - \Omega_\delta} \frac{\rho_e(\mathbf{r}')}{\varepsilon} \nabla g(\mathbf{r}, \mathbf{r}', \lambda)\, d\Omega'. \tag{6.22}$$

For the electrostatic problems considered here the Green's function satisfies (6.13) with $\lambda = 0$, i.e. (dropping the $\lambda$ designation),

$$\nabla^2 g(\mathbf{r}, \mathbf{r}') = -\delta(\mathbf{r} - \mathbf{r}') \tag{6.23}$$

subject to boundary condition $B(g) = 0$. The Green's function can be obtained using spectral methods as an eigenfunction expansion, which is discussed later in this section, or in some other form, perhaps using scattering superposition or image theory.

### Green's Function via Scattering Superposition

To obtain the Green's function for (6.21) using scattering superposition, we write (repeating from (1.45) and (1.46))

$$g(\mathbf{r},\mathbf{r}') = g^p(\mathbf{r},\mathbf{r}') + g^h(\mathbf{r},\mathbf{r}'),$$

where $g^p$ is the principal Green's function, satisfying

$$\nabla^2 g^p(\mathbf{r},\mathbf{r}') = -\delta(\mathbf{r}-\mathbf{r}')$$

and $g^h$ is the scattered Green's function, satisfying

$$\nabla^2 g^h(\mathbf{r},\mathbf{r}') = 0 \tag{6.24}$$

such that together $g = g^p + g^h$ satisfies the boundary condition $B(g) = 0$. In three dimensions (in fact, for $m \geq 3$) we have

$$g^p(\mathbf{r},\mathbf{r}') = \frac{1}{4\pi\left|\mathbf{r}-\mathbf{r}'\right|}, \tag{6.25}$$

and for the two-dimensional case ($m = 2$)

$$g^p(\mathbf{r},\mathbf{r}') = -\frac{1}{2\pi}\ln(\left|\mathbf{r}-\mathbf{r}'\right|). \tag{6.26}$$

In any event, one must be able to determine $g$ obeying $B(g) = 0$. Solution techniques for when such a Green's function cannot be found are discussed in Section 6.6.

**Example 6.1.**

Consider Figure 6.3, which depicts a charge density having support in a bounded domain $\Omega_c$ in the vicinity of an infinite grounded half-plane $\Gamma$. To determine the potential for $z \geq 0$ we need to solve Poisson's equation (6.4) subject to homogeneous Dirichlet conditions,

$$\begin{aligned} -\nabla^2\phi &= \frac{\rho_e}{\varepsilon}, \\ \phi\big|_{z=0} &= 0. \end{aligned} \tag{6.27}$$

The *Dirichlet Green's function* is

$$g^d(\mathbf{r},\mathbf{r}') = \frac{1}{4\pi\left|\mathbf{r}-\mathbf{r}'\right|} + g^h(\mathbf{r},\mathbf{r}') \tag{6.28}$$

with $g^h$ chosen so that $g^d(\mathbf{r},\mathbf{r}')\big|_{\mathbf{r}\in\Gamma} = 0$. With $\mathbf{r}' = \widehat{\mathbf{x}}x' + \widehat{\mathbf{y}}y' + \widehat{\mathbf{z}}z'$, from image theory it is clear that we should choose $g^h(\mathbf{r},\mathbf{r}') = -1/(4\pi\left|\mathbf{r}-\mathbf{r}'_i\right|)$, where $\mathbf{r}'_i = \widehat{\mathbf{x}}x' + \widehat{\mathbf{y}}y' - \widehat{\mathbf{z}}z'$. Therefore,

$$g^d(\mathbf{r},\mathbf{r}') = \frac{1}{4\pi\left|\mathbf{r}-\mathbf{r}'\right|} - \frac{1}{4\pi\left|\mathbf{r}-\mathbf{r}'_i\right|}, \tag{6.29}$$

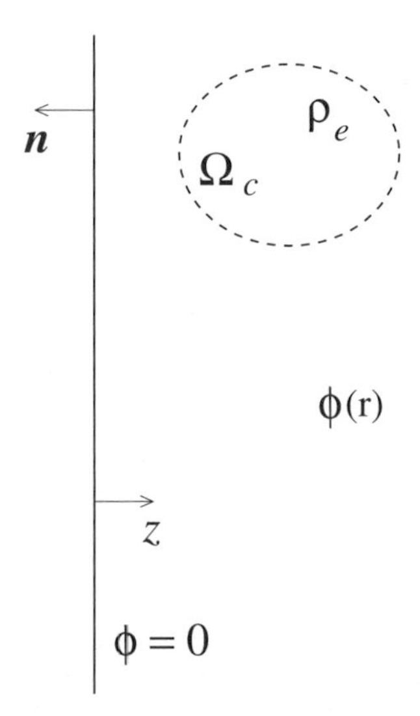


Figure 6.3: Charge density having support in a bounded domain $\Omega_c \subset \mathbf{R}^m$ in the vicinity of a grounded half-plane.

which also is regular at infinity for the half-space $z > 0$. Substituting into (6.21), the solution is

$$\phi(\mathbf{r}) = \lim_{\delta \to 0} \int_{\Omega_c - \Omega_\delta} \frac{\rho_e(\mathbf{r}')}{\varepsilon} \left[ \frac{1}{4\pi |\mathbf{r} - \mathbf{r}'|} - \frac{1}{4\pi |\mathbf{r} - \mathbf{r}_i'|} \right] d\Omega'. \tag{6.30}$$

### Eigenfunction Expansion Solutions

Continuing with the self-adjoint problem, the solution of

$$(L - \lambda) \phi = f$$

subject to $B(\phi) = 0$ by eigenfunction expansion proceeds as follows. The eigenfunctions are chosen such that $B(\phi_n) = 0$. With $\phi = \sum_{n=1}^{\infty} \langle \phi, \phi_n \rangle \phi_n$ and $L\phi = \sum_{n=1}^{\infty} \lambda_n \langle \phi, \phi_n \rangle \phi_n$, we have

$$(L - \lambda) \phi = \sum_{n=1}^{\infty} \lambda_n \langle \phi, \phi_n \rangle \phi_n - \lambda \sum_{n=1}^{\infty} \langle \phi, \phi_n \rangle \phi_n = f.$$

Taking the inner product of both sides with $\phi_m$ and exploiting orthonormality $\langle \phi_n, \phi_m \rangle = \delta_{nm}$ lead to $\langle \phi, \phi_m \rangle = (1/(\lambda_m - \lambda)) \langle f, \phi_m \rangle$, resulting in

$$\phi(\mathbf{r}) = \sum_{n=1}^{\infty} \frac{\langle f, \phi_n \rangle}{\lambda_n - \lambda} \phi_n(\mathbf{r}). \tag{6.31}$$

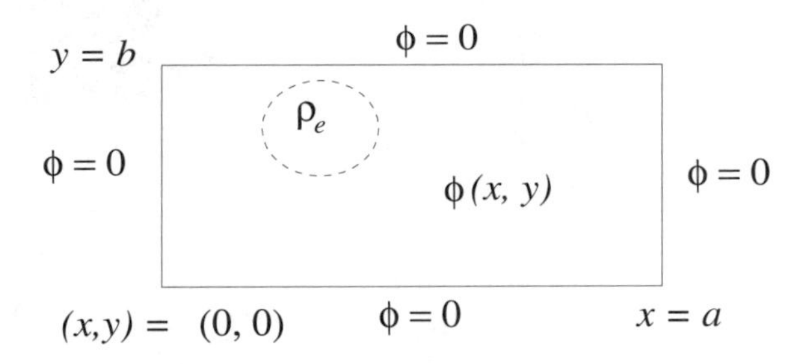


Figure 6.4: Charge density in a grounded rectangular region.

This is a generalized $m$-dimensional Fourier expansion solution with respect to the eigenfunctions of $L$, the negative Laplacian, and is analogous to the one-dimensional result (5.27) (recall that the summation in (6.31) is $m$-dimensional, representing an $m$-tuple of integers). Of course, the expansion does not exist if $\lambda = \lambda_n$, unless $\langle f, \phi_n \rangle = 0$ as discussed in Sections 3.11 and 4.5. In electrostatics we are interested in Poisson's equation ($\lambda = 0$), such that the potential is given as

$$\phi(\mathbf{r}) = \sum_{n=1}^{\infty} \frac{\left\langle \frac{\rho_e}{\varepsilon}, \phi_n \right\rangle}{\lambda_n} \phi_n(\mathbf{r}). \tag{6.32}$$

A similar form for the Green's function is also available. If we take the Green's function to be defined as

$$(L - \lambda)\, g(\mathbf{r}, \mathbf{r}', \lambda) = \delta(\mathbf{r} - \mathbf{r}')$$

with $B(g) = 0$, then formally substituting $\delta(\mathbf{r} - \mathbf{r}')$ for $f$ in (6.31) leads to

$$g(\mathbf{r}, \mathbf{r}', \lambda) = \sum_{n=1}^{\infty} \frac{\phi_n(\mathbf{r}) \overline{\phi}_n(\mathbf{r}')}{\lambda_n - \lambda}, \tag{6.33}$$

which is the multidimensional bilinear series form for $g$, analogous to (5.28). From (4.63) we also have the completeness relation

$$\frac{1}{2\pi i} \oint_C g(\mathbf{r}, \mathbf{r}', \lambda)\, d\lambda = -\delta(\mathbf{r} - \mathbf{r}'). \tag{6.34}$$

**Example 6.2.**

Consider the interior Dirichlet problem in two dimensions,

$$-\nabla^2 \phi = \frac{\rho_e}{\varepsilon},$$
$$\phi(\mathbf{r})\big|_{\mathbf{r} \in \Gamma} = 0,$$

with $\rho_e$ given, where the boundary $\Gamma$ is the closed rectangular path formed by $0 \leq x \leq a$ and $0 \leq y \leq b$ as shown in Figure 6.4. Eigenfunctions are defined by $-\nabla^2 \phi_n = \lambda_n \phi_n$ subject to $\phi_n(\mathbf{r})|_{\mathbf{r}\in\Gamma} = 0$. The eigenfunction problem is therefore

$$-\frac{\partial^2 \phi_n}{\partial x^2} - \frac{\partial^2 \phi_n}{\partial y^2} = \lambda \phi_n(x, y) \tag{6.35}$$

solvable by separation of variables, which converts (6.35) into two simple Sturm–Liouville problems. Proceeding, we replace the single-index notation $\phi_n$ by the double-indexed $\phi_{nm}$ and assume

$$\phi_{nm}(x, y) = X_n(x) Y_m(y).$$

Substituting into (6.35) yields

$$-\frac{1}{X_n}\frac{\partial^2 X_n}{\partial x^2} - \frac{1}{Y_m}\frac{\partial^2 Y_m}{\partial y^2} - \lambda_{nm} = 0.$$

With the identification of $-\frac{1}{X_n}\frac{d^2 X_n}{dx^2} = k_{n,x}^2$ subject to $X_n(0) = X_n(a) = 0$, and $-\frac{1}{Y_m}\frac{d^2 Y_m}{dy^2} = k_{m,y}^2$ subject to $Y_m(0) = Y_m(b) = 0$, and noting the separation equation $k_{n,x}^2 + k_{m,y}^2 = \lambda_{nm}$, we have

$$X_n(x) = \sqrt{\frac{2}{a}} \sin \frac{n\pi}{a} x,$$

$$Y_m(y) = \sqrt{\frac{2}{b}} \sin \frac{m\pi}{b} y,$$

where $k_{n,x} = n\pi/a$ and $k_{m,y} = m\pi/b$, such that eigenvalues are found to be

$$\lambda_{nm} = \left(\frac{n\pi}{a}\right)^2 + \left(\frac{m\pi}{b}\right)^2,$$

$n, m = 1, 2, 3, \ldots$. Therefore, the orthonormal eigenfunctions are

$$\phi_{nm}(x, y) = \sqrt{\frac{4}{ab}} \sin \frac{n\pi}{a} x \sin \frac{m\pi}{b} y,$$

which form a basis for $\mathbf{L}^2(\Omega)$, where $\Omega = \{x, y : x \in (0, a),\ y \in (0, b)\}$. Note that this product form is consistent with Theorem 2.16 and that, in this case, $\phi_{nm} \in \mathbf{C}^\infty(\widetilde{\Omega})$. The eigenfunction expansion solution to the given problem is, from (6.32),

$$\phi(\mathbf{r}) = \frac{4}{ab} \sum_{n=1}^{\infty} \sum_{m=1}^{\infty} \frac{\int_\Omega \rho_e(\mathbf{r}) \sin \frac{n\pi}{a} x \sin \frac{m\pi}{b} y \, d\Omega}{\left(\frac{n\pi}{a}\right)^2 + \left(\frac{m\pi}{b}\right)^2} \sin \frac{n\pi}{a} x \sin \frac{m\pi}{b} y. \tag{6.36}$$

An alternative view is to consider the Green's function from (6.33),

$$g(\mathbf{r}, \mathbf{r}', \lambda) = \frac{4}{ab} \sum_{n=1}^{\infty} \sum_{m=1}^{\infty} \frac{\sin \frac{n\pi}{a} x \sin \frac{m\pi}{b} y \sin \frac{n\pi}{a} x' \sin \frac{m\pi}{b} y'}{\left(\frac{n\pi}{a}\right)^2 + \left(\frac{m\pi}{b}\right)^2 - \lambda}, \tag{6.37}$$

with the solution for the potential given by (6.21). Furthermore, from (6.34) we have

$$\begin{aligned}\delta(\mathbf{r}-\mathbf{r}') &= \frac{-1}{2\pi i}\oint_C g(\mathbf{r},\mathbf{r}',\lambda)\,d\lambda\\ &= \sum_{n=1}^{\infty}\sum_{m=1}^{\infty}\sqrt{\frac{4}{ab}}\sin\frac{n\pi}{a}x\,\sin\frac{m\pi}{b}y\,\sqrt{\frac{4}{ab}}\sin\frac{n\pi}{a}x'\,\sin\frac{m\pi}{b}y'.\end{aligned}$$

**Partial Eigenfunction Expansion**

Note that the eigenfunction expansion in $m$ dimensions typically results in an $m$-fold summation. This purely spectral form can often be reduced to a lower-dimension summation using the concept of a *partial eigenfunction expansion*, which consists of separating the $m$-dimensional eigenvalue problem into $m$ one-dimensional Sturm–Liouville eigenvalue problems (other variations are also possible). All, or at least some, of the resulting Sturm–Liouville problems generates a complete set of functions in the corresponding space. The $m$-dimensional unknown is then expanded in one of these complete sets of one-dimensional eigenfunctions, with expansion coefficients depending on the remaining $(m-1)$ spatial coordinates and the $m$ primed coordinates. This generates an $(m-1)$ dimensional nonhomogeneous differential equation for the expansion coefficients. If this differential equation for the coefficients can be solved in closed form, then the result is a one-dimensional (rather than an $m$-dimensional) summation for the $m$-dimensional solution. If the $(m-1)$ dimensional differential equation for the coefficients cannot be solved in closed form, usually the process can be repeated. Since this procedure is fairly common, an example will not be included here, although the technique is used later (see, e.g., Chapter 8). Numerous other examples can be found in the literature, e.g. [7, pp. 154–164], [8].

## 6.5 Integral Methods and Separation of Variables Solutions for Nonself-Adjoint Problems

We now consider some generally nonself-adjoint problems, in particular those arising from nonhomogeneous Dirichlet or Neumann conditions. As stated earlier, we generally don't need to consider the Poisson problem with nonhomogeneous conditions, since this can be decomposed into a self-adjoint Poisson problem with homogeneous boundary conditions (6.19), with a solution of the form (6.21), and a Laplace problem with nonhomogeneous conditions (6.20). Having considered the self-adjoint Poisson problem, we now consider the nonself-adjoint Laplace problem

$$\begin{aligned} -\nabla^2 \phi &= 0, \\ B(\phi) &= \eta, \end{aligned} \tag{6.38}$$

where we will assume $\eta \in \mathbf{R}$.

### Integral Methods for Nonself-Adjoint Laplace Problems

The two principal methods commonly applied to the Laplace problem are the Green's function technique and separation of variables. Starting with the former, since $L\phi = 0$ and $Lg = \delta$, substitution into Green's second theorem leads to

$$\phi(\mathbf{r}') = -\oint_\Gamma [\phi(\mathbf{r})\nabla g(\mathbf{r},\mathbf{r}') - g(\mathbf{r},\mathbf{r}')\nabla\phi(\mathbf{r})] \cdot d\mathbf{\Gamma} \tag{6.39}$$

for $\mathbf{r}' \in \Omega$. Since $\nabla\phi(\mathbf{r}) \cdot \mathbf{n} = \partial\phi(\mathbf{r})/\partial n = -\mathbf{n} \cdot \mathbf{E}(\mathbf{r})$, the second term in the boundary integral is related to the normal component of electric field on the boundary (normal derivative of potential), which is associated with boundary charge, whereas the first term simply relates to the potential on the boundary.

Note that in (6.39) we enforce $B(\phi) = \eta$, but we have not yet specified any boundary conditions on $g$. Furthermore, (6.39) satisfies Laplace's equation for any $g$ a solution of (6.23) and represents a valid integral representation (assuming sufficient differentiability) for $\phi$. However, for (6.39) to provide an explicit solution of the given boundary value problem, one needs to choose $g$ accordingly. It is clear that for Dirichlet boundary conditions

$$\phi(\mathbf{r})\big|_{\mathbf{r}\in\Gamma} = \eta_d(\mathbf{r})$$

we need to choose $g = g^d(\mathbf{r},\mathbf{r}')$ such that

$$g^d(\mathbf{r},\mathbf{r}')\big|_{\mathbf{r}\in\Gamma} = 0.$$

The integral relation (6.39) then becomes

$$\phi(\mathbf{r}') = -\oint_\Gamma \eta_d(\mathbf{r})\nabla g^d(\mathbf{r},\mathbf{r}') \cdot d\mathbf{\Gamma}. \tag{6.40}$$

For Neumann conditions

$$\left.\frac{\partial\phi(\mathbf{r})}{\partial n}\right|_{\mathbf{r}\in\Gamma} = \eta_n(\mathbf{r})$$

it would seem that we need to choose $g = g^n(\mathbf{r},\mathbf{r}')$ such that $\left.\frac{\partial g^n(\mathbf{r},\mathbf{r}')}{\partial n}\right|_{\mathbf{r}\in\Gamma} = 0$; however, this is generally not consistent with the existence constraint

(6.8). Applying (6.8) to $\nabla^2 g^n(\mathbf{r},\mathbf{r}') = -\delta(\mathbf{r}-\mathbf{r}')$ provides $\int_\Gamma \partial g^n/\partial n \, d\Gamma =$ 1, so that, rather than $\left.\frac{\partial g^n(\mathbf{r},\mathbf{r}')}{\partial n}\right|_{\mathbf{r}\in\Gamma} = 0$, we can require (not uniquely)

$$\left.\frac{\partial g^n(\mathbf{r},\mathbf{r}')}{\partial n}\right|_{\mathbf{r}\in\Gamma} = \frac{1}{S},$$

which is known as a *Neumann Green's function*, where $S = \int_\Gamma d\Gamma$. With the Neumann Green's function, (6.39) becomes

$$\begin{aligned}\phi(\mathbf{r}') &= -\oint_\Gamma \phi(\mathbf{r})\nabla\, g(\mathbf{r},\mathbf{r}')\cdot d\mathbf{\Gamma} + \oint_\Gamma g^n(\mathbf{r},\mathbf{r}')\eta_n(\mathbf{r})\, d\Gamma \\ &= -\frac{1}{S}\oint_\Gamma \phi(\mathbf{r})\, d\Gamma + \oint_\Gamma g^n(\mathbf{r},\mathbf{r}')\eta_n(\mathbf{r})\, d\Gamma.\end{aligned} \tag{6.41}$$

Although the above is still not an explicit expression for the potential, if $S = \int_\Gamma d\Gamma \to \infty$, then, assuming $\oint_\Gamma \phi(\mathbf{r})\, d\Gamma < \infty$, the first term vanishes and we obtain

$$\phi(\mathbf{r}') = \oint_\Gamma g^n(\mathbf{r},\mathbf{r}')\eta_n(\mathbf{r})\, d\Gamma.$$

Note that for both the external and internal problems the left side of (6.39) is zero for $\mathbf{r}' \notin \Omega$.

By superposition, the solution of the Poisson problem $-\nabla^2\phi(\mathbf{r}) = \rho_e/\varepsilon$ subject to $B(\phi) = \eta$ is obtained by adding (6.21) and (6.39), consistent with (1.52). If one does not so decompose the Poisson problem as described, then (1.52) provides the desired form, where $g$ is any solution of (6.23). This provides an explicit solution to Poisson's problem if $g$ is chosen to have the appropriate boundary behavior.

**Example 6.3.**

Consider Figure 6.5, which depicts a charge density having support in a bounded domain $\Omega_c$ in the vicinity of a half-plane. On the half-plane the potential is $\eta(x,y)$. To determine the potential for $z \geq 0$ we need to solve Poisson's equation (6.4) subject to nonhomogeneous Dirichlet conditions,

$$\begin{aligned}-\nabla^2\phi &= \frac{\rho_e}{\varepsilon}, \\ \phi|_{z=0} &= \eta(x,y).\end{aligned} \tag{6.42}$$

From Example 6.1, the solution of

$$\begin{aligned}-\nabla^2\phi &= \frac{\rho_e}{\varepsilon}, \\ \phi|_{z=0} &= 0\end{aligned} \tag{6.43}$$

is

$$\phi(\mathbf{r}) = \lim_{\delta\to 0}\int_{\Omega_c-\Omega_\delta} \frac{\rho_e(\mathbf{r}')}{\varepsilon} g^d(\mathbf{r},\mathbf{r}')\, d\Omega',$$

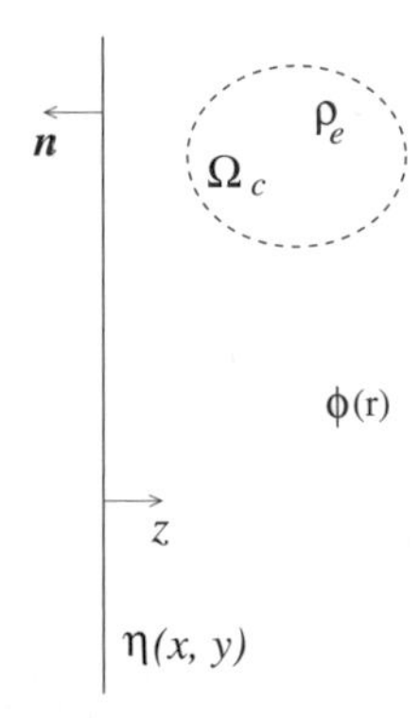


Figure 6.5: Charge density having support in a bounded domain $\Omega_c \subset \mathbf{R}^m$ in the vicinity of a half-plane having potential $\eta(x, y)$.

where the Dirichlet Green's function is (6.29)

$$g^d(\mathbf{r}, \mathbf{r}') = \frac{1}{4\pi |\mathbf{r} - \mathbf{r}'|} - \frac{1}{4\pi |\mathbf{r} - \mathbf{r}_i'|}$$

with $\mathbf{r}' = \widehat{\mathbf{x}}x' + \widehat{\mathbf{y}}y' + \widehat{\mathbf{z}}z'$ and $\mathbf{r}_i' = \widehat{\mathbf{x}}x' + \widehat{\mathbf{y}}y' - \widehat{\mathbf{z}}z'$. Therefore, by superposition the solution of (6.42) is, from (6.39) and noting that

$$\nabla g^d(\mathbf{r}, \mathbf{r}') \cdot \mathbf{n}\big|_{z=0} = -\frac{\partial g^d}{\partial z}\bigg|_{z=0} = -\frac{1}{2\pi} \frac{z'}{[(x - x')^2 + (y - y')^2 + z'^2]^{3/2}},$$

$$\begin{aligned} \phi(\mathbf{r}) &= \lim_{\delta \to 0} \int_{\Omega_c - \Omega_\delta} \frac{\rho_e(\mathbf{r}')}{\varepsilon} g^d(\mathbf{r}, \mathbf{r}')\, d\Omega' \\ &+ \frac{z}{2\pi} \int_{-\infty}^{\infty} \int_{-\infty}^{\infty} \frac{\eta(x', y')}{[(x - x')^2 + (y - y')^2 + z^2]^{3/2}} dx'\, dy'. \end{aligned} \tag{6.44}$$

For the special case of a constant potential $\eta(x, y) = \phi_0$ (perhaps $\phi_0 = 0$ as in Example 6.1, representing a grounded conductor), we have

$$\phi(\mathbf{r}) = \lim_{\delta \to 0} \int_{\Omega_c - \Omega_\delta} \frac{\rho_e(\mathbf{r}')}{\varepsilon} g^d(\mathbf{r}, \mathbf{r}')\, d\Omega' + \phi_0, \tag{6.45}$$

using $\int_{-\infty}^{\infty} \int_{-\infty}^{\infty} \left[(x - x')^2 + (y - y')^2 + z^2\right]^{-3/2} dx\, dy = 2\pi/z$.

Note also that for the half-plane problem, $S = \int_{-\infty}^{\infty} \int_{-\infty}^{\infty} d\Gamma \to \infty$. Assuming $\int_{-\infty}^{\infty} \int_{-\infty}^{\infty} \eta(x, y)\, dx\, dy < \infty$, we can also solve the Neumann problem where $\left.\frac{\partial \phi(\mathbf{r})}{\partial n}\right|_{\mathbf{r} \in \Gamma} = \eta(\mathbf{r})$ is specified, with the Neumann Green's function $\left.\frac{\partial g^n(\mathbf{r}, \mathbf{r}')}{\partial n}\right|_{\mathbf{r} \in \Gamma} = 0$. It can be easily found that

$$g^n(\mathbf{r}, \mathbf{r}') = \frac{1}{4\pi |\mathbf{r} - \mathbf{r}'|} + \frac{1}{4\pi |\mathbf{r} - \mathbf{r}_i'|}$$

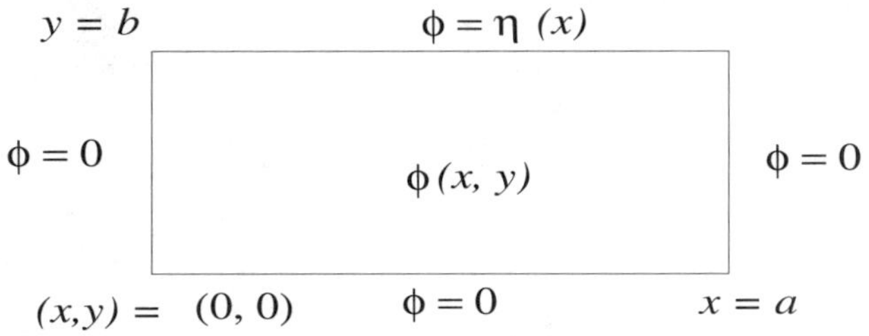

Figure 6.6: Rectangular region carrying potential $\eta$ on one side.

with the solution

$$\phi(\mathbf{r}) = \lim_{\delta \to 0} \int_{\Omega_c - \Omega_\delta} \frac{\rho_e(\mathbf{r}')}{\varepsilon} g^n(\mathbf{r}, \mathbf{r}')\, d\Omega' + \int_{-\infty}^{\infty} \int_{-\infty}^{\infty} g^n(\mathbf{r}, \mathbf{r}')\eta(\mathbf{r}')\, dx'\, dy', \tag{6.46}$$

which provides an explicit expression for $\phi$.

### Separation of Variables for Nonself-Adjoint Laplace Problems

The second method of attacking Laplace's equation is via separation of variables. For the case of nonhomogeneous boundary conditions this method depends on the principle of superposition.[14] The method is most easily illustrated by a simple example.

#### Example 6.4.

Consider the interior Dirichlet problem in two dimensions depicted in Figure 6.6,

$$-\nabla^2 \phi = 0, \tag{6.47}$$

where the boundary $\Gamma$ is the closed rectangular path formed by $0 \leq x \leq a$ and $0 \leq y \leq b$. The boundary conditions are

$$\begin{aligned} \phi(0, y) = \phi(a, y) &= 0, \qquad && 0 \leq y \leq b, \\ \phi(x, 0) &= 0, && 0 \leq x \leq a, \\ \phi(x, b) &= \eta(x), && 0 \leq x \leq a, \end{aligned} \tag{6.48}$$

where $\eta(x)$ is a given, real-valued piecewise-continuous function.

The solution of Laplace's equation (6.47) is separated as $\phi(x, y) = X(x)Y(y)$, leading to

$$-\frac{1}{X}\frac{\partial^2 X}{\partial x^2} - \frac{1}{Y}\frac{\partial^2 Y}{\partial y^2} = 0.$$

[14]Fourier series-type expansions can also be used, but are more cumbersome.

With the identification of $-\frac{1}{X}\frac{d^2X}{dx^2} = k_x^2$ subject to $X(0) = X(a) = 0$, we have

$$X(x) = X_n(x) = A_n \sin\frac{n\pi}{a}x,$$

where $k_x = k_{n,x} = n\pi/a$.

From $-\frac{1}{Y}\frac{d^2Y}{dy^2} = k_y^2$ subject to $Y(0) = 0$, and noting the separation equation $k_x^2 + k_y^2 = 0$, we obtain

$$Y(y) = Y_n(y) = B_n \sin\frac{in\pi}{a}y = iB_n \sinh\frac{n\pi}{a}y$$

such that the solution is, so far,

$$\phi = C_n \sin\frac{n\pi}{a}x \sinh\frac{n\pi}{a}y.$$

We have yet to impose the boundary condition $\phi(x, b) = \eta(x)$ for $0 \le x \le a$. In fact, we cannot satisfy this condition with a single term of the form $\sin(n\pi x/a)\sinh(n\pi y/a)$, and so, noting that a sum of homogeneous solutions is also a homogeneous solution, we form

$$\phi(x, y) = \sum_{n=1}^{\infty} C_n \sin\frac{n\pi}{a}x \ \sinh\frac{n\pi}{a}y$$

and enforce the remaining boundary condition. Proceeding, we have

$$\phi(x, b) = \sum_n C_n \sin\frac{n\pi}{a}x \sinh\frac{n\pi}{a}b = \eta(x),$$

which, when multiplied by $\sin(m\pi x/a)$ and integrated over $(0, a)$, leads to $C_n = \frac{2}{a}\frac{\int_0^a \eta(x)\sin\frac{n\pi}{a}x\,dx}{\sinh\frac{n\pi}{a}b}$. The final result is then

$$\phi(x, y) = \frac{2}{a}\sum_{n=1}^{\infty}\frac{\int_0^a \eta(x)\sin\frac{n\pi}{a}x\,dx}{\sinh\frac{n\pi}{a}b}\sin\frac{n\pi}{a}x \ \sinh\frac{n\pi}{a}y.$$

Because of the exponential increase of the hyperbolic sine function for moderately large arguments, the series converges rapidly as long as $y$ is not too close to $b$.

It is easy to see that the three null boundary conditions in (6.48) are identically satisfied due to the vanishing of the eigenfunctions at the boundary and that the boundary condition at $y = b$ is satisfied in the norm sense. Indeed,

$$\lim_{y\to b}\|\phi(x, y) - \eta(x)\| = \|\phi(x, b) - \eta(x)\|$$

by continuity of the norm, Section 2.5. Furthermore,

$$\begin{aligned}
\|\phi(x,b)-\eta(x)\|^2 &= \langle \phi(x,b)-\eta(x), \phi(x,b)-\eta(x)\rangle \\
&= \int_0^a (\phi(x,b)-\eta(x))^2\, dx \\
&= \int_0^a \left(\sqrt{\frac{2}{a}}\sum_{n=1}^{\infty} a_n \sin\frac{n\pi}{a}x - \eta(x)\right)^2 dx \\
&= \frac{2}{a}\sum_{n=1}^{\infty} a_n^2 \int_0^a \sin^2\frac{n\pi}{a}x\, dx - 2\sum_{n=1}^{\infty} a_n^2 + \int_0^a \eta^2(x)\, dx \\
&= -\sum_{n=1}^{\infty} a_n^2 + \|\eta\|^2,
\end{aligned}$$

where

$$a_n \equiv \sqrt{\frac{2}{a}}\int_0^a \eta(x)\sin\frac{n\pi}{a}x\, dx.$$

By Parseval's equality (2.16),

$$\sum_{n=1}^{\infty} |a_n|^2 = \|\eta\|^2$$

and therefore

$$\lim_{y\to b} \|\phi(x,y)-\eta(x)\| = -\|\eta\|^2 + \|\eta\|^2 = 0. \tag{6.49}$$

The success of any method for solving Laplace's equation depends on several factors. For the integral method, one must be able to determine an appropriate Dirichlet or Neumann Green's function, of course preferably in closed form. Various methods may be applied, including eigenfunction expansion, scattering superposition, and image theory. These methods will prove to be difficult for all but the simplest boundaries (canonical shapes like parallelepipeds and spheres, etc.) and will be impossible for many realistic geometries.

For the separation-of-variables method, one must have a surface adhering to a separable coordinate system. For three-dimensional problems 11 separable coordinate systems are identified in [9, p. 515], although of course many problems concerning quite simple geometries do not lead to separation of the Laplacian. For instance, the interior problem for both a parallelepiped and a spherical geometry is amenable to separation of variables, as is the exterior problem for a spherical structure, yet the exterior problem for the parallelepiped is not.

What can be done if the Dirichlet or Neumann's Green's function cannot be obtained in closed form, or if the problem is not separable? Several options exist. In the integral method one can formulate an integral equation, the solution of which leads to the harmonic term $g^h$ [7, pp. 146–147]. However, unless the integral equation can be solved analytically the computation of the potential must be obtained via a two-step numerical procedure. One can also chose a simple Green's function that satisfies (6.23), usually the free-space Green's function, and treat (6.39) or a similar integral relation as an integral equation to be solved, usually numerically, for the unknown $\phi$. This will be described in some detail below.

As an alternative, Laplace's or Poisson's equation can be solved directly using numerical differential equation methods, e.g., finite-difference techniques, although this topic is not discussed here.

## 6.6 Integral Equation Techniques for Potential Theory

One method to address the above-mentioned difficulties, when, for instance, an appropriate Green's function cannot be reasonably obtained, is via the formulation of integral equations. Although the resulting equations may need to be solved numerically, usually this can be accomplished using well-established techniques. Depending on the formulation, the associated integral equations can also be used to establish existence proofs and to ascertain solution properties. Moreover, integral equation formulations are often necessary for multiregion and scattering problems, especially those corresponding to complicated geometries. Here will we consider both the boundary value problem and the scattering problem, since they both can be analyzed using the same integral relations. Although not discussed here, the eigenvalue problem $-\nabla^2\phi_n = \lambda_n\phi_n$ can obviously be cast as an integral equation as well.

Consider the geometry depicted in Figure 6.7. The interior region $\Omega_1$, bounded by smooth boundary $\Gamma$, is characterized by constant permittivity $\varepsilon_1$ with source-charge density $\rho_1$. The region $\Omega_2$ is bounded by $\Gamma+\Gamma_\infty$ and characterized by constant permittivity $\varepsilon_2$, with source-charge density $\rho_2$. The potential is governed by

$$-\nabla^2\phi_i(\mathbf{r}) = \frac{\rho_i(\mathbf{r})}{\varepsilon_i}$$

for $\mathbf{r} \in \Omega_i$, $i = 1, 2$. The potential can be shown to satisfy the boundary conditions

$$(\phi_1 - \phi_2)|_\Gamma = 0, \tag{6.50}$$

$$\left.\left(\varepsilon_1 \frac{\partial \phi_1}{\partial n} - \varepsilon_2 \frac{\partial \phi_2}{\partial n}\right)\right|_\Gamma = \rho_s,$$

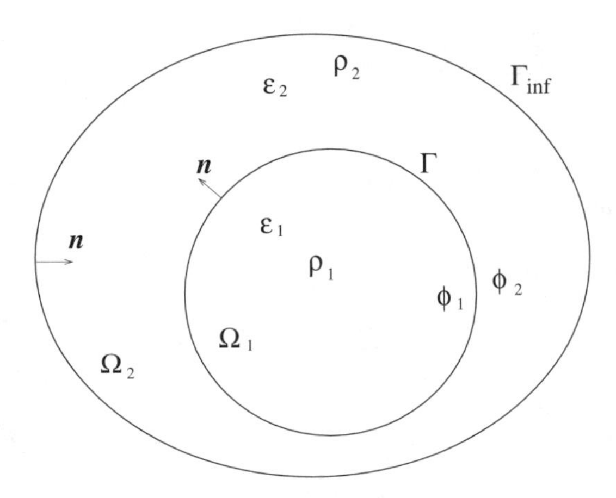


Figure 6.7: Two-region electrostatic problem. Charge $\rho_1$ inside region $\Omega_1$ characterized by $\varepsilon_1$ and bounded by $\Gamma$, and charge $\rho_2$ inside region $\Omega_2$ characterized by $\varepsilon_2$ and bounded by $\Gamma \cup \Gamma_\infty$.

where $\rho_s$ is a surface-charge density and $\partial/\partial n$ is the derivative normal to the boundary. Boundary conditions (6.50) can be obtained from (1.5) or by direct methods.

Green's functions are defined as

$$-\nabla^2 g_{1,2}(\mathbf{r}, \mathbf{r}') = \delta(\mathbf{r} - \mathbf{r}')$$

for $\mathbf{r} \in \Omega_{1,2}$. Inserting $\{\phi_1, g_1\}$ into Green's second theorem (see (1.52)) and noting that $\mathbf{n}$ points outward from $\Omega_1$, we have

$$\begin{array}{ll} \mathbf{r}' \in \Omega_1 & \phi_1(\mathbf{r}') \\ \mathbf{r}' \notin \Omega_1 & 0 \end{array} \Bigg\} = \lim_{\delta \to 0} \int_{\Omega_1 - \Omega_\delta} g_1(\mathbf{r}, \mathbf{r}') \, \frac{\rho_1(\mathbf{r})}{\varepsilon_1} \, d\Omega \tag{6.51}$$
$$- \oint_\Gamma \{\phi_1(\mathbf{r}) \nabla g_1(\mathbf{r}, \mathbf{r}') - g_1(\mathbf{r}, \mathbf{r}') \nabla \phi_1(\mathbf{r})\} \cdot d\mathbf{\Gamma}.$$

Repeating for region 2,

$$\begin{array}{ll} \mathbf{r}' \in \Omega_2 & \phi_2(\mathbf{r}') \\ \mathbf{r}' \notin \Omega_2 & 0 \end{array} \Bigg\} = \lim_{\delta \to 0} \int_{\Omega_2 - \Omega_\delta} g_2(\mathbf{r}, \mathbf{r}') \, \frac{\rho_2(\mathbf{r})}{\varepsilon_2} \, d\Omega \tag{6.52}$$
$$+ \oint_\Gamma \{\phi_2(\mathbf{r}) \nabla g_2(\mathbf{r}, \mathbf{r}') - g_2(\mathbf{r}, \mathbf{r}') \nabla \phi_2(\mathbf{r})\} \cdot d\mathbf{\Gamma},$$

where the positive sign in front on the boundary integral is due to $\mathbf{n}$ pointing inward to $\Omega_2$, and the boundary integral over $\Gamma_\infty$ has vanished assuming $\phi_2$ and $g_2$ are regular at infinity.

So far, integral relations (6.51) and (6.52) are not valid for $\mathbf{r}'$ on $\Gamma$. Because $\mathbf{r}$ ranges over $\Gamma$, if $\mathbf{r}'$ is some fixed point on $\Gamma$, then the principal part of the Green's function becomes singular, leading to an improper integral. It can be shown that $\oint_\Gamma g(\mathbf{r},\mathbf{r}')\nabla\phi(\mathbf{r})\cdot d\mathbf{\Gamma}$ is continuous as $\mathbf{r}' \to \mathbf{r}'_s \in \Gamma$ and exists as a convergent improper integral (for instance, in three dimensions $g^p = O(1/R)$ and $d\Gamma = O(R^2)$). The term $\oint_\Gamma \phi(\mathbf{r})\nabla g(\mathbf{r},\mathbf{r}')\cdot d\mathbf{\Gamma} = \oint_\Gamma \phi(\mathbf{r})\partial g(\mathbf{r},\mathbf{r}')/\partial n\, d\Gamma$, however, is discontinuous as $\mathbf{r}'$ approaches the boundary, such that

$$\begin{aligned}\lim_{\mathbf{r}'\to\mathbf{r}'_s}\oint_\Gamma \phi(\mathbf{r})\frac{\partial g(\mathbf{r},\mathbf{r}')}{\partial n}\,d\Gamma &= \mp\frac{\phi(\mathbf{r}'_s)}{2} + \oint_\Gamma \phi(\mathbf{r})\frac{\partial g(\mathbf{r},\mathbf{r}'_s)}{\partial n}\,d\Gamma,\\ \lim_{\mathbf{r}'\to\mathbf{r}'_s}\oint_\Gamma \phi(\mathbf{r})\frac{\partial g(\mathbf{r},\mathbf{r}')}{\partial n'}\,d\Gamma &= \pm\frac{\phi(\mathbf{r}'_s)}{2} + \oint_\Gamma \phi(\mathbf{r})\frac{\partial g(\mathbf{r},\mathbf{r}'_s)}{\partial n'}\,d\Gamma,\end{aligned} \tag{6.53}$$

where the upper sign is for the case of the limit being taken from the region into which $\mathbf{n}$ does not point, and the lower sign is for the case of the limit being taken from the region into which $\mathbf{n}$ points. This topic is considered in detail in many references; see, e.g., [7, Sec. 6.4] and [2, pp. 128–138]. Note that the continuity of $\oint_\Gamma g(\mathbf{r},\mathbf{r}')\nabla\phi(\mathbf{r})\cdot d\mathbf{\Gamma}$ and the discontinuity of $\oint_\Gamma \phi(\mathbf{r})\nabla g(\mathbf{r},\mathbf{r}')\cdot d\mathbf{\Gamma}$ as $\mathbf{r}' \to \mathbf{r}'_s$ are obtained for any $g$ that satisfies (6.23). This is because, from scattering superposition, we see that $g^p$ is the dominant term in $g = g^p + g^h$ near the source-point singularity, and so it is enough to consider $\nabla g^p$, the behavior of which leads to the above-described properties. If the boundary $\Gamma$ is not sufficiently smooth, then the term $\phi(\mathbf{r}'_s)/2$ needs to be modified. The appropriate expressions can be found in [2, pp. 222–228].

It is worthwhile to briefly describe the properties of the integral

$$f(\mathbf{r}'_s) = \oint_\Gamma \phi(\mathbf{r})\nabla' g(\mathbf{r},\mathbf{r}'_s)\cdot d\mathbf{\Gamma}$$

concentrating on the three-dimensional case. If $\phi \in \mathbf{C}(\Gamma)$, which is assumed everywhere in this section, then the improper integral $f(\mathbf{r}'_s)$ is uniformly Hölder continuous on $\Gamma$ (in fact, $f(\mathbf{r}'_s) \in H_{0,1/3}(\Gamma)$) and convergent [5, pp. 167, 300–301]. Recall from Section 1.3.4 that for a two-dimensional improper surface integral of the form considered here, a sufficient condition for convergence, assuming $\phi$ well behaved, is that the denominator is $O(R^n)$, $0 < n < 2$. The convergence of the surface integral[15]

$$\oint_\Gamma \phi(\mathbf{r})\frac{\partial g(\mathbf{r},\mathbf{r}'_s)}{\partial n'}\,d\Gamma = \oint_\Gamma \phi(\mathbf{r})\frac{\cos(\mathbf{r}-\mathbf{r}'_s,\mathbf{n})}{R^2(\mathbf{r},\mathbf{r}'_s)}\,d\Gamma$$

is obtained in this case even with the term $R^2$ in the denominator since as $\mathbf{r} \to \mathbf{r}'_s$ we have $\cos(\mathbf{r}-\mathbf{r}'_s,\mathbf{n}) \to 0$; therefore, the kernel is actually weakly singular [10, pp. 118–120],

$$\frac{|\cos(\mathbf{r}-\mathbf{r}'_s,\mathbf{n})|}{R^2(\mathbf{r},\mathbf{r}'_s)} \le \frac{C}{R^{(m-1-\alpha)}}$$

[15] $\cos(\mathbf{x},\mathbf{y})$ is the cos of the angle between the vectors $\mathbf{x}$ and $\mathbf{y}$.

in $m$-dimensional space for some constant $C$ and $\alpha > 0$. For the special case of $\mathbf{r}_s$ and $\mathbf{r}'_s$ lying on a planar surface $\Gamma$, $\cos(\mathbf{r}_s - \mathbf{r}'_s, \mathbf{n}) = 0$, and so in this case the source-point singularity is eliminated.

With the appropriate limiting procedure, (6.51) and (6.52) become[16]

$$\begin{array}{ll} \mathbf{r}' \in \Omega_1 & \phi_1(\mathbf{r}') \\ \mathbf{r}' = \mathbf{r}'_s \in \Gamma & \frac{\phi_1(\mathbf{r}')}{2} \\ \mathbf{r}' \in \Omega_2 & 0 \end{array} \left.\vphantom{\begin{array}{l} a \\ b \\ c \end{array}}\right\} = \lim_{\delta \to 0} \int_{\Omega_1 - \Omega_\delta} g_1(\mathbf{r}, \mathbf{r}') \frac{\rho_1(\mathbf{r})}{\varepsilon_1} \, d\Omega$$

$$- \oint_\Gamma \{\phi_1(\mathbf{r}) \nabla g_1(\mathbf{r}, \mathbf{r}') - g_1(\mathbf{r}, \mathbf{r}') \nabla \phi_1(\mathbf{r})\} \cdot d\mathbf{\Gamma} \quad (6.54)$$

and

$$\begin{array}{ll} \mathbf{r}' \in \Omega_2 & \phi_2(\mathbf{r}') \\ \mathbf{r}' = \mathbf{r}'_s \in \Gamma & \frac{\phi_2(\mathbf{r}')}{2} \\ \mathbf{r}' \in \Omega_1 & 0 \end{array} \left.\vphantom{\begin{array}{l} a \\ b \\ c \end{array}}\right\} = \lim_{\delta \to 0} \int_{\Omega_2 - \Omega_\delta} g_2(\mathbf{r}, \mathbf{r}') \frac{\rho_2(\mathbf{r})}{\varepsilon_2} \, d\Omega$$

$$+ \oint_\Gamma \{\phi_2(\mathbf{r}) \nabla g_2(\mathbf{r}, \mathbf{r}') - g_2(\mathbf{r}, \mathbf{r}') \nabla \phi_2(\mathbf{r})\} \cdot d\mathbf{\Gamma}. \quad (6.55)$$

Note that we have not imposed any boundary conditions on $g_{1,2}$. Assuming that a Dirichlet or Neumann Green's function is difficult to obtain, it is particularly convenient to take $g_1 = g_2 = g^p$, the principal Green's function corresponding to an unbounded region. Adding (6.54) and (6.55), and noting for the left sides that $\lim_{\mathbf{r}' \to \mathbf{r}'_s} \phi_1(\mathbf{r}') = \phi_s(\mathbf{r}'_s) = \lim_{\mathbf{r}' \to \mathbf{r}'_s} \phi_2(\mathbf{r}')$ from (6.50), where $\phi_s$ is the potential on the boundary, yield

$$\begin{array}{ll} \mathbf{r}' \in \Omega_1 & \phi_1(\mathbf{r}') \\ \mathbf{r}' = \mathbf{r}'_s \in \Gamma & \phi_s(\mathbf{r}') \\ \mathbf{r}' \in \Omega_2 & \phi_2(\mathbf{r}') \end{array} \left.\vphantom{\begin{array}{l} a \\ b \\ c \end{array}}\right\} = \lim_{\delta \to 0} \int_{\Omega_1 - \Omega_\delta} g^p(\mathbf{r}, \mathbf{r}') \frac{\rho_1(\mathbf{r})}{\varepsilon_1} \, d\Omega$$

$$+ \lim_{\delta \to 0} \int_{\Omega_2 - \Omega_\delta} g^p(\mathbf{r}, \mathbf{r}') \frac{\rho_2(\mathbf{r})}{\varepsilon_2} \, d\Omega$$

$$+ \oint_\Gamma g^p(\mathbf{r}, \mathbf{r}') \left[ \frac{\partial \phi_1(\mathbf{r})}{\partial n} - \frac{\partial \phi_2(\mathbf{r})}{\partial n} \right] d\Gamma.$$

For convenience we denote

$$\phi_{\text{inc}}^{(i)}(\mathbf{r}') \equiv \lim_{\delta \to 0} \int_{\Omega_i - \Omega_\delta} g^p(\mathbf{r}, \mathbf{r}') \frac{\rho_i(\mathbf{r})}{\varepsilon_i} \, d\Omega$$

[16]The terms $\oint_{\partial\Omega} g^p(\mathbf{r}, \mathbf{r}'_s) \nabla \phi(\mathbf{r}) \cdot d\mathbf{\Gamma}$ and $\oint_{\partial\Omega} \phi(\mathbf{r}) \nabla g^p(\mathbf{r}, \mathbf{r}'_s) \cdot d\mathbf{\Gamma}$ are known as the *single-layer* and *double-layer* potential, respectively.

for $i = 1, 2$; therefore,

$$\left.\begin{array}{ll} \mathbf{r}' \in \Omega_1 & \phi_1(\mathbf{r}') \\ \mathbf{r}' = \mathbf{r}'_s \in \Gamma & \phi_s(\mathbf{r}') \\ \mathbf{r}' \in \Omega_2 & \phi_2(\mathbf{r}') \end{array}\right\} = \phi_{\text{inc}}^{(1)}(\mathbf{r}') + \phi_{\text{inc}}^{(2)}(\mathbf{r}')$$

$$+ \oint_\Gamma g^p(\mathbf{r}, \mathbf{r}') \left[ \frac{\partial \phi_1(\mathbf{r})}{\partial n} - \frac{\partial \phi_2(\mathbf{r})}{\partial n} \right] d\Gamma. \quad (6.56)$$

Note also that $\phi_{\text{inc}}$ could be an externally imposed potential associated with an externally imposed electric field $\mathbf{E}_{\text{inc}}(\mathbf{r}') = -\nabla \phi_{\text{inc}}(\mathbf{r}')$.

### Boundary Value Problems Involving Conductors

At this point we need to be more specific, and so we will consider two distinct classes of problems of interest in electrostatics. The first is where the boundary $\Gamma$ is a perfect conductor (an equipotential surface), held at a known potential $\phi_s = \phi_0$. It is then natural to work with the boundary integral equation (6.56) for $\mathbf{r}' = \mathbf{r}'_s \in \Gamma$. We consider two cases.

### Region 1 Source-free ($\rho_1 = 0 \Rightarrow \phi_{\text{inc}}^{(1)} = 0$)

In this case $\phi_1 = \phi_0$ in $\Omega_1$ (we are assuming in this part that $\Omega_1$ is simply connected), and therefore $\left.\frac{\partial \phi_1}{\partial n}\right|_\Gamma = 0$ from continuity. The desired integral equation from (6.56) is, upon interchanging primed and unprimed coordinates,

$$\phi_0 - \phi_{\text{inc}}^{(2)}(\mathbf{r}_s) = -\oint_\Gamma g^p(\mathbf{r}_s, \mathbf{r}') \frac{\partial \phi_2(\mathbf{r}')}{\partial n'} \, d\Gamma'. \quad (6.57)$$

The problem of a conductor held at a given potential and immersed in an incident potential ($\phi_{\text{inc}}^{(2)} \neq 0$), and that of a conductor held at a given potential in a source-free space ($\phi_{\text{inc}}^{(2)} = 0$), can be treated by (6.57). In either case, if $\phi_0 - \phi_{\text{inc}}^{(2)}(\mathbf{r}'_s)$ is nonzero, then (6.57) is a nonhomogeneous first-kind integral equation with compact kernel. Once $\partial \phi_2 / \partial n$ is obtained from (6.57), the surface-charge density $\rho_s(\mathbf{r}_s)$ can be determined via $\rho_s(\mathbf{r}_s) = -\varepsilon_2 \partial \phi_2 / \partial n$, with the total charge residing on $\Gamma$ found as $Q = \int_\Gamma \rho_s(\mathbf{r}) \, d\Gamma$ and the potential in $\Omega_2$ found from inserting $\partial \phi_2 / \partial n$ in (6.56). Capacitance may be determined from

$$C = \frac{Q}{\phi_0} = \frac{1}{\phi_0} \int_\Gamma \rho_s(\mathbf{r}) \, d\Gamma.$$

Note also that for an open surface $\Gamma$ held at a constant potential $\phi_0$, as depicted in Figure 6.8, (6.57) becomes

$$\phi_0 - \phi_{\text{inc}}(\mathbf{r}_s) = \int_\Gamma g^p(\mathbf{r}_s, \mathbf{r}') \frac{\rho_s(\mathbf{r}')}{\varepsilon} \, d\Gamma', \quad (6.58)$$

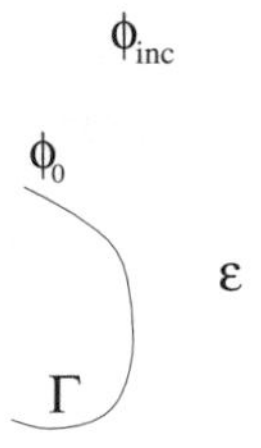


Figure 6.8: Open surface $\Gamma$ held at potential $\phi_0$ immersed in a homogeneous medium characterized by $\varepsilon$, where $\phi_{\text{inc}}$ represents an incident potential.

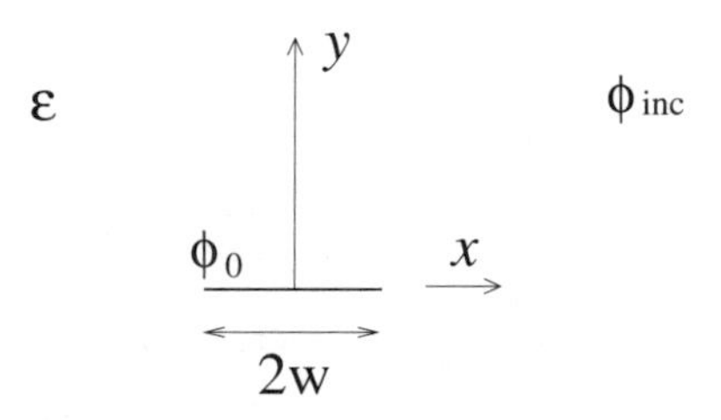


Figure 6.9: Perfectly conducting strip having width $2w$ held at potential $\phi_0$ in the presence of an incident potential $\phi_{\text{inc}}$.

where we have used $\rho_s = -\varepsilon \partial\phi/\partial n$.

Although in general (6.57) and (6.58) must be solved numerically, for some special cases exact analytical solutions can be found. For instance, if $\Gamma$ is a spherical or circular cylindrical surface, (6.57) can be solved using spherical harmonics (5.163) or Bessel functions, respectively. As another special case we have the following example.

**Example 6.5.**

Consider a thin, perfectly conducting strip of width $2w$ held at potential $\phi_0$, as depicted in Figure 6.9. The strip is infinite in the $z$-coordinate, and therefore (6.58) becomes, using (6.26),

$$\phi_0 - \phi_{\text{inc}}(x) = -\int_{-w}^{w} \ln|x - x'| \frac{\rho_s(x')}{2\pi\varepsilon}\, dx'. \tag{6.59}$$

The solution of (6.59) follows from considering the eigenfunctions of integral operators with logarithmic kernels, as detailed in Section 4.3.4. We assume that the charge density obeys the well-known edge condition[17]

$$\rho_s(x) = \frac{\rho(x)}{\sqrt{w^2 - x^2}}. \tag{6.60}$$

[17]Note that if we do not assume this form for the charge density, we simply expand the unknown in terms of weighted Chebyshev polynomials.

In this case, from (4.32) we have[18]

$$\rho(x) = \frac{2\pi\varepsilon}{\pi}\sum_{n=0}^{\infty}\frac{\varepsilon_n}{\lambda_n}\left\langle\phi_{\mathrm{inc}}(x) - \phi_0, T_n(x/w)\right\rangle T_n(x/w), \tag{6.61}$$

where $\lambda_0 = -\pi\ln(2/w)$ for $n = 0$ and $\lambda_n = -\pi/n$ for $n \neq 0$.

As a special case, if $\phi_{\mathrm{inc}} = 0$, then, upon noting that

$$\int_{-w}^{w}\frac{T_n(x/w)}{\sqrt{w^2 - x^2}}e^{i\xi x}\,dx = \pi i^n J_n(\xi w), \tag{6.62}$$

the charge density $\rho$ in (6.60) is found to be

$$\begin{aligned}\rho(x) &= 2\varepsilon\sum_{n=0}^{\infty}\frac{-\phi_0\varepsilon_n}{\lambda_n}\pi i^n J_n(0)\,T_n(x/w) \\ &= \frac{2\varepsilon\phi_0}{\ln(2/w)}\end{aligned} \tag{6.63}$$

since only the $n = 0$ term is nonzero for $J_n(0)$, and $T_0(x/w) = 1$. Therefore, the charge density on a strip in free space held at potential $\phi_0$ is

$$\rho_s(x) = \frac{2\varepsilon\phi_0}{\ln(2/w)\sqrt{w^2 - x^2}},$$

and the total charge is found to be

$$Q = \int_{-w}^{w}\rho_s(x)\,dx = \frac{2\pi\varepsilon\phi_0}{\ln(2/w)}.$$

A related problem is considered in Section 8.2.2.

Note that if $w = 2$ the above results don't hold. This is due to the eigenvalue $\lambda_0 = -\pi\ln(2/w) = 0$, leading to the presence of a nontrivial homogeneous solution of (6.59). In this case, by the discussion in Sections 4.5 and 3.11, a (nonunique) solution exists if $(\phi_{\mathrm{inc}}(x) - \phi_0) \perp N_A$, where $A$ is the integral operator

$$Af(x) = \int_{-w}^{w}\ln|x - x'|\frac{f(x')}{\sqrt{w^2 - x'^2}}\,dx'.$$

In this case the eigenfunction is a constant ($T_0 = 1$), and so we require $(\phi_{\mathrm{inc}}(x) - \phi_0)$ to be perpendicular to unity if a solution is to exist, i.e.,

$$\int_a^b\frac{(\phi_{\mathrm{inc}}(x) - \phi_0)(1)}{\sqrt{w^2 - x^2}}\,dx = 0.$$

[18]The correct inner product is

$$\langle f, g\rangle = \int_{-w}^{w} f(x)g(x)\frac{dx}{\sqrt{w^2 - x^2}}.$$

Therefore, if $\phi_{\text{inc}} = 0$, and $w = 2$, since $\int_{-w}^{w} (1/\sqrt{w^2 - x^2})\, dx = \pi$, no solution exists using this formulation.

One can also solve (6.59) using the general (*Carleman's*) formula for the solution of

$$\int_a^b \ln|x - x'|\, y(x')\, dx' = f(x), \tag{6.64}$$

which is, for $b - a \neq 4$ (i.e., in this case $w \neq 2$), [11, p. 217],[19]

$$y(x) = \frac{1}{\pi^2 \sqrt{(x-a)(b-x)}} \left\{ \int_a^b \frac{\sqrt{(x'-a)(b-x')} f'(x')}{x' - x}\, dx' + \frac{1}{\ln\left(\frac{1}{4}(b-a)\right)} \int_a^b \frac{f(x')}{\sqrt{(x'-a)(b-x')}}\, dx' \right\}.$$

If $b - a = 4$, then for (6.64) to be solvable we must have

$$\int_a^b f(x) \frac{dx}{\sqrt{x-a}\sqrt{b-x}} = 0,$$

in which case the solution is

$$y(x) = \frac{1}{\pi^2 \sqrt{(x-a)(b-x)}} \left\{ \int_a^b \frac{\sqrt{(x'-a)(b-x')} f'(x')}{x' - x}\, dx' + C \right\}$$

with $C$ an arbitrary constant.

Mathematical details concerning integral equations with logarithmic kernels in Hölder spaces are extensively discussed in [12].

### Region 1 Containing Sources ($\phi_{\text{inc}}^{(1)} \neq 0$)

In this case $\phi_1$ is not constant in $\Omega_1$. The solution can be obtained via the first-kind integral equation

$$\phi_0 - \left(\phi_{\text{inc}}^{(1)}(\mathbf{r}_s') + \phi_{\text{inc}}^{(2)}(\mathbf{r}_s')\right) = \oint_\Gamma g^p(\mathbf{r}, \mathbf{r}_s') \left[\frac{\partial \phi_1(\mathbf{r})}{\partial n} - \frac{\partial \phi_2(\mathbf{r})}{\partial n}\right] d\Gamma, \tag{6.65}$$

where, once $\partial\phi_1(\mathbf{r}_s)/\partial n - \partial\phi_2(\mathbf{r}_s)/\partial n$ is determined, insertion into (6.56) leads to the potential in both regions.

Note that in either case $Q$ must be conserved in the sense that for a conductor immersed in an incident field, the total charge on the conductor before being placed in the incident field must be the same as the total charge on the conductor after being placed in the incident field; the charge density only becomes rearranged by the action of the incident field.

[19] The formula in [11, p. 217] contains a misprint and erroneously contains a factor of $\pi$ in the second term.

One can also work directly with (6.54) or (6.55), generating boundary integral equations of the form

$$\begin{aligned}\frac{\phi_0}{2} &= \phi_{\text{inc}}^{(1)}(\mathbf{r}_s') - \oint_\Gamma \left( \phi_0 \frac{\partial g^p(\mathbf{r}, \mathbf{r}_s')}{\partial n} - g^p(\mathbf{r}, \mathbf{r}_s') \frac{\partial \phi_1(\mathbf{r})}{\partial n} \right) d\Gamma, \\ \frac{\phi_0}{2} &= \phi_{\text{inc}}^{(2)}(\mathbf{r}_s') + \oint_\Gamma \left( \phi_0 \frac{\partial g^p(\mathbf{r}, \mathbf{r}_s')}{\partial n} - g^p(\mathbf{r}, \mathbf{r}_s') \frac{\partial \phi_2(\mathbf{r})}{\partial n} \right) d\Gamma.\end{aligned} \tag{6.66}$$

The first can be solved for $\partial \phi_1 / \partial n$, and the second for $\partial \phi_2 / \partial n$. The principal difficulty with the form (6.66) is that the first term in the boundary integral is more singular than the term in (6.56) and leads to a more complicated integral equation compared to (6.57) or (6.65) (of course, if the conductor is grounded, $\phi_0 = 0$, and this problem is eliminated). For this reason the combined equation (6.56), leading to (6.57) or (6.65), is often preferred.

One can also derive corresponding integral equations of the second kind for this class of problem. For instance, assume the surface is a perfect conductor with $\phi_s = \phi_0$ given and $\phi_{\text{inc}}^{(1)} = 0$, such that $\phi_1 = \phi_0$. Then, taking $\lim_{\mathbf{r}' \to \mathbf{r}_s'} \partial \phi_1(\mathbf{r}') / \partial n'$ in (6.56), properly accounting for the discontinuity via (6.53), and noting that $\partial \phi_1(\mathbf{r}') / \partial n' = 0$, we obtain

$$\frac{1}{2} \frac{\partial \phi_2(\mathbf{r}_s')}{\partial n'} + \oint_\Gamma \frac{\partial g^p(\mathbf{r}, \mathbf{r}_s')}{\partial n'} \frac{\partial \phi_2(\mathbf{r})}{\partial n} d\Gamma = \frac{\partial}{\partial n'} \phi_{\text{inc}}^{(2)}(\mathbf{r}_s'). \tag{6.67}$$

One obtains the same result by taking $\lim_{\mathbf{r}' \to \mathbf{r}_s'} \partial \phi_2(\mathbf{r}') / \partial n'$ in (6.56). With the identification $\rho_s = -\varepsilon_2 \left. \frac{\partial \phi_2}{\partial n} \right|_\Gamma$ then we see that (6.67) is really an integral equation for the surface-charge density,

$$\frac{1}{2\varepsilon_2} \rho_s(\mathbf{r}_s') + \frac{1}{\varepsilon_2} \oint_\Gamma \rho_s(\mathbf{r}) \frac{\partial g^p(\mathbf{r}, \mathbf{r}_s')}{\partial n'} d\Gamma = -\frac{\partial}{\partial n'} \phi_{\text{inc}}^{(2)}(\mathbf{r}_s'). \tag{6.68}$$

The integral equation (6.68) is of the second kind with a weakly singular kernel, compact in $\mathbf{L}^2(\Gamma)$, so that the Fredholm alternative applies. We can rewrite (6.68) as an operator equation[20]

$$(\lambda K + I) \rho_s = \eta_s,$$

where

$$(K \rho_s)(\mathbf{r}_s') = \oint_\Gamma \rho_s(\mathbf{r}) \frac{\partial g^p(\mathbf{r}, \mathbf{r}_s')}{\partial n'} d\Gamma$$

is a weakly singular compact integral operator, $\lambda = 2$, and where $2\varepsilon_2$ multiplying the right side of (6.68) forms $\eta_s$.

[20]Because $K$ is an integral operator, it is standard to use the form where $\lambda$ multiplies $K$, rather than associate $\lambda$ with $I$.

The integral equation (6.68) has a nontrivial homogeneous solution,[21] i.e., $\lambda = 2$ is an eigenvalue of $K$, and therefore (6.68) has either no solution or a nonunique solution. In this case, from the discussion in Section 4.5 it follows that a (nonunique) solution exists if the right side of (6.68) is orthogonal to unity. The solution can be made unique [5, pp. 309–315] by specifying $Q = \int_\Gamma \rho_s(\mathbf{r})\, d\Gamma$ or by forcing the solution of (6.68) to satisfy (6.56) for $\mathbf{r}' = \mathbf{r}'_s$, i.e.,

$$\phi_0 - \phi_{\text{inc}}^{(2)}(\mathbf{r}'_s) = \frac{1}{\varepsilon_2} \oint_\Gamma g^p(\mathbf{r}, \mathbf{r}'_s) \rho_s(\mathbf{r})\, d\Gamma. \tag{6.69}$$

## Problems Involving Multi-Dielectric Regions

The second type of problem is where $\Gamma$ represents an interface between two dielectric regions. In this case we multiply (6.54) by $\varepsilon_1$ and (6.55) by $\varepsilon_2$, and add the resulting equations, yielding

$$\left.\begin{array}{ll} \mathbf{r}' \in \Omega_1 & \varepsilon_1 \phi_1(\mathbf{r}') \\ \mathbf{r}' = \mathbf{r}'_s \in \Gamma & \phi_s(\mathbf{r}') \frac{(\varepsilon_1+\varepsilon_2)}{2} \\ \mathbf{r}' \in \Omega_2 & \varepsilon_2 \phi_2(\mathbf{r}') \end{array}\right\} = \varepsilon_1 \phi_{\text{inc}}^{(1)}(\mathbf{r}') + \varepsilon_2 \phi_{\text{inc}}^{(2)}(\mathbf{r}') \tag{6.70}$$

$$+ (\varepsilon_2 - \varepsilon_1) \oint_\Gamma \phi_s(\mathbf{r}) \frac{\partial g^p(\mathbf{r}, \mathbf{r}')}{\partial n}\, d\Gamma,$$

where the boundary integral term involving $g^p\,[\varepsilon_1 \partial\phi_1/\partial n - \varepsilon_2 \partial\phi_2/\partial n]$ vanishes because of the condition (6.50) ($\rho_s = 0$), and we have also used $\phi_1|_\Gamma = \phi_2|_\Gamma = \phi_s$. The appropriate boundary integral equation is then

$$\phi_s(\mathbf{r}'_s) \frac{(\varepsilon_1 + \varepsilon_2)}{2} + (\varepsilon_1 - \varepsilon_2) \oint_\Gamma \phi_s(\mathbf{r}) \frac{\partial g^p(\mathbf{r}, \mathbf{r}'_s)}{\partial n}\, d\Gamma = \varepsilon_1 \phi_{\text{inc}}^{(1)}(\mathbf{r}'_s) + \varepsilon_2 \phi_{\text{inc}}^{(2)}(\mathbf{r}'_s). \tag{6.71}$$

This is a second-kind integral equation with a weakly singular kernel, compact in $\mathbf{L}^2(\Gamma)$, so that the Fredholm alternative applies. If we denote

$$\lambda = \frac{(\varepsilon_1 - \varepsilon_2)}{2\pi\,(\varepsilon_1 + \varepsilon_2)}$$

and

$$(K\phi_s)(\mathbf{r}'_s) = \oint_\Gamma \phi_s(\mathbf{r}) \frac{\partial}{\partial n'} \frac{1}{R(\mathbf{r}, \mathbf{r}'_s)}\, d\Gamma,$$

[21] It can be seen that $\rho_s$ being constant provides a homogeneous solution, since

$$\frac{\rho_s}{2\varepsilon_2} + \frac{\rho_s}{4\pi\varepsilon_2} \oint_\Gamma \frac{\partial}{\partial n'} \frac{1}{R(\mathbf{r}, \mathbf{r}'_s)}\, d\Gamma = \frac{\rho_s}{2\varepsilon_2} + \frac{\rho_s}{4\pi\varepsilon_2}(-2\pi) = 0,$$

where the factor $(-2\pi)$ comes from the formula in Appendix A.5 for the derivative $\frac{\partial}{\partial n'} \frac{1}{R(\mathbf{r}, \mathbf{r}'_s)}$ and the solid-angle formula (1.51), with the solid angle $4\pi$ becoming $2\pi$ for points on the surface.

we can write the above as

$$(\lambda K + I)\,\phi = \eta_s.$$

It is shown in [5, p. 310] that for this integral equation, eigenvalues $\lambda$ are such that $|\lambda| \geq 1$. Since $(\varepsilon_1 - \varepsilon_2)/(2\pi(\varepsilon_1 + \varepsilon_2)) < 1$, we see that the homogeneous form of (6.71) has only the trivial solution, and therefore (6.71) has a unique solution. The boundary potential $\phi_s$, once determined and inserted into (6.70), leads to the potential in both regions.

It is worthwhile to reiterate that integral equation methods are primarily utilized when the complexity of the problem is such that the appropriate Green's function cannot be determined, or other direct methods are not applicable. Some mathematical details concerning the integral equations considered in this section, along with iterative solution techniques, can be found in [13].

# 7

# Transmission-Line Analysis

Transmission lines are used to transfer electrical signals (information) or electrical power from one point to another in an electrical system. Transmission lines take a wide variety of forms, from simple wire pairs and cables to more complicated integrated structures for high-frequency applications. Several common transmission lines are shown in Figure 7.1.

The chapter begins with a discussion of general transmission-line analysis, both from the standpoint of impedance methods and from the theory of Sturm–Liouville operators. Arbitrarily terminated transmission lines are treated using Green's function techniques, and the spectral analysis of transmission-line operators is briefly discussed. Then, transmission-line resonators and unbounded transmission lines are treated using regular and singular Sturm–Liouville theory, respectively. The chapter concludes with a discussion of general multiconductor transmission lines.

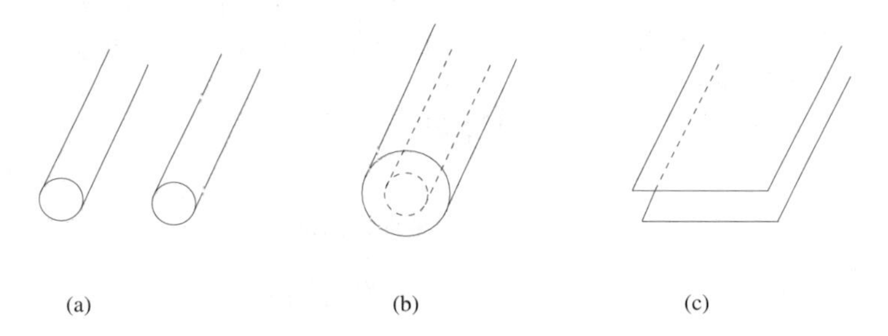


Figure 7.1: Several types of two-conductor TEM transmission lines: (a) parallel wires, (b) coaxial cable, (c) parallel-plate waveguide.

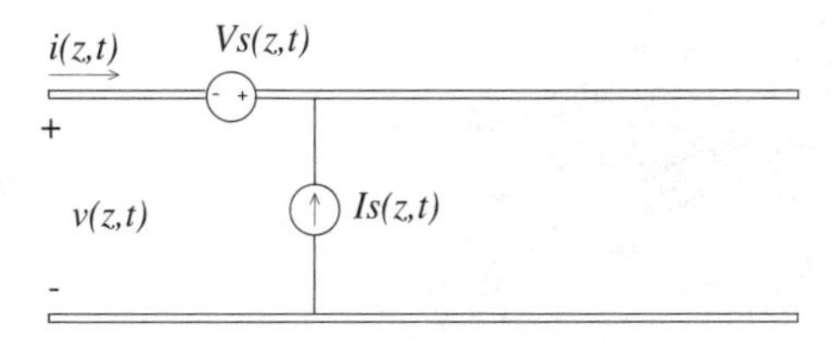


Figure 7.2: Two-conductor TEM transmission line with distributed sources $V_s(z,t)$ and $I_s(z,t)$.

## 7.1 General Analysis

To illustrate the analysis of transmission lines and transmission-line resonators, consider the two-conductor[1] TEM transmission line depicted in Figure 7.2, where $V_s$ and $I_s$ represent distributed sources. The lumped-element model for a small segment of the line is shown in Figure 7.3. The circuit elements are

- $R$: series resistance per unit length for both conductors, ohms/m
- $L$: series inductance per unit length for both conductors, H/m
- $G$: shunt conductance per unit length, S/m
- $C$: shunt capacitance per unit length, F/m
- $i_s$: shunt current source per unit length, A/m
- $v_s$: series voltage source per unit length, V/m

The distributed sources may represent, for instance, distributed currents and voltages induced on the transmission line by an external source (see, e.g., [14, pp. 429–436]). If desired, a localized source can be modeled by $v_s = v_0\delta(z - z')$, and similarly for $i_s$. With the exception of the distributed sources, the engineering analysis of this lumped-element circuit model is found in many books, [15, Ch. 2], [16, Ch. 3], and so here we will focus on considering the problem from the standpoint of Sturm–Liouville theory. We will assume $R, L, G, C \in \mathbf{R}$.

Applying Kirchhoff's voltage law and current law to the circuit of Figure 7.3 yields, respectively,

$$\begin{aligned} v(z,t) + v_s(z,t)\Delta z - R\,\Delta z\, i(z,t) \\ - L\,\Delta z\,\frac{\partial i(z,t)}{\partial t} - v(z+\Delta z, t) = 0 \end{aligned}$$

[1]Generally the term "conductor" is used in transmission-line analysis, and, in fact, the lines are usually conductive. However, other transmission systems, including those only involving dielectrics (e.g., optical fibers), can be modeled using these techniques.

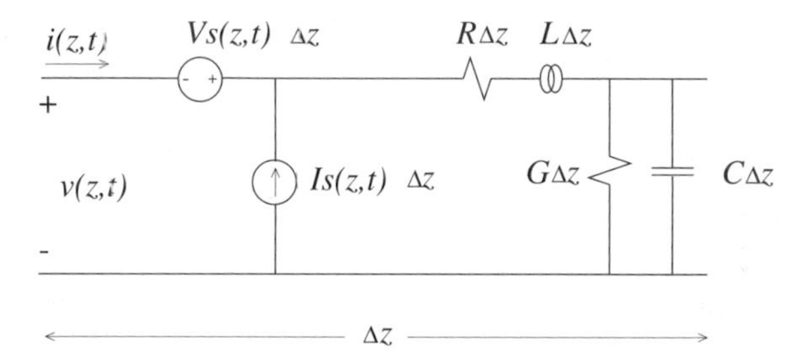


Figure 7.3: Lumped-element model for a small segment of a two-conductor transmission line with distributed sources.

and

$$
\begin{aligned}
i(z,t) + i_s(z,t)\Delta z - G\,\Delta z\, v(z+\Delta z, t)& \\
- C\,\Delta z\,\frac{\partial v(z+\Delta z,t)}{\partial t} - i(z+\Delta z, t) &= 0.
\end{aligned}
$$

Dividing these two equations by $\Delta z$ and taking the limit as $\Delta z \to 0$ lead to

$$
\begin{aligned}
\frac{\partial v(z,t)}{\partial z} &= -R\,i(z,t) - L\frac{\partial i(z,t)}{\partial t} + v_s(z,t), \\
\frac{\partial i(z,t)}{\partial z} &= -G\,v(z,t) - C\frac{\partial v(z,t)}{\partial t} + i_s(z,t),
\end{aligned}
\tag{7.1}
$$

and, upon assuming time-harmonic conditions or Fourier transformation,

$$
\begin{aligned}
\frac{dv(z)}{dz} &= -R\,i(z) - i\omega L\,i(z) + v_s(z), \\
\frac{di(z)}{dz} &= -G\,v(z) - i\omega C\,v(z) + i_s(z),
\end{aligned}
\tag{7.2}
$$

where $\omega \in \mathbf{R}$. These two coupled first-order differential equations can be easily decoupled by forming second-order differential equations

$$
\begin{aligned}
\frac{d^2 v(z)}{dz^2} - \gamma^2 v(z) &= -\left(R + i\omega L\right) i_s(z) + \frac{d\,v_s(z)}{dz}, \\
\frac{d^2 i(z)}{dz^2} - \gamma^2 i(z) &= -\left(G + i\omega C\right) v_s(z) + \frac{d\,i_s(z)}{dz},
\end{aligned}
\tag{7.3}
$$

where

$$
\gamma^2 = \left(R + i\omega L\right)\left(G + i\omega C\right),
$$

and $\gamma = \alpha + i\beta \in \mathbf{C}$ is called the *propagation constant* (1/m). The real and imaginary parts of the propagation constant are known as the *attenuation constant* ($\alpha$) and the *phase constant* ($\beta$), respectively. The equations (7.3) are nonhomogeneous Helmholtz equations or wave equations and are scalar, one-dimensional versions (for $v$ and $i$) of (1.25).

One can easily see that either equation can be cast in terms of the Sturm–Liouville operator (5.4),

$$(L - \lambda)\, u_{v,i} = f_{v,i} \tag{7.4}$$

with $L \equiv -d^2/dz^2$, $\lambda = -\gamma^2$, $u_v = v$, and $u_i = i$.

**Impedance Methods**

From (7.4) the analysis may proceed in several directions. Usually in microwave analysis one considers the homogeneous equations

$$(L - \lambda)\, u_{v,i} = 0$$

corresponding to the absence of any distributed source. General solutions[2] are found as (replacing $u_{v,i}$ with $v, i$)

$$\begin{aligned} v(z) &= v_0^+ e^{-i\sqrt{\lambda}z} + v_0^- e^{i\sqrt{\lambda}z} = v_0^+ e^{-\gamma z} + v_0^- e^{+\gamma z}, \\ i(z) &= i_0^+ e^{-i\sqrt{\lambda}z} + i_0^- e^{i\sqrt{\lambda}z} = i_0^+ e^{-\gamma z} + i_0^- e^{+\gamma z}, \end{aligned} \tag{7.5}$$

where $\gamma = i\sqrt{\lambda}$ with $\mathrm{Im}\,\sqrt{\lambda} < 0$, which represent voltage and current waves. The term $e^{-\gamma z}$ is a forward ($+z$-traveling) wave, while the term $e^{+\gamma z}$ is a backward ($-z$-traveling) wave, subject to the $e^{+i\omega t}$ time dependence. Exploiting (7.2) leads to the relationship between voltage and current as

$$i(z) = \frac{1}{Z_0}\left[v_0^+ e^{-\gamma z} - v_0^- e^{+\gamma z}\right],$$

where

$$Z_0 \equiv \frac{R + i\omega L}{\gamma}$$

is called the *characteristic impedance* (ohms) of the transmission line.

Properties of the voltage and current waves can be ascertained from a consideration of $\gamma$, which in turn depends on the parameters $R, G, C, L$ and frequency $\omega$. Terminated transmission lines, depicted schematically in Figure 7.4, can be analyzed by introducing the concept of a load reflection coefficient

$$\Gamma_L \equiv \frac{v_0^-}{v_0^+} = \frac{Z_L - Z_0}{Z_L + Z_0},$$

leading to

$$\begin{aligned} v(z) &= v_0^+ \left[e^{-\gamma z} + \Gamma_L e^{+\gamma z}\right], \\ i(z) &= \frac{v_0^+}{Z_0}\left[e^{-\gamma z} - \Gamma_L e^{+\gamma z}\right]. \end{aligned} \tag{7.6}$$

[2]In waveguide analysis one would consider these solutions as natural waveguide modes (see Chapter 8). In the usual transmission line analysis we only consider one mode, the so-called *transmission-line mode*, which corresponds to a transverse electromagnetic mode on an actual three-dimensional transmission line or waveguide.

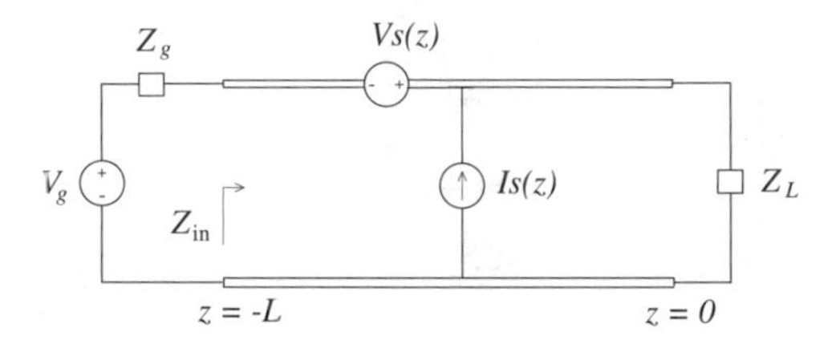


Figure 7.4: Source-driven transmission line terminated by load impedance $Z_L$.

Using this type of analysis one can determine engineering quantities of interest, such as input impedance,

$$Z_{\mathrm{in}} \equiv \frac{v(z=-l)}{i(z=-l)} = Z_0 \frac{Z_L + Z_0 \tanh \gamma l}{Z_0 + Z_L \tanh \gamma l}$$

and input reflection coefficient,

$$\Gamma_{\mathrm{in}} = \Gamma_L e^{-2\gamma l}.$$

If a generator is connected to the line as depicted in Figure 7.4, then $v_0^+$ can be determined from the voltage divider rule as $v(z=-l) = v_g Z_{in}/(Z_{in} + Z_g) = v_0^+ \left[e^{\gamma l} + \Gamma_L e^{-\gamma l}\right]$, leading to

$$v_0^+ = v_g \frac{Z_{\mathrm{in}}}{Z_{\mathrm{in}} + Z_g} \frac{1}{\left[e^{\gamma l} + \Gamma_L e^{-\gamma l}\right]}.$$

Using this expression the amount of power delivered to the line,

$$P_{\mathrm{in}} = \frac{1}{2} \operatorname{Re}\left\{v(-l)\bar{i}(-l)\right\},$$

and the power delivered to the load,

$$P_{\mathrm{L}} = \frac{1}{2} \operatorname{Re}\left\{v(0)\bar{i}(0)\right\},$$

can be determined. For details see, e.g., [15, Ch. 2].

If we consider the set of parameters $\{R, G, L, C, \omega\}$, which are, in turn, related to the physical structure of the transmission line, then we can study, for instance, the transmission-line voltage $v(z)$ as a function of these parameters. Thus, (7.3) are not usually solved as boundary value problems per se, but, in fact, very general solutions for typical problems of engineering interest are obtained using the impedance concept.

### Sturm–Liouville Analysis

If the transmission line exists over the finite interval $[a, b]$, and boundary conditions $B_a$ and $B_b$ of the form (5.2) are provided at the endpoints $a$

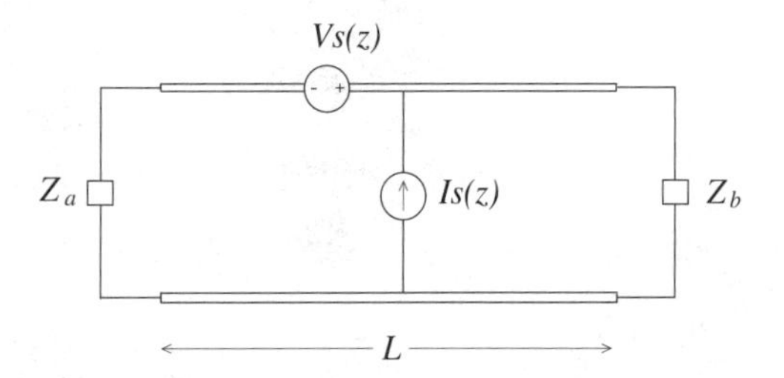


Figure 7.5: Transmission line with distributed sources terminated by impedances $Z_a$, $Z_b$.

and $b$,

$$\begin{aligned} B_a(u_{v,i}) &= \alpha_1 u_{v,i}(a) + \alpha_2 u'_{v,i}(a) = 0, \\ B_b(u_{v,i}) &= \beta_1 u_{v,i}(b) + \beta_2 u'_{v,i}(b) = 0, \end{aligned} \tag{7.7}$$

then we can treat (7.3) as a Sturm–Liouville boundary value problem

$$(L - \lambda)\, u_{v,i} = f_{v,i}$$

$$L \equiv -\frac{d^2}{dz^2}, \qquad \lambda = -\gamma^2$$

$$D_L \equiv \left\{ u : u' \text{ a.c., } u'' \in \mathbf{L}^2(a,b), B_a(u) = B_b(u) = 0 \right\}$$

From (7.2) note that if $v_s(z) = i_s(z) = 0$ for some particular point $z$, then $u'_{v,i}(z)$ is proportional to $u_{i,v}(z)$. If the endpoint at $z = a$ is terminated in a short (open) circuit, then for $u_v$ we have $\alpha_2 = 0$ ($\alpha_1 = 0$) and for $u_i$ we have $\alpha_1 = 0$ ($\alpha_2 = 0$). Similar comments apply for the endpoint $z = b$, and these cases will be discussed in the next section. If the endpoints are terminated with general impedances $Z_a, Z_b$ such as shown in Figure 7.5, and if the distributed sources do not exist at the endpoints, then

$$-\frac{\alpha_2}{\alpha_1} = u_v(a)/u'_v(a) = \frac{Z_a}{R + i\omega L}$$

assuming $u'_v(a) \neq 0$ (i.e., $i(a) \neq 0$). Similarly, at $z = b$ we have

$$-\frac{\beta_2}{\beta_1} = u_v(b)/u'_v(b) = -\frac{Z_b}{R + i\omega L}.$$

Note the sign difference between the conditions at endpoints $a$ and $b$; this is due to the sign convention chosen in Figure 7.2, which is used to obtain the governing differential equations. The case for $u_i$ leads to

$$-\frac{\alpha_2}{\alpha_1} = u_i(a)/u'_i(a) = \frac{Y_a}{G + i\omega C},$$

where $Y_a = Z_a^{-1}$, with a similar equation at $z = b$.

For arbitrary terminating impedances, $\alpha_{1,2}, \beta_{1,2} \in \mathbf{C}$ in (7.7), and the Sturm–Liouville problem is not self-adjoint[3] (see the discussion surrounding (5.11) and in Section 5.3). In this case (7.3) can be solved using the Green's function technique described in Section 5.3.

The problem of arbitrary terminating impedances is discussed in detail in [17, pp. 210–215], where the evaluation of the Green's function is simplified using a network approach. Following a similar method, let the Green's functions $g_{v,i}$, where $v, i$, represent voltage and current, respectively, satisfy

$$\left(-\frac{d^2}{dz^2} - \lambda\right) g_{v,i}(z, z', \lambda) = \delta(z - z') \tag{7.8}$$

subject to (5.17),

$$\begin{aligned} \left.\frac{dg_{v,i}(z, z', \lambda)}{dz}\right|_{z=a} - \eta_a^{v,i} g_{v,i}(a, z', \lambda) &= 0, \\ \left.\frac{dg_{v,i}(z, z', \lambda)}{dz}\right|_{z=b} - \eta_b^{v,i} g_{v,i}(b, z', \lambda) &= 0, \\ g_{v,i}(z, z', \lambda)\Big|_{z=z'-\varepsilon}^{z=z'+\varepsilon} &= 0, \\ \left.\frac{dg_{v,i}(z, z', \lambda)}{dz}\right|_{z=z'-\varepsilon}^{z=z'+\varepsilon} + 1 &= 0, \end{aligned} \tag{7.9}$$

where $\eta_{a,b}^v \equiv \pm Y_{a,b}\,(R + i\omega L)$ and $\eta_{a,b}^i \equiv \pm Z_{a,b}\,(G + i\omega C)$, with the positive sign for endpoint $a$ and the negative sign for endpoint $b$, and where $Z_{a,b}$ are the terminating impedances, with $Y_{a,b} = Z_{a,b}^{-1}$. The solution of (7.8) is obtained as

$$g_{v,i}(z, z', \lambda) = \begin{cases} A^{v,i}\left[\cos\sqrt{\lambda}z + X_a^{v,i}\sin\sqrt{\lambda}z\right], & z < z', \\ B^{v,i}\left[\cos\sqrt{\lambda}z - X_b^{v,i}\sin\sqrt{\lambda}z\right], & z > z', \end{cases} \tag{7.10}$$

where the form of (7.10) is motivated by (7.6), with $X_a^{v,i}$ chosen to satisfy the first two boundary conditions in (7.9). This leads to

$$X_{a(b)}^{v,i} = \pm\frac{\eta_{a(b)}^{v,i} + \sqrt{\lambda}\tan\sqrt{\lambda}a(b)}{\sqrt{\lambda} - \eta_{a(b)}^{v,i}\tan\sqrt{\lambda}a(b)},$$

where the positive sign is for $X_a^{v,i}$ and the negative sign is for $X_b^{v,i}$. The tangent indicates $\tan\sqrt{\lambda}a$ for $X_a^{v,i}$ and $\tan\sqrt{\lambda}b$ for $X_b^{v,i}$. The final two

[3] Some special cases exist in which the terminated transmission line is amenable to the self-adjoint Sturm–Liouville theory. For instance, a lossless line ($R = G = 0$) terminated by purely reactive elements $Z_{a,b} = iX_{a,b}$, $X_{a,b} \in \mathbf{R}$, results in $\alpha_{1,2}, \beta_{1,2} \in \mathbf{R}$.

conditions in (7.9) lead to the determination of $A^{v,i}$ and $B^{v,i}$, resulting in the solution

$$g_{v,i}(z,z',\lambda) = \frac{1}{\sqrt{\lambda}\left(X_b^{v,i}+X_a^{v,i}\right)} \cdot \begin{cases} \left[\cos\sqrt{\lambda}z' - X_b^{v,i}\sin\sqrt{\lambda}z'\right]\left[\cos\sqrt{\lambda}z + X_a^{v,i}\sin\sqrt{\lambda}z\right], & z<z', \\ \left[\cos\sqrt{\lambda}z - X_b^{v,i}\sin\sqrt{\lambda}z\right]\left[\cos\sqrt{\lambda}z' + X_a^{v,i}\sin\sqrt{\lambda}z'\right], & z>z'. \end{cases} \tag{7.11}$$

With the Green's function established, the solution of (7.4) is given by (5.101),

$$u_{v,i}(z') = \int_a^b f_{v,i}(z)g_{v,i}(z',z,\lambda)\,dz + \left[g_{v,i}(z',z,\lambda)\frac{du_{v,i}}{dz} - u_{v,i}(z)\frac{dg_{v,i}(z',z,\lambda)}{dz}\right]\Bigg|_{z=a}^{z=b}. \tag{7.12}$$

Since the boundary conditions at $z=a,b$ on $u$ are the same as those on $g$, the conjunct vanishes, leading to

$$u_{v,i}(z) = \int_a^b f_{v,i}(z')g_{v,i}(z,z',\lambda)\,dz'. \tag{7.13}$$

**Example 7.1.**

Consider a coaxial transmission line excited by a series voltage source

$$v_s(z) = v_0\delta(z-z_0),$$

where $z_0$ is the position of the source. For a coaxial line [15, p. 62],

$$L = \frac{\mu}{2\pi}\ln\frac{r_0}{r_i}, \qquad C = \frac{2\pi\varepsilon}{\ln\frac{r_0}{r_i}},$$

where $\mu$ and $\varepsilon$ are the parameters of the dielectric filling in the line, assumed real-valued, and $r_0$ and $r_i$ are the radii of the outer and inner conductors, respectively. Consider the simple example of a lossless line ($R=G=0$) with $a=0, b=1$ m, $\mu=\mu_0$, $\varepsilon=2\varepsilon_0$, $r_0=3$ mm, and $r_i=1$ mm. This results in $Z_0=\sqrt{L/C}=46.6$ ohms. The solution for the voltage induced on the line is

$$\begin{aligned} u_v(z) &= \int_0^1 v_0\delta'(z'-z_0)g_v(z,z',\lambda)\,dz' = -v_0\left.\frac{dg_v(z,z_0,\lambda)}{dz'}\right|_{z'=z_0} \\ &= \frac{-v_0}{(X_b+X_a)}\left(\cosh\gamma z + iX_b\sinh\gamma z\right)\left(i\sinh\gamma z_0 + X_a\cosh\gamma z_0\right) \end{aligned} \tag{7.14}$$

for $z > z_0$, where we have made the substitution $\sqrt{\lambda} = -i\gamma$ and used the distributional identity [18, p. 17]

$$\left\langle \frac{d\delta(z-z_0)}{dz}, f \right\rangle = -\left. \frac{df(z)}{dz} \right|_{z_0}.$$

With $v_0 = 1$ mV and $z_0 = 0.5$ m, the induced voltage at the load, $u_v(b)$, and the power dissipated by the load, $P_b = (1/2)\,\mathrm{Re}\,\{u_v(b)u_i^*(b)\}$, are given in Tables 7.1 and 7.2, where we use $u_i(b) = u_v(b)/Z_b$ to compute the current. In Table 7.1 we have $Z_a = 50$ ohms and $Z_b = 20$ ohms, whereas in Table 7.2 we have $Z_a = Z_0 = Z_b = 46.6$ ohms (matched conditions). Note that in Table 7.1 the power dissipated at the load increases with increasing frequency for a fixed $v_0$.

Table 7.1

| $f$ (MHz) | $u_v(b)$ (uv) | $P_b$ (nW) |
|---|---|---|
| 0.01 | $-285.72 + 0.0367i$ | 2.04 |
| 1 | $-285.70 + 3.667i$ | 2.04 |
| 100 | $-29.70 + 305.38i$ | 2.35 |
| 10000 | $346.98 + 42.47i$ | 3.02 |

Table 7.2

| $f$ (MHz) | $u_v(b)$ (uv) | $P_b$ (nW) |
|---|---|---|
| 0.01 | $-500.00 + 0.0741i$ | 2.69 |
| 1 | $-499.95 + 7.41i$ | 2.69 |
| 100 | $-45.70 + 497.08i$ | 2.69 |
| 10000 | $485.06 - 120.72i$ | 2.69 |

The solution (7.14) can account for the frequency-dependent losses due to $R$ and $G$, and for the possible frequency dependence of $v_0$.

The spectral analysis for the arbitrarily terminated line is difficult and won't be included here, although a brief discussion of the eigenvalue problem is in order. The discrete (proper) eigenfunctions are determined from

$$\left(-\frac{d^2}{dz^2} - \lambda_n\right) u_n^{v,i}(z) = 0 \tag{7.15}$$

subject to

$$\left.\frac{du_n^{v,i}(z)}{dz}\right|_{z=a} - \eta_a^{v,i} u_n^{v,i}(a) = 0, \tag{7.16}$$
$$\left.\frac{du_n^{v,i}(z)}{dz}\right|_{z=b} - \eta_b^{v,i} u_n^{v,i}(b) = 0,$$

where $\eta_{a,b}^{v,i}$ are generally complex-valued. A solution of (7.15) is

$$u_n^{v,i}(z) = A^{v,i}\left[\cos\sqrt{\lambda_n}z + X_a^{v,i}\sin\sqrt{\lambda_n}z\right], \tag{7.17}$$

which satisfies the boundary condition at endpoint $a$. Enforcement of the second boundary condition leads to a transcendental equation for the proper eigenvalues,

$$\tan\sqrt{\lambda_n}b = \frac{\sqrt{\lambda_n}\left[\eta_a^{v,i} + \sqrt{\lambda_n}\tan\sqrt{\lambda_n}a\right] - \eta_b^{v,i}\left[\sqrt{\lambda_n} - \eta_a^{v,i}\tan\sqrt{\lambda_n}a\right]}{\sqrt{\lambda_n}\left[\sqrt{\lambda_n} - \eta_a^{v,i}\tan\sqrt{\lambda_n}a\right] + \eta_b^{v,i}\left[\eta_a^{v,i} + \sqrt{\lambda_n}\tan\sqrt{\lambda_n}a\right]}. \tag{7.18}$$

Depending on the values of $Z_a$ and $Z_b$ the spectrum may be purely discrete, may be purely continuous, or may contain a combination of discrete and continuous elements. For instance, if $Z_a$ and $Z_b$ are reactive (i.e., $Z_{a,b}$ are imaginary), the problem is self-adjoint and the spectrum is discrete. If $Z_a = Z_b = Z_0$, the spectrum is purely continuous, and for general terminating impedances the spectrum will contain a continuous and a discrete part. The validity of these statements is clear from the analysis of several special cases in the following sections; in particular see Section 8.1.6 for the detailed analysis of an impedance plane structure where, in essence, $Z_a$ is arbitrary and $Z_b = Z_0$, leading to both a discrete and a continuous spectrum.

In the general case, the spectral analysis leading to an eigenfunction (proper or improper) representation of the Green's function and eigenfunction solution to (7.4) is considerably more difficult than the Green's function analysis described above. The eigenfunction approach using the completeness relation (5.108) with (7.11) as the Green's function is also somewhat difficult (although the analysis is similar to that detailed in Section 8.1.6 and can be preformed accordingly), and so we will not pursue the spectral problem further. As mentioned above, for certain line terminations these difficulties will not be encountered.

In the following sections we apply the self-adjoint Sturm–Liouville theory to the analysis of short- and open-circuited transmission lines, leading to the regular Sturm–Liouville problem and to the concept of resonance, and to unbounded lines, leading to the singular problem.

## 7.2 Transmission-Line Resonators

An important transmission-line problem amenable to analysis as a regular Sturm–Liouville problem is the forced transmission-line resonator. Consider the Sturm–Liouville equation (7.3) written as

$$\left(-\frac{d^2}{dz^2} + \gamma^2\right)u_{v,i}(z) = f_{v,i}(z), \tag{7.19}$$

where

$$f_v = (R + i\omega L)\, i_s(z) - \frac{d\,v_s(z)}{dz},$$
$$f_i = (G + i\omega C)\, v_s(z) - \frac{d\,i_s(z)}{dz}.$$

The Sturm–Liouville operator (5.4) is simply $L = -d^2/dz^2$, and so we have

$$(L - \lambda)\, u_{v,i} = f_{v,i},$$

where $\lambda = -\gamma^2$.

Assume the transmission line with distributed sources exists over a finite region $[0, b]$, with the boundary conditions $u_{v,i}(0) = u_{v,i}(b) = 0$. For $u_v$ this physically corresponds to short-circuit boundary conditions, whereas for $u_i$ this physically corresponds to an open-circuited transmission line. The physical situation is depicted in Figure 7.5 with $Z_\alpha = 0$ for short-circuit conditions and $Z_\alpha = \infty$ for open-circuit conditions, $\alpha = a, b$.

We may solve (7.19) in a number of ways, as described in Example 5.2. One form of the solution is (5.14), which, when specifically applied to (7.19), yields (see (5.22))

$$u_{v,i}(z) = \int_0^b f_{v,i}(z')g_{v,i}(z, z', \gamma)\, dz' \tag{7.20}$$

for $z \in [0, b]$ and $u_{v,i}(z) \in \mathbf{L}^2(0, b)$, where, repeating from (5.21),

$$g_{v,i}(z, z', \gamma) = \frac{\sinh \gamma(b - z_>)\, \sinh \gamma z_<}{\gamma \sinh \gamma b} \tag{7.21}$$

using the direct method or, equivalently, (5.26) using scattering superposition. The above Green's function can also be obtained as a special case of (7.11) with $Z_a = Z_b = a = 0$.

The solution does not exist if $\sinh \gamma b = 0$, such that $\gamma b = in\pi$. This condition represents a resonance of the structure, denoted as $\gamma_r b = in\pi$. Resonances can also be obtained by solving

$$\left(X_b^{v,i} + X_a^{v,i}\right) = 0,$$

i.e., at the poles of the Green's function (7.11). If $R, G \neq 0$, then $\gamma$ will be complex-valued such that a resonance will not occur for real frequencies $\omega$. Physically, this is because energy loss on the line prevents perfect constructive and destructive interference between waves traveling in different directions on the line, such that a true resonance cannot "build up." If $R = G = 0$ then $\gamma = i\beta = i\omega\sqrt{LC}$, and the resonance condition is $\beta_r = n\pi/b$. For fixed $L$ and $C$ this leads to the concept of a *resonance frequency*,

$$\omega_r = \frac{n\pi}{b\sqrt{LC}}.$$

For time-harmonic lossless problems, the solution does not exist at the real resonant frequency. The physical meaning of this is the following. Time-harmonic problems do not literally correspond to physical reality, since there must be a time when sources are turned on. A careful analysis of the temporal problem shows that, in the lossless case, a solution exists that tends toward infinitely large amplitudes as time increases. This "blowing-up" of the solution using a more realistic temporal model corresponds to nonexistence of the solution for the less-realistic yet far simpler time-harmonic model. Actually, the lossless model is not physically realistic itself, and the temporal solution for the lossy case shows that near the resonance frequency[4] the solution can become quite large (for small loss) as time progresses, yet the solution remains finite. The large yet finite time-harmonic solution for the lossy case then models this phenomena. Thus we see that time-harmonic and lossless conditions are, taken together, too simplistic for situations where the driving frequency is equal to a resonance frequency of the line.

As an alternative to the solution (7.20), the eigenfunction expansion method described in Section 5.1.3 leading to (5.28) and (5.27) may be used. Eigenfunctions of the operator $L = -d^2/dz^2$ are found from

$$(L - \lambda_n)\, u_n^{v,i} = -\left(\frac{d^2}{dz^2} + \lambda_n\right) u_n^{v,i}(z) = 0$$

subject to $u_n^{v,i}(0) = u_n^{v,i}(b) = 0$. The eigenfunctions are easily seen to be $u_n^{v,i} = A\sin(n\pi z/b)$, with associated eigenvalues $\lambda_n = (n\pi/b)^2$. Note that the eigenvalues are exactly the negative of the square of the resonance $\gamma_r = in\pi/b$; thus, in the lossless case the eigenvalues correspond to the square of the resonant phase constant $\beta_r$. When normalized[5] the eigenfunctions become

$$u_n^{v,i}(z) = \sqrt{\frac{2}{b}}\sin\frac{n\pi}{b}z.$$

The solution of (7.19) is then

$$u_{v,i}(z) = \sum_{n=1}^{\infty} \frac{\langle f_{v,i}, u_n^{v,i}\rangle}{\lambda_n - \lambda} u_n^{v,i} = \sum_{n=1}^{\infty} \frac{\frac{2}{b}\int_0^b f_{v,i}(z)\sin\frac{n\pi}{b}z\, dz}{\left(\frac{n\pi}{b}\right)^2 + \gamma^2}\sin\frac{n\pi}{b}z, \quad (7.22)$$

where again the solution does not exist if $\gamma = \gamma_r$ corresponding to a resonance. Alternately, one may use (7.20) with the Green's function repre-

---

[4]Recall $\gamma_r a = in\pi$ will never occur for real frequencies, yet $\gamma_r a$ may be very close to $in\pi$ for low-loss lines, making $\sinh \gamma_r a$ quite small.

[5]The normalization with respect to the inner product is

$$\langle u_n, u_n\rangle = A^2 \int_0^b \sin\frac{n\pi}{b}z \sin\frac{n\pi}{b}z\, dz = 1,$$

leading to $A = \sqrt{2/b}$.

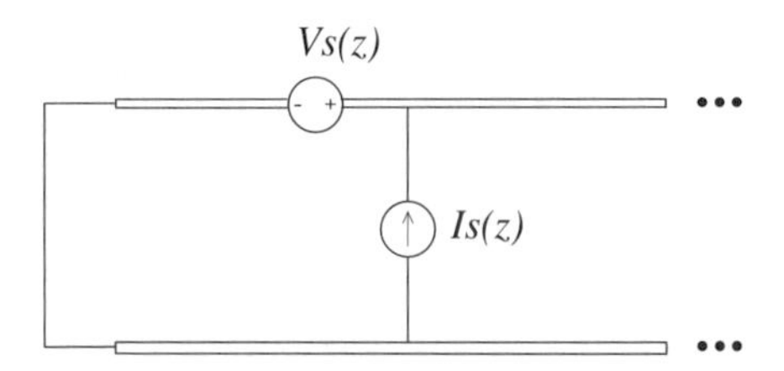


Figure 7.6: Semi-infinite transmission line with distributed sources, short-circuited at one end.

sented as an eigenfunction expansion (5.31),

$$g_{v,i}(z,z',\gamma) = \sum_{n=1}^{\infty} \frac{u_n^{v,i}(z)\overline{u^{v,i}}_n(z')}{\lambda_n - \lambda} = \sum_{n=1}^{\infty} \frac{\sqrt{\frac{2}{b}}\sin\frac{n\pi}{b}z\,\sqrt{\frac{2}{b}}\sin\frac{n\pi}{b}z'}{\left(\frac{n\pi}{b}\right)^2 + \gamma^2}. \quad (7.23)$$

If we consider a lossless line ($R = G = 0$), then $\gamma = i\beta$ with $\beta^2 = \omega^2 LC$. All of the above equations hold under the substitution $\gamma = i\beta$, with the hyperbolic circular functions becoming ordinary circular functions using $\sinh iy = i\sin y$. Similar results can be obtained from boundary conditions of the sort $\{u'(0) = 0,\ u'(b) = 0\}$, $\{u(0) = 0,\ u'(b) = 0\}$, and $\{u'(0) = 0, u(b) = 0\}$.

## 7.3 Semi-Infinite and Infinite Transmission Lines

If one or both of the transmission-line boundaries recede to infinity, the regular Sturm–Liouville problem considered in Section 7.2 becomes a singular problem. The governing Sturm–Liouville equation

$$\left(-\frac{d^2}{dz^2} + \gamma^2\right) u_{v,i}(z) = f_{v,i}(z) \quad (7.24)$$

over $[0,\infty)$, $(-\infty,0]$, or $(-\infty,\infty)$ is in the limit-point case at the singular endpoint(s). The semi-infinite interval $[0,\infty)$ with $u_{v,i}(0) = 0$ is treated in detail in Section 5.2, with the principal results repeated below. For the geometry depicted in Figure 7.6 the solution of (7.24) is (5.50),

$$u_{v,i}(z) = \int_0^\infty f_{v,i}(z')g_{v,i}(z,z',\gamma)\,dz' \quad (7.25)$$

for $z \in [0,\infty)$, where $u_{v,i}(z) \in \mathbf{L}^2(0,\infty)$. The Green's function is (5.49),

$$g_{v,i}(z,z',\gamma) = \frac{1}{\gamma}e^{-\gamma z_>}\sinh\gamma z_<,$$

which is again a special case of (7.11) with $a = 0$, $Z_a(Y_a) = 0$, and either $b \to \infty$ or $Z_b = Z_0$, leading to a purely continuous spectrum. The completeness relation is (5.60),

$$\delta(z - z') = \frac{2}{\pi} \int_0^\infty \sin kz' \sin kz \, dk. \tag{7.26}$$

An alternative solution based on an expansion in the continuous spectrum $\lambda = -\gamma^2 \in [0, \infty)$ (of improper eigenfunctions ) is (5.68),

$$u_{v,i}(z) = \frac{2}{\pi} \int_0^\infty \frac{\langle f_{v,i}, \sin kz) \rangle}{k^2 + \gamma^2} \sin kz \, dk. \tag{7.27}$$

Note that in the lossless case ($\alpha = 0$) $\gamma^2 = -\beta^2$ such that $k^2 + \gamma^2 = k^2 - \beta^2$, and since $\beta^2 \in [0, \infty)$ is part of the continuous spectrum the solution of (7.19) for $\beta^2 \in [0, \infty)$ does not exist unless $\langle f_{v,i}, \sin \beta z \rangle = 0$. Physically, one usually regards the lossless case as a limiting low-loss case, such that $\alpha \to 0$ through positive values. This avoids the problem with the phase constant being part of the continuous spectrum.

The case of an unbounded transmission line over $(-\infty, \infty)$, i.e.,

$$\left( -\frac{d^2}{dz^2} + \gamma^2 \right) u_{v,i}(z) = f(z) \tag{7.28}$$

on $(-\infty, \infty)$ with $\gamma = \alpha + i\beta$ and $\operatorname{Re} \gamma > 0$, is treated in Example 5.11. In summary, the solution of the defining equation for the Green's function,

$$\left( -\frac{d^2}{dz^2} + \gamma^2 \right) g_{v,i}(z, z', \gamma) = \delta(z - z')$$

is (5.77),

$$g_{v,i}(z, z', \gamma) = \frac{1}{2\gamma} e^{-\gamma |z - z'|}, \tag{7.29}$$

which is a special case of (7.11) under matched conditions, $Z_a = Z_b = Z_0 = (R + i\omega L)/\gamma$ (or removing the terminating impedances to $\pm\infty$). Upon defining a branch for $\gamma$, application of (5.66) leads to

$$\delta(z - z') = \frac{1}{2\pi} \int_{-\infty}^\infty e^{ik\,(z - z')} dk. \tag{7.30}$$

The solution of (7.28) is found to be

$$u_{v,i}(z) = \int_{-\infty}^\infty f_{v,i}(z') g_{v,i}(z, z', \gamma) \, dz', \tag{7.31}$$

for $z \in (-\infty, \infty)$, where $u_{v,i}(z) \in \mathbf{L}^2(-\infty, \infty)$. In the continuous expansion form the solution is

$$u_{v,i}(z) = \frac{1}{2\pi} \int_{-\infty}^\infty \frac{\langle f_{v,i}, e^{ikz}) \rangle}{k^2 + \gamma^2} e^{ikz} \, dk, \tag{7.32}$$

leading to the spectral form of the Green's function

$$g_{v,i}(z, z', \gamma) = \frac{1}{2\pi} \int_{-\infty}^{\infty} \frac{e^{ikz} e^{-ikz'}}{k^2 + \gamma^2} \, dk. \tag{7.33}$$

## 7.4 Multiconductor Transmission Lines

The analysis of the previous sections concerned a two-conductor transmission line. In this section we briefly describe methods for analyzing transmission-line systems involving more than two conductors. The general theory is developed in a variety of references; see, e.g., [14], [19].

We assume the multiconductor system has $N+1$ conductors, with the $N+1st$ conductor being the reference conductor. All conductors are parallel and uniform along the $z$-axis. By a generalization of the analysis presented in Section 7.1, each constant circuit element $R, L, G$, and $C$ becomes a real-valued $N \times N$ matrix of constants, denoted as $\mathrm{R}, \mathrm{G}, \mathrm{L}, \mathrm{C}$ (we will omit the bracket notation $[R]$, etc., used in Chapter 4).

The transmission-line equations (7.1) become

$$\begin{aligned} \frac{\partial \mathrm{v}(z,t)}{\partial z} &= -\mathrm{R}\,\mathrm{i}(z,t) - \mathrm{L}\frac{\partial \mathrm{i}(z,t)}{\partial t} + \mathrm{v_s}(z,t), \\ \frac{\partial \mathrm{i}(z,t)}{\partial z} &= -\mathrm{G}\,\mathrm{v}(z,t) - \mathrm{C}\frac{\partial \mathrm{v}(z,t)}{\partial t} + \mathrm{i_s}(z,t), \end{aligned} \tag{7.34}$$

where v and i are $N \times 1$ voltage and current vectors, such that the $j$th entry of v (i) is the voltage (current) on the $j$th conductor, and $\mathrm{v_s}$, $\mathrm{i_s}$ are $N \times 1$ distributed voltage and current source vectors, respectively. Upon assuming time-harmonic conditions or Fourier transformation,

$$\begin{aligned} \frac{d\mathrm{v}(z)}{dz} &= -\mathrm{Z}\,\mathrm{i}(z) + \mathrm{v_s}(z), \\ \frac{d\mathrm{i}(z)}{dz} &= -\mathrm{Y}\mathrm{v}(z) + \mathrm{i_s}(z), \end{aligned} \tag{7.35}$$

where

$$\begin{aligned} \mathrm{Z} &= (\mathrm{R} + i\omega \mathrm{L}), \\ \mathrm{Y} &= (\mathrm{G} + i\omega \mathrm{C}), \end{aligned} \tag{7.36}$$

with $\omega \in \mathbf{R}$. These two coupled first-order differential matrix equations can be easily decoupled by forming second-order differential equations

$$\begin{aligned} \frac{d^2\mathrm{v}(z)}{dz^2} - \mathrm{ZY}\mathrm{v}(z) &= -\mathrm{Z}\,\mathrm{i_s}(z) + \frac{d\,\mathrm{v_s}(z)}{dz}, \\ \frac{d^2\mathrm{i}(z)}{dz^2} - \mathrm{YZ}\mathrm{i}(z) &= -\mathrm{Y}\,\mathrm{v_s}(z) + \frac{d\,\mathrm{i_s}(z)}{dz}, \end{aligned} \tag{7.37}$$

where ZY and YZ are $N \times N$ matrices. Each matrix equation in (7.37) represents a system of coupled Sturm–Liouville equations. The general case of distributed-source driven lines is considered in [20] and [14] and references therein, and here we will concentrate on the solutions of the homogeneous forms

$$\begin{aligned}\frac{d^2\mathrm{v}(z)}{dz^2} - \mathrm{ZYv}(z) = 0,\\ \frac{d^2\mathrm{i}(z)}{dz^2} - \mathrm{YZi}(z) = 0.\end{aligned} \tag{7.38}$$

Determining the solutions of (7.38) is considerably simplified if one can diagonalize the matrices ZY or YZ. Toward this goal we introduce an invertible, complex-valued $N \times N$ transformation matrix $\mathrm{T_i}$ that relates the current matrix i and the *modal current* matrix $\mathrm{i_m}$ as

$$\mathrm{i} = \mathrm{T_i i_m}. \tag{7.39}$$

Substituting (7.39) into the second of (7.38) we have

$$\frac{d^2\mathrm{i_m}(z)}{dz^2} - \mathrm{T_i^{-1}YZT_i i_m}(z) = 0.$$

If YZ is diagonalizable, then $\mathrm{T_i}$ is such that

$$\mathrm{T_i^{-1}YZT_i} = \gamma^2$$

with $\gamma^2$ being a diagonal matrix ($\gamma^2 = \mathrm{diag}\left[\gamma_{jj}^2\right]$), and we have the diagonal (uncoupled) matrix system

$$\frac{d^2\mathrm{i_m}(z)}{dz^2} - \gamma^2\mathrm{i_m}(z) = 0 \tag{7.40}$$

with the solutions

$$\mathrm{i_m}\left(z\right) = e^{-\gamma z}\mathrm{i_m^+} - e^{\gamma z}\mathrm{i_m^-}. \tag{7.41}$$

In (7.41), $\gamma$ is the square root of the matrix $\gamma^2$, such that the entries of the diagonal matrix $\gamma$ are the square roots of the entries of the diagonal matrix $\gamma^2$ (i.e., $\gamma = \sqrt{\gamma^2} = \mathrm{diag}\left[\gamma_{jj}\right]$), and $\mathrm{i_m^{\mp}}$ are $N \times 1$ matrices. Furthermore, since $\gamma$ is diagonal, then by (4.45) the exponential of the matrix $\gamma$, $e^{-\gamma z}$, is a diagonal matrix with its entries being the values $e^{-\gamma_{jj} z}$. The solution for the current i is obtained as

$$\mathrm{i}\left(z\right) = \mathrm{T_i}\left[e^{-\gamma z}\mathrm{i_m^+} - e^{\gamma z}\mathrm{i_m^-}\right]. \tag{7.42}$$

An analogous procedure can be developed for the voltage matrix. Writing

$$\mathrm{v} = \mathrm{T_v v_m}, \tag{7.43}$$

and substituting (7.43) into the second of (7.38), we have

$$\frac{d^2\mathrm{v_m}(z)}{dz^2} - \mathrm{T_v^{-1}ZYT_v v_m}(z) = 0.$$

If $\mathrm{T_v}$ is such that

$$\mathrm{T_v^{-1} Z Y T_v} = \gamma^2,$$

then we have the diagonal matrix system

$$\frac{d^2 \mathrm{v_m}(z)}{dz^2} - \gamma^2 \mathrm{v_m}(z) = 0$$

with the solutions

$$\mathrm{v_m}(z) = e^{-\gamma z}\mathrm{v_m^+} + e^{\gamma z}\mathrm{v_m^-}. \tag{7.44}$$

The voltage vector is obtained as

$$\mathrm{v}(z) = \mathrm{T_v}\left[e^{-\gamma z}\mathrm{v_m^+} + e^{\gamma z}\mathrm{v_m^-}\right]. \tag{7.45}$$

From physical reasoning we assume the same propagation constant matrix $\gamma$ for both current and voltage waves, and therefore

$$\mathrm{ZY} = \mathrm{T_v T_i^{-1} Y Z T_i T_v^{-1}}.$$

From (7.35) we have

$$\begin{aligned}
\mathrm{v}(z) &= -\mathrm{Y}^{-1}\frac{\partial \mathrm{i}(z)}{\partial z} \\
&= \mathrm{Y^{-1} T_i}\gamma\left[e^{-\gamma z}\mathrm{i_m^+} + e^{\gamma z}\mathrm{i_m^-}\right] \\
&= \left(\mathrm{Y^{-1} T_i}\gamma\mathrm{T_i^{-1}}\right)\mathrm{T_i}\left[e^{-\gamma z}\mathrm{i_m^+} + e^{\gamma z}\mathrm{i_m^-}\right],
\end{aligned}$$

which prompts the definition of the characteristic impedance matrix [14, p. 513]

$$\begin{aligned}
\mathrm{Z_0} &\equiv \mathrm{Y^{-1} T_i}\gamma\mathrm{T_i^{-1}} \\
&= \mathrm{Z T_i}\gamma^{-1}\mathrm{T_i^{-1}}.
\end{aligned}$$

The amplitude vectors $\mathrm{i_m^\mp}$ and $\mathrm{v_m^\mp}$ can be determined by terminal conditions at the source and load ends of the line. In fact, once $\mathrm{i_m^\mp}$ are determined, the current is given by (7.42) and the voltage by

$$\mathrm{v}(z) = \mathrm{Z_0 T_i}\left[e^{-\gamma z}\mathrm{i_m^+} + e^{\gamma z}\mathrm{i_m^-}\right].$$

In the above analysis we assumed that the matrices YZ and ZY were diagonalizable, in which case, for instance, from Section 4.3.1 $\mathrm{T_i}$ would have as its columns the eigenvectors $\phi_m$ of YZ, and the entries of the diagonal matrix $\gamma^2$ would be the associated eigenvalues, i.e.,

$$\mathrm{YZ}\phi_m = \gamma^2\phi_m.$$

In this case, (7.40) has the form of a matrix eigenvalue equation for the differential matrix operator $(d^2/dz^2)\mathrm{I}_{N\times N}$, where $\mathrm{I}_{N\times N}$ is the $N \times N$ identity matrix, which explains why the current vector $\mathrm{i_m}$ is called the modal (eigenmode) current. The square roots of the associated eigenvalues are the propagation constants of the line. Similar comments obviously apply to ZY and the voltage vector $\mathrm{v_m}$. We also assume that $\gamma^2$ is a nonnegative matrix, so that its square root is well defined.

From Theorem 4.38, a complex matrix is diagonalizable if and only if it is normal, and a real-valued matrix is diagonalizable if and only if it is symmetric. It can be shown that the matrices L and C are symmetric [14], although their product is not necessarily so. Furthermore, from reciprocity one may expect that Z and Y are symmetric ($\mathrm{Z} = \mathrm{Z}^\top$, $\mathrm{Y} = \mathrm{Y}^\top$); however, this does not lead to the necessary condition for diagonalization. However, some special cases are of interest. For example, if the multiconductor line is immersed in lossless space, then $\mathrm{Z} = i\omega\mathrm{L}$ and $\mathrm{Y} = i\omega\mathrm{C}$, so that $\mathrm{YZ} = -\omega^2\mathrm{CL}$. Furthermore, if the lossless space is homogeneous and characterized by $\mu, \varepsilon$, then [14]

$$\mathrm{CL} = \mathrm{LC} = \mu\varepsilon\mathrm{I}_{N\times N}.$$

Therefore, for lossless homogeneous space, $\mathrm{YZ} = \mathrm{ZY} = \gamma^2$ is a diagonal matrix ($\mathrm{T_i} = \mathrm{T_v} = \mathrm{I}_{N\times N}$) and

$$\mathrm{Z}_0 = \mathrm{Z}\gamma^{-1} = \mathrm{Z}\left(\sqrt{\mathrm{YZ}}\right)^{-1} = \mathrm{L}\left(\sqrt{\mathrm{LC}}\right)^{-1}, \tag{7.46}$$

in agreement with the two-conductor lossless case, $Z_0 = L(\sqrt{LC})^{-1}$.

# 8

# Planarly Layered Media Problems

In the previous chapter we considered the analysis of an important electromagnetic problem that results in a one-dimensional mathematical model, where the solution was discussed within the framework of Sturm–Liouville theory. It was shown that a generally terminated transmission line leads to a nonself-adjoint problem due to the boundary conditions imposed at the terminating impedances. For certain special terminations the problem is self-adjoint, regardless of the presence or absence of loss on the line.

In this chapter we consider electromagnetic problems in planarly layered media that model physical structures of practical interest. For instance, multi-layer printed circuit boards have printed conducting traces embedded in insulating layers, with one or more conducting ground planes. Similar geometries are found in integrated circuits and in a variety of other electronic applications. Layered-media geometries are also used to model earthen layers and other geophysical structures, including aquatic and ionospheric environments.

The chapter begins with a treatment of two-dimensional problems. After a general formulation, parallel-plate, impedance-plane, and grounded dielectric structures are considered in detail. Since the basic electromagnetic analysis of these structures is well known, special emphasis is placed on the spectral properties of the governing differential operators and on the relationship between these spectral properties and physical electromagnetic phenomena.

We consider both the source-driven and waveguiding problems. For the waveguiding problem we examine the natural waveguide modes of the structure, which, for the propagating modes, are interpreted as eigenfunctions in the null space of the governing two-dimensional differential operator. For the source-driven problem Green's functions are derived, primarily via the method of partial eigenfunction expansion. Completeness relations and associated spectral field expansions (spectral transform pairs) are also discussed. The spectral expansions lead directly to a purely spectral representation of the Green's function, which, although usually not preferred for computational purposes, has a very simple form and can lead to other Green's function representations via complex-plane analysis. Then, the formulation of integral equations for scattering from dielectric cylinders and conducting strips embedded in a layered medium is discussed. Cylinders and strips in free space are obtained as special cases, and some examples of integral equations that admit exact solutions are presented.

Following the analysis of two-dimensional problems, the analysis of multilayered, three-dimensional planar structures is presented. Dyadic Green's functions are formulated in terms of Hertzian potentials, and special cases of interest are considered. The relationship between the three-dimensional Green's function components and the two-dimensional results previously obtained are discussed, and integral equations for three-dimensional objects embedded in a layered medium are developed.

## 8.1 Two-Dimensional Problems

### 8.1.1 General Analysis

#### Fundamental Relations

Consider the planar structure depicted in Figure 8.1, consisting of a planarly inhomogeneous medium characterized by scalar quantities $\varepsilon(x)$ and $\mu(x)$, which may be complex as described in Section 1.1.4. As shown, the structure is bounded by impenetrable walls at $x = -x_b, x_t$ and $z = -z_l, z_r$; any wall can be removed to infinity to implement an unbounded structure.

Working in the frequency domain, generally as temporal Fourier transforms with the $\omega$-dependence suppressed, Maxwell's equations (1.13) become

$$\begin{aligned}
\nabla \cdot \varepsilon(x)\mathbf{E}(\mathbf{r}) &= \rho_e(\mathbf{r}), \\
\nabla \cdot \mu(x)\mathbf{H}(\mathbf{r}) &= \rho_m(\mathbf{r},\omega), \\
\nabla \times \mathbf{E}(\mathbf{r}) &= -i\omega\mu(x)\mathbf{H}(\mathbf{r}) - \mathbf{J}_m(\mathbf{r}), \\
\nabla \times \mathbf{H}(\mathbf{r}) &= \ i\omega\varepsilon(x)\mathbf{E}(\mathbf{r}) + \mathbf{J}_e(\mathbf{r}), \\
\nabla \cdot \mathbf{J}_{e(m)}(\mathbf{r}) &= -i\omega\, \rho_{e(m)}(\mathbf{r}),
\end{aligned} \tag{8.1}$$

where the source quantities $\rho_{e(m)}$ and $\mathbf{J}_{e(m)}$ represent impressed sources, depicted in the figure by the symbol $f$. For simplicity, we will assume that all sources are line sources independent of $y$. Because the structure is also independent of $y$, there will be no field variation in the $y$-direction, and the problem is essentially two-dimensional, e.g., $\mathbf{E} = \widehat{\alpha}E_\alpha(x,z)$, $\alpha = \mathbf{x},\mathbf{y},\mathbf{z}$.

Several aspects of this problem will be discussed:

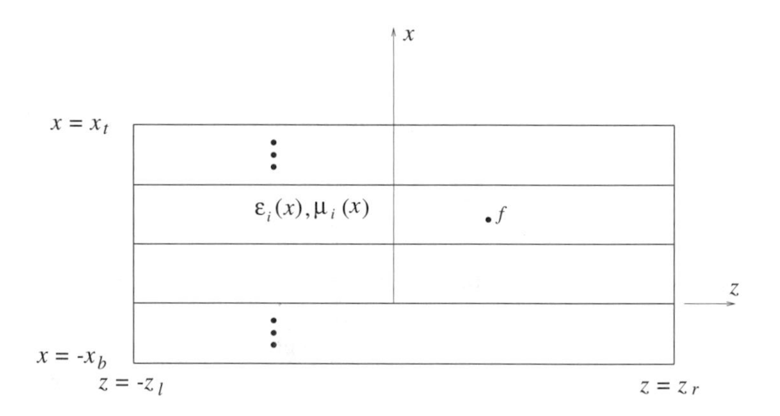


Figure 8.1: Planar waveguiding geometry. Any of the impenetrable walls at $x = -x_b, x_t$ and $z = -z_l, z_r$ can be removed to infinity to implement an unbounded structure. Material characteristics may be arbitrary functions of $x$, and source density is denoted by $f$.

- fundamental interrelationships among field quantities, and the development of appropriate Helmholtz equations;
- spectral properties (eigenfunctions and eigenvalues) and source-free solutions (natural modes);
- construction of Green's functions and field solutions;
- construction of completeness relations and associated spectral expansions.

To see the interrelationship among the various field components, we write the two curl equations in (8.1) as six scalar equations, where in all terms we set $\partial (\cdot) / \partial y = 0$. These equations can be grouped into two independent sets,

$$\frac{\partial E_y}{\partial z} = i\omega\mu H_x + J_{mx}, \tag{8.2}$$

$$\frac{\partial E_y}{\partial x} = -i\omega\mu H_z - J_{mz}, \tag{8.3}$$

$$\frac{\partial H_x}{\partial z} - \frac{\partial H_z}{\partial x} = i\omega\varepsilon E_y + J_{ey}, \tag{8.4}$$

and

$$\frac{\partial H_y}{\partial z} = -i\omega\varepsilon E_x - J_{ex}, \tag{8.5}$$

$$\frac{\partial H_y}{\partial x} = i\omega\varepsilon E_z + J_{ez}, \tag{8.6}$$

$$\frac{\partial E_x}{\partial z} - \frac{\partial E_z}{\partial x} = -i\omega\mu H_y - J_{my}. \tag{8.7}$$

The first group of equations describes modes *transverse-electric* to $z$ ($\text{TE}^z$), having field components $(E_y, H_x, H_z)$. The second group of equations describes modes *transverse-magnetic* to $z$ ($\text{TM}^z$), having field components $(H_y, E_x, E_z)$. Because the two groups of equations are not coupled, by an appropriate choice of sources $\text{TE}^z$-modes (also known as $H$-modes) and $\text{TM}^z$-modes ($E$-modes) may be excited individually, or a source may excite both mode types. The most general type of field that may exist on the structure can be decomposed into $\text{TE}^z$ and $\text{TM}^z$ components.

Helmholtz equations for the $\text{TE}^z$-modes are formed by applying $\partial/\partial z$ to (8.2) and $(\partial/\partial x)\mu^{-1}$ to (8.3), where, upon substitution into (8.4) and defining $k^2(x) \equiv \omega^2\mu(x)\varepsilon(x)$, we obtain

$$\left(\frac{\partial^2}{\partial z^2} + \mu\frac{\partial}{\partial x}\mu^{-1}\frac{\partial}{\partial x} + k^2(x)\right) E_y = i\omega\mu J_{ey} + \frac{\partial J_{mx}}{\partial z} - \mu\frac{\partial}{\partial x}\mu^{-1}J_{mz}. \tag{8.8}$$

Similar manipulations, or invoking duality (1.14), lead to

$$\left(\frac{\partial^2}{\partial z^2} + \varepsilon\frac{\partial}{\partial x}\varepsilon^{-1}\frac{\partial}{\partial x} + k^2(x)\right) H_y = i\omega\varepsilon J_{my} - \frac{\partial J_{ex}}{\partial z} + \varepsilon\frac{\partial}{\partial x}\varepsilon^{-1}J_{ez} \tag{8.9}$$

for the $\text{TM}^z$-modes.[1] The solution of the inhomogeneous Helmholtz equations, subject to appropriate boundary conditions, leads to the respective $y$-component of fields. The two remaining field components are obtained from (8.2) and (8.3) for the $\text{TE}^z$-modes, and from (8.5) and (8.6) for the $\text{TM}^z$-modes.

The Helmholtz equations can be written in generic form as ($w = s^{-1}$)

$$-\left(\frac{\partial^2}{\partial z^2} + s\frac{\partial}{\partial x}s^{-1}\frac{\partial}{\partial x} + k^2(x)\right)\Lambda(x,z) = f(x,z), \tag{8.10}$$

where $s = \mu(x)$ and $\Lambda = E_y$ for $\text{TE}^z$-modes, and $s = \varepsilon(x)$, $\Lambda = H_y$ for $\text{TM}^z$-modes, where $f$ represents the source current density. In the source-driven problem we assume $k(x)$ is a given function, corresponding to a certain physical structure and frequency.

Based on (8.10), defining $\Omega \equiv \{(x,z) : x \in (-x_b, x_t), z \in (-z_l, z_r)\}$ and $H \equiv \mathbf{L}^2_w(\Omega)$, we consider the general problem

$$(L - \lambda)\,\Lambda(x,z) = f(x,z),$$

[1] Note that both Helmholtz equations can be obtained from (1.23) under an appropriate restriction of field components, i.e., $E_y, H_x, H_z \neq 0$ for $\text{TE}^z$-modes, with the other field components absent, and similarly with $H_y, E_x, E_z \neq 0$ for the $\text{TM}^z$-modes, where, in both cases, $\partial(\cdot)/\partial y = 0$ is enforced in the $\nabla$ operator.

where

$$L \equiv -\left(\frac{\partial^2}{\partial z^2} + s\frac{\partial}{\partial x}s^{-1}\frac{\partial}{\partial x} + k^2(x)\right),$$
$$D_L \equiv \{\Lambda : \Lambda, L\Lambda \in H, B(\Lambda) = 0\},$$

and where $B(\Lambda)$ are boundary conditions or fitness conditions to be discussed later. In particular, if a boundary in the $\alpha$th coordinate is removed to infinity, then in that coordinate we require $\lim_{\alpha\to\infty} \Lambda = 0$. This fitness condition, which is usually associated with material loss, ensures that the function is in the appropriate $\mathbf{L}^2$ space as described in Chapter 5. An alternative is to consider lossless media and impose a radiation condition, although in this case one does not generally work in an $\mathbf{L}^2$ space, and, in addition, the operator is inherently nonself-adjoint.

In solving equations of the form (8.10) one usually seeks a Green's function that satisfies

$$\left(-\left(\frac{\partial^2}{\partial z^2} + s\frac{\partial}{\partial x}s^{-1}\frac{\partial}{\partial x} + k^2(x)\right) - \lambda\right) g(\rho, \rho', \lambda) = s(x)\delta(x - x')\delta(z - z'), \tag{8.11}$$

leading to the solution of (8.10) as

$$\Lambda(\rho) = \int_\Omega g(\rho, \rho', 0)\ f(\rho')s^{-1}(x')\ d\Omega', \tag{8.12}$$

where $\rho = \widehat{\mathbf{x}}x + \widehat{\mathbf{z}}z$.[2]

There are many well known methods of solving an equation of the form (8.11) and its associated extension to three dimensions, especially for the case where $s$ is a constant. In this section we will primarily be interested in solving (8.11) by reduction to coupled one-dimensional problems, making use of the one-dimensional Sturm–Liouville theory developed in Chapter 5.

### Eigenfunctions of the Planar Structure

Consideration of (8.10) leads to the eigenvalue equation $Lu = \lambda u$,

$$-\left(\frac{\partial^2}{\partial z^2} + s\frac{\partial}{\partial x}s^{-1}\frac{\partial}{\partial x} + k^2(x)\right) u_{\alpha,\beta}(x, z) = \lambda_{\alpha,\beta}u_{\alpha,\beta}(x, z). \tag{8.13}$$

The general form (8.13) encompasses both proper and improper eigenfunctions, with the four possible cases being

$$\begin{aligned} Lu_{n,m}(x, z) &= \lambda_{n,m}u_{n,m}(x, z), \\ Lu_{n,\nu}(x, z, \nu) &= \lambda_{n,\nu}u_{n,\nu}(x, z, \nu), \\ Lu_{\upsilon,m}(x, z, \upsilon) &= \lambda_{\upsilon,m}u_{\upsilon,m}(x, z, \upsilon), \\ Lu_{\upsilon,\nu}(x, z, \upsilon, \nu) &= \lambda_{\upsilon,\nu}u_{\upsilon,\nu}(x, z, \upsilon, \nu). \end{aligned} \tag{8.14}$$

---

[2] The parameter $\lambda$ is included in (8.11) for generality, and because it facilitates spectral analysis.

In the above equations $n, m$ $(\upsilon, \nu)$ represent discrete (continuous) indices, in particular,

- $u_{n,m}(x,z)$ represents eigenfunctions proper (discrete) in both spatial coordinates,
- $u_{n,\nu}(x,z,\nu)$ represents eigenfunctions proper in $x$ and improper (continuous) in $z$,
- $u_{\upsilon,m}(x,z,\upsilon)$ represents eigenfunctions proper in $z$ and improper in $x$, and
- $u_{\upsilon,\nu}(x,z,\upsilon,\nu)$ represents eigenfunctions improper in both coordinates.

It will be assumed that all eigenfunctions form a mutually orthonormal set (e.g., $\langle u_{\alpha,\beta}, u_{\gamma,\varsigma}\rangle = \delta_{\alpha,\gamma}\delta_{\beta,\varsigma}$ for proper eigenfunctions). The presence or absence of the different eigenfunction types depends on the physical structure of the specific problem of interest.

For nonself-adjoint problems, we also need to consider adjoint eigenfunctions satisfying $L^* u^* = \lambda^* u^*$,

$$-\left(\frac{\partial^2}{\partial z^2} + \overline{s}\frac{\partial}{\partial x}\overline{s}^{-1}\frac{\partial}{\partial x} + \overline{k}^2(x)\right) u^*_{\alpha,\beta}(x,z) = \lambda^*_{\alpha,\beta} u^*_{\alpha,\beta}(x,z)$$

where $\lambda^*_{\alpha,\beta} = \overline{\lambda}_{\alpha,\beta}$ and where $\{u, u^*\}$ satisfy a bi-orthogonality relation.

For proper eigenfunctions we impose homogeneous Dirichlet or Neumann conditions for bounded regions, whereas for unbounded regions we require the proper eigenfunctions to be in the appropriate $\mathbf{L}^2$ space. Improper eigenfunctions will be simply bounded at infinity.

### Natural Modes of the Planar Structure

In addition to the eigenvalue problem, a related problem of considerable interest is the concept of the *natural modes* of a structure. A natural mode is defined as a field configuration, including any constraints on parameter values, that exists in the absence of sources. The form of the natural-mode solution is dictated by the structure of the physical space in question, which, in turn, leads to the appropriate boundary conditions for the problem. The natural modes satisfy the homogeneous form $L\psi = 0$,

$$-\left(\frac{\partial^2}{\partial z^2} + s\frac{\partial}{\partial x}s^{-1}\frac{\partial}{\partial x} + k^2(x)\right)\psi(x,z) = 0. \tag{8.15}$$

Typical boundary conditions are homogeneous Dirichlet or Neumann conditions for bounded regions, and outgoing wave conditions for unbounded regions.

While the natural-mode solutions of (8.15) would appear to be simply eigenfunctions in the null space of $L$ (i.e., corresponding to $\lambda_{\alpha,\beta} = 0$), which is, in fact, often the case, it will become clear that there are important differences in the two concepts. The natural modes are perhaps the more physically meaningful quantities, representing the waveguide or cavity modes of the structure. The eigenfunctions and eigenvalues are useful in spectral representations, with eigenfunctions in the null space of $L$ being a subset of the natural modes, with their corresponding physical significance. In the following we denote eigenfunctions by $u$ and natural modes by $\psi$.

A two-dimensional resonator problem occurs when all of the conducting walls are present, as shown in Figure 8.1. We'll see later that for resonance problems all eigenfunctions in the null space of the operator $L$ (i.e., corresponding to $\lambda_{\alpha,\beta} = 0$) are natural modes (cavity resonances), and, conversely, that all natural modes are eigenfunctions in the null space of $L$. In this case frequency $\omega$ is a free parameter via $k^2(x) = \omega^2 \mu(x) \varepsilon(x)$, interpreted as a nonstandard eigenvalue (see (4.5)), which is determined such that the proper eigenvalue of the operator $L$ is zero, i.e., $\lambda_{\alpha,\beta}(\omega) = 0$. In this way the natural cavity fields are known as *eigenfields*, and the nonstandard eigenvalues $\omega$ such that $\lambda_{\alpha,\beta}(\omega) = 0$ are the resonate frequencies of the cavity.[3] Eigenfunctions that are not in the null space of $L$ are not directly related to physical resonances, yet are important in spectral expansions.

For waveguiding problems where some of the walls are removed to infinity, we assume that $k(x)$ is given. It will be shown then that all eigenfunctions in the null space of the operator $L$ are natural modes, representing propagating waves; however, natural modes corresponding to evanescent waves are not eigenfunctions of $L$ (they are not generally bounded on $\mathbf{R}$) and are, interestingly, not directly implicated in a purely spectral theory, their physical significance notwithstanding.

By invoking the necessary boundary conditions the natural-mode fields can be determined to within an arbitrary amplitude coefficient, which is determined by solving the excitation problem (i.e., the inhomogeneous Helmholtz equation). The actual field caused by a source in a waveguide or cavity can then be represented as a sum (continuous or discrete) of natural modes, or, alternatively, in a two-dimensional spectral form in terms of eigenfunctions.

The most common method for solving equations of the form (8.13) and (8.15) is separation-of-variables. For instance, we seek a solution to (8.13) as $u_{\alpha,\beta}(x,z) = u_{x,\alpha}(x) u_{z,\beta}(z)$, which, when substituted into (8.13), leads

[3] If the medium is homogeneous, then $k^2$ is a constant. Replacing $k^2$ with $\lambda_{n,m}$ leads to the ordinary eigenvalue equation

$$-\left(\frac{\partial^2}{\partial z^2} + \frac{\partial^2}{\partial x^2}\right) u_{n,m}(x,z) = \lambda_{n,m} u_{n,m}(x,z)$$

with the resonant (cavity) frequencies determined by $\omega_{n,m} = (1/\sqrt{\mu\varepsilon})\sqrt{\lambda_{n,m}}$. This is the usual method for homogeneous cavities, although the procedure described in the text is more general in the sense that it can accommodate planar stratifications.

to two one-dimensional eigenvalue equations

$$\left(-\frac{d^2}{dz^2} - \lambda_z\right) u_{z,\beta}(z) = 0, \tag{8.16}$$

$$\left(-s\frac{d}{dx}s^{-1}\frac{d}{dx} - \left(k^2(x) + \lambda_x\right)\right) u_{x,\alpha}(x) = 0 \tag{8.17}$$

with the separation equation $\lambda_z + \lambda_x = \lambda_{\alpha,\beta}$. For (8.15) we write $\psi(x,z) = \psi_x(x)\psi_z(z)$ and obtain the above forms for $\psi_x$ and $\psi_z$, where $\lambda_z + \lambda_x = 0$.

We assume that the eigenfunctions associated with (8.16) and (8.17) satisfy the bi-orthogonality relations

$$\begin{aligned} \left\langle u_{\alpha,n}, u^*_{\alpha,m}\right\rangle_{\alpha=x,z} &= \delta_{nm}, \\ \left\langle u_{\alpha,v}(v), u^*_{\alpha,p}(p)\right\rangle_{\alpha=x,z} &= \delta(v-p), \\ \left\langle u_{\alpha,n}, u^*_{\alpha,v_\alpha}(v_\alpha)\right\rangle_{\alpha=x,z} &= 0, \end{aligned} \tag{8.18}$$

where the above inner products are the appropriate one-dimensional inner products,

$$\begin{aligned} \langle f, g\rangle_x &= \int_{-x_b}^{x_t} f(x)\overline{g}(x)s^{-1}(x)\,dx, \\ \langle f, g\rangle_z &= \int_{-z_l}^{z_r} f(z)\overline{g}(z)\,dz. \end{aligned}$$

The differential operators in both above equations are seen to be of the Sturm–Liouville form (5.4), with $w = p = 1$, and $q = 0$ in (8.16), and $w = p = s^{-1}$, $q = -k^2$ in (8.17). If the medium is lossy, $s, k \in \mathbf{C}$ and the condition $w, p, q \in \mathbf{R}$ described in Section 5.1 is violated. Subsequently, completeness and other spectral relations obtained in Chapter 5 may not necessarily hold. Nevertheless, the presence of material loss does not lead to this problem if the medium is homogeneous with appropriate boundary conditions, as discussed in Chapter 7 and in Section 8.1.2. Moreover, by the discussion in Section 5.3.2, if the problem can be separated into equations of the Sturm–Liouville type, which are nonself-adjoint but satisfy the conditions of Theorem 5.3, then completeness of the eigenbasis is assured. Another method is to consider nonself-adjoint perturbations of self-adjoint operators, as briefly discussed on p. 435.

### Green's Functions by Partial Eigenfunction Expansion

Consideration of (8.16) and (8.17) lead us to study the associated one-dimensional Sturm–Liouville Green's function equations

$$\left(-\frac{d^2}{dz^2} - \lambda_z\right) g_z(z, z', \lambda_z) = \delta(z - z'), \tag{8.19}$$

$$\left(-s\frac{d}{dx}s^{-1}\frac{d}{dx} - \left(k^2(x) + \lambda_x\right)\right) g_x(x, x', \lambda_x) = s(x)\delta(x - x'). \tag{8.20}$$

While $u(x,z) = u_x(x)u_z(z)$ represents the solution of (8.13), it is *not* true that the product of $g_x$ and $g_z$ provides $g(x,z,x',z')$ which satisfies (8.11),

$$\left(-\left(\frac{\partial^2}{\partial z^2} + s\frac{\partial}{\partial x}s^{-1}\frac{\partial}{\partial x} + k^2(x)\right) - \lambda\right) g(\rho,\rho',\lambda) = s(x)\delta(x-x')\delta(z-z').$$

However, we can construct $g$ by using one of the Green's functions, i.e., $g_x$ (or $g_z$), and account for the remaining coordinate using eigenfunctions in $z$ (or $x$)[4]. This partial eigenfunction expansion procedure does not lead to a purely spectral form, yet the dimensionality of the resulting sums (continuous and discrete) are reduced from those encountered in a two-dimensional spectral representation, leading to a more practically useful expression.

To construct the two-dimensional Green's function $g(\rho,\rho',\lambda)$, consider the partial eigenfunction expansion

$$g(x,z,x',z',\lambda) = \sum_n a_n(z,x',z',n)u_{x,n}(x) + \int_v a(z,x',z',v)u_{x,v}(x,v)\,dv \tag{8.21}$$

with $u_{x,n}(x)$ and $u_{x,v}(x,v)$ proper and improper eigenfunctions, respectively, of the operator $L_x \equiv \left(-s\frac{d}{dx}s^{-1}\frac{d}{dx} - k^2(x)\right)$, i.e.,

$$\left(-s\frac{d}{dx}s^{-1}\frac{d}{dx} - k^2(x)\right) u_{x,n}(x) = \lambda_{x,n}u_{x,n}(x),$$

$$\left(-s\frac{d}{dx}s^{-1}\frac{d}{dx} - k^2(x)\right) u_{x,v}(x,v) = \lambda_{x,v}(v)\,u_{x,v}(x,v)$$

(in a slight abuse of notation, we replace $\lambda_x$ with $\lambda_{x,n}$ or $\lambda_{x,v}$, not to be confused with the two-dimensional eigenfunctions $\lambda_{\alpha,\beta}$). In the continuous case often $\lambda_{x,v}(v) = v^2$, or a similar relation exists, with the exact representation depending on the problem.

If we substitute (8.21) into (8.11) and take the inner product $\langle\cdot,\cdot\rangle_x$ of the resulting expression with $u^*_{x,m}(x')$, then, upon using (8.16) and (8.17), we obtain

$$\left(\frac{d^2}{dz^2} - (\lambda_{x,n} - \lambda)\right) a_m = -\overline{u}^*_{x,m}(x')\delta(z-z').$$

Subsequent to the identification

$$g_z(z,z',-(\lambda_{x,n}-\lambda)) = a_m/\overline{u}^*_{x,m}(x') \tag{8.22}$$

in the above we obtain

$$\left(\frac{d^2}{dz^2} - (\lambda_{x,n} - \lambda)\right) g_z(z,z',(\lambda_{x,n}-\lambda)) = -\delta(z-z'),$$

[4] Another method to construct the two-dimensional Green's function $g$ from the two one-dimensional Green's functions is shown in [25, pp. 273–274], [17, pp. 285–287], and [38, pp. 72–87].

the solution of which leads to the expansion coefficients $a_m$ via (8.22). Similar manipulations with $u_{x,\upsilon}(x', \upsilon)$ lead to

$$a(z, x', z', \upsilon) = g_z(z, z', -(\lambda_{x,\upsilon}(\upsilon) - \lambda))\overline{u}^*_{x,\upsilon}(x', \upsilon)$$

such that we obtain

$$\begin{aligned} g(x, z, x', z', \lambda) =& \sum_n u_{x,n}(x)\overline{u}^*_{x,n}(x') g_z(z, z', -(\lambda_{x,n} - \lambda)) \\ &+ \int_\upsilon u_{x,\upsilon}(x, \upsilon)\overline{u}^*_{x,\upsilon}(x', \upsilon) g_z(z, z', -(\lambda_{x,\upsilon}(\upsilon) - \lambda))\, d\upsilon \end{aligned} \tag{8.23}$$

where $g_z$ satisfies (8.19).

We can also construct an expansion like (8.21) but in the $z$-coordinate eigenfunctions with $x$-dependent coefficients, resulting in

$$\begin{aligned} g(x, z, x', z', \lambda) =& \sum_n u_{z,n}(z)\overline{u}^*_{z,n}(z') g_x(x, x', -(\lambda_{z,m} - \lambda)) \\ &+ \int_\nu u_{z,\nu}(z, \nu)\overline{u}^*_{z,\nu}(z', \nu) g_x(x, x', -(\lambda_{z,\nu}(\nu) - \lambda))\, d\nu. \end{aligned} \tag{8.24}$$

Of course, if either the discrete or continuous spectral components are missing for a given problem, the appropriate sum or integral term in (8.23) and (8.24) is omitted. It must be emphasized that we assume the appropriate one-dimensional eigenfunctions form an orthonormal basis for the corresponding $\mathbf{L}^2$ space in order for the expansion to be valid. This is discussed in more detail in Section 8.1.8. A two-dimensional spectral form is derived subsequent to the discussion of completeness relations below.

### Completeness Relations and Associated Spectral Expansions

The completeness relation (5.92) is used to form

$$\begin{aligned} s(x)\delta(x - x') =& \frac{-1}{2\pi i}\oint_{C_x} g_x(x, x', \lambda_x)\, d\lambda_x = \sum_n u_{x,n}(x)\overline{u}^*_{x,n}(x') \\ &+ \int_\upsilon u_{x,\upsilon}(x, \upsilon)\overline{u}^*_{x,\upsilon}(x', \upsilon)\, d\upsilon \\ \delta(z - z') =& \frac{-1}{2\pi i}\oint_{C_z} g_z(z, z', \lambda_z)\, d\lambda_z = \sum_n u_{z,n}(z)\overline{u}^*_{z,n}(z') \\ &+ \int_\nu u_{z,\nu}(z, \nu)\overline{u}^*_{z,\nu}(z', \nu)\, d\nu. \end{aligned} \tag{8.25}$$

If the associated one-dimensional boundary value problems are self-adjoint, then $u^* = u$. For generality, the possible presence of both a discrete ($u_{\alpha,n}$) and a continuous ($u_{\alpha,\upsilon_\alpha}(\alpha, \upsilon_\alpha)$) spectrum is accommodated in (8.25) for both coordinate problems. The discrete eigenfunctions $u_{\alpha,n}$ (if they exist) of the differential operators satisfy (8.16) and (8.17) subject to prescribed boundary conditions. If the structure is unbounded in one or more directions, then a continuous spectrum will be present. The continuous spectral components, represented by improper eigenfunctions $u_{\alpha,\upsilon_\alpha}(\alpha,\upsilon_\alpha)$, also satisfy the one-dimensional eigenvalue equations, though not the usual limiting or $\mathbf{L}^2$ inclusion condition in the $\alpha$-coordinate at infinity.

Rather than via (8.16) and (8.17), the spectrum, especially the continuous spectrum, is often identified indirectly via the integrals in (8.25) as discussed in Section 5.2. The advantage of obtaining the spectrum from the integration of $g_\alpha$ is that one automatically obtains the appropriate normalization, whereas if the spectral components are obtained from the eigenvalue equations they must be normalized accordingly. This normalization, especially in the continuous case, can be difficult. The possible disadvantage of obtaining the spectrum from the integration of $g_\alpha$ is that the form of $g_\alpha$ may be quite complicated; see, e.g., Section 8.1.7.

With the two-dimensional eigenfunctions given by the product of the one-dimensional eigenfunctions, forming the product $\delta(\rho-\rho')=\delta(x-x')\delta(z-z')$ leads to the two-dimensional completeness relation [17, p. 285]

$$\begin{aligned}
s(x)\delta(x-x')\delta(z-z') &= \left(\frac{-1}{2\pi i}\right)^2 \oint_{C_x}\oint_{C_z} g_x(x,x',\lambda_x)g_z(z,z',\lambda_z)\,d\lambda_z d\lambda_x \\
&= \sum_n\sum_m u_{n,m}(x,z)\,\overline{u}^*_{n,m}(x',z') + \sum_n \int_\nu u_{n,\nu}(x,z,\nu)\,\overline{u}^*_{n,\nu}(x',z',\nu)\,d\nu \\
&\quad + \sum_m \int_\upsilon u_{\upsilon,m}(x,z,\upsilon)\,\overline{u}^*_{\upsilon,m}(x',z',\upsilon)\,d\upsilon \\
&\quad + \int_\upsilon\int_\nu u_{\upsilon,\nu}(x,z,\upsilon,\nu)\,\overline{u}^*_{\upsilon,\nu}(x',z',\upsilon,\nu)\,d\nu\,d\upsilon
\end{aligned} \tag{8.26}$$

which will be valid if the eigenfunctions form a basis for the space. We also have a completeness relation of the type (4.63),

$$\frac{1}{2\pi i}\oint_C g(\rho,\rho',\lambda)\,d\lambda = -s(x)\,\delta(\rho-\rho'), \tag{8.27}$$

which will hold if the root system (including improper eigenfunctions) of the governing operator forms a basis for the space. However, one must be careful to provide a valid spectral form for the Green's function, especially in the case of a nonself-adjoint operator (e.g., it may happen that (8.27) is applicable, but that (8.23) and (8.24) may not provide a valid form for $g$).

### Spectral Expansions in $\mathbf{L}^2_w(\Omega)$ in Terms of Eigenfunctions

To determine a modal expansion of an arbitrary function $f(x,z)\in\mathbf{L}^2_w(\Omega)$, we take the inner product of (8.26) with $f(x,z)$, where we define the two-dimensional inner product by

$$\langle f, g \rangle \equiv \int_\Omega f(x,z)\overline{g}(x,x)s^{-1}(x)\, d\Omega.$$

This leads to the expansion

$$f(x,z) = \sum_n \sum_m a_{n,m} u_{n,m}(x,z) + \sum_n \int_\nu a^z_{x,n}(\nu) u_{n,\nu}(x,z,\nu)\, d\nu \qquad (8.28)$$
$$+ \sum_m \int_\upsilon a^x_{z,m}(\upsilon) u_{m,\upsilon}(x,z,\upsilon)\, d\upsilon$$
$$+ \int_\upsilon \int_\nu a(\upsilon,\nu) u_{\upsilon,\nu}(x,z,\upsilon,\nu)\, d\nu\, d\upsilon,$$

where $a_{n,m} = \langle f, u^*_{n,m} \rangle$, $a^z_{x,n}(\nu) = \langle f, u^*_{n,\nu} \rangle$, $a^x_{z,m}(\upsilon) = \langle f, u^*_{m,\upsilon} \rangle$, and $a(\upsilon,\nu) = \langle f, u^*_{\upsilon,\nu} \rangle$, assuming that the eigenfunctions form a basis for $\mathbf{L}^2_w(\Omega)$.

The above expansion and associated coefficients form a transform pair for the space in question. For example, if the space is bounded by impenetrable walls, the continuous spectrum will be absent, and (8.28) takes the form

$$f(x,z) = \sum_n \sum_m a_{n,m} u_{n,m}(x,z),$$

i.e., a two-dimensional generalized Fourier series. If the geometry is unbounded in all directions, the discrete spectrum may be absent, in which case (8.28) will have the form of a two-dimensional generalized Fourier transform,

$$f(x,z) = \int_\upsilon \int_\nu a(\upsilon,\nu) u_{\upsilon,\nu}(x,z,\upsilon,\nu)\, d\nu\, d\upsilon.$$

Of course, other combinations are also possible, as the examples to follow illustrate.

The expansion (8.28) has two principal uses. First, recalling that usually a function can be represented in a given space by many equivalent representations, (8.28) is the representation in terms of the complete set of two-dimensional eigenfunctions. While these are not waveguide or cavity "modes" until coupled by the separation equation $\lambda_x + \lambda_z = 0$ (i.e., for the eigenvalue $\lambda_{\alpha,\beta} = 0$), it is still a useful representation for expanding a given source, allowing orthonormality to be utilized in solving for an unknown field response. The second principal use of the expansion is for the representation of an unknown quantity in the form of (8.28), with coefficients to be determined by a mode-matching procedure.

### Spectral Solution of the Helmholtz Equation and Spectral Green's Function

The expansion (8.28) also leads to an eigenfunction expansion solution of

$$\left(-\left(\frac{\partial^2}{\partial z^2}+s\frac{\partial}{\partial x}s^{-1}\frac{\partial}{\partial x}+k^2(x)\right)-\lambda\right)\Lambda(x,z)=f \tag{8.29}$$

and to a purely spectral Green's function representation. A spectral solution for a one-dimensional problem involving both a discrete and a continuous spectrum is detailed beginning on p. 326 (see also Section 4.5.2), and the procedure in the two-dimensional case is analogous. The expansion (8.28) leads to

$$\begin{aligned}\Lambda(x,z)=&\sum_n\sum_m\left\langle\Lambda,u^*_{n,m}\right\rangle u_{n,m}(x,z)+\sum_n\int_\nu\left\langle\Lambda,u^*_{n,\nu}\right\rangle u_{n,\nu}(x,z,\nu)\,d\nu\\&+\sum_m\int_\upsilon\left\langle\Lambda,u^*_{\upsilon,m}\right\rangle u_{\upsilon,m}(x,z,\upsilon)\,d\upsilon\\&+\int_\upsilon\int_\nu\left\langle\Lambda,u^*_{\upsilon,\nu}\right\rangle u_{\upsilon,\nu}(x,z,\upsilon,\nu)\,d\nu\,d\upsilon.\end{aligned}$$

Exploiting bi-orthonormality and following the general procedure for eigenfunction expansions, we obtain the spectral solution of (8.29) as

$$\begin{aligned}\Lambda(x,z)=&\sum_n\sum_m\frac{\left\langle f,u^*_{n,m}\right\rangle}{\lambda_{n,m}-\lambda}u_{n,m}(x,z)+\sum_n\int_\nu\frac{\left\langle f,u^*_{n,\nu}\right\rangle}{\lambda_{n,\nu}-\lambda}u_{n,\nu}(x,z,\nu)\,d\nu\\&+\sum_m\int_\upsilon\frac{\left\langle f,u^*_{\upsilon,m}\right\rangle}{\lambda_{\upsilon,m}-\lambda}u_{\upsilon,m}(x,z,\upsilon)\,d\upsilon\\&+\int_\upsilon\int_\nu\frac{\left\langle f,u^*_{\upsilon,\nu}\right\rangle}{\lambda_{\upsilon,\nu}-\lambda}u_{\upsilon,\nu}(x,z,\upsilon,\nu)\,d\nu\,d\upsilon.\end{aligned} \tag{8.30}$$

The eigenvalue subscript indicates a proper (discrete; $n,m$) or improper (continuous; $\upsilon,\nu$) eigenvalue parameter as described subsequent to (8.14). Alternatively, following the procedure detailed in Section 5.1.3 we obtain from (8.30) the spectral form of the Green's function:

$$\begin{aligned}g(\rho,\rho',\lambda)=&\sum_n\sum_m\frac{\overline{u}^*_{n,m}(x',z')}{\lambda_{n,m}-\lambda}u_{n,m}(x,z)\\&+\sum_n\int_\nu\frac{\overline{u}^*_{n,\nu}(x',z',\nu)}{\lambda_{n,\nu}-\lambda}u_{n,\nu}(x,z,\nu)\,d\nu\\&+\int_\upsilon\sum_m\frac{\overline{u}^*_{\upsilon,m}(x',z',\upsilon)}{\lambda_{\upsilon,m}-\lambda}u_{\upsilon,m}(x,z,\upsilon)\,d\upsilon\\&+\int_\upsilon\int_\nu\frac{\overline{u}^*_{\upsilon,\nu}(x',z',\upsilon,\nu)}{\lambda_{\upsilon,\nu}-\lambda}u_{\upsilon,\nu}(x,z,\upsilon,\nu)\,d\nu\,d\upsilon.\end{aligned} \tag{8.31}$$

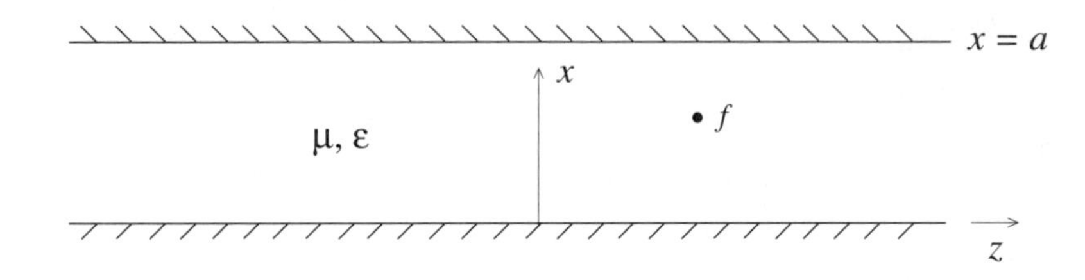


Figure 8.2: Homogeneously filled parallel-plate waveguide structure bounded by perfectly conducting infinite plates at $x = 0, a$.

We remark that the forms (8.30) and (8.31) assume the existence of a complete basis of ordinary eigenfunctions (no associated eigenfunctions exist). As in previous chapters, we are most directly interested in the case $\lambda = 0$, although the variable-$\lambda$ case allows for spectral integrations utilized in completeness relations.

The above formulations will be applied to several specific waveguiding and cavity geometries in the following sections.

### 8.1.2 Homogeneously Filled Parallel-Plates

Consider the homogeneously filled parallel-plate waveguide structure shown in Figure 8.2, consisting of a homogeneous, laterally unbounded region characterized by constants $\mu$ and $\varepsilon$ and bounded by perfectly conducting infinite plates at $x = 0, a$. The appropriate Helmholtz equation, from (8.8) and (8.9), is

$$-\left(\frac{\partial^2}{\partial z^2} + \frac{\partial^2}{\partial x^2} + k^2\right) \Lambda(x, z) = f(x, z) \tag{8.32}$$

with $\Lambda = E_y$ for TE$^z$-modes and $\Lambda = H_y$ for TM$^z$-modes, and where the term $f$ represents a component of source. For both types of modes we will enforce

$$\lim_{z \to \pm\infty} \Lambda(x, z) = 0,$$

whereas the boundary condition for TE$^z$-modes is

$$B(\Lambda) = \Lambda(0, z) = \Lambda(a, z) = 0, \tag{8.33}$$

and for TM$^z$-modes we have

$$B(\Lambda) = \frac{\partial \Lambda(0, z)}{\partial x} = \frac{\partial \Lambda(a, z)}{\partial x} = 0. \tag{8.34}$$

Letting $\Omega \equiv \{(x, z) : x \in [0, a], z \in (-\infty, \infty)\}$ and $H \equiv \mathbf{L}^2(\Omega)$, we consider in the following the operator $L : \mathbf{L}^2(\Omega) \to \mathbf{L}^2(\Omega)$ defined by

$$L \equiv -\left(\frac{\partial^2}{\partial z^2} + \frac{\partial^2}{\partial x^2} + k^2\right), \tag{8.35}$$

$$D_L \equiv \left\{\Lambda : \Lambda, L\Lambda \in H, B(\Lambda) = 0, \lim_{z\to\pm\infty} \Lambda(x,z) = 0\right\}$$

with the inner product

$$\langle f, g\rangle = \int_{-\infty}^{\infty}\int_0^a f(x,z)\,\overline{g(x,z)}\,dx\,dz.$$

### Spectral Properties and Eigenfunctions of the Parallel-Plate Structure

We first consider the spectral properties of the differential operator for the parallel-plate structure, i.e., solutions of

$$-\left(\frac{\partial^2}{\partial z^2} + \frac{\partial^2}{\partial x^2} + k^2\right) u_{n,\nu}(x,z) = \lambda_{n,\nu} u_{n,\nu}(x,z) \tag{8.36}$$

subject to the appropriate boundary conditions. We seek a solution as $u(x,z) = u_x(x)u_z(z)$, which, from (8.16) and (8.17), lead to

$$\left(-\frac{d^2}{dz^2} - \lambda_{z,\nu}\right) u_{z,\nu}(z) = 0, \tag{8.37}$$

$$\left(-\frac{d^2}{dx^2} - \lambda_{x,n}\right) u_{x,n}(x) = 0 \tag{8.38}$$

subject to the separation equation[5] $\lambda_{z,\nu} + \lambda_{x,n} - k^2 = \lambda_{n,\nu}$. The differential operators in both of the above equations are seen to be of the self-adjoint Sturm–Liouville form (5.4), which is extensively studied in Chapter 5. This is true even for a lossy medium.[6] Boundary conditions in the vertical direction, (8.33) and (8.34), are dictated by the conducting walls at $x = 0, a$, whereas in the $z$-coordinate we look for solutions bounded at infinity.

[5] Because $k$ is constant, the natural form of the separation equation is $\lambda_{x,n} + \lambda_{z,\nu} - k^2 = \lambda_{n,\nu}$. In this section we consider $k^2$ as part of the operator, to be consistent with the preceding general development. In this case the presence of $k^2$ simply shifts the eigenvalues of the transverse Laplacian, which is the natural operator for the homogeneous medium case (see Theorem 4.4).

[6] For TE$^z$-modes one can also consider an inhomogeneous lossy medium $(\varepsilon(x), \mu_0)$ between the plates, leading to

$$-\left(\frac{\partial^2}{\partial x^2} + k^2(x)\right) u_{x,n} = \lambda_{x,n} u_{x,n}.$$

By an example in Kato [24, p. 297], assuming that $k^2(x)$ is merely bounded on $(0,a)$, we treat $k^2(x)$ as a nonself-adjoint perturbation of the self-adjoint operator $-\partial^2/\partial x^2$ with boundary conditions $u(0) = u(a) = 0$. The eigenfunctions form a basis in $\mathbf{L}^2(0,a)$.

In the vertical direction the eigenvalue problem (8.38) leads to

$$u_{x,n}(x) = A_x \sin\sqrt{\lambda_x}x + B_x \cos\sqrt{\lambda_x}x.$$

Application of the boundary conditions for $TE^z$-modes ($H$-modes) provides $B_x = 0$ and $\sqrt{\lambda_{x,n}}a = n\pi$ (see Example 5.3), such that the normalized eigenfunctions and eigenvalues are

$$\begin{aligned} u^H_{x,n}(x) &= \sqrt{\frac{2}{a}}\sin\frac{n\pi}{a}x, \\ \lambda^H_{x,n} &= \left(\frac{n\pi}{a}\right)^2. \end{aligned}$$

In a similar manner, application of the boundary conditions for $TM^z$-modes ($E$-modes) provides $A_x = 0$ and $\sqrt{\lambda_{x,n}}a = n\pi$ (see Example 5.5), such that the normalized eigenfunctions and eigenvalues are

$$\begin{aligned} u^E_{x,n}(x) &= \sqrt{\frac{\varepsilon_n}{a}}\cos\frac{n\pi}{a}x, \\ \lambda^E_{x,n} &= \lambda^H_{x,n} = \left(\frac{n\pi}{a}\right)^2, \end{aligned}$$

where $\varepsilon_n$ is Neumann's number. This constitutes the spectral problem in the $x$-coordinate, where it is easily seen that the eigenfunctions form an eigenbasis in $\mathbf{L}^2(0,a)$.

From the analysis in Example 5.11, we see that (8.37) leads to the improper eigenfunctions and eigenvalues

$$\begin{aligned} u_{z,\nu}(z,\nu) &= \frac{1}{\sqrt{2\pi}}e^{i\nu z}, \\ \lambda_z(\nu) &= \nu^2, \qquad \nu \in (-\infty,\infty). \end{aligned}$$

The two-dimensional eigenfunctions are $u_{n,\nu}(x,z) = u_{x,n}(x)\,u_{z,\nu}(z,\nu)$, forming a basis in $\mathbf{L}^2(\Omega)$. Specifically, for the $H$- and $E$-modes we have

$$u^H_{n,\nu}(x,z,\nu) = \frac{1}{\sqrt{a\pi}}\sin\frac{n\pi}{a}x\, e^{i\nu z}, \tag{8.39}$$

$$u^E_{n,\nu}(x,z,\nu) = \sqrt{\frac{\varepsilon_n}{2\pi a}}\cos\frac{n\pi}{a}x\, e^{i\nu z}, \tag{8.40}$$

$$\lambda_{n,\nu} = \nu^2 + \left(\frac{n\pi}{a}\right)^2 - k^2, \tag{8.41}$$

where $k$ is given and $\nu$ is an infinite continuum. Orthonormality is expressed as

$$\left\langle u^\alpha_{n,\nu}, u^\alpha_{m,p}\right\rangle = \delta_{nm}\delta(\nu - p)$$

for $\alpha = E, H$.

### Natural Waveguide Modes of the Parallel-Plate Structure

The natural modes of the parallel-plate structure are solutions of

$$-\left(\frac{\partial^2}{\partial z^2}+\frac{\partial^2}{\partial x^2}+k^2\right)\psi(x,z)=0, \tag{8.42}$$

leading to

$$\left(-\frac{d^2}{dx^2}-\lambda_x\right)\psi_x(x)=0, \tag{8.43}$$

$$\left(-\frac{d^2}{dz^2}-\lambda_z\right)\psi_z(z)=0, \tag{8.44}$$

where $\lambda_x+\lambda_z-k^2=0$, and which are solved subject to the prescribed boundary conditions and an outgoing-wave condition. In this way the natural modes are obtained as $\psi_n(x,z)=\psi_x(x)\,\psi_z(z)$.

As with the $x$-component eigenfunctions, (8.43) is solved as

$$\psi^H_{x,n}(x)=\sqrt{\frac{2}{a}}\sin\frac{n\pi}{a}x,$$

$$\psi^E_{x,n}(x)=\sqrt{\frac{\varepsilon_n}{a}}\cos\frac{n\pi}{a}x,$$

where $\lambda^E_{x,n}=\lambda^H_{x,n}=(n\pi/a)^2$. Considering the solution of (8.44), if $\operatorname{Im}\lambda_z\neq 0$ from Theorem 5.2, we know there is exactly one solution in $\mathbf{L}^2(0,\infty)$, $\psi_z(z)=Ae^{-i\sqrt{\lambda_z}z}$, and one solution in $\mathbf{L}^2(-\infty,0)$, $\psi_z(z)=Be^{i\sqrt{\lambda_z}z}$, where we assume $\operatorname{Im}\sqrt{\lambda_z}<0$. We will take the solution as

$$\psi_z(z)=A\,e^{\mp i\sqrt{\lambda_z}z}$$

with the sign chosen according to the sign of $z$, and where $A$ is an arbitrary constant. Since $\lambda_z+\lambda_x=k^2$, then $\lambda_z=\lambda_{z,n}=k^2-(n\pi/a)^2$ and the propagation constant in the $z$-direction is determined as

$$\sqrt{\lambda_{z,n}}=\sqrt{k^2-(n\pi/a)^2}.$$

We therefore have

$$\begin{aligned}\psi^H_n(x,z)&=A\sqrt{\frac{2}{a}}\sin\frac{n\pi}{a}x\,e^{\mp i\sqrt{k^2-\left(\frac{n\pi}{a}\right)^2}z},\\ \psi^E_n(x,z)&=A\sqrt{\frac{\varepsilon_n}{a}}\cos\frac{n\pi}{a}x\,e^{\mp i\sqrt{k^2-\left(\frac{n\pi}{a}\right)^2}z}\end{aligned} \tag{8.45}$$

as the $n$th modal (natural mode) solution of (8.42).[7] For $n=0$, $\psi^E_0$ represents a TEM$^z$-mode since $E_z=0$ from (8.6), leaving only $H_y, E_x\neq 0$.

[7] With $k^2=\omega^2\mu\varepsilon\in\mathbf{R}$ and $k^2>(n\pi/a)^2$, the solution (8.45) is not in $\mathbf{L}^2(\Omega^\pm)$, where $\Omega^\pm=\{x,z: x\in[0,a],\ z\in(0,\pm\infty)\}$, but we can examine the limit of small material loss, $\sigma_{e,m}\to 0$.

The solution (8.45) represents traveling waves in the $z$-direction, with a standing-wave pattern in the $x$-direction. For simplicity assume the limitingly lossless case. Then, if $k^2 > (n\pi/a)^2$, the $n$th mode is purely propagating in the $\pm z$-direction with a phase factor $e^{\mp i\beta_n z}$, where

$$\beta_n(\omega) = \sqrt{k^2 - \left(\frac{n\pi}{a}\right)^2}.$$

If $k^2 < (n\pi/a)^2$, then the $n$th mode undergoes pure attenuation according to $e^{\mp\alpha_n z}$, where

$$\alpha_n(\omega) = \sqrt{\left(\frac{n\pi}{a}\right)^2 - k^2}.$$

In this range the mode is said to be *evanescent.* As frequency is lowered, the point at which the mode transitions from propagating to attenuating is called the cutoff frequency and is given by

$$\omega_c = \frac{1}{\sqrt{\mu\varepsilon}}\left(\frac{n\pi}{a}\right).$$

For a parallel-plate waveguide $\pm\omega_c$ represent branch points in the complex frequency plane [22], which may impact the transient analysis of the parallel-plate problem.

### Comparison of Eigenfunctions and Natural Modes

Comparing (8.39), (8.40), and (8.45), we see that the eigenfunctions can be interpreted as propagating modes ($k^2 > (n\pi/a)^2$) and can be cast in the form (8.45) by requiring that the improper eigenvalue $\lambda_{n,\nu}$ in (8.41) vanish, leading to

$$\nu = \pm\sqrt{k^2 - \left(\frac{n\pi}{a}\right)^2}.$$

This isolates one component of the continuum, associated with the propagation constant of the mode. Every point in the continuum $\nu$ can be so isolated by varying $k$. Note that the cutoff point $k^2 = (n\pi/a)^2$ corresponds to $\nu = 0$. As such, the propagating modes are eigenfunctions corresponding to null eigenvalues of the governing two-dimensional differential operator.

For evanescent modes ($k^2 < (n\pi/a)^2$), no point $\nu^2 \in [0,\infty)$ can result in the vanishing of the eigenvalue $\lambda_{n,\nu}$. As shown above, in this case the natural-mode fields have the dependence $e^{\mp\alpha_n z}$, with $\alpha_n(\omega) = \sqrt{(n\pi/a)^2 - k^2}$. Because this term is not bounded on $\mathbf{R}$, we do not regard evanescent modes as improper eigenfunctions of the governing operator, even though they satisfy (8.36) with $\lambda_{n,\nu} = 0$. Thus, we have the interesting result that a purely spectral theory would not implicate the evanescent modes, although in physical terms they have considerable importance and significance. The evanescent modes can, however, be obtained from the spectral eigenfunctions by analytic continuation into the complex $\nu$-plane.

### Green's Functions for the Parallel-Plate Structure

The Green's function equation for the parallel-plate structure is ($g(\rho,\rho') \equiv g(\rho,\rho',\lambda=0)$)

$$-\left(\frac{\partial^2}{\partial z^2}+\frac{\partial^2}{\partial x^2}+k^2\right)g(\rho,\rho')=\delta(x-x')\delta(z-z'), \tag{8.46}$$

and the associated one-dimensional Green's function equations are

$$\left(-\frac{d^2}{dz^2}-\lambda_z\right)g_z(z,z',\lambda_z)=\delta(z-z'), \tag{8.47}$$

$$\left(-\frac{d^2}{dx^2}-\lambda_x\right)g_x(x,x',\lambda_x)=\delta(x-x'). \tag{8.48}$$

For the singular (limit-point) Sturm–Liouville problem (8.47) we impose the condition

$$\lim_{z\to\pm\infty} g_z(z,z',\lambda_z)=0.$$

As considered in Example 5.11, the solution is (5.78),

$$g_z(z,z',\lambda_z)=\frac{1}{2i\sqrt{\lambda_z}}e^{-i\sqrt{\lambda_z}|z-z'|}. \tag{8.49}$$

For the regular Sturm–Liouville problem (8.48), which is the same as (5.19) considered previously, we impose the boundary conditions $g_x^H(0,x')=g_x^H(a,x')=0$ for the $H$-modes, and $\partial g_x^E(0,x')/\partial x=\partial g_x^E(a,x')/\partial x=0$ for the $E$-modes, leading to (see the derivation preceding (5.21) and also Section 5.1.3),

$$\begin{aligned} g_x^H(x,x',\lambda_x)&=\sum_{n=1}^{\infty}\frac{\sqrt{\frac{2}{a}}\sin\frac{n\pi}{a}x\,\sqrt{\frac{2}{a}}\sin\frac{n\pi}{a}x'}{\left(\frac{n\pi}{a}\right)^2-\lambda_x}\\ &=\frac{\sin\sqrt{\lambda_x}(a-x_>)\,\sin\sqrt{\lambda_x}x_<}{\sqrt{\lambda_x}\sin\sqrt{\lambda_x}a}, \end{aligned} \tag{8.50}$$

$$\begin{aligned} g_x^E(x,x',\lambda_x)&=\sum_{n=0}^{\infty}\frac{\sqrt{\frac{\varepsilon_n}{a}}\cos\frac{n\pi}{a}x\,\sqrt{\frac{\varepsilon_n}{a}}\cos\frac{n\pi}{a}x'}{\left(\frac{n\pi}{a}\right)^2-\lambda_x}\\ &=\frac{-\cos\sqrt{\lambda_x}(a-x_>)\,\cos\sqrt{\lambda_x}x_<}{\sqrt{\lambda_x}\sin\sqrt{\lambda_x}a} \end{aligned} \tag{8.51}$$

with $x_{<,>}$ defined on p. 301.

The two-dimensional Green's functions for this problem are, from (8.23) and using (8.49) with $\lambda_x+\lambda_z=k^2$,

$$g^{H,E}(\rho,\rho')=\sum_n u_{x,n}^{H,E}(x)\overline{u}_{x,n}^{H,E}(x')g_z(z,z',-\lambda_{x,n}+k^2)$$

$$g^H(\rho, \rho') = \sum_{n=1}^{\infty} \frac{1}{ia} \sin \frac{n\pi}{a} x \sin \frac{n\pi}{a} x' \cdot \frac{1}{\sqrt{k^2 - \left(\frac{n\pi}{a}\right)^2}} e^{-i\sqrt{k^2-\left(\frac{n\pi}{a}\right)^2}|z-z'|} \tag{8.52}$$

$$g^E(\rho, \rho') = \sum_{n=0}^{\infty} \frac{\varepsilon_n}{2ia} \cos \frac{n\pi}{a} x \cos \frac{n\pi}{a} x' \cdot \frac{1}{\sqrt{k^2 - \left(\frac{n\pi}{a}\right)^2}} e^{-i\sqrt{k^2-\left(\frac{n\pi}{a}\right)^2}|z-z'|}. \tag{8.53}$$

Comparing with (8.45) we see that this form of the Green's function is an expansion over the natural waveguide modes, rather than a pure eigenfunction expansion.

Because we require $g \to 0$ as $|z| \to \infty$, we must have (perhaps small) material loss via $k^2 = w^2\mu\varepsilon$. Therefore, the condition $\lim_{z\to\pm\infty} \Lambda(x, z) = 0$ in (8.35) is associated with nonvanishing material loss. In this case the operator $L$ in (8.35) is not self-adjoint, or even symmetric (consistent with the eigenvalue (8.41) being complex-valued). However, this consequence is due to the definition of the operator $L$, where we included $k^2$ as part of $L$ to be consistent with the general development for an inhomogeneous medium. If we had made the more natural definition (for the homogeneous medium case)

$$L \equiv -\left(\frac{\partial^2}{\partial z^2} + \frac{\partial^2}{\partial x^2}\right), \tag{8.54}$$

$$D_L \equiv \left\{\Lambda : \Lambda, L\Lambda \in H, B(\Lambda) = 0, \lim_{z\to\pm\infty} \Lambda(x, z) = 0\right\},$$

then we would have a self-adjoint operator[8] even though $k \in \mathbf{C}$. In this case the eigenfunctions are given by (8.39)–(8.40), with eigenvalues $\lambda_{n,\nu} = \nu^2 + \left(\frac{n\pi}{a}\right)^2$, forming an eigenbasis for $\mathbf{L}^2(\Omega)$. Thus, the natural formulation of the homogeneous-space problem is to consider (perhaps limitingly-small) material loss, and the self-adjoint operator (8.54). Because the homogeneous parallel-plate case with small dielectric loss leads to an inherently self-adjoint boundary value problem, the spectral expansions developed in this section are justified.

The above Green's function representation is in terms of natural modes guided in the $z$-direction, both propagating and evanescent, utilizing cross-sectional eigenfunctions. Another form of the Green's functions is (8.24),

[8] If one considers a perfectly lossless medium the condition $\lim_{z\to\pm\infty} \Lambda(x, z) = 0$ is not reasonable, and a radiation condition must be formulated. In this case, regardless of the formal definition of $L$, a nonself-adjoint problem results which is not defined within an $\mathbf{L}^2$ space.

obtained by first expanding over the (continuous) $z$-dependent spectrum, leading to

$$g^H(\rho,\rho') = \int_\nu u_z(z,\nu)\overline{u}_z(z',\nu) g_x^H(x,x',-\lambda_{z,\nu}(\nu)+k^2)\,d\nu \tag{8.55}$$

$$= \frac{1}{2\pi}\int_{-\infty}^{\infty} \frac{\sin\sqrt{\lambda_x}(a-x_>)\,\sin\sqrt{\lambda_x}x_<}{\sqrt{\lambda_x}\sin\sqrt{\lambda_x}a} e^{i\nu\,(z-z')}\,d\nu, \tag{8.56}$$

$$g^E(\rho,\rho') = \frac{-1}{2\pi}\int_{-\infty}^{\infty} \frac{\cos\sqrt{\lambda_x}(a-x_>)\,\cos\sqrt{\lambda_x}x_<}{\sqrt{\lambda_x}\sin\sqrt{\lambda_x}a} e^{i\nu\,(z-z')}\,d\nu \tag{8.57}$$

with $\sqrt{\lambda_x} = \sqrt{k^2-\lambda_z} = \sqrt{k^2-\nu^2}$. This form of the Green's function is an expansion over all possible plane waves traveling in the $z$-direction, modified by the vertical eigenfunctions. The equivalence of the Green's functions (8.52) and (8.56) can easily be checked via complex plane analysis, and similarly for the $E$-mode Green's functions. Both forms for $g^{H,E}$ are seen to be special cases of the general three-dimensional Green's function considered in Section 8.3.2.

### Completeness Relations and Associated Spectral Expansions

The completeness relations (8.25) are obtained by integrating $g_\alpha$ in the $\lambda_\alpha$-plane, as shown in Section 5.2, or directly from the discrete and continuous eigenfunctions, leading to

$$\delta^H(x-x') = \frac{2}{a}\sum_{n=1}^{\infty} \sin\frac{n\pi}{a}x\,\sin\frac{n\pi}{a}x', \tag{8.58}$$

$$\delta^E(x-x') = \sum_{n=0}^{\infty} \frac{\varepsilon_n}{a}\cos\frac{n\pi}{a}x\,\cos\frac{n\pi}{a}x', \tag{8.59}$$

$$\delta(z-z') = \frac{1}{2\pi}\int_{-\infty}^{\infty} e^{i\nu\,(z-z')}\,d\nu. \tag{8.60}$$

Forming the product $\delta(\rho-\rho') = \delta(x-x')\delta(z-z')$ results in

$$\delta^H(x-x')\delta(z-z') = \sum_{n=1}^{\infty}\int_{-\infty}^{\infty} \frac{1}{\pi a}\sin\frac{n\pi}{a}x\,\sin\frac{n\pi}{a}x'\,e^{i\nu(z-z')}\,d\nu,$$

$$\delta^E(x-x')\delta(z-z') = \sum_{n=0}^{\infty}\int_{-\infty}^{\infty} \frac{\varepsilon_n}{2\pi a}\cos\frac{n\pi}{a}x\,\cos\frac{n\pi}{a}x'\,e^{i\nu(z-z')}\,d\nu. \tag{8.61}$$

From (8.28) we can determine a modal expansion of an arbitrary function $f(x, z) \in \mathbf{L}^2(\Omega)$ in terms of the $H$-mode or $E$-mode eigenfunctions as

$$f(x, z) = \sum_{n=1}^{\infty} \int_{-\infty}^{\infty} a_{x,n}^{z}(\nu) u_{n,\nu}^{H}(x, z, \nu)\, d\nu, \tag{8.62}$$

$$f(x, z) = \sum_{n=1}^{\infty} \int_{-\infty}^{\infty} a_{x,n}^{z}(\nu)\, u_{n,\nu}^{E}(x, z, \nu)\, d\nu, \tag{8.63}$$

where $a_{x,n}^{z}(\nu) = \langle f, u_{n,\nu}^{H,E} \rangle$. It can be seen that (8.62) is nothing more than a Fourier sine series/Fourier transform, whereas (8.63) utilizes the Fourier cosine series.

## Field Due to a Source in a Parallel-Plate Waveguide

To obtain an expression for the field caused by a source in the parallel-plate waveguide geometry, i.e., a solution of (8.32), we start from the two-dimensional form[9] of (1.52),

$$\Lambda(\rho') = \lim_{\delta \to 0} \int_{\Omega - \Omega_\delta} g(\rho, \rho')\, f(\rho)\, d\Omega - \oint_{\Gamma} (g(\rho, \rho') \nabla \Lambda(\rho) - \Lambda(\rho) \nabla g(\rho, \rho')) \cdot d\mathbf{\Gamma},$$

where $\rho = \widehat{\mathbf{x}}x + \widehat{\mathbf{z}}z$. The cross-sectional line integral consists of a path along the plates at $x = 0, a$ and a path at $z = \pm\infty$. Since we assume $\lim_{z \to \pm\infty} \Lambda, g = 0$, the line integral term for those sections of the path vanishes. Along the plates at $x = 0, a$ we have $\Lambda(x_p, z) = g(x_p, z)|_{x_p = 0, a} = 0$ for the TE$^z$-modes, and $\frac{\partial \Lambda(x,z)}{\partial x} = \frac{\partial g(x,z)}{\partial x}\big|_{x = x_p = 0, a} = 0$ for the TM$^z$-modes, resulting in no contribution from those sections. Therefore, the line integral vanishes and, interchanging spatial coordinates and noting that $g(\rho, \rho') = g(\rho', \rho)$, we obtain the solution of (8.32) as

$$\Lambda(\rho) = \int_{\Omega} g(\rho, \rho')\, f(\rho')\, d\Omega', \tag{8.64}$$

in agreement with (8.12), where the limiting notation has been suppressed and $\Omega$ represents the area of the two-dimensional source.

As an example, if the source consists of a single component $J_{ey}(x, z)$, then only TE$^z$-modes are excited. In this case $\Lambda = E_y$ and $-f = i\omega\mu J_{ey}$, so that

$$E_y(x, z) = -i\omega\mu \int_{\Omega} g^H(\rho, \rho')\; J_{ey}(x', z')\; d\Omega', \tag{8.65}$$

where $g^H$ is either (8.52) or (8.56).

[9] Alternatively, one could have at the onset considered taking a spatial Fourier transform in the $y$-coordinate of the $y$-invariant equations (8.1), resulting in a term $\delta(y)$ being associated with each field quantity. If one takes this view, the three-dimensional result (1.52) can be used directly.

**Spectral Solution and Spectral Green's Function**

A two-dimensional spectral solution of (8.32),

$$-\left(\frac{\partial^2}{\partial z^2}+\frac{\partial^2}{\partial x^2}+k^2\right)\Lambda(x,z)=f, \tag{8.66}$$

and a corresponding Green's function can also be obtained. For example, let $u_{n,\nu}(x,z,\nu)$ represent either of the orthonormal improper eigenfunctions (8.39) or (8.40), where $\lambda_{n,\nu}$ is an improper eigenvalue

$$\lambda_{n,\nu}=\nu^2+\left(\frac{n\pi}{a}\right)^2-k^2.$$

Then, from (8.61) we have the expansion

$$\Lambda(x,z)=\sum_{n=1}^{\infty}\int_{-\infty}^{\infty}\langle\Lambda,u_{n,\nu}\rangle\,u_{n,\nu}(x,z,\nu)\,d\nu.$$

Following the usual procedure for eigenfunction expansions we obtain the solution of (8.32) in spectral form as a special case of (8.30),

$$\Lambda(x,z)=\sum_{n=1}^{\infty}\int_{-\infty}^{\infty}\frac{\langle f,u_{n,\nu}\rangle}{\nu^2+\left(\frac{n\pi}{a}\right)^2-k^2}u_{n,\nu}(x,z,\nu)\,d\nu, \tag{8.67}$$

(for $E$-modes the lower summation limit is $n=0$). For example, using $H$-modes we obtain the solution of (8.32) as

$$\Lambda(x,z)=\frac{1}{a\pi}\sum_{n=1}^{\infty}\int_{-\infty}^{\infty}\frac{\int_{-\infty}^{\infty}\int_0^a f(x,z)\sin\frac{n\pi}{a}x\;e^{-i\nu z}dx\,dz}{\nu^2+\left(\frac{n\pi}{a}\right)^2-k^2}\sin\frac{n\pi}{a}x\;e^{i\nu z}d\nu. \tag{8.68}$$

As a comparison with (8.65), if the source consists of a single component $J_{ey}(x,z)$, then $\Lambda=E_y$ and $-f=i\omega\mu J_{ey}$, so that

$$E_y(x,z)=\frac{-i\omega\mu}{a\pi}\sum_{n=1}^{\infty}\int_{-\infty}^{\infty}\frac{\int_{-\infty}^{\infty}\int_0^a J_{ey}(x,z)\sin\frac{n\pi}{a}x\;e^{-i\nu z}dx\,dz}{\nu^2+\left(\frac{n\pi}{a}\right)^2-k^2}\sin\frac{n\pi}{a}x\;e^{i\nu z}d\nu. \tag{8.69}$$

Complex plane analysis leads to the discrete form

$$E_y(x,z)=\frac{-\omega\mu}{a}\sum_{n=1}^{\infty}a_m^{\pm}\sin\frac{n\pi}{a}x\,\frac{e^{\mp i\sqrt{k^2-\left(\frac{n\pi}{a}\right)^2}\,z}}{\sqrt{k^2-\left(\frac{n\pi}{a}\right)^2}} \tag{8.70}$$

for $z\gtrless 0$, where $a_m^{\pm}=\int_{-\infty}^{\infty}\int_0^a J_{ey}(x,z)\sin(n\pi x/a)\;e^{\pm i\sqrt{k^2-(n\pi/a)^2}\,z}dx\,dz$.

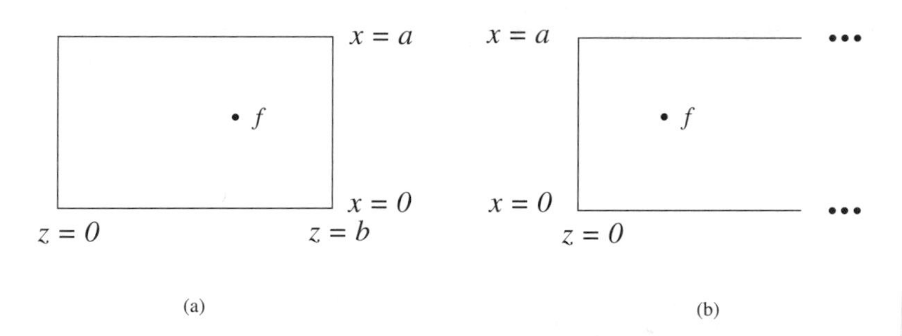


Figure 8.3: (a) Two-dimensional resonator structure, (b) semi-infinite waveguide structure.

Alternatively, we can consider the solution to be given by (8.64), with the Green's function $g = g^{H,E}$ given by

$$g(x,z,x',z') = \sum_{n=1}^{\infty} \int_{-\infty}^{\infty} \frac{\overline{u}_{n,\nu}(x',z',\nu)}{\nu^2 + \left(\frac{n\pi}{a}\right)^2 - k^2} u_{n,\nu}(x,z,\nu)\, d\nu \tag{8.71}$$

which is a special case of (8.31) and which can be shown to lead to (8.52) or (8.53) by complex plane analysis. Note, however, that while the spectral form is an aid in understanding the relationship between spectral operator properties and properties of solutions, the purely spectral form of the Green's function is not computationally convenient compared to the form obtained via partial eigenfunction expansion (which may, however, be obtained by complex plane analysis of the spectral form).

### 8.1.3 Two-Dimensional Resonator

For each of the structures depicted in Figure 8.3, the analysis is similar to that in the previous section. All of the $x$-dependent results remain the same, while the $z$-dependent Green's function (8.49) and completeness relation (8.60) differ due to the possible presence of a perfectly conducting wall(s).

Specifically, consider $H$-modes of the two-dimensional resonator depicted in Figure 8.3(a). With the boundary condition

$$g_z^H(0,z') = g_z^H(b,z') = 0,$$

(8.49) becomes (see Example 5.3)

$$\begin{aligned} g_z^H(z,z',\lambda_z) &= \sum_{n=1}^{\infty} \frac{\sqrt{\frac{2}{b}}\sin\frac{n\pi}{b}z\,\sqrt{\frac{2}{b}}\sin\frac{n\pi}{b}z'}{\left(\frac{n\pi}{b}\right)^2 - \lambda_z} \\ &= \frac{\sin\sqrt{\lambda_z}(b-z_>)\,\sin\sqrt{\lambda_z}z_<}{\sqrt{\lambda_z}\sin\sqrt{\lambda_z}b}. \end{aligned} \tag{8.72}$$

The two-dimensional Green's function satisfying (8.11) for this problem is

$$
\begin{aligned}
g^H(\rho,\rho') &= \sum_n u^H_{x,n}(x)\overline{u}^H_{x,n}(x')g^H_z(z,z',\lambda_{zn}) \\
&= \sum_{n=1}^{\infty} \frac{2}{a}\sin\frac{n\pi}{a}x\,\sin\frac{n\pi}{a}x'\frac{\sin\sqrt{\lambda_{z,n}}(b-z_>)\,\sin\sqrt{\lambda_{z,n}}z_<}{\sqrt{\lambda_{z,n}}\sin\sqrt{\lambda_{z,n}}b}
\end{aligned}
\tag{8.73}
$$

where $\sqrt{\lambda_{z,n}} = \sqrt{k^2-(n\pi/a)^2}$. Poles of the Green's function given by

$$\sin\sqrt{\lambda_{z,n}}b = 0$$

lead to the two-dimensional resonance frequencies of the cavity, resulting in

$$\lambda_{z,n} = \left(\frac{m\pi}{b}\right)^2 = k^2 - \left(\frac{n\pi}{a}\right)^2$$

or

$$\omega = \frac{1}{\sqrt{\mu\varepsilon}}\sqrt{\left(\frac{m\pi}{b}\right)^2 + \left(\frac{n\pi}{a}\right)^2}.$$

The delta-function in the $z$-coordinate is seen to be

$$\delta(z-z') = \frac{2}{b}\sum_{m=1}^{\infty}\sin\frac{m\pi}{b}z\sin\frac{m\pi}{b}z',$$

leading to the two-dimensional delta-function

$$\delta(x-x')\delta(z-z') = \frac{4}{ab}\sum_{n=1}^{\infty}\sum_{m=1}^{\infty}\sin\frac{n\pi}{a}x\sin\frac{n\pi}{a}x'\sin\frac{m\pi}{b}z\sin\frac{m\pi}{b}z'$$

and the associated spectral expansions.

The spectral solution of (8.32) for this structure is modified from (8.67) as

$$\Lambda(x,z) = \sum_{n=1}^{\infty}\sum_{m=1}^{\infty}\frac{\langle f, u^H_{n,m}\rangle}{\left(\frac{n\pi}{a}\right)^2+\left(\frac{m\pi}{b}\right)^2-k^2}u^H_{n,m}(x,z), \tag{8.74}$$

where

$$u^H_{n,m}(x,z) = \frac{2}{\sqrt{ab}}\sin\frac{n\pi}{a}x\sin\frac{m\pi}{b}z$$

are $H$-mode eigenfunctions of the operator $-\left(\frac{\partial^2}{\partial z^2}+\frac{\partial^2}{\partial x^2}+k^2\right)$, with eigenvalues identified as

$$\lambda_{n,m} = \left(\frac{n\pi}{a}\right)^2 + \left(\frac{m\pi}{b}\right)^2 - k^2.$$

The spectral form of the Green's function is obtained as

$$g^H(x,z,x',z') = \sum_{n=1}^{\infty}\sum_{m=1}^{\infty} \frac{\overline{u}_{n,m}^H(x',z')}{\left(\frac{n\pi}{a}\right)^2 + \left(\frac{m\pi}{b}\right)^2 - k^2} u_{n,m}^H(x,z). \tag{8.75}$$

As discussed previously, frequencies $\omega$ such that $\lambda(\omega) = 0$ are the natural cavity resonance frequencies. Similar results are obtained for $E$-modes (see Example 5.5).

### 8.1.4 Semi-Open Structure

For the situation depicted in Figure 8.3(b), with $g_z^H(0,z') = 0$ and a limit point at $z = \infty$, (8.49) becomes (see Example 5.9)

$$g_z^H(z,z',\lambda_z) = \frac{2}{\pi}\int_0^{\infty} \frac{\sin \upsilon z \sin \upsilon z'}{\upsilon^2 - \lambda_z} d\upsilon = \frac{1}{\sqrt{\lambda_z}} e^{-i\sqrt{\lambda_z}z_>} \sin\sqrt{\lambda_z} z_<$$

and

$$\delta(z-z') = \frac{2}{\pi}\int_0^{\infty} \sin \upsilon z' \sin \upsilon z \, d\upsilon.$$

With the above, modifications to (8.61)–(8.57) are straightforward, and similar results follow for the $E$-modes (see Example 5.7).

The walls could also be impedance planes. If both walls in an opposing pair of walls are impedance planes, then the Green's function in that coordinate becomes (7.11). Other combinations of impedance and perfectly conducting walls, and with certain walls removed to infinity, can be analyzed in a straightforward manner.

### 8.1.5 Free Space

We may also consider free space as a modification to the parallel-plate geometry. If both plates recede to infinity, we have

$$\begin{aligned} g_\alpha(\alpha,\alpha',\lambda_\alpha) &= \frac{1}{2i\sqrt{\lambda_\alpha}} e^{-i\sqrt{\lambda_\alpha}\left|\alpha-\alpha'\right|}, \\ \delta(\alpha-\alpha') &= \frac{1}{2\pi}\int_{-\infty}^{\infty} e^{i\upsilon_\alpha(\alpha-\alpha')} d\upsilon_\alpha \end{aligned} \tag{8.76}$$

for $\alpha = x, z$ and $\lambda_\alpha = \upsilon_\alpha^2$, where $\upsilon_\alpha \in (-\infty,\infty)$ with $\upsilon_x = \upsilon$ and $\upsilon_z = \nu$.

The two-dimensional delta-function is

$$\begin{aligned} \delta(x-x')\delta(z-z') &= \int_{-\infty}^{\infty}\int_{-\infty}^{\infty} u_{x,\upsilon}(x,\upsilon)\overline{u}_{x,\upsilon}(x',\upsilon)u_{z,\nu}(z,\nu)\overline{u}_{z,\nu}(z',\nu)\, d\upsilon\, d\nu \\ &= \frac{1}{(2\pi)^2}\int_{-\infty}^{\infty}\int_{-\infty}^{\infty} e^{i\upsilon(x-x')} e^{i\nu(z-z')} d\upsilon\, d\nu. \end{aligned} \tag{8.77}$$

To determine a modal expansion of an arbitrary function $f(x,z) \in \mathbf{L}^2(\Omega)$, where $\Omega = \{(x,z) : x \in (-\infty,\infty),\ z \in (-\infty,\infty)\}$, we take the inner product of the above expression with $f(x,z)$, where we define the two-dimensional inner product by

$$\langle f, g \rangle = \int_{\mathbf{R}^2} f(x,z)\overline{g}(x,x)\, d\Omega.$$

This leads to the expansion

$$f(x,z) = \frac{1}{2\pi}\int_{-\infty}^{\infty}\int_{-\infty}^{\infty} a(\upsilon,\nu)e^{i\upsilon\, x}e^{i\nu z}d\upsilon\, d\nu,$$

where

$$a(\upsilon,\nu) = \frac{1}{2\pi}\int_{-\infty}^{\infty}\int_{-\infty}^{\infty} f(x,z)e^{-i\upsilon\, x}e^{-i\nu z}dx\, dz = \left\langle f, u_{x,\upsilon}(\upsilon)u_{z,\nu}(\nu)\right\rangle,$$

which is easily recognized as simply a two-dimensional Fourier transform. The associated Green's function is, from (8.23) and (8.24),

$$\begin{aligned} g(x,z,x',z') &= \frac{1}{2\pi}\int_{-\infty}^{\infty} e^{i\upsilon\,(x-x')}\frac{1}{2i\sqrt{\lambda_z}}e^{-i\sqrt{\lambda_z}|z-z'|}d\upsilon \\ &= \frac{1}{2\pi}\int_{-\infty}^{\infty} e^{i\nu\,(z-z')}\frac{1}{2i\sqrt{\lambda_x}}e^{-i\sqrt{\lambda_x}|x-x'|}d\nu \\ &= \frac{1}{4i}H_0^{(2)}(k\,|\rho-\rho'|), \end{aligned} \tag{8.78}$$

with $\upsilon^2+\lambda_z = k^2$ in the first expression and $\lambda_x+\nu^2 = k^2$ in the second form, both leading to the Hankel function with $\rho-\rho' = \widehat{\mathbf{x}}\,(x-x') + \widehat{\mathbf{z}}\,(z-z')$, in agreement with (1.43) for two-dimensional free space.

### 8.1.6 Impedance Plane Structure

As an example of a vertically open structure, consider an infinite impedance plane bounding a semi-infinite half-space with material parameters $\varepsilon_1$ and $\mu_1$, as shown in Figure 8.4. This structure is useful as a canonical problem, exhibiting the intricate spectral characteristics of more complicated vertically unbounded planar structures in as simple a manner as possible. It is also of practical interest, representing, for instance, the surface of an imperfectly conducting sheet, and as a simple model for a surface-wave antenna. This structure is extensively studied in [23, pp. 22–92].

An impedance plane is most generally characterized by a dyadic surface impedance $\underline{\mathbf{Z}}_s$, each component of which has units of ohms. On the impedance surface we must have

$$\mathbf{E}|_{x=0} = \underline{\mathbf{Z}}_s \cdot \mathbf{H}|_{x=0}\,.$$

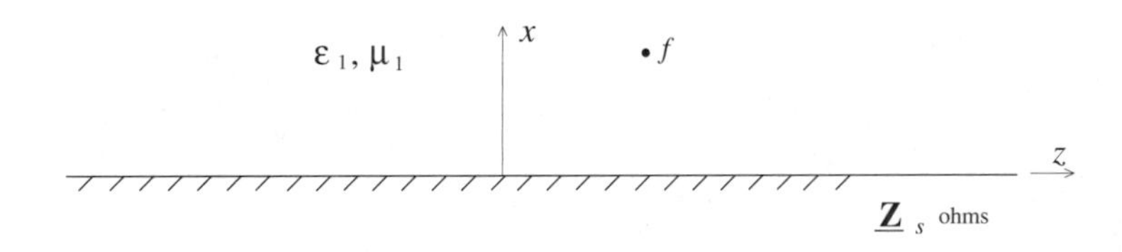


Figure 8.4: Source $f$ over an infinite plane characterized by surface impedance $\underline{\mathbf{Z}}_s$.

For the analysis of TE$^z$- and TM$^z$-fields considered here, each of which has a single tangential electric- and magnetic field component, we simply use a scalar surface impedance

$$Z_s = R_s + iX_s$$

that must equal the ratio of the tangential electric field to the tangential magnetic field.

If we assume the surface is passive, then we have $R_s \geq 0$, where, in addition, it is noted that the surface impedance is generally frequency-dependent. If $Z_s = 0$, then the plane implements a perfectly conducting electric ground plane, whereas if $1/Z_s = 0$, the plane implements a perfectly conducting magnetic ground plane.

The appropriate Helmholtz equation from (8.8) and (8.9) is

$$-\left(\frac{\partial^2}{\partial z^2} + \frac{\partial^2}{\partial x^2} + k^2\right) \Lambda(x, z) = f(x, z), \tag{8.79}$$

with $\Lambda = E_y$ for TE$^z$-modes and $\Lambda = H_y$ for TM$^z$-modes, where $f$ represents the source density. For both types of modes we will enforce the fitness conditions

$$\lim_{z \to \pm\infty} \Lambda(x, z) = 0,$$
$$\lim_{x \to \infty} \Lambda(x, z) = 0$$

or that the field component belongs to the appropriate $\mathbf{L}^2$ space. The boundary condition for TM$^z$-modes is $\left(\frac{E_z}{H_y}\right)\Big|_{x=0} = Z_s$, leading to

$$\frac{\partial \Lambda(0, z)}{\partial x} = i\omega\varepsilon Z_s \, \Lambda(0, z),$$

whereas the boundary condition for TE$^z$-modes is $\left(-\frac{E_y}{H_z}\right)\Big|_{x=0} = Z_s$, leading to

$$\frac{\partial \Lambda(0, z)}{\partial x} = \frac{i\omega\mu}{Z_s} \Lambda(0, z).$$

Both cases can be subsumed under the condition

$$B(\Lambda) = \frac{\partial \Lambda(0,z)}{\partial x} - \eta\, \Lambda(0,z) = 0, \tag{8.80}$$

where $\eta \in \mathbf{C}$.

For this problem we define $\Omega \equiv \{(x,z) : x \in [0,\infty), z \in (-\infty,\infty)\}$ and $H \equiv \mathbf{L}^2(\Omega)$, and consider in the following the operator $L : \mathbf{L}^2(\Omega) \to \mathbf{L}^2(\Omega)$ defined by

$$L \equiv -\left(\frac{\partial^2}{\partial z^2} + \frac{\partial^2}{\partial x^2} + k^2\right),$$

$$D_L \equiv \left\{\Lambda : \Lambda, L\Lambda \in H, B(\Lambda) = 0, \lim_{x,|z|\to\infty} \Lambda(x,z) = 0\right\},$$

with the inner product

$$\langle f, g \rangle = \int_{-\infty}^{\infty}\int_{0}^{\infty} f(x,z)\,\overline{g(x,z)}\,dx\,dz.$$

Subsequent to performing separation of variables, we will see that the $z$-dependent equations form a self-adjoint problem, but those in the $x$-direction do not since $\eta \in \mathbf{C}$ (see Chapter 7 and Section 5.3), from which it is clear that the adjoint boundary condition is

$$\frac{\partial \Lambda^*(0,z)}{\partial x} = \overline{\eta}\Lambda^*(0,z).$$

The special case $\eta \in \mathbf{R}$ leads to a self-adjoint problem.

This problem is considered in [25, pp. 230–233] and [8, pp. 128–132], and we utilize the described approach in the following analysis. In the vertical direction this structure is analogous to a transmission line terminated with an impedance $Z_s$ at one end and match terminated at the other end. The resulting Green's function $g_x$ is the same as (7.11) with $a = 0$, $Z_a = Z_s$, and $Z_b = Z_0$.

### Spectral Properties and Eigenfunctions of the Impedance Plane

We first consider the eigenfunctions of the structure, governed by

$$-\left(\frac{\partial^2}{\partial z^2} + \frac{\partial^2}{\partial x^2} + k^2\right) u_{\alpha,\beta}(x,z) = \lambda_{\alpha,\beta} u_{\alpha,\beta}(x,z) \tag{8.81}$$

subject to the appropriate boundary conditions. We seek a solution as $u_{\alpha,\beta}(x,z) = u_x(x)u_z(z)$, leading to

$$\left(-\frac{d^2}{dz^2} - \lambda_z\right) u_z(z) = 0, \tag{8.82}$$

$$\left(-\frac{d^2}{dx^2} - \lambda_x\right) u_x(x) = 0, \tag{8.83}$$

with the separation equation $\lambda_z + \lambda_x - k^2 = \lambda_{\alpha,\beta}$. Separated boundary conditions are

$$\lim_{z\to\pm\infty} u_z(z) = 0, \qquad \lim_{x\to\infty} u_x(x) = 0,$$
$$\frac{\partial u_x(0)}{\partial x} = \eta\, u_x(0).$$

The differential operators in both of these equations are seen to be of the Sturm–Liouville form (5.4).

In the vertical direction the eigenvalue problem (8.83), taken as a spectral problem with $\lambda_x$ unspecified, has both discrete (proper) and continuous (improper) eigenfunctions.

**Proper Eigenfunctions in the $x$-Coordinate**

The discrete eigenfunctions satisfy

$$\frac{\partial u_x(0)}{\partial x} = \eta\, u_x(0), \qquad u_x \in \mathbf{L}^2(0,\infty).$$

Taking the solution of (8.83) as

$$u_x(x) = Ae^{-i\sqrt{\lambda_x}x} + Be^{i\sqrt{\lambda_x}x},$$

with $\operatorname{Im}\sqrt{\lambda_x} < 0$, and applying the aforementioned two conditions lead to $B = 0$ and $-i\sqrt{\lambda_x} = \eta$ (which can also be obtained from (7.18) under $a = 0$ and $b \to \infty$, leading to $\sqrt{\lambda_n} = i\eta_a$), so that the lone discrete eigenfunction and eigenvalue are

$$u_{x,1}(x) = Ae^{\eta x},$$
$$\lambda_x = -\eta^2,$$

where $\operatorname{Re}\eta < 0$, as discussed later. Consideration of the adjoint problem

$$\left(-\frac{d^2}{dx^2} - \overline{\lambda}_x\right) u_x^*(x) = 0$$

leads to

$$u_x^*(x) = A^* e^{-i\sqrt{\overline{\lambda}_x}x} + B^* e^{i\sqrt{\overline{\lambda}_x}x}$$

with $\operatorname{Im}\sqrt{\overline{\lambda}_x} < 0$. Applying the aforementioned two conditions leads to $B^* = 0$ and $-i\sqrt{\overline{\lambda}_x} = \overline{\eta}$, so that the lone discrete adjoint eigenfunction and eigenvalue are

$$u_{x,1}^*(x) = A^* e^{\overline{\eta}x},$$
$$\lambda_x^* = \overline{\lambda}_x = \overline{\eta},$$

consistent with Theorem 4.7. Eigenfunctions are normalized so that

$$\langle u_{x,1}, u_{x,1}^* \rangle = \int_0^\infty u_{x,1}(x)\overline{u_{x,1}^*}(x)\, dx = 1,$$

leading to

$$\begin{aligned} u_{x,1}(x) &= i\sqrt{2\eta}\, e^{\eta x}, \\ u_{x,1}^*(x) &= -i\sqrt{2\overline{\eta}}\, e^{\overline{\eta} x}, \end{aligned} \tag{8.84}$$

where it is noted that $u_{x,1}^* = \overline{u}_{x,1}$.

### Improper Eigenfunctions in the $x$-Coordinate

The improper eigenfunctions are determined as

$$u_x(x) = A \sin \sqrt{\lambda_x} x + B \cos \sqrt{\lambda_x} x.$$

Application of the boundary condition $\partial u_x(0)/\partial x = \eta\, u_x(0)$ yields

$$u_x(x, \lambda_x) = B \left[ \cos \sqrt{\lambda_x} x + \frac{\eta}{\sqrt{\lambda_x}} \sin \sqrt{\lambda_x} x \right], \tag{8.85}$$

where the adjoint improper eigenfunctions are easily seen to be

$$u_x^*(x, \lambda_x) = B^* \left( \cos \sqrt{\overline{\lambda}_x} x + \frac{\overline{\eta}}{\sqrt{\overline{\lambda}_x}} \sin \sqrt{\overline{\lambda}_x} x \right).$$

It will be shown that these improper eigenfunctions, which represent the continuous spectrum, are taken over the continuous variable $\lambda_x \in [0, \infty)$, in which case $\overline{\lambda}_x = \lambda_x$. Recall that the improper eigenfunctions are not in $\mathbf{L}^2(0, \infty)$ in the $x$-coordinate yet are bounded at $x = \infty$.

The improper eigenfunctions can be normalized as

$$\langle u_x, u_x^* \rangle = \int_0^\infty u_x(x, \upsilon)\overline{u_x^*}(x, p)\, dx = \delta(\upsilon - p),$$

where $\upsilon^2 = \lambda_x$, leading to

$$B\overline{B^*} \int_0^\infty \left( \cos \upsilon x + \frac{\eta}{\upsilon} \sin \upsilon x \right) \left( \cos px + \frac{\eta}{p} \sin px \right) dx = \delta(\upsilon - p).$$

Using (5.60) and (5.72),

$$\frac{2}{\pi} \int_0^\infty \sin \upsilon x \sin px\, dx = \delta(\upsilon - p), \tag{8.86}$$

$$\frac{2}{\pi} \int_0^\infty \cos \upsilon x \cos px\, dx = \delta(\upsilon - p), \tag{8.87}$$

and[10]

$$\frac{2}{\pi}\int_0^\infty \sin \upsilon x \cos px \, dx = \frac{2}{\pi}\frac{\upsilon}{\upsilon^2 - p^2} \tag{8.88}$$

leads to $B\overline{B^*} = \frac{2}{\pi}\frac{\upsilon^2}{\eta^2+\upsilon^2}$ for $\upsilon = p$, so that

$$u_{x,\upsilon}(x,\upsilon) = \sqrt{\frac{2}{\pi}}\frac{\upsilon}{\sqrt{\eta^2+\upsilon^2}}\left(\cos \upsilon x + \frac{\eta}{\upsilon}\sin \upsilon x\right),$$

$$u^*_{x,\upsilon}(x,\upsilon) = \overline{u}_{x,\upsilon}(x,\upsilon) = \sqrt{\frac{2}{\pi}}\frac{\upsilon}{\sqrt{\overline{\eta}^2+\upsilon^2}}\left(\cos \upsilon x + \frac{\overline{\eta}}{\upsilon}\sin \upsilon x\right)$$

represent the properly normalized improper eigenfunctions for the original and adjoint problems, $\upsilon \in [0, \infty)$ (see also Example 5.13, which concerns the case of two identical impedance plates separated by unit distance).

In the $z$-coordinate, from Example 5.11 we obtain the improper eigenfunctions and eigenvalues

$$u_{z,\nu}(z,\nu) = \frac{1}{\sqrt{2\pi}}e^{i\nu z},$$
$$\lambda_z(\nu) = \nu^2, \qquad \nu \in (-\infty, \infty).$$

The resulting two-dimensional eigenfunctions and eigenvalues are then

$$\begin{aligned} u_{1,\nu}(x,z,\nu) &= u_{x,1}(x)u_{z,\nu}(z,\nu) \\ &= i\sqrt{\frac{\eta}{\pi}}e^{\eta x}e^{i\nu z}, \\ \lambda_{1,\nu} &= \lambda_x + \lambda_z - k^2 = -\eta^2 + \nu^2 - k^2, \end{aligned}$$

and

$$\begin{aligned} u_{\upsilon,\nu}(x,z,\upsilon,\nu) &= u_{x,\upsilon}(x,\upsilon)u_{z,\nu}(z,\nu) \\ &= \frac{1}{\pi}\frac{\upsilon}{\sqrt{\eta^2+\upsilon^2}}\left(\cos \upsilon x + \frac{\eta}{\upsilon}\sin \upsilon x\right)e^{i\nu z}, \\ \lambda_{\upsilon,\nu} &= \lambda_x + \lambda_z - k^2 = \upsilon^2 + \nu^2 - k^2. \end{aligned}$$

Adjoint eigenfunctions are found from $u^*_{1,\nu} = \overline{u}_{1,\nu}$ and $u^*_{\upsilon,\nu} = \overline{u}_{\upsilon,\nu}$. Bi-orthonormality is expressed as

$$\langle u_{1,\nu}, u^*_{1,p}\rangle = \delta(\nu - p),$$
$$\langle u_{\upsilon,\nu}, u^*_{p,q}\rangle = \delta(\upsilon - p)\,\delta(\nu - q).$$

---

[10] This result follows simply from noting that

$$\int_0^\infty \sin \upsilon x \cos px \, dx = \frac{1}{2}\int_0^\infty [\sin(\upsilon + p)x + \sin(\upsilon - p)x]\, dx$$

and that $\lim_{L\to\infty}\int_0^L \sin \upsilon x \, dx = 1/\upsilon$.

### Natural Waveguide Modes of the Impedance Plane

The natural waveguide modes of the impedance plane structure satisfy

$$-\left(\frac{\partial^2}{\partial z^2}+\frac{\partial^2}{\partial x^2}+k^2\right)\psi(x,z)=0 \tag{8.89}$$

subject to the appropriate boundary conditions. We seek a solution as $\psi(x,z)=\psi_x(x)\psi_z(z)$, leading to

$$\left(-\frac{d^2}{dz^2}-\lambda_z\right)\psi_z(z)=0, \tag{8.90}$$

$$\left(-\frac{d^2}{dx^2}-\lambda_x\right)\psi_x(x)=0, \tag{8.91}$$

with the separation equation $\lambda_z+\lambda_x-k^2=0$, where for the $z$-dependent problem we enforce an outgoing-wave condition. The solution of (8.91) follows from that of (8.83). For (8.90), as with the parallel-plate waveguide, we take the solution as

$$\psi_z(z)=A\,e^{\mp i\sqrt{\lambda_z}z},$$

where $\operatorname{Im}\sqrt{\lambda_z}<0$, with the appropriate sign chosen according to the sign of $z$, and where $A$ is an arbitrary constant. Because $\lambda_z+\lambda_x=k^2$, then the discrete propagation constant is determined as

$$\sqrt{\lambda_z}=\sqrt{k^2+\eta^2}.$$

We therefore have for the lone ($n=1$) discrete natural mode

$$\psi_1(x,z)=Ai\sqrt{2\eta}e^{\eta x}e^{\mp i\sqrt{k^2+\eta^2}z}, \tag{8.92}$$

and for the continuous (improper) natural mode

$$\psi(x,z,v)=A\sqrt{\frac{2}{\pi}}\frac{v}{\sqrt{\eta^2+v^2}}\left[\cos vx+\frac{\eta}{v}\sin vx\right]e^{\mp i\sqrt{k^2-v^2}z}$$

for $v^2=\lambda_x\in[0,\infty)$.

The discrete mode (8.92) represents a wave bound to the surface by the exponential decay factor $e^{\eta x}$, where $\operatorname{Re}(\eta)<0$, and propagating in the $z$-direction with the dispersion equation given as

$$\beta(\omega)=\sqrt{\omega^2\mu\varepsilon+\eta^2}.$$

If $\operatorname{Im}\eta=0$ ($R_s=0$), the wave propagates without attenuation at all frequencies. Similarly to the parallel-plate waveguide, for $\eta\neq 0$ the dispersion relation indicates branch points in the complex frequency plane. For the case of $\eta$, $\mu$, and $\varepsilon$ independent of frequency, these branch points occur at $\omega_b=\pm i\eta/\sqrt{\mu\varepsilon}$.

The improper modes have a physical meaning quite different from that of the discrete mode, with the term $e^{\mp i\sqrt{k^2-\upsilon^2}z}$ representing two distinct types of behavior. For $\lambda_x = \upsilon^2 \in (0, k^2)$, the term inside the radical is positive, such that we have oscillatory behavior (propagation) along the $z$-axis, again only purely so in the lossless limit. For $\upsilon^2 \in (k^2, \infty)$ we have $e^{\mp i\sqrt{-(\upsilon^2-k^2)}z} = e^{\mp\sqrt{(\upsilon^2-k^2)}z}$, exhibiting pure exponential decay along the $z$-axis. It can be considered that for $\upsilon^2 \in (0, k^2)$ the improper eigenfunctions represent *propagating radiation modes* (above cutoff) of the structure, whereas for $\upsilon^2 \in (k^2, \infty)$ the improper eigenfunctions represent *evanescent radiation modes* (below cutoff). As was the case for the parallel-plate waveguide, only the propagating modes are eigenfunctions (in the null space) of the governing two-dimensional differential operator.

Neither the propagating nor the evanescent radiation modes decay as $x \to \infty$, yet both remain bounded at infinity. This would seem to be at odds with our solution being in $\mathbf{L}^2(0, \infty)$ in the vertical direction. The resolution of this difficulty is the observation that a single radiation mode cannot be physically excited by an electromagnetic source; rather a real source can only excite a continuum of such modes (for instance, see the form of the Green's function (8.98)). The continuous summation (integration) over this continuum must vanish as $x \to \infty$, which indeed occurs as discussed later. This observation was first made in [23, p. 14] for the physical problem of open-boundary waveguides. This fact is also evident from the theory of Fourier transforms, where the exponential kernel $e^{i\upsilon x}$ of the transform represents an improper eigenfunction of the operator $-d^2/dx^2$ with improper eigenvalue $\lambda = \upsilon^2$.

### Green's Functions for the Impedance Plane Structure

Green's functions for the impedance plane structure satisfy

$$-\left(\frac{\partial^2}{\partial z^2} + \frac{\partial^2}{\partial x^2} + k^2\right) g(\rho, \rho') = \delta(x - x')\delta(z - z'). \tag{8.93}$$

The associated one-dimensional Green's function problems are

$$\left(-\frac{d^2}{dz^2} - \lambda_z\right) g_z(z, z', \lambda_z) = \delta(z - z'), \tag{8.94}$$

$$\left(-\frac{d^2}{dx^2} - \lambda_x\right) g_x(x, x', \lambda_x) = \delta(x - x'), \tag{8.95}$$

where $g_{x,z}$ are subject to the same conditions as $u_{x,z}$. To obtain the Green's functions for this structure, we first consider the singular (limit-point) Sturm–Liouville problem (8.94), where we impose the condition $\lim_{z\to\pm\infty} g_z(z, z', \lambda_z) = 0$. As considered in Example 5.11, the solution is (5.78),

$$g_z(z, z', \lambda_z) = \frac{1}{2i\sqrt{\lambda_z}} e^{-i\sqrt{\lambda_z}|z-z'|}. \tag{8.96}$$

For the singular (limit-point) Sturm–Liouville problem (8.95) we have

$$g_x(x,x',\lambda_x) = \begin{cases} A\cos\sqrt{\lambda_x}x + B\sin\sqrt{\lambda_x}x, & x < x', \\ Ce^{-i\sqrt{\lambda_x}x} + De^{i\sqrt{\lambda_x}x}, & x > x', \end{cases}$$

where we take $\operatorname{Im}\sqrt{\lambda_x} < 0$. Applying the condition that $g_x(x,x',\lambda_x) \in \mathbf{L}^2(0,\infty)$ leads to $D = 0$, and the boundary condition $\partial g_x(0,x,\lambda_x)/\partial x = \eta g_x(0,x',\lambda_x)$, yields $B = (\eta/\sqrt{\lambda_x})A$, resulting in

$$g_x(x,x',\lambda_x) = \begin{cases} A\left(\cos\sqrt{\lambda_x}x + \frac{\eta}{\sqrt{\lambda_x}}\sin\sqrt{\lambda_x}x\right), & x < x', \\ Ce^{-i\sqrt{\lambda_x}x}, & x > x'. \end{cases}$$

Applying the continuity and jump conditions (5.17),

$$g_x(x,x',\lambda_x)\Big|_{x=x'-\varepsilon}^{x=x'+\varepsilon} = 0,$$

$$\left.\frac{dg_x(x,x',\lambda_x)}{dx}\right|_{x=x'-\varepsilon}^{x=x'+\varepsilon} + 1 = 0,$$

yields

$$A = \frac{e^{-i\sqrt{\lambda_x}x'}}{i\sqrt{\lambda_x}+\eta},$$

$$C = \frac{\cos\sqrt{\lambda_x}x' + \frac{\eta}{\sqrt{\lambda_x}}\sin\sqrt{\lambda_x}x'}{i\sqrt{\lambda_x}+\eta},$$

such that the Green's function is

$$g_x(x,x',\lambda_x) = \frac{e^{-i\sqrt{\lambda_x}x_>}}{i\sqrt{\lambda_x}+\eta}\left(\cos\sqrt{\lambda_x}x_< + \frac{\eta}{\sqrt{\lambda_x}}\sin\sqrt{\lambda_x}x_<\right), \tag{8.97}$$

which is a special case of (7.11). The adjoint Green's function is obtained as $g_x^*(x,x',\lambda_x) = \overline{g}_x(x',x,\lambda_x)$ from (5.100), although it is not needed here.

Note that (8.97) has a branch point in the $\lambda_x$-plane at $\lambda_x = 0$, resulting in a branch cut along the positive real axis, consistent with $\operatorname{Im}\sqrt{\lambda_x} < 0$. The Green's function also has a simple pole singularity at $\lambda_{xp} = -\eta^2$. Because a first-order branch point exists, the $\lambda_x$-plane is two-sheeted. As such, it remains to check that the pole is on the Riemann sheet of interest. It is easy to see that $\operatorname{Im}\sqrt{\lambda_{xp}} = \operatorname{Re}\eta < 0$, so that $\operatorname{Re}\eta < 0$ *must* occur for the pole to reside on the sheet of interest, upon which the completeness relation integration (discussed later) is performed. For TM$^z$-modes,

$$\eta = i\omega\varepsilon Z_s = i\omega\varepsilon\left(R_s + iX_s\right),$$

implying that the surface impedance must have $X_s > 0$ (i.e., the surface must be inductive) for $E$-modes to propagate along and bound to the surface, with a spatial dependence $e^{i\omega\varepsilon R_s}e^{-\omega\varepsilon X_s x}$. For TE$^z$-modes,

$$\eta = \frac{i\omega\mu}{Z_s} = \frac{i\omega\mu}{R_s^2 + X_s^2}\left(R_s - iX_s\right),$$

so that $X_s < 0$ must occur (i.e., the surface must be capacitive) for $H$-modes to propagate, where the spatial dependence is $e^{\frac{i\omega\mu}{R_s^2+X_s^2} R_s\, x}\, e^{\frac{\omega\mu}{R_s^2+X_s^2} X_s\, x}$. In either case the wave decays exponentially away from the surface and for a constant frequency is more tightly bound to the surface as $X_s$ increases in magnitude.

The two-dimensional Green's function satisfying (8.11) for this problem is, from (8.23) and using (8.96),

$$\begin{aligned} g(x,z,x',z') &= \sum_n u_{x,n}(x)\overline{u^*_{x,n}}(x')g_z(z,z',-\lambda_{x,n}+k^2) \\ &\quad + \int_{\upsilon} u_{x,\upsilon}(x,\upsilon)\overline{u^*_{x,\upsilon}}(x',\upsilon)g_z(z,z',-\lambda_{x,\upsilon}(\upsilon)+k^2)\,d\upsilon \\ &= i\frac{\eta}{\sqrt{k^2+\eta^2}}e^{\eta(x+x')}e^{-i\sqrt{k^2+\eta^2}|z-z'|} - \frac{i}{\pi}\int_{\upsilon=0}^{\infty} d\upsilon \frac{\upsilon^2}{\eta^2+\upsilon^2} \\ &\quad \cdot \left(\cos\upsilon x + \frac{\eta}{\upsilon}\sin\upsilon x\right)\left(\cos\upsilon x' + \frac{\eta}{\upsilon}\sin\upsilon x'\right)\frac{1}{\sqrt{\lambda_z}}e^{-i\sqrt{\lambda_z}|z-z'|}, \end{aligned} \tag{8.98}$$

where $\sqrt{\lambda_z} = \sqrt{k^2-\lambda_x^2} = \sqrt{k^2-\upsilon^2}$ for the continuous spectrum term. We see that this form of the Green's function involves the discrete mode as well as the propagating and evanescent radiation modes.

If we first expand over the (continuous) $z$-dependent spectrum we have, from (8.24), (8.50), and (8.51),

$$\begin{aligned} &g(x,z,x',z') \\ &\quad = \int_{\nu} u_{z,\nu}(z,\nu)\overline{u}_{z,\nu}(z',\nu)g_x(x,x',-\lambda_{z,\nu}(\nu)+k^2)\,d\nu \\ &\quad = \frac{1}{2\pi}\int_{-\infty}^{\infty}\frac{e^{-i\sqrt{\lambda_x}x_>}}{i\sqrt{\lambda_x}+\eta}\left(\cos\sqrt{\lambda_x}x_< + \frac{\eta}{\sqrt{\lambda_x}}\sin\sqrt{\lambda_x}x_<\right)e^{i\nu(z-z')}d\nu \end{aligned} \tag{8.99}$$

with $\sqrt{\lambda_x} = \sqrt{k^2-\nu^2}$. Note that we have $g(\rho,\rho') = g(\rho',\rho)$. A purely spectral form can also be obtained from (8.30).

It is interesting to compare the two equivalent forms for the Green's function. The form (8.98) explicitly has the surface-wave term shown, which provides the dominant contribution to the Green's function for $\mathrm{Re}\,(\eta(x+x'))$ sufficiently small. This term is in the form of a plane wave traveling along the $z$-axis, exponentially damped in the vertical direction. As $\mathrm{Re}\,(\eta(x+x'))$ becomes large, the continuous spectral components, representing a radiation field, dominate the response. The spectral integral can be evaluated asymptotically for $(x,z)\gg(x',z')$ to obtain a cylindrical wave form for the radiation (far) field [25, pp. 286–290]. Furthermore, it is straightforward to obtain the surface-wave term from (8.99) via complex plane analysis, and the remaining integral term will be equivalent to the integral term in (8.98). This second form is also convenient for asymptotic analysis for large $(z - z')$.

The earlier statement that the continuum of improper eigenfunctions excited by a source must vanish at $x \to \infty$ is readily apparent from (8.98). Indeed, as $x \to \infty$ the integral term has the form

$$\begin{aligned}
&\lim_{x\to\infty} \int_\alpha^\beta u_x(x,v)\overline{u}_x(x',v) f(v)\, dv \\
&\quad = \lim_{x\to\infty} \int_\alpha^\beta [\cos vx + g_1(v)\sin vx]\,[\cos vx' + g_1(v)\sin vx']\, g_2(v)\, dv \\
&\quad = \lim_{x\to\infty} \int_\alpha^\beta (h_1(v)\cos vx + h_2(v)\sin vx)\ dv = 0,
\end{aligned}$$

where the last equality is valid since

$$\lim_{x\to\infty} \int_\alpha^\beta h_1(v)\cos vx\, dv = \lim_{x\to\infty} \int_\alpha^\beta h_1(v)\sin vx\, dv = 0$$

by the Riemann–Lebesgue lemma [26, p. 278]. According to the statement of the lemma we must have $h_{1,2} \in \mathbf{L}^1(\alpha,\beta)$, which is easily verified. Another method of proof [23, p. 26] is via the asymptotic evaluation of the continuous spectrum integral, resulting in a closed-form expression for the radiation field as a cylindrical wave.

The behavior of the individual radiation modes is given a physical interpretation as follows. The normalized improper eigenfunctions (8.85) can be written as

$$u_{x,v}(x,v) = \frac{1}{\sqrt{2\pi}}\sqrt{\frac{v - i\eta}{v + i\eta}}\left(e^{ivx} + \frac{v + i\eta}{v - i\eta}e^{-ivx}\right). \tag{8.100}$$

The first term represents a plane wave incident from $x = \infty$, while the second term represents the reflection of this incident plane wave, with reflection coefficient $\Gamma = (v + i\eta)/(v - i\eta)$. In fact, this viewpoint can be utilized as the basis of a normalization procedure [25, pp. 245–248], although this will not be detailed here. We simply remark that the improper spectral components are physically related to plane waves, which individually do not vanish at infinity. The superposition of such waves forms a cylindrical wave that does have the proper asymptotic behavior.

### Completeness Relations and Associated Spectral Expansions

The completeness relation (8.25) is obtained by integrating the one-dimensional Green's function in the appropriate $\lambda$-plane, as shown in Chapter 5 (see Examples 5.8 and 5.11), or directly from the appropriately normalized discrete and continuous eigenfunctions.

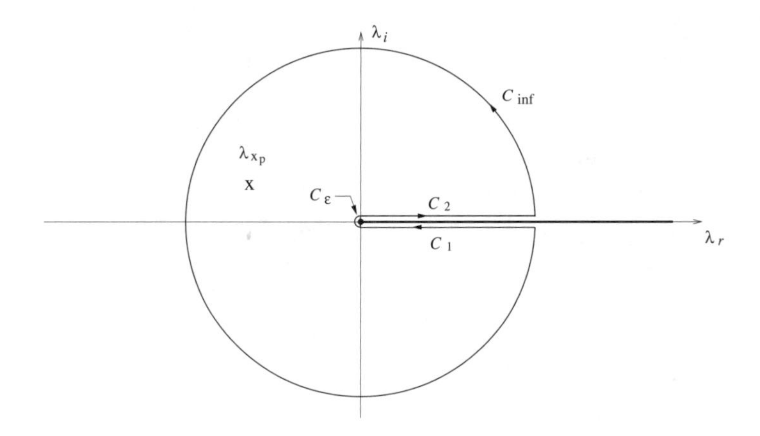


Figure 8.5: Complex plane analysis for evaluation of (8.102). Branch point at $\lambda = 0$ (denoted by "•") is shown with an associated branch cut along the positive real axis. Location of pole, shown in second quadrant, will depend on the impedance parameter and may occur in the first quadrant. Circle $C_{\text{inf}}$ has infinite radius.

From (8.96), the spectral representation of the delta-function in the $z$-coordinate is (5.81),

$$\delta(z - z') = \int_{-\infty}^{\infty} u_{z,\nu}(z,\nu)\overline{u}_{z,\nu}(z',\nu)\,d\nu = \frac{1}{2\pi}\int_{-\infty}^{\infty} e^{i\nu\,(z-z')}d\nu. \quad (8.101)$$

To determine the spectral representation of the delta-function in the $x$-coordinate, consider the integration path in Figure 8.5, leading to

$$\oint_{C_\infty + C_1 + C_2 + C_\varepsilon} g_x(x, x', \lambda_x)\,d\lambda_x = 2\pi i \operatorname{Res}\left\{g_x(x, x', \lambda_{xp})\right\}, \quad (8.102)$$

where $\operatorname{Res}\{g_x(x, x', \lambda_{xp})\}$ is the residue at the pole

$$\lambda_{x,p} = -\eta^2 = -\left(\eta_r^2 - \eta_i^2\right) - 2i\eta_r\eta_i.$$

Because $\eta_r < 0$ due to $\operatorname{Im}\sqrt{\lambda_x} < 0$, and $\eta_i \geq 0$ because $R_s \geq 0$, we see that $\operatorname{Im}(\lambda_{xp}) \geq 0$. If $\eta_r^2 > \eta_i^2$ ($|X_s| > R_s$), the pole will reside in the second quadrant of the $\lambda_x$-plane, and if $\eta_r^2 < \eta_i^2$ ($|X_s| < R_s$), the pole will lie in the first quadrant of the $\lambda_x$-plane. The residue is easily evaluated as

$$\operatorname{Res}\left\{g_x(x, x', \lambda_{xp})\right\} = \frac{e^{-i\sqrt{\lambda_x}x}\left(\cos\sqrt{\lambda_x}x' + \frac{\eta}{\sqrt{\lambda_x}}\sin\sqrt{\lambda_x}x'\right)\Big|_{\lambda_x = \lambda_{xp} = -\eta^2}}{\frac{d}{d\lambda_x}\left(i\sqrt{\lambda_x} + \eta\right)\Big|_{\lambda_x = \lambda_{xp} = -\eta^2}}$$

$$= 2\eta e^{\eta(x+x')}.$$

From (5.66) we have

$$\int_{C_\infty} g_x(x,x',\lambda_x)\,d\lambda_x = -2\pi i\,\delta(x-x'). \tag{8.103}$$

The contribution from $C_\varepsilon$ vanishes as $\varepsilon \to 0$, and the contributions from $C_1$ and $C_2$ are obtained as in Section 5.2 (see (5.59)). As $\varepsilon \to 0$ on $C_1$, $\lambda_x = re^{i4\pi^-}$, leading to $\sqrt{\lambda_x} = \sqrt{r}e^{i2\pi^-} = \sqrt{r}$; on $C_2$, $\lambda_x = re^{i2\pi^+}$ such that $\sqrt{\lambda_x} = \sqrt{r}e^{i\pi^+} = -\sqrt{r}$, where $r = |\lambda_x|$. Therefore, for $x > x'$,

$$\begin{aligned}
\lim_{\varepsilon\to 0}\int_{C_1+C_2} g_x(x,x',\lambda_x)\,d\lambda_x &= \int_\infty^0 \frac{e^{-i\sqrt{r}x}}{i\sqrt{r}+\eta}\left(\cos\sqrt{r}x' + \frac{\eta}{\sqrt{r}}\sin\sqrt{r}x'\right)dr \\
&\quad + \int_0^\infty \frac{e^{i\sqrt{r}x}}{-i\sqrt{r}+\eta}\left(\cos\sqrt{r}x' + \frac{\eta}{\sqrt{r}}\sin\sqrt{r}x'\right)dr \\
&= \int_0^\infty \left(\cos\sqrt{r}x' + \frac{\eta}{\sqrt{r}}\sin\sqrt{r}x'\right)\left[\frac{e^{i\sqrt{r}x}}{-i\sqrt{r}+\eta} - \frac{e^{-i\sqrt{r}x}}{i\sqrt{r}+\eta}\right]dr \\
&= 2i\int_0^\infty \left(\cos\sqrt{r}x' + \frac{\eta}{\sqrt{r}}\sin\sqrt{r}x'\right)\left(\cos\sqrt{r}x + \frac{\eta}{\sqrt{r}}\sin\sqrt{r}x\right)\frac{\sqrt{r}}{r+\eta^2}dr.
\end{aligned}$$

Letting $v = \sqrt{r}$ and $dv = dr/(2\sqrt{r})$, we obtain

$$\begin{aligned}
&\lim_{\varepsilon\to 0}\int_{C_1+C_2} g_x(x,x',\lambda_x)\,d\lambda_x \\
&\quad = 4i\int_0^\infty \left(\cos vx + \frac{\eta}{v}\sin vx\right)\left(\cos vx' + \frac{\eta}{v}\sin vx'\right)\frac{v^2}{v^2+\eta^2}dv,
\end{aligned} \tag{8.104}$$

which, when combined with (8.102)–(8.104), leads to

$$\begin{aligned}
\delta(x-x') &= -2\eta e^{\eta(x+x')} \\
&\quad + \frac{2}{\pi}\int_0^\infty \left(\cos vx + \frac{\eta}{v}\sin vx\right)\left(\cos vx' + \frac{\eta}{v}\sin vx'\right)\frac{v^2}{\eta^2+v^2}dv.
\end{aligned} \tag{8.105}$$

The same result is obtained for $x < x'$. If $\operatorname{Re}\eta > 0$, then the discrete term in (8.105) is omitted, since the pole is not captured by the spectral integration.

Comparing the last expression with (5.108), one sees that the proper and improper eigenfunctions and those of the adjoint problem, obtained from the spectral integration of the Green's function, are in agreement with (8.84) and (8.85).

Forming the product $\delta(\rho-\rho')=\delta(x-x')\delta(z-z')$ results in

$$\begin{aligned}
&\delta(x-x')\delta(z-z') \\
&\quad=\sum_n \int_\nu u_{x,n}(x)\overline{u}^*_{x,n}(x')u_{z,\nu}(z,\nu)\overline{u}_{z,\nu}(z',\nu)\,d\nu \\
&\qquad+\int_\upsilon\int_\nu u_{x,\upsilon}(x,\upsilon)\overline{u}^*_{x,\upsilon}(x',\upsilon)u_{z,\nu}(z,\nu)\overline{u}_{z,\nu}(z',\nu)\,d\nu\,d\upsilon \qquad (8.106)\\
&\quad=-\frac{\eta}{\pi}e^{\eta(x+x')}\int_{-\infty}^{\infty}d\nu\, e^{i\nu\,(z-z')}+\frac{1}{\pi^2}\int_0^\infty d\upsilon\int_{-\infty}^{\infty}d\nu\, e^{i\nu\,(z-z')}\\
&\qquad\cdot\left(\cos\upsilon x+\frac{\eta}{\upsilon}\sin\upsilon x\right)\left(\cos\upsilon x'+\frac{\eta}{\upsilon}\sin\upsilon x'\right)\frac{\upsilon^2}{\eta^2+\upsilon^2}.
\end{aligned}$$

From (8.28) we can determine a modal expansion of an arbitrary function $f(x,z)\in\mathbf{L}^2(\Omega)$, where $\Omega=\{(x,z):x\in[0,\infty),\ z\in(-\infty,\infty)\}$, as

$$\begin{aligned}
f(x,z)=&\int_{-\infty}^{\infty}a^z_{x1}(\nu)u_{x,1}(x)u_{z,\nu}(z,\nu)\,d\nu \\
&+\int_0^\infty\int_{-\infty}^{\infty}a(\upsilon,\nu)u_{x,\upsilon}(x,\upsilon)u_{z,\nu}(z,\nu)\,d\nu\,d\upsilon \\
=&i\sqrt{\frac{\eta}{\pi}}e^{\eta x}\int_{-\infty}^{\infty}a^z_{x1}(\nu)\,e^{i\nu\,z}d\nu \\
&+\frac{1}{\pi}\int_0^\infty\int_{-\infty}^{\infty}a(\upsilon,\nu)e^{i\nu\,z}\frac{\upsilon}{\sqrt{\eta^2+\upsilon^2}}\left(\cos\upsilon x+\frac{\eta}{\upsilon}\sin\upsilon x\right)d\nu\,d\upsilon,
\end{aligned}$$

where

$$\begin{aligned}
a^z_{x1}(\nu)&=\left\langle f,u^*_{x,n}(x)u_{z,\nu}(z,\nu)\right\rangle\\
a(\upsilon,\nu)&=\left\langle f,u^*_{x,\upsilon}(x,\upsilon)u_{z,\nu}(z,\nu)\right\rangle.
\end{aligned}$$

Note that the various spectral expansions are justified by Theorem 5.3 if we consider this structure as having a perfectly conducting top cover at $x=a$, and letting $a\to\infty$.

### Field Caused by a Source in an Impedance-Plane Waveguide

To obtain an expression for the field caused by a source in the impedance-plane waveguide geometry, we start with the two-dimensional form of (1.52),

$$\Lambda(\rho')=\lim_{\delta\to 0}\int_{\Omega-\Omega_\delta}g(\rho,\rho')\,f(\rho)\,d\Omega-\oint_\Gamma\left(g(\rho,\rho')\nabla\Lambda(\rho)-\Lambda(\rho)\nabla g(\rho,\rho')\right)\cdot d\mathbf{\Gamma}. \tag{8.107}$$

The cross-sectional line integral consists of a path along the impedance boundary at $x=0$, a path at $z=\pm\infty$, and a path at $x=\infty$. Since we assume $\lim_{z\to\pm\infty}\Lambda,g=0$ and $\lim_{x\to\infty}\Lambda,g=0$, the line-integral term for those sections of the path vanishes. Along the impedance plane at $x = 0$ we have $\partial\Lambda(0,z)/\partial x = \eta\,\Lambda(0,z)$, and the same for $g$, so that there is no contribution from this section of the line integral. Therefore, the line integral vanishes and, noting that $g(\rho,\rho') = g(\rho',\rho)$,

$$\Lambda(\rho) = \int_\Omega g(\rho,\rho')\; f(\rho')\; d\Omega', \tag{8.108}$$

in agreement with (8.12), where the limiting notation has been suppressed and $\Omega$ represents the area of the two-dimensional source $f$. As with the parallel-plate example, a purely spectral form of the solution can also be obtained.

We conclude this section with two specific cases.

### Imperfectly Conducting Plane

A common occurrence is the impedance-plane's being a good, yet imperfect, conductor, having electrical conductivity $\sigma < \infty$. It can be shown that [16, p. 200]

$$Z_s = (1+i)\sqrt{\frac{\omega\varepsilon}{2\sigma}},$$

and the analysis for general $\eta$ described above can be used to obtain field behavior in this case.

### Perfectly Conducting Plane

If the impedance plane is a perfect conductor, $Z_s = 0$. The boundary conditions reduce to

$$\frac{\partial u(0,z)}{\partial x} = 0$$

($\eta = 0$) for TM$^z$-modes and

$$u(0,z) = 0$$

($\eta = \infty$) for TE$^z$-modes. Either way the discrete term in (8.98) vanishes (for $\eta \to \infty$ recall $\mathrm{Re}\,\eta < 0$). This is in agreement with the fact that a perfect electrical conductor cannot support a surface wave. For $\eta = 0$ the Green's function (8.97) reduces to (5.71), the spectral delta-function representation (8.105) reduces to (5.72), and, rather than (8.106), we have

$$\delta(x-x')\delta(z-z') = \frac{1}{\pi^2}\int_0^\infty \int_{-\infty}^\infty \cos\upsilon x \cos\upsilon x'\, e^{i\nu\,(z-z')}\, d\nu\, d\upsilon$$

for the TM$^z$-modes; the corresponding formula for the TE$^z$-modes ($\eta \to \infty$) involves sine terms replacing the cosine terms. The modal expansion for TM$^z$-modes becomes simply a combination of a Fourier exponential and cosine transform. The two-dimensional Green's function for the $E$-modes is

$$\begin{aligned} g^E(x,z,x',z') &= \frac{1}{i\pi}\int_{\upsilon}^{\infty} \cos \upsilon x \cos \upsilon x' \frac{1}{\sqrt{\lambda_z}} e^{-i\sqrt{\lambda_z}|z-z'|} d\upsilon \\ &= \frac{1}{2\pi}\int_{-\infty}^{\infty} \frac{e^{-i\sqrt{\lambda_x}x_>}}{i\sqrt{\lambda_x}} \cos \sqrt{\lambda_x} x_< e^{i\nu(z-z')} d\nu, \end{aligned} \tag{8.109}$$

with $\sqrt{\lambda_z} = \sqrt{k^2 - \upsilon^2}$ and $\sqrt{\lambda_x} = \sqrt{k^2 - \nu^2}$, with a corresponding form for the $H$-modes.

### 8.1.7 Grounded Dielectric Layer

Consider the grounded dielectric-layer structure shown in Figure 8.6, consisting of two homogeneous regions characterized by constants $\mu_1$ and $\varepsilon_1$ for $-a < x < 0$, and $\mu_2$ and $\varepsilon_2$ for $x > 0$, with an infinite, perfectly conducting plate located at $x = -a$. The appropriate Helmholtz equation is, from (8.8) and (8.9),

$$-\left(\frac{\partial^2}{\partial z^2} + s\frac{\partial}{\partial x}s^{-1}\frac{\partial}{\partial x} + k^2(x)\right)\Lambda(x,z) = f(x,z), \tag{8.110}$$

where $s = \mu(x)$ and $\Lambda = E_y$ for TE$^z$-modes, and $s = \varepsilon(x)$, $\Lambda = H_y$ for TM$^z$-modes, with $f$ a component relating to the source density. For both types of modes we impose limiting conditions

$$\begin{aligned} \lim_{z\to\pm\infty} \Lambda(x,z) &= 0, \\ \lim_{x\to\infty} \Lambda(x,z) &= 0. \end{aligned}$$

In addition, we have $\Lambda(-a,z) = 0$ for TE$^z$-modes and $\left.\frac{\partial\Lambda(x,z)}{\partial x}\right|_{x=-a} = 0$ for TM$^z$-modes. These conditions, along with the requirement that $\Lambda, L\Lambda \in \mathbf{L}^2(\Omega)$, where $\Omega \equiv \{(x,z) : x \in [-a,\infty), z \in (-\infty,\infty)\}$, form the domain of $L : \mathbf{L}^2(\Omega) \to \mathbf{L}^2(\Omega)$,

$$L \equiv -\left(\frac{\partial^2}{\partial z^2} + s\frac{\partial}{\partial x}s^{-1}\frac{\partial}{\partial x} + k^2(x)\right)$$

with the inner product

$$\langle f,g\rangle = \int_{-\infty}^{\infty}\int_{-a}^{\infty} f(x,z)\,\overline{g(x,z)}\,dx\,dz.$$

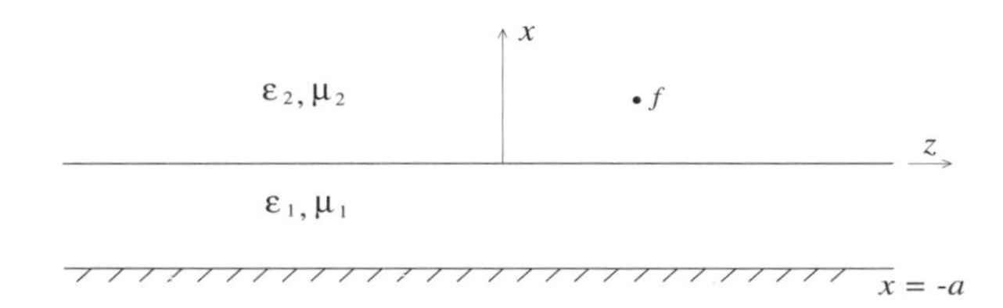


Figure 8.6: Grounded dielectric-layer structure. Source $f$ resides in the region $x > 0$.

### Spectral Properties and Eigenfunctions of the Grounded Dielectric Layer

We first consider the spectral properties of the operator for the grounded dielectric-layer waveguide, with eigenfunctions satisfying

$$-\left(\frac{\partial^2}{\partial z^2} + s\frac{\partial}{\partial x}s^{-1}\frac{\partial}{\partial x} + k^2(x)\right)u_{\alpha,\beta}(x,z) = \lambda_{\alpha,\beta}u_{\alpha,\beta}(x,z) \tag{8.111}$$

subject to the appropriate boundary conditions. We seek a solution as $u_{\alpha,\beta}(x,z) = u_{x,\alpha}(x)u_{z,\beta}(z)$, which, from (8.15), leads to

$$\left(-\frac{d^2}{dz^2} - \lambda_z\right)u_{z,\beta} = 0, \tag{8.112}$$

$$\left(-s\frac{d}{dx}s^{-1}\frac{d}{dx} - \left(k^2(x) + \lambda_x\right)\right)u_{x,\alpha} = 0, \tag{8.113}$$

where $k^2(x) = \omega^2\mu(x)\varepsilon(x)$, with the separation equation $\lambda_z + \lambda_x = \lambda_{\alpha,\beta}$. The differential operators in both of the above equations are seen to be of the Sturm–Liouville form (5.4), with $w = p = 1$ and $q = 0$ in (8.112), and $w = p = s^{-1}$, $q = -k^2$ in (8.113). In the horizontal direction (8.112) presents no new difficulties, as this problem was examined in several preceding sections and in Section 5.2. Orthonormal improper eigenfunctions and corresponding eigenvalues are found to be

$$u_{z,\nu}(z,\nu) = \frac{1}{\sqrt{2\pi}}e^{i\nu z},$$
$$\lambda_z(\nu) = \nu^2, \qquad \nu \in (-\infty,\infty),$$

forming a basis in $\mathbf{L}^2(-\infty,\infty)$. However, in the vertical coordinate we are confronted with two problems, both associated with the violation of the regular Sturm–Liouville conditions, described in the beginning of Chapter 5, that $w, p,$ and $q$ in (5.4) should be real-valued and continuous.

First, as noted in the general development, if the medium is lossy (consistent with the fitness condition $\lim_{x\to\infty}\Lambda(x,z) = 0$) the condition

$w, p, q \in \mathbf{R}$ is violated for the $x$-coordinate problem, such that the problem is not self-adjoint. However, Theorem 5.3 can be applied to this structure if we assume that a top plate exists over the structure, and then the plate is removed to infinity. Therefore, we assume some material loss, however small, and the eigenfunctions will form a basis for the space $\mathbf{L}^2(\Omega)$ except possibly at a set of parameters where modal degeneracies exist (see the discussion in Section 8.1.8). The most serious consequence of this fact is that associated eigenfunctions (rank $> 1$) will be implicated at certain combinations of parameters, thereby affecting modal expansions. In the following we consider the limitingly low-loss case, and we assume $\mu_1\varepsilon_1 \geq \mu_2\varepsilon_2$ in order to draw conclusions concerning the modal eigenvalue equation; other conditions are easily analyzed.

The second difficulty is that in the vertical coordinate the eigenvalue problem (8.113) encompasses a stratified inhomogeneous medium. Note that in Chapter 5 we require $p, p', q$, and $w$ to be continuous. Because $s$ ($s = \mu$ or $s = \varepsilon$) is discontinuous in the problem considered here, it would seem to violate the stated conditions for the applicability of the Sturm–Liouville theory developed in Chapter 5.

The procedure is to relax the conditions concerning continuity of $w, p$, and $q$, yet require $Lu$ to be at least piecewise continuous [25, pp. 174–177]. If this is the case, then we find that $s^{-1}(d/dx)u_x$ and $u_x$ must be continuous at $x = 0$.[11] These two conditions are called *discontinuity conditions* and allow the application of Sturm–Liouville theory to problems with discontinuous media. Assuming the source terms in (8.2)–(8.7) vanish at, or are continuous across, the boundary at $x = 0$, then for TE$^z$-modes this implies $E_y = u_x$ and $H_z = -(1/(i\omega\mu))(\partial u_x/\partial x)$ are continuous, and for TM$^z$-modes this implies $H_y = u_x$ and $E_z = (1/(i\omega\varepsilon))(\partial u_x/\partial x)$ are continuous, consistent with (1.5).

One method to solve (8.113) is to divide the interval $-a \leq x < \infty$ into sections wherein the material parameters are constant. We therefore write (8.113) as

$$\left(-\frac{d^2}{dx^2} - \left(k_1^2 + \lambda_x\right)\right) u_x^1(x) = 0, \tag{8.114}$$
$$\left(-\frac{d^2}{dx^2} - \left(k_2^2 + \lambda_x\right)\right) u_x^2(x) = 0$$

for regions 1 and 2, respectively, with region 1 defined as $-a \leq x \leq 0$ and region 2 as $0 \leq x < \infty$. We then solve the above equations in each region, apply the boundary condition at $x = -a$, tangential field continuity at

[11] To see this, consider $u_x$ to be discontinuous across the boundary. One easily sees that $Lu_x$ would contain the derivative of a delta-function, which is not allowed. Hence, $u_x$ must be continuous. Similarly, if $s^{-1}(d/dx)u_x$ is discontinuous, application of the operator $d/dx$ as in (8.113) would generate a delta-function discontinuity in $Lu_x$, and so $s^{-1}(d/dx)u_x$ must be continuous.

$x = 0$, and the limiting condition at $x = \infty$. Since the problem is similar to one considered in Section 8.1.6, it is not surprising that in the vertical coordinate we encounter both a discrete and a continuous spectrum. We consider the TE$^z$-modes first.

### Proper TE$^z$ Eigenfunctions in the $x$-Coordinate

For the discrete spectrum we solve (8.114) as

$$\begin{aligned} u_x^1(x) &= A\sin[\sqrt{\gamma_1}\,(x+a)] + B\cos[\sqrt{\gamma_1}\,(x+a)], \\ u_x^2(x) &= Ce^{-i\sqrt{\gamma_2}x} + De^{i\sqrt{\gamma_2}x}, \end{aligned} \tag{8.115}$$

where the superscript indicates the region, and $\gamma_j = k_j^2 + \lambda_x$ for the $j$th region, $j = 1, 2$. The above form is chosen because it facilitates evaluation of the unknown coefficients. Assuming $\text{Im}\sqrt{\gamma_2} < 0$ with the condition $u_x^2(x) \in \mathbf{L}^2(0,\infty)$ leads to $D = 0$. The boundary condition at the perfectly conducting plate, $u_x^1(x = -a) = 0$, results in $B = 0$. Enforcing $E_y$- and $H_z$- field continuity at $x = 0$,

$$\begin{aligned} u_x^1(x) - u_x^2(x)\Big|_{x=0} &= 0, \\ \frac{1}{\mu_1}\frac{du_x^1(x)}{dx} - \frac{1}{\mu_2}\frac{du_x^2(x)}{dx}\Bigg|_{x=0} &= 0, \end{aligned}$$

respectively, leads to $C = A\sin\sqrt{\gamma_1}a$ and

$$\cot\sqrt{\gamma_1}a = -i\frac{\mu_1}{\mu_2}\frac{\sqrt{\gamma_2}}{\sqrt{\gamma_1}}. \tag{8.116}$$

The TE$^z$ proper eigenmodes are then

$$u_{x,n}^1(x) = C_n\frac{\sin[\sqrt{\gamma_1}\,(x+a)]}{\sin\sqrt{\gamma_1}a}, \qquad -a < x < 0, \tag{8.117}$$

$$u_{x,n}^2(x) = C_n e^{-i\sqrt{\gamma_2}x}, \qquad x > 0, \tag{8.118}$$

where each term is evaluated at the eigenvalue $\lambda_x = \lambda_{x,n}$ through $\gamma_j = k_j^2 + \lambda_{x,n}$, with $\lambda_{x,n}$ a discrete solution to the TE$^z$-mode *dispersion equation* (8.116). Such solutions must generally be found numerically, although a well known graphical procedure [16, pp. 712–716] can be utilized to obtain information concerning the roots of (8.116). Below we take an approach suggested in [17, p. 292] for a similar problem.

Since (8.113) is self-adjoint in the limitingly low-loss case, $\lambda_{x,n}$ must be real-valued; together with $k_j^2 \in \mathbf{R}$, this leads us to see that we have either $\gamma_j > 0$ or $\gamma_j < 0$. For both $\gamma_{1,2}$ real and positive, (8.116) has no solution. Therefore, at least one of $\gamma_1$ or $\gamma_2$, or both, must be negative, such that either $-k_1^2 < \lambda_{x,n} < -k_2^2$, leading to $\gamma_2 < 0$ and $\gamma_1 > 0$, or $\lambda_{x,n} < -k_1^2$, leading to $\gamma_1, \gamma_2 < 0$ (recall $\mu_1\varepsilon_1 > \mu_2\varepsilon_2$). The latter condition is impossible because, with $\text{Im}\sqrt{\gamma_2} < 0$, the left and right sides of (8.116) are real but with differing signs, such that (8.116) would have no solution. Therefore, if (8.116) is to have solutions, we must have $\gamma_2 < 0$ and $\gamma_1 > 0$, so that

$$-k_1^2 < \lambda_{x,n} < -k_2^2.$$

With $\sqrt{\gamma_2} = -i\sqrt{-\lambda_{x,n} - k_2^2} = -i\alpha$, where $\alpha > 0$, (8.118) becomes $C_n e^{-\alpha x}$, which displays the proper asymptotic behavior at $x = \infty$.

With these observations we can normalize the discrete modes as $\langle u_{x,n}, u_{x,n}\rangle = 1$, where orthogonality follows from self-adjointness of the operator. Therefore, for the TE$^z$-modes the normalization is

$$C_n^2 \left[ \int_{-a}^{0} \left( \frac{\sin[\sqrt{\gamma_1}\,(x+a)]}{\sin\sqrt{\gamma_1}a} \right)^2 \frac{1}{\mu_1} dx + \int_0^\infty \left( e^{-i\sqrt{\gamma_2}x} \right)^2 \frac{1}{\mu_2} dx \right] = 1,$$

leading to

$$C_n^2 = \frac{2\mu_2}{\frac{\mu_2}{\mu_1}\frac{a}{\sin^2\sqrt{\gamma_1}a} + i\frac{\gamma_2-\gamma_1}{\gamma_1\sqrt{\gamma_2}}}. \tag{8.119}$$

### Improper TE$^z$ Eigenfunctions in the $x$-Coordinate

For the improper (continuous) transverse eigenfunctions we solve (8.113) as

$$\begin{aligned} u_x^1(x,\lambda_x) &= A\sin[\sqrt{\gamma_1}\,(x+a)] + B\cos[\sqrt{\gamma_1}\,(x+a)], \\ u_x^2(x,\lambda_x) &= C\sin\sqrt{\gamma_2}x + D\cos\sqrt{\gamma_2}x, \end{aligned} \tag{8.120}$$

where the form for $u_x^2(x)$ is motivated by the fact that we don't require the improper eigenfunctions to belong to $\mathbf{L}^2\,(0,\infty)$. The boundary condition at the perfectly conducting plate, $u_x^1(x=-a) = 0$, leads to $B = 0$. Enforcing $E_y$- and $H_z$-field continuity at $x = 0$ leads to $A = C(\mu_1/\mu_2)(\sqrt{\gamma_2}/\sqrt{\gamma_1})$ $(1/\cos\sqrt{\gamma_1}a)$ and $D = C(\mu_1/\mu_2)(\sqrt{\gamma_2}/\sqrt{\gamma_1})(\tan\sqrt{\gamma_1}a)$. A rearrangement of constants leads to the TE$^z$ improper eigenmodes as

$$\begin{aligned} u_x^1(x,\lambda_x) &= E_v \frac{\sin[\sqrt{\gamma_1}\,(x+a)]}{\sin\sqrt{\gamma_1}a}, & -a < x < 0, \\ u_x^2(x,\lambda_x) &= E_v \left[ \cos\sqrt{\gamma_2}x + \frac{\alpha^H}{\sqrt{\gamma_2}}\sin\sqrt{\gamma_2}x \right], & x > 0, \end{aligned} \tag{8.121}$$

where

$$\alpha^H = \frac{\mu_2}{\mu_1}\sqrt{\gamma_1}\cot\sqrt{\gamma_1}a.$$

In the above, $\gamma_j = k_j^2 + \lambda_x$ with the spectral parameter $\lambda_x$ continuous rather than discrete. We will see later that as a continuous parameter, $\lambda_x \in (-k_2^2, \infty)$, leading to $\gamma_{1,2} \geq 0$. Therefore, $u_x^{1,2}(x,\lambda_x)$ are real-valued such that the improper modes can be accordingly normalized. The normalization is usually given as

$$\langle u_x, \overline{u}_x \rangle = \int_{-a}^{\infty} u_{x,\upsilon}(x,\upsilon) u_{x,p}(x,p) \mu(x)^{-1} dx = \delta(\upsilon - p),$$

where $\upsilon = \sqrt{\gamma_2} = \sqrt{k_2^2 + \lambda_x}$ and $p = \sqrt{\widetilde{\gamma}_2} = \sqrt{k_2^2 + \widetilde{\lambda}_x}$ for different values $\lambda_x, \widetilde{\lambda}_x$ (the appropriate eigenfunction is used for $\int_{-a}^{0} (\cdot)\, dx$). Using (8.86)–(8.88), we obtain

$$E_\upsilon = \left[ \frac{\pi}{2} \frac{1}{\mu_2} \left( 1 + \left( \frac{\alpha^H}{\sqrt{\gamma_2}} \right)^2 \right) \right]^{-1/2}. \tag{8.122}$$

### Proper TM$^z$ Eigenfunctions in the $x$-Coordinate

For the TM$^z$-modes, similar considerations result in the proper eigenmodes as

$$\begin{aligned} u_{x,n}^1(x) &= C_n \frac{\cos[\sqrt{\gamma_1}\,(x+a)]}{\cos\sqrt{\gamma_1}a}, & -a < x < 0, \\ u_{x,n}^2(x) &= C_n e^{-i\sqrt{\gamma_2}x}, & x > 0, \end{aligned} \tag{8.123}$$

where each term is evaluated at $\lambda_x = \lambda_{x,n}$ through $\gamma_j = k_j^2 + \lambda_{x,n}$, and where $\lambda_{x,n}$ is a discrete solution to the TM$^z$ *dispersion equation*

$$\tan\sqrt{\gamma_1}a = i\frac{\varepsilon_1}{\varepsilon_2}\frac{\sqrt{\gamma_2}}{\sqrt{\gamma_1}}. \tag{8.124}$$

As with the TE$^z$-modes, we find that if (8.124) is to have solutions we must have $\gamma_2 < 0$ and $\gamma_1 > 0$, so that

$$-k_1^2 < \lambda_{x,n} < -k_2^2,$$

leading to $\sqrt{\gamma_2} = -i\sqrt{-\lambda_{x,n} - k_2^2} = -i\alpha$, where $\alpha > 0$, and the proper asymptotic behavior at $x = \infty$. For the TM$^z$ discrete modes the normalization $\langle u_{x,n}, u_{x,n} \rangle = 1$ leads to

$$C_n^2 = \frac{2\varepsilon_2}{\frac{\varepsilon_2}{\varepsilon_1}\frac{a}{\cos^2\sqrt{\gamma_1}a} + i\frac{\gamma_2 - \gamma_1}{\gamma_1\sqrt{\gamma_2}}}. \tag{8.125}$$

### Improper TM$^z$ Eigenfunctions in the $x$-Coordinate

The TM$^z$ improper transverse eigenfunctions are found to be

$$\begin{aligned} u_x^1(x,\lambda_x) &= E_\upsilon \frac{\cos[\sqrt{\gamma_1}\,(x+a)]}{\cos\sqrt{\gamma_1}a}, & -a < x < 0, \\ u_x^2(x,\lambda_x) &= E_\upsilon \left[ \cos\sqrt{\gamma_2}x - \frac{\alpha^E}{\sqrt{\gamma_2}} \sin\sqrt{\gamma_2}x \right], & x > 0, \end{aligned} \tag{8.126}$$

where

$$\alpha^E = \frac{\varepsilon_2}{\varepsilon_1}\sqrt{\gamma_1}\tan\sqrt{\gamma_1}a$$

and $\gamma_j = k_j^2 + \lambda_x$, with the spectral continuous parameter $\lambda_x \in (-k_2^2, \infty)$. As with the TE$^z$-modes, (8.126) can be normalized as

$$\langle u_x, \overline{u}_x \rangle = \int_{-a}^{\infty} u_x(x, \upsilon)u_x(x, p)\varepsilon(x)^{-1}dx = \delta(\upsilon - p).$$

Using (8.86)–(8.88), we obtain

$$E_\upsilon = \left[\frac{\pi}{2}\frac{1}{\varepsilon_2}\left(1 + \left(\frac{\alpha^E}{\sqrt{\gamma_2}}\right)^2\right)\right]^{-1/2}. \tag{8.127}$$

**Summary of Two-Dimensional Eigenfunctions**

It is easy to see from the above that

$$\begin{aligned} u_{n,\nu}^1(x, z, \nu) &= \frac{C_n}{\sqrt{2\pi}} \frac{\binom{\cos}{\sin}\left[\sqrt{\gamma_1}\,(x + a)\right]}{\binom{\sin}{\cos}\sqrt{\gamma_1}a} e^{i\nu z}, \qquad -a < x < 0, \\ u_{n,\nu}^2(x, z, \nu) &= \frac{C_n}{\sqrt{2\pi}} e^{-i\sqrt{\gamma_2}x} e^{i\nu z}, \qquad x > 0, \end{aligned} \tag{8.128}$$

are eigenfunctions of $L$ for the $\binom{\text{TE}}{\text{TM}}$ cases, proper in the $x$-coordinate and improper in $z$, where $\gamma_j = k_j^2 + \lambda_{x,n}$, and where $\lambda_{x,n}$ are solutions of the corresponding dispersion equation (8.116) or (8.124), with $C_n$ given by (8.119) or (8.125). The corresponding eigenvalues are

$$\lambda_{n,\nu} = \lambda_x + \lambda_z = \lambda_{x,n} + \nu^2, \qquad \nu \in (-\infty, \infty).$$

Orthonormality is expressed as

$$\langle u_{n,\nu}, u_{m,p} \rangle = \delta_{n,m}\delta\left(\nu - p\right).$$

Moreover,

$$\begin{aligned} u_{\upsilon,\nu}^1(x, z, \upsilon, \nu) &= \frac{E_\upsilon}{\sqrt{2\pi}} \frac{\binom{\sin}{\cos}\left[\sqrt{\gamma_1}\,(x + a)\right]}{\binom{\sin}{\cos}\sqrt{\gamma_1}a} e^{i\nu z}, \qquad -a < x < 0, \\ u_{\upsilon,\nu}^2(x, z, \upsilon, \nu) &= \frac{E_\upsilon}{\sqrt{2\pi}} \left[\cos \upsilon x \pm \frac{\alpha^{H,E}}{\upsilon}\sin \upsilon x\right] e^{i\nu z}, \qquad 0 < x, \end{aligned} \tag{8.129}$$

are eigenfunctions of $L$ for the $\binom{\text{TE}}{\text{TM}}$ modes, improper in both coordinates, where $E_\upsilon$ is given by (8.122) or (8.127), $\upsilon = \sqrt{\gamma_2}$, and $\gamma_j = k_j^2 + \lambda_x$, and where $\lambda_x \in (-k_2^2, \infty)$ is a continuous variable ($\upsilon \in [0, \infty)$) and $\nu \in$

$(-\infty, \infty)$. The corresponding eigenvalues are $\lambda_{\upsilon,\nu} = \lambda_x + \lambda_z = \lambda_x + \nu^2$; however, because $\upsilon = \sqrt{\gamma_2}$, we have $\lambda_x = \upsilon^2 - k_2^2$ such that

$$\lambda_{\upsilon,\nu} = \upsilon^2 + \nu^2 - k_2^2, \qquad \upsilon \in [0, \infty), \quad \nu \in (-\infty, \infty).$$

Orthonormality is expressed as

$$\langle u_{\upsilon,\nu}, u_{p,q} \rangle = \delta(\upsilon - p)\, \delta(\nu - q).$$

### Natural Waveguide Modes of the Grounded Dielectric Layer

The natural modes of the grounded dielectric-layer structure satisfy

$$-\left( \frac{\partial^2}{\partial z^2} + s\frac{\partial}{\partial x}s^{-1}\frac{\partial}{\partial x} + k^2(x) \right) \psi(x, z) = 0 \tag{8.130}$$

subject to the appropriate boundary conditions. We seek a solution as $\psi(x, z) = \psi_x(x)\psi_z(z)$, which, from (8.15), leads to

$$\left( -\frac{d^2}{dz^2} - \lambda_z \right) \psi_z = 0, \tag{8.131}$$

$$\left( -s\frac{d}{dx}s^{-1}\frac{d}{dx} - \left(k^2(x) + \lambda_x\right) \right) \psi_x = 0, \tag{8.132}$$

where $k^2(x) = \omega^2 \mu(x) \varepsilon(x)$, with the separation equation $\lambda_z + \lambda_x = 0$. In the $x$-coordinate the solution is the same as the solution of (8.113), and in the $z$-coordinate we follow the method described for the parallel-plate structure and obtain

$$\psi_z(z) = A\, e^{\mp i\sqrt{\lambda_z} z},$$

where $\operatorname{Im}\sqrt{\lambda_z} < 0$ and $A$ is arbitrary. Because $\lambda_z + \lambda_x = 0$, we have for the discrete modes $\sqrt{\lambda_{z,n}} = \sqrt{-\lambda_{x,n}}$; recall that $-k_1^2 < \lambda_{x,n} < -k_2^2$; then the term $e^{\mp i\sqrt{-\lambda_{x,n}} z}$ represents purely oscillatory behavior along the $z$-axis, although we take $\lambda_{x,n}$ to have a very small imaginary part due to slight material loss (consistent with $\operatorname{Im}\sqrt{\lambda_z} < 0$) so that the function exhibits the proper decay at $z = \pm\infty$. We then have

$$\begin{aligned} \psi_n^1(x, z) &= A \cdot C_n \frac{\binom{\sin}{\cos}\left[\sqrt{\gamma_1}\,(x + a)\right]}{\binom{\sin}{\cos}\sqrt{\gamma_1}a} e^{\mp i\sqrt{-\lambda_{x,n}} z}, & -a < x < 0, \\ \psi_n^2(x, z) &= A \cdot C_n e^{-i\sqrt{\gamma_2}x} e^{\mp i\sqrt{-\lambda_{x,n}} z}, & x > 0, \end{aligned} \tag{8.133}$$

for the proper discrete $\binom{\text{TE}}{\text{TM}}$ modes, where $\gamma_j = k_j^2 + \lambda_{x,n}$, $-k_1^2 < \lambda_{x,n} < -k_2^2$, and $\lambda_{zn} = -\lambda_{x,n}$, with $\lambda_{x,n}$ a solution of the corresponding dispersion equation (8.116) or (8.124), and where $C_n$ is given by (8.119) or (8.125). Because of the attenuation in the vertical direction, these modes are bound to and propagate along the dielectric surface and are known as *surface waves*.

The improper modes are

$$\psi_x^1(x,\lambda_x) = A\cdot E_\upsilon \frac{\binom{\sin}{\cos}\left[\sqrt{\gamma_1}\,(x+a)\right]}{\binom{\sin}{\cos}\sqrt{\gamma_1}a} e^{\mp i\sqrt{-\lambda_x}z}, \qquad -a<x<0,$$

$$\psi_x^2(x,\lambda_x) = A\cdot E_\upsilon \left[\cos \upsilon x \pm \frac{\alpha^{H,E}}{\upsilon}\sin \upsilon x\right] e^{\mp i\sqrt{-\lambda_x}z}, \qquad 0<x,$$

$$(8.134)$$

for the $\binom{\text{TE}}{\text{TM}}$ modes, where $\gamma_j = k_j^2 + \lambda_x$ and $\lambda_z = -\lambda_x$, with $\lambda_x \in (-k_2^2, \infty)$ a continuous variable and $\upsilon = \sqrt{\gamma_2}$.

The term $e^{\mp i\sqrt{-\lambda_x}z}$ encompasses two distinct types of modes. For $\lambda_x \in (-k_2^2, 0)$ we have $e^{\mp i\sqrt{|\lambda_x|}z}$, which represents oscillatory behavior along the $z$-axis (again only purely so in the lossless limit). For $\lambda_x \in (0,\infty)$ we have $e^{\mp i\sqrt{-\lambda_x}z} = e^{\mp\sqrt{\lambda_x}z}$, exhibiting pure exponential decay along the $z$-axis. It can be considered that for $\lambda_x \in (-k_2^2, 0)$ the improper modes represent propagating radiation modes of the structure, whereas for $\lambda_x \in (0,\infty)$ the improper modes represent evanescent radiation modes of the structure. For a single spectral component, neither the propagating nor evanescent radiation modes decay as $x \to \infty$, yet both remain bounded at infinity, as was the case for the impedance plane waveguide. Note also that the discrete modes (8.133) and the propagating radiation modes ((8.134) with $\lambda_x \in (-k_2^2, 0)$) are eigenfunctions in the null space of the two-dimensional operator $L$. The evanescent radiation modes are homogeneous solutions of $L$, yet are not improper eigenfunctions of $L$ since they are not bounded on $\mathbf{R}$.

### Leaky Modes

For the discrete modes we found $-k_1^2 < \lambda_{x,n} < -k_2^2$, with $\lambda_{x,n}$ a solution of (8.116) or (8.124). These equations also admit complex solutions $\widetilde{\lambda}_x$ that are *not* eigenvalues of (8.113) yet nevertheless have physical significance. It can be shown graphically using the procedure in [16, pp. 712–716] that for very large, real-valued frequency $\omega$ the eigenvalue $\lambda_{x,n}$ tends toward $-k_1^2$, and that as frequency is lowered $\lambda_{x,n}$ tends toward $-k_2^2$. At some sufficiently low frequency we have $\lambda_{x,n} = -k_2^2$, such that $\gamma_2 = 0$. This frequency, known as the cutoff frequency, is given by

$$\omega_c = \frac{n\pi}{a\sqrt{\varepsilon_1\mu_1 - \varepsilon_2\mu_2}}$$

for TM$^z$-modes, with $n$ replaced by $n+1/2$ for TE$^z$-modes, $n = 0,1,2,\ldots$. As frequency is further lowered, the eigenvalue equations fail to have real solutions, yet complex solutions $\widetilde{\lambda}_{x,n}$ exist that represent the analytic continuation of the real eigenvalues $\lambda_{x,n}$ below their cutoff frequency. This is associated with $\lambda_{x,n}$ passing through the branch point at $\lambda_x = -k_2^2$ in the complex $\lambda$-plane as described later, such that the analytic continuation of $\lambda_{x,n}$ below cutoff resides on the improper Riemann sheet $\operatorname{Im}(\sqrt{\lambda_x}) > 0$. Inserting this condition into (8.133) yields the behavior

$$e^{-i\sqrt{\gamma_2}x} = e^{-i\operatorname{Re}\{\sqrt{\gamma_2}\}x} e^{\operatorname{Im}\{\sqrt{\gamma_2}\}x}$$

such that the discrete mode is exponentially increasing in the vertical direction. We will call (8.133) with $\widetilde{\lambda}_{x,n}$ a *leaky mode*, representing the analytical continuation of the proper discrete mode below its cutoff frequency. The nonphysical behavior of this mode can represent, in some limited spatial region, the physical wave associated with the radiation field. This is discussed in [17, pp. 538–543] for a point-source excitation. Some details of the behavior of $\lambda_{x,n}$ (and $\widetilde{\lambda}_{x,n}$) in the complex $\omega$-plane are provided in [22] and [27], including the identification of $\omega$-plane branch points of the dispersion function.

### Green's Functions for the Grounded Dielectric-Layer Waveguide

The Green's function for the grounded dielectric layer satisfies

$$\left(\frac{\partial^2}{\partial z^2} + s\frac{\partial}{\partial x}s^{-1}\frac{\partial}{\partial x} + k^2(x)\right) g(\rho, \rho') = -s(x)\delta(x - x')\delta(z - z'). \quad (8.135)$$

The associated one-dimensional Green's function equations are, from (5.13),

$$\left(-\frac{d^2}{dz^2} - \lambda_z\right) g_z(z, z', \lambda_z) = \delta(z - z'), \quad (8.136)$$

$$\begin{aligned}\left(-s\frac{d}{dx}s^{-1}\frac{d}{dx} - \left(k^2(x) + \lambda_x\right)\right) g_x(x, x', \lambda_x) &= \frac{\delta(x - x')}{w(x)} \\ &= s(x)\delta(x - x'),\end{aligned} \quad (8.137)$$

subject to the appropriate boundary conditions. For the singular (limit-point) Sturm–Liouville problem (8.136) we impose the condition

$$\lim_{z\to\pm\infty} g_z(z, z') = 0.$$

As considered in (5.11), the solution is (5.78)

$$g_z(z, z', \lambda_z) = \frac{1}{2i\sqrt{\lambda_z}} e^{-i\sqrt{\lambda_z}|z-z'|}. \quad (8.138)$$

We divide the vertical problem into two regions, $-a \le x \le 0$ and $x \ge 0$, such that (8.136) becomes

$$\left(-\frac{d^2}{dx^2} - \gamma_j\right) g_{xj}(x, x', \gamma_j) = \begin{cases} 0, & j = 1, \\ s(x)\delta(x - x'), & j = 2, \end{cases}$$

where $\gamma_j = k_j + \lambda_x$ since the source[12] is in region 2. The resulting equations can be solved by the direct method (see Section 5.1.1 and (5.17)). We write the solution of $(d^2/dx^2 + \gamma_j)\, g_{xj}(x, x', \gamma_j) = 0$ as

$$\begin{aligned} g_{x1}(x, x', \gamma_1) &= A_1 \sin\sqrt{\gamma_1}\,(x+a) + B_1 \cos\sqrt{\gamma_1}\,(x+a)\,, & -a < x < 0, \\ g_{x2}(x, x', \gamma_2) &= A_2 e^{i\sqrt{\gamma_2} x} + B_2 e^{-i\sqrt{\gamma_2} x}, & 0 < x < x', \\ g_{x3}(x, x', \gamma_2) &= A_3 e^{i\sqrt{\gamma_2} x} + B_3 e^{-i\sqrt{\gamma_2} x}, & x' < x, \end{aligned} \tag{8.139}$$

where $g_{x2}$ and $g_{x3}$ are both in region 2, below and above the source, respectively. The boundary conditions are

$$g_{x1}(-a, x', \gamma_1) = 0, \tag{8.140}$$

$$g_{x2}(0, x', \gamma_2) = g_{x1}(0, x', \gamma_1), \tag{8.141}$$

$$\frac{1}{\mu_2} \left.\frac{dg_{x2}(x, x', \gamma_2)}{dx}\right|_{x=0} = \frac{1}{\mu_1} \left.\frac{dg_{x1}(x, x', \gamma_1)}{dx}\right|_{x=0}, \tag{8.142}$$

$$g_{x2}(x', x', \gamma_2) = g_{x3}(x', x', \gamma_2), \tag{8.143}$$

$$\left.\frac{dg_{x3}(x, x', \gamma_2)}{dx}\right|_{x=x'} - \left.\frac{dg_{x2}(x, x', \gamma_2)}{dx}\right|_{x=x'} = -\frac{1}{\mu^{-1}(x')}, \tag{8.144}$$

$$g_{x3}(x, x', \gamma_2) \in \mathbf{L}^2(x', \infty), \tag{8.145}$$

for $TE^z$-modes, and

$$\left.\frac{dg_{x1}(x, x', \gamma_1)}{dx}\right|_{x=-a} = 0, \tag{8.146}$$

$$g_{x2}(0, x', \gamma_2) = g_{x1}(0, x', \gamma_1), \tag{8.147}$$

$$\frac{1}{\varepsilon_2} \left.\frac{dg_{x2}(x, x', \gamma_2)}{dx}\right|_{x=0} = \frac{1}{\varepsilon_1} \left.\frac{dg_{x1}(x, x', \gamma_1)}{dx}\right|_{x=0}, \tag{8.148}$$

$$g_{x2}(x', x', \gamma_2) = g_{x3}(x', x', \gamma_2), \tag{8.149}$$

$$\left.\frac{dg_{x3}(x, x', \gamma_2)}{dx}\right|_{x=x'} - \left.\frac{dg_{x2}(x, x', \gamma_2)}{dx}\right|_{x=x'} = -\frac{1}{\varepsilon^{-1}(x')}, \tag{8.150}$$

$$g_{x3}(x, x', \gamma_2) \in \mathbf{L}^2(x', \infty), \tag{8.151}$$

for $TM^z$-modes. The boundary conditions arise by noting that, in addition to (5.17), for the $TE^z$-modes the Green's function represents the tangential electric field $E_y$, and its derivative is proportional $H_z$ through (8.3). The condition (8.140) is due to the vanishing of the tangential electric field at the perfectly conducting plate. The conditions (8.141) and (8.142) arise from the continuity of tangential electric and magnetic fields at $x = 0$. The

[12] The problem can be analyzed for a source in region 1 in a similar manner.

remaining three conditions come from (5.17). Similar comments apply to (8.146)–(8.151), where the Green's function represents the magnetic field $H_y$.

Application of the boundary conditions lead to

$$A_1 = \mu_2 \frac{1}{2\sqrt{\gamma_1}} \frac{\mu_1}{\mu_2} \frac{e^{-i\sqrt{\gamma_2}x'}}{\cos\sqrt{\gamma_1}a} \frac{\left(Z^H - N^H\right)}{Z^H},$$

$$A_2 = \mu_2 \frac{e^{-i\sqrt{\gamma_2}x'}}{2i\sqrt{\gamma_2}},$$

$$B_2 = \mu_2 \frac{e^{-i\sqrt{\gamma_2}x'}}{2i\sqrt{\gamma_2}} \frac{N^H}{Z^H},$$

$$B_3 = \mu_2 \frac{1}{2i\sqrt{\gamma_2}} \left[e^{i\sqrt{\gamma_2}x'} + e^{-i\sqrt{\gamma_2}x'} \frac{N^H}{Z^H}\right],$$

and $B_1 = A_3 = 0$ for the TE$^z$-modes, where

$$Z^H = 1 - i\frac{\mu_2}{\mu_1} \frac{\sqrt{\gamma_1}}{\sqrt{\gamma_2}} \cot\sqrt{\gamma_1}a,$$

$$N^H = 1 + i\frac{\mu_2}{\mu_1} \frac{\sqrt{\gamma_1}}{\sqrt{\gamma_2}} \cot\sqrt{\gamma_1}a.$$

The final Green's function is, combining $g_2$ and $g_3$ into a single expression,

$$g_x^H(x, x', \lambda_x) = \begin{cases} \mu_2 \frac{1}{2\sqrt{\gamma_1}} \frac{\mu_1}{\mu_2} \frac{e^{-i\sqrt{\gamma_2}x'}}{\cos\sqrt{\gamma_1}a} \frac{\left(Z^H - N^H\right)}{Z^H} \sin\sqrt{\gamma_1}(x+a), & -a < x < 0, \\ \mu_2 \frac{1}{2i\sqrt{\gamma_2}} \left[e^{-i\sqrt{\gamma_2}|x-x'|} + \frac{N^H}{Z^H} e^{-i\sqrt{\gamma_2}(x+x')}\right], & 0 < x. \end{cases} \tag{8.152}$$

Note that $Z^H = 0$ provides the same equation as (8.116), such that the poles of the Green's function correspond to discrete modes (eigenvalues).

For the TM$^z$-modes we obtain

$$B_1 = -\varepsilon_2 \frac{1}{2\sqrt{\gamma_1}} \frac{\varepsilon_1}{\varepsilon_2} \frac{e^{-i\sqrt{\gamma_2}x'}}{\sin\sqrt{\gamma_1}a} \frac{\left(Z^E - N^E\right)}{Z^E},$$

$$B_2 = \varepsilon_2 \frac{e^{-i\sqrt{\gamma_2}x'}}{2i\sqrt{\gamma_2}} \frac{N^E}{Z^E},$$

$$B_3 = \varepsilon_2 \frac{1}{2i\sqrt{\gamma_2}} \left[e^{i\sqrt{\gamma_2}x'} + e^{-i\sqrt{\gamma_2}x'} \frac{N^E}{Z^E}\right],$$

$$A_2 = \varepsilon_2 \frac{e^{-i\sqrt{\gamma_2}x'}}{2i\sqrt{\gamma_2}},$$

and $A_1 = A_3 = 0$, where

$$Z^E = 1 + i\frac{\varepsilon_2}{\varepsilon_1}\frac{\sqrt{\gamma_1}}{\sqrt{\gamma_2}}\tan\sqrt{\gamma_1}a,$$
$$N^E = 1 - i\frac{\varepsilon_2}{\varepsilon_1}\frac{\sqrt{\gamma_1}}{\sqrt{\gamma_2}}\tan\sqrt{\gamma_1}a,$$

and $Z^E = 0$ is the same as (8.124), leading to

$$g_x^E(x, x', \lambda_x) = \begin{cases} -\varepsilon_2 \frac{1}{2\sqrt{\gamma_1}} \frac{\varepsilon_1}{\varepsilon_2} \frac{e^{-i\sqrt{\gamma_2}x'}}{\sin\sqrt{\gamma_1}a} \frac{\left(Z^E - N^E\right)}{Z^E} \cos\sqrt{\gamma_1}(x+a), & -a < x < 0, \\ \varepsilon_2 \frac{1}{2i\sqrt{\gamma_2}} \left[e^{-i\sqrt{\gamma_2}|x-x'|} + \frac{N^E}{Z^E} e^{-i\sqrt{\gamma_2}(x+x')}\right], & 0 < x. \end{cases} \tag{8.153}$$

The two-dimensional Green's function can be obtained from (8.23) or (8.24). From (8.23) we obtain

$$\begin{aligned} g(x, z, x', z') = &\sum_{n=1}^{N} u_{x,n}(x)\overline{u}_{x,n}(x') \frac{1}{2i\sqrt{-\lambda_{x,n}}} e^{-i\sqrt{-\lambda_{x,n}}|z-z'|} \\ &+ \int_0^\infty u_{x,\upsilon}(x,\upsilon)\overline{u}_{x,\upsilon}(x',\upsilon) \frac{1}{2i\sqrt{k_2^2 - \upsilon^2}} e^{-i\sqrt{k_2^2-\upsilon^2}|z-z'|} d\upsilon \end{aligned} \tag{8.154}$$

with the proper and improper eigenfunctions appropriately chosen from (8.117), (8.121), (8.123), and (8.126). $N$ is the number of proper modes above cutoff, $\upsilon = \sqrt{\gamma_2}$, and in all cases $\text{Im}\sqrt{\lambda_z} < 0$.

If the medium has nontrivial loss, we need to replace $\overline{u}_{x,n}$ with $\overline{u}^*_{x,n}$, the adjoint eigenfunctions. The resulting modal expansions will fail at points of modal degeneracy where associated eigenfunctions (rank $> 1$) occur. One can see from the form of the eigenfunctions that at a modal degeneracy (where two or more roots of the same dispersion equation coincide) among TE modes, or among TM modes, the ordinary eigenfunctions become equal.[13] As discussed in [27] for $\text{TM}_n$- and $\text{TM}_{n+2}$-modes, this can occur for certain values of material loss.[14] At such points the modal expansions need to be augmented with associated eigenfunctions. This is discussed in more detail in Section 8.1.8.

[13] In the lossless case this cannot occur.

[14] If $H(\omega, \varepsilon_1, \varepsilon_2, a, \lambda_{x,n}) = 0$ represents either of the dispersion equations (8.116) or (8.124), then at a modal degeneracy of rank 2,

$$H(\omega, \varepsilon_1, \varepsilon_2, a, \lambda_{x,n}) = 0,$$
$$\frac{\partial}{\partial\lambda_x} H(\omega, \varepsilon_1, \varepsilon_2, a, \lambda_{x,n}) = 0,$$

indicating a second-order root of $H$ at frequency $\omega$.

The form of the Green's function (8.24) is more readily utilized for the grounded dielectric waveguide, leading to

$$g^{H,E}(x,z,x',z') = \int_\nu u_{z,\nu}(z,\nu)\overline{u}_{z,\nu}(z',\nu)g_x^{H,E}(x,x',-\lambda_z(\nu))\,d\nu, \quad (8.155)$$

$$g^H(x,z,x',z') = \frac{1}{2\pi}\int_{-\infty}^{\infty} d\nu\, e^{i\nu\,(z-z')}$$

$$\cdot\begin{cases} \frac{\mu_2}{2\sqrt{\gamma_1}}\frac{\mu_1}{\mu_2}\frac{e^{-i\sqrt{\gamma_2}x'}}{\cos\sqrt{\gamma_1}a}\frac{(Z^H-N^H)}{Z^H}\sin\sqrt{\gamma_1}\,(x+a)\,, & -a<x<0, \\ \frac{\mu_2}{2i\sqrt{\gamma_2}}\left[e^{-i\sqrt{\gamma_2}|x-x'|}+\frac{N^H}{Z^H}e^{-i\sqrt{\gamma_2}(x+x')}\right], & 0<x, \end{cases}$$

$$g^E(x,z,x',z') = \frac{1}{2\pi}\int_{-\infty}^{\infty} d\nu\, e^{i\nu\,(z-z')}$$

$$\cdot\begin{cases} -\frac{\varepsilon_2}{2\sqrt{\gamma_1}}\frac{\varepsilon_1}{\varepsilon_2}\frac{e^{-i\sqrt{\gamma_2}x'}}{\sin\sqrt{\gamma_1}a}\frac{(Z^E-N^E)}{Z^E}\cos\sqrt{\gamma_1}\,(x+a)\,, & -a<x<0, \\ \frac{\varepsilon_2}{2i\sqrt{\gamma_2}}\left[e^{-i\sqrt{\gamma_2}|x-x'|}+\frac{N^E}{Z^E}e^{-i\sqrt{\gamma_2}(x+x')}\right], & 0<x. \end{cases}$$

Because $\lambda_z = \nu^2$, we have $\gamma_j = k_j^2 - \nu^2$ in the above expressions. The first terms for $x > 0$ can be analytically evaluated to obtain a Hankel function (see (8.78)). Dyadic Green's functions for grounded dielectric layers are considered in Section 8.3.

In comparing the two forms of the Green's function we see that (8.154) is a modal representation, which is physically significant, yet it can be burdensome to compute because the eigenvalues $\lambda_{x,n}$ must be determined numerically before the summation term can be evaluated. A convenient aspect of this form, though, is that sufficiently close to the dielectric surface the surface-wave term should dominate the response. Often only one or a few discrete modes are above cutoff; in this case once the corresponding eigenvalues $\lambda_{x,n}$ are determined, the associated terms in the summation in (8.154) provide a simple approximation to the total field. On the other hand, (8.155) does not require knowledge of the various discrete $\lambda_{x,n}$-values, although the integral, which is clearly interpreted as an inverse Fourier transform integral,[15] is of the *Sommerfeld form* and can be difficult to evaluate. However, this representation is usually the preferred form and is amenable to asymptotic analysis. Moreover, this form is clearly valid in the event of a lossy medium. In this case poles of the Green's function may be of the order $m > 1$, corresponding to eigenvalues of multiplicity $m$. A

[15] In all of the $z$-invariant waveguide examples investigated here (parallel-plate, impedance-boundary, and dielectric-slab), the two-dimensional Green's function, plane-wave spectral form $(1/2\pi)\int_{-\infty}^{\infty} d\nu\, e^{i\nu\,(z-z')}\,\{\cdot\}$ could also have been obtained by taking a Fourier transform of the governing two-dimensional wave equation and then solving the resulting $x$-dependent problem. Although this is the usual method of analysis, here we have purposely avoided direct Fourier methods and instead concentrate on the interpretation as a continuous summation of improper eigenfunctions.

two-dimensional spectral form for the Green's function is provided at the end of this section.

### Field Caused by a Source in a Grounded Dielectric Waveguide

The solution of (8.110),

$$-\left(\frac{\partial^2}{\partial z^2}+s\frac{\partial}{\partial x}s^{-1}\frac{\partial}{\partial x}+k^2\left(x\right)\right)\Lambda(x,z)=f\left(x,z\right), \tag{8.156}$$

is, from (8.12),

$$\Lambda(x,z)=\int_S g^{H,E}(x,z,x',z')\; f(x',z')s^{-1}(x')\; dS'. \tag{8.157}$$

For example, if the source consists of a single component $J_{ey}(x,z)$, then only $\mathrm{TE}^z$-modes are excited, $\Lambda=E_y$ and $-f=i\omega\mu J_{ey}$, so that

$$E_y(x,z)=-i\omega\int_S g^H(x,z,x',z')\; J_{ey}(x',z')\; dS'. \tag{8.158}$$

### Completeness Relations and Associated Spectral Expansions

The completeness relation in the horizontal direction is

$$\begin{aligned}\delta(z-z')&=\frac{-1}{2\pi i}\oint_{C_z} g_z(z,z',\lambda_z)\,d\lambda_z=\int_\nu u_{z,\nu}(z,\nu)\overline{u}_{z,\nu}(z',\nu)\,d\nu\\&=\frac{1}{2\pi}\int_{-\infty}^{\infty}e^{i\nu\,(z-z')}d\nu.\end{aligned}$$

One method to form the completeness relation in the $x$-coordinate and to identify the discrete (proper) and continuous (improper) eigenfunctions is via integration of the Green's function as in (5.92),

$$\begin{aligned}\frac{\delta(x-x')}{w_x(x')}&=\frac{-1}{2\pi i}\oint_{C_x} g(x,x',\lambda_x)\,d\lambda_x\\&=\sum_n u_{x,n}(x)\overline{u}_{x,n}(x')+\int_\upsilon u_{x,\upsilon}(x,\upsilon)\overline{u}_{x,\upsilon}(x',\upsilon)\,d\upsilon,\end{aligned} \tag{8.159}$$

where $\upsilon=\sqrt{\gamma_2}$ and $w_x(x)=s(x)^{-1}$.

For example, consider the Green's function (8.139), which involves the terms $\sqrt{\gamma_2}$ and $\sqrt{\gamma_1}$. It is straightforward to see that the Green's function is even in the term $\sqrt{\gamma_1}$, so that the branch point at $\lambda_x=-k_1^2$ due to $\sqrt{\gamma_1}$ is removable and not of further interest. However, the Green's function is not even in $\sqrt{\gamma_2}$, and therefore $\sqrt{\gamma_2}$ induces a branch point at $\lambda_x=-k_2^2$. The condition $\operatorname{Im}\sqrt{\lambda_x}<0$ defines a branch cut in the complex $\lambda_x$-plane along the real axis $[-k_2^2,\infty)$, which will lead to the continuous spectral

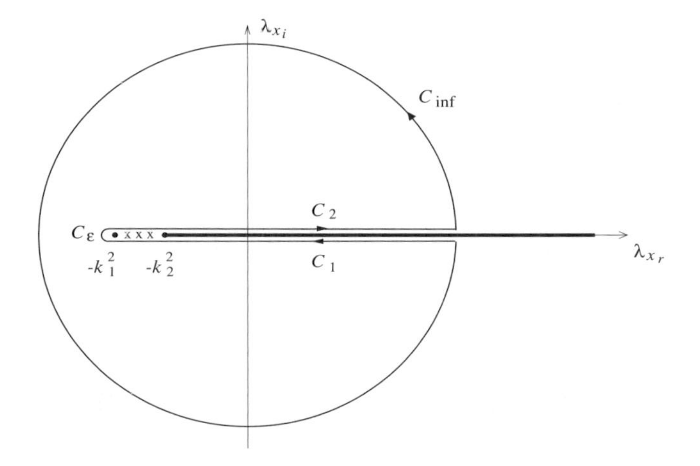


Figure 8.7: Complex plane analysis for evaluation of (8.160) showing the branch point at $\lambda_x = -k_2^2$ (denoted by "•") and the associated branch cut along the real $\lambda_x$-axis. Poles are denoted by "**x**" (three shown), and the branch point at $\lambda_x = -k_1^2$ is removable.

integral in (8.159) and to the continuous spectrum as $\lambda_x \in [-k_2^2, \infty)$. The Green's function (8.139) also has pole singularities where $Z_1^{E,H} = 0$, which, as previously mentioned, correspond to the discrete eigenvalues $\lambda_{x,n}$ from (8.116) and (8.124), with $-k_1^2 < \lambda_{x,n} < -k_2^2$.

Following the method detailed in Section 8.1.6 we consider the integration path in Figure 8.7, leading to

$$\oint_{C_\infty + C_1 + C_2 + C_\varepsilon} g_x(x, x', \lambda_x)\, d\lambda_x = 2\pi i \sum_n \text{Res}\,\{g_x(x, x', \lambda_{x,n})\}\,, \quad (8.160)$$

where $\text{Res}\{g_x(x, x', \lambda_{x,n})\}$ is the residue at the $n$th pole, $\lambda_{x,n}$. From (5.92) we also have

$$\int_{C_\infty} g_x(x, x', \lambda_x)\, d\lambda_x = -2\pi i\, \frac{\delta(x - x')}{w_x(x)}, \quad (8.161)$$

and, as in previous analyses, the contribution from $C_\varepsilon$ vanishes as $\varepsilon \to 0$. The contributions from $C_1$ and $C_2$ are obtained as in Section 5.2 (see (5.59)). As $\varepsilon \to 0$ on $C_1$, $\lambda_x = re^{i4\pi^-}$, and on $C_2$, $\lambda_x = re^{i2\pi^+}$, leading to

$$\lim_{\varepsilon\to 0}\int_{C_1+C_2} g_x(x,x',\lambda_x)\,d\lambda_x$$
$$= \int_{\infty}^{-k_2^2} g_x(x,x',\lambda_x)|_{\lambda_x=re^{i4\pi^-}}\,d\lambda_x + \int_{-k_2^2}^{\infty} g_x(x,x',\lambda_x)|_{\lambda_x=re^{i2\pi^+}}\,d\lambda_x$$
$$= \int_{-k_2^2}^{\infty}\left(g_x(x,x',\lambda_x)|_{\lambda_x=re^{i2\pi^+}} - g_x(x,x',\lambda_x)|_{\lambda_x=re^{i4\pi^-}}\right)d\lambda_x$$
$$= 2i\,\mathrm{Im}\int_{-k_2^2}^{\infty}\left(g_x(x,x',\lambda_x)|_{\lambda_x=re^{i2\pi^+}}\right)d\lambda_x,$$

where the last equality is valid since $\lambda_x = re^{i4\pi^-}$ is complex conjugate to $\lambda_x = re^{i2\pi^+}$, and $g_x(x,x',\overline{\lambda_x}) = \overline{g_x}(x,x',\lambda_x)$, again, assuming a lossless medium. With $\upsilon = \sqrt{\gamma_2}$, from (8.160) we therefore have[16]

$$\frac{\delta(x-x')}{w_x(x)} = -\frac{2}{\pi}\,\mathrm{Im}\int_0^{\infty}\left\{\upsilon g_x(x,x',\upsilon^2-k_2^2)\right\}d\upsilon - \sum_n \mathrm{Res}\left\{g_x(x,x',\lambda_{x,n})\right\}.$$

Comparing with (8.159) we see that the integral will provide the properly normalized improper eigenfunctions $u_x(x,\upsilon)$, consistent with (8.121) and (8.122), or (8.126) and (8.127). The negative of the residue terms will provide the properly normalized discrete eigenfunctions $u_{x,n}(x)$, consistent with (8.117) and (8.119), or (8.123) and (8.125). The evaluation for both the discrete and continuous terms is straightforward.

To determine a spectral expansion (spectral transform pair) for an arbitrary function $f(x,z) \in \mathbf{L}^2_w(\Omega)$, where $\Omega = \{(x,z) : x \in (-a,\infty),\ z \in (-\infty,\infty)\}$, let $u_{n,\nu}(x,z,\nu)$ be the eigenfunctions (8.128) and $u_{\upsilon,\nu}(x,z,\upsilon,\nu)$ be the eigenfunctions (8.129). The relevant eigenvalue equations are

$$-\left(\frac{\partial^2}{\partial z^2} + s\frac{\partial}{\partial x}s^{-1}\frac{\partial}{\partial x} + k^2(x)\right)u_{n,\nu}(x,z,\nu) = \lambda_{n,\upsilon}u_{n,\nu}(x,z,\nu),$$
$$-\left(\frac{\partial^2}{\partial z^2} + s\frac{\partial}{\partial x}s^{-1}\frac{\partial}{\partial x} + k^2(x)\right)u_{\upsilon,\nu}(x,z,\upsilon,\nu) = \lambda_{\upsilon,\upsilon}u_{\upsilon,\nu}(x,z,\upsilon,\nu),$$

where

$$\lambda_{n,\nu} = \lambda_{x,n} + \nu^2, \qquad \nu \in (-\infty,\infty),$$

with $\lambda_{x,n}$ a solution of (8.116) or (8.124) as appropriate, and where

$$\lambda_{\upsilon,\nu} = \upsilon^2 + \nu^2 - k_2^2, \qquad \upsilon \in (0,\infty), \quad \nu \in (-\infty,\infty).$$

[16]Note that because $\lambda_x \in \mathbf{R}$, the substitution $\lambda_x = \varsigma^2$ and subsequent integration over $\varsigma$ will convert the integral into the familiar integration along the hyperbolic branch cuts in the $\varsigma$-plane.

From (8.28) we have

$$f(x,z) = \sum_{n=1}^{N} \int_{-\infty}^{\infty} a_{x,n}^{z}(\nu) u_{n,\nu}(x,z,\nu)\, d\nu + \int_{0}^{\infty} \int_{-\infty}^{\infty} a(\upsilon,\nu) u_{\upsilon,\nu}(x,z,\upsilon,\nu)\, d\nu\, d\upsilon,$$

where $a_{x,n}^{z}(\nu) = \langle f, u_{n,\nu} \rangle$ and $a(\upsilon,\nu) = \langle f, u_{\upsilon,\nu} \rangle$, with

$$\langle f, g \rangle = \int_{-\infty}^{\infty} \int_{-a}^{\infty} f(x,z) \overline{g}(x,z)\, s^{-1}(x)\, dx\, dz.$$

### Spectral Solution and Spectral Green's Function

From the spectral expansion we have

$$\Lambda(x,z) = \sum_{n=1}^{N} \int_{-\infty}^{\infty} \langle \Lambda, u_{n,\nu} \rangle\, u_{n,\nu}(x,z,\nu)\, d\nu + \int_{0}^{\infty} \int_{-\infty}^{\infty} \langle \Lambda, u_{\upsilon,\nu} \rangle\, u_{\upsilon,\nu}(x,z,\upsilon,\nu)\, d\nu\, d\upsilon.$$

Following the usual eigenfunction expansion procedure, and using a two-dimensional version of the orthogonality relations (8.18), we obtain the solution of (8.110) in spectral form as

$$\Lambda(x,z) = \sum_{n=1}^{N} \int_{-\infty}^{\infty} \frac{\langle f, u_{n,\nu} \rangle}{\lambda_{x,n} + \nu^2} u_{n,\nu}(x,z,\nu)\, d\nu + \int_{0}^{\infty} \int_{-\infty}^{\infty} \frac{\langle f, u_{\upsilon,\nu} \rangle}{\upsilon^2 + \nu^2 - k_2^2} u_{\upsilon,\nu}(x,z,\upsilon,\nu)\, d\nu\, d\upsilon, \tag{8.162}$$

which is a special case of (8.30). Alternatively, we can consider the solution to be given by (8.157), with the Green's function obtained as a special case of (8.31),

$$g(x,z,x',z') = \sum_{n=1}^{N} \int_{-\infty}^{\infty} \frac{\overline{u}_{n,\nu}(x',z',\nu)}{\lambda_{x,n} + \nu^2} u_{n,\nu}(x,z,\nu)\, d\nu + \int_{0}^{\infty} \int_{-\infty}^{\infty} \frac{\overline{u}_{\upsilon,\nu}(x',z',\upsilon,\nu)}{\upsilon^2 + \nu^2 - k_2^2} u_{\upsilon,\nu}(x,z,\upsilon,\nu)\, d\nu\, d\upsilon, \tag{8.163}$$

which can be shown to be equivalent to (8.154) using complex plane analysis.

As mentioned previously, (8.162) and (8.163) are purely spectral forms and most directly utilize the eigenfunctions of the governing operator, whereas (8.154) and (8.155) arise from partial eigenfunction expansions and are more computationally convenient.

### 8.1.8 Comments on General Multilayered Media Problems, Completeness, and Associated Eigenfunctions

In Section 8.1.1 we assumed a planar multilayered medium, as depicted in Figure 8.1, and in subsequent sections several special cases were considered. Here we provide some brief comments concerning the validity of spectral methods for the class of structures shown in Figure 8.1.

If the medium is lossless and the four perfectly conducting bounding walls depicted in the figure are present, the governing differential operator $L : \mathbf{L}^2(\Omega) \to \mathbf{L}^2(\Omega)$,

$$L \equiv -\left(\frac{\partial^2}{\partial z^2} + s\frac{\partial}{\partial x}s^{-1}\frac{\partial}{\partial x} + k^2(x)\right),$$
$$D_L \equiv \left\{\Lambda : \Lambda, L\Lambda \in \mathbf{L}^2(\Omega), B(\Lambda) = 0\right\}, \tag{8.164}$$

subject to physically appropriate boundary conditions $B(\Lambda)$ is self-adjoint with compact inverse.[17] By Theorem 4.41 and the discussion in Section 4.3.3, the discrete eigenfunctions form a basis for $\mathbf{L}^2(\Omega)$, where $\Omega$ corresponds to the geometry of the structure.

If the medium is lossy and homogeneous, then we consider the operator $L : \mathbf{L}^2(\Omega) \to \mathbf{L}^2(\Omega)$

$$L \equiv -\left(\frac{\partial^2}{\partial z^2} + \frac{\partial^2}{\partial x^2}\right),$$
$$D_L \equiv \left\{\Lambda : \Lambda, L\Lambda \in \mathbf{L}^2(\Omega), B(\Lambda) = 0\right\}, \tag{8.165}$$

and draw the same conclusions. However, if the medium is lossy and inhomogeneous, then we must consider the operator (8.164), which in this case is not self-adjoint because $k \in \mathbf{C}$. By Theorem 5.3, expansions utilizing eigenfunctions and adjoint eigenfunctions will be valid at all points in parameter space (frequency, permittivity, etc.) except at the set of parameters where eigenvalues have multiplicity greater than 1, as discussed below. It may also be possible to establish completeness of the eigenfunctions by other methods; see, e.g., the footnote on p. 435.

Now for generality assume that the medium is inhomogeneous, and consider the special case where the two horizontal walls are retained and the two vertical sidewalls ($z$-coordinate) are removed to infinity, where we impose the condition that functions should vanish at infinity.[18] Physically, this corresponds to enforcing outgoing wave behavior in the presence of (perhaps small) material loss, and in this case the governing

[17] For example, $B(\Lambda) = 0$ may indicate that $\Lambda = 0$ or $\partial\Lambda / \partial n = 0$ at the bounding walls.

[18] Alternatively, we can consider the medium to be lossless and consider a set of functions obeying an outgoing radiation condition in the $z$-coordinate. In this case one can define a formally self-adjoint (symmetric) operator outside an $\mathbf{L}^2$ space, although

operator is

$$L \equiv -\left(\frac{\partial^2}{\partial z^2} + s\frac{\partial}{\partial x}s^{-1}\frac{\partial}{\partial x} + k^2(x)\right),$$

$$D_L \equiv \left\{\Lambda : \Lambda, L\Lambda \in \mathbf{L}^2(\Omega), B_{x=x_b,x_t}(\Lambda) = 0, \lim_{z\to\pm\infty} \Lambda(x,z) = 0\right\}.$$

The operator is nonself-adjoint because $k$ is complex-valued.[19]

By Theorem 5.3, expansions utilizing eigenfunctions and adjoint eigenfunctions will be valid at all points in parameter space except where eigenvalues have multiplicity greater than 1. Even in the event of modal degeneracies, the eigenfunctions may still form a basis for $H = \mathbf{L}^2(\Omega)$ (biorthonormal with the set of adjoint eigenfunctions), however at these points Theorem 5.3 is not applicable. In this case, if the ordinary eigenfunctions are linearly independent (corresponding to different physical field configurations), then the ordinary eigenfunctions will form a complete set in the appropriate $\mathbf{L}^2$ space.[20] Physically, this is a trivial degeneracy since the modes do not couple, which is associated with the linear independence of the eigenfunctions (although small structural perturbations may result in modal coupling).

Nontrivial modal degeneracies may occur that relate to a degeneracy in the eigenvalue (propagation constant) and field configuration that is associated with mode coupling. In this case eigenvalues of higher multiplicity may not have corresponding linearly independent ordinary eigenfunctions, and associated (rank $> 1$) eigenfunctions may be implicated. This is analogous to the consideration of the Jordan canonical form discussed in Chapter 4. In this case the operator cannot be diagonalized, which is reflected in the fact that an eigenbasis of ordinary eigenfunctions does not exist. The associated eigenfunctions correspond to poles of the characteristic Green's function having order $m$, $m > 1$, as shown in Example 5.14.

In this case, characteristic propagation behavior $z^{m-1}e^{-i\gamma z}$ may be obtained, arising from the residue formula for an $m$th-order pole, (5.115). Some general discussion of the role of associated eigenfunctions in electromagnetic waveguiding problems can be found in [45].

---

the operator will not be self-adjoint due to the imposition of the radiation condition (the adjoint operator acts on functions obeying an incoming radiation condition, and so $D_L \neq D_{L^*}$).

[19]However, if the medium is homogeneous we can consider (8.165) and work with a self-adjoint operator, which is exactly the case considered in Section 8.1.2 (see the discussion on p. 440)

[20]This is analogous to an $n \times n$ matrix having $n$ linearly independent eigenvectors (thus forming an eigenbasis), even if some eigenvalues have multiplicity greater than unity, as described in Section 4.3.1.

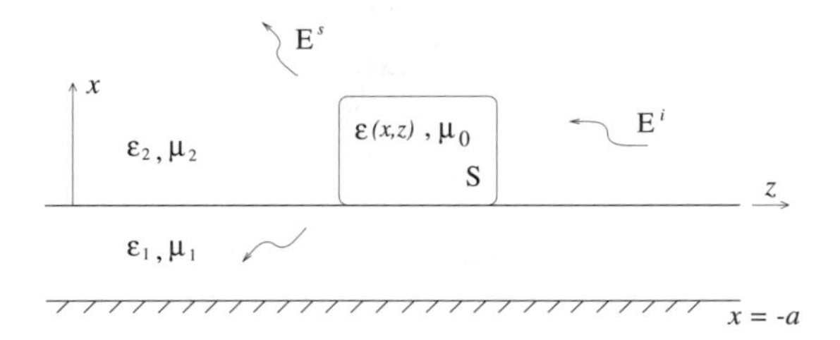


Figure 8.8: Scattering from an inhomogeneous dielectric cylinder on a grounded dielectric layer.

# 8.2 Two-Dimensional Scattering Problems in Planar Media

In this section we consider scattering from two-dimensional objects embedded in a planar medium using the Green's functions developed in the previous sections. While no physical structure is actually two-dimensional, this serves as a model of cylindrical objects with cross-sections that are $y$-invariant over sufficiently long distances. For simplicity, $\mathrm{TE}^z$ scattering will be considered.

## 8.2.1 TE Scattering from Inhomogeneous Dielectric Cylinders in Layered Media

Figure 8.8 depicts scattering from an inhomogeneous cylinder residing on top of a grounded dielectric layer. In electronics applications, the cylinder could represent, for instance, an integrated dielectric waveguide. Alternatively, as a geophysical application, the cylinder could model plastic piping in a layered-earth model, where the bottom region has sufficiently large conductivity.

Most importantly, the presence or absence of various material layers, or the ground plane, is incidental to the scattering formulation. If a different "background environment" for the cylinder is required, the scattering formulation and resulting integral equations remain valid, only the Green's functions need to be changed. While the derivation of the Green's function becomes more cumbersome as the number of layers increases, the methods of the previous sections are applicable and no new mathematical difficulties are encountered. For example, the problem depicted in Figure 8.9 of a plastic pipe embedded in a layer of sand over a rock-bottomed stream bed is easily analyzed by deriving a Green's function for a four-layer (three-interface) dielectric medium. This Green's function can be obtained in a straightforward manner using the methods of Section 8.1.7. Furthermore, the three-dimensional dyadic Green's function for this environment is formulated in Section 8.3, from which the two-dimensional Green's functions

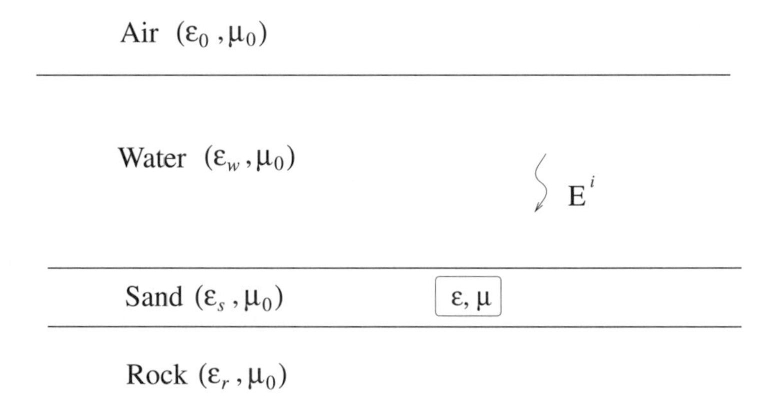


Figure 8.9: Scattering from an inhomogeneous dielectric cylinder embedded in the sand layer of a stream bed.

may be easily obtained.

To analyze the problem shown in Figure 8.8, we use the volume equivalence principle and the domain integral equations derived in Section 1.4.1. We assume the excitation is $y$-invariant, $\mathbf{E}^i = \widehat{\mathbf{y}} E_y^i(x,z)$, such that only TE$^z$-fields are excited. For instance, the excitation could be an above-cutoff surface-wave mode of the background structure,

$$E_y^i(x,z) = e^{-i\sqrt{k_2^2+\lambda_{x,n}}x} e^{-i\sqrt{-\lambda_{x,n}}z},$$

where $-k_1^2 < \lambda_{x,n} < -k_2^2$, with $\lambda_{x,n}$ a solution of the TE$^z$ dispersion equation (8.116). Alternatively, the excitation may be an incident plane wave

$$E_y^i(x,z) = e^{-i\mathbf{k}\cdot\rho},$$

where $\mathbf{k} = \widehat{\mathbf{k}} k_2$, with $\widehat{\mathbf{k}}$ giving the direction of propagation and $\rho = \widehat{\mathbf{x}}x + \widehat{\mathbf{z}}z$.

As the problem is two-dimensional, the volume integral in the domain integral equation (1.109) is reduced to a surface integral,

$$\mathbf{E}(\rho) - \int_S \underline{\mathbf{G}}_{e,ea}(\rho,\rho') \cdot \mathbf{E}(\rho')\, dS' = \mathbf{E}^i(\rho). \tag{8.166}$$

The Green's function is, from (1.108),

$$\underline{\mathbf{G}}_{e,ea}(\rho,\rho') \equiv i\omega \underline{\mathbf{G}}_{ee}^{\wedge}(\rho,\rho') \left[\varepsilon(\rho') - \varepsilon_2\right],$$

where $\underline{\mathbf{G}}_{ee}^{\wedge}$ is the dyadic Green's function that directly relates fields and currents, i.e., from (1.88) $\underline{\mathbf{G}}_{ee}^{\wedge}$ is such that the relationship between fields and currents in the layered medium is

$$\mathbf{E}(\rho) = \int_S \underline{\mathbf{G}}_{ee}^{\wedge}(\rho,\rho') \cdot \mathbf{J}_e(\rho')\, dS'. \tag{8.167}$$

Because in this TE$^z$ case $\mathbf{J}_e = \widehat{\mathbf{y}} J_y$ and $\mathbf{E} = \widehat{\mathbf{y}} E_y$, (8.167) becomes simply

$$E_y(\rho) = \int_S g_{yy}(\rho, \rho') \, J_y(\rho') \; dS',$$

where $g_{yy} = \widehat{\mathbf{y}} \cdot \underline{\mathbf{G}}_{ee}^{\wedge} \cdot \widehat{\mathbf{y}}$. From (8.158) we have the relationship between $E_y$ and $J_y$ as

$$E_y(\rho) = -i\omega \int_S g^H(\rho, \rho') \; J_y(\rho') \; dS' \tag{8.168}$$

and therefore the desired Green's function that directly relates $E_y$ and $J_y$ is

$$g_{yy}(\rho, \rho') = -i\omega g^H(\rho, \rho'),$$

where $g^H$ is any of the TE$^z$ Green's functions (8.154), (8.155), or (8.163). The domain integral equation (8.166) becomes

$$E_y(\rho) - \int_S G_{e,ea}^{yy}(\rho, \rho') E_y(\rho') \; dS' = E_y^i(\rho),$$

where

$$G_{e,ea}^{yy}(\rho, \rho') \equiv \omega^2 g^H(\rho, \rho') \left[\varepsilon(\rho') - \varepsilon_2\right].$$

The integral equation to be solved is therefore

$$E_y(\rho) - \omega^2 \int_S g^H(\rho, \rho') \left[\varepsilon(\rho') - \varepsilon_2\right] E_y(\rho') \; dS' = E_y^i(\rho) \tag{8.169}$$

for $\rho \in S$.

In (8.169), $g^H$ is the Green's function accounting for the background environment, i.e., the grounded planar layer. Any of the forms (8.154), (8.155), or (8.163) are applicable, although (8.155) is perhaps the easiest for computation. Therefore, from (8.155) for $x \geq 0$ we have

$$g^H(\rho, \rho') = \frac{\mu_0}{2\pi} \int_{-\infty}^{\infty} e^{i\nu(z-z')} \frac{1}{2i\sqrt{\gamma_2}} \left[e^{-i\sqrt{\gamma_2}|x-x'|} + R^H e^{-i\sqrt{\gamma_2}(x+x')}\right] d\nu, \tag{8.170}$$

where

$$R^H(\nu) \equiv \frac{N^H}{Z^H} = \frac{\sqrt{\gamma_2} + i\sqrt{\gamma_1}\cot\sqrt{\gamma_1}a}{\sqrt{\gamma_2} - i\sqrt{\gamma_1}\cot\sqrt{\gamma_1}a} \tag{8.171}$$

with $\gamma_j = k_j^2 - \nu^2$. If desired, from (8.78) the Green's function can be written in scattering superposition form as

$$g^H(\rho, \rho') = \frac{\mu_0}{4i} H_0^{(2)}(k|\rho - \rho'|) + \frac{\mu_0}{2\pi} \int_{-\infty}^{\infty} e^{i\nu(z-z')} \frac{1}{2i\sqrt{\gamma_2}} R^H e^{-i\sqrt{\gamma_2}(x+x')} d\nu. \tag{8.172}$$

The advantage of the scattering superposition form (8.172) is that the second term accounts for the layered medium, while the first term is the

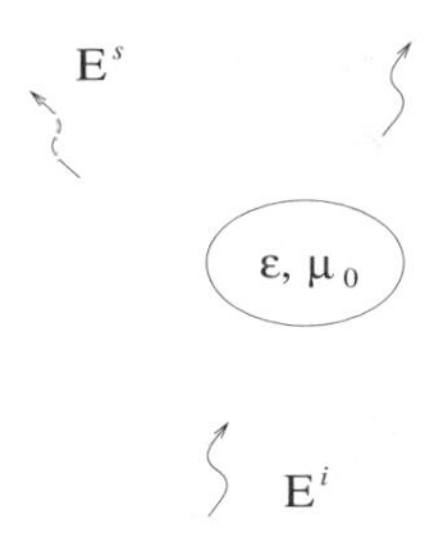


Figure 8.10: Scattering from a dielectric cylinder in free space.

response maintained by a source in free space. Therefore, for scattering from an inhomogeneous dielectric cylinder in free space, as depicted in Figure 8.10, the desired integral equation is

$$E_y(\rho) - \frac{\omega^2 \mu_0}{4i} \int_S H_0^{(2)}(k|\rho - \rho'|) [\varepsilon(\rho') - \varepsilon_2] E_y(\rho') \, dS' = E_y^i(\rho) \quad (8.173)$$

for $\rho \in S$. With mild restrictions on the surface $S$ and permittivity $\varepsilon$, the integral operator in (8.173) is compact in $\mathbf{L}^2(S)$, making (8.173) a second-kind Fredholm integral equation. The existence of a solution is governed by the Fredholm alternative, Theorem 3.44, and by Theorem 3.49 one can obtain a norm-convergent method-of-moments solution that converges to the exact solution.[21]

As described above, if the dielectric cylinder is located in a different planar environment, the appropriate Green's function is substituted for (8.170), or, equivalently, the second term in (8.172) would be modified accordingly. For example, if the cylinder is located in a parallel-plate waveguide as considered in Section 8.1.2, then for $g^H$ we can use (8.52) or (8.56).[22] Using the form (8.56), we have

$$\mu_0 g^H(\rho, \rho') = \frac{\mu_0}{2\pi} \int_{-\infty}^{\infty} \frac{\sin\sqrt{\lambda_x}(a - x_>) \sin\sqrt{\lambda_x} x_<}{\sqrt{\lambda_x} \sin\sqrt{\lambda_x} a} e^{i\nu(z-z')} d\nu \quad (8.174)$$

with $\sqrt{\lambda_x} = \sqrt{k^2 - \nu^2}$, leading to the integral equation

$$E_y(\rho) - \frac{\omega^2 \mu_0}{2\pi} \int_S \int_{-\infty}^{\infty} \left\{ \frac{e^{i\nu(z-z')} \sin\sqrt{\lambda_x}(a - x_>) \sin\sqrt{\lambda_x} x_<}{\sqrt{\lambda_x} \sin\sqrt{\lambda_x} a} \right. \quad (8.175)$$
$$\left. [\varepsilon(\rho') - \varepsilon_2] E_y(\rho') \right\} d\nu \, dS' = E_y^i(\rho)$$

for all $\rho \in S$.

[21] If the second term in (8.172) can be shown to generate a square-integrable kernel, then a compact operator results and the same comments apply to (8.169).

[22] Note that in this case $g^H$ needs to be multiplied by a factor of $\mu_0$ due to the source-field relationship (8.65) instead of (8.168).

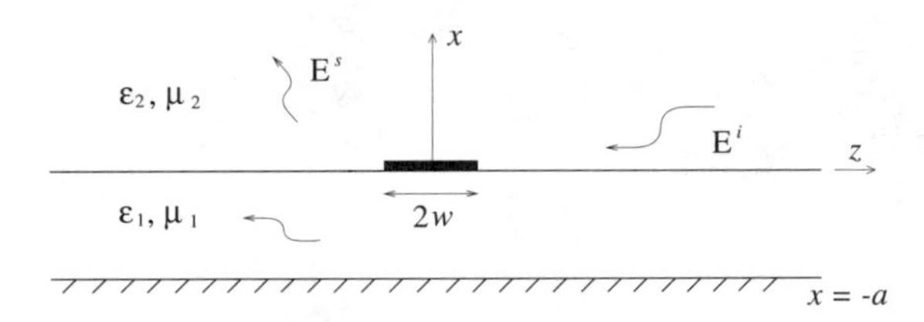


Figure 8.11: Scattering from a perfectly conducting thin strip (microstrip line) on a grounded dielectric layer.

Numerical results for scattering from dielectric cylinders in planar layers are available in the literature; see, e.g., [28]–[31]. $\mathrm{TM}^z$ scattering can be analyzed in a similar manner, although the formulation is more complicated and will be omitted here; see, for example, [1, Sec. 2.6].

### 8.2.2 TE Scattering from Thin Conducting Strips in Layered Media

As another example, consider a $y$-invariant excitation $\mathbf{E}^i = \widehat{\mathbf{y}} E_y^i(x,z)$ incident on a perfectly conducting, infinitely thin strip on a grounded dielectric layer (microstrip transmission line), as shown in Figure 8.11. With the source-field relationship

$$E_y(\rho) = -i\omega \int_S g^H(\rho,\rho')\, J_y(\rho')\, dS', \tag{8.176}$$

where $g^H$ is any of the $\mathrm{TE}^z$ Green's functions (8.154), (8.155), or (8.163), the electric field integral equation (1.115) becomes

$$i\omega \int_{-w}^{w} g^H(z,z')\big|_{x=x'=0}\, J_y(z')\, dz' = E_y^i(0,z) \tag{8.177}$$

for $z \in (-w,w)$. Because the integral equation is enforced at $x = x' = 0$, the Green's function is simply

$$g^H(z,z') = \frac{\mu_0}{2\pi} \int_{-\infty}^{\infty} \frac{e^{i\nu(z-z')}}{2i\sqrt{k_2^2-\nu^2}} \left[1 + R^H(\nu)\right] d\nu \tag{8.178}$$

with $R^H$ given by (8.171), so that the integral equation to be solved is

$$\int_{-w}^{w} \int_{-\infty}^{\infty} \frac{e^{i\nu(z-z')}}{2i\sqrt{k_2^2-\nu^2}} \left[1 + R^H(\nu)\right] J_y(z')\, d\nu\, dz' = \frac{2\pi}{i\omega\mu_0} E_y^i(0,z) \tag{8.179}$$

for $z \in (-w,w)$. Alternatively, one can use the Green's function written in scattering superposition form,

$$g^H(z,z') = \frac{\mu_0}{4i} H_0^{(2)}(k|z-z'|) + \frac{\mu_0}{2\pi} \int_{-\infty}^{\infty} \frac{e^{i\nu(z-z')}}{2i\sqrt{k_2^2-\nu^2}} R^H(\nu)\, d\nu. \tag{8.180}$$

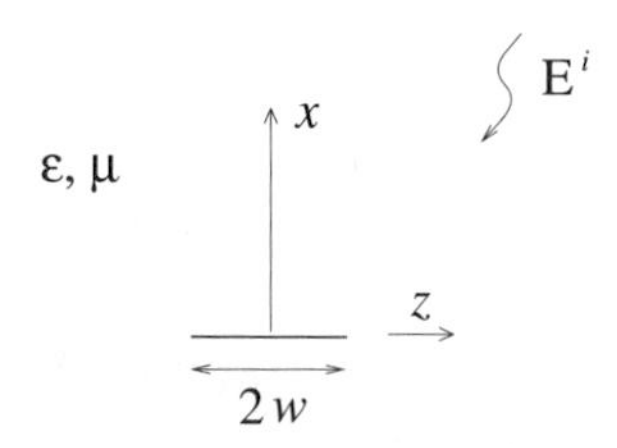


Figure 8.12: Perfectly conducting strip in a homogeneous medium, in the presence of an incident field $\mathbf{E}^i$.

As before, for a strip embedded in a different layered-medium environment, the integral equation (8.177) holds, with the Green's function appropriately modified. For instance, for a strip in a homogeneous space characterized by $\mu$ and $\varepsilon$, the appropriate integral equation is

$$\int_{-w}^{w} H_0^{(2)}(k\,|z-z'|)J_y(z')\,dz' = \frac{4}{\omega\mu}E_y^i\,(0,z)\,, \qquad z\in(-w,w)\,. \tag{8.181}$$

The above electric field integral equation is a first-kind integral equation with compact kernel in $\mathbf{L}^2\,(-w,w)$. For the electrically narrow-strip ($kw \ll 1$) problem the solution is discussed in [34] and in Example 8.1.

For arbitrary strips in a layered medium, the solution is detailed in [32] and references therein. The rigorous solution procedure detailed in this reference is related to the method of analytic regularization [33] and is based on (i) efficient treatment of the second term in (8.180), and (ii) utilization of Neumann's expansion for $H_0^{(2)}$,

$$\begin{aligned} H_0^{(2)}(k\,|z-z'|) &= \left[1 - i\frac{2}{\pi}\left(\ln\frac{k\,|z-z'|}{2} + \gamma\right)\right] J_0\left(k\,|z-z'|\right) \\ &\quad + i\frac{4}{\pi}\sum_{n=1}^{\infty}(-1)^n\,\frac{J_{2n}\left(k\,|z-z'|\right)}{n}, \end{aligned} \tag{8.182}$$

where $\gamma$ is Euler's constant ($\gamma \simeq 0.5772$).

**Example 8.1.**

We revisit the problem of a thin strip in free space considered in Example 6.5 for the electrostatic potential. As depicted in Figure 8.12, the strip has width $2w$ and is excited by an incident $TE^z$ source $\mathbf{E}^i$. We assume that the strip is narrow, $kw \ll 1$, such that we approximate the Hankel function from (8.182) (or more directly from the small-argument limit of the Bessel functions $J_0$ and $Y_0$ on p. 345) as

$$H_0^{(2)}(k|z-z'|) \simeq 1 - i\frac{2}{\pi}\left[\ln\left(\frac{k|z-z'|}{2}\right)+\gamma\right]$$
$$= 1 - i\frac{2}{\pi}\left(\gamma + \ln\left(\frac{k}{2}\right)\right) - i\frac{2}{\pi}\ln(|z-z'|),$$

where $\gamma = 0.5772$. The integral equation (8.181) becomes

$$-i\frac{\omega\mu}{2\pi}\int_{-w}^{w}\ln(|z-z'|)\,J_y(z')\,dz' + C\int_{-w}^{w}J_y(z')\,dz' = E_y^i(0,z) \qquad (8.183)$$

for $z \in (-w,w)$, where $C = -(i\omega\mu)/(2\pi)\{i\pi/2+\gamma+\ln(k/2)\}$. The solution [34] $J_y \in L^2(-w,w)$ is obtained by substituting the expansion

$$J_y(z) = \frac{1}{\sqrt{w^2-z^2}}\sum_{n=0}^{\infty}a_n\sqrt{\frac{\varepsilon_n}{\pi}}T_n(z/w)$$

into (8.183). Exploiting orthonormality (4.31), the spectral property (4.30), and relation (6.62), we find that

$$a_0 = \frac{I_0}{-i\frac{\omega\mu}{2}\sqrt{\pi}\left(\frac{i\pi}{2}+\gamma+\ln\left(\frac{kw}{4}\right)\right)}, \qquad (8.184)$$

$$a_n = \frac{n\sqrt{\frac{2}{\pi}}}{i\frac{\omega\mu}{2}}I_n, \qquad n \neq 0, \qquad (8.185)$$

where

$$I_n = \int_{-w}^{w}T_n(z/w)\frac{1}{\sqrt{w^2-z^2}}E_y^i(0,z)\,dz.$$

As a special case, if the incident wave is a plane wave

$$E_y^i(0,z) = E_0e^{ikz\sin\theta_i},$$

where $\theta_i$ is the incidence angle measured relative to the $x$-axis, then, from (6.62),

$$I_0 = \int_{-w}^{w}\frac{E_0e^{ikz\sin\theta_i}}{\sqrt{w^2-z^2}}dz = \pi E_0J_0(kw\sin\theta_i),$$
$$I_n = \int_{-w}^{w}\frac{T_n(z/w)}{\sqrt{w^2-z^2}}E_0e^{ikz\sin\theta_i}\,dz = \pi E_0i^nJ_n(kw\sin\theta_i)$$
$$\simeq \pi E_0i^n\frac{1}{n!}\left(\frac{kw\sin\theta_i}{2}\right)^n,$$

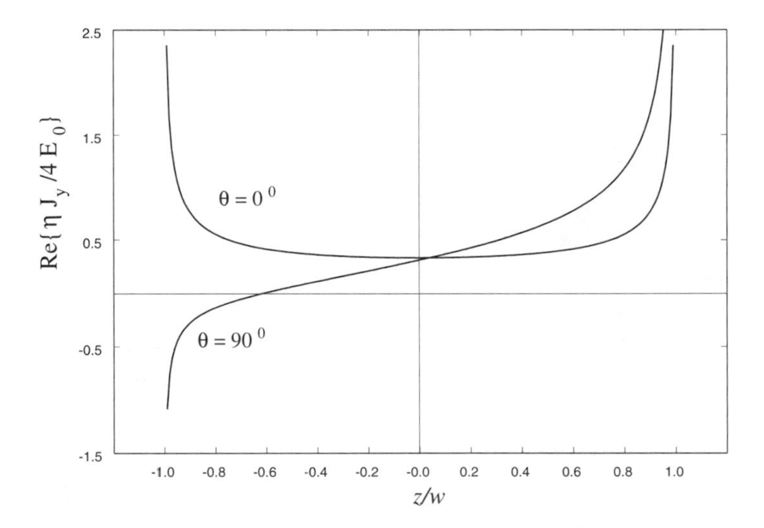


Figure 8.13: Normalized current on a perfectly conducting strip in free space due to an incident plane wave having amplitude $E_0$. The angle between the strip normal and the direction of the incident field is $\theta$.

leading to the strip current

$$J_y(z) = \frac{4E_0 i}{kw\eta\sqrt{1-(z/w)^2}} \left( \frac{J_0(kw\sin\theta_i)}{2\left(\frac{i\pi}{2}+\gamma+\ln\left(\frac{kw}{4}\right)\right)} + \sum_{n=1}^{\infty} -ni^n J_n(kw\sin\theta_i)\, T_n(z/w) \right),$$

where $\eta = \sqrt{\mu_0/\varepsilon_0}$.

It is easy to see that the summation is rapidly convergent, since with $kw \ll 1$,

$$J_n(kw\sin\theta_i) \sim \frac{(kw\sin\theta_i)^n}{n!}.$$

Furthermore, for normal incidence ($\theta_i = 0$) the summation term does not contribute to the current. The real part of the normalized current, $\mathrm{Re}(\eta J_y/4E_0)$, versus $z/w$ is shown in Figure 8.13 for $kw = 0.5$.

## 8.3 Three-Dimensional Planar Problems

In the previous sections we considered several planar structures invariant along the $y$-axis. We also took the source terms to be $y$-invariant, leading to a two-dimensional problem. In this section we consider planar structures with arbitrary sources, resulting in a three-dimensional problem.

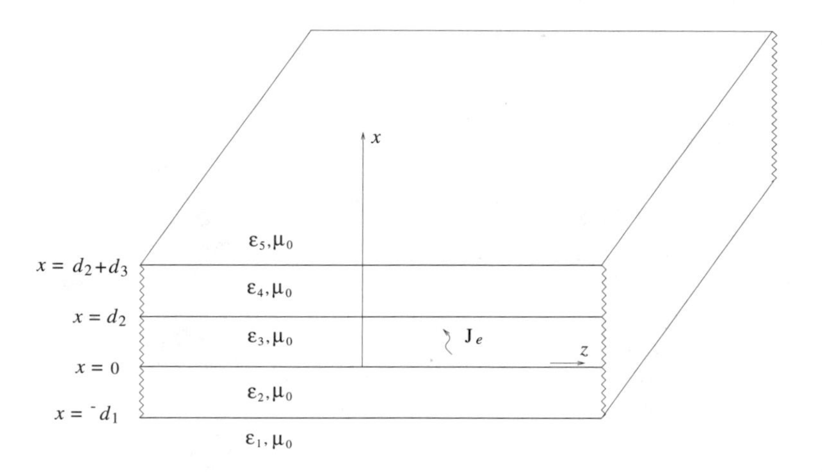


Figure 8.14: Multilayered planar medium with a source $\mathbf{J}_e$ in one of the layers.

### 8.3.1 General Analysis

Consider the planar structure depicted in Figure 8.14, consisting of a planarly inhomogeneous medium characterized by piecewise constant permittivity $\varepsilon$ and constant permeability $\mu_0$ and excited by an electric current source $\mathbf{J}_e$. As shown, the structure is vertically unbounded, although the limit $\varepsilon_{1,5} \to -i\infty$ can be used to implement a perfect electrical conductor in either outer region. The structure is assumed laterally infinite.

While there are many different methods for determining dyadic Green's functions for layered media (see, e.g., [35], [36]), here we will use Hertzian potential functions [37]. The relationship between fields and Hertzian potentials in each region is given by [38]

$$\begin{aligned} \mathbf{E}(\mathbf{r}) &= \left(k^2 + \nabla\nabla\cdot\right)\pi(\mathbf{r}), \\ \mathbf{H}(\mathbf{r}) &= i\omega\varepsilon\nabla\times\pi(\mathbf{r}), \end{aligned} \tag{8.186}$$

where $k^2 = \omega^2\mu_0\varepsilon$. The (electric) Hertzian potential $\pi(\mathbf{r})$ satisfies

$$\left(\nabla^2 + k^2\right)\pi(\mathbf{r}) = -\frac{\mathbf{J}_e(\mathbf{r})}{i\omega\varepsilon_j}, \tag{8.187}$$

where $\mathbf{J}_e$ is an electric-type current density having compact support in the $j$th region. In this section we will assume that the fields, sources, and potentials exist in the function space $\mathbf{L}^2(\Omega)^3$.

Obviously, comparing with the magnetic vector potential discussed in Section 1.3.2, the relationship between potentials in the time domain is

$$\mathbf{A}(\mathbf{r},t) = \varepsilon\mu\frac{\partial\pi(\mathbf{r},t)}{\partial t}$$

or simply

$$\mathbf{A}(\mathbf{r},\omega) = i\omega\varepsilon\mu\pi(\mathbf{r})$$

in the frequency domain. While in the time domain the Hertzian potential is well behaved compared to the magnetic vector potential (derivatives tending to make functions less smooth), in the frequency domain they exhibit similar behavior. Therefore, the fundamental theory and development of Hertzian potentials in the frequency domain closely follows the methods detailed in Section 1.3.2 and is not repeated here.[23]

From the form of (8.187) it would seem that a current $\widehat{\alpha}J_\alpha$ would maintain only a potential $\widehat{\alpha}\pi_\alpha$, such that (8.187) could be solved using the scalar methods developed in the previous sections. This is true for a homogeneous medium; however, for a layered medium the boundary conditions associated with the Hertzian potential cannot be sufficiently decoupled [37]. It turns out that a vertical component of current results in a vertical potential, although horizontal currents maintain both horizontal and vertical potentials. The resulting relationship between currents and potentials is

$$\pi(\mathbf{r}) = \int_\Omega \underline{\mathbf{G}}_\pi(\mathbf{r},\mathbf{r}') \cdot \frac{\mathbf{J}_e(\mathbf{r}')}{i\omega\varepsilon_j}\, d\Omega',$$

where $\underline{\mathbf{G}}_\pi$ is not, in general, diagonal.

Although a purely spectral representation of $\underline{\mathbf{G}}_\pi(\mathbf{r},\mathbf{r}')$ as an eigenfunction expansion is possible, representing a three-dimensional, vector generalization of the formulation in Section 8.1, it is easier to determine the Green's dyadic using other methods. As we will see, the resulting Green's dyadic components have the form of the partial eigenfunction expansions (8.23) and (8.24), generalized to three dimensions. The scalar theory developed in the previous sections is useful as an aid in interpreting the three-dimensional results.

To determine the Hertzian-potential Green's dyadic, the potential is decomposed via scattering superposition into a primary potential $\pi^p$ and a scattered potential $\pi^s$. We then have

$$\begin{aligned}\pi(\mathbf{r}) &= \pi^p(\mathbf{r}) + \pi^s(\mathbf{r})\\ &= \int_\Omega \left\{\underline{\mathbf{G}}_\pi^p(\mathbf{r},\mathbf{r}') + \underline{\mathbf{G}}_\pi^s(\mathbf{r},\mathbf{r}')\right\} \cdot \frac{\mathbf{J}_e(\mathbf{r}')}{i\omega\varepsilon_j}\, d\Omega',\end{aligned}$$

where the two potentials satisfy

$$\begin{aligned}\nabla^2\pi^p(\mathbf{r}) + k_j^2\pi^p(\mathbf{r}) &= -\frac{\mathbf{J}_e(\mathbf{r})}{i\omega\varepsilon_j},\\ \nabla^2\pi^s(\mathbf{r}) + k^2\pi^s(\mathbf{r}) &= \mathbf{0}.\end{aligned}$$

[23] Magnetic Hertzian potentials are also useful and are developed in a manner similar to the electric vector potential $\mathbf{F}$ considered in Section 1.3.2.

The decomposition $\underline{\mathbf{G}}_\pi(\mathbf{r},\mathbf{r}') = \underline{\mathbf{G}}_\pi^p(\mathbf{r},\mathbf{r}') + \underline{\mathbf{G}}_\pi^s(\mathbf{r},\mathbf{r}')$ allows for the separation of the source-point singularity associated with $\underline{\mathbf{G}}_\pi^p$ from the rest of the Green's dyadic. Furthermore, if $\|\underline{\mathbf{G}}^s\| \ll \|\underline{\mathbf{G}}^p\|$, then the planar layering outside the $j$th region has a small influence on the electromagnetic response of an object embedded in the $j$th region, and perturbation methods may be applicable [39].

Using the methods of Section 1.3, the primary potential is obtained as

$$\pi^p(\mathbf{r}) = \int_\Omega \underline{\mathbf{G}}_\pi^p(\mathbf{r},\mathbf{r}') \cdot \frac{\mathbf{J}_e(\mathbf{r}')}{i\omega\varepsilon_j}\, d\Omega',$$

where

$$\underline{\mathbf{G}}_\pi^p(\mathbf{r},\mathbf{r}') = \underline{\mathbf{I}}\, G_\pi^p = \underline{\mathbf{I}}\, \frac{e^{-ik_jR}}{4\pi R}$$

with $R = |\mathbf{r}-\mathbf{r}'|$. As described in Chapter 3, the operator $A : \mathbf{L}^2(\Omega)^3 \to \mathbf{L}^2(\Omega)^3$ defined by

$$(A\mathbf{x})(\mathbf{r}) \equiv \int_\Omega \underline{\mathbf{G}}_\pi^p(\mathbf{r},\mathbf{r}') \cdot \mathbf{x}(\mathbf{r}')\, d\Omega'$$

is a weakly singular integral operator, compact for $\mathbf{r}$ restricted to a finite region $\Omega \subset \mathbf{R}^3$. Differentiability properties of this operator are discussed in Section 1.3.

In order to facilitate evaluation of the scattered potential, we make use of the equality [35, p. 64]

$$\frac{e^{-ik_jR}}{4\pi R} = \frac{1}{(2\pi)^2} \int_{-\infty}^{\infty}\int_{-\infty}^{\infty} \frac{e^{i\lambda\cdot(\mathbf{r}-\mathbf{r}')}e^{-p_j|x-x'|}}{2p_j}\, d^2\lambda, \tag{8.188}$$

where $p_j = \sqrt{\lambda^2 - k_j^2}$, $k_j = \omega\sqrt{\mu_0\varepsilon_j}$ for $j = 1,2,\ldots,5$, $\lambda = \widehat{\mathbf{y}}k_y + \widehat{\mathbf{z}}k_z$, $\lambda^2 = k_y^2 + k_z^2$, and $d^2\lambda = dk_y\, dk_z$. Because

$$e^{i\lambda\cdot(\mathbf{r}-\mathbf{r}')} = e^{ik_y(y-y')}e^{ik_z(z-z')},$$

it can be seen that (8.188) has the form of an expansion over lateral eigenfunctions, as in (8.24), although it is more generally considered as an inverse spatial Fourier transform of the one-dimensional Green's function $e^{-p_j|x-x'|}/(2p_j)$. For convergence of the integral (8.188), and to ensure the proper field behavior at infinity, we require $\mathrm{Re}\{p_j\} > 0$, which determines branch cuts in the complex $\lambda$-plane.

By enforcing the Hertzian-potential boundary conditions[24] on the total potential $\pi = \pi^p + \pi^s$, one obtains the scattered potential $\pi^s$. Although

[24] These are derived from the electric- and magnetic field boundary conditions and are given in [37].

the details are omitted here, the resulting Hertzian-potential Green's dyadic has the form [37]

$$\begin{aligned}\underline{\mathbf{G}}_{\pi}\left(\mathbf{r},\mathbf{r}'\right) =& \underline{\mathbf{G}}_{\pi}^{p}\left(\mathbf{r},\mathbf{r}'\right)+\underline{\mathbf{G}}_{\pi}^{s}\left(\mathbf{r},\mathbf{r}'\right)\\ =& \underline{\mathbf{G}}_{\pi}^{p}\left(\mathbf{r},\mathbf{r}'\right)+\widehat{\mathbf{x}}\widehat{\mathbf{x}}G_{n}^{s}\left(\mathbf{r},\mathbf{r}'\right)+\left(\widehat{\mathbf{x}}\widehat{\mathbf{y}}\frac{\partial}{\partial y}+\widehat{\mathbf{x}}\widehat{\mathbf{z}}\frac{\partial}{\partial z}\right)G_{c}^{s}\left(\mathbf{r},\mathbf{r}'\right)\\ &+\left(\widehat{\mathbf{y}}\widehat{\mathbf{y}}+\widehat{\mathbf{z}}\widehat{\mathbf{z}}\right)G_{t}^{s}\left(\mathbf{r},\mathbf{r}'\right),\end{aligned} \tag{8.189}$$

where

$$\underline{\mathbf{G}}_{\pi}^{p}\left(\mathbf{r},\mathbf{r}'\right)=\underline{\mathbf{I}}\,\frac{e^{-ik_{j}R}}{4\pi R}=\underline{\mathbf{I}}\,\frac{1}{(2\pi)^{2}}\int_{-\infty}^{\infty}\int_{-\infty}^{\infty}e^{-p_{j}|x-x'|}\frac{e^{i\lambda\cdot(\mathbf{r}-\mathbf{r}')}}{2p_{j}}\,d^{2}\lambda,$$

and the scalar components of the scattered dyadic are

$$\left.\begin{array}{c}G_{t}^{s}\left(\mathbf{r},\mathbf{r}'\right)\\ G_{n}^{s}\left(\mathbf{r},\mathbf{r}'\right)\\ G_{c}^{s}\left(\mathbf{r},\mathbf{r}'\right)\end{array}\right\}=\frac{1}{(2\pi)^{2}}\int_{-\infty}^{\infty}\int_{-\infty}^{\infty}\left\{\begin{array}{c}R_{t}\left(x,x',\lambda\right)\\ R_{n}\left(x,x',\lambda\right)\\ R_{c}\left(x,x',\lambda\right)\end{array}\right\}\frac{e^{i\lambda\cdot(\mathbf{r}-\mathbf{r}')}}{2p_{j}\left(\lambda\right)}\,d^{2}\lambda. \tag{8.190}$$

The coefficients $R_{\alpha}$ are determined by the specific structure of the layered medium. In fact, the form (8.189) with (8.190) is valid for any planarly layered medium, for any observation point $\mathbf{r}$, and source point $\mathbf{r}'$ in the $j$th region. For the five-layer structure shown in Figure 8.12, and for the case where both the source point and observation point are in region 3, the coefficients are given in Appendix D along with several special cases.

In addition to the condition $\operatorname{Re}\{p_j\}>0$ in (8.188), the coefficients $R_{\alpha}$ in (8.190) involve the parameters $p_j$, $j=1,2,\ldots,5$. As described in [35, Ch. 2], the coefficients $R_{\alpha}$ are even in all $p_j$ except those for the outermost regions, i.e., $j=1,5$, which physically relates to the branch cuts being associated with radiation into the unbounded outer regions. Therefore, (8.188) implicates branch points associated with the source region via $p_j$ (these are removable unless $j=1,5$), and (8.190) implicates branch points associated with the outermost regions via $p_{1,5}$. The coefficients $R_{\alpha}$ also have pole singularities via the denominator terms in (D.1), which are associated with surface-wave propagation as described in Section 8.1.7.

Green's functions for a planar medium in lower dimensions may be easily obtained from the above spectral form. Let a component of Green's function be $g^{3d}\left(\mathbf{r},\mathbf{r}'\right)$, such that

$$g^{3d}\left(\mathbf{r},\mathbf{r}'\right)=\frac{1}{(2\pi)^{2}}\int_{-\infty}^{\infty}\int_{-\infty}^{\infty}g\left(k_{y},k_{z},x,x'\right)e^{ik_{y}(y-y')}e^{ik_{z}(z-z')}dk_{y}\,dk_{z}.$$

Then,

$$\begin{aligned}g^{2d}\left(\rho,\rho'\right)&=\frac{1}{2\pi}\int_{-\infty}^{\infty}g\left(k_{y},k_{z},x,x'\right)e^{ik_{z}(z-z')}dk_{z},\\ g^{1d}\left(z,z'\right)&=g\left(k_{y},k_{z},x,x'\right),\end{aligned}$$

where in two dimensions $k_y$ is assumed constant, and $k_y = 0$ for structures and sources invariant along $y$. Similarly, in one dimension both $k_y$ and $k_z$ are constant, and $(k_y, k_z) = (0, 0)$ indicates no variation along the planar layers, such as occurs for a uniform plane wave normally incident upon the layered medium.

### Sommerfeld's Representation

Components of the scattered Green's dyadic (8.190) are interpreted as a continuum of plane waves. This form naturally occurs in solving layered-media boundary value problems using Fourier transforms. Alternative forms in terms of cylindrical waves are possible [35, Ch. 2] and provide further insight into the nature of waves in a planarly layered medium.

In order to obtain a cylindrical wave expansion, we transform both the spatial and spectral coordinates into polar form,

$$\begin{aligned} y - y' &= \rho \cos\theta, & z - z' &= \rho \sin\theta, \\ k_y &= \lambda \cos\phi, & k_z &= \lambda \sin\phi. \end{aligned}$$

Then, with $e^{i\boldsymbol{\lambda}\cdot(\mathbf{r}-\mathbf{r}')} = e^{i\lambda\rho\cos(\phi-\theta)}$ and noting that the coefficients $R_\alpha$ only depend on $\{k_y, k_z\}$ via $\lambda^2 = k_y^2 + k_z^2$, (8.190) becomes

$$\left.\begin{array}{l} G_t^s(\mathbf{r},\mathbf{r}') \\ G_n^s(\mathbf{r},\mathbf{r}') \\ G_c^s(\mathbf{r},\mathbf{r}') \end{array}\right\} = \frac{1}{(2\pi)^2}\int_0^\infty \int_{-\pi}^{\pi} \left\{\begin{array}{l} R_t(x,x',\lambda) \\ R_n(x,x',\lambda) \\ R_c(x,x',\lambda) \end{array}\right\} \frac{e^{i\lambda\rho\cos(\phi-\theta)}}{2p_j(\lambda)}\lambda\, d\phi\, d\lambda$$

$$= \frac{1}{(2\pi)^2}\int_0^\infty \frac{\lambda\, d\lambda}{2p_j(\lambda)} \left\{\begin{array}{l} R_t(x,x',\lambda) \\ R_n(x,x',\lambda) \\ R_c(x,x',\lambda) \end{array}\right\} \int_{-\pi}^{\pi} e^{i\lambda\rho\cos\alpha} d\alpha.$$

Because [40]

$$J_0(\lambda\rho) = \frac{1}{2\pi}\int_{-\pi}^{\pi} e^{i\lambda\rho\cos\alpha} d\alpha,$$

we have

$$\left.\begin{array}{l} G_t^s(\mathbf{r},\mathbf{r}') \\ G_n^s(\mathbf{r},\mathbf{r}') \\ G_c^s(\mathbf{r},\mathbf{r}') \end{array}\right\} = \frac{1}{2\pi}\int_0^\infty \left\{\begin{array}{l} R_t(x,x',\lambda) \\ R_n(x,x',\lambda) \\ R_c(x,x',\lambda) \end{array}\right\} \frac{J_0(\lambda\rho)}{2p_j(\lambda)}\lambda\, d\lambda, \tag{8.191}$$

where $\rho = \sqrt{(y-y')^2 + (z-z')^2}$. From (8.188) we also have

$$\underline{\mathbf{G}}_\pi^p(\mathbf{r},\mathbf{r}') = \underline{\mathbf{I}}\,\frac{e^{-ik_j R}}{4\pi R} = \underline{\mathbf{I}}\,\frac{1}{2\pi}\int_0^\infty e^{-p_j|x-x'|}\frac{J_0(\lambda\rho)}{2p_j(\lambda)}\lambda\, d\lambda. \tag{8.192}$$

This form is interpreted as a continuous summation of cylindrical waves in the $\rho$-direction, multiplied by a weighted plane wave in the vertical direction. Another form, which is particularly useful for complex plane analysis, is obtained by noting [40]

$$J_0(\alpha) = \frac{1}{2}\left[H_0^{(1)}(\alpha) + H_0^{(2)}(\alpha)\right],$$
$$H_0^{(2)}(-\alpha) = -H_0^{(1)}(\alpha),$$

resulting in the form

$$\left.\begin{array}{c} G_t^s(\mathbf{r},\mathbf{r}') \\ G_n^s(\mathbf{r},\mathbf{r}') \\ G_c^s(\mathbf{r},\mathbf{r}') \end{array}\right\} = \frac{1}{2\pi}\int_{-\infty}^{\infty}\left\{\begin{array}{c} R_t(x,x',\lambda) \\ R_n(x,x',\lambda) \\ R_c(x,x',\lambda) \end{array}\right\}\frac{H_0^{(2)}(\lambda\rho)}{4p_j(\lambda)}\lambda\, d\lambda \tag{8.193}$$

and

$$\underline{\mathbf{G}}_\pi^p(\mathbf{r},\mathbf{r}') = \underline{\mathbf{I}}\,\frac{e^{-ik_jR}}{4\pi R} = \underline{\mathbf{I}}\,\frac{1}{2\pi}\int_{-\infty}^{\infty} e^{-p_j|x-x'|}\frac{H_0^{(2)}(\lambda\rho)}{4p_j(\lambda)}\lambda\, d\lambda. \tag{8.194}$$

In the Hankel function form, the inversion path must be below the logarithmic branch point associated with $H_0^{(2)}$, such that the above integrals are actually $\int_{-\infty-ja}^{0^+-ja}(\cdot)\,\lambda\, d\lambda + \int_{0^++ja}^{\infty+ja}(\cdot)\,\lambda\, d\lambda$ for $a$ a small, positive constant.

Using any of the possible forms for $\underline{\mathbf{G}}_\pi(\mathbf{r},\mathbf{r}')$ the fields are given by (8.186),

$$\begin{aligned}\mathbf{E}(\mathbf{r}) &= \left(k^2 + \nabla\nabla\cdot\right)\int_\Omega \underline{\mathbf{G}}_\pi(\mathbf{r},\mathbf{r}')\cdot\frac{\mathbf{J}_e(\mathbf{r}')}{i\omega\varepsilon_j}\, d\Omega', \\ \mathbf{H}(\mathbf{r}) &= i\omega\varepsilon\nabla\times\int_\Omega \underline{\mathbf{G}}_\pi(\mathbf{r},\mathbf{r}')\cdot\frac{\mathbf{J}_e(\mathbf{r}')}{i\omega\varepsilon_j}\, d\Omega',\end{aligned} \tag{8.195}$$

where in the expression for $\mathbf{H}$ the operator $\nabla\times$ may be brought under the integral symbol; that is also the case for the divergence operator in the expression for $\mathbf{E}$. Alternatively, field/source relations in the form of (1.88) may be obtained that incorporate the depolarizing dyadic,

$$\begin{aligned}\mathbf{E}(\mathbf{r}) &= \int_\Omega \underline{\mathbf{G}}_{ee}^{\wedge}(\mathbf{r},\mathbf{r}')\cdot\mathbf{J}_e(\mathbf{r}')\, d\Omega', \\ \mathbf{H}(\mathbf{r}) &= \int_\Omega \underline{\mathbf{G}}_{me}(\mathbf{r},\mathbf{r}')\cdot\mathbf{J}_e(\mathbf{r}')\, d\Omega',\end{aligned} \tag{8.196}$$

where

$$\begin{aligned}\underline{\mathbf{G}}_{ee}^{\wedge}(\mathbf{r},\mathbf{r}') &\equiv \text{P.V.}\left\{-i\omega\mu_0\frac{\varepsilon}{\varepsilon_j}\left(\underline{\mathbf{I}} + \frac{\nabla\nabla}{k^2}\right)\cdot\underline{\mathbf{G}}_\pi(\mathbf{r},\mathbf{r}')\right\} - \frac{\widehat{\mathbf{x}}\widehat{\mathbf{x}}\delta(\mathbf{r}-\mathbf{r}')}{i\omega\varepsilon_j}, \\ \underline{\mathbf{G}}_{me}(\mathbf{r},\mathbf{r}') &\equiv \frac{\varepsilon}{\varepsilon_j}\nabla\times\underline{\mathbf{G}}_\pi(\mathbf{r},\mathbf{r}').\end{aligned} \tag{8.197}$$

This form of the Green's dyadic easily allows one to utilize the integral equations developed in Section 1.4 for scattering and resonance problems

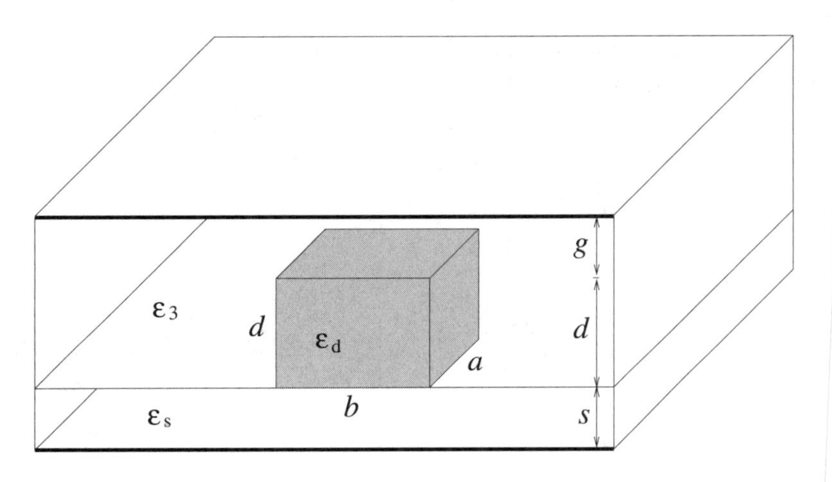


Figure 8.15: Rectangular dielectric resonator on a grounded dielectric layer with top and bottom perfectly conducting plates.

in layered media. For example, if an inhomogeneous dielectric region $\Omega$ is immersed in layer $j$, the second-kind integral equation (1.110),

$$\mathbf{E}(\mathbf{r}) - \int_{\Omega} \underline{\mathbf{G}}_{e,eq}(\mathbf{r},\mathbf{r}') \cdot \mathbf{E}(\mathbf{r}')\, d\Omega' = \mathbf{E}^{i}(\mathbf{r}), \qquad \mathbf{r} \in \Omega, \tag{8.198}$$

can be used to find the field $\mathbf{E}$ in $\Omega$ caused by an excitation $\mathbf{E}^{i}$, where

$$\underline{\mathbf{G}}_{e,ea}(\mathbf{r},\mathbf{r}') \equiv i\omega \underline{\mathbf{G}}_{ee}^{\wedge}(\mathbf{r},\mathbf{r}') \cdot \left[ \underline{\varepsilon}(\mathbf{r}') - \underline{\varepsilon}_j \right].$$

Note, however, that the integral operator in (8.198) is not compact due to the strong singularity of the electric dyadic Green's function.

This formulation can also be used to determine the resonance characteristics of dielectric resonators immersed in a planar medium, as the next example illustrates.

**Example 8.2.**

Consider a dielectric resonator having the shape of a parallelepiped immersed in a microwave integrated circuit environment, as depicted in Figure 8.15. Similar structures are used in filtering and coupling applications. In seeking the natural resonances of the structure, we consider the spectral problem

$$(I - A(\omega_n, \tau))\, \mathbf{E}_n = \lambda_n(\omega_n)\, \mathbf{E}_n \tag{8.199}$$

and seek the value of the nonstandard eigenvalue $\omega_n$ such that $\lambda_n(\omega_n) = 0$ (see Section 4.1), corresponding to the homogeneous solution of (8.198). In (8.199), $\tau$ is a vector of material and geometrical parameters associated with the Green's function, $\omega_n$ is the desired complex-valued resonant

Table 8.1: Computed real part of the fundamental resonant frequency $\omega_0$ (in GHz) for the dielectric resonator depicted in Figure 8.15. Permittivity values are relative to $\varepsilon_0$, and all dimensions are in mm.

| Case | $\varepsilon_d/\varepsilon_0$ | $\varepsilon_s/\varepsilon_0$ | $\varepsilon_3/\varepsilon_0$ | $a=b$ | $d$ | $s$ | $g$ | $\omega_0$ |
|---|---|---|---|---|---|---|---|---|
| 1 | 34.19 | 9.6 | 1.0 | 14.98 | 7.48 | 0.7 | 0.72 | 26.45 |
| 2 | 34.21 | 9.6 | 1.0 | 13.99 | 6.95 | 0.7 | 1.25 | 27.49 |
| 3 | 34.02 | 9.6 | 1.0 | 11.99 | 5.98 | 0.7 | 2.215 | 30.68 |
| 4 | 36.13 | 9.6 | 1.0 | 6.03 | 4.21 | 0.7 | 10.10 | 49.22 |
| 5 | 36.20 | 1.0 | 1.0 | 4.06 | 5.15 | 2.93 | 2.93 | 61.84 |
| 6 | 36.20 | 1.0 | 1.0 | 8.00 | 2.14 | 4.43 | 4.43 | 46.48 |

frequency, and the operator $A(\omega_n, \tau)$ is

$$(A\mathbf{E}_n)(\mathbf{r}) = i\omega(\varepsilon_d - \varepsilon_3) \int_\Omega \underline{\mathbf{G}}_{ee}^{\wedge}(\mathbf{r}, \mathbf{r}') \cdot \mathbf{E}_n(\mathbf{r}') \, d\Omega'$$

for $\mathbf{E}_n \in \mathbf{L}^2(\Omega)^3$, where $\Omega$ is the volume of the resonator. The nonstandard eigenvalues $\omega_n$ are obtained from the necessary condition[25]

$$\det\left[(I - A(\omega_n, \tau))\right] = 0. \tag{8.200}$$

Since the Green's dyadic is pseudo-symmetric from reciprocity (see the footnote on p. 162), then, by the discussion in Section 4.2.11, (8.200) is a stationary form that leads to second-order errors in $\omega_n$ due to first-order errors in approximating $\mathbf{E}_n$. A discussion of the numerical solution, along with various numerical results for several related geometries, can be found in [41]. Table 8.1 lists some representative results for the structure depicted in Figure 8.15.

In the above example the existence of the complex $\omega$-plane resonances (exterior resonances) was not established theoretically. To do this it is much simpler to consider a homogeneous resonator. Then, second-kind surface integral equations involving compact operators may be derived [44, pp. 385–387]. Furthermore, if the resonator is immersed in a homogeneous lossless space, the resulting operators will be analytic operator-valued functions in the finite $\omega$-plane, and Theorem 4.34 is applicable. For homogeneous resonators immersed in a layered medium having infinite extent, the resulting operators will not be analytic in $\omega$ because of $\omega$-plane branch-point singularities associated with the background surface waves (e.g., see the discussion on pp. 438 and 471).

[25] Determinants of operators on infinite-dimensional spaces are discussed extensively in [42]. While one can proceed in this direction, here the function det is merely symbolic and indicates that one take the determinant of the matrix generated by a numerical discretization of the homogeneous form of (8.199).

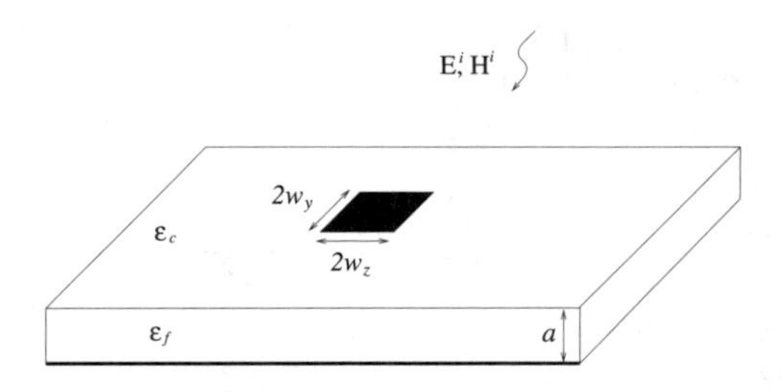


Figure 8.16: Microstrip patch on a grounded dielectric layer, excited by incident field $\mathbf{E}^i, \mathbf{H}^i$.

For a perfectly conducting surface $S$ immersed in layer $j$, and oriented parallel to the planar layers (such as a microstrip patch), the electric field integral equation (1.115),

$$\mathbf{n} \times \int_S \underline{\mathbf{G}}_{ee}^{\wedge}(\mathbf{r}, \mathbf{r}') \cdot \mathbf{J}_e^s(\mathbf{r}')\, dS' = -\mathbf{n} \times \mathbf{E}^i(\mathbf{r}), \qquad \mathbf{r} \in S, \tag{8.201}$$

is applicable, where $\mathbf{J}_e^s$ is the unknown surface current caused by the excitation $\mathbf{E}^i$. In this case of planar currents, the depolarizing dyadic is not needed.

**Example 8.3.**

Consider a microstrip patch located on the surface of a grounded dielectric layer, depicted in Figure 8.16. For this structure we have

$$\mathbf{J}_e^s(\mathbf{r}) = \widehat{\mathbf{y}} J_y(y, z) + \widehat{\mathbf{z}} J_z(y, z)$$

and, noting that $\mathbf{n} = \widehat{\mathbf{x}}$ and $x = x' = 0$, we obtain the coupled system of first-kind integral equations

$$\int_{-w_y}^{w_y} \int_{-w_z}^{w_z} \left[ G_{yy}(\mathbf{r}, \mathbf{r}') J_y(y', z') + G_{yz}(\mathbf{r}, \mathbf{r}') J_z(y', z') \right] dz' dy' = -E_y^i(y, z),$$
$$\int_{-w_y}^{w_y} \int_{-w_z}^{w_z} \left[ G_{zy}(\mathbf{r}, \mathbf{r}') J_y(y', z') + G_{zz}(\mathbf{r}, \mathbf{r}') J_z(y', z') \right] dz' dy' = -E_z^i(y, z),$$

which are enforced for all $y, z \in [-w_y, w_y] \times [-w_z, w_z]$, where $G_{\alpha\beta} = \widehat{\alpha} \cdot \underline{\mathbf{G}}_{ee}^{\wedge} \cdot \widehat{\beta}$. In this case

$$G_{\alpha\alpha} = \int_{-\infty}^{\infty} \int_{-\infty}^{\infty} \left[ (1 + R_t) \left( k_c^2 - k_\alpha^2 \right) + k_\alpha^2 p_c R_c \right] \frac{e^{ik_y(y-y')} e^{ik_z(z-z')}}{2(2\pi)^2 j\omega\varepsilon_c p_c}\, dk_y\, dk_z,$$
$$G_{yz} = \int_{-\infty}^{\infty} \int_{-\infty}^{\infty} k_y k_z \left[ p_c R_c - (1 + R_t) \right] \frac{e^{ik_y(y-y')} e^{ik_z(z-z')}}{2(2\pi)^2 j\omega\varepsilon_c p_c}\, dk_y\, dk_z$$
$$= G_{zy},$$

for $\alpha = y, z$, where $R_t$ and $R_c$ are given as (D.4) in Appendix D. Upon interchanging the order of integrations, the final system of integral equations becomes

$$\int_{-\infty}^{\infty}\int_{-\infty}^{\infty}[f_{yy}(k_y,k_z)J_y(k_y,k_z)+f_{yz}(k_y,k_z)J_z(k_y,k_z)]\,dk_y\,dk_z = -E_y^i(y,z),$$

$$\int_{-\infty}^{\infty}\int_{-\infty}^{\infty}[f_{zy}(k_y,k_z)J_y(k_y,k_z)+f_{zz}(k_y,k_z)J_z(k_y,k_z)]\,dk_y\,dk_z = -E_z^i(y,z),$$

for all $y, z \in [-w_y, w_y] \times [-w_z, w_z]$, where

$$J_{y,z}(k_y,k_z) = \frac{1}{(2\pi)^2}\int_{-w_y}^{w_y}\int_{-w_z}^{w_z} J_{y,z}(y',z')\, e^{-ik_y y'} e^{-ik_z z'}\,dz'dy',$$

$$f_{\alpha\alpha}(y,z,k_y,k_z) = \left\{(1+R_t)\left(k_c^2-k_\alpha^2\right)+k_\alpha^2 p_c R_c\right\}\frac{e^{ik_y y}e^{ik_z z}}{2j\omega\varepsilon_c p_c},$$

$$f_{yz}(y,z,k_y,k_z) = k_y k_z\left\{p_c R_c - (1+R_t)\right\}\frac{e^{ik_y y}e^{ik_z z}}{2j\omega\varepsilon_c p_c}.$$

Convergence properties of the integrals depend, in part, on the transform-domain current $J_{y,z}(k_y,k_z)$.

The system of equations has been numerically solved in a number of references (e.g., [43]). For a narrow patch with $w_z \ll w_y$ we may assume $\mathbf{J}_e^s(\mathbf{r}) \simeq \widehat{\mathbf{y}} J_y(y,z)$ and obtain the single integral equation

$$\int_{-\infty}^{\infty}\int_{-\infty}^{\infty} f_{yy}(k_y,k_z)J_y(k_y,k_z)\;dk_y\,dk_z = -E_y^i(y,z).$$

In the next section we consider a homogeneous parallel-plate structure, which was considered for the two-dimensional case in Section 8.1.2.

### 8.3.2 Parallel-Plate Structure

Consider the homogeneously filled parallel-plate structure depicted in Figure 8.2 with material parameters $\varepsilon, \mu_0$, where in this section we consider a general three-dimensional source $\mathbf{J}_e$. As provided in Appendix D, the Hertzian potential is given as

$$\pi(\mathbf{r}) = \int_\Omega \underline{\mathbf{G}}_\pi(\mathbf{r},\mathbf{r}')\cdot\frac{\mathbf{J}_e(\mathbf{r}')}{i\omega\varepsilon}\,d\Omega',$$

where

$$\begin{aligned}\underline{\mathbf{G}}_\pi(\mathbf{r},\mathbf{r}') &= \underline{\mathbf{G}}_\pi^p(\mathbf{r},\mathbf{r}') + \widehat{\mathbf{x}}\widehat{\mathbf{x}}G_n^s(\mathbf{r},\mathbf{r}') + (\widehat{\mathbf{y}}\widehat{\mathbf{y}}+\widehat{\mathbf{z}}\widehat{\mathbf{z}})\,G_t^s(\mathbf{r},\mathbf{r}')\\ &= \widehat{\mathbf{x}}\widehat{\mathbf{x}}\left[G_\pi^p(\mathbf{r},\mathbf{r}')+G_n^s(\mathbf{r},\mathbf{r}')\right] + (\widehat{\mathbf{y}}\widehat{\mathbf{y}}+\widehat{\mathbf{z}}\widehat{\mathbf{z}})\left[G_\pi^p(\mathbf{r},\mathbf{r}')+G_t^s(\mathbf{r},\mathbf{r}')\right]\end{aligned} \tag{8.202}$$

with the coefficients in the equivalent forms (8.190), (8.191), or (8.193) given by (D.2).

It is sometimes convenient to combine the primary Green's dyadic with the scattered Green's dyadic to yield, using the spectral form for $G_\pi^p$ and (D.2),

$$\underline{\mathbf{G}}_\pi(\mathbf{r},\mathbf{r}') = \widehat{\mathbf{x}}\widehat{\mathbf{x}} G_\pi^E(\mathbf{r},\mathbf{r}') + (\widehat{\mathbf{y}}\widehat{\mathbf{y}} + \widehat{\mathbf{z}}\widehat{\mathbf{z}})\, G_\pi^H(\mathbf{r},\mathbf{r}'), \tag{8.203}$$

where

$$\begin{aligned}
&G_\pi^H(\mathbf{r},\mathbf{r}') \\
&\equiv G_\pi^p(\mathbf{r},\mathbf{r}') + G_t^s(\mathbf{r},\mathbf{r}') \\
&= \frac{1}{(2\pi)^2}\int_{-\infty}^{\infty}\int_{-\infty}^{\infty} \frac{\cosh[p(x-x'\mp a)] - \cosh[p(x+x'-a)]}{\sinh pa}\,\frac{e^{i\lambda\cdot(\mathbf{r}-\mathbf{r}')}}{2p(\lambda)}\,d^2\lambda \\
&= \frac{1}{(2\pi)^2}\int_{-\infty}^{\infty}\int_{-\infty}^{\infty} \sinh px_< \frac{\sinh[p(a-x_>)]}{p\sinh pa} e^{i\lambda\cdot(\mathbf{r}-\mathbf{r}')}\,d^2\lambda,
\end{aligned} \tag{8.204}$$

$$\begin{aligned}
&G_\pi^E(\mathbf{r},\mathbf{r}') \\
&\equiv G_\pi^p(\mathbf{r},\mathbf{r}') + G_n^s(\mathbf{r},\mathbf{r}') \\
&= \frac{1}{(2\pi)^2}\int_{-\infty}^{\infty}\int_{-\infty}^{\infty} \frac{\cosh[p(x-x'\mp a)] + \cosh[p(x+x'-a)]}{\sinh pa}\,\frac{e^{i\lambda\cdot(\mathbf{r}-\mathbf{r}')}}{2p(\lambda)}\,d^2\lambda \\
&= \frac{1}{(2\pi)^2}\int_{-\infty}^{\infty}\int_{-\infty}^{\infty} \cosh px_< \frac{\cosh[p(a-x_>)]}{p\sinh pa} e^{i\lambda\cdot(\mathbf{r}-\mathbf{r}')}\,d^2\lambda
\end{aligned} \tag{8.205}$$

for $x \gtrless x'$ in (8.204) and (8.205), where $p = \sqrt{\lambda^2 - k^2}$ and $\lambda^2 = k_y^2 + k_z^2$. It is clear from the form of (8.203) that vertical currents excite vertical potentials, whereas horizontal currents excite parallel, horizontal potentials.

Note that

$$G_\pi^{H,E}(\mathbf{r},\mathbf{r}') = \frac{1}{2\pi}\int_{-\infty}^{\infty} g^{H,E}(x,z,x',z')\Big|_{\substack{\nu\to k_z \\ \sqrt{k^2-\nu^2}\to\sqrt{k^2-k_y^2-k_z^2}}} e^{ik_y(y-y')}\,dk_y, \tag{8.206}$$

where $g^{H,E}$ are the two-dimensional parallel-plate Green's functions developed in Section 8.1.2, given by (8.56) and (8.57), respectively. The substitution $\nu \to k_z$ merely indicates renaming a variable to be consistent with the notation used in this section, whereas $\sqrt{k^2-\nu^2} \to \sqrt{k^2-k_y^2-k_z^2}$ accounts for the spatial variation encountered in the three-dimensional case considered here.

The Green's components (8.204) and (8.205) may be converted to a discrete summation form by utilizing the Hankel function representation (8.193). Indeed, starting from

$$G_\pi^H(\mathbf{r},\mathbf{r}') = \frac{1}{2\pi}\int_{-\infty}^{\infty} \sinh px_< \frac{\sinh[p(a-x_>)]}{2p\sinh pa} H_0^{(2)}(\lambda\rho)\,\lambda\,d\lambda,$$

$$G_\pi^E(\mathbf{r},\mathbf{r}') = \frac{1}{2\pi}\int_{-\infty}^{\infty} \cosh px_< \frac{\cosh[p(a-x_>)]}{2p\sinh pa} H_0^{(2)}(\lambda\rho)\,\lambda\,d\lambda,$$

note that the integrands are meromorphic in the lower-half complex $\lambda$-plane, with pole singularities at

$$\sinh p_n a = 0.$$

Therefore, poles occur at $p_n = \pm in\pi/a$, $n = 0, 1, 2, \ldots$, such that

$$\lambda_n = \sqrt{k^2 - \left(\frac{n\pi}{a}\right)^2}. \tag{8.207}$$

Closing the integration contour with a semicircle of infinite radius in the lower-half $\lambda$-plane and invoking Cauchy's theorem (see Section 5.2), we obtain the Green's components as a sum of residues,

$$\begin{aligned} G_\pi^H(\mathbf{r}, \mathbf{r}') &= \frac{1}{2ia} \sum_{n=1}^{\infty} \sin\left(\frac{n\pi}{a}x\right) \sin\left(\frac{n\pi}{a}x'\right) H_0^{(2)}(\lambda_n \rho), \\ G_\pi^E(\mathbf{r}, \mathbf{r}') &= \frac{1}{4ia} \sum_{n=0}^{\infty} \epsilon_n \cos\left(\frac{n\pi}{a}x\right) \cos\left(\frac{n\pi}{a}x'\right) H_0^{(2)}(\lambda_n \rho). \end{aligned} \tag{8.208}$$

The various forms of the Green's components represent partial eigenfunction expansions. The integral forms (8.204) and (8.205) represent an expansion over the improper eigenfunctions in $\{y, z\}$, multiplied by a Green's function for the vertical coordinate, as in (8.24). The discrete summation form (8.208) represents an expansion over the proper eigenfunctions in the vertical coordinate, multiplied by a Green's function for the radial direction, analogous to (8.23). Using a procedure similar to (8.206), and noting the integral definition of the Hankel function (8.78), the Green's components (8.208) can also be obtained by an integration of the two-dimensional Green's functions $g^{H,E}$ (8.52) and (8.53).

# 9

# Cylindrical Waveguide Problems

In this chapter we consider various problems associated with scattering within a cylindrical waveguide environment. We first discuss some general concepts relating to magnetic potential and electric Green's dyadics of the first and second kinds. Then, a variety of integral equations are formulated for waveguide scattering problems, including the problem of an infinite waveguide containing perfectly conducting obstacles, an infinite waveguide with apertures that couple energy to the region outside the waveguide, semi-infinite waveguides coupled through apertures in a common ground plane, and semi-infinite waveguides containing perfectly conducting obstacles and coupled through apertures in a common ground plane. In these sections the background waveguide (i.e., the waveguide with all apertures and obstacles removed) contains a homogeneous medium, and all formulations utilize the corresponding dyadic Green's function for the background waveguide. These Green's dyadics are developed via a scalar partial eigenfunction expansion method, and explicit forms are provided for the special case of rectangular waveguides.

Next, integral equations are developed for the analysis of waveguide-based electric- (patch, strip) and magnetic- (slot, aperture) type antennas for spatial power combining, and a general approach to construct the corresponding electric Green's dyadics of the third kind is demonstrated for the background waveguide containing a layered medium. The generalized scattering matrix is discussed for interacting strip and slot layers in a layered waveguide and for an aperture-coupled patch array in an $N$-port waveguide transition.

Finally, the method of integral representations for overlapping regions is demonstrated for the analysis of a shielded microstrip line, again using a rectangular waveguide as an example. It is shown that the method results in Fredholm-type integral equations of the second kind with a compact operator in the sequence space $\mathbf{l}^1$.

# 9.1 Scattering Problems for Waveguides Filled with a Homogeneous Medium

In this section we discuss electric- and magnetic field integral equation formulations for scattering problems inside a cylindrical waveguide filled with a homogeneous medium. Various geometries are considered, followed by the derivation of the associated Green's functions for the background waveguide. We begin with some general material on dyadic Green's functions.

## 9.1.1 Green's Dyadics—General Concepts and Definitions

Consider Maxwell's equations (1.13) for a region filled with a linear, isotropic, homogeneous medium. As discussed in Section 1.3.1, the system of Maxwell's curl equations can be decoupled, resulting in two independent vector wave equations for the electric and magnetic fields generated by an impressed electric current $\mathbf{J}_{\text{imp}}(\mathbf{r})$. Repeating from (1.24), we have

$$\nabla \times \nabla \times \mathbf{E}(\mathbf{r}) - k^2\mathbf{E}(\mathbf{r}) = -i\omega\mu\mathbf{J}_{\text{imp}}(\mathbf{r}), \tag{9.1}$$

$$\nabla \times \nabla \times \mathbf{H}(\mathbf{r}) - k^2\mathbf{H}(\mathbf{r}) = \nabla \times \mathbf{J}_{\text{imp}}(\mathbf{r}), \tag{9.2}$$

where $k = \omega\sqrt{\varepsilon\mu}$, $\varepsilon = \varepsilon_0\varepsilon_r$, and $\mu = \mu_0\mu_r$, with $\varepsilon_r$ and $\mu_r$ the relative dielectric permittivity and magnetic permeability, respectively, of the medium. Also of interest is the magnetic vector potential $\mathbf{A}(\mathbf{r})$, considered in Section 1.3.2 and defined via the relationship $\mathbf{H}(\mathbf{r}) = \frac{1}{\mu}\nabla \times \mathbf{A}(\mathbf{r})$, which satisfies the vector Helmholtz equation (1.30),

$$\nabla^2\mathbf{A}(\mathbf{r}) + k^2\mathbf{A}(\mathbf{r}) = -\mu\mathbf{J}_{\text{imp}}(\mathbf{r}). \tag{9.3}$$

Corresponding to $\mathbf{E}, \mathbf{H}$, and $\mathbf{A}$ we introduce dyadic Green's functions that have physical meaning analogous to the corresponding fields maintained by a unit point source, but elevated to dyadic level. Corresponding to the electric field associated with an electric source we have the electric Green's dyadic $\underline{\mathbf{G}}_e(\mathbf{r}, \mathbf{r}')$; for the magnetic field caused by an electric source we have the magnetic Green's dyadic $\underline{\mathbf{G}}_m(\mathbf{r}, \mathbf{r}')$ and corresponding to the magnetic vector potential we have the magnetic potential Green's dyadic $\underline{\mathbf{G}}_A(\mathbf{r}, \mathbf{r}')$.[1]

[1] Because we consider only electric-type driving current sources, we abandon the double-subscript notation followed in Chapter 1.

Analogous to (9.1)–(9.3), the Green's dyadics satisfy the dyadic wave equations

$$\nabla \times \nabla \times \underline{\mathbf{G}}_e(\mathbf{r},\mathbf{r}') - k^2 \underline{\mathbf{G}}_e(\mathbf{r},\mathbf{r}') = \underline{\mathbf{I}}\,\delta(\mathbf{r}-\mathbf{r}'), \tag{9.4}$$

$$\nabla \times \nabla \times \underline{\mathbf{G}}_m(\mathbf{r},\mathbf{r}') - k^2 \underline{\mathbf{G}}_m(\mathbf{r},\mathbf{r}') = \nabla\delta(\mathbf{r}-\mathbf{r}') \times \underline{\mathbf{I}}, \tag{9.5}$$

$$\nabla^2 \underline{\mathbf{G}}_A(\mathbf{r},\mathbf{r}') + k^2 \underline{\mathbf{G}}_A(\mathbf{r},\mathbf{r}') = -\underline{\mathbf{I}}\,\delta(\mathbf{r}-\mathbf{r}'). \tag{9.6}$$

Note that $\underline{\mathbf{G}}_e$ and $\underline{\mathbf{G}}_m$ are related as

$$\underline{\mathbf{G}}_e(\mathbf{r},\mathbf{r}') = \frac{1}{k^2}\left(\nabla \times \underline{\mathbf{G}}_m(\mathbf{r},\mathbf{r}') - \underline{\mathbf{I}}\,\delta(\mathbf{r}-\mathbf{r}')\right),$$
$$\underline{\mathbf{G}}_m(\mathbf{r},\mathbf{r}') = \nabla \times \underline{\mathbf{G}}_e(\mathbf{r},\mathbf{r}'),$$

which are analogous to the equations for electric and magnetic fields

$$\mathbf{E}(\mathbf{r}) = -\frac{i}{\omega\varepsilon}\left(\nabla \times \mathbf{H}(\mathbf{r}) - \mathbf{J}_{\text{imp}}(\mathbf{r})\right),$$
$$\mathbf{H}(\mathbf{r}) = \frac{i}{\omega\mu}\nabla \times \mathbf{E}(\mathbf{r}).$$

Also, $\underline{\mathbf{G}}_A$ is related to $\underline{\mathbf{G}}_e$ and $\underline{\mathbf{G}}_m$ as

$$\underline{\mathbf{G}}_e(\mathbf{r},\mathbf{r}') = \left(\underline{\mathbf{I}} + \frac{1}{k^2}\nabla\nabla\right) \cdot \underline{\mathbf{G}}_A(\mathbf{r},\mathbf{r}'),$$
$$\underline{\mathbf{G}}_m(\mathbf{r},\mathbf{r}') = \nabla \times \underline{\mathbf{G}}_A(\mathbf{r},\mathbf{r}'),$$

where we note the analogous relations from Section 1.3.2,

$$\mathbf{E}(\mathbf{r}) = -i\omega\left(\underline{\mathbf{I}} + \frac{1}{k^2}\nabla\nabla\right) \cdot \mathbf{A}(\mathbf{r}),$$
$$\mathbf{H}(\mathbf{r}) = \frac{1}{\mu}\nabla \times \mathbf{A}(\mathbf{r}).$$

Next, we define boundary conditions that separate the Green's dyadics into two classes. Assume that on a boundary $S_{e(m)}$ we have homogeneous Dirichlet conditions

$$\mathbf{n} \times \mathbf{E}(\mathbf{r})|_{S_e} = \mathbf{0}, \qquad \mathbf{n} \times \mathbf{H}(\mathbf{r})|_{S_m} = \mathbf{0},$$
$$\mathbf{n} \cdot \mathbf{H}(\mathbf{r})|_{S_e} = 0, \qquad \mathbf{n} \cdot \mathbf{E}(\mathbf{r})|_{S_m} = 0,$$

where $\mathbf{n}$ is the unit normal vector on $S$. The above boundary conditions physically relate to $S_e$ being a perfect electrical conductor and $S_m$ being a perfect magnetic conductor (see (1.5)). Using Maxwell's curl equations we can also write the equations containing $\mathbf{n}\times$ as homogeneous Neumann conditions,

$$\mathbf{n} \times \nabla \times \mathbf{H}(\mathbf{r})|_{S_e} = \mathbf{0}, \qquad \mathbf{n} \times \nabla \times \mathbf{E}(\mathbf{r})|_{S_m} = \mathbf{0}.$$

Analogous to the above, we introduce boundary conditions for the electric and magnetic Green's dyadics,

$$\mathbf{n} \times \underline{\mathbf{G}}_e^{(1)}(\mathbf{r}, \mathbf{r}') = \underline{\mathbf{0}}, \qquad \mathbf{n} \times \underline{\mathbf{G}}_m^{(1)}(\mathbf{r}, \mathbf{r}') = \underline{\mathbf{0}}, \tag{9.7}$$

$$\mathbf{n} \times \nabla \times \underline{\mathbf{G}}_m^{(2)}(\mathbf{r}, \mathbf{r}') = \underline{\mathbf{0}}, \qquad \mathbf{n} \times \nabla \times \underline{\mathbf{G}}_e^{(2)}(\mathbf{r}, \mathbf{r}') = \underline{\mathbf{0}}, \tag{9.8}$$

$$\mathbf{n} \cdot \underline{\mathbf{G}}_m^{(2)}(\mathbf{r}, \mathbf{r}') = \mathbf{0}, \qquad \mathbf{n} \cdot \underline{\mathbf{G}}_e^{(2)}(\mathbf{r}, \mathbf{r}') = \mathbf{0}, \tag{9.9}$$

where the superscript corresponds to the Green's dyadic of the first or second kind [36]. The first-kind Green's dyadic satisfies a homogeneous Dirichlet condition, whereas the second-kind dyadic satisfies a homogeneous Neumann condition. The electric, magnetic, and magnetic potential Green's dyadics of the first and second kinds are related as

$$\begin{aligned} \underline{\mathbf{G}}_e^{(1,2)}(\mathbf{r}, \mathbf{r}') &= \left( \underline{\mathbf{I}} + \frac{1}{k^2} \nabla\nabla \right) \cdot \underline{\mathbf{G}}_A^{(1,2)}(\mathbf{r}, \mathbf{r}') \\ &= \frac{1}{k^2} \left( \nabla \times \underline{\mathbf{G}}_m^{(2,1)}(\mathbf{r}, \mathbf{r}') - \underline{\mathbf{I}}\, \delta(\mathbf{r} - \mathbf{r}') \right), \end{aligned} \tag{9.10}$$

$$\underline{\mathbf{G}}_m^{(1,2)}(\mathbf{r}, \mathbf{r}') = \nabla \times \underline{\mathbf{G}}_e^{(2,1)}(\mathbf{r}, \mathbf{r}') = \nabla \times \underline{\mathbf{G}}_A^{(2,1)}(\mathbf{r}, \mathbf{r}'). \tag{9.11}$$

For the magnetic potential Green's dyadic we have the boundary conditions[2]

$$\begin{aligned} \mathbf{n} \times \underline{\mathbf{G}}_A^{(1)}(\mathbf{r}, \mathbf{r}') &= \underline{\mathbf{0}}, \\ \nabla \cdot \underline{\mathbf{G}}_A^{(1)}(\mathbf{r}, \mathbf{r}') &= \mathbf{0}, \\ \mathbf{n} \times \nabla \times \underline{\mathbf{G}}_A^{(2)}(\mathbf{r}, \mathbf{r}') &= \underline{\mathbf{0}}, \\ \mathbf{n} \cdot \left( \underline{\mathbf{I}} + \frac{1}{k^2} \nabla\nabla \right) \cdot \underline{\mathbf{G}}_A^{(2)}(\mathbf{r}, \mathbf{r}') &= \mathbf{0}, \end{aligned} \tag{9.12}$$

[2]Note that $\mathbf{n} \times \underline{\mathbf{G}}_A^{(1)}(\mathbf{r}, \mathbf{r}') = \underline{\mathbf{0}}$ and $\nabla \cdot \underline{\mathbf{G}}_A^{(1)}(\mathbf{r}, \mathbf{r}') = \mathbf{0}$ on a perfectly conducting boundary are equivalent to the boundary condition

$$\mathbf{n} \times \left( \underline{\mathbf{I}} + (1/k^2) \nabla\nabla \right) \cdot \underline{\mathbf{G}}_A^{(1)}\left(\mathbf{r}, \mathbf{r}'\right) = \underline{\mathbf{0}},$$

which corresponds to a homogeneous Dirichlet condition for

$$\underline{\mathbf{G}}_e^{(1)}\left(\mathbf{r}, \mathbf{r}'\right) : \mathbf{n} \times \underline{\mathbf{G}}_e^{(1)}\left(\mathbf{r}, \mathbf{r}'\right) = \underline{\mathbf{0}}.$$

Noting that $\nabla \cdot \mathbf{E}(\mathbf{r}) = 0$ on the boundary (assuming that there is no source charge-density on the boundary) results in $\nabla \cdot \mathbf{A}(\mathbf{r}) = 0$,

$$\nabla \cdot \mathbf{E}(\mathbf{r}) = -i\omega \nabla \cdot \mathbf{A}(\mathbf{r}) + \frac{1}{i\omega\varepsilon\mu} \nabla \cdot (\nabla\nabla \cdot \mathbf{A}(\mathbf{r})) = -i\omega \nabla \cdot \mathbf{A}(\mathbf{r}) = 0;$$

hence, $\nabla \cdot \mathbf{A}(\mathbf{r}) = 0$, which is analogous to $\nabla \cdot \underline{\mathbf{G}}_A^{(1)}(\mathbf{r}, \mathbf{r}') = \mathbf{0}$. Also note that $\nabla \cdot \underline{\mathbf{G}}_A^{(2)}(\mathbf{r}, \mathbf{r}') \neq \mathbf{0}$ on the boundary.

and

$$\begin{aligned}\mathbf{n} \times \nabla \times \underline{\mathbf{G}}_A^{(2)}(\mathbf{r}, \mathbf{r}') &= \underline{\mathbf{0}}, \\ \mathbf{n} \times \nabla \times \nabla \times \underline{\mathbf{G}}_A^{(1)}(\mathbf{r}, \mathbf{r}') &= \underline{\mathbf{0}}, \\ \mathbf{n} \cdot \nabla \times \underline{\mathbf{G}}_A^{(1)}(\mathbf{r}, \mathbf{r}') &= \mathbf{0}.\end{aligned} \tag{9.13}$$

These two sets of boundary conditions are obtained using the boundary conditions (9.7)–(9.9) for electric and magnetic Green's dyadics of the first and second kinds, and relations (9.10) and (9.11), respectively. In fact, the set (9.12) corresponds to the boundary conditions for electric Green's dyadics and the set (9.13) to those obtained for magnetic Green's dyadics.

Symmetry properties of Green's functions for a reciprocal medium are understood in the sense of interchanging the position of the source point $\mathbf{r}'$ and the observation point $\mathbf{r}$. This becomes useful in formulating integral representations for fields and potentials in terms of Green's dyadics and impressed or induced sources. Here we summarize results for the symmetry of electric, magnetic, and potential Green's dyadics [36],[46],

$$\begin{aligned}
&\underline{\mathbf{G}}_e^{(1,2)}(\mathbf{r}, \mathbf{r}') = \left[\underline{\mathbf{G}}_e^{(1,2)}(\mathbf{r}', \mathbf{r})\right]^\top, \quad \nabla \times \underline{\mathbf{G}}_e^{(1,2)}(\mathbf{r}, \mathbf{r}') = \left[\nabla' \times \underline{\mathbf{G}}_e^{(2,1)}(\mathbf{r}', \mathbf{r})\right]^\top, \\
&\underline{\mathbf{G}}_m^{(1,2)}(\mathbf{r}, \mathbf{r}') = \left[\underline{\mathbf{G}}_m^{(2,1)}(\mathbf{r}', \mathbf{r})\right]^\top, \quad \nabla \times \underline{\mathbf{G}}_m^{(1,2)}(\mathbf{r}, \mathbf{r}') = \left[\nabla' \times \underline{\mathbf{G}}_m^{(1,2)}(\mathbf{r}', \mathbf{r})\right]^\top, \\
&\underline{\mathbf{G}}_A^{(1,2)}(\mathbf{r}, \mathbf{r}') = \underline{\mathbf{G}}_A^{(1,2)}(\mathbf{r}', \mathbf{r}), \qquad \nabla \times \underline{\mathbf{G}}_A^{(1,2)}(\mathbf{r}, \mathbf{r}') = \left[\nabla' \times \underline{\mathbf{G}}_A^{(2,1)}(\mathbf{r}', \mathbf{r})\right]^\top.
\end{aligned}$$

Having stated preliminary concepts associated with dyadic Green's functions, we next develop integral equations for scattering in a cylindrical waveguide environment. In the following section we consider waveguides filled with a homogeneous medium, and in Section 9.3 we consider waveguides filled with a planarly inhomogeneous medium.

## 9.1.2 Infinite Waveguide Containing Metal Obstacles

Consider an infinite waveguide consisting of a volume $V$ enclosed by a perfectly conducting surface $S_M$ and filled with a linear homogeneous medium having material parameters $\varepsilon$ and $\mu$. Furthermore, assume that the waveguide contains arbitrarily shaped, perfectly conducting objects having surfaces $S_m^i$, as depicted in Figure 9.1 for the special case of a rectangular waveguide. An impressed electric current $\mathbf{J}_{\text{imp}}(\mathbf{r})$, $\mathbf{r} \in V_{\text{imp}} \subset V$, generates an incident electric and magnetic field, and scattered fields are generated by the electric surface current induced on the perfectly conducting metallization $S_m^i$ (the Green's dyadics to be introduced will account for the waveguide surface $S_M$).

The electric field is determined as the solution of the vector wave equation

$$\nabla \times \nabla \times \mathbf{E}(\mathbf{r}) - k^2\mathbf{E}(\mathbf{r}) = -i\omega\mu\mathbf{J}_{\text{imp}}(\mathbf{r}), \qquad \mathbf{r} \in V, \tag{9.14}$$

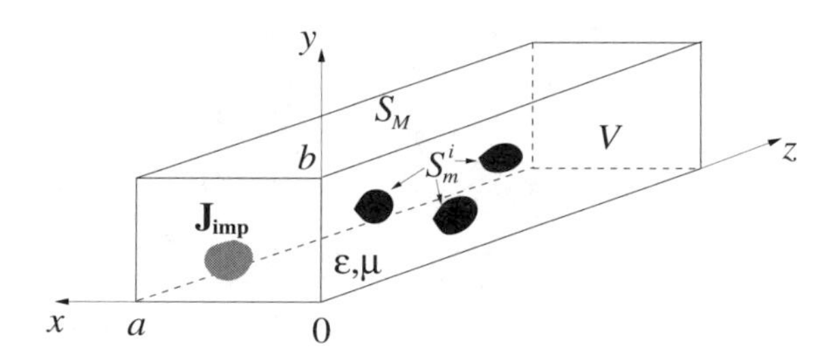


Figure 9.1: Infinite waveguide filled with a homogeneous medium and containing arbitrarily shaped, perfectly conducting objects.

subject to the boundary condition

$$\mathbf{n} \times \mathbf{E}(\mathbf{r})|_S = \mathbf{0}, \tag{9.15}$$

where $S = S_M \cup \left(\cup_{i=1}^n S_m^i\right)$ and $\mathbf{n}$ is the unit normal vector, outward to $S_M$ and inward to $S_m^i$.[3] As $z \to \pm\infty$ we require the fitness condition that the field vanishes (limiting absorption principle [47]), which is associated with small material loss, and also guarantees that there is no incoming wave from infinity in the scattered field. Assuming the field dependence $e^{i\omega t - \gamma z}$ in the positive $z$-direction, where $\gamma = \alpha + i\beta$ is the propagation constant, we require that $\mathrm{Re}\{\gamma\} \geq 0$ and $\mathrm{Im}\{\gamma\} \geq 0$.

Note that the operator $L : \mathbf{L}^2(V)^3 \to \mathbf{L}^2(V)^3$ defined as

$$LE \equiv \nabla \times \nabla \times \mathbf{E}(\mathbf{r}), \tag{9.16}$$

$$D_L \equiv \left\{ \mathbf{E} : \mathbf{E}(\mathbf{r}), \nabla \times \nabla \times \mathbf{E}(\mathbf{r}) \in \mathbf{L}^2(V)^3, \ \ \mathbf{n} \times \mathbf{E}(\mathbf{r})|_S = \mathbf{0}, \lim_{z \to \pm\infty} \mathbf{E} = \mathbf{0} \right\}$$

is self-adjoint, such that (9.14) and (9.15) constitute a self-adjoint boundary value problem. Then, by the discussion in Section 4.3.3, we can expand any $\mathbf{E} \in \mathbf{L}^2(V)^3$ in terms of the complete set of waveguide eigenfunctions (see also Section 10.2).[4]

In order to solve (9.14) and (9.15) we introduce the dyadic wave equation

$$\nabla \times \nabla \times \underline{\mathbf{G}}_e^{(1)}(\mathbf{r}, \mathbf{r}') - k^2 \underline{\mathbf{G}}_e^{(1)}(\mathbf{r}, \mathbf{r}') = \underline{\mathbf{I}}\, \delta(\mathbf{r} - \mathbf{r}'), \qquad \mathbf{r}, \mathbf{r}' \in V, \tag{9.17}$$

[3] In addition to (9.15) we also have a boundary condition for the normal component of the electric field on a perfectly conducting boundary, $\nabla \cdot \mathbf{E}(\mathbf{r})|_S = 0$ (assuming that there is no source-charge density on the boundary $S$). This condition, in conjunction with (9.15), results in $\partial E_n / \partial n = 0$ (excluding the edge points).

[4] As discussed in Section 8.1.8, if the waveguide is lossy and filled with an inhomogeneous medium, than one considers

$$LE \equiv \nabla \times \nabla \times \mathbf{E}(\mathbf{r}) - k^2(\mathbf{r})\, \mathbf{E}(\mathbf{r})$$

which is nonself-adjoint because $k$ is complex-valued. Nontrivial modal degeneracies may exists at which points the resulting eigenfunction expansions may not be valid.

subject to the boundary condition of the first kind on the waveguide surface $S_M$,

$$\mathbf{n} \times \underline{\mathbf{G}}_e^{(1)}(\mathbf{r}, \mathbf{r}')\Big|_{\mathbf{r} \in S_M} = \underline{\mathbf{0}}, \tag{9.18}$$

and a fitness condition (limiting absorption) at infinity. Note that the Green's function will account for the waveguide walls $S_M$ but not the scatterers $S_m^i$.

The vector-dyadic Green's second theorem (see Appendix A.4) applied to $\mathbf{E}(\mathbf{r})$ and $\underline{\mathbf{G}}_e^{(1)}(\mathbf{r}, \mathbf{r}')$, with the boundary conditions (9.15) and (9.18), results in the integral representation[5]

$$\begin{aligned}\mathbf{E}(\mathbf{r}') = &- i\omega\mu \int_{V_{\text{imp}}} \mathbf{J}_{\text{imp}}(\mathbf{r}) \cdot \underline{\mathbf{G}}_e^{(1)}(\mathbf{r}, \mathbf{r}')\, dV \\ &+ \sum_{i=1}^{n} \int_{S_m^i} [\mathbf{n}_i \times \nabla \times \mathbf{E}(\mathbf{r})] \cdot \underline{\mathbf{G}}_e^{(1)}(\mathbf{r}, \mathbf{r}')\, dS\end{aligned} \tag{9.19}$$

with the unit normal vector $\mathbf{n}_i$ pointing outward on $S_m^i$. Defining induced scatterer currents

$$\mathbf{J}_i(\mathbf{r}) \equiv \mathbf{n}_i \times \mathbf{H}(\mathbf{r}) = \frac{i}{\omega\mu} \mathbf{n}_i \times \nabla \times \mathbf{E}(\mathbf{r})$$

for $\mathbf{r} \in S_m^i$, interchanging $\mathbf{r}'$ and $\mathbf{r}$, and using the identity

$$\mathbf{J}(\mathbf{r}') \cdot \underline{\mathbf{G}}_e^{(1)}(\mathbf{r}', \mathbf{r}) = [\underline{\mathbf{G}}_e^{(1)}(\mathbf{r}', \mathbf{r})]^{\top} \cdot \mathbf{J}(\mathbf{r}') = \underline{\mathbf{G}}_e^{(1)}(\mathbf{r}, \mathbf{r}') \cdot \mathbf{J}(\mathbf{r}'), \tag{9.20}$$

we obtain the total electric field at $\mathbf{r} \in V$ caused by a known impressed current $\mathbf{J}_{\text{imp}}(\mathbf{r}')$ and induced currents $\mathbf{J}_i(\mathbf{r}')$ as[6]

$$\begin{aligned}\mathbf{E}(\mathbf{r}) = &- i\omega\mu \int_{V_{\text{imp}}} \underline{\mathbf{G}}_e^{(1)}(\mathbf{r}, \mathbf{r}') \cdot \mathbf{J}_{\text{imp}}(\mathbf{r}')\, dV' \\ &- i\omega\mu \sum_{i=1}^{n} \int_{S_m^i} \underline{\mathbf{G}}_e^{(1)}(\mathbf{r}, \mathbf{r}') \cdot \mathbf{J}_i(\mathbf{r}')\, dS'.\end{aligned} \tag{9.21}$$

---

[5]Indeed, multiplying (9.14) by $\underline{\mathbf{G}}_e^{(1)}(\mathbf{r}, \mathbf{r}')$ from the right (dot product), and (9.17) by $\mathbf{E}(\mathbf{r})$ from the left (dot product), and subtracting the left and right sides we obtain

$$\begin{aligned}&[\nabla \times \nabla \times \mathbf{E}(\mathbf{r})] \cdot \underline{\mathbf{G}}_e^{(1)}(\mathbf{r}, \mathbf{r}') - \mathbf{E}(\mathbf{r}) \cdot [\nabla \times \nabla \times \underline{\mathbf{G}}_e^{(1)}(\mathbf{r}, \mathbf{r}')] \\ &\qquad = -i\omega\mu \mathbf{J}_{\text{imp}}(\mathbf{r}) \cdot \underline{\mathbf{G}}_e^{(1)}(\mathbf{r}, \mathbf{r}') - \mathbf{E}(\mathbf{r})\delta(\mathbf{r} - \mathbf{r}').\end{aligned}$$

Applying the vector-dyadic Green's second theorem results in

$$\begin{aligned}\mathbf{E}(\mathbf{r}') = &-i\omega\mu \int_{V_{\text{imp}}} \mathbf{J}_{\text{imp}}(\mathbf{r}) \cdot \underline{\mathbf{G}}_e^{(1)}(\mathbf{r}, \mathbf{r}')\, dV \\ &- \oint_S \mathbf{n} \cdot \left[\mathbf{E}(\mathbf{r}) \times \nabla \times \underline{\mathbf{G}}_e^{(1)}(\mathbf{r}, \mathbf{r}') + \nabla \times \mathbf{E}(\mathbf{r}) \times \underline{\mathbf{G}}_e^{(1)}(\mathbf{r}, \mathbf{r}')\right] dS.\end{aligned}$$

Imposing boundary conditions (9.15) and (9.18) reduces the surface integral to $\sum_{i=1}^{n} \int_{S_m^i} [\mathbf{n}_i \times \nabla \times \mathbf{E}(\mathbf{r})] \cdot \underline{\mathbf{G}}_e^{(1)}(\mathbf{r}, \mathbf{r}')\, dS$.

[6]Note that in the source region ($\mathbf{r} \in V_{\text{imp}}$) the volume integral must be evaluated

In order to determine the unknown induced scatterer currents we impose the boundary condition for tangential components of the total electric field on the metal surfaces $S_m^p$,

$$\mathbf{n}_p \times \mathbf{E}(\mathbf{r})|_{S_m^p} = \mathbf{0}$$

for $p = 1, \ldots, n$, resulting in a coupled system of $n$ integral equations of the first kind for the induced electric currents $\mathbf{J}_i(\mathbf{r}')$,

$$\mathbf{n}_p \times \sum_{i=1}^{n} \int_{S_m^i} \underline{\mathbf{G}}_e^{(1)}(\mathbf{r}, \mathbf{r}') \cdot \mathbf{J}_i(\mathbf{r}')\, dS' = -\,\mathbf{n}_p \times \int_{V_{\mathrm{imp}}} \underline{\mathbf{G}}_e^{(1)}(\mathbf{r}, \mathbf{r}') \cdot \mathbf{J}_{\mathrm{imp}}(\mathbf{r}')\, dV' \tag{9.22}$$

for $\mathbf{r} \in S_m^p$. Once the induced currents are found, (9.21) provides the total electric field in the waveguide, and the magnetic field can be found from $\mathbf{E}$ using Faraday's law. If $S_m$ denotes a closed surface, (9.22) may exhibit problems associated with interior resonances, as described in Section 1.4.2.

Alternatively, the magnetic field can be determined directly from the currents induced on the scatterer. The magnetic field is such that

$$\begin{aligned} \nabla \times \nabla \times \mathbf{H}(\mathbf{r}) - k^2 \mathbf{H}(\mathbf{r}) &= \nabla \times \mathbf{J}_{\mathrm{imp}}(\mathbf{r}), & \mathbf{r} \in V, \\ \mathbf{n} \times \nabla \times H(r) = 0 \quad \mathbf{n} \cdot \mathbf{H}(\mathbf{r}) &= 0, & \mathbf{r} \in S, \end{aligned} \tag{9.23}$$

and the analogous problem for the electric Green's dyadic of the second kind, which accounts for the background waveguide, is

$$\begin{aligned} \nabla \times \nabla \times \underline{\mathbf{G}}_e^{(2)}(\mathbf{r}, \mathbf{r}') - k^2 \underline{\mathbf{G}}_e^{(2)}(\mathbf{r}, \mathbf{r}') &= \underline{\mathbf{I}}\, \delta(\mathbf{r} - \mathbf{r}'), & \mathbf{r}, \mathbf{r}' \in V, \\ \mathbf{n} \times \nabla \times \underline{\mathbf{G}}_e^{(2)}(\mathbf{r}, \mathbf{r}') = \underline{\mathbf{0}}, \quad \mathbf{n} \cdot \underline{\mathbf{G}}_e^{(2)}(\mathbf{r}, \mathbf{r}') &= \mathbf{0}, & \mathbf{r} \in S_M, \end{aligned} \tag{9.24}$$

subject to the fitness condition (limiting absorption) at infinity.

From (9.23), (9.24), and the vector-dyadic Green's second theorem we obtain the total magnetic field caused by impressed and induced electric currents as

$$\begin{aligned} \mathbf{H}(\mathbf{r}') = &\int_{V_{\mathrm{imp}}} [\nabla \times \mathbf{J}_{\mathrm{imp}}(\mathbf{r})] \cdot \underline{\mathbf{G}}_e^{(2)}(\mathbf{r}, \mathbf{r}')\, dV \\ &+ \sum_{i=1}^{n} \int_{S_m^i} \mathbf{J}_i(\mathbf{r}) \cdot [\nabla \times \underline{\mathbf{G}}_e^{(2)}(\mathbf{r}, \mathbf{r}')]\, dS. \end{aligned} \tag{9.25}$$

---

using a limiting procedure (see Section 1.3.5), resulting in

$$\begin{aligned} \mathbf{E}(\mathbf{r}) = &-i\omega\mu \lim_{\delta \to 0} \int_{V_{imp} - V_\delta} \underline{\mathbf{G}}_e^{(1)}\left(\mathbf{r}, \mathbf{r}'\right) \cdot \mathbf{J}_{\mathrm{imp}}(\mathbf{r}')\, dV' - \frac{\underline{\mathbf{L}}(\mathbf{r}) \cdot \mathbf{J}_{\mathrm{imp}}(\mathbf{r})}{i\omega\varepsilon} \\ &- i\omega\mu \sum_{i=1}^{n} \int_{S_m^i} \underline{\mathbf{G}}_e^{(1)}\left(\mathbf{r}, \mathbf{r}'\right) \cdot \mathbf{J}_i(\mathbf{r}')\, dS', \end{aligned}$$

where $\underline{\mathbf{L}}(\mathbf{r})$ is the depolarizing dyadic (1.79).

By interchanging $\mathbf{r}'$ and $\mathbf{r}$ and using the identities

$$\begin{aligned}[\nabla' \times \mathbf{J}(\mathbf{r}')] \cdot \underline{\mathbf{G}}_e^{(2)}(\mathbf{r}', \mathbf{r}) &= [\underline{\mathbf{G}}_e^{(2)}(\mathbf{r}', \mathbf{r})]^\top \cdot [\nabla' \times \mathbf{J}(\mathbf{r}')] \\ &= \underline{\mathbf{G}}_e^{(2)}(\mathbf{r}, \mathbf{r}') \cdot [\nabla' \times \mathbf{J}(\mathbf{r}')],\end{aligned} \tag{9.26}$$

$$\begin{aligned}\mathbf{J}(\mathbf{r}') \cdot [\nabla' \times \underline{\mathbf{G}}_e^{(2)}(\mathbf{r}', \mathbf{r})] &= [\nabla' \times \underline{\mathbf{G}}_e^{(2)}(\mathbf{r}', \mathbf{r})]^\top \cdot \mathbf{J}(\mathbf{r}') \\ &= [\nabla \times \underline{\mathbf{G}}_e^{(1)}(\mathbf{r}, \mathbf{r}')] \cdot \mathbf{J}(\mathbf{r}'),\end{aligned} \tag{9.27}$$

we obtain

$$\begin{aligned}\mathbf{H}(\mathbf{r}) = &\int_{V_{\text{imp}}} \underline{\mathbf{G}}_e^{(2)}(\mathbf{r}, \mathbf{r}') \cdot [\nabla' \times \mathbf{J}_{\text{imp}}(\mathbf{r}')] \, dV' \\ &+ \sum_{i=1}^{n} \int_{S_m^i} [\nabla \times \underline{\mathbf{G}}_A^{(1)}(\mathbf{r}, \mathbf{r}')] \cdot \mathbf{J}_i(\mathbf{r}') \, dS',\end{aligned} \tag{9.28}$$

where $\nabla \times \underline{\mathbf{G}}_e^{(1)}(\mathbf{r}, \mathbf{r}') = \nabla \times \underline{\mathbf{G}}_A^{(1)}(\mathbf{r}, \mathbf{r}')$ by (9.11). Once the induced currents are found (perhaps by solving (9.22)), then (9.28) provides the total magnetic field within the waveguide. Moreover, if the obstacle surfaces are closed, a third alternative is to form a magnetic field integral equation using (9.28), as described in Section 1.4.2. We provide explicit expressions for the Green's dyadics $\underline{\mathbf{G}}_A^{(1)}$ and $\underline{\mathbf{G}}_e^{(1,2)}$ in Section 9.2.

When the metal scatterers contain edges,[7] we have to restrict the solution by enforcing a finite energy condition on the total field in the vicinity of the edge,

$$\lim_{V_e \to 0} \int_{V_e} \left( |\mathbf{E}(\mathbf{r})|^2 + \frac{1}{k^2} |\nabla \times \mathbf{E}(\mathbf{r})|^2 \right) dV = 0, \tag{9.29}$$

which guarantees that there is no source in $V_e$. The condition (9.29) defines a subspace of $\mathbf{L}^2(V)^3$ as the Sobolev space $\mathbf{W}_2^1(V)^3 \subset \mathbf{L}^2(V)^3$ and, in particular, as the Hilbert space $H(\text{curl}, V)$ defined as (see Section 2.1.2)

$$H(\text{curl}, V) \equiv \left\{ \mathbf{E} : \mathbf{E}(\mathbf{r}), \nabla \times \mathbf{E}(\mathbf{r}) \in \mathbf{L}^2(V)^3 \right\}$$

with $\mathbf{n} \times \mathbf{E}(\mathbf{r})|_S \in H^{1/2}(S)$ [48]. In addition, in a source-free region we also have

$$H(\text{div}, V) \equiv \left\{ \mathbf{E} : \mathbf{E}(\mathbf{r}) \in \mathbf{L}^2(V)^3, \nabla \cdot \mathbf{E}(\mathbf{r}) \in \mathbf{L}^2(V) \right\},$$

and the solution is sought in the subspace $H(\text{div}, \text{curl}, V) = H(\text{curl}, V) \cap H(\text{div}, V)$. In general, these conditions mostly concern scattered fields calculated away from the source region. The incident and scattered electric fields have square-integrable curls and divergences with the boundary values of their tangential components defined within the space $H^{1/2}(S_m^i) \subset \mathbf{L}^2(S_m^i)$.

[7] In the vicinity of an edge the boundary value of the normal derivative of the normal component of the electric (magnetic) field, $\partial E_n/\partial n$ ($\partial H_n/\partial n$), is defined within the Sobolev space $H^{-1/2}(S)$, and boundary values of the normal components $E_n$ ($H_n$) are defined within the space $H^{1/2}(S)$ (see Appendix E).

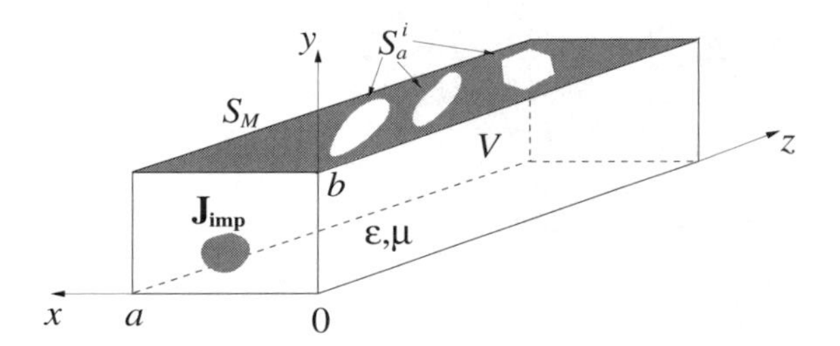


Figure 9.2: Infinite waveguide filled with a homogeneous medium and having arbitrarily shaped apertures located on the waveguide surface.

### 9.1.3 Infinite Waveguide with Apertures

Consider an infinite cylindrical waveguide characterized by material parameters $\varepsilon$ and $\mu$ and having arbitrarily shaped apertures (slots) $S_a^i$ located on the waveguide surface $S_M$, as depicted in Figure 9.2 for the special case of a rectangular waveguide. An impressed electric current $\mathbf{J}_{\text{imp}}(\mathbf{r})$, $\mathbf{r} \in V_{\text{imp}} \subset V$, generates an impressed electric and magnetic field. In this section we obtain integral representations for the total electric and magnetic field inside the waveguide as the superposition of incident and scattered fields caused by impressed and induced currents, respectively.

The magnetic field is such that

$$\begin{aligned} \nabla \times \nabla \times \mathbf{H}(\mathbf{r}) - k^2\mathbf{H}(\mathbf{r}) = \nabla \times \mathbf{J}_{\text{imp}}(\mathbf{r}), \qquad & \mathbf{r} \in V, \\ \mathbf{n} \times \nabla \times \mathbf{H}(\mathbf{r}) = \mathbf{0}, \quad \mathbf{n} \cdot \mathbf{H}(\mathbf{r}) = 0, \qquad & \mathbf{r} \in S_M, \end{aligned} \tag{9.30}$$

and so we formulate the boundary value problem for the electric Green's dyadic of the second kind as

$$\begin{aligned} \nabla \times \nabla \times \underline{\mathbf{G}}_e^{(2)}(\mathbf{r},\mathbf{r}') - k^2\underline{\mathbf{G}}_e^{(2)}(\mathbf{r},\mathbf{r}') = \underline{\mathbf{I}}\,\delta(\mathbf{r}-\mathbf{r}'), \qquad & \mathbf{r},\mathbf{r}' \in V, \\ \mathbf{n} \times \nabla \times \underline{\mathbf{G}}_e^{(2)}(\mathbf{r},\mathbf{r}') = \underline{\mathbf{0}}, \quad \mathbf{n} \cdot \underline{\mathbf{G}}_e^{(2)}(\mathbf{r},\mathbf{r}') = \mathbf{0}, \qquad & \mathbf{r} \in \widetilde{S}, \end{aligned} \tag{9.31}$$

where $\widetilde{S} = S_M \cup \left(\cup_{i=1}^n S_a^i\right)$, subject to the fitness condition (limiting absorption) at infinity. The vector-dyadic Green's second theorem applied to (9.30) and (9.31) provides the total magnetic field inside the waveguide,

$$\mathbf{H}(\mathbf{r}') = \int_{V_{\text{imp}}} [\nabla \times \mathbf{J}_{\text{imp}}(\mathbf{r})] \cdot \underline{\mathbf{G}}_e^{(2)}(\mathbf{r},\mathbf{r}')\, dV - \sum_{i=1}^{n} \int_{S_a^i} [\mathbf{n}_i \times \nabla \times \mathbf{H}(\mathbf{r})] \cdot \underline{\mathbf{G}}_e^{(2)}(\mathbf{r},\mathbf{r}')\, dS.$$

By interchanging $\mathbf{r}'$ and $\mathbf{r}$ and using (9.26) we obtain

$$\mathbf{H}(\mathbf{r}) = \int_{V_{\text{imp}}} \underline{\mathbf{G}}_e^{(2)}(\mathbf{r},\mathbf{r}') \cdot [\nabla' \times \mathbf{J}_{\text{imp}}(\mathbf{r}')]\, dV' + i\omega\varepsilon \sum_{i=1}^{n} \int_{S_a^i} \underline{\mathbf{G}}_e^{(2)}(\mathbf{r},\mathbf{r}') \cdot \mathbf{K}_i(\mathbf{r}')\, dS' \quad (9.32)$$

where we have introduced magnetic currents on the apertures,

$$\mathbf{K}_i(\mathbf{r}) \equiv -\mathbf{n}_i \times \mathbf{E}(\mathbf{r}) = \frac{i}{\omega\varepsilon}\mathbf{n}_i \times \nabla \times \mathbf{H}(\mathbf{r})$$

for $\mathbf{r} \in S_a^i$.

The unknown aperture currents may be determined by enforcing continuity of the tangential components of magnetic field across the aperture. On the waveguide side of the aperture we have $\mathbf{H}_{wg}(\mathbf{r})$, given by (9.32), and letting $\mathbf{H}_0(\mathbf{r})$ be the magnetic field in the region outside the waveguide, we have

$$\mathbf{n}_p \times \mathbf{H}_{wg}(\mathbf{r})|_{S_a^p} = \mathbf{n}_p \times \mathbf{H}_0(\mathbf{r})|_{S_a^p}$$

for $p = 1, \ldots, n$. Determining a representation for the magnetic field $\mathbf{H}_0(\mathbf{r})$ in terms of the aperture currents is difficult but can be done for some canonical geometries, although the details are omitted here.

Once the aperture currents are determined, (9.32) provides the total magnetic field inside the waveguide, and the electric field may be determined from Ampère's law. Alternatively, the electric field inside the waveguide may be determined directly from the aperture currents. The electric field satisfies the vector wave equation

$$\nabla \times \nabla \times \mathbf{E}(\mathbf{r}) - k^2\mathbf{E}(\mathbf{r}) = -i\omega\mu\mathbf{J}_{\text{imp}}(\mathbf{r}), \qquad \mathbf{r} \in V, \quad (9.33)$$

and the boundary condition on the waveguide surface[8] $S_M$

$$\mathbf{n} \times \mathbf{E}(\mathbf{r})|_{S_M} = \mathbf{0}, \quad (9.34)$$

where $\mathbf{n}$ is an outward unit normal vector to the surface $S_M$. The Green's function problem is formulated as

$$\nabla \times \nabla \times \underline{\mathbf{G}}_e^{(1)}(\mathbf{r},\mathbf{r}') - k^2\underline{\mathbf{G}}_e^{(1)}(\mathbf{r},\mathbf{r}') = \underline{\mathbf{I}}\,\delta(\mathbf{r}-\mathbf{r}'), \qquad \mathbf{r},\mathbf{r}' \in V, \quad (9.35)$$

$$\mathbf{n} \times \underline{\mathbf{G}}_e^{(1)}(\mathbf{r},\mathbf{r}') = \underline{\mathbf{0}}, \qquad \mathbf{r} \in \widetilde{S}, \quad (9.36)$$

where $\widetilde{S} = S_M \cup \left(\cup_{i=1}^n S_a^i\right)$. Therefore, the Green's dyadic corresponds to an infinite waveguide filled with a homogeneous medium.

[8] Note that the electric field is unknown on the apertures $S_a^i$.

The vector-dyadic Green's second theorem applied to $\mathbf{E}$ and $\underline{\mathbf{G}}_e^{(1)}$, with boundary conditions (9.34) and (9.36), results in the integral representation for the electric field

$$\mathbf{E}(\mathbf{r}') = -i\omega\mu \int_{V_{\text{imp}}} \mathbf{J}_{\text{imp}}(\mathbf{r}) \cdot \underline{\mathbf{G}}_e^{(1)}(\mathbf{r}, \mathbf{r}')\, dV + \sum_{i=1}^{n} \int_{S_a^i} \mathbf{K}_i(\mathbf{r}) \cdot [\nabla \times \underline{\mathbf{G}}_e^{(1)}(\mathbf{r}, \mathbf{r}')]\, dS$$

where $\mathbf{K}_i(\mathbf{r}) \equiv -\mathbf{n}_i \times \mathbf{E}(\mathbf{r})$ on the aperture. By interchanging $\mathbf{r}'$ and $\mathbf{r}$ and using (9.11) and the identity

$$\begin{aligned}[\mathbf{n}_i \times \mathbf{E}(\mathbf{r}')] \cdot [\nabla' \times \underline{\mathbf{G}}_e^{(1)}(\mathbf{r}', \mathbf{r})] &= [\nabla' \times \underline{\mathbf{G}}_e^{(1)}(\mathbf{r}', \mathbf{r})]^{\top} \cdot [\mathbf{n}_i \times \mathbf{E}(\mathbf{r}')] \\ &= [\nabla \times \underline{\mathbf{G}}_e^{(2)}(\mathbf{r}, \mathbf{r}')] \cdot [\mathbf{n}_i \times \mathbf{E}(\mathbf{r}')]\end{aligned} \tag{9.37}$$

we obtain[9]

$$\mathbf{E}(\mathbf{r}) = -i\omega\mu \int_{V_{\text{imp}}} \underline{\mathbf{G}}_e^{(1)}(\mathbf{r}, \mathbf{r}') \cdot \mathbf{J}_{\text{imp}}(\mathbf{r}')\, dV' + \sum_{i=1}^{n} \int_{S_a^i} [\nabla \times \underline{\mathbf{G}}_A^{(2)}(\mathbf{r}, \mathbf{r}')] \cdot \mathbf{K}_i(\mathbf{r}')\, dS'. \tag{9.42}$$

In Section 9.2 we provide explicit representations for the Green's dyadics $\underline{\mathbf{G}}_e^{(1,2)}$ and $\underline{\mathbf{G}}_A^{(2)}$.

---

[9] Equation (9.42), written in terms of the electric field, has been applied in the method of overlapping regions [46], [53]–[55], where $\mathbf{r}$ and $\mathbf{r}'$ are geometrically separated, which allows for the interchange of curl and integral operators,

$$\mathbf{E}(\mathbf{r}) = -j\omega\mu \int_{V_{\text{imp}}} \underline{\mathbf{G}}_e^{(1)}\left(\mathbf{r}, \mathbf{r}'\right) \cdot \mathbf{J}_{\text{imp}}(\mathbf{r}')\, dV' - \sum_{i=1}^{n} \nabla \times \int_{S_a^i} \underline{\mathbf{G}}_A^{(2)}\left(\mathbf{r}, \mathbf{r}'\right) \cdot \left[\mathbf{n}_i \times \mathbf{E}(\mathbf{r}')\right] dS'. \tag{9.38}$$

The surface integral is weakly singular, and it has been shown in the analysis of microstrip lines [55] that this leads to a matrix equation of the second kind with a compact operator in the sequence space $\mathbf{l}^1$ (see Section 9.5). The magnetic potential Green's dyadic of the second kind, $\underline{\mathbf{G}}_A^{(2)}(\mathbf{r}, \mathbf{r}')$, used in the representation of the scattered electric field, is obtained as the solution of

$$\nabla^2 \underline{\mathbf{G}}_A^{(2)}\left(\mathbf{r}, \mathbf{r}'\right) + k^2 \underline{\mathbf{G}}_A^{(2)}\left(\mathbf{r}, \mathbf{r}'\right) = -\underline{\mathbf{I}}\delta(\mathbf{r} - \mathbf{r}'), \qquad \mathbf{r}, \mathbf{r}' \in V, \tag{9.39}$$

$$\mathbf{n} \times \nabla \times \underline{\mathbf{G}}_A^{(2)}\left(\mathbf{r}, \mathbf{r}'\right) = \underline{\mathbf{0}}, \qquad \mathbf{r} \in \widetilde{S}, \tag{9.40}$$

$$\mathbf{n} \cdot \left(\underline{\mathbf{I}} + \frac{1}{k^2}\nabla\nabla\right) \cdot \underline{\mathbf{G}}_A^{(2)}\left(\mathbf{r}, \mathbf{r}'\right) = \mathbf{0}, \qquad r \in \widetilde{S}. \tag{9.41}$$

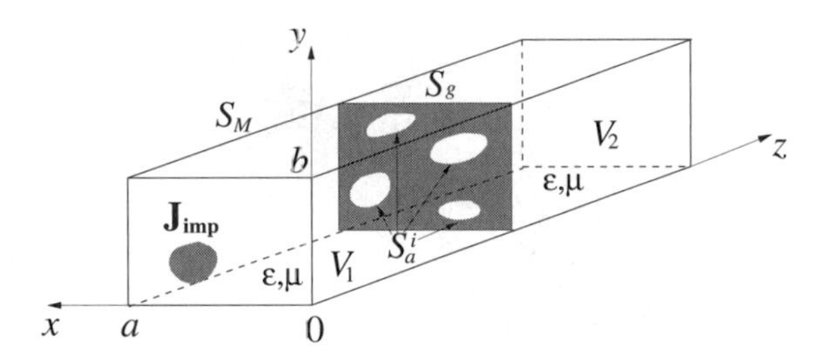


Figure 9.3: Semi-infinite waveguides coupled through apertures in a common ground plane.

## 9.1.4 Semi-Infinite Waveguides Coupled through Apertures in a Common Ground Plane

Consider an infinite cylindrical waveguide filled with a homogeneous medium and containing a ground plane perpendicular to the waveguiding axis having arbitrarily shaped apertures, as depicted in Figure 9.3 for the special case of a rectangular waveguide. The ground plane $S_g$ separates the waveguide of volume $V$ into two semi-infinite waveguides, $V_1$ and $V_2$, coupled together by field penetration through the apertures $S_a^i$. An impressed electric current in region 1, $\mathbf{J}_{\text{imp}}(\mathbf{r})$ where $\mathbf{r} \in V_{\text{imp}} \subset V_1$, generates an impressed electric and magnetic field.[10] The total electric and magnetic field caused by impressed and induced currents can be written in an integral form using the previously obtained representations (9.42) and (9.32),

$$\begin{aligned}\mathbf{E}_l(\mathbf{r}) = &-\delta_{1l} i\omega\mu \int_{V_{\text{imp}}} \underline{\mathbf{G}}_{e1}^{(1)}(\mathbf{r},\mathbf{r}') \cdot \mathbf{J}_{\text{imp}}(\mathbf{r}')\, dV' \\ &- (-1)^l \sum_{i=1}^{n} \int_{S_a^i} [\nabla \times \underline{\mathbf{G}}_{Al}^{(2)}(\mathbf{r},\mathbf{r}')] \cdot \mathbf{K}_i(\mathbf{r}')\, dS',\end{aligned} \tag{9.43}$$

$$\begin{aligned}\mathbf{H}_l(\mathbf{r}) = &\,\delta_{1l} \int_{V_{\text{imp}}} \underline{\mathbf{G}}_{e1}^{(2)}(\mathbf{r},\mathbf{r}') \cdot [\nabla' \times \mathbf{J}_{\text{imp}}(\mathbf{r}')]\, dV' \\ &- (-1)^l i\omega\varepsilon \sum_{i=1}^{n} \int_{S_a^i} \underline{\mathbf{G}}_{el}^{(2)}(\mathbf{r},\mathbf{r}') \cdot \mathbf{K}_i(\mathbf{r}')\, dS',\end{aligned} \tag{9.44}$$

$l = 1, 2$, where $\delta_{1l} = 1$ for $l = 1$ and 0 for $l \neq 1$. The Green's dyadics for semi-infinite rectangular waveguides, $\underline{\mathbf{G}}_{el}^{(2)}(\mathbf{r},\mathbf{r}')$, $\underline{\mathbf{G}}_{e1}^{(1)}(\mathbf{r},\mathbf{r}')$, and $\underline{\mathbf{G}}_{Al}^{(2)}(\mathbf{r},\mathbf{r}')$ satisfy dyadic differential equations (9.31), (9.35), and (9.39), respectively, and boundary conditions (9.31), (9.36), and (9.40), (9.41) for $\mathbf{r} \in S_M \cup S_g \cup \left(\cup_{i=1}^n S_a^i\right)$.

[10] A similar formulation can also be obtained for an impressed electric current located in region $V_2$.

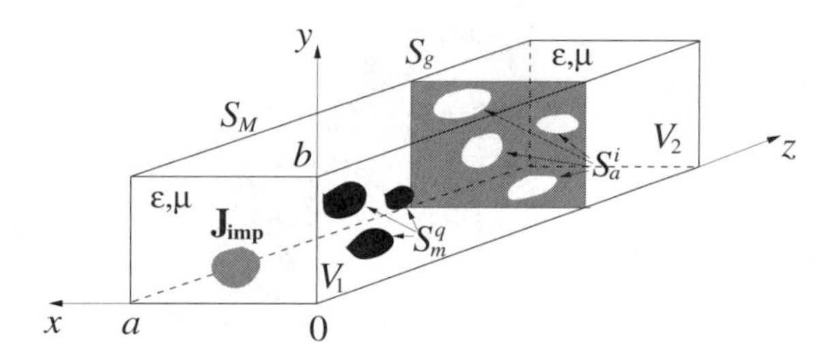


Figure 9.4: Semi-infinite waveguides containing perfectly conducting objects and coupled through apertures in a common ground plane.

A coupled system of $n$ first-kind integral equations is obtained by enforcing the continuity condition for the tangential components of the magnetic field across the apertures $S_a^p$,

$$\mathbf{n}_p \times \mathbf{H}_1(\mathbf{r})|_{S_a^p} = \mathbf{n}_p \times \mathbf{H}_2(\mathbf{r})|_{S_a^p}$$

for $p = 1, \ldots, n$, resulting in the coupled set of $n$ first-kind integral equations

$$\begin{aligned} i\omega\varepsilon\, \mathbf{n}_p \times \sum_{i=1}^{n} \int_{S_a^i} (\underline{\mathbf{G}}_{e1}^{(2)}(\mathbf{r},\mathbf{r}') + \underline{\mathbf{G}}_{e2}^{(2)}(\mathbf{r},\mathbf{r}')) \cdot \mathbf{K}_i(\mathbf{r}')\, dS' \\ = -\mathbf{n}_p \times \int_{V_{\text{imp}}} \underline{\mathbf{G}}_{e1}^{(2)}(\mathbf{r},\mathbf{r}') \cdot [\nabla' \times \mathbf{J}_{\text{imp}}(\mathbf{r}')]\, dV', \qquad \mathbf{r} \in S_a^p. \quad (9.45) \end{aligned}$$

### 9.1.5 Semi-Infinite Waveguides with Metal Obstacles Coupled through Apertures in a Common Ground Plane

In this section we consider the composite example of perfectly conducting obstacles in region $V_1$ of semi-infinite waveguides coupled through apertures in a common ground plane, as depicted in Figure 9.4 for the special case of a rectangular waveguide. The incident electric and magnetic fields caused by the impressed electric current $\mathbf{J}_{\text{imp}}(\mathbf{r})$, $\mathbf{r} \in V_{\text{imp}} \subset V_1$, will simultaneously induce electric currents on the surface of the conducting scatterers $S_m^q$ and magnetic currents on the apertures $S_a^i$. In this case, we obtain scattered electric and magnetic fields caused by both induced electric and magnetic currents. The total electric and magnetic fields in region $V_1$ can be written in integral form as a superposition of incident and scattered electric and magnetic fields, respectively. By (9.21), (9.28), (9.43), and (9.44) we obtain

$$\begin{aligned}
\mathbf{E}_1(\mathbf{r}) = &-i\omega\mu \int_{V_{\text{imp}}} \underline{\mathbf{G}}_{e1}^{(1)}(\mathbf{r}, \mathbf{r}') \cdot \mathbf{J}_{\text{imp}}(\mathbf{r}')\, dV' \\
&- i\omega\mu \sum_{q=1}^{N} \int_{S_m^q} \underline{\mathbf{G}}_{e1}^{(1)}(\mathbf{r}, \mathbf{r}') \cdot \mathbf{J}_q(\mathbf{r}')\, dS' \\
&+ \sum_{i=1}^{M} \int_{S_a^i} [\nabla \times \underline{\mathbf{G}}_{A1}^{(2)}(\mathbf{r}, \mathbf{r}')] \cdot \mathbf{K}_i(\mathbf{r}')\, dS'
\end{aligned} \tag{9.46}$$

$$\begin{aligned}
\mathbf{H}_1(\mathbf{r}) = &\int_{V_{\text{imp}}} \underline{\mathbf{G}}_{e1}^{(2)}(\mathbf{r}, \mathbf{r}') \cdot [\nabla' \times \mathbf{J}_{\text{imp}}(\mathbf{r}')]\, dV' \\
&+ \sum_{q=1}^{N} \int_{S_m^q} [\nabla \times \underline{\mathbf{G}}_{A1}^{(1)}(\mathbf{r}, \mathbf{r}')] \cdot \mathbf{J}_q(\mathbf{r}')\, dS' \\
&+ i\omega\varepsilon \sum_{i=1}^{M} \int_{S_a^i} \underline{\mathbf{G}}_{e1}^{(2)}(\mathbf{r}, \mathbf{r}') \cdot \mathbf{K}_i(\mathbf{r}')\, dS'.
\end{aligned} \tag{9.47}$$

The representation for $\mathbf{E}_2(\mathbf{r})$ and $\mathbf{H}_2(\mathbf{r})$ is the same as that obtained in the previous example (see (9.43) and (9.44) for $l = 2$). The presence of metal obstacles in $V_1$ does not affect the formulation for the Green's dyadics introduced in (9.46) and (9.47), which have been discussed above.

We can formulate a coupled set of integral equations by enforcing the boundary condition for tangential components of electric field on the obstacles $S_m^r$, and continuity of the tangential components of magnetic field across the apertures $S_a^p$. From (9.46) and

$$\mathbf{n}_r \times \mathbf{E}_1(\mathbf{r})|_{S_m^r} = \mathbf{0}$$

for $r = 1, \ldots, N$, we obtain

$$\begin{aligned}
&i\omega\mu\mathbf{n}_r \times \sum_{q=1}^{N} \int_{S_m^q} \underline{\mathbf{G}}_{e1}^{(1)}(\mathbf{r}, \mathbf{r}') \cdot \mathbf{J}_q(\mathbf{r}')\, dS' \\
&\qquad -\mathbf{n}_r \times \sum_{i=1}^{M} \int_{S_a^i} [\nabla \times \underline{\mathbf{G}}_{A1}^{(2)}(\mathbf{r}, \mathbf{r}')] \cdot \mathbf{K}_i(\mathbf{r}')\, dS' \\
&= -i\omega\mu\mathbf{n}_r \times \int_{V_{\text{imp}}} \underline{\mathbf{G}}_{e1}^{(1)}(\mathbf{r}, \mathbf{r}') \cdot \mathbf{J}_{\text{imp}}(\mathbf{r}')\, dV', \qquad \mathbf{r} \in S_m^r,
\end{aligned} \tag{9.48}$$

and from (9.44), (9.47), and

$$\mathbf{n}_p \times \mathbf{H}_1(\mathbf{r})|_{S_a^p} = \mathbf{n}_p \times \mathbf{H}_2(\mathbf{r})|_{S_a^p}$$

we obtain

$$\mathbf{n}_p \times \sum_{q=1}^{N} \int_{S_m^q} [\nabla \times \underline{\mathbf{G}}_{A1}^{(1)}(\mathbf{r}, \mathbf{r}')] \cdot \mathbf{J}_q(\mathbf{r}')\, dS'$$

$$+ i\omega\varepsilon \mathbf{n}_p \times \sum_{i=1}^{M} \int_{S_a^i} (\underline{\mathbf{G}}_{e1}^{(2)}(\mathbf{r}, \mathbf{r}') + \underline{\mathbf{G}}_{e2}^{(2)}(\mathbf{r}, \mathbf{r}')) \cdot \mathbf{K}_i(\mathbf{r}')\, dS'$$

$$= -\mathbf{n}_p \times \int_{V_{\text{imp}}} \underline{\mathbf{G}}_{e1}^{(2)}(\mathbf{r}, \mathbf{r}') \cdot [\nabla' \times \mathbf{J}_{\text{imp}}(\mathbf{r}')]\, dV', \qquad \mathbf{r} \in S_a^p. \quad (9.49)$$

The integral equations represent a coupled system of $N + M$ integral equations with respect to the unknown induced electric currents $\mathbf{J}_q(\mathbf{r}')$ and magnetic currents $\mathbf{K}_i(\mathbf{r}')$.

## 9.2 Green's Dyadics for Waveguides Filled with a Homogeneous Medium

In the previous section we developed integral equation formulations that utilize various dyadic Green's functions. In this section we obtain explicit representations for these Green's functions for the special case of a rectangular waveguide. We consider arbitrarily oriented, three-dimensional sources.

A traditional and general way to construct Green's functions for closed-boundary, guided-wave structures, semi-infinite waveguides, and cavities is to use the Hansen vector wave functions $\mathbf{M}$, $\mathbf{N}$, and $\mathbf{L}$ (see Section 10.2) in a double-series expansion [82], [36]. Electric and magnetic dyadic Green's functions for uniform infinite and semi-infinite rectangular waveguides were obtained in [38], [36], and [49], and for a rectangular cavity in [38], [36], [49], and [50]. Also, dyadic Green's functions for a magnetic current in a rectangular waveguide were obtained in the form of a Fourier integral in the direction along the waveguiding axis, and a Fourier series in the transverse direction in [51].

An alternative representation of the Green's function for cylindrical waveguides and cavities is a partial eigenfunction expansion involving the complete system of eigenfunctions of the transverse Laplacian operator and a one-dimensional characteristic Green's function [17]. The properties of completeness and orthogonality of the transverse eigenfunctions allow for the formulation of a Sturm–Liouville problem for the characteristic Green's function in the waveguiding coordinate. This method has been applied in [52] for the derivation of the magnetic potential dyadic Green's function (diagonal tensor) for rectangular waveguides and cavities using scalar eigenfunctions of the Laplacian operator. Important developments in this approach have been reported in Russian and Ukrainian literature,

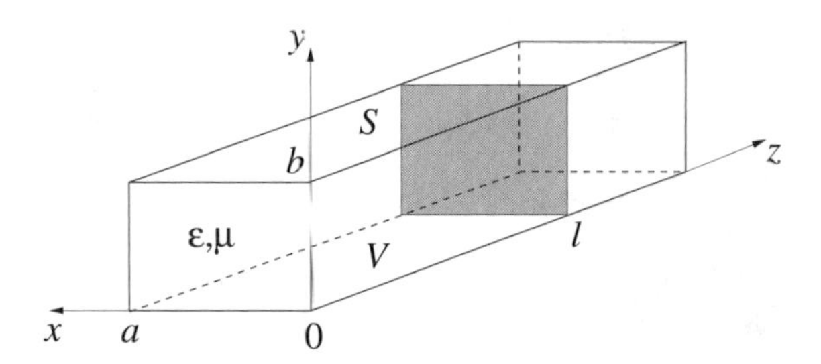


Figure 9.5: Infinite rectangular waveguide filled with a linear homogeneous medium with material parameters $\varepsilon$ and $\mu$. Transverse perfectly conducting plane is shown at $z = l$ to obtain boundary conditions for the Green's function components for the case of semi-infinite waveguide and cavity.

for example, in [46] for applications to three-dimensional waveguide discontinuities, in [53] and [54] for antenna problems, and in [55] and [56] for the electric-field analysis of shielded printed-circuit transmission lines.

## 9.2.1 Magnetic Potential Green's Dyadics

Here we will derive the magnetic potential Green's dyadics of the first and second kinds, $\underline{\mathbf{G}}_A^{(1,2)}(\mathbf{r}, \mathbf{r}')$, for an infinite and semi-infinite rectangular waveguide, and for a rectangular cavity. The geometry of a rectangular waveguide of volume $V$ enclosed by a perfectly conducting surface $S$ and filled with a linear homogeneous medium with material parameters $\varepsilon$ and $\mu$ is shown in Figure 9.5. It can be shown that the magnetic potential Green's dyadic for a rectangular waveguide is a diagonal tensor of rank 2, such that [50], [52]

$$\underline{\mathbf{G}}_A(\mathbf{r}, \mathbf{r}') = \widehat{\mathbf{x}}\widehat{\mathbf{x}}G_{Axx}(\mathbf{r}, \mathbf{r}') + \widehat{\mathbf{y}}\widehat{\mathbf{y}}G_{Ayy}(\mathbf{r}, \mathbf{r}') + \widehat{\mathbf{z}}\widehat{\mathbf{z}}G_{Azz}(\mathbf{r}, \mathbf{r}').$$

### Magnetic Potential Green's Dyadic of the First Kind

We seek the first-kind magnetic potential Green's dyadic $\underline{\mathbf{G}}_A^{(1)}(\mathbf{r}, \mathbf{r}')$ as the solution of the dyadic Helmholtz equation

$$\nabla^2 \underline{\mathbf{G}}_A^{(1)}(\mathbf{r}, \mathbf{r}') + k^2 \underline{\mathbf{G}}_A^{(1)}(\mathbf{r}, \mathbf{r}') = -\underline{\mathbf{I}}\,\delta(\mathbf{r} - \mathbf{r}'), \qquad \mathbf{r}, \mathbf{r}' \in V, \tag{9.50}$$

subject to the boundary condition of the first kind on the surface $S$,

$$\begin{aligned} \mathbf{n} \times \underline{\mathbf{G}}_A^{(1)}(\mathbf{r}, \mathbf{r}') &= \underline{\mathbf{0}}, \qquad \mathbf{r} \in S, \\ \nabla \cdot \underline{\mathbf{G}}_A^{(1)}(\mathbf{r}, \mathbf{r}') &= \mathbf{0}, \qquad \mathbf{r} \in S, \end{aligned} \tag{9.51}$$

where $\mathbf{n}$ is an outward unit normal vector on $S$. The dyadic equation (9.50) for a diagonal tensor $\underline{\mathbf{G}}_A^{(1)}(\mathbf{r}, \mathbf{r}')$ is equivalent to three independent scalar

Helmholtz equations[11]

$$\nabla^2 G^{(1)}_{A\nu\nu}(\mathbf{r},\mathbf{r}') + k^2 G^{(1)}_{A\nu\nu}(\mathbf{r},\mathbf{r}') = -\delta(\mathbf{r}-\mathbf{r}'), \qquad \nu = x, y, z. \tag{9.52}$$

The boundary condition (9.51) can be written as

$$\begin{aligned}
\mathbf{n}=\widehat{\mathbf{x}}: &\quad \frac{\partial G^{(1)}_{Axx}}{\partial x}=0, \quad G^{(1)}_{Ayy}=0, \quad G^{(1)}_{Azz}=0 \quad \text{at } x=0,a,\\
\mathbf{n}=\widehat{\mathbf{y}}: &\quad G^{(1)}_{Axx}=0, \quad \frac{\partial G^{(1)}_{Ayy}}{\partial y}=0, \quad G^{(1)}_{Azz}=0 \quad \text{at } y=0,b,\\
\mathbf{n}=\widehat{\mathbf{z}}: &\quad G^{(1)}_{Axx}=0, \quad G^{(1)}_{Ayy}=0, \quad \frac{\partial G^{(1)}_{Azz}}{\partial z}=0 \quad \text{at } z=l,
\end{aligned} \tag{9.53}$$

where for the infinite or semi-infinite waveguide in the $z$-direction the fitness condition at infinity (limiting absorption) is satisfied.

### Magnetic Potential Green's Dyadic of the Second Kind

The magnetic potential Green's dyadic $\underline{\mathbf{G}}^{(2)}_A(\mathbf{r},\mathbf{r}')$ is obtained as the solution of the dyadic Helmholtz equation

$$\nabla^2 \underline{\mathbf{G}}^{(2)}_A(\mathbf{r},\mathbf{r}') + k^2 \underline{\mathbf{G}}^{(2)}_A(\mathbf{r},\mathbf{r}') = -\underline{\mathbf{I}}\,\delta(\mathbf{r}-\mathbf{r}'), \qquad \mathbf{r},\mathbf{r}' \in V, \tag{9.54}$$

with boundary conditions of the second kind on the metal surface $S$,

$$\mathbf{n}\times\nabla\times\underline{\mathbf{G}}^{(2)}_A(\mathbf{r},\mathbf{r}') = \underline{\mathbf{0}}, \qquad \mathbf{r}\in S, \tag{9.55}$$

$$\mathbf{n}\cdot\left(\underline{\mathbf{I}}+\frac{1}{k^2}\nabla\nabla\right)\cdot\underline{\mathbf{G}}^{(2)}_A(\mathbf{r},\mathbf{r}') = \mathbf{0}, \qquad \mathbf{r}\in S. \tag{9.56}$$

The scalar form of the boundary condition can be obtained using either (9.55) or (9.56) as

$$\begin{aligned}
\mathbf{n}=\widehat{\mathbf{x}}: &\quad G^{(2)}_{Axx}=0, \quad \frac{\partial G^{(2)}_{Ayy}}{\partial x}=0, \quad \frac{\partial G^{(2)}_{Azz}}{\partial x}=0 \quad \text{at } x=0,a,\\
\mathbf{n}=\widehat{\mathbf{y}}: &\quad \frac{\partial G^{(2)}_{Axx}}{\partial y}=0, \quad G^{(2)}_{Ayy}=0, \quad \frac{\partial G^{(2)}_{Azz}}{\partial y}=0 \quad \text{at } y=0,b,\\
\mathbf{n}=\widehat{\mathbf{z}}: &\quad \frac{\partial G^{(2)}_{Axx}}{\partial z}=0, \quad \frac{\partial G^{(2)}_{Ayy}}{\partial z}=0, \quad G^{(2)}_{Azz}=0 \quad \text{at } z=l.
\end{aligned} \tag{9.57}$$

[11] Note that, in general, the dyadic equation (9.50) is equivalent to nine independent scalar Helmholtz equations [46],

$$\nabla^2 G^{(1)}_{A\nu\upsilon}\left(\mathbf{r},\mathbf{r}'\right) + k^2 G^{(1)}_{A\nu\upsilon}\left(\mathbf{r},\mathbf{r}'\right) = -\delta_{\nu\upsilon}\delta(\mathbf{r}-\mathbf{r}'), \qquad \nu,\upsilon = x, y, z,$$

where $\delta_{\nu\upsilon} = 1$ for $\nu = \upsilon$ and 0 for $\nu \neq \upsilon$. However, for $\nu \neq \upsilon$, $G^{(1)}_{A\nu\upsilon}(\mathbf{r},\mathbf{r}')$ satisfy independent homogeneous equations and therefore do not represent Green's functions in the classical sense.

**Partial Eigenfunction Expansion Form**

The Green's dyadic components $G_{A\nu\nu}^{(1,2)}(\mathbf{r}, \mathbf{r}')$, satisfying Helmholtz equations (9.52) and boundary conditions (9.53) and (9.57), can be expressed as a double-series expansion over the complete system of eigenfunctions of the transverse Laplacian operator in the waveguide cross-section (see, e.g., Section 8.1.1),

$$G_{A\nu\nu}^{(1,2)}(\mathbf{r}, \mathbf{r}') = \sum_{m=0}^{\infty} \sum_{n=0}^{\infty} \phi_{mn}^{(1,2)\nu}(x, y) \phi_{mn}^{(1,2)\nu}(x', y') f_{mn}^{(1,2)\nu}(z, z'), \tag{9.58}$$

where $\phi_{mn}^{(1,2)\nu}(x, y)$ form an orthonormal

$$\int_0^a \int_0^b \phi_{mn}^{(1,2)\nu}(x, y) \phi_{ps}^{(1,2)\nu}(x, y)\, dy\, dx = \delta_{mp} \delta_{ns} \tag{9.59}$$

and complete

$$\sum_{m=0}^{\infty} \sum_{n=0}^{\infty} \phi_{mn}^{(1,2)\nu}(x, y) \phi_{mn}^{(1,2)\nu}(x', y') = \delta(x - x') \delta(y - y') \tag{9.60}$$

set of eigenfunctions of the transverse Laplacian operator

$$L = \nabla_{xy}^2 = \frac{\partial^2}{\partial x^2} + \frac{\partial^2}{\partial y^2}.$$

The eigenfunctions are determined as the solution of

$$(\nabla_{xy}^2 + \kappa_{mn}^2) \phi_{mn}^{(1,2)\nu}(x, y) = 0 \tag{9.61}$$

subject to boundary conditions of the first kind,

$$\begin{aligned} &\frac{\partial \phi_{mn}^{(1)x}}{\partial x} = 0, \quad \phi_{mn}^{(1)y} = 0, \quad \phi_{mn}^{(1)z} = 0 \quad \text{at } x = 0, a, \\ &\phi_{mn}^{(1)x} = 0, \quad \frac{\partial \phi_{mn}^{(1)y}}{\partial y} = 0, \quad \phi_{mn}^{(1)z} = 0 \quad \text{at } y = 0, b \end{aligned} \tag{9.62}$$

and of the second kind,

$$\begin{aligned} &\phi_{mn}^{(2)x} = 0, \quad \frac{\partial \phi_{mn}^{(2)y}}{\partial x} = 0, \quad \frac{\partial \phi_{mn}^{(2)z}}{\partial x} = 0 \quad \text{at } x = 0, a, \\ &\frac{\partial \phi_{mn}^{(2)x}}{\partial y} = 0, \quad \phi_{mn}^{(2)y} = 0, \quad \frac{\partial \phi_{mn}^{(2)z}}{\partial y} = 0 \quad \text{at } y = 0, b. \end{aligned} \tag{9.63}$$

Solving the eigenvalue problem (9.61) and (9.62), we obtain

$$\begin{aligned}
\phi_{mn}^{(1)x}(x,y) &= \sqrt{\frac{\varepsilon_{0m}\varepsilon_{0n}}{ab}}\cos\left(\frac{m\pi x}{a}\right)\sin\left(\frac{n\pi y}{b}\right),\\
\phi_{mn}^{(1)y}(x,y) &= \sqrt{\frac{\varepsilon_{0m}\varepsilon_{0n}}{ab}}\sin\left(\frac{m\pi x}{a}\right)\cos\left(\frac{n\pi y}{b}\right),\\
\phi_{mn}^{(1)z}(x,y) &= \sqrt{\frac{\varepsilon_{0m}\varepsilon_{0n}}{ab}}\sin\left(\frac{m\pi x}{a}\right)\sin\left(\frac{n\pi y}{b}\right),
\end{aligned} \tag{9.64}$$

and for the eigenvalue problem (9.61) and (9.63),

$$\begin{aligned}
\phi_{mn}^{(2)x}(x,y) &= \sqrt{\frac{\varepsilon_{0m}\varepsilon_{0n}}{ab}}\sin\left(\frac{m\pi x}{a}\right)\cos\left(\frac{n\pi y}{b}\right),\\
\phi_{mn}^{(2)y}(x,y) &= \sqrt{\frac{\varepsilon_{0m}\varepsilon_{0n}}{ab}}\cos\left(\frac{m\pi x}{a}\right)\sin\left(\frac{n\pi y}{b}\right),\\
\phi_{mn}^{(2)z}(x,y) &= \sqrt{\frac{\varepsilon_{0m}\varepsilon_{0n}}{ab}}\cos\left(\frac{m\pi x}{a}\right)\cos\left(\frac{n\pi y}{b}\right),
\end{aligned} \tag{9.65}$$

where $\varepsilon_{0m}$, $\varepsilon_{0n}$ are Neumann indexes, such that $\varepsilon_{00} = 1$ and $\varepsilon_{0m} = 2$, $m \neq 0$, and where the eigenvalue in either case is

$$\kappa_{mn}^2 = \left(\frac{m\pi}{a}\right)^2 + \left(\frac{n\pi}{b}\right)^2.$$

Because of the form (9.54), we identify $\kappa_{mn}^2 = k^2 + \gamma_{mn}^2$, where $\gamma$ is the propagation constant associated with the field dependence $e^{\pm\gamma_{mn}z}$ in the positive $(-)$ and negative $(+)$ $z$-directions.

The coefficients $f_{mn}^{(1,2)\nu}(z,z')$ in (9.58) represent one-dimensional $z$-coordinate characteristic Green's functions. From (9.52) and (9.58) we obtain a boundary value problem for the characteristic Green's function $f_{mn}^{(1,2)\nu}(z,z')$,

$$\left(\frac{\partial^2}{\partial z^2} - \gamma_{mn}^2\right) f_{mn}^{(1,2)\nu}(z,z') = -\delta(z-z'), \tag{9.66}$$

where $\gamma_{mn} = \sqrt{\kappa_{mn}^2 - k^2}$, subject to boundary conditions of the first and second kinds, respectively,

$$f_{mn}^{(1)x} = 0, \qquad f_{mn}^{(1)y} = 0, \qquad \frac{\partial f_{mn}^{(1)z}}{\partial z} = 0 \qquad \text{at } z = l, \tag{9.67}$$

$$\frac{\partial f_{mn}^{(2)x}}{\partial z} = 0, \qquad \frac{df_{mn}^{(2)y}}{dz} = 0, \qquad f_{mn}^{(2)z} = 0 \qquad \text{at } z = l. \tag{9.68}$$

The magnetic potential dyadic Green's function is completely specified by (9.58), using (9.64) and (9.65), subject to determination of the characteristic Green's function $f_{mn}^{(1,2)\nu}$.

We next provide the characteristic one-dimensional Green's functions $f_{mn}^{(1,2)\nu}$ for several geometries of interest.

### Infinite Waveguide

For an infinite waveguide the characteristic Green's functions $f_{mn}^{(1,2)\nu}(z,z')$ are sought as the solution of the differential equation (9.66) on $(-\infty,+\infty)$. The problem can be solved using a variety of methods as detailed in Chapter 5, including direct and spectral expansion methods.[12]

A closed-form solution of (9.66) for an infinite waveguide can be obtained as in Example 5.11, leading to the primary[13] or principal Green's function

$$f_{mn}^{(1,2)\nu}(z,z') = f_{mn}^{(1,2)}(z,z') = \frac{1}{2\gamma_{mn}} e^{-\gamma_{mn}|z-z'|} \tag{9.69}$$

$$= \frac{1}{2}\left\{ \begin{array}{ll} \dfrac{e^{-i|z-z'|\sqrt{k^2-\kappa_{mn}^2}}}{i\sqrt{k^2-\kappa_{mn}^2}}, & k^2 > \kappa_{mn}^2, \\ \dfrac{e^{-|z-z'|\sqrt{\kappa_{mn}^2-k^2}}}{\sqrt{\kappa_{mn}^2-k^2}}, & k^2 < \kappa_{mn}^2 \end{array} \right\}. \tag{9.70}$$

Assuming an $e^{i\omega t}$ time dependence, we require $\mathrm{Re}\{\gamma\}, \mathrm{Im}\{\gamma\} > 0$, where the "$\pm$" sign corresponds to waves propagating in the negative and positive $z$-directions.

From (9.58) we obtain a double-series expansion for the magnetic potential Green's dyadics of the first and second kinds for an infinite rectangular waveguide as

$$G_{A\nu\nu}^{(1,2)}(\mathbf{r},\mathbf{r}') = \sum_{m=0}^{\infty}\sum_{n=0}^{\infty} \phi_{mn}^{(1,2)\nu}(x,y)\phi_{mn}^{(1,2)\nu}(x',y')\frac{1}{2\gamma_{mn}}e^{-\gamma_{mn}|z-z'|}, \tag{9.71}$$

where $\gamma_{mn} = \sqrt{\kappa_{mn}^2 - k^2}$ and $\nu = x, y, z$.

### Semi-Infinite Waveguide

For a waveguide that extends to infinity in the positive $z$-coordinate and is terminated with a perfectly conducting ground plane at $z = 0$, the characteristic Green's functions $f_{mn}^{(1,2)\nu}(z,z')$ are obtained as the solution of (9.66)

---

[12]Spectral expansion methods would result in an integral expansion for the one-dimensional characteristic Green's functions $f_{mn}^{(1,2)\nu}(z,z')$, and, therefore, the three-dimensional potential Green's dyadics $G_{A\nu\nu}^{(1,2)}(\mathbf{r},\mathbf{r}')$ would be obtained as a triple expansion

$$G_{A\nu\nu}^{(1,2)}(\mathbf{r},\mathbf{r}') = \frac{1}{2\pi}\int_{-\infty}^{\infty}\sum_{m=0}^{\infty}\sum_{n=0}^{\infty}\frac{\phi_{mn}^{(1,2)\nu}(x,y)\phi_{mn}^{(1,2)\nu}(x',y')e^{iv(z-z')}}{v^2+\kappa_{mn}^2-k^2}dv.$$

This is a purely spectral form, although it is not practical from a computational point of view.

[13]Note that the primary Green's function does not satisfy any boundary conditions (only the limiting absorption condition at infinity) and that it has the same representation for all components of the magnetic potential Green's dyadics of the first and second kinds.

with either (9.67) or (9.68) satisfied at $z = 0$, and the limiting absorption condition at $z = +\infty$. This problem was considered in Example 5.7, resulting in (5.53) and (5.54),

$$\begin{aligned} f_{mn}^{(1)x}(z,z') = f_{mn}^{(1)y}(z,z') &\equiv f_{mn}(z,z') \\ &= \frac{1}{\gamma_{mn}} \left\{ \begin{array}{ll} e^{-\gamma_{mn}z}\sinh\gamma_{mn}z', & z \geq z', \\ e^{-\gamma_{mn}z'}\sinh\gamma_{mn}z, & z \leq z' \end{array} \right\}, \end{aligned} \tag{9.72}$$

$$f_{mn}^{(1)z}(z,z') \equiv g_{mn}(z,z') = \frac{1}{\gamma_{mn}} \left\{ \begin{array}{ll} e^{-\gamma_{mn}z}\cosh\gamma_{mn}z', & z \geq z', \\ e^{-\gamma_{mn}z'}\cosh\gamma_{mn}z, & z \leq z' \end{array} \right\} \tag{9.73}$$

for the first-kind Green's functions, where for the second-kind functions we have

$$\begin{aligned} f_{mn}^{(2)x}(z,z') = f_{mn}^{(2)y}(z,z') = f_{mn}^{(1)z}(z,z') = g_{mn}(z,z'), \\ f_{mn}^{(2)z}(z,z') = f_{mn}^{(1)x}(z,z') = f_{mn}^{(1)y}(z,z') = f_{mn}(z,z'). \end{aligned} \tag{9.74}$$

Alternately, as described in Chapter 5, the method of scattering superposition can be applied to obtain representations for $f_{mn}^{(1,2)\nu}(z,z')$ in terms of the primary Green's function (9.69) and scattered Green's functions. The scattered parts of the Green's functions have the physical meaning of waves reflected from the ground plane and traveling along the waveguide (and decaying at infinity in the presence of small loss). This approach results in the alternative representations for $f_{mn}^{(1)\nu}(z,z')$ as

$$\begin{aligned} f_{mn}^{(1)x}(z,z') = f_{mn}^{(1)y}(z,z') &= f_{mn}(z,z') \\ &= \frac{1}{2\gamma_{mn}}e^{-\gamma_{mn}|z-z'|} - \frac{1}{2\gamma_{mn}}e^{-\gamma_{mn}(z+z')}, \end{aligned} \tag{9.75}$$

$$f_{mn}^{(1)z}(z,z') = g_{mn}(z,z') = \frac{1}{2\gamma_{mn}}e^{-\gamma_{mn}|z-z'|} + \frac{1}{2\gamma_{mn}}e^{-\gamma_{mn}(z+z')}, \tag{9.76}$$

with the expressions for $f_{mn}^{(2)\nu}(z,z')$ given by (9.74). It can be seen that (9.75) and (9.76) are easily reduced to (9.72) and (9.73), respectively.

From (9.58) we obtain a double-series expansion for the magnetic potential Green's dyadics of the first and second kinds for a semi-infinite rectangular waveguide as (see also [46])

$$\begin{aligned} G_{Axx}^{(1)}(\mathbf{r},\mathbf{r}') &= \sum_{m=0}^{\infty}\sum_{n=0}^{\infty} \phi_{mn}^{(1)x}(x,y)\phi_{mn}^{(1)x}(x',y')f_{mn}(z,z'), \\ G_{Ayy}^{(1)}(\mathbf{r},\mathbf{r}') &= \sum_{m=0}^{\infty}\sum_{n=0}^{\infty} \phi_{mn}^{(1)y}(x,y)\phi_{mn}^{(1)y}(x',y')f_{mn}(z,z'), \\ G_{Azz}^{(1)}(\mathbf{r},\mathbf{r}') &= \sum_{m=0}^{\infty}\sum_{n=0}^{\infty} \phi_{mn}^{(1)z}(x,y)\phi_{mn}^{(1)z}(x',y')g_{mn}(z,z') \end{aligned} \tag{9.77}$$

and

$$
\begin{aligned}
G_{Axx}^{(2)}(\mathbf{r},\mathbf{r}') &= \sum_{m=0}^{\infty}\sum_{n=0}^{\infty} \phi_{mn}^{(2)x}(x,y)\phi_{mn}^{(2)x}(x',y')g_{mn}(z,z'), \\
G_{Ayy}^{(2)}(\mathbf{r},\mathbf{r}') &= \sum_{m=0}^{\infty}\sum_{n=0}^{\infty} \phi_{mn}^{(2)y}(x,y)\phi_{mn}^{(2)y}(x',y')g_{mn}(z,z'), \qquad (9.78) \\
G_{Azz}^{(2)}(\mathbf{r},\mathbf{r}') &= \sum_{m=0}^{\infty}\sum_{n=0}^{\infty} \phi_{mn}^{(2)z}(x,y)\phi_{mn}^{(2)z}(x',y')f_{mn}(z,z'),
\end{aligned}
$$

where the eigenfunctions $\phi_{mn}^{(1,2)\nu}(x,y)$ for $\nu = x, y, z$ are given by (9.64) and (9.65), and the characteristic Green's functions $f_{mn}(z,z')$ and $g_{mn}(z,z')$ are given by (9.72) and (9.73), or (9.75) and (9.76).

The termination of a rectangular waveguide by a perfectly conducting ground plane at $z = 0$ is analogous to the case of a transmission line short-circuited at $z = 0$, and in this case we may define $\underline{\mathbf{G}}_A^{(1,2)\mathrm{sc}}(\mathbf{r},\mathbf{r}') \equiv \underline{\mathbf{G}}_A^{(1,2)}(\mathbf{r},\mathbf{r}')$. It can be shown that for an open-circuit termination at $z = 0$ we have

$$
\underline{\mathbf{G}}_A^{(1,2)\mathrm{oc}}(\mathbf{r},\mathbf{r}') = \underline{\mathbf{G}}_A^{(2,1)\mathrm{sc}}(\mathbf{r},\mathbf{r}').
$$

### Rectangular Cavity

For a rectangular waveguide terminated with perfectly conducting ground planes at $z = 0$ and $z = l$ (rectangular cavity), the characteristic Green's functions $f_{mn}^{(1,2)\nu}(z,z')$ are obtained as the solution of (9.66) subject to (9.67) or (9.68) applied at $z = 0, l$. As shown in Example 5.2 we obtain the solution as

$$
\begin{aligned}
f_{mn}^{(1)x}(z,z') &= f_{mn}^{(1)y}(z,z') \equiv f_{mn}(z,z') \\
&= \frac{1}{\gamma_{mn}\sinh\gamma_{mn}l}\left\{\begin{array}{ll} \sinh\gamma_{mn}(l-z)\sinh\gamma_{mn}z', & z \geq z', \\ \sinh\gamma_{mn}(l-z')\sinh\gamma_{mn}z, & z \leq z' \end{array}\right\},
\end{aligned} \qquad (9.79)
$$

$$
\begin{aligned}
f_{mn}^{(1)z}(z,z') &\equiv g_{mn}(z,z') \\
&= \frac{1}{\gamma_{mn}\sinh\gamma_{mn}l}\left\{\begin{array}{ll} \cosh\gamma_{mn}(l-z)\cosh\gamma_{mn}z', & z \geq z', \\ \cosh\gamma_{mn}(l-z')\cosh\gamma_{mn}z, & z \leq z' \end{array}\right\}.
\end{aligned} \qquad (9.80)
$$

The expressions for $f_{mn}^{(2)\nu}(z,z')$ are then obtained from (9.74).

We can also determine these Green's functions via scattering superposition, leading to

$$\begin{aligned} f_{mn}^{(1)x}(z,z') &= f_{mn}^{(1)y}(z,z') = f_{mn}(z,z') \\ &= \frac{1}{2\gamma_{mn}} e^{-\gamma_{mn}|z-z'|} - \frac{1}{2\gamma_{mn}} e^{\gamma_{mn}(z+z'-2l)} \\ &\quad - \frac{e^{-\gamma_{mn}l}}{\gamma_{mn}\sinh\gamma_{mn}l} \sinh\gamma_{mn}(l-z)\sinh\gamma_{mn}(l-z'), \end{aligned} \tag{9.81}$$

$$\begin{aligned} f_{mn}^{(1)z}(z,z') &= g_{mn}(z,z') = \frac{1}{2\gamma_{mn}} e^{-\gamma_{mn}|z-z'|} + \frac{1}{2\gamma_{mn}} e^{\gamma_{mn}(z+z'-2l)} \\ &\quad + \frac{e^{-\gamma_{mn}l}}{\gamma_{mn}\sinh\gamma_{mn}l} \cosh\gamma_{mn}(l-z)\cosh\gamma_{mn}(l-z'). \end{aligned} \tag{9.82}$$

The component form of $\underline{\mathbf{G}}_A^{(1,2)}(\mathbf{r},\mathbf{r}')$ for the rectangular cavity is identical to that obtained for the semi-infinite rectangular waveguide, (9.77) and (9.78), with the characteristic Green's functions determined by (9.79) and (9.80), or (9.81) and (9.82). Dyadic Green's functions for cavities are also briefly considered in Section 10.2.

The case of terminating a rectangular waveguide by perfectly conducting ground planes at $z = 0, l$ corresponds to short-circuit terminations. For open-circuit terminations at $z = 0, l$ we have

$$\underline{\mathbf{G}}_A^{(1,2)\mathrm{oc}}(\mathbf{r},\mathbf{r}') = \underline{\mathbf{G}}_A^{(2,1)\mathrm{sc}}(\mathbf{r},\mathbf{r}').$$

Of course, one can easily derive, using the same approach, characteristic Green's functions for a cavity short-circuited at $z = 0$ and open-circuited at $z = l$, or vice versa.

### 9.2.2 Electric Green's Dyadics

Electric and magnetic Green's dyadics $\underline{\mathbf{G}}_e^{(1,2)}(\mathbf{r},\mathbf{r}')$ and $\underline{\mathbf{G}}_m^{(1,2)}(\mathbf{r},\mathbf{r}')$ for a waveguide or cavity filled with a homogeneous medium can be obtained from the magnetic potential Green's dyadics derived in the previous section via the relations

$$\underline{\mathbf{G}}_e^{(1,2)}(\mathbf{r},\mathbf{r}') = \left(\underline{\mathbf{I}} + \frac{1}{k^2}\nabla\nabla\right) \cdot \underline{\mathbf{G}}_A^{(1,2)}(\mathbf{r},\mathbf{r}'), \tag{9.83}$$

$$\underline{\mathbf{G}}_m^{(1,2)}(\mathbf{r},\mathbf{r}') = \nabla \times \underline{\mathbf{G}}_A^{(2,1)}(\mathbf{r},\mathbf{r}'). \tag{9.84}$$

Alternatively, in this section we demonstrate a general procedure for deriving electric Green's dyadics of the first and second kinds directly from the dyadic wave equation (9.4),

$$\nabla \times \nabla \times \underline{\mathbf{G}}_e^{(1,2)}(\mathbf{r},\mathbf{r}') - k^2 \underline{\mathbf{G}}_e^{(1,2)}(\mathbf{r},\mathbf{r}') = \underline{\mathbf{I}}\,\delta(\mathbf{r}-\mathbf{r}'), \qquad \mathbf{r},\mathbf{r}' \in V, \tag{9.85}$$

subject to boundary conditions of the first and second kinds on the perfectly conducting surface $S$,

$$\mathbf{n} \times \underline{\mathbf{G}}_e^{(1)}(\mathbf{r}, \mathbf{r}') = \underline{\mathbf{0}}, \qquad \mathbf{r} \in S, \tag{9.86}$$

$$\mathbf{n} \times \nabla \times \underline{\mathbf{G}}_e^{(2)}(\mathbf{r}, \mathbf{r}') = \underline{\mathbf{0}}, \qquad \mathbf{n} \cdot \underline{\mathbf{G}}_e^{(2)}(\mathbf{r}, \mathbf{r}') = \mathbf{0}, \qquad \mathbf{r} \in S. \tag{9.87}$$

The component form of the dyadic equation (9.85) represents nine second-order differential equations grouped into three subsystems of equations with respect to the components $G_{e\nu x}^{(1,2)}, G_{e\nu y}^{(1,2)}$, and $G_{e\nu z}^{(1,2)}$ for $\nu = x, y, z$.[14] These components are expressed as double-series expansions over the complete system of eigenfunctions of the transverse Laplacian operator, leading to

$$G_{e\nu\upsilon}^{(1,2)}(\mathbf{r}, \mathbf{r}') = \sum_{m=0}^{\infty} \sum_{n=0}^{\infty} \phi_{mn}^{(1,2)\nu}(x, y) \phi_{mn}^{(1,2)\upsilon}(x', y') f_{mn}^{(1,2)\nu\upsilon}(z, z') \tag{9.88}$$

for $\nu, \upsilon = x, y, z$, where $\phi_{mn}^{(1,2)\nu,\upsilon}(x, y)$ are the complete set of eigenfunctions (9.64) and (9.65) in the waveguide cross-section and $f_{mn}^{(1,2)\nu\upsilon}(z, z')$ are one-dimensional characteristic Green's functions. The properties of orthogonality (9.59) and completeness (9.60) of the eigenfunctions allow for the reduction of the three-dimensional boundary value problem (9.85), with (9.86) or (9.87), to a Sturm–Liouville problem for the one-dimensional characteristic Green's functions,

$$\left(\frac{\partial^2}{\partial z^2} - \gamma_{mn}^2\right) f_{mn}^{(1,2)pq}(z, z') = -\xi_{mn}^{pq} \delta(z - z'), \qquad p, q = x, y, \tag{9.89}$$

$$\left(\frac{\partial^2}{\partial z^2} - \gamma_{mn}^2\right) f_{mn}^{(1,2)pz}(z, z') = -\xi_{mn}^{(1,2)pz} \frac{\partial}{\partial z} \delta(z - z'), \qquad p = x, y, \tag{9.90}$$

with boundary conditions of the first and second kinds at $z = l$,

$$\begin{aligned} f_{mn}^{(1)xx} &= 0, & f_{mn}^{(1)xy} &= 0, & f_{mn}^{(1)yx} &= 0, \\ f_{mn}^{(1)yy} &= 0, & f_{mn}^{(1)xz} &= 0, & f_{mn}^{(1)yz} &= 0, \end{aligned} \tag{9.91}$$

$$\begin{aligned} \frac{\partial f_{mn}^{(2)xx}}{\partial z} &= 0, & \frac{\partial f_{mn}^{(2)xy}}{\partial z} &= 0, & \frac{\partial f_{mn}^{(2)xz}}{\partial z} &= 0, \\ \frac{\partial f_{mn}^{(2)yx}}{\partial z} &= 0, & \frac{\partial f_{mn}^{(2)yy}}{\partial z} &= 0, & \frac{\partial f_{mn}^{(2)yz}}{\partial z} &= 0, \end{aligned} \tag{9.92}$$

[14] All of the components $G_{e\nu x}^{(1,2)}, G_{e\nu y}^{(1,2)}$, and $G_{e\nu z}^{(1,2)}$ represent Green's functions even though six out of nine differential equations are homogeneous equations. Within each of three subsystems of three differential equations the components $G_{e\nu x}^{(1,2)}, G_{e\nu y}^{(1,2)}$, and $G_{e\nu z}^{(1,2)}$ are coupled by the curl-curl operator.

where $\xi_{mn}^{pq}$, $\xi_{mn}^{pz}$ are defined as

$$\xi_{mn}^{xx} = \frac{k^2 - \left(\frac{m\pi}{a}\right)^2}{k^2}, \quad \xi_{mn}^{xy} = \xi_{mn}^{yx} = -\frac{\left(\frac{m\pi}{a}\right)\left(\frac{n\pi}{b}\right)}{k^2}, \quad \xi_{mn}^{yy} = \frac{k^2 - \left(\frac{n\pi}{b}\right)^2}{k^2}$$
$$\xi_{mn}^{(1)xz} = \frac{\left(\frac{m\pi}{a}\right)}{k^2}, \qquad \xi_{mn}^{(1)yz} = \frac{\left(\frac{n\pi}{b}\right)}{k^2},$$
$$\xi_{mn}^{(2)xz} = -\xi_{mn}^{(1)xz}, \qquad \xi_{mn}^{(2)yz} = -\xi_{mn}^{(1)yz}. \tag{9.93}$$

The longitudinal components $f_{mn}^{(1,2)zi}(z,z')$ for $i = x, y, z$ are expressed in terms of the transverse components as

$$f_{mn}^{(1)zx}(z,z') = \frac{1}{\gamma_{mn}^2}\frac{\partial}{\partial z}\left(\frac{m\pi}{a}f_{mn}^{(1)xx}(z,z') + \frac{n\pi}{b}f_{mn}^{(1)yx}(z,z')\right),$$
$$f_{mn}^{(1)zy}(z,z') = \frac{1}{\gamma_{mn}^2}\frac{\partial}{\partial z}\left(\frac{m\pi}{a}f_{mn}^{(1)xy}(z,z') + \frac{n\pi}{b}f_{mn}^{(1)yy}(z,z')\right),$$
$$f_{mn}^{(1)zz}(z,z') = \frac{1}{\gamma_{mn}^2}\left[\delta(z-z') + \frac{\partial}{\partial z}\left(\frac{m\pi}{a}f_{mn}^{(1)xz}(z,z') + \frac{n\pi}{b}f_{mn}^{(1)yz}(z,z')\right)\right], \tag{9.94}$$

$$f_{mn}^{(2)zx}(z,z') = -\frac{1}{\gamma_{mn}^2}\frac{\partial}{\partial z}\left(\frac{m\pi}{a}f_{mn}^{(2)xx}(z,z') + \frac{n\pi}{b}f_{mn}^{(2)yx}(z,z')\right),$$
$$f_{mn}^{(2)zy}(z,z') = -\frac{1}{\gamma_{mn}^2}\frac{\partial}{\partial z}\left(\frac{m\pi}{a}f_{mn}^{(2)xy}(z,z') + \frac{n\pi}{b}f_{mn}^{(2)yy}(z,z')\right),$$
$$f_{mn}^{(2)zz}(z,z') = \frac{1}{\gamma_{mn}^2}\left[\delta(z-z') - \frac{\partial}{\partial z}\left(\frac{m\pi}{a}f_{mn}^{(2)xz}(z,z') + \frac{n\pi}{b}f_{mn}^{(2)yz}(z,z')\right)\right]. \tag{9.95}$$

The electric Green's dyadics are completely specified by (9.88) subsequent to determination of the one-dimensional characteristic Green's functions $f_{mn}^{(1,2)\nu\upsilon}$. We next provide these functions for several geometries of interest.

## Infinite Waveguide

For an infinite rectangular waveguide the solution of Sturm–Liouville equations (9.89) and (9.90) for transverse components of characteristic Green's functions of the first and second kinds can be obtained as

$$f_{mn}^{(1,2)pq}(z,z') = f_{mn}^{pq}(z,z') = \xi_{mn}^{pq}\frac{1}{2\gamma_{mn}}e^{-\gamma_{mn}|z-z'|}, \qquad p,q = x,y, \tag{9.96}$$
$$\begin{aligned} f_{mn}^{(1)pz}(z,z') &= -f_{mn}^{(2)pz}(z,z') = f_{mn}^{pz}(z,z') \\ &= -\frac{1}{2}\xi_{mn}^{(1)pz}\operatorname{sgn}(z-z')e^{-\gamma_{mn}|z-z'|}, \qquad p = x,y, \end{aligned} \tag{9.97}$$

where $\xi^{pq}_{mn}$ and $\xi^{(1)pz}_{mn}$ are defined by (9.93) and $\operatorname{sgn}(z-z') = 1$ for $z > z'$ and $-1$ for $z < z'$. Using representations (9.94) and (9.95), we obtain expressions for the longitudinal components as

$$\begin{aligned} f^{(1)zp}_{mn}(z,z') &= -f^{(1)pz}_{mn}(z,z') = -f^{(2)zp}_{mn}(z,z') \\ &= f^{(2)pz}_{mn}(z,z') = -f^{pz}_{mn}(z,z'), \qquad p = x, y, \end{aligned} \tag{9.98}$$

$$\begin{aligned} f^{(1,2)zz}_{mn}(z,z') &= f^{zz}_{mn}(z,z') \\ &= \left(1 + \frac{\gamma^2_{mn} - 2\gamma_{mn}\delta(z-z')}{k^2}\right)\frac{1}{2\gamma_{mn}} e^{-\gamma_{mn}|z-z'|}. \end{aligned} \tag{9.99}$$

Note that $f^{(1,2)pq}_{mn}(z,z')$ for $p, q = x, y$ are continuous at $z = z'$ with a discontinuous derivative and that components $f^{(1,2)pz}_{mn}(z,z')$, $f^{(1,2)zp}_{mn}(z,z')$, and $f^{(1,2)zz}_{mn}(z,z')$ are discontinuous at $z = z'$. Furthermore, components $f^{pz}_{mn}(z,z')$ and $f^{zp}_{mn}(z,z')$ of the first and second kinds differ in sign.

The double-series expansion (9.88) with the eigenfunctions defined by (9.64) and (9.65) and the characteristic Green's functions (9.96)–(9.99) provide a component form for the electric Green's dyadics of the first and second kinds for an infinite homogeneous rectangular waveguide.

### Semi-Infinite Waveguide

For a waveguide that extends to infinity in the positive $z$-direction and is terminated with a ground plane at $z = 0$, the solution of the boundary value problems (9.89)–(9.92) with limiting absorption at infinity for the characteristic Green's functions $f^{(1,2)pq}_{mn}(z,z')$ and $f^{(1,2)pz}_{mn}(z,z')$, $p, q = x, y$, can be obtained by the method of scattering superposition. In this representation the solution is expressed in terms of the primary, (9.96) and (9.97), and scattered parts, such that

$$f^{(1,2)pq}_{mn}(z,z') = \xi^{pq}_{mn}\frac{1}{2\gamma_{mn}} e^{-\gamma_{mn}|z-z'|} + \eta^{(1,2)}_{pq}(z') e^{-\gamma_{mn} z}, \tag{9.100}$$

$$\begin{aligned} f^{(1,2)pz}_{mn}(z,z') = &-\frac{1}{2}\xi^{(1,2)pz}_{mn}\operatorname{sgn}(z-z') e^{-\gamma_{mn}|z-z'|} \\ &+ \eta^{(1,2)}_{pz}(z') e^{-\gamma_{mn} z}, \end{aligned} \tag{9.101}$$

where the $\eta$-coefficients are the unknown amplitudes to be determined subject to the boundary conditions (9.91) and (9.92), respectively. This results in closed-form expressions for $f^{(1,2)pq}_{mn}(z,z')$ and $f^{(1,2)pz}_{mn}(z,z')$,

$$\begin{aligned} f^{(1)pq}_{mn}(z,z') &= \xi^{pq}_{mn}\frac{1}{2\gamma_{mn}}\left(e^{-\gamma_{mn}|z-z'|} - e^{-\gamma_{mn}(z+z')}\right), \\ f^{(1)pz}_{mn}(z,z') &= -\frac{1}{2}\xi^{(1)pz}_{mn}\left(\operatorname{sgn}(z-z') e^{-\gamma_{mn}|z-z'|} + e^{-\gamma_{mn}(z+z')}\right), \end{aligned} \tag{9.102}$$

$$f_{mn}^{(2)pq}(z,z') = \xi_{mn}^{pq}\frac{1}{2\gamma_{mn}}\left(e^{-\gamma_{mn}|z-z'|} + e^{-\gamma_{mn}(z+z')}\right),$$
$$f_{mn}^{(2)pz}(z,z') = -\frac{1}{2}\xi_{mn}^{(2)pz}\left(\operatorname{sgn}(z-z')e^{-\gamma_{mn}|z-z'|} - e^{-\gamma_{mn}(z+z')}\right). \tag{9.103}$$

The longitudinal components $f_{mn}^{(1,2)zi}(z,z')$ for $i = x, y, z$ are obtained from (9.94) and (9.95), together with (9.102) and (9.103), leading to

$$f_{mn}^{(1)zp}(z,z') = \frac{1}{2}\xi_{mn}^{(1)pz}\left(\operatorname{sgn}(z-z')e^{-\gamma_{mn}|z-z'|} - e^{-\gamma_{mn}(z+z')}\right),$$
$$f_{mn}^{(1)zz}(z,z') = \left(1 + \frac{\gamma_{mn}^2 - 2\gamma_{mn}\delta(z-z')}{k^2}\right)\frac{1}{2\gamma_{mn}}e^{-\gamma_{mn}|z-z'|} \tag{9.104}$$
$$+ \left(1 + \frac{\gamma_{mn}^2}{k^2}\right)\frac{1}{2\gamma_{mn}}e^{\gamma_{mn}(z+z')},$$

$$f_{mn}^{(2)zp}(z,z') = \frac{1}{2}\xi_{mn}^{(2)pz}\left(\operatorname{sgn}(z-z')e^{-\gamma_{mn}|z-z'|} + e^{-\gamma_{mn}(z+z')}\right),$$
$$f_{mn}^{(2)zz}(z,z') = \left(1 + \frac{\gamma_{mn}^2 - 2\gamma_{mn}\delta(z-z')}{k^2}\right)\frac{1}{2\gamma_{mn}}e^{-\gamma_{mn}|z-z'|} \tag{9.105}$$
$$- \left(1 + \frac{\gamma_{mn}^2}{k^2}\right)\frac{1}{2\gamma_{mn}}e^{\gamma_{mn}(z+z')}.$$

The component form of the electric Green's dyadics for a semi-infinite rectangular waveguide is determined as the double-series expansion (9.88) over the complete system of eigenfunctions (9.64) and (9.65), with one-dimensional characteristic Green's functions obtained by expressions (9.102)–(9.105).

### Rectangular Cavity

For a rectangular waveguide terminated by perfectly conducting ground planes at $z = 0, l$, we apply the method of scattering superposition to obtain the solution of the Sturm–Liouville boundary value problems (9.89)–(9.92). The components $f_{mn}^{(1,2)pq}(z,z')$ and $f_{mn}^{(1,2)pz}(z,z')$ for $p, q = x, y$ are expressed in terms of the primary, (9.96) and (9.97), and scattered Green's functions as forward and backward traveling waves in the cavity region,

$$f_{mn}^{(1,2)pq}(z,z') = \xi_{mn}^{pq}\frac{1}{2\gamma_{mn}}e^{-\gamma_{mn}|z-z'|} + \eta_{pq}^{(1,2)}(z')e^{-\gamma_{mn}z} \tag{9.106}$$
$$+ \zeta_{pq}^{(1,2)}(z')e^{\gamma_{mn}(z-l)},$$
$$f_{mn}^{(1,2)pz}(z,z') = -\frac{1}{2}\xi_{mn}^{(1,2)pz}\operatorname{sgn}(z-z')e^{-\gamma_{mn}|z-z'|} \tag{9.107}$$
$$+ \eta_{pz}^{(1,2)}(z')e^{-\gamma_{mn}z} + \zeta_{pz}^{(1,2)}(z')e^{\gamma_{mn}(z-l)},$$

with the unknown $\eta$- and $\zeta$-coefficients to be determined by satisfying the boundary conditions (9.91) and (9.92) at $z = 0, l$. The solutions for the components $f_{mn}^{(1,2)pq}(z,z')$ and $f_{mn}^{(1,2)pz}(z,z')$ are

$$
\begin{aligned}
f_{mn}^{(1)pq}(z,z') &= \xi_{mn}^{pq}\Big(\frac{1}{2\gamma_{mn}}e^{-\gamma_{mn}|z-z'|} - \frac{1}{2\gamma_{mn}}e^{\gamma_{mn}(z+z'-2l)} \\
&\quad - \frac{e^{-\gamma_{mn}l}}{\gamma_{mn}\sinh\gamma_{mn}l}\sinh\gamma_{mn}(l-z)\sinh\gamma_{mn}(l-z')\Big), \\
f_{mn}^{(1)pz}(z,z') &= -\frac{1}{2}\xi_{mn}^{(1)pz}\Big(\mathrm{sgn}(z-z')e^{-\gamma_{mn}|z-z'|} - e^{\gamma_{mn}(z+z'-2l)} \\
&\quad + \frac{2e^{-\gamma_{mn}l}}{\sinh\gamma_{mn}l}\sinh\gamma_{mn}(l-z)\cosh\gamma_{mn}(l-z')\Big),
\end{aligned} \tag{9.108}
$$

$$
\begin{aligned}
f_{mn}^{(2)pq}(z,z') &= \xi_{mn}^{pq}\Big(\frac{1}{2\gamma_{mn}}e^{-\gamma_{mn}|z-z'|} + \frac{1}{2\gamma_{mn}}e^{\gamma_{mn}(z+z'-2l)} \\
&\quad + \frac{e^{-\gamma_{mn}l}}{\gamma_{mn}\sinh\gamma_{mn}l}\cosh\gamma_{mn}(l-z)\cosh\gamma_{mn}(l-z')\Big), \\
f_{mn}^{(2)pz}(z,z') &= -\frac{1}{2}\xi_{mn}^{(2)pz}\Big(\mathrm{sgn}(z-z')e^{-\gamma_{mn}|z-z'|} + e^{-\gamma_{mn}(z+z'-2l)} \\
&\quad - \frac{2e^{-\gamma_{mn}l}}{\sinh\gamma_{mn}l}\cosh\gamma_{mn}(l-z)\sinh\gamma_{mn}(l-z')\Big).
\end{aligned} \tag{9.109}
$$

The longitudinal $z$-directed components are obtained from (9.94) and (9.95) in conjunction with (9.108) and (9.109), resulting in

$$
\begin{aligned}
f_{mn}^{(1)zp}(z,z') &= \frac{1}{2}\xi_{mn}^{(1)pz}\Big(\mathrm{sgn}(z-z')e^{-\gamma_{mn}|z-z'|} + e^{\gamma_{mn}(z+z'-2l)} \\
&\quad - \frac{2e^{-\gamma_{mn}l}}{\sinh\gamma_{mn}l}\cosh\gamma_{mn}(l-z)\sinh\gamma_{mn}(l-z')\Big), \\
f_{mn}^{(1)zz}(z,z') &= \left(1+\frac{\gamma_{mn}^2 - 2\gamma_{mn}\delta(z-z')}{k^2}\right)\frac{1}{2\gamma_{mn}}e^{-\gamma_{mn}|z-z'|} \\
&\quad + \Big(1+\frac{\gamma_{mn}^2}{k^2}\Big)\Big(\frac{e^{\gamma_{mn}(z+z'-2l)}}{2\gamma_{mn}} \\
&\quad + \frac{e^{-\gamma_{mn}l}}{\gamma_{mn}\sinh\gamma_{mn}l}\cosh\gamma_{mn}(l-z)\cosh\gamma_{mn}(l-z')\Big),
\end{aligned} \tag{9.110}
$$

$$
\begin{aligned}
f_{mn}^{(2)zp}(z,z') &= \frac{1}{2}\xi_{mn}^{(2)pz}\Big(\mathrm{sgn}(z-z')e^{-\gamma_{mn}|z-z'|} - e^{\gamma_{mn}(z+z'-2l)} \\
&\quad + \frac{2e^{-\gamma_{mn}l}}{\sinh\gamma_{mn}l}\sinh\gamma_{mn}(l-z)\cosh\gamma_{mn}(l-z')\Big), \\
f_{mn}^{(2)zz}(z,z') &= \left(1+\frac{\gamma_{mn}^2 - 2\gamma_{mn}\delta(z-z')}{k^2}\right)\frac{1}{2\gamma_{mn}}e^{-\gamma_{mn}|z-z'|} \\
&\quad - \Big(1+\frac{\gamma_{mn}^2}{k^2}\Big)\Big(\frac{e^{\gamma_{mn}(z+z'-2l)}}{2\gamma_{mn}} \\
&\quad + \frac{e^{-\gamma_{mn}l}}{\gamma_{mn}\sinh\gamma_{mn}l}\cosh\gamma_{mn}(l-z)\cosh\gamma_{mn}(l-z')\Big).
\end{aligned} \tag{9.111}
$$

The double-series expansion (9.88) provides a component form for the electric Green's dyadics of the first and second kinds for a rectangular cavity, with the eigenfunctions determined by (9.64) and (9.65), and the one-dimensional characteristic Green's functions obtained as (9.108)–(9.111).

## 9.3 Scattering Problems for Waveguides Filled with a Planarly Layered Medium

In this section we present an integral equation formulation for the full-wave analysis of arbitrarily shaped interacting electric (strip, patch) and magnetic (slot, aperture) transversely located discontinuities in a layered waveguide. The work described here was motivated by the need to develop a modeling environment for quasi-optical and spatial power combining circuits [57], [58]. Spatial power amplifiers are used to combine power from an array of solid-state devices at millimeter-wave frequencies, resulting in increased output power and power combining efficiencies. An array of antennas that contain active devices (amplifiers) is excited with the same magnitude and phase (uniform field excitation), and power carried by the incident field is amplified and reradiated by the use of output antennas. This power is combined in space due to coherent conditions (the same magnitude and phase) satisfied for each output antenna. Spatially distributed power combining systems generally consist of a number of transverse electric and magnetic layers (for example, a patch antenna array coupled to a slot antenna array) separated by waveguide sections. The idea of a *generalized scattering matrix* (GSM) is utilized for the simulation of large waveguide-based electromagnetic and quasi-optical systems [59], wherein a whole system is decomposed into individual modules (electric and magnetic layers). The GSM of each layer is obtained based on a full-wave integral equation formulation, and the overall system response is obtained by cascading GSMs of individual modules. The GSM is constructed for all propagating and evanescent TE- and TM-modes and provides an accurate account of the interactions between neighboring modules.

### 9.3.1 Interacting Electric- and Magnetic-Type Discontinuities in a Layered-Medium Waveguide

Consider a cylindrical waveguide that contains arbitrarily shaped metal surfaces $S_m$ and slot apertures $S_a$, as shown for the special case of a rectangular waveguide in Figure 9.6, where the $S_m$ are located on the interface $S_d$ of adjacent dielectric layers with permittivities $\varepsilon_1$ and $\varepsilon_2$. The apertures $S_a$ in the ground plane $S_g$ separate two dielectrics with permittivities[15] $\varepsilon_2$

[15] Here we assume that the dielectric filling regions $V_1$, $V_2$, and $V_3$ is nonmagnetic, with $\mu_1 = \mu_2 = \mu_3 = \mu_0$.

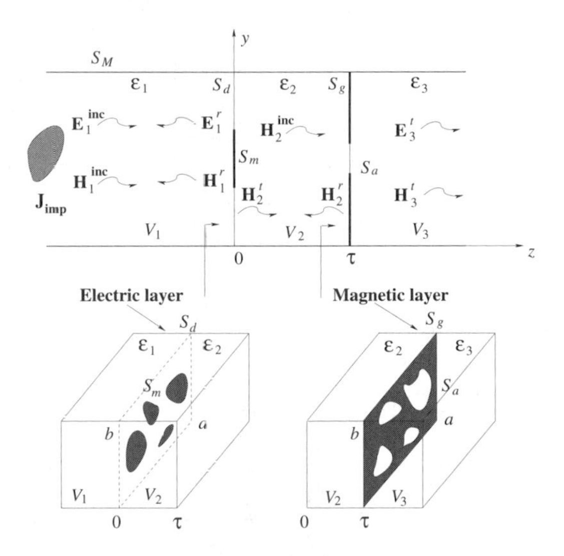


Figure 9.6: Waveguide-based transition module of arbitrarily shaped interacting electric- (patch, strip) and magnetic-type (slot, aperture) discontinuities.

and $\varepsilon_3$. The incident electric and magnetic fields in region $V_1$ are generated by an impressed electric current $\mathbf{J}_{\text{imp}}(\mathbf{r})$, $\mathbf{r} \in V_{\text{imp}} \subset V_1$.[16] The scattered electric and magnetic fields in regions $V_1$ and $V_2$ are generated by the electric currents induced on the metal surface $S_m$ and by the magnetic currents induced on the surface of apertures $S_a$. The scattered electric and magnetic fields in region $V_3$ are caused by induced magnetic currents on the surface $S_a$.

In the formulation detailed here we obtain integral representations for the total electric and magnetic fields in regions $V_1$, $V_2$, and $V_3$ by implementing boundary conditions for tangential components of the electric field on perfectly conducting surfaces $S_m$ and enforcing the continuity of tangential components of the magnetic field across apertures $S_a$, which results in a coupled system of integral equations.

Our goal is to obtain a GSM for interacting electric- (strip, patch) and magnetic- (slot, aperture) type discontinuities located in a rectangular waveguide containing a layered medium, which, in general, allows for the propagation of higher-order modes.

[16]The incident fields due to an impressed electric current in region $V_3$ are similarly handled.

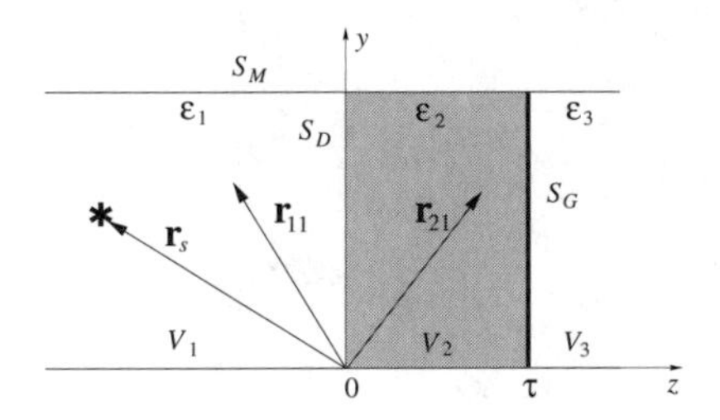


Figure 9.7: Two-layered, semi-infinite waveguide with a point source arbitrarily located in region $V_1$ used in the electric-field formulation.

### Electric-Field Integral Representation

The electric-field vector at each interior point of regions $V_1$ and $V_2$ is determined as the solution of the vector wave equations

$$\nabla \times \nabla \times \mathbf{E}_1(\mathbf{r}) - k_1^2 \mathbf{E}_1(\mathbf{r}) = -i\omega\mu_0 \mathbf{J}_{\text{imp}}(\mathbf{r}), \qquad \mathbf{r} \in V_1, \tag{9.112}$$

$$\nabla \times \nabla \times \mathbf{E}_2(\mathbf{r}) - k_2^2 \mathbf{E}_2(\mathbf{r}) = \mathbf{0}, \qquad \mathbf{r} \in V_2, \tag{9.113}$$

subject to boundary conditions for tangential components of electric field on the waveguide surface $S_M$, on the metal surface $S_m$, and on the surface of the ground plane $S_g$ (Figure 9.6),

$$\mathbf{n} \times \mathbf{E}_1(\mathbf{r})|_{S_1} = \mathbf{0}, \tag{9.114}$$

$$\mathbf{n} \times \mathbf{E}_2(\mathbf{r})|_{S_2} = \mathbf{0}, \tag{9.115}$$

where $S_1 = S_M \cup S_m$ and $S_2 = S_1 \cup S_g$, with $\mathbf{n}$ an outward normal unit vector to the surface enclosing volumes $V_1$ and $V_2$. Furthermore, we impose continuity conditions across the dielectric interface $S_d$,

$$\begin{aligned} \widehat{\mathbf{z}} \times \mathbf{E}_1(\mathbf{r})|_{S_d} &= \widehat{\mathbf{z}} \times \mathbf{E}_2(\mathbf{r})|_{S_d}, \\ \widehat{\mathbf{z}} \times \nabla \times \mathbf{E}_1(\mathbf{r})|_{S_d} &= \widehat{\mathbf{z}} \times \nabla \times \mathbf{E}_2(\mathbf{r})|_{S_d}, \end{aligned} \tag{9.116}$$

where $k_i = k_0\sqrt{\varepsilon_i}$ for $i = 1, 2$, with $k_0 = 2\pi/\lambda_0$, and the fitness condition (limiting absorption) at infinity.

Analogous to (9.112)–(9.116), we formulate a boundary value problem for the electric Green's dyadic of the third kind [36] for a two-layered, semi-infinite waveguide in the absence of a metal surface at $z = 0$ and apertures at $z = \tau$ (Figure 9.7). Electric dyadic Green's functions in regions $V_1$ and $V_2$ associated with a point source arbitrarily located in $V_1$ are obtained as the solution of the system of dyadic differential equations

$$\nabla \times \nabla \times \underline{\mathbf{G}}_{e11}^{(1)}(\mathbf{r}, \mathbf{r}') - k_1^2 \underline{\mathbf{G}}_{e11}^{(1)}(\mathbf{r}, \mathbf{r}') = \underline{\mathbf{I}}\,\delta(\mathbf{r} - \mathbf{r}'), \qquad \mathbf{r}, \mathbf{r}' \in V_1, \tag{9.117}$$

$$\nabla \times \nabla \times \underline{\mathbf{G}}_{e21}^{(1)}(\mathbf{r}, \mathbf{r}') - k_2^2 \underline{\mathbf{G}}_{e21}^{(1)}(\mathbf{r}, \mathbf{r}') = \underline{\mathbf{0}}, \qquad \mathbf{r} \in V_2,\ \mathbf{r}' \in V_1, \tag{9.118}$$

subject to boundary conditions of the first kind on the waveguide surface $S_M$ and on the surface of the ground plane $S_G = S_g \cup S_a$ (Figure 9.7),

$$\mathbf{n} \times \underline{\mathbf{G}}_{e11}^{(1)}(\mathbf{r}, \mathbf{r}') = \underline{\mathbf{0}}, \qquad \mathbf{r} \in S_M, \tag{9.119}$$

$$\mathbf{n} \times \underline{\mathbf{G}}_{e21}^{(1)}(\mathbf{r}, \mathbf{r}') = \underline{\mathbf{0}}, \qquad \mathbf{r} \in S_M \cup S_G, \tag{9.120}$$

and mixed continuity conditions for the electric Green's dyadics of the third kind (analogous to the field continuity conditions (9.116) expressed in terms of the electric field) across the dielectric interface $S_D = S_d \cup S_m$,

$$\begin{aligned} \widehat{\mathbf{z}} \times \underline{\mathbf{G}}_{e11}^{(1)}(\mathbf{r}, \mathbf{r}') &= \widehat{\mathbf{z}} \times \underline{\mathbf{G}}_{e21}^{(1)}(\mathbf{r}, \mathbf{r}'), \qquad && \mathbf{r} \in S_D, \\ \widehat{\mathbf{z}} \times \nabla \times \underline{\mathbf{G}}_{e11}^{(1)}(\mathbf{r}, \mathbf{r}') &= \widehat{\mathbf{z}} \times \nabla \times \underline{\mathbf{G}}_{e21}^{(1)}(\mathbf{r}, \mathbf{r}'), \qquad && \mathbf{r} \in S_D. \end{aligned} \tag{9.121}$$

The fitness condition (limiting absorption) at infinity for $\underline{\mathbf{G}}_{e11}^{(1)}(\mathbf{r}, \mathbf{r}')$ is accounted for in a similar manner as for the scattered field.

The vector-dyadic Green's second theorem applied to $\mathbf{E}_1$ and $\underline{\mathbf{G}}_{e11}^{(1)}$, with the boundary conditions (9.114) and (9.119), results in an integral representation for the electric field in region $V_1$,

$$\begin{aligned} \mathbf{E}_1(\mathbf{r}') = & \\ & - i\omega\mu_0 \int_{V_{\text{imp}}} \mathbf{J}_{\text{imp}}(\mathbf{r}) \cdot \underline{\mathbf{G}}_{e11}^{(1)}(\mathbf{r}, \mathbf{r}')\, dV \\ & - \int_{S_d} \widehat{\mathbf{z}} \cdot \left( \mathbf{E}_1(\mathbf{r}) \times [\nabla \times \underline{\mathbf{G}}_{e11}^{(1)}(\mathbf{r}, \mathbf{r}')] + [\nabla \times \mathbf{E}_1(\mathbf{r})] \times \underline{\mathbf{G}}_{e11}^{(1)}(\mathbf{r}, \mathbf{r}') \right) dS \\ & - \int_{S_m} \widehat{\mathbf{z}} \times [\nabla \times \mathbf{E}_1(\mathbf{r})] \cdot \underline{\mathbf{G}}_{e11}^{(1)}(\mathbf{r}, \mathbf{r}')\, dS. \end{aligned} \tag{9.122}$$

An integral representation for the electric field in region $V_2$ is obtained as a result of the vector-dyadic Green's second theorem applied to the source-free equations (9.113) and (9.118) with the boundary conditions (9.115) and (9.120),

$$\begin{aligned} & \int_{S_d} \widehat{\mathbf{z}} \cdot \left( \mathbf{E}_2(\mathbf{r}) \times [\nabla \times \underline{\mathbf{G}}_{e21}^{(1)}(\mathbf{r}, \mathbf{r}')] + [\nabla \times \mathbf{E}_2(\mathbf{r})] \times \underline{\mathbf{G}}_{e21}^{(1)}(\mathbf{r}, \mathbf{r}') \right) dS \\ & \quad = - \int_{S_m} \widehat{\mathbf{z}} \times [\nabla \times \mathbf{E}_2(\mathbf{r})] \cdot \underline{\mathbf{G}}_{e21}^{(1)}(\mathbf{r}, \mathbf{r}')\, dS \\ & \quad + \int_{S_a} [\widehat{\mathbf{z}} \times \mathbf{E}_2(\mathbf{r})] \cdot [\nabla \times \underline{\mathbf{G}}_{e21}^{(1)}(\mathbf{r}, \mathbf{r}')]\, dS. \end{aligned} \tag{9.123}$$

It can be seen that the integrands of the surface integrals over the dielectric interface $S_d$ in (9.122) and (9.123) are coupled by means of continuity conditions (9.116) and (9.121). This allows us to represent the total electric field in region $V_1$ in terms of the incident field caused by an impressed electric current $\mathbf{J}_{\text{imp}}(\mathbf{r})$ and a scattered field caused by induced electric current

$\mathbf{J}(\mathbf{r})$,

$$\mathbf{J}(\mathbf{r}) \equiv \frac{i}{\omega\mu_0}\widehat{\mathbf{z}} \times \nabla \times [\mathbf{E}_2(\mathbf{r}) - \mathbf{E}_1(\mathbf{r})] = \widehat{\mathbf{z}} \times [\mathbf{H}_2(\mathbf{r}) - \mathbf{H}_1(\mathbf{r})], \qquad \mathbf{r} \in S_m,$$

and an induced magnetic current $\mathbf{K}(\mathbf{r}) \equiv -\widehat{\mathbf{z}} \times \mathbf{E}_2(\mathbf{r})$ for $\mathbf{r} \in S_a$, such that

$$\begin{aligned}\mathbf{E}_1(\mathbf{r}') = &- i\omega\mu_0 \int_{V_{\text{imp}}} \mathbf{J}_{\text{imp}}(\mathbf{r}) \cdot \underline{\mathbf{G}}_{e11}^{(1)}(\mathbf{r}, \mathbf{r}')\, dV \\ &- i\omega\mu_0 \int_{S_m} \mathbf{J}(\mathbf{r}) \cdot \underline{\mathbf{G}}_{e11}^{(1)}(\mathbf{r}, \mathbf{r}')\, dS + \int_{S_a} \mathbf{K}(\mathbf{r}) \cdot [\nabla \times \underline{\mathbf{G}}_{e21}^{(1)}(\mathbf{r}, \mathbf{r}')]\, dS.\end{aligned} \tag{9.124}$$

An electric-field integral equation is obtained by enforcing on tangential components of the total electric field the boundary condition on the perfectly conducting surface $S_m$ at $z = 0$,

$$\widehat{\mathbf{z}} \times (\mathbf{E}_1^{\text{inc}}(\mathbf{r}) + \mathbf{E}_1^{\text{scat}}(\mathbf{r}))\Big|_{S_m} = \mathbf{0},$$

resulting in

$$\begin{aligned}&- i\omega\mu_0 \widehat{\mathbf{z}} \times \int_{V_{\text{imp}}} \mathbf{J}_{\text{imp}}(\mathbf{r}') \cdot \underline{\mathbf{G}}_{e11}^{(1)}(\mathbf{r}', \mathbf{r})\, dV' \\ &= i\omega\mu_0 \widehat{\mathbf{z}} \times \int_{S_m} \mathbf{J}(\mathbf{r}') \cdot \underline{\mathbf{G}}_{e11}^{(1)}(\mathbf{r}', \mathbf{r})\, dS' - \widehat{\mathbf{z}} \times \int_{S_a} \mathbf{K}(\mathbf{r}') \cdot [\nabla' \times \underline{\mathbf{G}}_{e21}^{(1)}(\mathbf{r}', \mathbf{r})]\, dS',\end{aligned} \tag{9.125}$$

where $\mathbf{r}'$ has been interchanged with $\mathbf{r}$. The electric Green's dyadics $\underline{\mathbf{G}}_{e11}^{(1)}$ and $\underline{\mathbf{G}}_{e21}^{(1)}$ (provided in Section 9.4) are obtained as the solution of the boundary value problem (9.117)–(9.121) for a semi-infinite two-layered waveguide terminated by a ground plane at $z = \tau$. Note that the integral equation (9.125) itself does not provide a complete formulation of the problem (we have two unknown currents, $\mathbf{J}(\mathbf{r}')$ and $\mathbf{K}(\mathbf{r}')$). An additional integral equation for $\mathbf{J}(\mathbf{r}')$ and $\mathbf{K}(\mathbf{r}')$ is developed next by enforcing a continuity condition for tangential components of magnetic field across apertures $S_a$. This condition, in conjunction with (9.125), will result in a coupled system of integral equations in terms of induced electric and magnetic currents.

### Magnetic-Field Integral Representation

The magnetic field vector satisfies a vector wave equation at each interior point of regions $V_1$ and $V_2$,

$$\nabla \times \nabla \times \mathbf{H}_1(\mathbf{r}) - k_1^2 \mathbf{H}_1(\mathbf{r}) = \nabla \times \mathbf{J}_{\text{imp}}(\mathbf{r}), \qquad \mathbf{r} \in V_1, \tag{9.126}$$

$$\nabla \times \nabla \times \mathbf{H}_2(\mathbf{r}) - k_2^2 \mathbf{H}_2(\mathbf{r}) = \mathbf{0}, \qquad \mathbf{r} \in V_2, \tag{9.127}$$

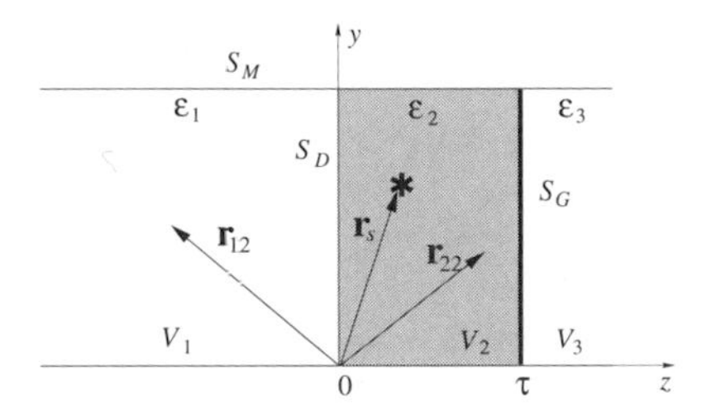


Figure 9.8: Two-layered, semi-infinite waveguide with a point source arbitrarily located in region $V_2$ used in the magnetic field formulation.

subject to boundary conditions on the waveguide surface $S_M$, on the metal surface $S_m$, and on the surface of the ground plane $S_g$ (Figure 9.6),

$$\mathbf{n} \times \nabla \times \mathbf{H}_1(\mathbf{r}) = \mathbf{0}, \qquad \mathbf{n} \cdot \mathbf{H}_1(\mathbf{r}) = 0, \qquad \mathbf{r} \in S_1, \tag{9.128}$$

$$\mathbf{n} \times \nabla \times \mathbf{H}_2(\mathbf{r}) = \mathbf{0}, \qquad \mathbf{n} \cdot \mathbf{H}_2(\mathbf{r}) = 0, \qquad \mathbf{r} \in S_2, \tag{9.129}$$

and continuity conditions across the dielectric interface $S_d$,

$$\begin{aligned} \widehat{\mathbf{z}} \times \mathbf{H}_1(\mathbf{r}) &= \widehat{\mathbf{z}} \times \mathbf{H}_2(\mathbf{r}), & \mathbf{r} \in S_d, \\ \frac{1}{\varepsilon_1}\widehat{\mathbf{z}} \times \nabla \times \mathbf{H}_1(\mathbf{r}) &= \frac{1}{\varepsilon_2}\widehat{\mathbf{z}} \times \nabla \times \mathbf{H}_2(\mathbf{r}), & \mathbf{r} \in S_d. \end{aligned} \tag{9.130}$$

The fitness condition (limiting absorption) is assumed at infinity.

Next, we formulate a boundary value problem for the electric Green's dyadics with a point source arbitrarily positioned in region $V_2$ (Figure 9.8) in order to obtain an integral representation for the magnetic field $\mathbf{H}_2(\mathbf{r})$. We determine the electric Green's dyadics of the third kind,[17] $\underline{\mathbf{G}}_{e12}^{(2)}(\mathbf{r}, \mathbf{r}')$ and $\underline{\mathbf{G}}_{e22}^{(2)}(\mathbf{r}, \mathbf{r}')$, as the solution of a coupled set of dyadic differential equations

$$\nabla \times \nabla \times \underline{\mathbf{G}}_{e12}^{(2)}(\mathbf{r}, \mathbf{r}') - k_1^2 \underline{\mathbf{G}}_{e12}^{(2)}(\mathbf{r}, \mathbf{r}') = \underline{\mathbf{0}}, \qquad \mathbf{r} \in V_1,\ \mathbf{r}' \in V_2, \tag{9.131}$$

$$\nabla \times \nabla \times \underline{\mathbf{G}}_{e22}^{(2)}(\mathbf{r}, \mathbf{r}') - k_2^2 \underline{\mathbf{G}}_{e22}^{(2)}(\mathbf{r}, \mathbf{r}') = \underline{\mathbf{I}}\,\delta(\mathbf{r} - \mathbf{r}'), \qquad \mathbf{r}, \mathbf{r}' \in V_2, \tag{9.132}$$

subject to boundary conditions of the second kind on the surface of the conducting shield $S_M$ and on the ground plane $S_G$ (Figure 9.8),

$$\mathbf{n} \times \nabla \times \underline{\mathbf{G}}_{e12}^{(2)}(\mathbf{r}, \mathbf{r}') = \underline{\mathbf{0}}, \quad \mathbf{n} \cdot \underline{\mathbf{G}}_{e12}^{(2)}(\mathbf{r}, \mathbf{r}') = \mathbf{0}, \quad \mathbf{r} \in S_M, \tag{9.133}$$

$$\mathbf{n} \times \nabla \times \underline{\mathbf{G}}_{e22}^{(2)}(\mathbf{r}, \mathbf{r}') = \underline{\mathbf{0}}, \quad \mathbf{n} \cdot \underline{\mathbf{G}}_{e22}^{(2)}(\mathbf{r}, \mathbf{r}') = \mathbf{0}, \quad \mathbf{r} \in S_M \cup S_G, \tag{9.134}$$

and mixed continuity conditions for the electric Green's dyadics of the third kind across the dielectric interface $S_D$ (analogous to the field continuity

[17] The superscript indicates that in this case the electric Green's dyadics of the third kind represent the analog of the magnetic field.

conditions (9.130) expressed in terms of the magnetic field),

$$\begin{aligned} \widehat{\mathbf{z}} \times \underline{\mathbf{G}}_{e12}^{(2)}(\mathbf{r},\mathbf{r}') &= \widehat{\mathbf{z}} \times \underline{\mathbf{G}}_{e22}^{(2)}(\mathbf{r},\mathbf{r}'), && \mathbf{r} \in S_D, \\ \frac{1}{\varepsilon_1}\widehat{\mathbf{z}} \times \nabla \times \underline{\mathbf{G}}_{e12}^{(2)}(\mathbf{r},\mathbf{r}') &= \frac{1}{\varepsilon_2}\widehat{\mathbf{z}} \times \nabla \times \underline{\mathbf{G}}_{e22}^{(2)}(\mathbf{r},\mathbf{r}'), && \mathbf{r} \in S_D, \end{aligned} \tag{9.135}$$

and the fitness condition (limiting absorption) at infinity.

The vector-dyadic Green's second theorem applied to the magnetic fields and Green's dyadics governed by equations (9.126), (9.127), (9.131), and (9.132), with boundary and continuity conditions (9.128)–(9.130) and (9.133)–(9.135), results in the integral representation for the total magnetic field in region $V_2$ as

$$\begin{aligned} \mathbf{H}_2(\mathbf{r}') =& \frac{\varepsilon_2}{\varepsilon_1}\int_{V_{\text{imp}}} [\nabla \times \mathbf{J}_{\text{imp}}(\mathbf{r})] \cdot \underline{\mathbf{G}}_{e12}^{(2)}(\mathbf{r},\mathbf{r}')\, dV \\ &+ \frac{\varepsilon_2}{\varepsilon_1}\int_{S_m} \mathbf{J}(\mathbf{r}) \cdot [\nabla \times \underline{\mathbf{G}}_{e12}^{(2)}(\mathbf{r},\mathbf{r}')]\, dS \\ &+ i\omega\varepsilon_0\varepsilon_2 \int_{S_a} \mathbf{K}(\mathbf{r}) \cdot \underline{\mathbf{G}}_{e22}^{(2)}(\mathbf{r},\mathbf{r}')\, dS. \end{aligned} \tag{9.136}$$

Note that the volume integral represents part of the incident magnetic field transmitted from region $V_1$ through the dielectric interface at $z = 0$, and the surface integrals determine the scattered magnetic field caused by induced electric and magnetic currents.

Following the above procedure, an integral representation for the scattered magnetic field in region $V_3$ caused by induced magnetic current $\mathbf{K}(\mathbf{r})$ is obtained as

$$\mathbf{H}_3(\mathbf{r}') = -i\omega\varepsilon_0\varepsilon_3 \int_{S_a} \mathbf{K}(\mathbf{r}) \cdot \underline{\mathbf{G}}_{e3}^{(2)}(\mathbf{r},\mathbf{r}')\, dS, \tag{9.137}$$

where $\underline{\mathbf{G}}_{e3}^{(2)}(\mathbf{r},\mathbf{r}')$ is the electric Green's dyadic of the second kind, obtained for a semi-infinite rectangular waveguide (region $V_3$) in Section 9.2.

The continuity condition for tangential components of the magnetic field across the surface of apertures $S_a$ at $z = \tau$,

$$\widehat{\mathbf{z}} \times (\mathbf{H}_2^{\text{inc}}(\mathbf{r}) + \mathbf{H}_2^{\text{scat}}(\mathbf{r})) = \widehat{\mathbf{z}} \times \mathbf{H}_3^{\text{scat}}(\mathbf{r}), \qquad \mathbf{r} \in S_a,$$

provides an integral equation for the unknown electric and magnetic currents,

$$\begin{aligned} \frac{\varepsilon_2}{\varepsilon_1}\widehat{\mathbf{z}} \times \int_{V_{\text{imp}}} & [\nabla' \times \mathbf{J}_{\text{imp}}(\mathbf{r}')] \cdot \underline{\mathbf{G}}_{e12}^{(2)}(\mathbf{r}',\mathbf{r})\, dV' \\ = &- \frac{\varepsilon_2}{\varepsilon_1}\widehat{\mathbf{z}} \times \int_{S_m} \mathbf{J}(\mathbf{r}') \cdot [\nabla' \times \underline{\mathbf{G}}_{e12}^{(2)}(\mathbf{r}',\mathbf{r})]\, dS' \\ &- i\omega\varepsilon_0 \widehat{\mathbf{z}} \times \int_{S_a} \mathbf{K}(\mathbf{r}') \cdot [\varepsilon_2 \underline{\mathbf{G}}_{e22}^{(2)}(\mathbf{r}',\mathbf{r}) + \varepsilon_3 \underline{\mathbf{G}}_{e3}^{(2)}(\mathbf{r}',\mathbf{r})]\, dS', \end{aligned} \tag{9.138}$$

where $\mathbf{r}'$ is interchanged with $\mathbf{r}$. Together, (9.125) and (9.138) form a coupled system of integral equations that may be solved for the unknown currents $\mathbf{J}$ and $\mathbf{K}$. The electric Green's dyadics $\underline{\mathbf{G}}_{e12}^{(2)}(\mathbf{r},\mathbf{r}')$ and $\underline{\mathbf{G}}_{e22}^{(2)}(\mathbf{r},\mathbf{r}')$ are developed in Section 9.4.

In the vicinity of an edge of the metal surfaces $S_m$ and apertures $S_a$ we enforce a finite energy condition on the total field

$$\lim_{V_i \to 0} \int_{V_i} \left( |\mathbf{H}_i(\mathbf{r})|^2 + \frac{1}{k_i^2} |\nabla \times \mathbf{H}_i(\mathbf{r})|^2 \right) dV = 0. \tag{9.139}$$

The condition (9.139) defines a subspace of $\mathbf{L}^2(V_i)^3$ for the solution as the Sobolev space $\mathbf{W}_2^1(V_i)^3 \subset \mathbf{L}^2(V_i)^3$, and, in particular, in a source free region we have the Hilbert space $H(\text{div}, \text{curl}, V_i)$ with the boundary values defined in $H^{1/2}(\widetilde{S}_i)$, where $\widetilde{S}_i = S_i \cup S_d$. The incident and scattered magnetic fields have square-integrable curls and divergences, with the boundary values of their tangential components defined within the space $H^{1/2}(S_a) \subset \mathbf{L}^2(S_a)$.

### Generalized Scattering Matrix for Waveguide-Based Antennas

Here we outline a procedure for developing a generalized scattering matrix for waveguide-based electric- and magnetic-type antennas [60]. The goal is to obtain a matrix representation that relates magnitudes of incident and scattered modes, including propagating and evanescent TE- and TM-modes.

For this procedure it is useful to obtain a discrete form for the incident fields. The form introduced in (9.125) and (9.138) requires an explicit representation for the impressed electric current $\mathbf{J}_{\text{imp}}(\mathbf{r}')$, and here we use an alternative development in terms of a series eigenmode expansion that includes both propagating and evanescent TE- and TM-modes.

The total incident electric field at $z = 0$ can be written as

$$\mathbf{E}_1^{\text{inc}}(\mathbf{r})\big|_{z=0} = \sum_{m=0}^{\infty} \sum_{m \neq n=0}^{\infty} a_{mn}^{\text{TE}} \mathbf{e}_{mn}^{+\text{TE}}(x,y)(1 + R_{mn}^{\text{TE}}) + \sum_{m=1}^{\infty} \sum_{n=1}^{\infty} a_{mn}^{\text{TM}} \mathbf{e}_{mn}^{+\text{TM}}(x,y)(1 + R_{mn}^{\text{TM}}), \tag{9.140}$$

where the first term is the direct incident wave and the second term (containing the reflection coefficient) is the wave reflected from the dielectric layer.[18] The coefficients $a_{mn}^{\text{TE}}$ and $a_{mn}^{\text{TM}}$ are the magnitudes of propagating

[18]Note that $R_{mn}^{\text{TE}}$ and $R_{mn}^{\text{TM}}$ represent total reflection coefficients for TE- and TM-modes. These coefficients account for the total reflection at $z = 0$, which also includes the reflection from the ground plane at $z = \tau$.

and evanescent modes. The total incident magnetic field at $z = \tau$ is

$$\mathbf{H}_2^{\rm inc}(\mathbf{r})\big|_{z=\tau} = \sum_{m=0}^{\infty} \sum_{m \neq n=0}^{\infty} 2a_{mn}^{\rm TE} \mathbf{h}_{mn}^{+\rm TE}(x,y) T_{mn}^{\rm TE} e^{-\gamma_{mn}^{(2)}\tau} + \sum_{m=1}^{\infty} \sum_{n=1}^{\infty} 2a_{mn}^{\rm TM} \mathbf{h}_{mn}^{+\rm TM}(x,y) T_{mn}^{\rm TM} e^{-\gamma_{mn}^{(2)}\tau}. \quad (9.141)$$

The modal vectors $\mathbf{e}_{mn}(x,y)$ and $\mathbf{h}_{mn}(x,y)$, normalized by the unity power condition

$$\int_{S_w} [\mathbf{e}_{mn}^{\pm}(x,y) \times \mathbf{h}_{mn}^{\pm}(x,y)] \cdot (\pm\widehat{\mathbf{z}})\, dS = 1 \quad (9.142)$$

are obtained using a procedure described in [38], leading to the expressions for TE- and TM-modes. In (9.142), $S_w$ is the waveguide cross-section, and the "$\pm$" sign corresponds to waves propagating in the positive (+) and negative (−) $z$-directions. These modal vectors are given by the following expressions.

**TE-modes:**

$$e_{xmn}^{+} = \sqrt{\frac{\varepsilon_{0m}\varepsilon_{0n}}{ab}} \frac{\left(\frac{n\pi}{b}\right)\sqrt{Z_h}}{\sqrt{\left(\frac{n\pi}{b}\right)^2 + \left(\frac{m\pi}{a}\right)^2}} \cos\left(\frac{m\pi x}{a}\right) \sin\left(\frac{n\pi y}{b}\right),$$

$$e_{ymn}^{+} = -\sqrt{\frac{\varepsilon_{0m}\varepsilon_{0n}}{ab}} \frac{\left(\frac{m\pi}{a}\right)\sqrt{Z_h}}{\sqrt{\left(\frac{n\pi}{b}\right)^2 + \left(\frac{m\pi}{a}\right)^2}} \sin\left(\frac{m\pi x}{a}\right) \cos\left(\frac{n\pi y}{b}\right),$$

$$h_{xmn}^{+} = \sqrt{\frac{\varepsilon_{0m}\varepsilon_{0n}}{ab}} \frac{\left(\frac{m\pi}{a}\right)}{\sqrt{Z_h}\sqrt{\left(\frac{n\pi}{b}\right)^2 + \left(\frac{m\pi}{a}\right)^2}} \sin\left(\frac{m\pi x}{a}\right) \cos\left(\frac{n\pi y}{b}\right),$$

$$h_{ymn}^{+} = \sqrt{\frac{\varepsilon_{0m}\varepsilon_{0n}}{ab}} \frac{\left(\frac{n\pi}{b}\right)}{\sqrt{Z_h}\sqrt{\left(\frac{n\pi}{b}\right)^2 + \left(\frac{m\pi}{a}\right)^2}} \cos\left(\frac{m\pi x}{a}\right) \sin\left(\frac{n\pi y}{b}\right). \quad (9.143)$$

Here, $\varepsilon_{0m}$ and $\varepsilon_{0n}$ are Neumann indexes, such that $\varepsilon_{00} = 1$ and $\varepsilon_{0m}, \varepsilon_{0n} = 2$, $m, n \neq 0$, and $Z_h = i\omega\mu_0/\gamma_{mn}^{(1)}$ is the wave impedance for TE-modes.

**TM-modes:**

$$e_{xmn}^{+} = -\sqrt{\frac{\varepsilon_{0m}\varepsilon_{0n}}{ab}} \frac{\left(\frac{m\pi}{a}\right)\sqrt{Z_e}}{\sqrt{\left(\frac{n\pi}{b}\right)^2 + \left(\frac{m\pi}{a}\right)^2}} \cos\left(\frac{m\pi x}{a}\right) \sin\left(\frac{n\pi y}{b}\right),$$

$$e_{ymn}^{+} = -\sqrt{\frac{\varepsilon_{0m}\varepsilon_{0n}}{ab}} \frac{\left(\frac{n\pi}{b}\right)\sqrt{Z_e}}{\sqrt{\left(\frac{n\pi}{b}\right)^2 + \left(\frac{m\pi}{a}\right)^2}} \sin\left(\frac{m\pi x}{a}\right) \cos\left(\frac{n\pi y}{b}\right),$$

$$h_{xmn}^{+} = \sqrt{\frac{\varepsilon_{0m}\varepsilon_{0n}}{ab}} \frac{\left(\frac{n\pi}{b}\right)}{\sqrt{Z_e}\sqrt{\left(\frac{n\pi}{b}\right)^2 + \left(\frac{m\pi}{a}\right)^2}} \sin\left(\frac{m\pi x}{a}\right) \cos\left(\frac{n\pi y}{b}\right),$$

$$h_{ymn}^{+} = -\sqrt{\frac{\varepsilon_{0m}\varepsilon_{0n}}{ab}} \frac{\left(\frac{m\pi}{a}\right)}{\sqrt{Z_e}\sqrt{\left(\frac{n\pi}{b}\right)^2 + \left(\frac{m\pi}{a}\right)^2}} \cos\left(\frac{m\pi x}{a}\right) \sin\left(\frac{n\pi y}{b}\right), \tag{9.144}$$

where $Z_e = \gamma_{mn}^{(2)}/i\omega\varepsilon_0\varepsilon_2$ is the wave impedance for TM-modes.

For backward TE and TM waves (in the negative $z$-direction) defined by (9.143) and (9.144), respectively, we have

$$e_{xmn}^{-} = e_{xmn}^{+}, \qquad e_{ymn}^{-} = e_{ymn}^{+},$$
$$h_{xmn}^{-} = -h_{xmn}^{+}, \qquad h_{ymn}^{-} = -h_{ymn}^{+}.$$

The total reflection ($R_{mn}$) and transmission ($T_{mn}$) coefficients for the TE and TM incident modes at $z = 0$ are obtained in the absence of metal obstacles and aperture, but account for the dielectric layer and the ground plane. Note that in integral equations (9.125) and (9.138) the electric and magnetic currents induced on the metal obstacles and apertures, respectively, are caused by the total incident electric and magnetic fields, which include the direct and reflected terms.

Solving for the TE- and TM-mode amplitudes by matching tangential components on the dielectric interface at $z = 0$, and satisfying boundary conditions on the ground plane at $z = \tau$, we obtain[19]

$$R_{mn}^{\mathrm{TE,TM}} = \frac{R_{12}^{\mathrm{TE,TM}} - e^{-2\gamma_{mn}^{(2)}\tau}}{1 - R_{12}^{\mathrm{TE,TM}} e^{-2\gamma_{mn}^{(2)}\tau}}, \qquad T_{mn}^{\mathrm{TE,TM}} = \frac{T_{12}^{\mathrm{TE,TM}}}{1 - R_{12}^{\mathrm{TE,TM}} e^{-2\gamma_{mn}^{(2)}\tau}}, \tag{9.145}$$

where

$$R_{12}^{\mathrm{TE}} = \frac{\gamma_{mn}^{(1)} - \gamma_{mn}^{(2)}}{\gamma_{mn}^{(1)} + \gamma_{mn}^{(2)}}, \qquad T_{12}^{\mathrm{TE}} = \frac{2\sqrt{\gamma_{mn}^{(1)}\gamma_{mn}^{(2)}}}{\gamma_{mn}^{(1)} + \gamma_{mn}^{(2)}},$$
$$R_{12}^{\mathrm{TM}} = \frac{\varepsilon_1\gamma_{mn}^{(2)} - \varepsilon_2\gamma_{mn}^{(1)}}{\varepsilon_1\gamma_{mn}^{(2)} + \varepsilon_2\gamma_{mn}^{(1)}}, \qquad T_{12}^{\mathrm{TM}} = \frac{2\sqrt{\varepsilon_1\gamma_{mn}^{(2)}\varepsilon_2\gamma_{mn}^{(1)}}}{\varepsilon_1\gamma_{mn}^{(2)} + \varepsilon_2\gamma_{mn}^{(1)}}, \tag{9.146}$$

with the propagation constant

$$\gamma_{mn}^{(i)} = \sqrt{\kappa_{mn}^2 - k_i^2} \tag{9.147}$$

[19] The transverse components of electric and magnetic fields in regions $V_1$ and $V_2$ are expressed as a superposition of forward and backward traveling with unknown magnitudes, to be determined subject to the boundary condition on the ground plane at $z = \tau$ and the continuity condition across the dielectric interface at $z = 0$.

for $i = 1, 2$, where $\kappa_{mn}^2 = (m\pi/a)^2 + (n\pi/b)^2$ with $k_i = (2\pi)/(\lambda_0)\sqrt{\varepsilon_i}$.

The coupled system of integral equations (9.125) and (9.138), with $\mathbf{E}_1^{\text{inc}}$ and $\mathbf{H}_2^{\text{inc}}$ given by (9.140) and (9.141), are then given by

$$\begin{aligned}\widehat{\mathbf{z}} \times \mathbf{E}_1^{\text{inc}}(\mathbf{r}) =& i\omega\mu_0 \widehat{\mathbf{z}} \times \int_{S_m} \mathbf{J}(\mathbf{r}') \cdot \underline{\mathbf{G}}_{e11}^{(1)}(\mathbf{r}', \mathbf{r})\, dS', \qquad \mathbf{r} \in S_m \\ &- \widehat{\mathbf{z}} \times \int_{S_a} \mathbf{K}(\mathbf{r}') \cdot [\nabla' \times \underline{\mathbf{G}}_{e21}^{(1)}(\mathbf{r}', \mathbf{r})]\, dS',\end{aligned} \tag{9.148}$$

$$\begin{aligned}\widehat{\mathbf{z}} \times \mathbf{H}_2^{\text{inc}}(\mathbf{r}) =& -\frac{\varepsilon_2}{\varepsilon_1} \widehat{\mathbf{z}} \times \int_{S_m} \mathbf{J}(\mathbf{r}') \cdot [\nabla' \times \underline{\mathbf{G}}_{e12}^{(2)}(\mathbf{r}', \mathbf{r})]\, dS', \qquad \mathbf{r} \in S_a \\ &- i\omega\varepsilon_0 \widehat{\mathbf{z}} \times \int_{S_a} \mathbf{K}(\mathbf{r}') \cdot [\varepsilon_2 \underline{\mathbf{G}}_{e22}^{(2)}(\mathbf{r}', \mathbf{r}) + \varepsilon_3 \underline{\mathbf{G}}_{e3}^{(2)}(\mathbf{r}', \mathbf{r})]\, dS',\end{aligned} \tag{9.149}$$

and can be discretized via a Galerkin-type projection technique, whereby the unknown induced currents are expanded in a set of known functions with unknown coefficients. This results in a matrix system for the unknown coefficients,

$$Ax = f, \tag{9.150}$$

where $A$ is the matrix of all self- and mutual interactions of electric- and magnetic field components caused by electric and magnetic currents, $x$ is the vector of unknown current coefficients, and $f$ represents the "tested" incident electric and magnetic fields expressed in terms of the magnitudes $a_{mn}$,

$$x = \begin{bmatrix} J \\ K \end{bmatrix}, \qquad f = \begin{bmatrix} V \\ I \end{bmatrix}. \tag{9.151}$$

The vector $f$ can be written as

$$\begin{bmatrix} V \\ I \end{bmatrix} = BCa, \tag{9.152}$$

where

$$B = \begin{bmatrix} \langle e_x^{\text{TE}}, \Phi_x \rangle & \langle e_x^{\text{TM}}, \Phi_x \rangle & 0 & 0 \\ \langle e_y^{\text{TE}}, \Phi_y \rangle & \langle e_y^{\text{TM}}, \Phi_y \rangle & 0 & 0 \\ 0 & 0 & \langle h_x^{\text{TE}}, \Psi_x \rangle & \langle h_x^{\text{TM}}, \Psi_x \rangle \\ 0 & 0 & \langle h_y^{\text{TE}}, \Psi_y \rangle & \langle h_y^{\text{TM}}, \Psi_y \rangle \end{bmatrix},$$

$$C = \begin{bmatrix} E + R^{\text{TE}} & 0 \\ 0 & E + R^{\text{TM}} \\ T^{\text{TE}} & 0 \\ 0 & T^{\text{TM}} \end{bmatrix}, \qquad a = \begin{bmatrix} a^{\text{TE}} \\ a^{\text{TM}} \end{bmatrix}.$$

Here, $\langle e, \Phi \rangle$ and $\langle h, \Psi \rangle$ represent the inner product of the vector components with the testing functions $\Phi$ and $\Psi$, corresponding to electric and magnetic current expansions,[20] $R^{\mathrm{TE}}$ and $R^{\mathrm{TM}}$ are the total reflection coefficients for incident modes defined by expressions (9.145) and (9.146), $E$ is the identity matrix, and $T^{\mathrm{TE}}$ and $T^{\mathrm{TM}}$ are the total transmission coefficients for incident modes,

$$T^{\mathrm{TE}} = 2e^{-\gamma_{mn}^{(2)}\tau}T_{mn}^{\mathrm{TE}}, \qquad T^{\mathrm{TM}} = 2e^{-\gamma_{mn}^{(2)}\tau}T_{mn}^{\mathrm{TM}},$$

where $T_{mn}^{\mathrm{TE}}$ and $T_{mn}^{\mathrm{TM}}$ are determined at $z = 0$ by (9.145) and (9.146). We can obtain the vector $x$ in terms of the magnitudes of the incident modes $a_{mn}$ as the matrix product

$$x = \begin{bmatrix} J \\ K \end{bmatrix} = A^{-1}BCa. \tag{9.153}$$

Next, consider the representation for the scattered electric field (reflected at $z = 0$), which can be obtained either in integral form or as an eigenmode expansion. The integral form immediately follows from (9.124), resulting in

$$\begin{aligned}\mathbf{E}_1^{\mathrm{ref}}(\mathbf{r}) = &- i\omega\mu_0 \int_{S_m} \mathbf{J}(\mathbf{r}') \cdot \underline{\mathbf{G}}_{e11}^{(1)}(\mathbf{r}', \mathbf{r})\, dS' \\ &+ \int_{S_a} \mathbf{K}(\mathbf{r}') \cdot [\nabla' \times \underline{\mathbf{G}}_{e21}^{(1)}(\mathbf{r}', \mathbf{r})]\, dS' \\ &+ \sum_{m=0}^{\infty} \sum_{m \neq n=0}^{\infty} a_{mn}^{\mathrm{TE}} R_{mn}^{\mathrm{TE}} \mathbf{e}_{mn}^{+\mathrm{TE}}(x, y) + \sum_{m=1}^{\infty} \sum_{n=1}^{\infty} a_{mn}^{\mathrm{TM}} R_{mn}^{\mathrm{TM}} \mathbf{e}_{mn}^{+\mathrm{TM}}(x, y).\end{aligned} \tag{9.154}$$

The reflected part of the total incident field is included in (9.154). The eigenmode series expansion for the reflected electric field at $z = 0$ can be written in the form

$$\mathbf{E}_1^{\mathrm{ref}}(\mathbf{r}) = \sum_{m=0}^{\infty} \sum_{m \neq n=0}^{\infty} b_{mn}^{\mathrm{TE}} \mathbf{e}_{mn}^{-\mathrm{TE}}(x, y) + \sum_{m=1}^{\infty} \sum_{n=1}^{\infty} b_{mn}^{\mathrm{TM}} \mathbf{e}_{mn}^{-\mathrm{TM}}(x, y). \tag{9.155}$$

Equating (9.154) and (9.155) and using the unity power condition (9.142) allow us to express the magnitudes $b_{mn}$ of reflected TE- and TM-modes

[20] In [60], overlapping piecewise-sinusoidal basis and testing functions have been used in the method-of-moments discretization for electric and magnetic currents. For canonical shapes (for example, rectangular patches or slots) we could expand currents in terms of entire-domain basis functions, for example, Chebyshev polynomials, which form a complete set and adequately model the current behavior in the vicinity of an edge. Moreover, as described in Section 4.3.4, Chebyshev polynomials are eigenfunctions of a singular logarithmic-type operator and may be useful in regularizing the singular part of the operator (where possible), resulting in Fredholm equations of the second kind.

as[21]

$$\left\{ \begin{array}{c} b_{mn}^{\mathrm{TE}} \\ b_{mn}^{\mathrm{TM}} \end{array} \right\} = -\, i\omega\mu_0 \int_{S_w} \int_{S_m} \left[ \mathbf{J} \cdot \underline{\mathbf{G}}_{e11}^{(1)} \times \left\{ \begin{array}{c} \mathbf{h}_{mn}^{-\mathrm{TE}} \\ \mathbf{h}_{mn}^{-\mathrm{TM}} \end{array} \right\} \right] \cdot (-\widehat{\mathbf{z}})\, dS'\, dS$$
$$+ \int_{S_w} \int_{S_a} \left[ \mathbf{K} \cdot [\nabla' \times \underline{\mathbf{G}}_{e21}^{(1)}] \times \left\{ \begin{array}{c} \mathbf{h}_{mn}^{-\mathrm{TE}} \\ \mathbf{h}_{mn}^{-\mathrm{TM}} \end{array} \right\} \right] \cdot (-\widehat{\mathbf{z}})\, dS'\, dS$$
$$+ \left\{ \begin{array}{c} a_{mn}^{\mathrm{TE}} \\ a_{mn}^{\mathrm{TM}} \end{array} \right\} \left\{ \begin{array}{c} R_{mn}^{\mathrm{TE}} \\ R_{mn}^{\mathrm{TM}} \end{array} \right\}. \tag{9.156}$$

The system of equations (9.156) can be written in the matrix form

$$b = D \left[ \begin{array}{c} J \\ K \end{array} \right] + Ra, \tag{9.157}$$

where $D$ is a $2 \times 2$ block matrix obtained as a result of integration over $S_w, S_m$ and $S_w, S_a$ in (9.156), and $R$ is a diagonal matrix with elements $R_{mn}^{\mathrm{TE}}$ and $R_{mn}^{\mathrm{TM}}$ determined by (9.145) and (9.146). Substituting the vector of electric and magnetic current magnitudes determined as the matrix product (9.153) into the matrix equation (9.157), we obtain a matrix relationship for the magnitudes of reflected and incident modes,

$$b = S_{11} a,$$

where

$$S_{11} = DA^{-1}BC + R, \tag{9.158}$$

which represents the reflection coefficient part of the GSM for the whole structure.

The transmission coefficient part of the GSM is obtained by relating magnitudes of incident and transmitted modes. The transmitted electric field caused by the magnetic current induced on the surface of apertures $S_a$ at $z = \tau$ is obtained in the integral form

$$\mathbf{E}_3^{\mathrm{tr}}(\mathbf{r}) = - \int_{S_a} \mathbf{K}(\mathbf{r}') \cdot [\nabla' \times \underline{\mathbf{G}}_{e3}^{(1)}(\mathbf{r}', \mathbf{r})]\, dS', \tag{9.159}$$

where $\underline{\mathbf{G}}_{e3}^{(1)}$ is the electric Green's dyadic of the first kind developed in Section 9.2 for a semi-infinite waveguide terminated by a ground plane at

[21] Note that TE- and TM-modes are orthogonal, such that

$$\int_{S_w} [\mathbf{e}_{mn}^{\pm\mathrm{TE}}(x,y) \times \mathbf{h}_{mn}^{\pm\mathrm{TM}}(x,y)] \cdot (\pm\widehat{\mathbf{z}})\, dS = 0,$$
$$\int_{S_w} [\mathbf{e}_{mn}^{\pm\mathrm{TM}}(x,y) \times \mathbf{h}_{mn}^{\pm\mathrm{TE}}(x,y)] \cdot (\pm\widehat{\mathbf{z}})\, dS = 0.$$

$z = \tau$.[22] The electric field transmitted into region $V_3$ is also expressed as an eigenmode series expansion with magnitudes $c_{mn}$ of transmitted TE- and TM-modes,

$$\mathbf{E}_3^{\mathrm{tr}}(\mathbf{r}) = \sum_{m=0}^{\infty} \sum_{m \neq n=0}^{\infty} c_{mn}^{\mathrm{TE}} \mathbf{e}_{mn}^{+\mathrm{TE}}(x,y) + \sum_{m=1}^{\infty} \sum_{n=1}^{\infty} c_{mn}^{\mathrm{TM}} \mathbf{e}_{mn}^{+\mathrm{TM}}(x,y). \quad (9.160)$$

The unity power condition (9.142) applied to (9.159) and (9.160) results in the expression for the magnitudes of transmitted modes in terms of induced magnetic currents,

$$\left\{ \begin{array}{c} c_{mn}^{\mathrm{TE}} \\ c_{mn}^{\mathrm{TM}} \end{array} \right\} = -\int_{S_w} \int_{S_a} \left[ \mathbf{K} \cdot [\nabla' \times \underline{\mathbf{G}}_{A3}^{(1)}] \times \left\{ \begin{array}{c} \mathbf{h}_{mn}^{+\mathrm{TE}} \\ \mathbf{h}_{mn}^{+\mathrm{TM}} \end{array} \right\} \right] \cdot \widehat{\mathbf{z}} \, dS' \, dS. \quad (9.161)$$

Following a similar procedure to that developed for $S_{11}$ we obtain a matrix relationship between magnitudes of transmitted and incident modes,

$$c = S_{21} a$$

with

$$S_{21} = L A^{-1} B C, \quad (9.162)$$

which is the transmission coefficient part of the GSM for the structure. Here, the matrix $L$ is analogous to the matrix $D$ in the matrix product (9.158). The $S_{22}$ and $S_{12}$ coefficients of the GSM are obtained similarly with the impressed electric current (incident fields) positioned in region $V_3$.

It should be noted that the integral equation formulation in conjunction with the GSM method does not require the calculation of electric and magnetic currents.[23] Instead, those currents are used to relate magnitudes of scattered (reflected and transmitted) modes to magnitudes of incident modes by means of matrix transformations. Numerical results for the scattering parameters of various waveguide-based strip-to-slot transition modules are presented in [60].

As an example, consider determining the $S$-parameters of an overmoded rectangular waveguide-based strip-to-slot transition module, as depicted in Figure 9.9. Numerical results generated using the formulation presented here are compared with those generated by the GSM cascading scheme [59] and the *Agilent* HFSS program. Figures 9.10 and 9.11 demonstrate dispersion characteristics (magnitude and phase) for the transmission coefficient $S_{21}$ for the dominant $\mathrm{TE}_{10}$-mode in the resonance frequency range (18.5–20.3 GHz). It can be seen that for electrically large substrates ($\tau = 2.5$ mm) the GSM technique discussed here and presented in [60], and the GSM modeling scheme proposed in [59], can both be used to generate an accurate solution for waveguide-based interacting discontinuities.

[22] Note that $\nabla' \times \underline{\mathbf{G}}_{e3}^{(1)}(\mathbf{r}', \mathbf{r}) = \nabla' \times \underline{\mathbf{G}}_{A3}^{(1)}(\mathbf{r}', \mathbf{r})$, where $\underline{\mathbf{G}}_{A3}^{(1)}$ is the magnetic potential Green's dyadic of the first kind obtained as a diagonal dyadic for rectangular waveguides and cavities.

[23] Of course, we can solve the matrix equation (9.153) for $J$ and $K$ if the magnitudes of the incident modes, $a_{mn}^{\mathrm{TE}}$ and $a_{mn}^{\mathrm{TM}}$, are known.

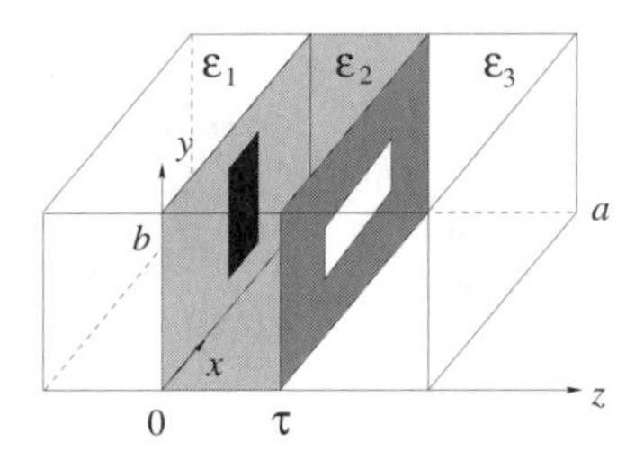


Figure 9.9: Rectangular waveguide-based strip-to-slot transition module: Strip is 0.6 mm × 5.4 mm, slot is 5.4 mm × 0.6 mm, $a$ = 22.86 mm, $b$ = 10.16 mm, $\tau$ = 2.5 mm, $\varepsilon_1 = 1.0$, $\varepsilon_2 = 6.0$, $\varepsilon_3 = 1.0$.

## 9.3.2 A Waveguide-Based Aperture-Coupled Patch Array

In this section the previously presented technique is applied to the full-wave analysis of a waveguide-based aperture-coupled patch amplifier array. In [61] a perpendicularly fed patch array is proposed for quasi-optical power combining, and Figure 9.12 depicts a waveguide-based analog of the array with a hard-horn excitation. In this system, the incident field from the horn couples to an array of aperture-coupled patch antennas, where each antenna is coupled to a dielectric-filled waveguide through the antenna aperture. The dielectric-filled waveguide is connected to a microstrip circuit [62] that contains amplifier networks. The amplified signal is coupled back to an aperture-coupled patch antenna array and then collected by the receive horn. The amplifier networks are isolated from the receive and transmit antennas by the use of dielectric-filled waveguides and the specific antenna-feed configuration developed in [62]. This allows us to separate the circuit part of the problem from the part requiring electromagnetic analysis. Passive structures in the waveguide-based amplifier array can be modeled using different full-wave methods, and the overall response of the system is obtained by cascading responses of individual elements, including the amplifiers.

In this section we present an integral equation formulation for the analysis of a waveguide-to-aperture-coupled patch array, resulting in the GSM of an $N$-port spatial divider (combiner) [63].

Consider the $N$-port waveguide-based aperture-coupled patch array transition shown in Figure 9.13. Rectangular patch antennas $S_m^p$ are located on the interface $S_d$ of two adjacent dielectric layers with permittivities $\varepsilon_1$ and $\varepsilon_2$, and rectangular slot apertures $S_a^q$ are located in a ground plane $S_g$ connected to $N-1$ single-mode waveguides (regions $V_a^q$) filled by the same dielectric material with permittivity $\varepsilon_3$. The incident electric and magnetic fields in the large waveguide (region $V_1$) are generated by an impressed electric current $\mathbf{J}_{\text{imp}}(\mathbf{r}) \in V_{\text{imp}} \subset V_1$.[24]

[24]Note that similar formulations can be obtained for excitation of the transition from

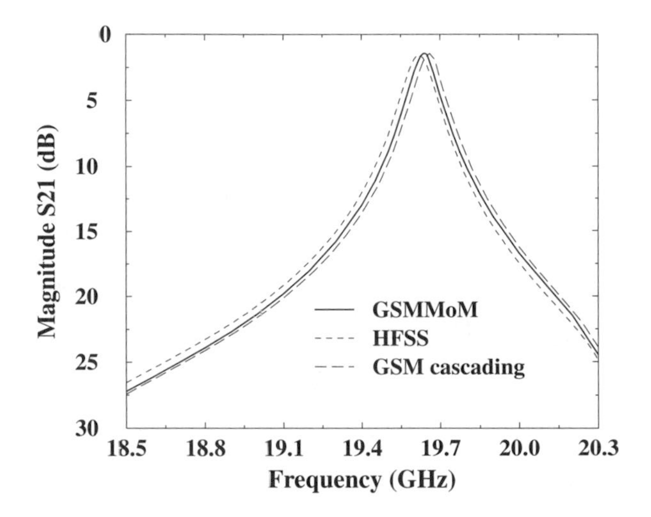


Figure 9.10: Magnitude of the transmission coefficient $S_{21}$ of the dominant mode $TE_{10}$ against frequency for a strip-to-slot transition in an overmoded rectangular waveguide. Numerical results are compared with data obtained using a GSM cascading modeling scheme [59] and a 3D commercial FEM program (*Agilent* HFSS).

We obtain a coupled system of integral equations with respect to the unknown induced electric and magnetic currents $\mathbf{J}_p$ and $\mathbf{K}_q$ by enforcing boundary conditions on the tangential components of the electric field on the perfectly conductive surfaces $S_m^p$ of the patch array at $z = 0$, and continuity conditions on the tangential components of the magnetic field on the magnetic (aperture) surfaces $S_a^q$ of the slot array at $z = \tau$. The integral form of the electric-field boundary condition is obtained using the vector-dyadic Green's second theorem with appropriate boundary and continuity conditions for fields and Green's dyadics (a similar procedure is discussed in the previous example; see (9.148)), resulting in the integral equation

$$\begin{aligned}\widehat{\mathbf{z}} \times \mathbf{E}_1^{\text{inc}}(\mathbf{r}) =& i\omega\mu_0 \widehat{\mathbf{z}} \times \sum_{p=2}^{N} \int_{S_m^p} \mathbf{J}_p(\mathbf{r}') \cdot \underline{\mathbf{G}}_{e11}^{(1)}(\mathbf{r}', \mathbf{r})\, dS' \\ &- \widehat{\mathbf{z}} \times \sum_{q=2}^{N} \int_{S_a^q} \mathbf{K}_q(\mathbf{r}') \cdot [\nabla' \times \underline{\mathbf{G}}_{e21}^{(1)}(\mathbf{r}', \mathbf{r})]\, dS'. \end{aligned} \tag{9.163}$$

each of the single-mode waveguides $V_a^q$. Positioning an impressed current in either of the waveguides does not affect the scattered field representation due to induced electric and magnetic currents. The consideration of the excitation at each particular waveguide ($V_1$ and $V_a^q$) becomes important when obtaining reflection and transmission parts of the GSM associated with each port.

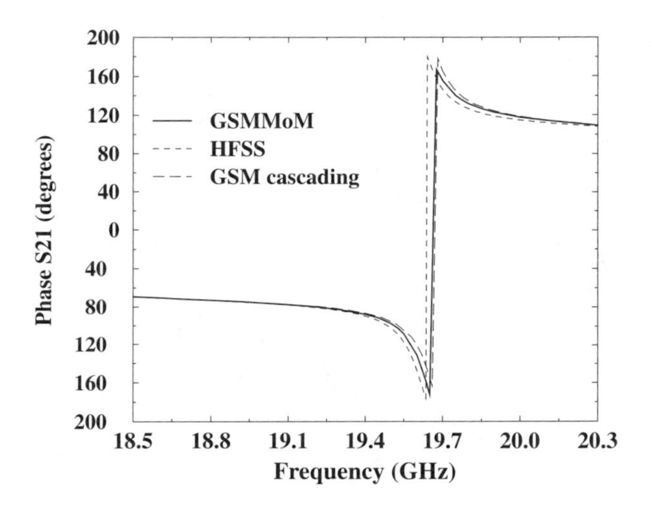


Figure 9.11: Phase of the transmission coefficient $S_{21}$ against frequency for a strip-to-slot transition module.

The electric Green's dyadics of the third kind, $\underline{\mathbf{G}}_{e11}^{(1)}$ and $\underline{\mathbf{G}}_{e21}^{(1)}$, are obtained as the solution of the boundary value problem (9.117)–(9.121) for a semi-infinite, partially filled waveguide (regions $V_1$ and $V_2$) terminated by a ground plane at $z = \tau$ (see Section 9.4). The incident electric field, $\mathbf{E}_1^{\text{inc}}$, is expressed as a series eigenmode expansion (9.140) for all propagating and evanescent TE- and TM-modes. The electric vector functions in this expansion satisfy the unity power condition (9.142).

The continuity condition for tangential components of the magnetic field at $z = \tau$ results in an integral equation for the unknown currents (a similar integral equation, (9.149), is obtained in the previous example of coupled electric- and magnetic-type discontinuities in a two-port waveguide transition),

$$\widehat{\mathbf{z}} \times \mathbf{H}_2^{\text{inc}}(\mathbf{r}) = -\frac{\varepsilon_2}{\varepsilon_1}\widehat{\mathbf{z}} \times \sum_{p=2}^{N} \int_{S_m^p} \mathbf{J}_p(\mathbf{r}') \cdot [\nabla' \times \underline{\mathbf{G}}_{e12}^{(2)}(\mathbf{r}', \mathbf{r})]\, dS'$$
$$- i\omega\varepsilon_0 \widehat{\mathbf{z}} \times \sum_{q=2}^{N} \int_{S_a^q} \mathbf{K}_q(\mathbf{r}') \cdot [\varepsilon_2 \underline{\mathbf{G}}_{e22}^{(2)}(\mathbf{r}', \mathbf{r}) + \varepsilon_3 \underline{\mathbf{G}}_{\text{eq}}^{(2)}(\mathbf{r}', \mathbf{r})]\, dS'. \quad (9.164)$$

The electric Green's dyadics of the third kind, $\underline{\mathbf{G}}_{e12}^{(2)}$ and $\underline{\mathbf{G}}_{e22}^{(2)}$, have been obtained for a semi-infinite, partially filled waveguide with termination at $z = \tau$ as the solution of the boundary value problem (9.131)–(9.135) (see Section 9.4 for details). The electric Green's dyadics of the second kind, $\underline{\mathbf{G}}_{\text{eq}}^{(2)}$, are obtained for the semi-infinite homogeneous waveguides (regions $V_a^q$) terminated by a ground plane at $z = \tau$ and are derived in Section 9.2.2.

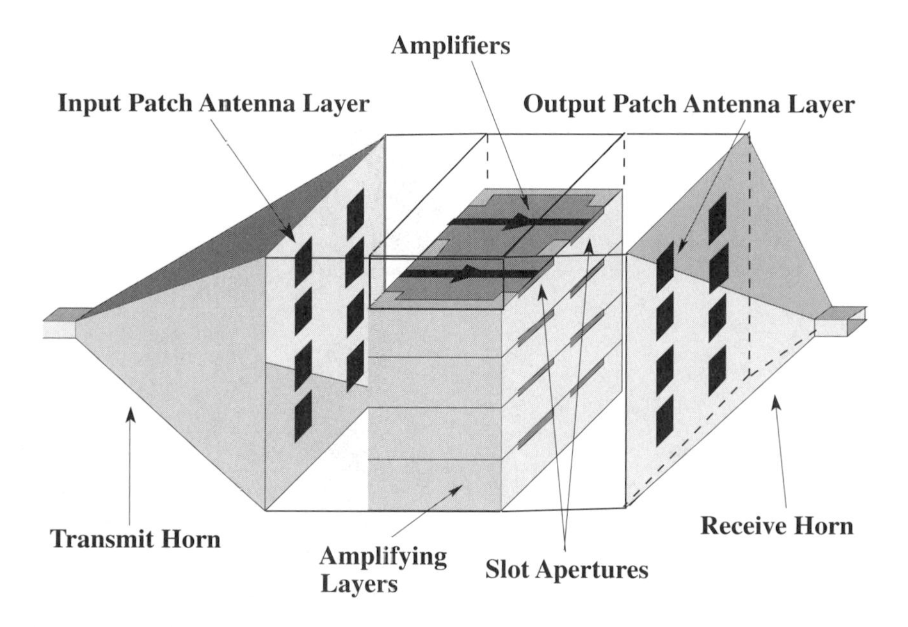


Figure 9.12: Waveguide-based aperture-coupled patch amplifier array.

The slot-to-waveguide interface is shown in Figure 9.14, where each slot is associated with the corresponding single-mode waveguide. The incident magnetic field, $\mathbf{H}_2^{\text{inc}}$, represents that part of the total incident magnetic field transmitted into region $V_2$ from region $V_1$ across the dielectric interface $S_d$. It is obtained as a series eigenmode expansion (9.141) for all propagating and evanescent TE- and TM-modes with the magnetic vector functions and transmission coefficients defined in region $V_2$.

The coupled system of integral equations (9.163) and (9.164) can be discretized via Galerkin's technique, which results in a matrix equation of the first kind (9.150) with respect to the unknown coefficients in the current expansion. Inverting the matrix $A$, we obtain the solution of (9.150) for the magnitudes of electric and magnetic currents $J$ and $K$,

$$\begin{bmatrix} J \\ K \end{bmatrix} = A^{-1} \begin{bmatrix} V \\ I \end{bmatrix}, \tag{9.165}$$

where $V$ and $I$ are defined in the previous section.

The GSM of the $N$-port waveguide transition relates magnitudes of incident and scattered (reflected and transmitted) propagating and evanescent modes at each port of the transition, and it includes all possible interactions among modes and ports. Following the same procedure shown in the previous example for the magnitudes $b_{mn}$ of reflected TE- and TM-modes

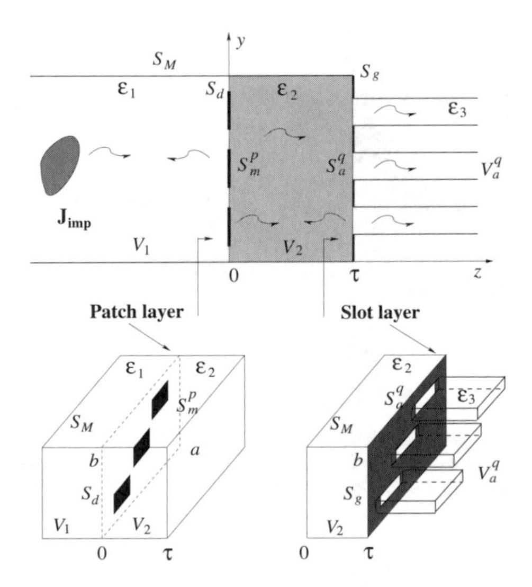


Figure 9.13: $N$-port waveguide-based aperture-coupled patch array transition.

in the waveguide $V_1$ at the interface $z = 0$, we obtain

$$\left\{ \begin{array}{c} b_{mn}^{-\text{TE}} \\ b_{mn}^{-\text{TM}} \end{array} \right\} = -\, i\omega\mu_0 \sum_{p=2}^{N} \int_{S_w} \int_{S_m^p} \left[ \mathbf{J}_p \cdot \underline{\mathbf{G}}_{e11}^{(1)} \times \left\{ \begin{array}{c} \mathbf{h}_{mn}^{-\text{TE}} \\ \mathbf{h}_{mn}^{-\text{TM}} \end{array} \right\} \right] \cdot (-\widehat{\mathbf{z}})\, dS'\, dS$$
$$+ \sum_{q=2}^{N} \int_{S_w} \int_{S_a^q} \left[ \mathbf{K}_q \cdot [\nabla' \times \underline{\mathbf{G}}_{e21}^{(1)}] \times \left\{ \begin{array}{c} \mathbf{h}_{mn}^{-\text{TE}} \\ \mathbf{h}_{mn}^{-\text{TM}} \end{array} \right\} \right] \cdot (-\widehat{\mathbf{z}})\, dS'\, dS$$
$$+ \left\{ \begin{array}{c} a_{mn}^{+\text{TE}} \\ a_{mn}^{+\text{TM}} \end{array} \right\} \left\{ \begin{array}{c} R_{mn}^{\text{TE}} \\ R_{mn}^{\text{TM}} \end{array} \right\}. \tag{9.166}$$

Here "±" corresponds to waves propagating in the positive (+) and negative (−) $z$-directions, $S_w$ is the cross-section of $V_1$, $\mathbf{h}_{mn}$ are magnetic vector functions normalized by the unity power condition (9.142), $R_{mn}$ are the total reflection coefficients of incident TE- and TM-modes (in the absence of patches and slots) (9.145) and (9.146), and $a_{mn}$ are the magnitudes of the incident modes.

The magnitudes $c_{mnq}$ of TE- and TM-modes transmitted into the waveguides $V_a^q$ with the excitation in region $V_1$ are expressed in terms of induced magnetic currents at the interface $z = \tau$,

$$\left\{ \begin{array}{c} c_{mnq}^{+\text{TE}} \\ c_{mnq}^{+\text{TM}} \end{array} \right\} = -\int_{S_w^q} \int_{S_a^q} \left[ \mathbf{K}_q \cdot [\nabla' \times \underline{\mathbf{G}}_{Aq}^{(1)}] \times \left\{ \begin{array}{c} \mathbf{h}_{mnq}^{+\text{TE}} \\ \mathbf{h}_{mnq}^{+\text{TM}} \end{array} \right\} \right] \cdot \widehat{\mathbf{z}}\, dS'\, dS, \tag{9.167}$$

where $S_w^q$ is the cross-section and $\mathbf{h}_{mnq}$ are the magnetic vector functions associated with waveguides $V_a^q$. The magnetic potential Green's dyadics of

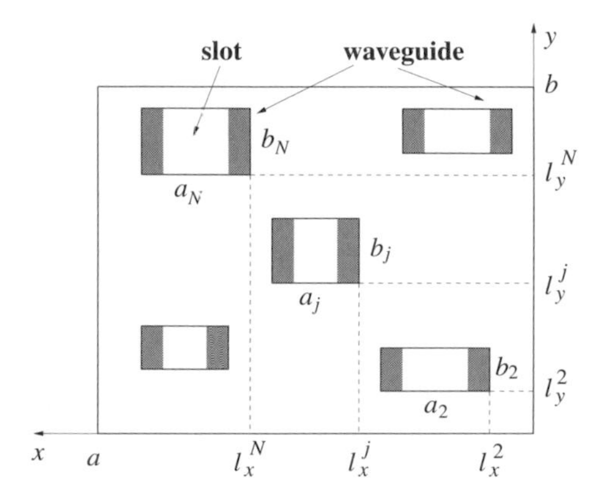


Figure 9.14: Cross-section of a slot-to-waveguide interface at $z = \tau$ (ground plane). Note that each slot in the ground plane is placed in front of the corresponding waveguide (regions $V_a^q$), and it has the same height as the waveguide.

the first kind, $\underline{\mathbf{G}}_{Aq}^{(1)}$, are obtained for the semi-infinite waveguides $V_a^q$ terminated by a ground plane at $z = \tau$ (see Section 9.2.1). The expressions (9.166) and (9.167), in conjunction with (9.165), represent a matrix form that relates the magnitudes $a_{mn}$ of incident modes and the magnitudes $b_{mn}$ and $c_{mnq}$ of reflected and transmitted modes, respectively. This corresponds to the reflected and transmitted parts of the GSM of the $N$-port transition with the excitation from waveguide $V_1$.

Next, we obtain the reflected and transmitted parts of the GSM with the excitation at each of the waveguides $V_a^q$. The integral representation for scattered fields, in conjunction with the eigenmode expansion, yields the expression for the magnitudes $b_{mnq}$ of modes reflected back in the $q$th waveguide $V_a^q$ and modes transmitted (coupled back) to $N-2$ waveguides $V_a^q$,

$$\left\{ \begin{array}{c} b_{mnq}^{+\mathrm{TE}} \\ b_{mnq}^{+\mathrm{TM}} \end{array} \right\} = -\, i\omega\varepsilon_0\varepsilon_3 \int_{S_w^q} \int_{S_a^q} \left[ \left\{ \begin{array}{c} \mathbf{e}_{mnq}^{+\mathrm{TE}} \\ \mathbf{e}_{mnq}^{+\mathrm{TM}} \end{array} \right\} \times [\mathbf{K}_q \cdot \underline{\mathbf{G}}_{\mathrm{eq}}^{(2)}] \right] \cdot \widehat{\mathbf{z}}\, dS'\, dS$$
$$- \,\delta_{qs} \left\{ \begin{array}{c} a_{mns}^{-\mathrm{TE}} \\ a_{mns}^{-\mathrm{TM}} \end{array} \right\}, \qquad (9.168)$$

where $\delta_{qs}$ is the Kronecker delta, $\mathbf{e}_{mnq}$ are the electric vector functions of the waveguides $V_a^q$, and $a_{mns}$ are the magnitudes of incident TE- and TM-modes in the $s$th waveguide, $s = 1, \ldots, N-1$. The electric Green's dyadics of the second kind, $\underline{\mathbf{G}}_{\mathrm{eq}}^{(2)}$, are obtained for the semi-infinite waveguides $V_a^q$ terminated by a ground plane at $z = \tau$ (see Section 9.2.2).

Finally, the magnitudes $c_{mns}$ of TE- and TM-modes transmitted into waveguide $V_1$ with the excitation in the $q$th waveguide $V_a^q$ are obtained at

$z = 0$ as

$$\left\{ \begin{array}{c} c_{mns}^{-\mathrm{TE}} \\ c_{mns}^{-\mathrm{TM}} \end{array} \right\} = \sum_{p=2}^{N} \int_{S_w} \int_{S_m^p} \left[ \left\{ \begin{array}{c} \mathbf{e}_{mn}^{-\mathrm{TE}} \\ \mathbf{e}_{mn}^{-\mathrm{TM}} \end{array} \right\} \times \mathbf{J}_p \cdot [\nabla' \times \underline{\mathbf{G}}_{e11}^{(2)}] \right] \cdot (-\widehat{\mathbf{z}})\, dS'\, dS$$
$$+ i\omega\varepsilon_0\varepsilon_1 \sum_{q=2}^{N} \delta_{qs} \int_{S_w} \int_{S_a^q} \left[ \left\{ \begin{array}{c} \mathbf{e}_{mn}^{-\mathrm{TE}} \\ \mathbf{e}_{mn}^{-\mathrm{TM}} \end{array} \right\} \times [\mathbf{K}_q \cdot \underline{\mathbf{G}}_{e21}^{(2)}] \right] \cdot (-\widehat{\mathbf{z}})\, dS'\, dS, \tag{9.169}$$

where $\mathbf{e}_{mn}$ are the electric vector functions of waveguide $V_1$. The electric Green's dyadics of the third kind, $\underline{\mathbf{G}}_{e11}^{(2)}$ and $\underline{\mathbf{G}}_{e21}^{(2)}$, are obtained for a semi-infinite, partially filled waveguide (regions $V_1$ and $V_2$ with a point source positioned in region $V_1$) satisfying boundary and continuity conditions for the magnetic field (similar Green's dyadics are considered in Section 9.4).

In the case of excitation in the $q$th waveguide $V_a^q$, Galerkin's method allows for the representation of the magnitudes of induced electric and magnetic currents in the matrix form

$$\left[ \begin{array}{c} J \\ K \end{array} \right] = A^{-1} \left[ \begin{array}{c} 0 \\ I \end{array} \right]. \tag{9.170}$$

This result, in conjunction with (9.168) and (9.169), leads to the GSM representation for reflected and transmitted modes for the case when the incident field is generated in the $q$th waveguide $V_a^q$, taking into account the coupling of all propagating and evanescent TE- and TM-modes from all ports of the transition.

As an example, numerical results for the $S$-parameters (magnitude and phase) of the dominant $\mathrm{TE}_{10}$-mode in the $2\times3$ aperture-coupled patch array waveguide transition operating at X-band (similar to the $N$-port waveguide transition shown in Figure 9.13) are shown in Figures 9.15 and 9.16. The results are obtained for the following parameters: The patch antennas have 340 mil width and 320 mil height; the slots are of 250 mil width and 15 mil height;[25] and the substrate thickness is 31 mil with permittivity of 2.2. The large waveguide (regions $V_1$ and $V_2$) has 1200 mil height and 1811 mil width, and the small waveguides (regions $V_a^q$) have 450 mil width and 15 mil height. Unit cells in the array are separated by a distance of 600 mil. In Figures 9.15 and 9.16, $S_{11}$ is the reflection coefficient at the interface $z = 0$ in region $V_1$ (with the excitation from $V_1$), and $S_{22}$ and $S_{33}$ are the reflection coefficients at $z = \tau$ (ground plane) in regions $V_a^2$ and $V_a^3$ (with the excitation from $V_a^2$ and $V_a^3$, respectively).[26] The transmission coefficients from $V_1$ into $V_a^2$ and $V_a^3$ are denoted as $S_{21}$ and $S_{31}$, respectively. The sharp resonance obtained at approximately 10.9 GHz corresponds to the occurrence of a transverse resonance and is associated with the coupling

[25] One *mil* is 25.4 $\mu m$.

[26] Other waveguides $V_a^q$ in the $2\times3$ array are symmetric to $V_a^2$ and $V_a^3$.

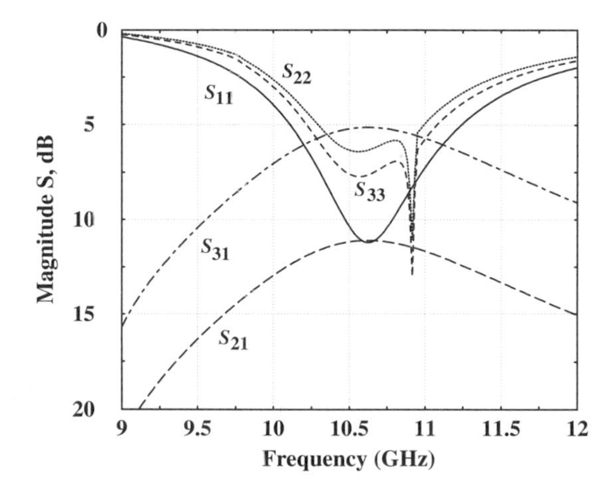


Figure 9.15: Magnitude of the $S$-parameters against frequency for the 2 × 3 aperture-coupled patch array in the seven-port rectangular waveguide transition.

of a mode propagating along the waveguide to the surface wave $TM_0$ of a dielectric slab (substrate) propagating in the transverse direction (waveguide cross-section). This phenomenon is discussed in Section 9.4.1 in the analysis of Green's function components.

## 9.4 Electric Green's Dyadics for Waveguides Filled with a Planarly Layered Medium

The previous section presented integral equation formulations that utilize electric Green's dyadics of the third kind. In this section we derive these Green's functions for a semi-infinite, partially filled rectangular waveguide terminated by a perfectly conducting ground plane.

Electric and magnetic Green's dyadics for an infinite waveguide with a transverse dielectric slab are presented in [64] for the full-wave analysis of antenna radiation in a layered rectangular waveguide. The method of mode expansion and scattering superposition is applied in [65] to construct electric Green's dyadics for a multilayered waveguide. A general method of constructing electric and magnetic Green's dyadics for a multilayered, semi-infinite rectangular waveguide is discussed in [66], where the coefficients of the Green's functions are in double-series expansion form, obtained in terms of recurrent transmission matrices.

In the approach described here, the electric Green's dyadics are obtained as solutions of the electric- and magnetic-type boundary value prob-

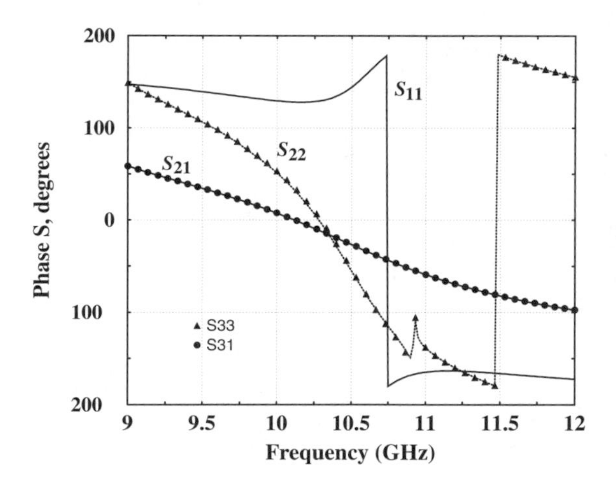


Figure 9.16: Phase of the $S$-parameters against frequency for the $2 \times 3$ aperture-coupled patch array in the seven-port rectangular waveguide transition.

lems (9.117)–(9.121) and (9.131)–(9.135), respectively. Components of the Green's dyadics are expressed in a double infinite series expansion over the complete system of orthonormal eigenfunctions of the transverse Laplacian operator. The unknown coefficients in these expansions represent one-dimensional characteristic Green's functions in the waveguiding coordinate. In this representation, transverse and longitudinal coordinates are functionally separated, which allows one to immediately reduce the three-dimensional problem to a one-dimensional Sturm–Liouville boundary value problem for the unknown characteristic Green's functions. The technique presented is general and straightforward and is also applicable to multilayered waveguides with applications to transmission-line problems [55], [56].

## 9.4.1 Electric Green's Dyadics of the Third Kind

### Electric-Type Boundary Value Problem

The electric Green's dyadics $\underline{\mathbf{G}}_{e11}^{(1)}(\mathbf{r}, \mathbf{r}')$ and $\underline{\mathbf{G}}_{e21}^{(1)}(\mathbf{r}, \mathbf{r}')$ utilized in (9.124) lead to the integral equation (9.125) for the induced currents on metal surfaces and slot apertures and to (9.163) in the example of an aperture-coupled patch array in an $N$-port waveguide transition. To determine the Green's dyadics we formulate the boundary value problem (9.117)–(9.121) for a semi-infinite rectangular waveguide with a ground plane at $z = \tau$ in the absence of metal surfaces and apertures, as depicted in Figure 9.7. The solution of the boundary value problem (9.117)–(9.121) yields nine components of the electric Green's dyadics, which can be expressed as

$$\left\{ \begin{array}{c} G_{e11}^{(1)\nu\upsilon} \\ G_{e21}^{(1)\nu\upsilon} \end{array} \right\} = \sum_{m=0}^{\infty} \sum_{n=0}^{\infty} \phi_{mn}^{(1)\nu}(x, y) \phi_{mn}^{(1)\upsilon}(x', y') \left\{ \begin{array}{c} f_{mn}^{(11)\nu\upsilon}(z, z') \\ f_{mn}^{(21)\nu\upsilon}(z, z') \end{array} \right\} \tag{9.171}$$

for $\nu, \upsilon = x, y, z$, where $\phi_{mn}^{(1)\nu}(x, y)$ are the orthonormal eigenfunctions of the transverse Laplacian operator determined as (9.64), and $f_{mn}^{(11)\nu\upsilon}(z, z')$, $f_{mn}^{(21)\nu\upsilon}(z, z')$ are one-dimensional characteristic Green's functions.[27]

Orthogonality (9.59) and completeness (9.60) of the eigenfunctions $\phi_{mn}^{(1)\nu}$ in the double-series expansion (9.171) allow for the reduction of the system of dyadic differential equations (9.117) and (9.118) to a system of second-order differential equations for the one-dimensional characteristic Green's functions $f_{mn}^{(11)\nu\upsilon}(z, z')$ and $f_{mn}^{(21)\nu\upsilon}(z, z')$,

$$\begin{aligned} \left( \frac{\partial^2}{\partial z^2} - \gamma_{mn}^{(1)2} \right) f_{mn}^{(11)\nu\upsilon}(z, z') &= -\xi_{mn}^{\nu\upsilon} \delta(z - z'), \\ \left( \frac{\partial^2}{\partial z^2} - \gamma_{mn}^{(2)2} \right) f_{mn}^{(21)\nu\upsilon}(z, z') &= 0, \qquad \nu, \upsilon = x, y, \end{aligned} \tag{9.172}$$

where

$$\xi_{mn}^{xx} = \frac{k_1^2 - \left(\frac{m\pi}{a}\right)^2}{k_1^2}, \quad \xi_{mn}^{xy} = \xi_{mn}^{yx} = -\frac{\left(\frac{m\pi}{a}\right)\left(\frac{n\pi}{b}\right)}{k_1^2}, \quad \xi_{mn}^{yy} = \frac{k_1^2 - \left(\frac{n\pi}{b}\right)^2}{k_1^2},$$

and $\gamma_{mn}^{(1,2)}$ is given by (9.147).

The boundary condition (9.120) on the ground plane $S_G$ (Figure 9.7), written in terms of the Green's function components, is reduced to the boundary condition for the one-dimensional characteristic Green's functions at $z = \tau$ and $z > z'$,

$$f_{mn}^{(21)\nu\upsilon}(\tau, z') = 0. \tag{9.173}$$

The continuity conditions (9.121) on the dielectric interface at $z = 0$ and $z > z'$ are obtained as

$$f_{mn}^{(11)\nu\upsilon}(0, z') = f_{mn}^{(21)\nu\upsilon}(0, z'), \tag{9.174}$$

$$\frac{\partial}{\partial \nu} f_{mn}^{(11)z\upsilon}(0, z') - \frac{\partial}{\partial z} f_{mn}^{(11)\nu\upsilon}(0, z') = \frac{\partial}{\partial \nu} f_{mn}^{(21)z\upsilon}(0, z') - \frac{\partial}{\partial z} f_{mn}^{(21)\nu\upsilon}(0, z'),$$

and the $z$-directed Green's functions $f_{mn}^{(11)z\upsilon}$ and $f_{mn}^{(21)z\upsilon}$ introduced in (9.174) can be expressed in terms of transverse components,

$$f_{mn}^{(i1)z\upsilon}(z, z') = \frac{1}{\gamma_{mn}^{(i)2}} \frac{\partial}{\partial z} \left( \frac{m\pi}{a} f_{mn}^{(i1)x\upsilon}(z, z') + \frac{n\pi}{b} f_{mn}^{(i1)y\upsilon}(z, z') \right) \tag{9.175}$$

[27] Note that the Green's dyadics developed here are used in the integral equation formulations for transverse planar metal conductors and slot apertures. Thus, $z$-directed current sources are not considered, which eliminates the need to consider Green's function components $G_{e11}^{(1)\nu z}$ and $G_{e21}^{(1)\nu z}$, $\nu = x, y, z$, in the analysis to follow.

for $i = 1, 2$.

The solution of the system of differential equations (9.172) is obtained as a superposition of primary and scattered parts,

$$f_{mn}^{(11)\nu\upsilon}(z,z') = \xi_{mn}^{\nu\upsilon}\frac{e^{-\gamma_{mn}^{(1)}|z-z'|}}{2\gamma_{mn}^{(1)}} + \eta_{\nu\upsilon}^{-(11)}(z')e^{\gamma_{mn}^{(1)}z}, \tag{9.176}$$

$$f_{mn}^{(21)\nu\upsilon}(z,z') = \eta_{\nu\upsilon}^{+(21)}(z')e^{-\gamma_{mn}^{(2)}z} + \eta_{\nu\upsilon}^{-(21)}(z')e^{\gamma_{mn}^{(2)}(z-\tau)}. \tag{9.177}$$

In (9.176) the primary part is due to a point source positioned in region $V_1$, and the scattered part is reflected from the interface $z = 0$, traveling in the negative $z$-direction. In (9.177) we have only the scattered part, which represents traveling backward and forward waves propagating in region $V_2$ (between the interface at $z = 0$ and the ground plane at $z = \tau$). The unknown $\eta$-coefficients to be determined are subject to the boundary and continuity conditions (9.173) and (9.174), respectively, and can be obtained in closed form.

Finally, this procedure results in the representation of the one-dimensional transverse characteristic Green's functions $f_{mn}^{(11)\nu\upsilon}(z,z')$ in terms of the primary and scattered parts,

$$\begin{aligned}
&f_{mn}^{(11)xx}(z,z')\\
&\quad= \xi_{mn}^{xx}\frac{e^{-\gamma_{mn}^{(1)}|z-z'|}}{2\gamma_{mn}^{(1)}} - e^{\gamma_{mn}^{(1)}(z+z')}\left(\frac{\xi_{mn}^{xx}}{2\gamma_{mn}^{(1)}} - \frac{1}{Z_o^{\mathrm{TE}}} + \frac{(\frac{m\pi}{a})^2 Z_e^{\mathrm{TE}}}{k_1^2 Z_o^{\mathrm{TE}} Z_e^{\mathrm{TM}}}\right),\\
&f_{mn}^{(11)xy}(z,z')\\
&\quad= f_{mn}^{(11)yx}(z,z')\\
&\quad= \xi_{mn}^{xy}\left(\frac{e^{-\gamma_{mn}^{(1)}|z-z'|}}{2\gamma_{mn}^{(1)}} - e^{\gamma_{mn}^{(1)}(z+z')}\left(\frac{1}{2\gamma_{mn}^{(1)}} - \frac{Z_e^{\mathrm{TE}}}{Z_o^{\mathrm{TE}} Z_e^{\mathrm{TM}}}\right)\right),\\
&f_{mn}^{(11)yy}(z,z')\\
&\quad= \xi_{mn}^{yy}\frac{e^{-\gamma_{mn}^{(1)}|z-z'|}}{2\gamma_{mn}^{(1)}} - e^{\gamma_{mn}^{(1)}(z+z')}\left(\frac{\xi_{mn}^{yy}}{2\gamma_{mn}^{(1)}} - \frac{1}{Z_o^{\mathrm{TE}}} + \frac{(\frac{n\pi}{b})^2 Z_e^{\mathrm{TE}}}{k_1^2 Z_o^{\mathrm{TE}} Z_e^{\mathrm{TM}}}\right),
\end{aligned} \tag{9.178}$$

and functions $f_{mn}^{(21)\nu\upsilon}(z,z')$ are obtained in terms of scattered waves,

$$\begin{aligned}
f_{mn}^{(21)xx}(z,z') &= -\left(\frac{1}{Z_o^{\mathrm{TE}}} - \frac{(\frac{m\pi}{a})^2 Z_e^{\mathrm{TE}}}{k_1^2 Z_o^{\mathrm{TE}} Z_e^{\mathrm{TM}}}\right)\frac{e^{\gamma_{mn}^{(1)}z'}\sinh\gamma_{mn}^{(2)}(z-\tau)}{\sinh\gamma_{mn}^{(2)}\tau},\\
f_{mn}^{(21)xy}(z,z') = f_{mn}^{(21)yx}(z,z') &= -\xi_{mn}^{xy}\frac{Z_e^{\mathrm{TE}}}{Z_o^{\mathrm{TE}} Z_e^{\mathrm{TM}}}\frac{e^{\gamma_{mn}^{(1)}z'}\sinh\gamma_{mn}^{(2)}(z-\tau)}{\sinh\gamma_{mn}^{(2)}\tau},\\
f_{mn}^{(21)yy}(z,z') &= -\left(\frac{1}{Z_o^{\mathrm{TE}}} - \frac{(\frac{n\pi}{b})^2 Z_e^{\mathrm{TE}}}{k_1^2 Z_o^{\mathrm{TE}} Z_e^{\mathrm{TM}}}\right)\frac{e^{\gamma_{mn}^{(1)}z'}\sinh\gamma_{mn}^{(2)}(z-\tau)}{\sinh\gamma_{mn}^{(2)}\tau}.
\end{aligned} \tag{9.179}$$

The $z$-coordinate characteristic Green's functions $f_{mn}^{(i1)zx}$ and $f_{mn}^{(i1)zy}$ for $i = 1, 2$ are determined by (9.175) with (9.178) and (9.179).

Here, $Z_e^{\mathrm{TE}}, Z_o^{\mathrm{TE}}$, and $Z_e^{\mathrm{TM}}$ are the characteristic functions of even and odd TE- and TM-modes, respectively, of a grounded dielectric slab of thickness $\tau$ bounded with electric walls at $x = 0, a$ and $y = 0, b$ (region $V_2$), given as

$$
\begin{aligned}
Z_e^{\mathrm{TE}} &= \gamma_{mn}^{(1)} + \gamma_{mn}^{(2)} \tanh \gamma_{mn}^{(2)} \tau, \\
Z_o^{\mathrm{TE}} &= \gamma_{mn}^{(1)} + \gamma_{mn}^{(2)} \coth \gamma_{mn}^{(2)} \tau, \\
Z_e^{\mathrm{TM}} &= \frac{\varepsilon_2}{\varepsilon_1} \gamma_{mn}^{(1)} + \gamma_{mn}^{(2)} \tanh \gamma_{mn}^{(2)} \tau,
\end{aligned}
\tag{9.180}
$$

where the propagation constant $\gamma_{mn}^{(i)}$ for $i = 1, 2$ is given by (9.147) in terms of transverse wave numbers $\kappa_{mn} = \sqrt{(m\pi/a)^2 + (n\pi/b)^2}$. Zeros of the characteristic functions (9.180) represent resonance frequencies of TE and TM oscillations in the waveguide cross-section. Moreover, it can be seen that zeros of $Z_o^{\mathrm{TE}}$ and $Z_e^{\mathrm{TM}}$ for TE-odd and TM-even modes correspond to poles of the characteristic Green's functions (9.178) and (9.179). The characteristic functions (9.180) for an infinite grounded dielectric slab of thickness $\tau$ define even and odd TE and TM surface waves with the propagation constant $\kappa_{mn}$, as discussed in Section 8.1.7.

At a resonance frequency corresponding to the transverse wavenumber $\kappa_{mn}$ of the shielded grounded dielectric slab, the value of $\kappa_{mn}$ is equal to the value of the propagation constant of a surface wave associated with an infinite dielectric slab. This is related to coupling of waves propagating along the waveguide in the $z$-direction with propagation constants $\gamma_{mn}^{(i)}$ to TE and TM surface waves propagating in an infinite grounded dielectric slab (associated with resonance wavenumbers $\kappa_{mn}$ in the waveguide cross-section). This becomes important in the analysis of waveguide-based aperture-coupled patch amplifier arrays, where coupling to surface waves results in a loss of power and undesirable cross-talk between neighboring antennas.

### Magnetic-Type Boundary Value Problem

The electric Green's dyadics of the third kind, $\underline{\mathbf{G}}_{e12}^{(2)}(\mathbf{r}, \mathbf{r}')$ and $\underline{\mathbf{G}}_{e22}^{(2)}(\mathbf{r}, \mathbf{r}')$, appear in the magnetic field integral equation formulation (9.136), resulting in integral equations (9.138) and (9.164). The boundary value problem (9.131)–(9.135) is formulated for a semi-infinite rectangular waveguide with a ground plane at $z = \tau$ (Figure 9.8) with a point source positioned in region $V_2$. The solution of the problem can be expressed in the form of a partial eigenfunction expansion,

$$
\left\{ \begin{array}{c} G_{e12}^{(2)\nu\upsilon} \\ G_{e22}^{(2)\nu\upsilon} \end{array} \right\} = \sum_{m=0}^{\infty} \sum_{n=0}^{\infty} \phi_{mn}^{(2)\nu}(x, y) \phi_{mn}^{(2)\upsilon}(x', y') \left\{ \begin{array}{c} g_{mn}^{(12)\nu\upsilon}(z, z') \\ g_{mn}^{(22)\nu\upsilon}(z, z') \end{array} \right\}
\tag{9.181}
$$

for $\nu, \upsilon = x, y, z$, where $\phi_{mn}^{(2)\nu}(x,y)$ are the orthonormal eigenfunctions of the transverse Laplacian operator given by (9.65), and $g_{mn}^{(12)\nu\upsilon}(z,z')$ and $g_{mn}^{(22)\nu\upsilon}(z,z')$ are the one-dimensional characteristic Green's functions to be determined as the solution of a Sturm–Liouville problem.[28]

A system of second-order differential equations for the one-dimensional characteristic Green's functions $g_{mn}^{(12)\nu\upsilon}(z,z')$ and $g_{mn}^{(22)\nu\upsilon}(z,z')$ is obtained from (9.131) and (9.132) using properties of orthogonality (9.59) and completeness (9.60) of the eigenfunctions $\phi_{mn}^{(2)\nu}(x,y)$,

$$\begin{aligned} &\left(\frac{\partial^2}{\partial z^2} - \gamma_{mn}^{(1)2}\right) g_{mn}^{(12)\nu\upsilon}(z,z') = 0, \qquad \nu,\upsilon = x,y, \\ &\left(\frac{\partial^2}{\partial z^2} - \gamma_{mn}^{(2)2}\right) g_{mn}^{(22)\nu\upsilon}(z,z') = -\zeta_{mn}^{\nu\upsilon}\delta(z-z'), \end{aligned} \tag{9.182}$$

where

$$\begin{aligned} \zeta_{mn}^{xx} &= \frac{k_2^2 - \left(\frac{m\pi}{a}\right)^2}{k_2^2}, \\ \zeta_{mn}^{xy} = \zeta_{mn}^{yx} &= -\frac{\left(\frac{m\pi}{a}\right)\left(\frac{n\pi}{b}\right)}{k_2^2}, \\ \zeta_{mn}^{yy} &= \frac{k_2^2 - \left(\frac{n\pi}{b}\right)^2}{k_2^2}. \end{aligned}$$

The boundary conditions for the transverse characteristic Green's functions at $z = \tau$ and $z > z'$ (see Figure 9.8) are obtained directly from (9.134) using the eigenfunction expansion (9.181),

$$\frac{\partial}{\partial z} g_{mn}^{(22)\nu\upsilon}(\tau, z') = 0, \tag{9.183}$$

and the continuity conditions (9.135) are reduced to the continuity conditions for the one-dimensional Green's functions determined on the dielectric interface at $z = 0$ and $z < z'$,

$$\begin{aligned} g_{mn}^{(12)\nu\upsilon}(0,z') &= g_{mn}^{(22)\nu\upsilon}(0,z') \\ \frac{1}{\varepsilon_1}\left(\frac{\partial}{\partial \nu} g_{mn}^{(12)z\upsilon}(0,z') - \frac{\partial}{\partial z} g_{mn}^{(12)\nu\upsilon}(0,z')\right) & \\ &= \frac{1}{\varepsilon_2}\left(\frac{\partial}{\partial \nu} g_{mn}^{(22)z\upsilon}(0,z') - \frac{\partial}{\partial z} g_{mn}^{(22)\nu\upsilon}(0,z')\right). \end{aligned} \tag{9.184}$$

[28] As in the electric-type boundary value problem, we will consider only those components of Green's dyadics that are associated with transverse electric and magnetic currents (planar transverse metal conductors and slot apertures in waveguide-based antenna problems). The longitudinal components $G_{e12}^{(2)\nu z}$ and $G_{e22}^{(2)\nu z}$ for $\nu = x, y, z$ do not appear in the formulation for the magnetic field (9.136).

The $z$-directed characteristic Green's functions $g_{mn}^{(12)zv}$ and $g_{mn}^{(22)zv}$ introduced in (9.184) are obtained in terms of transverse Green's functions as

$$g_{mn}^{(i2)zv}(z,z') = -\frac{1}{\gamma_{mn}^{(i)2}}\frac{\partial}{\partial z}\left(\frac{m\pi}{a}g_{mn}^{(i2)xv}(z,z') + \frac{n\pi}{b}g_{mn}^{(i2)yv}(z,z')\right) \tag{9.185}$$

for $i = 1, 2$.

The solution of the system of second-order differential equations (9.182) is given by a superposition of primary and scattered parts,

$$g_{mn}^{(12)\nu v}(z,z') = \theta_{\nu v}^{-(12)}(z')e^{\gamma_{mn}^{(1)}z}, \tag{9.186}$$

$$g_{mn}^{(22)\nu v}(z,z') = \zeta_{mn}^{\nu v}\frac{e^{-\gamma_{mn}^{(2)}|z-z'|}}{2\gamma_{mn}^{(2)}} + \theta_{\nu v}^{+(22)}(z')e^{-\gamma_{mn}^{(2)}z} + \theta_{\nu v}^{-(22)}(z')e^{\gamma_{mn}^{(2)}(z-\tau)}. \tag{9.187}$$

In (9.186) we have only backward traveling waves propagating in the negative $z$-direction in region $V_1$. In (9.187) the solution is represented by the primary Green's function maintained by a point source positioned in region $V_2$, and the scattered part is made up of backward and forward traveling waves propagating in region $V_2$ (between the interface at $z = 0$ and the ground plane at $z = \tau$). The unknown $\theta$-coefficients are determined subject to the boundary and continuity conditions (9.183) and (9.184), respectively.

The one-dimensional transverse characteristic Green's functions $g_{mn}^{(12)\nu v}$ are obtained as

$$g_{mn}^{(12)xx}(z,z') = \left(\frac{1}{Z_e^{\mathrm{TM}}} - \frac{(\frac{m\pi}{a})^2 Z}{k_2^2 Z_o^{\mathrm{TE}} Z_e^{\mathrm{TM}}}\right)\frac{e^{\gamma_{mn}^{(1)}z}\cosh\gamma_{mn}^{(2)}(z'-\tau)}{\cosh\gamma_{mn}^{(2)}\tau}, \tag{9.188}$$

$$g_{mn}^{(12)xy}(z,z') = g_{mn}^{(12)yx}(z,z') = \zeta_{mn}^{xy}\frac{Z}{Z_o^{\mathrm{TE}} Z_e^{\mathrm{TM}}}\frac{e^{\gamma_{mn}^{(1)}z}\cosh\gamma_{mn}^{(2)}(z'-\tau)}{\cosh\gamma_{mn}^{(2)}\tau},$$

$$g_{mn}^{(12)yy}(z,z') = \left(\frac{1}{Z_e^{\mathrm{TM}}} - \frac{(\frac{n\pi}{b})^2 Z}{k_2^2 Z_o^{\mathrm{TE}} Z_e^{\mathrm{TM}}}\right)\frac{e^{\gamma_{mn}^{(1)}z}\cosh\gamma_{mn}^{(2)}(z'-\tau)}{\cosh\gamma_{mn}^{(2)}\tau},$$

and transverse Green's functions $g_{mn}^{(22)\nu v}(z,z')$ are determined in terms of the primary and scattered parts,

$$g_{mn}^{(22)xx}(z,z') = \zeta_{mn}^{xx}\frac{e^{-\gamma_{mn}^{(2)}|z-z'|}}{2\gamma_{mn}^{(2)}} + \zeta_{mn}^{xx}\frac{e^{\gamma_{mn}^{(2)}(z+z'-2\tau)}}{2\gamma_{mn}^{(2)}} + \left(\frac{1}{Z_e^{\mathrm{TM}}\cosh\gamma_{mn}^{(2)}\tau} - \frac{(\frac{m\pi}{a})^2 Z}{k_2^2 Z_o^{\mathrm{TE}} Z_e^{\mathrm{TM}}\cosh\gamma_{mn}^{(2)}\tau} - \frac{\zeta_{mn}^{xx}e^{-\gamma_{mn}^{(2)}\tau}}{\gamma_{mn}^{(2)}}\right) \times \frac{\cosh\gamma_{mn}^{(2)}(z-\tau)\cosh\gamma_{mn}^{(2)}(z'-\tau)}{\cosh\gamma_{mn}^{(2)}\tau},$$

$$g_{mn}^{(22)xy}(z,z') = g_{mn}^{(22)yx}(z,z') = \zeta_{mn}^{xy}\frac{e^{-\gamma_{mn}^{(2)}|z-z'|}}{2\gamma_{mn}^{(2)}} + \zeta_{mn}^{xy}\frac{e^{\gamma_{mn}^{(2)}(z+z'-2\tau)}}{2\gamma_{mn}^{(2)}}$$
$$+ \zeta_{mn}^{xy}\left(\frac{Z}{Z_o^{\mathrm{TE}}Z_e^{\mathrm{TM}}\cosh\gamma_{mn}^{(2)}\tau} - \frac{e^{-\gamma_{mn}^{(2)}\tau}}{\gamma_{mn}^{(2)}}\right)$$
$$\times\ \frac{\cosh\gamma_{mn}^{(2)}(z-\tau)\cosh\gamma_{mn}^{(2)}(z'-\tau)}{\cosh\gamma_{mn}^{(2)}\tau},$$
$$g_{mn}^{(22)yy}(z,z') = \zeta_{mn}^{yy}\frac{e^{-\gamma_{mn}^{(2)}|z-z'|}}{2\gamma_{mn}^{(2)}} + \zeta_{mn}^{yy}\frac{e^{\gamma_{mn}^{(2)}(z+z'-2\tau)}}{2\gamma_{mn}^{(2)}}$$
$$+\left(\frac{1}{Z_e^{\mathrm{TM}}\cosh\gamma_{mn}^{(2)}\tau} - \frac{(\frac{n\pi}{b})^2 Z}{k_2^2 Z_o^{\mathrm{TE}}Z_e^{\mathrm{TM}}\cosh\gamma_{mn}^{(2)}\tau} - \frac{\zeta_{mn}^{yy}e^{-\gamma_{mn}^{(2)}\tau}}{\gamma_{mn}^{(2)}}\right)$$
$$\times\ \frac{\cosh\gamma_{mn}^{(2)}(z-\tau)\cosh\gamma_{mn}^{(2)}(z'-\tau)}{\cosh\gamma_{mn}^{(2)}\tau}, \tag{9.189}$$

where $Z = \gamma_{mn}^{(1)} + (\varepsilon_2/\varepsilon_1)\gamma_{mn}^{(2)}\coth\gamma_{mn}^{(2)}\tau$. Poles of the Green's functions $g_{mn}^{(12)\upsilon\upsilon}(z,z')$ and $g_{mn}^{(22)\upsilon\upsilon}(z,z')$ represent zeros of characteristic functions corresponding to the resonance frequencies of TE-odd and TM-even oscillations in the waveguide cross-section.

The $z$-directed characteristic Green's functions $g_{mn}^{(i2)zx}$ and $g_{mn}^{(i2)zy}$, $i = 1, 2$, can be obtained from (9.185) using the expressions for the transverse Green's functions (9.188) and (9.189).

## 9.5 The Method of Overlapping Regions

The method of overlapping regions introduced in [67] for the analysis of noncoordinate waveguide discontinuity problems is an alternative to mode-matching techniques. In a mode-matching analysis, a geometrical domain is subdivided into regions with common boundaries, within which a general solution of the Helmholtz equation can be obtained by the method of separation of variables. Unknown fields are expanded in these solutions and matched across the boundary of neighboring regions, leading to a system of equations that yield the field magnitudes.

There is a variety of noncoordinate problems (for example, junctions of cylindrical and rectangular waveguides) where a complex geometry can be divided into overlapping regions with a common subregion (called a *Zwischenmedium* according to the terminology of [67]), for which eigenfunctions of the Laplacian operator are known. This method was proposed to avoid intermediate regions with noncoordinate boundaries in the mode-matching procedure. Various waveguide discontinuity problems, including junctions of circular and rectangular waveguides, junctions between canonical horns, and different types of waveguide bends, have been investigated using the idea of a Zwischenmedium and presented in [68]. The problem of radiation from a flanged waveguide has been studied in [69] using a cylindrical Zwischenmedium common for a waveguide and a sector horn region.

Furthermore, the method of overlapping regions has been utilized in [70] using integral representations for overlapping regions. In this formulation, a complex geometrical domain is divided into simple overlapping regions for which the solution of the Laplacian equation exists as the solution of a Green's function problem. Based on the continuity of fields in common subregions, the method results in a coupled set of Fredholm integral equations of the second kind, which is discretized using Galerkin-type projection techniques. The matrix equation obtained is of the second kind with a compact operator in the sequence space $\mathbf{l}^1$. This guarantees the existence of a bounded inverse operator (see Theorem 3.40) and allows for the truncation of the infinite system of linear algebraic equations (see Section 3.11.2). The method has been applied to the analysis of waveguide discontinuity and antenna problems [46], [53], [54], [71], microstrip discontinuities [72], and for the study of natural waves in shielded printed-circuit transmission lines [55], [56].

Another variant is Schwarz's iterative method of overlapping regions [73], which has been applied to the analysis of triangular waveguides [74], scattering by a polygonal cylinder [75], and waveguide discontinuity problems [76], [77]. This method involves dividing the geometry of a structure into several simple overlapping subregions, wherein the wave equation can be solved by separation of variables, leading to an analytical solution for the Green's function. The method requires assumptions for the field component determined on nonmetallic boundaries of overlapping subregions for the first iteration, but usually converges in a small number of iterations.

### 9.5.1 Integral Equation Formulations for Overlapping Regions

In this section we present a general integral equation formulation for overlapping regions in a complex geometrical domain [46].

Consider a complex domain having volume $V$ enclosed by a perfectly conducting surface $S$, as shown in Figure 9.17. Within $V$ we introduce two overlapping regions, $V_1$ and $V_2$, enclosed by surfaces $S_1$ and $S_2$, respectively, such that $S_1 = S_m^{(1)} \cup S_a^{(1)}$ and $S_2 = S_m^{(2)} \cup S_a^{(2)}$. Here $S_m^{(1)}$ and $S_m^{(2)}$ are metal surfaces that are part of the surface $S = S_m^{(1)} \cup S_m^{(2)}$, and $S_a^{(1)}$ and $S_a^{(2)}$ represent coupling apertures (artificial surfaces) corresponding to regions $V_1$ and $V_2$. An impressed electric current $\mathbf{J}_{\text{imp}}(\mathbf{r})$, $\mathbf{r} \in V_{\text{imp}} \subset V_1$, generates an incident electric and magnetic field, and scattered fields are caused by secondary sources induced on the surface of apertures $S_a^{(1)}$ and $S_a^{(2)}$. We first obtain integral representations for fields in regions $V_1$ and $V_2$ and then couple the obtained system of equations resulting in Fredholm integral equations of the second kind.

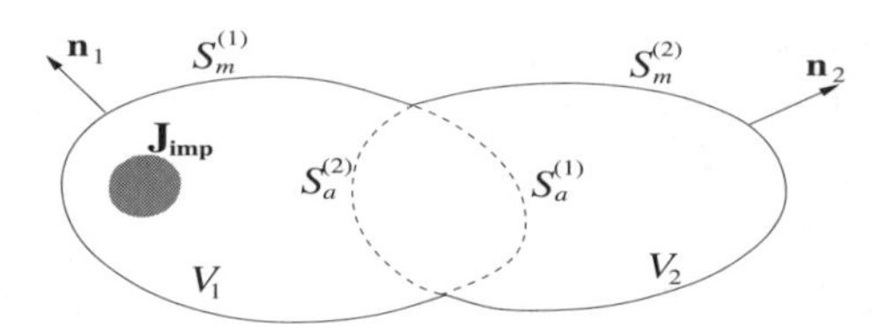


Figure 9.17: Complex domain having volume $V$ enclosed by a perfectly conducting surface $S$. Two overlapping subregions are introduced within $V$ coupled through the apertures $S_a^{(1)}$ and $S_a^{(2)}$.

## Electric-Field Formulation

A boundary value problem can be formulated for the electric field in regions $V_1$ and $V_2$ such that

$$\nabla \times \nabla \times \mathbf{E}_i(\mathbf{r}) - k^2 \mathbf{E}_i(\mathbf{r}) = -\delta_{i1} i\omega\mu \mathbf{J}_{\text{imp}}(\mathbf{r}), \quad \mathbf{r} \in V_i, \; i = 1, 2, \tag{9.190}$$

$$\mathbf{n}_i \times \mathbf{E}_i(\mathbf{r})\big|_{S_m^{(i)}} = \mathbf{0}, \tag{9.191}$$

where $\delta_{i1} = 1$ for $i = 1$ and 0 for $i = 2$, $\mathbf{n}_i$ is the outward unit normal vector on surface $S_i$, and $k = \omega\sqrt{\varepsilon\mu}$.[29] In the vicinity of the conducting edge formed by contour $\Gamma = S_1 \cap S_2$ we restrict the solution by enforcing the finite energy condition

$$\lim_{V_i \to 0} \int_{V_i} \left( |\mathbf{E}_i(\mathbf{r})|^2 + \frac{1}{k^2} |\nabla \times \mathbf{E}_i(\mathbf{r})|^2 \right) dV = 0. \tag{9.192}$$

The electric Green's dyadics can be obtained for regions $V_1$ and $V_2$ as the solution of the dyadic wave equations

$$\nabla \times \nabla \times \underline{\mathbf{G}}_{ei}^{(1)}(\mathbf{r}, \mathbf{r}') - k^2 \underline{\mathbf{G}}_{ei}^{(1)}(\mathbf{r}, \mathbf{r}') = \underline{\mathbf{I}}\, \delta(\mathbf{r} - \mathbf{r}'), \qquad \mathbf{r}, \mathbf{r}' \in V_i, \tag{9.193}$$

subject to the boundary condition of the first kind on the closed surface $S_i$,

$$\mathbf{n}_i \times \underline{\mathbf{G}}_{ei}^{(1)}(\mathbf{r}, \mathbf{r}')\Big|_{\mathbf{r} \in S_i} = \underline{\mathbf{0}}. \tag{9.194}$$

The vector-dyadic Green's second theorem applied to $\mathbf{E}_i$ and $\underline{\mathbf{G}}_{ei}^{(1)}$, along with the boundary conditions (9.191) and (9.194), results in the electric-

[29] If region $V$ and, therefore, regions $V_1$ and/or $V_2$ extend to infinity, then we also impose the limiting absorption condition.

field representation

$$\mathbf{E}_1(\mathbf{r}') = -\,i\omega\mu \int_{V_{\text{imp}}} \mathbf{J}_{\text{imp}}(\mathbf{r}) \cdot \underline{\mathbf{G}}_{e1}^{(1)}(\mathbf{r},\mathbf{r}')\, dV \\ - \int_{S_a^{(1)}} [\mathbf{n}_1 \times \mathbf{E}_1(\mathbf{r})] \cdot [\nabla\times\underline{\mathbf{G}}_{e1}^{(1)}(\mathbf{r},\mathbf{r}')]\, dS, \tag{9.195}$$

$$\mathbf{E}_2(\mathbf{r}') = -\int_{S_a^{(2)}} [\mathbf{n}_2 \times \mathbf{E}_2(\mathbf{r})] \cdot [\nabla\times\underline{\mathbf{G}}_{e2}^{(1)}(\mathbf{r},\mathbf{r}')]\, dS, \tag{9.196}$$

and, by interchanging $\mathbf{r}'$ and $\mathbf{r}$ and using the identities (9.20) and (9.37) for transposition of $\underline{\mathbf{G}}_{e1}^{(1)}$ and $\nabla\times\underline{\mathbf{G}}_{e1}^{(1)}$, we obtain

$$\mathbf{E}_1(\mathbf{r}) = -\,i\omega\mu \int_{V_{\text{imp}}} \underline{\mathbf{G}}_{e1}^{(1)}(\mathbf{r},\mathbf{r}') \cdot \mathbf{J}_{\text{imp}}(\mathbf{r}')\, dV' \\ - \nabla\times \int_{S_a^{(1)}} \underline{\mathbf{G}}_{e1}^{(2)}(\mathbf{r},\mathbf{r}') \cdot \left[\mathbf{n}_1 \times \mathbf{E}_1(\mathbf{r}')\right]\, dS', \tag{9.197}$$

$$\mathbf{E}_2(\mathbf{r}) = -\,\nabla\times \int_{S_a^{(2)}} \underline{\mathbf{G}}_{e2}^{(2)}(\mathbf{r},\mathbf{r}') \cdot \left[\mathbf{n}_2 \times \mathbf{E}_2(\mathbf{r}')\right]\, dS'. \tag{9.198}$$

The curl and integral operators can be interchanged because $\mathbf{r}'$ and $\mathbf{r}$ are geometrically separated in this formulation.[30]

As can be seen from Figure 9.17, $S_a^{(1)} \subset V_2$ and $S_a^{(2)} \subset V_1$, such that

$$\mathbf{n}_1 \times \mathbf{E}_1(\mathbf{r})\big|_{S_a^{(1)}} = \mathbf{n}_1 \times \mathbf{E}_2(\mathbf{r})\big|_{S_a^{(1)}}, \\ \mathbf{n}_2 \times \mathbf{E}_2(\mathbf{r})\big|_{S_a^{(2)}} = \mathbf{n}_2 \times \mathbf{E}_1(\mathbf{r})\big|_{S_a^{(2)}}.$$

This observation allows us to interchange $\mathbf{E}_1(\mathbf{r}')$ with $\mathbf{E}_2(\mathbf{r}')$ in the integrand over $S_a^{(1)}$ in (9.197) and interchange $\mathbf{E}_2(\mathbf{r}')$ with $\mathbf{E}_1(\mathbf{r}')$ in the integrand over $S_a^{(2)}$ in (9.198), resulting in a coupled system of integral

[30]If the regions $V_1$ and $V_2$ represent rectangular waveguides or cavities, then $\nabla\times\underline{\mathbf{G}}_{ei}^{(2)}(\mathbf{r},\mathbf{r}') = \nabla\times\underline{\mathbf{G}}_{Ai}^{(2)}(\mathbf{r},\mathbf{r}')$, where the magnetic potential Green's dyadics $\underline{\mathbf{G}}_{Ai}^{(2)}$ of the second kind satisfy the boundary value problem

$$\nabla^2\underline{\mathbf{G}}_{Ai}^{(2)}\left(\mathbf{r},\mathbf{r}'\right) + k^2\underline{\mathbf{G}}_{Ai}^{(2)}\left(\mathbf{r},\mathbf{r}'\right) = -\underline{\mathbf{I}}\,\delta(\mathbf{r}-\mathbf{r}'), \qquad \mathbf{r},\mathbf{r}' \in V_i,$$

$$\mathbf{n}_i\times\nabla\times\underline{\mathbf{G}}_{Ai}^{(2)}\left(\mathbf{r},\mathbf{r}'\right) = \underline{\mathbf{0}}, \qquad \mathbf{r} \in S_i,$$

$$\mathbf{n}_i \cdot \left(\underline{\mathbf{I}} + \frac{1}{k^2}\nabla\nabla\right) \cdot \underline{\mathbf{G}}_{Ai}^{(2)}\left(\mathbf{r},\mathbf{r}'\right) = \mathbf{0}, \qquad \mathbf{r} \in S_i.$$

equations with respect to the field values on apertures $S_a^{(1)}$ and $S_a^{(2)}$,

$$\mathbf{E}_1(\mathbf{r}) = -\,i\omega\mu \int_{V_{\mathrm{imp}}} \underline{\mathbf{G}}_{e1}^{(1)}(\mathbf{r},\mathbf{r}') \cdot \mathbf{J}_{\mathrm{imp}}(\mathbf{r}')\, dV' - \nabla\times \int_{S_a^{(1)}} \underline{\mathbf{G}}_{e1}^{(2)}(\mathbf{r},\mathbf{r}') \cdot \left[\mathbf{n}_1 \times \mathbf{E}_2(\mathbf{r}')\right] dS', \tag{9.199}$$

$$\mathbf{E}_2(\mathbf{r}) = -\,\nabla\times \int_{S_a^{(2)}} \underline{\mathbf{G}}_{e2}^{(2)}(\mathbf{r},\mathbf{r}') \cdot \left[\mathbf{n}_2 \times \mathbf{E}_1(\mathbf{r}')\right] dS'. \tag{9.200}$$

Note that in (9.200) $\mathbf{r} \in V_2$, and, therefore, this representation is also defined for $\mathbf{r} \in S_a^{(1)} \subset V_2$. This allows us to substitute $\mathbf{E}_2(\mathbf{r})$ given by (9.200) in the integral over $S_a^{(1)}$ in (9.199), leading to

$$\begin{aligned}\mathbf{E}_1(\mathbf{r}) = &-\,i\omega\mu \int_{V_{\mathrm{imp}}} \underline{\mathbf{G}}_{e1}^{(1)}(\mathbf{r},\mathbf{r}') \cdot \mathbf{J}_{\mathrm{imp}}(\mathbf{r}')\, dV' \\ &- \nabla\times \int_{S_a^{(2)}} \int_{S_a^{(1)}} \underline{\mathbf{G}}_{e1}^{(2)}(\mathbf{r},\mathbf{r}') \cdot \left[\mathbf{n}_1 \times \nabla' \times \underline{\mathbf{G}}_{e2}^{(2)}(\mathbf{r}',\mathbf{r}'')\right] \\ &\cdot \left[\mathbf{n}_2 \times \mathbf{E}_1(\mathbf{r}'')\right] dS'\, dS''.\end{aligned} \tag{9.201}$$

This is observed to be a Fredholm integral equation of the second kind

$$\mathbf{E}_1(\mathbf{r}) + \int_{S_a^{(2)}} \underline{\mathbf{L}}_E(\mathbf{r},\mathbf{r}'') \cdot \left[\mathbf{n}_2 \times \mathbf{E}_1(\mathbf{r}'')\right] dS'' = \mathbf{E}_{\mathrm{inc}}(\mathbf{r}), \tag{9.202}$$

where $\underline{\mathbf{L}}_E(\mathbf{r},\mathbf{r}'')$ is the integral dyadic kernel defined as an integral product operator,

$$\underline{\mathbf{L}}_E(\mathbf{r},\mathbf{r}'') \equiv \nabla\times \int_{S_a^{(1)}} \underline{\mathbf{G}}_{e1}^{(2)}(\mathbf{r},\mathbf{r}') \cdot \left[\mathbf{n}_1 \times \nabla' \times \underline{\mathbf{G}}_{e2}^{(2)}(\mathbf{r}',\mathbf{r}'')\right] dS', \tag{9.203}$$

and $\mathbf{E}_{\mathrm{inc}}(\mathbf{r})$ is the incident electric field caused by $\mathbf{J}_{\mathrm{imp}}$,

$$\mathbf{E}_{\mathrm{inc}}(\mathbf{r}) = -\,i\omega\mu \int_{V_{\mathrm{imp}}} \underline{\mathbf{G}}_{e1}^{(1)}(\mathbf{r},\mathbf{r}') \cdot \mathbf{J}_{\mathrm{imp}}(\mathbf{r}')\, dV'. \tag{9.204}$$

To prove that (9.202) is Fredholm integral equation of the second kind, we must show that the integral operator $A : \mathbf{L}^2(S_a^{(2)})^3 \to \mathbf{L}^2(S_a^{(2)})^3$ defined by

$$(A\mathbf{E}_1)(\mathbf{r}) \equiv \int_{S_a^{(2)}} \underline{\mathbf{L}}_E(\mathbf{r},\mathbf{r}'') \cdot \left[\mathbf{n}_2 \times \mathbf{E}_1(\mathbf{r}'')\right] dS'', \qquad \mathbf{r},\mathbf{r}'' \in S_a^{(2)}, \tag{9.205}$$

is compact (see Example 4, Section 3.6.2), where[31] $\underline{\mathbf{L}}_E(\mathbf{r},\mathbf{r}'') \in \mathbf{L}^2(S_a^{(2)} \times S_a^{(2)})^{3\times 3}$ is bounded, assuming

$$\|\underline{\mathbf{L}}_E\|^2 = \int_{S_a^{(2)}} \int_{S_a^{(2)}} |(\underline{\mathbf{L}}_E(\mathbf{r},\mathbf{r}''))_{mn}|^2\, dS''\, dS < \infty.$$

[31]That is, each component of the $(3 \times 3)$-tuple dyadic kernel $\underline{\mathbf{L}}_E(\mathbf{r},\mathbf{r}'')$ determined by (9.203) belongs to $\mathbf{L}^2(S_a^{(2)} \times S_a^{(2)})$.

To prove that the integral operator $A$ is compact in $\mathbf{L}^2(S_a^{(2)})^3$ we have to show that for each component of the operator $A$ there is a sequence of finite-rank operators that converges in the $\mathbf{L}^2$-norm to the appropriate component. This is based on the approximation in the $\mathbf{L}^2$-norm of components of the dyadic kernel $\underline{\mathbf{L}}_E$ by a sequence of continuous degenerate kernels. The proof of compactness of the integral operator $A$ is usually done for a specific problem or for a class of problems that results in the same operator. For example, to prove that the matrix operator $A$ is compact in $\mathbf{l}^2$, it suffices to show that the condition

$$\sum_{i=1}^{\infty}\sum_{j=1}^{\infty} |a_{ij}|^2 < \infty$$

is satisfied (see Example 2, Section 3.6.2), which represents a sufficient condition for $A$ to be compact in $\mathbf{l}^2$.

The integral equation (9.202) can be discretized using a Galerkin-type projection technique, leading to a matrix equation of the second kind

$$(I - A)x = f,$$

where $I$ is the identity operator, $A$ is the matrix operator corresponding to the infinite matrix $[a_{ij}]$ (the matrix analog of the integral operator (9.205)), $x$ is the vector of unknown coefficients in the field expansion, and $f$ is the matrix analog of the incident field.

### Magnetic Field Formulation

The magnetic field in regions $V_1$ and $V_2$ satisfies

$$\nabla \times \nabla \times \mathbf{H}_i(\mathbf{r}) - k^2\mathbf{H}_i(\mathbf{r}) = \delta_{i1}\nabla \times \mathbf{J}_{\text{imp}}(\mathbf{r}), \quad \mathbf{r} \in V_i, \quad i = 1, 2, \tag{9.206}$$

$$\mathbf{n}_i \times \nabla \times \mathbf{H}_i(\mathbf{r}) = \mathbf{0}, \qquad \mathbf{n}_i \cdot \mathbf{H}_i(\mathbf{r}) = 0, \quad \mathbf{r} \in S_m^{(i)}, \tag{9.207}$$

with the fitness condition (limiting absorption) for infinite domains. Similarly to the electric field, we enforce a finite energy condition in the vicinity of a conducting edge formed by contour $\Gamma = S_1 \cap S_2$ as

$$\lim_{V_i \to 0} \int_{V_i} \left( |\mathbf{H}_i(\mathbf{r})|^2 + \frac{1}{k^2} |\nabla \times \mathbf{H}_i(\mathbf{r})|^2 \right) dV = 0. \tag{9.208}$$

The corresponding boundary value problem for the electric Green's dyadics of the second kind can be formulated as[32]

$$\nabla \times \nabla \times \underline{\mathbf{G}}_{ei}^{(2)}(\mathbf{r}, \mathbf{r}') - k^2\underline{\mathbf{G}}_{ei}^{(2)}(\mathbf{r}, \mathbf{r}') = \underline{\mathbf{I}}\,\delta(\mathbf{r} - \mathbf{r}'), \quad \mathbf{r}, \mathbf{r}' \in V_i, \tag{9.209}$$

$$\mathbf{n}_i \times \nabla \times \underline{\mathbf{G}}_{ei}^{(2)}(\mathbf{r}, \mathbf{r}') = \underline{\mathbf{0}}, \quad \mathbf{r} \in S_i \tag{9.210}$$

$$\mathbf{n}_i \cdot \underline{\mathbf{G}}_{ei}^{(2)}(\mathbf{r}, \mathbf{r}') = \mathbf{0}, \quad \mathbf{r} \in S_i.$$

[32]For rectangular waveguides and cavities the electric Green's dyadics of the second kind are given in Section 9.2.2.

The integral representations for the magnetic field are obtained using the vector-dyadic Green's second theorem applied to $\mathbf{H}_i(\mathbf{r})$ and $\underline{\mathbf{G}}_{ei}^{(2)}(\mathbf{r},\mathbf{r}')$ satisfying (9.206) and (9.209), with the boundary conditions (9.207) and (9.210), leading to

$$\mathbf{H}_1(\mathbf{r}') = \int_{V_{\text{imp}}} [\nabla \times \mathbf{J}_{\text{imp}}(\mathbf{r})] \cdot \underline{\mathbf{G}}_{e1}^{(2)}(\mathbf{r},\mathbf{r}')\, dV - \int_{S_a^{(1)}} [\mathbf{n}_1 \times \nabla \times \mathbf{H}_1(\mathbf{r})] \cdot \underline{\mathbf{G}}_{e1}^{(2)}(\mathbf{r},\mathbf{r}')\, dS, \tag{9.211}$$

$$\mathbf{H}_2(\mathbf{r}') = -\int_{S_a^{(2)}} [\mathbf{n}_2 \times \nabla \times \mathbf{H}_2(\mathbf{r})] \cdot \underline{\mathbf{G}}_{e2}^{(2)}(\mathbf{r},\mathbf{r}')\, dS, \tag{9.212}$$

and, by interchanging $\mathbf{r}'$ and $\mathbf{r}$ and using the identity (9.26) for transposition of $\underline{\mathbf{G}}_{ei}^{(2)}(\mathbf{r},\mathbf{r}')$, we have

$$\mathbf{H}_1(\mathbf{r}) = \int_{V_{\text{imp}}} \underline{\mathbf{G}}_{e1}^{(2)}(\mathbf{r},\mathbf{r}') \cdot [\nabla' \times \mathbf{J}_{\text{imp}}(\mathbf{r}')]\, dV' - \int_{S_a^{(1)}} \underline{\mathbf{G}}_{e1}^{(2)}(\mathbf{r},\mathbf{r}') \cdot \left[\mathbf{n}_1 \times \nabla' \times \mathbf{H}_1(\mathbf{r}')\right] dS', \tag{9.213}$$

$$\mathbf{H}_2(\mathbf{r}) = -\int_{S_a^{(2)}} \underline{\mathbf{G}}_{e2}^{(2)}(\mathbf{r},\mathbf{r}') \cdot \left[\mathbf{n}_2 \times \nabla' \times \mathbf{H}_2(\mathbf{r}')\right] dS'. \tag{9.214}$$

The continuity of the magnetic field across the apertures $S_a^{(1)}$ and $S_a^{(2)}$ results in a coupled system of integral equations

$$\mathbf{H}_1(\mathbf{r}) = \int_{V_{\text{imp}}} \underline{\mathbf{G}}_{e1}^{(2)}(\mathbf{r},\mathbf{r}') \cdot [\nabla' \times \mathbf{J}_{\text{imp}}(\mathbf{r}')]\, dV' - \int_{S_a^{(1)}} \underline{\mathbf{G}}_{e1}^{(2)}(\mathbf{r},\mathbf{r}') \cdot \left[\mathbf{n}_1 \times \nabla' \times \mathbf{H}_2(\mathbf{r}')\right] dS', \tag{9.215}$$

$$\mathbf{H}_2(\mathbf{r}) = -\int_{S_a^{(2)}} \underline{\mathbf{G}}_{e2}^{(2)}(\mathbf{r},\mathbf{r}') \cdot \left[\mathbf{n}_2 \times \nabla' \times \mathbf{H}_1(\mathbf{r}')\right] dS'. \tag{9.216}$$

Substituting (9.216) into (9.215), we obtain an integral equation for $\mathbf{H}_1$,

$$\begin{aligned}\mathbf{H}_1(\mathbf{r}) = &\int_{V_{\text{imp}}} \underline{\mathbf{G}}_{e1}^{(2)}(\mathbf{r},\mathbf{r}') \cdot [\nabla' \times \mathbf{J}_{\text{imp}}(\mathbf{r}')]\, dV' \\ &+ \int_{S_a^{(2)}} \int_{S_a^{(1)}} \underline{\mathbf{G}}_{e1}^{(2)}(\mathbf{r},\mathbf{r}') \cdot \left[\mathbf{n}_1 \times \nabla' \times \underline{\mathbf{G}}_{e2}^{(2)}(\mathbf{r}',\mathbf{r}'')\right] \\ &\cdot \left[\mathbf{n}_2 \times \nabla'' \times \mathbf{H}_1(\mathbf{r}'')\right] dS'\, dS'',\end{aligned} \tag{9.217}$$

which is recognized as a Fredholm integral equation of the second kind,

$$\mathbf{H}_1(\mathbf{r}) - \int_{S_a^{(2)}} \underline{\mathbf{L}}_H(\mathbf{r},\mathbf{r}'') \cdot \left[\mathbf{n}_2 \times \nabla'' \times \mathbf{H}_1(\mathbf{r}'')\right] dS'' = \mathbf{H}_{\text{inc}}(\mathbf{r}), \tag{9.218}$$

where $\underline{\mathbf{L}}_H(\mathbf{r}, \mathbf{r}'')$ is the integral dyadic kernel defined as the integral product operator

$$\underline{\mathbf{L}}_H(\mathbf{r}, \mathbf{r}'') \equiv \int_{S_a^{(1)}} \underline{\mathbf{G}}_{e1}^{(2)}(\mathbf{r}, \mathbf{r}') \cdot \left[\mathbf{n}_1 \times \nabla' \times \underline{\mathbf{G}}_{e2}^{(2)}(\mathbf{r}', \mathbf{r}'')\right] dS' \tag{9.219}$$

and $\mathbf{H}_{\mathrm{inc}}(\mathbf{r})$ is the incident magnetic field caused by $\mathbf{J}_{\mathrm{imp}}$,

$$\mathbf{H}_{\mathrm{inc}}(\mathbf{r}) = \int_{V_{\mathrm{imp}}} \underline{\mathbf{G}}_{e1}^{(2)}(\mathbf{r}, \mathbf{r}') \cdot [\nabla' \times \mathbf{J}_{\mathrm{imp}}(\mathbf{r}')]\, dV'. \tag{9.220}$$

Similarly to the electric field, we have to show that the integral operator $A : \mathbf{L}^2(S_a^{(2)})^3 \to \mathbf{L}^2(S_a^{(2)})^3$ defined by

$$(A\mathbf{H}_1)(\mathbf{r}) \equiv \int_{S_a^{(2)}} \underline{\mathbf{L}}_H(\mathbf{r}, \mathbf{r}'') \cdot [\mathbf{n}_2 \times \nabla'' \times \mathbf{H}_1(\mathbf{r}'')]\, dS'', \qquad \mathbf{r}, \mathbf{r}'' \in S_a^{(2)}, \tag{9.221}$$

where $\underline{\mathbf{L}}_H(\mathbf{r}, \mathbf{r}'') \in \mathbf{L}^2(S_a^{(2)} \times S_a^{(2)})^{3\times 3}$ is compact, such that each component of (9.219) is a Fredholm kernel and each component of the operator $A$ is compact in $\mathbf{L}^2(S_a^{(2)})$. The resulting matrix equation, obtained by discretization of the integral equation (9.218), is of the second kind, where it has to be shown that the matrix operator $A$ is compact in an appropriate sequence space (see Section 3.7).

In the next section we will demonstrate the application of the method to the analysis of natural modes in a shielded microstrip line.

### 9.5.2 Shielded Microstrip Line

To illustrate the method discussed above, a shielded microstrip line with a finite thickness conductor will be considered, as depicted in Figure 9.18. The structure is uniform in the $z$-coordinate, and it has symmetry with respect to the $y$-coordinate. A perfectly conducting strip of finite thickness $t$ ($t = t_2 - t_1$) is partially buried in the substrate material, which has thickness $d$ and relative dielectric permittivity $\varepsilon_r$. Electric and magnetic walls are placed in the plane of symmetry of the structure at $x = 0$, as shown in Figure 9.19, which lead to the characterization of odd and even natural waveguide modes, respectively.

We consider the natural modes having symmetry about $x = 0$ (imposed by the electric wall) and apply the method of overlapping regions to the geometry shown in Figure 9.19. The problem can be formulated completely using representations only for the electric-field vector. Dependence of the electric field on $e^{i(\omega t + k_z z)}$ is assumed, where $k_z = i\alpha - \beta$ ($\mathrm{Re}\{k_z\} \leq 0$, $\mathrm{Im}\{k_z\} \geq 0$) is the propagation constant in the $z$-direction. The structure is invariant along $z$, and therefore the problem can be formulated in the axial-Fourier transform domain ($z \longleftrightarrow k_z$) using the Fourier transform

$$\mathbf{E}(x, y, z) = \frac{1}{2\pi} \int_{-\infty}^{\infty} \widetilde{\mathbf{E}}(x, y, k_z) e^{ik_z z}\, dk_z. \tag{9.222}$$

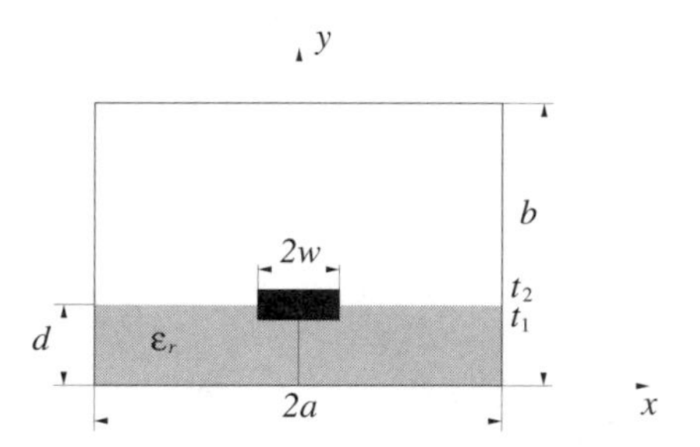


Figure 9.18: Shielded microstrip line with a finite thickness strip conductor.

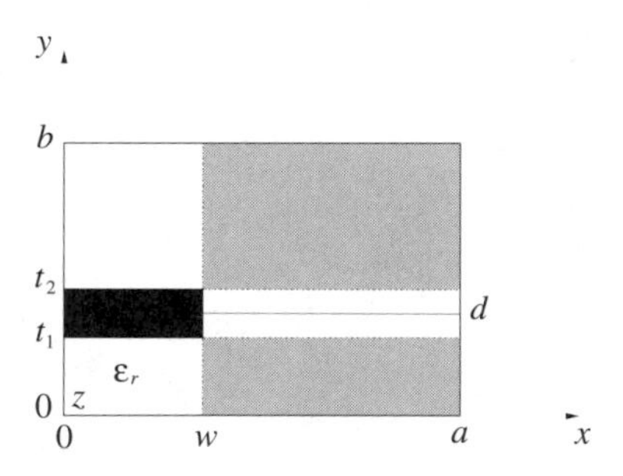


Figure 9.19: Symmetry of the structure shown in Figure 9.18 by the electric wall about $x = 0$. A discrete set of odd natural modes is investigated using the method of overlapping regions.

The geometrical domain surrounding the metal strip in the cross-section can be divided into simple one-layer and two-layer rectangular subregions that overlap in the shaded areas, as shown in Figure 9.19. We identify each region with the corresponding electric field $\widetilde{\mathbf{E}}(x, y, k_z)$, such that

$$\text{Region 1:} \qquad \left\{ \begin{array}{lll} \text{1st layer, } \widetilde{\mathbf{E}}_1^1(x,y,k_z): & w \le x \le a, & 0 \le y \le d \\ \text{2nd layer, } \widetilde{\mathbf{E}}_1^2(x,y,k_z): & w \le x \le a, & d \le y \le b \end{array} \right\},$$

$$\text{Region 2:} \qquad \widetilde{\mathbf{E}}_2(x,y,k_z): \quad 0 \le x \le a, \quad 0 \le y \le t_1,$$

$$\text{Region 3:} \qquad \widetilde{\mathbf{E}}_3(x,y,k_z): \quad 0 \le x \le a, \quad t_2 \le y \le b.$$

The electric field inside each region is obtained in terms of field values at the apertures. The continuity of the electric field across the apertures of the corresponding regions,

$$\begin{aligned}
\widehat{\mathbf{x}} \times \widetilde{\mathbf{E}}_1^1(w,y,k_z) &= \widehat{\mathbf{x}} \times \widetilde{\mathbf{E}}_2(w,y,k_z), && 0 \le y \le t_1, \\
\widehat{\mathbf{x}} \times \widetilde{\mathbf{E}}_1^2(w,y,k_z) &= \widehat{\mathbf{x}} \times \widetilde{\mathbf{E}}_3(w,y,k_z), && t_2 \le y \le b, \\
\widehat{\mathbf{y}} \times \widetilde{\mathbf{E}}_1^1(x,t_1,k_z) &= \widehat{\mathbf{y}} \times \widetilde{\mathbf{E}}_2(x,t_1,k_z), && w \le x \le a, \\
\widehat{\mathbf{y}} \times \widetilde{\mathbf{E}}_1^2(x,t_2,k_z) &= \widehat{\mathbf{y}} \times \widetilde{\mathbf{E}}_3(x,t_2,k_z), && w \le x \le a,
\end{aligned} \tag{9.223}$$

is used to couple the integral representations. Below we develop integral representations for one- and two-layer overlapping regions.

### One-Layer Region

For the one-layer regions (regions 2 and 3), we can immediately write integral representations for electric fields $\widetilde{\mathbf{E}}_2$ and $\widetilde{\mathbf{E}}_3$ based on the general formulation developed in the previous section, (9.199) and (9.200), in the axial-Fourier transform domain ($z \longleftrightarrow k_z$),

$$\widetilde{\mathbf{E}}_2(x, y, k_z) = -\left(\nabla_{xy} + ik_z\widehat{\mathbf{z}}\right) \times \int_w^a \widetilde{\underline{\mathbf{G}}}_{A2}^{(2)}(x, y, x', y', k_z) \cdot \left[\widehat{\mathbf{y}} \times \widetilde{\mathbf{E}}_1^1(x', y', k_z)\right]\Big|_{y'=t_1} dx', \tag{9.224}$$

$$\widetilde{\mathbf{E}}_3(x, y, k_z) = \left(\nabla_{xy} + ik_z\widehat{\mathbf{z}}\right) \times \int_w^a \widetilde{\underline{\mathbf{G}}}_{A3}^{(2)}(x, y, x', y', k_z) \cdot \left[\widehat{\mathbf{y}} \times \widetilde{\mathbf{E}}_1^2(x', y', k_z)\right]\Big|_{y'=t_2} dx', \tag{9.225}$$

where $\nabla_{xy} = \widehat{\mathbf{x}}\partial/\partial x + \widehat{\mathbf{y}}\partial/\partial y$, and $\widetilde{\underline{\mathbf{G}}}_{A2}^{(2)}$ and $\widetilde{\underline{\mathbf{G}}}_{A3}^{(2)}$ are magnetic potential transform-domain Green's dyadics of the second kind for the two-dimensional cavity regions 2 and 3, respectively, obtained using the Fourier transform

$$\underline{\mathbf{G}}_A^{(2)}(x, y, z, x', y', z') = \frac{1}{2\pi} \int_{-\infty}^{\infty} \widetilde{\underline{\mathbf{G}}}_A^{(2)}(x, y, x', y', k_z)\, e^{ik_z(z-z')}\, dk_z. \tag{9.226}$$

In Section 9.2 we developed magnetic potential Green's dyadics for a three-dimensional rectangular cavity with Green's function components expressed as a double-series (partial eigenfunction) expansion over the complete system of eigenfunctions in the cavity cross-section and the characteristic Green's functions along the cavity axis ($z$-coordinate). In the microstrip-line problem (for the one-layer regions 2 and 3) we use the same magnetic potential Green's dyadics, with the difference being that they are obtained in the Fourier transform domain. In this case, the components of the Green's dyadics are expressed as a single-series expansion over the eigenfunctions in the $y$-coordinate (normalized over the intervals $[0, t_1]$ and $[t_2, b]$ for Green's dyadics in regions 2 and 3, respectively) and characteristic Green's functions in the $x$-coordinate satisfying boundary conditions on the electric walls at $x = 0, a$.

### Two-Layer Region

For the two-layer region (region 1) we use the methodology described in Section 9.3 for the analysis of interacting electric- and magnetic-type discontinuities in layered-media waveguides, with the electric Green's dyadics of the third kind obtained for a two-layer, semi-infinite rectangular waveguide, as described in Section 9.4. Details concerning the boundary value problems for the electric field in a multilayered rectangular cavity region, the corresponding electric Green's dyadics, and integral representations for the method of overlapping regions are discussed in [55] and [56]. The electric fields $\widetilde{\mathbf{E}}_1^1$ and $\widetilde{\mathbf{E}}_1^2$ in the first and second layers of region 1, respectively, are obtained in terms of the field values on the corresponding apertures ($x = w,\ y \in [0, t_1]$ and $x = w,\ y \in [t_2, b]$) and on the dielectric interface at $y = d,\ x \in [w, a]$. Matching tangential components of electric fields and the corresponding electric Green's dyadics across the interface, we obtain integral representations for $\widetilde{\mathbf{E}}_1^1$ and $\widetilde{\mathbf{E}}_1^2$ in terms of integrals over the apertures,

$$\begin{aligned}\widetilde{\mathbf{E}}_1^1(x, y, k_z) = &\int_0^{t_1} \left[\widehat{\mathbf{x}} \times \widetilde{\mathbf{E}}_2(x', y', k_z)\right] \\ &\cdot \left[(\nabla'_{xy} - ik_z\widehat{\mathbf{z}}) \times \widetilde{\underline{\mathbf{G}}}_{e11}^{(1)}(x', y', x, y, -k_z)\right]\Big|_{x'=w} dy' \\ &+ \int_{t_2}^{b} \left[\widehat{\mathbf{x}} \times \widetilde{\mathbf{E}}_3(x', y', k_z)\right] \\ &\cdot \left[(\nabla'_{xy} - ik_z\widehat{\mathbf{z}}) \times \widetilde{\underline{\mathbf{G}}}_{e21}^{(1)}(x', y', x, y, -k_z)\right]\Big|_{x'=w} dy',\end{aligned} \tag{9.227}$$

$$\begin{aligned}\widetilde{\mathbf{E}}_1^2(x, y, k_z) = &\int_0^{t_1} \left[\widehat{\mathbf{x}} \times \widetilde{\mathbf{E}}_2(x', y', k_z)\right] \\ &\cdot \left[(\nabla'_{xy} - ik_z\widehat{\mathbf{z}}) \times \widetilde{\underline{\mathbf{G}}}_{e12}^{(1)}(x', y', x, y, -k_z)\right]\Big|_{x'=w} dy' \\ &+ \int_{t_2}^{b} \left[\widehat{\mathbf{x}} \times \widetilde{\mathbf{E}}_3(x', y', k_z)\right] \\ &\cdot \left[(\nabla'_{xy} - ik_z\widehat{\mathbf{z}}) \times \widetilde{\underline{\mathbf{G}}}_{e22}^{(1)}(x', y', x, y, -k_z)\right]\Big|_{x'=w} dy',\end{aligned} \tag{9.228}$$

where $\widetilde{\underline{\mathbf{G}}}_{eij}^{(1)}$ for $i, j = 1, 2$ represent electric transform-domain Green's dyadics of the third kind, obtained for the two-dimensional two-layer rectangular cavity (region 1) [55], [56]. The components of $\widetilde{\underline{\mathbf{G}}}_{eij}^{(1)}$ are expressed as a single-series expansion over the complete system of eigenfunctions in the $x$-coordinate (normalized over the interval $[w, a]$) and characteristic Green's functions in the $y$-coordinate satisfying boundary conditions on the electric walls at $y = 0, b$ and continuity conditions across the dielectric interface at $y = d$.

### System of Integral Equations of the Second Kind

Following the method of overlapping regions, we substitute integral representations for $\widetilde{\mathbf{E}}_1^1$ and $\widetilde{\mathbf{E}}_1^2$, (9.227) and (9.228), into the representations for $\widetilde{\mathbf{E}}_2$ and $\widetilde{\mathbf{E}}_3$, (9.224) and (9.225), respectively. This results in a coupled system of integral equations of the second kind for the field values $\widetilde{\mathbf{E}}_2$ and $\widetilde{\mathbf{E}}_3$ on the apertures $x = w$, $y \in [0, t_1]$ and $x = w$, $y \in [t_2, b]$,

$$\begin{aligned} \widetilde{\mathbf{E}}_2(x, y, k_z) & \qquad\qquad (9.229) \\ = \int_0^{t_1} \underline{\mathbf{L}}_{11}\,(x, y, x'', y'', k_z) \cdot & \left[\widehat{\mathbf{x}} \times \widetilde{\mathbf{E}}_2(x'', y'', k_z)\right]\Big|_{x''=w} dy'' \\ + \int_{t_2}^{b} \underline{\mathbf{L}}_{21}\,(x, y, x'', y'', k_z) \cdot & \left[\widehat{\mathbf{x}} \times \widetilde{\mathbf{E}}_3(x'', y'', k_z)\right]\Big|_{x''=w} dy'', \end{aligned}$$

$$\begin{aligned} \widetilde{\mathbf{E}}_3(x, y, k_z) & \qquad\qquad (9.230) \\ = \int_0^{t_1} \underline{\mathbf{L}}_{12}\,(x, y, x'', y'', k_z) \cdot & \left[\widehat{\mathbf{x}} \times \widetilde{\mathbf{E}}_2(x'', y'', k_z)\right]\Big|_{x''=w} dy'' \\ + \int_{t_2}^{b} \underline{\mathbf{L}}_{22}\,(x, y, x'', y'', k_z) \cdot & \left[\widehat{\mathbf{x}} \times \widetilde{\mathbf{E}}_3(x'', y'', k_z)\right]\Big|_{x''=w} dy'', \end{aligned}$$

where $\underline{\mathbf{L}}_{ij}$ $(i, j = 1, 2)$ represent integral dyadic kernels defined as product operators,

$$\begin{aligned} \underline{\mathbf{L}}_{i1}\,(x, y, x'', y'', k_z) = -(\nabla_{xy} + jk_z\widehat{\mathbf{z}}) \times \int_w^a \underline{\widetilde{\mathbf{G}}}_{A2}^{(2)}\,(x, y, x', y', k_z) \\ \cdot \left[(\nabla''_{xy} - jk_z\widehat{\mathbf{z}}) \times \underline{\widetilde{\mathbf{G}}}_{ei1}^{(1)}\,(x'', y'', x', y', -k_z)\right]^{\top}\Bigg|_{y'=t_1} dx', \end{aligned} \qquad (9.231)$$

$$\begin{aligned} \underline{\mathbf{L}}_{i2}\,(x, y, x'', y'', k_z) = (\nabla_{xy} + jk_z\widehat{\mathbf{z}}) \times \int_w^a \underline{\widetilde{\mathbf{G}}}_{A3}^{(2)}\,(x, y, x', y', k_z) \\ \cdot \left[(\nabla''_{xy} - jk_z\widehat{\mathbf{z}}) \times \underline{\widetilde{\mathbf{G}}}_{ei2}^{(1)}\,(x'', y'', x', y', -k_z)\right]^{\top}\Bigg|_{y'=t_2} dx'. \end{aligned} \qquad (9.232)$$

For the system of the second kind, it has to be shown that each integral operator $A_{1j} : \mathbf{L}^2(0, t_1)^3 \to \mathbf{L}^2(0, t_1)^3$ and $A_{2j} : \mathbf{L}^2(t_2, b)^3 \to \mathbf{L}^2(t_2, b)^3$, defined by

$$\begin{aligned} (A\widetilde{\mathbf{E}}_2)_{1j}(w, y) \equiv \int_0^{t_1} \underline{\mathbf{L}}_{1j}\,(w, y, x'', y'', k_z) \\ \cdot \left[\widehat{\mathbf{x}} \times \widetilde{\mathbf{E}}_2(x'', y'', k_z)\right]\Big|_{x''=w} dy'', \end{aligned} \qquad (9.233)$$

$$\begin{aligned} (A\widetilde{\mathbf{E}}_3)_{2j}(w, y) \equiv \int_{t_2}^{b} \underline{\mathbf{L}}_{2j}\,(w, y, x'', y'', k_z) \\ \cdot \left[\widehat{\mathbf{x}} \times \widetilde{\mathbf{E}}_3(x'', y'', k_z)\right]\Big|_{x''=w} dy'', \end{aligned} \qquad (9.234)$$

is bounded (each component of the dyadic kernels $\underline{\mathbf{L}}_{ij}$ is of Fredholm type) and compact in the sense that each component of the integral operators $A_{1j}$ and $A_{2j}$ is compact in $\mathbf{L}^2(0, t_1)$ and $\mathbf{L}^2(t_2, b)$, respectively.

The system of coupled functional equations can be discretized using Galerkin's method, resulting in a system of coupled matrix equations of the second kind,

$$(I - A)X = 0, \tag{9.235}$$

where $I$ is the identity matrix, $A$ is the $2 \times 2$ block-matrix operator corresponding to the infinite matrix $[a_{mn}]$ (i.e., the matrix analog of the integral operators $A_{1j}$ and $A_{2j}$), and $X$ is the vector of unknown coefficients in the series expansion for the field components.

It can be shown that in the vicinity of an edge of the perfectly conducting strip of finite thickness (rectangular edge), the asymptotic behavior of the normal to the edge field components defines the asymptotics for the coefficients $X$, such that $X = \{x_n\} = O(n^{-5/3})$, $n \to \infty$. This result defines a sequence space for $X : \{x_n\} \subset \mathbf{l}^1$ as discussed in the examples in Section 2.1.3.

The asymptotic behavior of the matrix elements in the diagonal blocks of the matrix operator $A$ can be shown to be [55][33]

$$a_{mn} \simeq \sum_{s=1}^{\infty} \frac{sn}{(s^2 + \xi^2 m^2)(\eta^2 s^2 + n^2)}, \qquad m, n \gg 1, \tag{9.236}$$

where $\xi$, $\eta > 0$ are real constants. It can be seen that (9.236) can be written as a product of two similar matrices, $a_{mn} = \sum_{s=1}^{\infty} b_{ms} c_{sn}$ with the elements of each matrix $d_{ms} \equiv b_{ms}(c_{ms})$ defined as

$$d_{ms} = \frac{s}{s^2 + \zeta^2 m^2}, \qquad \zeta \equiv \xi(\eta). \tag{9.237}$$

To show that the matrix operator $A$ (diagonal blocks) is compact with the matrix elements given by (9.236), we use conditions for continuity

$$\sup_{1 \le s < \infty} \left( \sum_{m=1}^{\infty} |d_{ms}|^p \right)^{1/p} < \infty \tag{9.238}$$

and $\omega$-continuity

$$\lim_{k \to \infty} \sup_{k \le s < \infty} \sum_{m=1}^{\infty} |d_{ms}|^p = 0 \tag{9.239}$$

of matrices $[d_{ms}]$ from $\mathbf{l}^1$ into $\mathbf{l}^p$, $p \ge 1$ (see Section 3.7.2, Theorem 3.20)

---

[33]Note that the same asymptotics for matrix elements has been obtained in [72] in the analysis of microstrip junctions (T-junction, right-angled bend) using the method of overlapping regions in an approximation due to a planar dispersive waveguide model.

[78]. It can be shown [55] that estimates can be obtained for $[d_{ms}]$ as

$$\sum_{m=1}^{\infty} \frac{s}{s^2+\zeta^2 m^2} = \frac{\pi}{2\zeta} \coth\left(\frac{s\pi}{\zeta}\right) - \frac{1}{2s}, \tag{9.240}$$

$$\sum_{m=1}^{\infty} \left(\frac{s}{s^2+\zeta^2 m^2}\right)^{1+\varepsilon} \leq \frac{\text{const}}{s^{\varepsilon}} + O\left(\frac{1}{s^{1+\varepsilon}}\right), \qquad \varepsilon > 0, \tag{9.241}$$

which proves the continuity of $[d_{ms}] : \mathbf{l}^1 \rightarrow \mathbf{l}^1$ and $\omega$-continuity of $[d_{ms}] : \mathbf{l}^1 \rightarrow \mathbf{l}^{1+\varepsilon}$. The $\omega$-continuity of $[d_{ms}] : \mathbf{l}^1 \rightarrow \mathbf{l}^{1+\varepsilon}$ implies the complete continuity (compactness) of $[d_{ms}] : \mathbf{l}^1 \rightarrow \mathbf{l}^{1+\varepsilon}$. The product of a continuous and a completely continuous operator results in a completely continuous operator. It can also be shown that nondiagonal matrix operators defined by the matrix elements

$$a_{mn} \simeq \sum_{s=1}^{\infty} \frac{n}{m^2 s(s^2+\nu^2 n^2)} \simeq O\left(\frac{\ln(\nu n)}{m^2 n}\right), \qquad m, n \gg 1,$$

$$a_{mn} \simeq \sum_{s=1}^{\infty} \frac{n s e^{-\sigma s}}{(s^2+\nu^2 m^2)(\upsilon^2 s^2 + n^2)}, \qquad m, n \gg 1,$$

where $\nu, \upsilon, \sigma > 0$ are real constants, are completely continuous operators in $\mathbf{l}^1$, and, therefore, the entire matrix operator $A$ is a completely continuous mapping $\mathbf{l}^1 \rightarrow \mathbf{l}^{1+\varepsilon}$.

Based on this result we conclude that the system of coupled matrix equations of the second kind (9.235) generates a Fredholm second-kind operator equation. By the discussion in Section 4.5.3 we may approximate the natural mode propagation constant as the root of the truncated eigenvalue equation.

Numerical results of dispersion characteristics for the dominant, $E_z$-even–$H_z$-odd (symmetry by the magnetic wall), and $E_z$-odd–$H_z$-even (symmetry by the electric wall) cases are shown in Figures 9.20 and 9.21 for ridge and buried shielded microstrip lines [55]. The results are obtained for a dielectric substrate thickness of $d = 1.27$ mm, strip width $2w = 1.27$ mm, shield $2a \times b = 12.7 \times 12.7$ mm, and $\varepsilon_r = 8.875$. Dispersion characteristics for the dominant and $E_z$-even–$H_z$-odd higher-order modes of a ridge microstrip line are presented in Figure 9.20. An increase in strip thickness causes large changes in the propagation constant behavior and in the cutoff frequency values of higher-order modes. A redistribution of coupling regions (circled) occurs, and an increase in mode coupling is observed. An increase of strip thickness in a buried microstrip line leads to a loss of mode coupling. Dispersion characteristics and coupling regions for $E_z$-odd–$H_z$-even higher-order modes are shown in Figure 9.21. Mode-coupling regions in a variety of guided-wave structures have been investigated using concepts of critical and singular points from catastrophe and bifurcation theories [22], [27], [79]–[81].

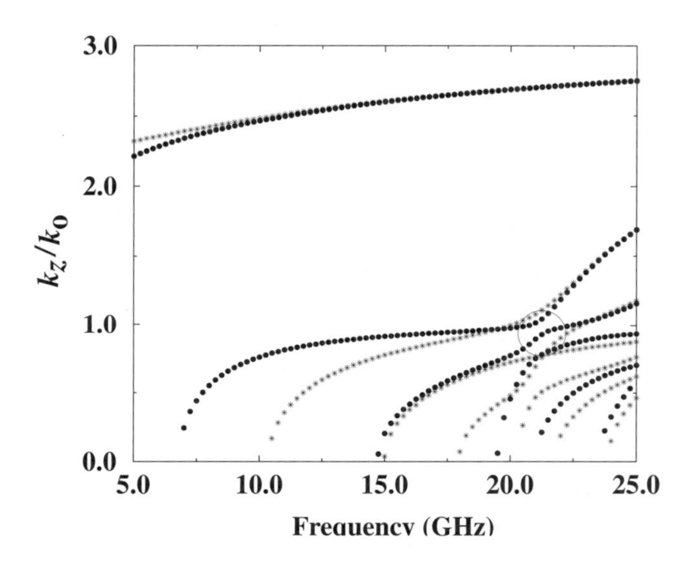


Figure 9.20: Dispersion characteristics for $E_z$-even–$H_z$-odd dominant and higher-order modes of ridge microstrip line: * $t_1$ = 1.2699 mm, $t_2$ = 2.54 mm; • $t_1$ = 1.2699 mm, $t_2$ = 7.62 mm.

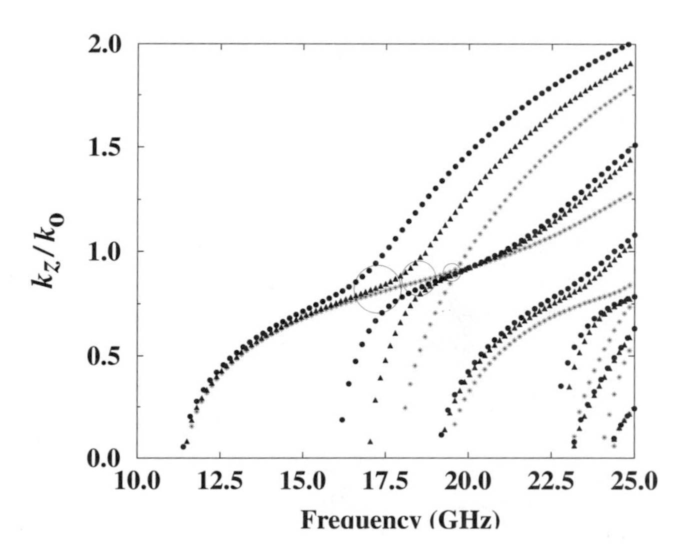


Figure 9.21: Dispersion characteristics for $E_z$-odd–$H_z$-even higher-order modes of buried microstrip line: • $t_1$ = 0.635 mm, $t_2$ = 1.271 mm; ▲ $t_1$ = 0.3175 mm, $t_2$ = 1.271 mm; * $t_1$ = 0.1 mm, $t_2$ = 1.271 mm; circles represent mode-coupling regions.

# 10

# Electromagnetic Cavities

Electromagnetic cavities are encountered in many applications, such as in forming shielding structures for high-speed/high-frequency electronic circuits, and as high quality factor resonators. Since the analysis of electromagnetic cavities is standard in many texts, the treatment here is somewhat brief and directed toward highlighting aspects of cavity analysis associated with the mathematical topics previously covered.

An electromagnetic cavity is formed by a closed surface surrounding a material medium. Generally the cavity surface is made of a conducting material, which is usually modeled as a perfect conductor. The consideration of imperfect conductivity, if necessary, is often accomplished via a perturbational approach as a secondary calculation. The cavity may, however, have one or more apertures that allow for coupling to the external environment.

The chapter begins with the identification of vector wave equations for fields inside a cavity, and associated eigenvalue problems. Next, spectral properties of the governing vector differential operators are detailed, and eigenfunction expansion solutions are considered. Vector wave equations are briefly discussed, and an example is included for a spherical cavity. Integral equation techniques are then considered, and the chapter concludes with a discussion of vector potentials for source-free cavities.

## 10.1 Problem Formulation

Consider a three-dimensional, simply connected volume $\Omega$ with sufficiently smooth boundary surface $\Gamma$. Inside $\Omega$ the medium is anisotropic and inhomogeneous, such that electric and magnetic fields satisfy the vector wave equations (1.21) and (1.22), repeated below as

$$\begin{aligned}\nabla \times \underline{\mu}(\mathbf{r})^{-1} \cdot \nabla \times \mathbf{E}(\mathbf{r}) - \omega^2 \underline{\varepsilon}(\mathbf{r}) \cdot \mathbf{E}(\mathbf{r}) \\ = -i\omega \mathbf{J}_e(\mathbf{r}) - \nabla \times \underline{\mu}(\mathbf{r})^{-1} \cdot \mathbf{J}_m(\mathbf{r}), \\ \nabla \times \underline{\varepsilon}(\mathbf{r})^{-1} \cdot \nabla \times \mathbf{H}(\mathbf{r}) - \omega^2 \underline{\mu}(\mathbf{r}) \cdot \mathbf{H}(\mathbf{r}) \\ = -i\omega \mathbf{J}_m(\mathbf{r}) + \nabla \times \underline{\varepsilon}(\mathbf{r})^{-1} \cdot \mathbf{J}_e(\mathbf{r}).\end{aligned} \tag{10.1}$$

It is convenient to define the operators $L_{E(H)} : \mathbf{L}^2(\Omega)^3 \to \mathbf{L}^2(\Omega)^3$ as

$$\begin{aligned}L_E \mathbf{x} &\equiv \nabla \times \underline{\mu}(\mathbf{r})^{-1} \cdot \nabla \times \mathbf{x} - \omega^2 \underline{\varepsilon}(\mathbf{r}) \cdot \mathbf{x}, \\ D_{L_E} &\equiv \left\{ \mathbf{x} : \mathbf{x}, \nabla \times \underline{\mu}(\mathbf{r})^{-1} \cdot \nabla \times \mathbf{x} \in \mathbf{L}^2(\Omega)^3, B(\mathbf{x}) = 0 \right\},\end{aligned} \tag{10.2}$$

and

$$\begin{aligned}L_H \mathbf{x} &\equiv \nabla \times \underline{\varepsilon}(\mathbf{r})^{-1} \cdot \nabla \times \mathbf{x} - \omega^2 \underline{\mu}(\mathbf{r}) \cdot \mathbf{x}, \\ D_{L_H} &\equiv \left\{ \mathbf{x} : \mathbf{x}, \nabla \times \underline{\varepsilon}(\mathbf{r})^{-1} \cdot \nabla \times \mathbf{x} \in \mathbf{L}^2(\Omega)^3, B(\mathbf{x}) = 0 \right\},\end{aligned}$$

where either

$$B(\mathbf{x}) = \mathbf{n} \times \mathbf{x}|_\Gamma = \mathbf{0}$$

or

$$B(\mathbf{x}) = \mathbf{n} \times \nabla \times \mathbf{x}|_\Gamma = \mathbf{0}.$$

For the source-driven problem we assume frequency $\omega$ is given, and the goal is to determine $\mathbf{E}$ and $\mathbf{H}$ maintained in $\Omega$ (and possibly external to $\Omega$ if coupling apertures are present) by the sources $\mathbf{J}_{e(m)} \neq \mathbf{0}$ inside the cavity. Often this is approached as a Green's function problem, and formally the solution is

$$\begin{aligned}\mathbf{E}(\mathbf{r}) &= L_E^{-1} \left( -i\omega \mathbf{J}_e(\mathbf{r}) - \nabla \times \underline{\mu}(\mathbf{r})^{-1} \cdot \mathbf{J}_m(\mathbf{r}) \right), \\ \mathbf{H}(\mathbf{r}) &= L_H^{-1} \left( -i\omega \mathbf{J}_m(\mathbf{r}) + \nabla \times \underline{\varepsilon}(\mathbf{r})^{-1} \cdot \mathbf{J}_e(\mathbf{r}) \right),\end{aligned}$$

where $L_{E(H)}^{-1}$ are obviously Green's operators as described in Section 3.9.3.

A related problem of interest on its own, and because it affects the source-driven problem, is the determination of the natural cavity fields and natural cavity resonance frequencies, which is inherently a spectral problem. In this case one is interested in the eigenvalue problems

$$\begin{aligned}L_E(\omega_n)\mathbf{x}_n &= \lambda_n(\omega_n)\mathbf{x}_n, \\ L_H(\omega_n)\mathbf{x}_n &= \lambda_n(\omega_n)\mathbf{x}_n,\end{aligned} \tag{10.3}$$

where the eigenfunctions $\mathbf{x}_n$ are subject to some appropriate vector boundary conditions as described below. A common form of the problem is to consider $\omega$ to be a free spectral parameter (a nonstandard eigenvalue), such that the cavity resonances are determined as those values $\omega_n$ that result in vanishing of the standard eigenvalue, i.e.,

$$\lambda_n(\omega_n) = 0.$$

Eigenfunctions corresponding to these eigenvalues are in the null space of $L_{E(H)}$ and represent the natural cavity modes of the structure. For the cavity problem each natural mode (see the discussion in Section 8.1.1) is an eigenfunction in the null space of $L_{E(H)}$, with the converse also true. Alternatively, one can consider the generalized eigenvalue problems

$$\begin{aligned} \nabla \times \underline{\mu}(\mathbf{r})^{-1} \cdot \nabla \times \mathbf{x}_n(\mathbf{r}) &= \lambda_n \underline{\varepsilon}(\mathbf{r}) \cdot \mathbf{x}_n(\mathbf{r}), \\ \nabla \times \underline{\varepsilon}(\mathbf{r})^{-1} \cdot \nabla \times \mathbf{x}_n(\mathbf{r}) &= \lambda_n \underline{\mu}(\mathbf{r}) \cdot \mathbf{x}_n(\mathbf{r}) \end{aligned} \tag{10.4}$$

with eigenvalues $\lambda_n = \omega_n^2$, where $\underline{\varepsilon}$ and $\underline{\mu}$ provide weighting factors.

In the special case of a homogeneous isotropic medium we have

$$\nabla \times \nabla \times \mathbf{E}(\mathbf{r}) - k^2\mathbf{E}(\mathbf{r}) = -i\omega\mu\mathbf{J}_e(\mathbf{r}) - \nabla \times \mathbf{J}_m(\mathbf{r}), \tag{10.5}$$

$$\nabla \times \nabla \times \mathbf{H}(\mathbf{r}) - k^2\mathbf{H}(\mathbf{r}) = -i\omega\varepsilon\mathbf{J}_m(\mathbf{r}) + \nabla \times \mathbf{J}_e(\mathbf{r}), \tag{10.6}$$

where $k^2 = \omega^2\mu\varepsilon$ such that the relevant operators become

$$L_E\mathbf{x} = L_H\mathbf{x} = \nabla \times \nabla \times \mathbf{x} - k^2\mathbf{x}.$$

In this case one can discuss the eigenvalue problem (10.3), although it is more useful to consider the generalized eigenvalue problem

$$L\mathbf{x} = \nabla \times \nabla \times \mathbf{x}_n = \lambda_n \mu\varepsilon \mathbf{x}_n$$

with $\lambda_n = \omega_n^2$. If the medium is lossless, it is also convenient to consider $k^2$ as an eigenvalue, such that the relevant eigenvalue equation is

$$\nabla \times \nabla \times \mathbf{x}_n = \lambda_n \mathbf{x}_n \tag{10.7}$$

with $\lambda_n = k_n^2 = \omega_n^2\mu\varepsilon$. In the following section we consider spectral properties of the operators $L$, $L_E$, and $L_H$.

## 10.2 Operator Properties and Eigenfunction Methods

### Self-Adjointness

We first consider the eigenvalue problems (10.3) and (10.7). As shown in Example 5, p. 151, the operator

$$\begin{aligned} L\mathbf{x} &\equiv \nabla \times \nabla \times \mathbf{x}, \\ D_L &\equiv \left\{ \mathbf{x} : \mathbf{x}(\mathbf{t}), \nabla \times \nabla \times \mathbf{x}(\mathbf{t}) \in \mathbf{L}^2(\Omega)^3, \mathbf{n} \times \mathbf{x}(\mathbf{t})|_\Gamma = \mathbf{0} \right\} \end{aligned}$$

is self-adjoint ($L = L^*$). For the general case of an anisotropic inhomogeneous medium, $L_{E(H)}$ can be shown to be self-adjoint if $\underline{\mu}$ and $\underline{\varepsilon}$ are Hermitian and $\mathbf{n} \times \mathbf{x}(\mathbf{t})|_\Gamma = \mathbf{0}$ or $\mathbf{n} \times \nabla \times \mathbf{x}(\mathbf{t})|_\Gamma = \mathbf{0}$.

To prove $L_E = (L_E)^*$ for the case of an isotropic inhomogeneous medium, we need to show that

$$\left\langle \mathbf{x}, \nabla\times\mu(\mathbf{r})^{-1}\nabla \times \mathbf{y} - \omega^2\varepsilon(\mathbf{r})\mathbf{y}\right\rangle = \left\langle \nabla\times\mu(\mathbf{r})^{-1}\nabla \times \mathbf{x} - \omega^2\varepsilon(\mathbf{r})\mathbf{x}, \mathbf{y}\right\rangle \tag{10.8}$$

since it will be obvious that the domains are identical. Assuming $\omega \in \mathbf{R}$, we have

$$\begin{aligned} &\left\langle \mathbf{x}, \nabla \times \mu(\mathbf{r})^{-1}\nabla \times \mathbf{y} - \omega^2\varepsilon(\mathbf{r})\mathbf{y}\right\rangle \\ &\quad = \int_\Omega \mathbf{x}\cdot\overline{(\nabla \times \mu(\mathbf{r})^{-1}\nabla \times \mathbf{y} - \omega^2\varepsilon(\mathbf{r})\mathbf{y})}\, d\Omega \qquad (10.9) \\ &\quad = \int_\Omega \mathbf{x}\cdot \left(\nabla \times \overline{\mu}(\mathbf{r})^{-1}\nabla \times \overline{\mathbf{y}} - \omega^2\overline{\varepsilon}(\mathbf{r})\overline{\mathbf{y}}\right)\, d\Omega. \end{aligned}$$

Using the vector identity $\nabla\cdot(\mathbf{A} \times \mathbf{B}) = \mathbf{B} \cdot \nabla \times \mathbf{A} - \mathbf{A} \cdot \nabla \times \mathbf{B}$ leads to the right side of (10.9) becoming

$$\int_\Omega \left[\left(\overline{\mu}(\mathbf{r})^{-1}\nabla \times \overline{\mathbf{y}}\right) \cdot (\nabla \times \mathbf{x}) - \omega^2\overline{\varepsilon}(\mathbf{r})\mathbf{x} \cdot \overline{\mathbf{y}}\right]\, d\Omega, \tag{10.10}$$

where we have used the divergence theorem and the boundary condition $\mathbf{n} \times \mathbf{x}(\mathbf{t})|_\Gamma = \mathbf{0}$. Using again the same vector identity as before, and the divergence theorem a second time, along with $\mathbf{n} \times \mathbf{y}(\mathbf{t})|_\Gamma = \mathbf{0}$, (10.10) becomes

$$\int_\Omega \left[\left(\nabla \times \overline{\mu}(\mathbf{r})^{-1}\nabla \times \mathbf{x}\right) - \omega^2\overline{\varepsilon}(\mathbf{r})\mathbf{x}\right] \cdot \overline{\mathbf{y}}\, d\Omega.$$

The same result is obtained with the boundary condition $\mathbf{n}\times\nabla\times\mathbf{x}(\mathbf{t})|_\Gamma = \mathbf{0}$. If $\overline{\mu}^{-1} = \mu^{-1}$ and $\overline{\varepsilon} = \varepsilon$, we have the desired result, (10.8). Therefore, in the case of a lossless isotropic, inhomogeneous medium (as the natural reduction of the Hermitian requirement to scalars), $L_E$ is self-adjoint; the same conclusion is similarly reached for $L_H$. The proof for an anisotropic medium is more involved and will be omitted here.

For a homogeneous isotropic medium one simply uses the eigenfunctions of $L = \nabla \times \nabla\times$, where self-adjointness follows from the boundary condition $\mathbf{n} \times \mathbf{x}(\mathbf{t})|_\Gamma = \mathbf{0}$ or $\mathbf{n} \times \nabla \times \mathbf{x}(\mathbf{t})|_\Gamma = \mathbf{0}$, regardless of the presence or absence of loss in the material.[1]

[1]As seen in several previous instances, this is a general feature encountered in most electromagnetic problems governed by a second-order differential operator (e.g., $\partial/\partial x^2$, $\nabla^2$, $\nabla \times \nabla\times$, etc.) in a bounded region of space. If homogeneous boundary conditions are specified, then the governing differential operator will be self-adjoint if the interior medium is homogeneous, regardless of the presence or absence of loss. If the interior medium is inhomogeneous, then the governing operator will usually only be self-adjoint if the interior medium is Hermitian or lossless.

If $L_E$ is self-adjoint, then (10.3) constitutes a self-adjoint boundary value problem, and by a generalization of Theorem 4.46 the orthonormal eigenfunctions defined by

$$L_E \mathbf{x}_n = \left(\nabla\times\underline{\mu}(\mathbf{r})^{-1}\cdot\nabla\times -\omega^2\underline{\varepsilon}(\mathbf{r})\cdot\right)\mathbf{x}_n = \lambda_n\mathbf{x}_n$$

(or using $L_H$ or $L$) subject to appropriate boundary conditions (as discussed next) form a complete set in $H = \mathbf{L}^2(\Omega)^3$ and lead to the expansion

$$\mathbf{F} = \sum_n \langle \mathbf{F}, \mathbf{x}_n\rangle \mathbf{x}_n$$

for any $\mathbf{F} \in H$. We also know that $\lambda_n \in \mathbf{R}$, and

$$\begin{aligned}\sigma\left(L_{E(H)}\right) &= \sigma_p\left(L_{E(H)}\right),\\ \sigma(L) &= \sigma_p(L).\end{aligned}$$

Also note that the summation in the eigenfunction expansion is three-dimensional, such that $n$ represents a triplet of integers.

### Boundary Conditions

We need to impose appropriate boundary conditions to uniquely determine the eigenfunctions $\mathbf{x}_n$ of $L_E$, $L_H$, or $L$. Due to the vector nature of the problem, the boundary conditions $\mathbf{n}\times\mathbf{x}_n(\mathbf{t})|_\Gamma = \mathbf{0}$ or $\mathbf{n}\times\nabla\times\mathbf{x}_n(\mathbf{t})|_\Gamma = \mathbf{0}$ are not generally sufficient[2] to determine $\mathbf{x}_n$. An immediate generalization of the scalar homogeneous Dirichlet conditions to the vector case is

$$\mathbf{x}_n|_\Gamma = \mathbf{0},$$

i.e.,

$$\begin{aligned}\mathbf{n}\times\mathbf{x}_n(\mathbf{t})|_\Gamma &= \mathbf{0},\\ \mathbf{n}\cdot\mathbf{x}_n|_\Gamma &= 0,\end{aligned} \tag{10.11}$$

which may be used to uniquely determine the eigenfunction $\mathbf{x}_n$. However, these are often not the most useful conditions in practice.

Of more practical interest are the boundary conditions

$$\begin{aligned}\mathbf{n}\times\mathbf{x}_n|_\Gamma &= \mathbf{0},\\ \nabla\cdot\mathbf{x}_n|_\Gamma &= 0,\end{aligned} \tag{10.12}$$

which lead to the *short-circuit electric eigenfunctions*, denoted as $\mathbf{x}_n = \mathbf{E}_n$. The condition $\mathbf{n}\times\mathbf{E}_n|_\Gamma = \mathbf{0}$ is consistent with the boundary condition

[2]Note that the specification of tangential boundary conditions is sufficient to uniquely determine vector fields satisfying Maxwell's equations (see Section 1.5). If vector eigenfunctions $\mathbf{x}_n$ are only required to satisfy the single vector eigenvalue equation $L\mathbf{x}_n = \lambda\mathbf{x}_n$ (or using $L_E$ or $L_H$), three scalar boundary conditions need to be specified to uniquely determine the three scalar components of the eigenfunction.

obeyed by the electric field at a perfectly conducting cavity wall, and $\nabla \cdot \mathbf{E}_n|_\Gamma = 0$ is physically motivated from Gauss' law if one assumes the absence of a source charge density on the boundary $\Gamma$. Although from a mathematical standpoint we could also choose the Dirichlet conditions (10.11), or some other appropriate conditions, for electric-field expansions the short-circuit conditions are generally preferred on physical and mathematical grounds, especially for perfectly conducting cavities.[3]

Yet another set of eigenfunctions can be obtained from the boundary conditions

$$\begin{aligned} \mathbf{n} \times \nabla \times \mathbf{x}_n|_\Gamma &= \mathbf{0}, \\ \mathbf{n} \cdot \mathbf{x}_n|_\Gamma &= 0, \end{aligned} \tag{10.13}$$

leading to the *short-circuit magnetic eigenfunctions*, denoted by $\mathbf{x}_n = \mathbf{H}_n$. These are particularly convenient for expanding the magnetic field, as these are the conditions that occur for the magnetic field on the surface of a perfectly conducting cavity.

Any of the three described sets of conditions uniquely determines the complete set of vector eigenfunctions, orthogonal for distinct eigenvalues. Degenerate eigenfunctions (corresponding to the same eigenvalues) can be turned into an orthonormal set via the Gram–Schmidt procedure. Other boundary conditions can also be considered; for instance, those corresponding to a perfect magnetic boundary.

For an inhomogeneous medium, if the material dyadics are not Hermitian (lossless in the scalar case), then the governing operators are not self-adjoint, and eigenfunction methods may not be valid. However, it is reasonable to assume that for small material loss eigenfunction solution techniques would still lead to accurate results.[4] A much more difficult problem is the actual determination of the eigenfunctions themselves. Due to the vector nature of the problem, this is extremely difficult or impossible for all but the simplest separable geometries. Even in these cases the presence of inhomogeneous anisotropic media generally eliminates the possibility of determining the desired vector eigenfunctions in closed form.

---

[3]Note that the Dirichlet eigenfunctions are such that their tangential and normal components vanish at the boundary. In cases where the electric field has nonvanishing tangential or normal components (e.g., the normal component of electric field at a perfectly conducting cavity wall), one has nonuniform convergence of those components of the resulting eigenfunction expansion for the electric field, which is one reason why the short-circuit eigenfunctions are generally preferred. Similar situations arise whenever a field component satisfies a boundary condition that differs from that satisfied by the corresponding eigenfunction component. One further consequence is that termwise differentiation of the series will not generally be possible.

[4]This may be treated rigorously as a nonself-adjoint perturbation of a self-adjoint operator, which, for a suitably weak perturbation, may still lead to an eigenbasis.

### Simple Cavities Filled with a Homogeneous Medium

Even for cavities filled with a homogeneous isotropic medium, eigenfunctions cannot generally be found in closed form, except simple separable geometries such as a parallelepiped, cylinder, or sphere. However, these shapes are important in practice, and, furthermore, for these problems loss associated with the medium filling the cavity is irrelevant to self-adjointness of the governing operator

$$\begin{aligned} L\mathbf{x} &= \nabla \times \nabla \times \mathbf{x}, \\ D_L &= \left\{ \mathbf{x} : \mathbf{x}, \nabla \times \nabla \times \mathbf{x} \in \mathbf{L}^2(\Omega)^3, \mathbf{n} \times \mathbf{x}|_\Gamma = \mathbf{0} \right\} \end{aligned}$$

(or with $\mathbf{n} \times \nabla \times \mathbf{x}|_\Gamma = \mathbf{0}$). Therefore, for these geometries eigenfunction methods have been found to be quite useful in practice. We consider here the case of an isotropic homogeneous medium filling a cavity having a simple separable geometry, leading to the vector wave equations (10.5) and associated eigenvalue problem (10.7).

### Vector Wavefunctions

The vector eigenfunctions $\mathbf{x}_n$ of the operator $L$, also called *vector wavefunctions,* can be separated into irrotational (lamellar) eigenfunctions $\mathbf{L}$ with

$$\begin{aligned} \nabla \times \mathbf{L} &= \mathbf{0}, \\ \nabla \cdot \mathbf{L} &\neq 0 \end{aligned}$$

and rotational (solenoidal) eigenfunctions $\mathbf{F}$ with

$$\begin{aligned} \nabla \times \mathbf{F} &\neq \mathbf{0}, \\ \nabla \cdot \mathbf{F} &= 0. \end{aligned}$$

Because $\nabla \times \mathbf{L} = \mathbf{0}$, the irrotational eigenfunctions must be in the null space of the curl–curl operator. The solenoidal eigenfunctions $\mathbf{F}$ are often further divided into vector wavefunctions $\mathbf{M}$ and $\mathbf{N}$. See [36] for a comprehensive analysis of vector wavefunctions.

The usefulness of the vector wavefunctions is that they are obtained in terms of scalar wavefunctions that satisfy a simple scalar Helmholtz equation of the form

$$\left(\nabla^2 + \kappa^2\right) \psi_i = 0$$

for $i = 1, 2, 3$. The solenoidal eigenfunctions $\mathbf{F}$ are obtained as [82]

$$\begin{aligned} \mathbf{M} &= \nabla \times (\psi_1 \mathbf{c}), \\ \mathbf{N} &= \frac{1}{\kappa} \nabla \times \nabla \times (\psi_2 \mathbf{c}), \end{aligned} \tag{10.14}$$

where the vector $\mathbf{c}$ is known as the *piloting vector*. There is considerable flexibility in choosing $\mathbf{c}$, although certain choices result in the desirable decomposition of $\mathbf{M}$ representing $TM$ $(TE)$ fields and $\mathbf{N}$ representing $TE$ $(TM)$ fields with respect to some preferred coordinate.

The solenoidal eigenfunctions do not form a complete set in the space $\mathbf{L}^2(\Omega)^3$ (they do, however, form a complete set in the space of solenoidal vector fields, corresponding to source-free cavity fields), and one must consider the irrotational eigenfunctions $\mathbf{L}$, which are obtained as

$$\mathbf{L} = \nabla(\psi_3). \tag{10.15}$$

Often a scale factor is included in (10.15).

It is easy to see that $\mathbf{M}, \mathbf{N}, \mathbf{L}$ as defined above satisfy (10.7), and, in fact, all such eigenfunctions of $\nabla \times \nabla \times$ can be so obtained. Taken together, the solenoidal and irrotational eigenfunctions form a basis for $\mathbf{L}^2(\Omega)^3$.

### Eigenfunction Expansions

Let $\{\mathbf{x}_n\}$ represent the complete, orthonormal set of eigenfunctions in $\mathbf{L}^2(\Omega)^3$ of the operator $\nabla \times \nabla \times$, i.e., let $\mathbf{x}_n$ satisfy

$$\nabla \times \nabla \times \mathbf{x}_n = \lambda_n \mathbf{x}_n$$

subject to one of the boundary condition sets (10.11), (10.12), or (10.13). The quantities $\mathbf{E}$, $\mathbf{J}_e$, and $\nabla \times \mathbf{J}_m$ are written as expansions[5] in $\mathbf{x}_n$,

$$\begin{aligned}\mathbf{E}(\mathbf{r}) &= \sum_n \langle \mathbf{E}, \mathbf{x}_n \rangle \mathbf{x}_n(\mathbf{r}), \\ \mathbf{J}_e(\mathbf{r}) &= \sum_n \langle \mathbf{J}_e, \mathbf{x}_n \rangle \mathbf{x}_n(\mathbf{r}), \\ \nabla \times \mathbf{J}_m(\mathbf{r}) &= \sum_n \langle \nabla \times \mathbf{J}_m, \mathbf{x}_n \rangle \mathbf{x}_n(\mathbf{r}),\end{aligned}$$

where the summations are three-dimensional with $n$ representing a triplet of integers. Exploiting orthonormality of the eigenfunctions leads to the solution of the vector wave equation (10.5) as

$$\mathbf{E}(\mathbf{r}) = \sum_n \frac{-i\omega\mu \langle \mathbf{J}_e, \mathbf{x}_n \rangle - \langle \nabla \times \mathbf{J}_m, \mathbf{x}_n \rangle}{\lambda_n - k^2} \mathbf{x}_n. \tag{10.16}$$

The expansion (10.16) can also be written as

$$\mathbf{E}(\mathbf{r}) = -i\omega\mu \int_\Omega \underline{\mathbf{G}}_{ee}(\mathbf{r}, \mathbf{r}') \cdot \mathbf{J}_e(\mathbf{r}') \, d\Omega' - \int_\Omega \underline{\mathbf{G}}_{em}(\mathbf{r}, \mathbf{r}') \cdot \mathbf{J}_m(\mathbf{r}') \, d\Omega' \tag{10.17}$$

[5] If $\mathbf{x}_n$ are separated into vector wavefunctions $\mathbf{M}, \mathbf{N}$, and $\mathbf{L}$, then the expansion of $\mathbf{F} \in \mathbf{L}^2(\Omega)^3$ has the form (10.22).

with the Green's function

$$\underline{\mathbf{G}}_{ee}(\mathbf{r}, \mathbf{r}', k^2) = \sum_n \frac{\mathbf{x}_n(\mathbf{r})\, \overline{\mathbf{x}_n}(\mathbf{r}')}{\lambda_n - k^2}, \tag{10.18}$$

$$\underline{\mathbf{G}}_{em}(\mathbf{r}, \mathbf{r}', k^2) = \sum_n \frac{\mathbf{x}_n(\mathbf{r})\, \nabla \times \overline{\mathbf{x}_n}(\mathbf{r}')}{\lambda_n - k^2}. \tag{10.19}$$

Of course, different convergence properties will be obtained from different eigenfunction expansions, corresponding to different sets of vector boundary conditions. As usual, poles of the Green's function correspond to resonances of the structure.[6]

**Example: Spherical Cavity**

For illustrative purposes, the eigenfunctions that satisfy

$$\nabla \times \nabla \times \mathbf{E}_{nmp} = \lambda_{nmp} \mathbf{E}_{nmp},$$

where $\lambda_{nmp} = \omega_{nmp}^2 \mu\varepsilon$, subject to

$$\begin{aligned} \mathbf{n} \times \mathbf{E}_{nmp}|_\Gamma &= \mathbf{0}, \\ \nabla \cdot \mathbf{E}_{nmp}|_\Gamma &= 0 \end{aligned} \tag{10.20}$$

(short-circuit electric eigenfunctions) for a spherical cavity of radius $a$ are listed below [83, pp. 306–307]. The irrotational eigenfunctions are[7]

$$\begin{aligned} \mathbf{L}_{nmp} = \mathbf{E}^{ir}_{nmp} = Q^L_{nmp} \Bigg\{ & \widehat{\mathbf{r}} \left\{ \begin{array}{c} \cos m\phi \\ \sin m\phi \end{array} \right\} P_n^m(\cos\theta) \frac{dj_n(k_{np} r)}{dr} \\ & + \widehat{\theta} \frac{1}{r} \left\{ \begin{array}{c} \cos m\phi \\ \sin m\phi \end{array} \right\} \frac{dP_n^m(\cos\theta)}{d\theta} j_n(k_{np} r) \\ & + \widehat{\phi} \frac{m}{r \sin\theta} \left\{ \begin{array}{c} -\sin m\phi \\ \cos m\phi \end{array} \right\} P_n^m(\cos\theta)\, j_n(k_{np} r) \Bigg\}, \end{aligned}$$

where $k_{np}$ are roots of $j_n(k_{np} a) = 0$, with corresponding eigenvalues

$$\lambda_{nmp} = 0.$$

One can easily check that, although in general $\nabla \cdot \mathbf{L} \neq 0$, the boundary condition $\nabla \cdot \mathbf{L}|_\Gamma = 0$ is satisfied in accordance with the prescribed short-circuit boundary conditions. The normalization factor $Q^L_{nmp}$ is

$$Q^L_{nmp} = \left\{ \frac{\varepsilon_m (2n+1)(n-m)!}{2\pi k_{np} (m+n)! a^3} j_{n+1}^{-2}(k_{np} a) \right\}^{1/2},$$

[6] In [38] a method is described to convert the purely spectral, three-dimensional series representation into a two-dimensional modal form by extracting the source-point singularity (depolarizing dyadic) contribution.

[7] The spherical Bessel functions $j_n$ and associated Legendre functions $P_n^m$ are defined in Section 5.4.2.

where $\varepsilon_m$ is Neumann's number.

The solenoidal eigenfunctions are divided into two groups, one corresponding to the modes $TM^r$ ($E$-modes), and the other to modes $TE^r$ ($H$-modes). For the $E$-modes we have

$$\begin{aligned}\mathbf{N}_{nmp} =& \mathbf{E}^{s,E}_{nmp} = Q^N_{nmp} \left\{ \widehat{\mathbf{r}} \frac{n(n+1)}{k'_{np} r} \left\{ \begin{array}{c} \sin m\phi \\ \cos m\phi \end{array} \right\} P^m_n (\cos\theta)\, j_n (k'_{np} r) \right. \\ &+ \widehat{\theta} \frac{1}{k'_{np} r} \left\{ \begin{array}{c} \sin m\phi \\ \cos m\phi \end{array} \right\} \frac{d(r j_n (k'_{np} r))}{dr} \frac{dP^m_n (\cos\theta)}{d\theta} \\ &\left. + \widehat{\phi} \frac{m}{k'_{np} r \sin\theta} \left\{ \begin{array}{c} \cos m\phi \\ -\sin m\phi \end{array} \right\} P^m_n (\cos\theta) \frac{d(r j_n (k'_{np} r))}{dr} \right\},\end{aligned}$$

where $k'_{np} a = x_{np}$, with $x_{np}$ the roots of $d(r j_n (r))/dr = 0$, and

$$\lambda_{nmp} = k'^2_{np}$$

are the associated eigenvalues. The normalization factor is

$$Q^N_{nmp} = \left\{ \frac{\varepsilon_m (2n+1)(n-m)!}{2\pi n (n+1)(n+m)! a^3} \left( 1 - \frac{n(n+1)}{k'_{np} a} \right)^{-1} \frac{1}{j^2_n (k'_{np} a)} \right\}^{1/2}.$$

For the $H$-modes,

$$\begin{aligned}\mathbf{M}_{nmp} =& \mathbf{E}^{s,H}_{nmp} = Q^M_{nmp} \left\{ \widehat{\theta} \frac{m}{\sin\theta} \left\{ \begin{array}{c} \cos m\phi \\ -\sin m\phi \end{array} \right\} j_n (k_{np} r) P^m_n (\cos\theta) \right. \\ &\left. - \widehat{\phi} \left\{ \begin{array}{c} \sin m\phi \\ \cos m\phi \end{array} \right\} \frac{dP^m_n (\cos\theta)}{d\theta} j_n (k_{np} r) \right\},\end{aligned}$$

where $k_{np}$ are roots of $j_n (k_{np} a) = 0$ and

$$\lambda_{nmp} = k^2_{np}$$

are the associated eigenvalues. The normalization factor is

$$Q^M_{nmp} = \left\{ \frac{\varepsilon_m (2n+1)(n-m)!}{2\pi n (n+1)(n+m)! a^3} j^{-2}_{n+1} (k_{np} a) \right\}^{1/2}.$$

In summary, the functions $\mathbf{M}, \mathbf{N}$, and $\mathbf{L}$ satisfy

$$\begin{aligned}\nabla \times \nabla \times \mathbf{L}_{nmp} &= \mathbf{0}, \\ \nabla \times \nabla \times \mathbf{N}_{nmp} &= k'^2_{np} \mathbf{N}_{nmp}, \\ \nabla \times \nabla \times \mathbf{M}_{nmp} &= k^2_{np} \mathbf{M}_{nmp},\end{aligned}$$

subject to the boundary conditions (10.20) at $r = a$, and form a complete set in $\mathbf{L}^2 (\Omega)^3$ where $\Omega$ is a sphere of radius $a$.

The solution of

$$\nabla \times \nabla \times \mathbf{E}(\mathbf{r}) - k^2 \mathbf{E}(\mathbf{r}) = -i\omega\mu \mathbf{J}_e(\mathbf{r}) \tag{10.21}$$

via eigenfunction expansion follows from

$$\begin{aligned}\mathbf{E}(\mathbf{r}) = & \sum_n \sum_m \sum_p \{ \langle \mathbf{E}, \mathbf{M}_{nmp} \rangle \mathbf{M}_{nmp}(\mathbf{r}) + \langle \mathbf{E}, \mathbf{N}_{nmp} \rangle \mathbf{N}_{nmp}(\mathbf{r}) \\ & + \langle \mathbf{E}, \mathbf{L}_{nmp} \rangle \mathbf{L}_{nmp}(\mathbf{r}) \}, \qquad (10.22) \\ \mathbf{J}_e(\mathbf{r}) = & \sum_n \sum_m \sum_p \{ \langle \mathbf{J}_e, \mathbf{M}_{nmp} \rangle \mathbf{M}_{nmp}(\mathbf{r}) + \langle \mathbf{J}_e, \mathbf{N}_{nmp} \rangle \mathbf{N}_{nmp}(\mathbf{r}) \\ & + \langle \mathbf{J}_e, \mathbf{L}_{nmp} \rangle \mathbf{L}_{nmp}(\mathbf{r}) \}.\end{aligned}$$

Substituting into (10.21), and invoking orthonormality, lead to

$$\begin{aligned}\mathbf{E}(\mathbf{r}) = -i\omega\mu \sum_n \sum_m \sum_p & \frac{\langle \mathbf{J}_e, \mathbf{M}_{nmp} \rangle}{k_{np}^2 - k^2} \mathbf{M}_{nmp}(\mathbf{r}) \\ & + \frac{\langle \mathbf{J}_e, \mathbf{N}_{nmp} \rangle}{k_{np}'^2 - k^2} \mathbf{N}_{nmp}(\mathbf{r}) - \frac{\langle \mathbf{J}_e, \mathbf{L}_{nmp} \rangle}{k^2} \mathbf{L}_{nmp}(\mathbf{r}) \qquad (10.23)\end{aligned}$$

or the form

$$\mathbf{E}(\mathbf{r}) = -i\omega\mu \int_\Omega \underline{\mathbf{G}}_{ee}(\mathbf{r}, \mathbf{r}', k^2) \cdot \mathbf{J}_e(\mathbf{r}') \, d\Omega', \tag{10.24}$$

where

$$\begin{aligned}\underline{\mathbf{G}}_{ee}(\mathbf{r}, \mathbf{r}', k^2) = \sum_n \sum_m \sum_p & \frac{\mathbf{M}_{nmp}(\mathbf{r}) \mathbf{M}_{nmp}(\mathbf{r}')}{k_{np}^2 - k^2} \\ & + \frac{\mathbf{N}_{nmp}(\mathbf{r}) \mathbf{N}_{nmp}(\mathbf{r}')}{k_{np}'^2 - k^2} - \frac{\mathbf{L}_{nmp}(\mathbf{r}) \mathbf{L}_{nmp}(\mathbf{r}')}{k^2}. \qquad (10.25)\end{aligned}$$

From (4.63) we have the completeness relation

$$\begin{aligned}\underline{\mathbf{I}}\,\delta(\mathbf{r} - \mathbf{r}') = & \frac{-1}{2\pi i} \oint_C \underline{\mathbf{G}}_{ee}(\mathbf{r}, \mathbf{r}', \lambda) \, d\lambda \qquad (10.26) \\ = & \sum_n \sum_m \sum_p \{ \mathbf{M}_{nmp}(\mathbf{r}) \mathbf{M}_{nmp}(\mathbf{r}') + \mathbf{N}_{nmp}(\mathbf{r}) \mathbf{N}_{nmp}(\mathbf{r}') \\ & + \mathbf{L}_{nmp}(\mathbf{r}) \mathbf{L}_{nmp}(\mathbf{r}') \},\end{aligned}$$

where the usual conjugate on the terms involving $\mathbf{r}'$ is not necessary since the eigenfunctions are real-valued, and therefore we may use a pseudo inner product. Dyadic Green's functions for rectangular cavities are extensively discussed in Section 9.2.

The interested reader can consult [36], [38], and [83] for a thorough treatment of vector wavefunctions for rectangular, cylindrical, and spherical cavities.

## 10.3 Integral Equation Methods for Cavities

For generally shaped cavities, and for cavities filled with an inhomogeneous isotropic lossless or anisotropic Hermitian medium, the above eigenfunction techniques continue to hold in theory. However, vector eigenfunctions will be extremely difficult to determine in closed form. In these instances it is preferable to utilize noneigenfunction techniques, usually numerical in nature, such as finite-element or finite-difference methods, or integral-equation methods. Here we will briefly discuss the formulation of integral equations for generally shaped cavities, although we assume for simplicity that the cavity is filled with a homogeneous isotropic medium. The interior problem is of primary interest, although the cavity may have apertures that couple energy into the exterior region.

Consider an arbitrarily shaped, three-dimensional simply connected volume $\Omega$ bounded by a sufficiently smooth surface $\Gamma$. The medium interior to $\Omega$ is characterized by homogeneous isotropic permittivity $\varepsilon_1$ and permeability $\mu_1$, whereas the medium external to $\Omega$ is characterized by homogeneous isotropic permittivity $\varepsilon_2$ and permeability $\mu_2$. The wavenumber in either region is $k_i = \omega\sqrt{\mu_i \varepsilon_i}$ for $i = 1, 2$. Electromagnetic sources $\mathbf{J}_e$ exist interior to $\Omega$, and the unit normal points inward to $\Omega$.

Referring to the material in Section 1.3.5, we assume the electric field in region $i$ satisfies the vector wave equation

$$\nabla \times \nabla \times \mathbf{E}_i(\mathbf{r}) - k^2 \mathbf{E}_i(\mathbf{r}) = -i\omega\mu_i \mathbf{J}_e(\mathbf{r})\delta_{i1} \tag{10.27}$$

for $\mathbf{r}$ in region $i$, where $\delta_{i1}$ is the Kronecker delta-function. Defining an electric dyadic Green's function by

$$\nabla \times \nabla \times \underline{\mathbf{G}}_i(\mathbf{r}, \mathbf{r}') - k_i^2 \underline{\mathbf{G}}_i(\mathbf{r}, \mathbf{r}') = \underline{\mathbf{I}}\,\delta(\mathbf{r} - \mathbf{r}'), \tag{10.28}$$

where $\mathbf{r}$ is in region $i$ and $\mathbf{r}'$ is arbitrary, and using the vector-dyadic Green's second theorem from the appendix, lead to the solution of (10.27) as

$$\left\{ \begin{array}{ll} \mathbf{r}' \in \Omega & \mathbf{E}_1(\mathbf{r}') \\ \mathbf{r}' \notin \Omega & 0 \end{array} \right\} = -i\omega\mu_1 \int_\Omega \mathbf{J}_e(\mathbf{r}) \cdot \underline{\mathbf{G}}_1(\mathbf{r}, \mathbf{r}')\, d\Omega \tag{10.29}$$

$$+ \oint_\Gamma \left\{ (\mathbf{n} \times \mathbf{E}_1(\mathbf{r})) \cdot \nabla \times \underline{\mathbf{G}}_1(\mathbf{r}, \mathbf{r}') + (\mathbf{n} \times \nabla \times \mathbf{E}_1(\mathbf{r})) \cdot \underline{\mathbf{G}}_1(\mathbf{r}, \mathbf{r}') \right\} d\Gamma$$

and

$$\left\{ \begin{array}{ll} \mathbf{r}' \notin \Omega & \mathbf{E}_2(\mathbf{r}') \\ \mathbf{r}' \in \Omega & 0 \end{array} \right\} = -\oint_\Gamma \left\{ (\mathbf{n} \times \mathbf{E}_2(\mathbf{r})) \cdot \nabla \times \underline{\mathbf{G}}_2(\mathbf{r}, \mathbf{r}') \right.$$

$$\left. + (\mathbf{n} \times \nabla \times \mathbf{E}_2(\mathbf{r})) \cdot \underline{\mathbf{G}}_2(\mathbf{r}, \mathbf{r}') \right\} d\Gamma. \tag{10.30}$$

For $\mathbf{r}' \notin \Omega$ we assume that the Green's function satisfies a dyadic radiation condition or fitness condition.

In (10.29) the volume term includes the depolarizing dyadic as discussed in Section 1.3.5, although the limiting notation is suppressed for simplicity. For instance, in the case of the free-space dyadic Green's function we have

$$\underline{\mathbf{G}}_1(\mathbf{r},\mathbf{r}') = \text{P.V.}\left[\underline{\mathbf{I}} + \frac{\nabla\nabla}{k_1^2}\right]\frac{e^{-i\,k_1\,|\mathbf{r}-\mathbf{r}'|}}{4\pi\,|\mathbf{r}-\mathbf{r}'|} - \frac{\underline{\mathbf{L}}\,(\mathbf{r}')\,\delta\,(\mathbf{r}-\mathbf{r}')}{k_1^2}, \tag{10.31}$$

where P.V. indicates that the integral is to be interpreted in the principal-value sense, i.e., as $\lim_{\delta\to 0}\int_{\Omega-\Omega_\delta}(\cdot)$, and $\underline{\mathbf{L}}$ is the depolarizing dyadic (1.79). The usual situation for cavities having a complex shape is to take $\underline{\mathbf{G}}_1 = \underline{\mathbf{G}}_2$ as the free-space dyadic Green's function, which will be assumed here; however, if other Green's functions can be found, then the surface integral terms in (10.29) and (10.30) may be simplified.

For arbitrary bodies it is usually convenient to take the null equations in (10.29) and (10.30) as the desired set of integral equations. These are coupled via the boundary conditions

$$\begin{aligned}\mathbf{n}\times\mathbf{E}_1(\mathbf{r})|_\Gamma &= \mathbf{n}\times\mathbf{E}_2(\mathbf{r})|_\Gamma\,,\\ \mathbf{n}\times\mathbf{H}_1(\mathbf{r})|_\Gamma &= \mathbf{n}\times\mathbf{H}_2(\mathbf{r})|_\Gamma\,,\end{aligned} \tag{10.32}$$

which can be solved for the unknown surface currents $\mathbf{n}\times\mathbf{E}$ and $\mathbf{n}\times\nabla\times\mathbf{E} = -i\omega\mu\mathbf{n}\times\mathbf{H}$.

In general, these surface integral equations hold for $\Omega$ an arbitrary body, including the case where $\Gamma$ is the interface between two differing dielectrics in the absence of any conductors. However, two special cases are of particular interest for the cavity problem; their descriptions follow.

### Perfectly Conducting Cavities without Apertures

First, assume that the cavity surface $\Gamma$ is a perfect conductor with no apertures. Then we are interested in the case $\mathbf{r}'\in\Omega$, and, since $\mathbf{n}\times\mathbf{E}_1(\mathbf{r})|_\Gamma = 0$, (10.29) becomes

$$\mathbf{E}_1(\mathbf{r}') = -i\omega\mu_1\int_\Omega \mathbf{J}_e(\mathbf{r})\cdot\underline{\mathbf{G}}_1(\mathbf{r},\mathbf{r}')\,d\Omega + \oint_\Gamma(\mathbf{n}\times\nabla\times\mathbf{E}_1(\mathbf{r}))\cdot\underline{\mathbf{G}}_1(\mathbf{r},\mathbf{r}')\,d\Gamma,$$

where we assume $\underline{\mathbf{G}}_1$ is the free-space dyadic Green's function (10.31). In particular, taking the cross-product with the unit normal vector $\mathbf{n}$ and evaluating the resulting equation for $\mathbf{r}'\in\Gamma$ lead to

$$\begin{aligned}\mathbf{0} = -i\omega\mu_1 &\int_\Omega \left(\mathbf{n}\times\underline{\mathbf{G}}_1(\mathbf{r},\mathbf{r}')\right)\cdot\mathbf{J}_e(\mathbf{r})\cdot\,d\Omega\\ &+\oint_\Gamma\left(\mathbf{n}\times\underline{\mathbf{G}}_1(\mathbf{r},\mathbf{r}')\right)\cdot(\mathbf{n}\times\nabla\times\mathbf{E}_1(\mathbf{r}))\,d\Gamma,\end{aligned} \tag{10.33}$$

where we have used the symmetry property (1.85),

$$[\underline{\mathbf{G}}(\mathbf{r},\mathbf{r}')]^\top = \underline{\mathbf{G}}(\mathbf{r}',\mathbf{r}).$$

We assume $\mathbf{J}_e$ is known. With

$$\mathbf{n} \times \nabla \times \mathbf{E}_1|_\Gamma = -i\omega\mu_1 \mathbf{n} \times \mathbf{H}_1|_\Gamma \equiv -i\omega\mu_1 \mathbf{J}_{es},$$

where $\mathbf{J}_{es}$ is an unknown surface current, we have

$$\oint_\Gamma \left(\mathbf{n} \times \underline{\mathbf{G}}_1(\mathbf{r},\mathbf{r}')\right) \cdot \mathbf{J}_{es}(\mathbf{r})\, d\Gamma = -\int_\Omega \left(\mathbf{n} \times \underline{\mathbf{G}}_1(\mathbf{r},\mathbf{r}')\right) \cdot \mathbf{J}_e(\mathbf{r})\, d\Omega, \quad (10.34)$$

interpreted as a nonhomogeneous first-kind integral equation for the unknown surface current caused by the known source $\mathbf{J}_e$.

The natural resonances of the cavity are found by setting $\mathbf{J}_e = \mathbf{0}$, leading to the homogeneous form

$$\oint_\Gamma \left(\mathbf{n} \times \underline{\mathbf{G}}_1(\mathbf{r},\mathbf{r}')\right) \cdot \mathbf{J}_{es}^r(\mathbf{r})\, d\Gamma = \mathbf{0}, \quad (10.35)$$

which can be interpreted as a nonstandard eigenvalue problem for the resonant frequency $\omega$. The resonant cavity mode current $\mathbf{J}_{es}^r$ maintains a nontrivial resonant cavity eigenfield $\mathbf{E}^r$ interior to $\Omega$ such that $\mathbf{n} \times \mathbf{E}^r(\mathbf{r})|_\Gamma = 0$, and the associated exterior field for $\mathbf{r}' \notin \Omega$ is null.

### Perfectly Conducting Cavities with Apertures

Another special case of interest is when the surface $\Gamma$ is a perfect conductor with an aperture. Let $\Gamma = \Gamma_{na} \cup \Gamma_a$, where $\Gamma_{na}$ is the perfectly conducting portion of the surface and $\Gamma_a$ is the aperture surface. In this case $\mathbf{n} \times \mathbf{E}(\mathbf{r})|_{\Gamma_{na}} = 0$ and $\mathbf{n} \times \mathbf{E}(\mathbf{r})|_{\Gamma_a} \neq 0$. The integral equations become

$$\begin{aligned} 0 = &- i\omega\mu_1 \int_\Omega \mathbf{J}_e(\mathbf{r}) \cdot \underline{\mathbf{G}}_1(\mathbf{r},\mathbf{r}')\, d\Omega \quad (10.36) \\ &+ \oint_{\Gamma_a} \left[(\mathbf{n} \times \mathbf{E}_1(\mathbf{r})) \cdot \nabla \times \underline{\mathbf{G}}_1(\mathbf{r},\mathbf{r}') + (\mathbf{n} \times \nabla \times \mathbf{E}_1(\mathbf{r})) \cdot \underline{\mathbf{G}}_1(\mathbf{r},\mathbf{r}')\right] d\Gamma \\ &+ \oint_{\Gamma_{na}} (\mathbf{n} \times \nabla \times \mathbf{E}_1(\mathbf{r})) \cdot \underline{\mathbf{G}}_1(\mathbf{r},\mathbf{r}')\, d\Gamma \end{aligned}$$

for $\mathbf{r}' \notin \Omega$ and

$$\begin{aligned} 0 = &- \oint_{\Gamma_a} \left[(\mathbf{n} \times \mathbf{E}_2(\mathbf{r})) \cdot \nabla \times \underline{\mathbf{G}}_2(\mathbf{r},\mathbf{r}') + (\mathbf{n} \times \nabla \times \mathbf{E}_2(\mathbf{r})) \cdot \underline{\mathbf{G}}_2(\mathbf{r},\mathbf{r}')\right] d\Gamma \\ &- \oint_{\Gamma_{na}} (\mathbf{n} \times \nabla \times \mathbf{E}_2(\mathbf{r})) \cdot \underline{\mathbf{G}}_2(\mathbf{r},\mathbf{r}')\, d\Gamma \quad (10.37) \end{aligned}$$

for $\mathbf{r}' \in \Omega$, where $\underline{\mathbf{G}}_{1,2}$ is the free-space dyadic Green's function for each medium. The integral equations (10.36) and (10.37) are coupled by the boundary conditions (10.32) and can be solved for the surface currents $\mathbf{n} \times \mathbf{E}_{1,2}$ and $\mathbf{n} \times \nabla \times \mathbf{E}_{1,2}$. It is also possible to formulate second-kind volume integral equations for the cavity problem, although this topic will be omitted here.

## 10.4 Cavity Resonances from Vector Potentials

It is often very important to determine the modal field patterns and modal resonance frequencies of a cavity. For instance, if a circuit is housed in a cavity for shielding purposes, knowledge of these modal parameters can be used to ensure that the circuit does not strongly interact with the cavity and affect circuit function. Cavities are also used as resonant devices, where knowledge of modal frequencies and modal field patterns is obviously of paramount importance. For the analysis of complicated cavity geometries, the integral equation formulation, solved numerically, or some other numerical method is usually necessary. These techniques can be used either for the source-driven problem or for resonance problems. For cavities having a simple shape corresponding to a separable coordinate system, and filled with a relatively simple medium, vector eigenfunction techniques are often practical and useful. However, the analysis of resonance (i.e., nonsource-driven) problems for simple cavities filled with an isotropic homogeneous medium is particularly straightforward using a scalar approach [84], which is considered in this section.

Assume that $\Omega$ forms a three-dimensional, simply connected cavity, where $\Gamma$ is its perfectly conducting surface. The medium in the interior of the cavity is assumed to be homogeneous and isotropic.

The simplest formulation is to introduce the source-free vector potentials from (1.37),

$$(\nabla^2 + k^2)\begin{pmatrix} \mathbf{A} \\ \mathbf{F} \end{pmatrix} = \begin{pmatrix} \mathbf{0} \\ \mathbf{0} \end{pmatrix}, \tag{10.38}$$

and construct the fields from (1.38),

$$\begin{aligned} \mathbf{B} &= \nabla \times \mathbf{A} - i\omega\mu\varepsilon\mathbf{F} + \frac{1}{i\omega}\nabla\nabla\cdot\mathbf{F}, \\ \mathbf{E} &= -i\omega\mathbf{A} + \frac{1}{i\omega\mu\varepsilon}\nabla\nabla\cdot\mathbf{A} - \nabla\times\mathbf{F}, \end{aligned} \tag{10.39}$$

although in the spherical case we need to choose a slightly different formulation as described later. Note that the electric and magnetic fields satisfy the source-free vector wave equation

$$\begin{aligned} \nabla\times\nabla\times\mathbf{E}(\mathbf{r}) - k^2\mathbf{E}(\mathbf{r}) &= \mathbf{0}, \\ \nabla\times\nabla\times\mathbf{H}(\mathbf{r}) - k^2\mathbf{H}(\mathbf{r}) &= \mathbf{0}, \end{aligned}$$

and, because $\nabla\cdot\mathbf{E} = 0$ and $\nabla\cdot\mathbf{H} = 0$, using the vector identity $\nabla\times\nabla\times\mathbf{A} = \nabla\nabla\cdot\mathbf{A} - \nabla^2\mathbf{A}$ leads to

$$(\nabla^2 + k^2)\begin{pmatrix} \mathbf{E} \\ \mathbf{H} \end{pmatrix} = \begin{pmatrix} \mathbf{0} \\ \mathbf{0} \end{pmatrix}. \tag{10.40}$$

Therefore, comparing (10.38) and (10.40), it may seem that the method utilizing the intermediate step of the vector potentials is unduly complicated, compared to solving directly for the fields. However, it is well known that the complete vector cavity fields may be generated by a single scalar component of vector potential, and so the potential method is usually preferred.

### Rectangular Coordinate Cavities

Consider $\Omega \subset \mathbf{R}^3$ to represent the region interior to a parallelepiped with $0 \leq x \leq a$, $0 \leq y \leq b$, and $0 \leq z \leq c$. The cavity modes can be decomposed [84] into $E$-modes ($\mathrm{TM}^\alpha$) and $H$-modes ($\mathrm{TE}^\alpha$), with $\alpha$ any coordinate variable $x$, $y$, or $z$. In this case $E$-modes are generated by

$$\mathbf{A} = \widehat{\alpha} A_\alpha, \qquad \mathbf{F} = \mathbf{0},$$

and $H$-modes are generated by

$$\mathbf{A} = \mathbf{0}, \qquad \mathbf{F} = \widehat{\alpha} F_\alpha.$$

Because

$$\nabla^2 \mathbf{C} = \widehat{\mathbf{x}}\, \nabla^2 C_x + \widehat{\mathbf{y}}\, \nabla^2 C_y + \widehat{\mathbf{z}}\, \nabla^2 C_z,$$

(10.38) becomes

$$\nabla^2 \psi + k^2 \psi = 0 \tag{10.41}$$

for $\psi = A_\alpha$ or $\psi = F_\alpha$. Mathematically, it makes no difference which coordinate is chosen, although often the $z$-coordinate is preferred since the resulting cavity modes have some similarities with the usual $z$-invariant rectangular waveguide modes. The homogeneous Helmholtz equation (10.41) can be thought of as an eigenvalue equation

$$\nabla^2 \psi_n = \lambda_n \psi_n \tag{10.42}$$

with $\lambda_n = -k^2 = -k_n^2$, $n$ representing a triplet of integers. The appropriate boundary conditions on $\psi_n$ are determined from the electric field. Using $z$ as the preferred coordinate, from (10.39) we have

$$\begin{aligned}
E_x &= \frac{1}{i\omega\mu\varepsilon} \frac{\partial^2 A_z}{\partial x \partial z} - \frac{\partial F_z}{\partial y}, \\
E_y &= \frac{1}{i\omega\mu\varepsilon} \frac{\partial^2 A_z}{\partial y \partial z} + \frac{\partial F_z}{\partial x}, \\
E_z &= -i\omega A_z + \frac{1}{i\omega\mu\varepsilon} \frac{\partial^2 A_z}{\partial z^2},
\end{aligned} \tag{10.43}$$

with the boundary conditions $\mathbf{n} \times \mathbf{E} = \mathbf{0}$ on all six sides of the cavity. The eigenvalue problem (10.42) is recognized as being self-adjoint in $\mathbf{L}^2(\Omega)$, involving the negative operator $\nabla^2$. Therefore, from Theorems 4.16 and 4.19 we expect the eigenvalues to be real-valued, negative quantities.

Using separation of variables, (10.42) is solved for $E$-modes as

$$A_z = \psi^E_{mnp}(\mathbf{r}) = \sin\frac{m\pi x}{a}\sin\frac{n\pi y}{b}\cos\frac{p\pi z}{c}$$

with $m,n = 1,2,3,\ldots$ and $p = 0,1,2,\ldots$, and for $H$-modes as

$$F_z = \psi^H_{mnp}(\mathbf{r}) = \cos\frac{m\pi x}{a}\cos\frac{n\pi y}{b}\sin\frac{p\pi z}{c}$$

with $m,n = 0,1,2,\ldots$, except $m = n = 0$, and $p = 1,2,3,\ldots$.

In either case eigenvalues are clearly

$$\lambda^{E,H}_{nmp} = -k^2_{nmp} = -\left[\left(\frac{m\pi}{a}\right)^2 + \left(\frac{n\pi}{b}\right)^2 + \left(\frac{p\pi}{c}\right)^2\right],$$

leading to the cavity resonances

$$\omega^{E,H}_{nmp} = \frac{1}{\sqrt{\mu\varepsilon}}\sqrt{\left(\frac{m\pi}{a}\right)^2 + \left(\frac{n\pi}{b}\right)^2 + \left(\frac{p\pi}{c}\right)^2}.$$

It is obvious that modal degeneracies will occur among $E$- and $H$-modes and within $E$- and $H$-mode sets. Note that (10.43) can generate the solenoidal vector eigenfunctions $\mathbf{M}$ and $\mathbf{N}$, but not the irrotational eigenfunctions $\mathbf{L}$.

### Cylindrical Coordinate Cavities

Consider $\Omega \subset \mathbf{R}^3$ to represent the region interior to a finite-length cylinder having circular cross-section with $0 \leq \rho \leq a$, $0 \leq \phi \leq 2\pi$, and $0 \leq z \leq c$. Although other decompositions are possible, cavity modes are most conveniently decomposed into $E$-modes ($\mathrm{TM}^z$) and $H$-modes ($\mathrm{TE}^z$). As in the rectangular coordinate case, $E$-modes are generated by

$$\mathbf{A} = \widehat{\mathbf{z}}A_z, \qquad \mathbf{F} = \mathbf{0},$$

and $H$-modes are generated by

$$\mathbf{A} = \mathbf{0}, \qquad \mathbf{F} = \widehat{\mathbf{z}}F_z.$$

Because

$$\nabla^2\mathbf{C} = \widehat{\rho}\left(\nabla^2 C_\rho - \frac{C_\rho}{\rho^2} - \frac{2}{\rho^2}\frac{\partial C_\phi}{\partial\phi}\right) + \widehat{\phi}\left(\nabla^2 C_\phi - \frac{C_\phi}{\rho^2} + \frac{2}{\rho^2}\frac{\partial C_\rho}{\partial\phi}\right) + \widehat{\mathbf{z}}\nabla^2 C_z,$$

(10.38) becomes

$$\nabla^2\psi + k^2\psi = 0 \tag{10.44}$$

for $\psi = A_z$ or $\psi = F_z$, again interpreted as an eigenvalue equation

$$\nabla^2 \psi_n = \lambda_n \psi_n \tag{10.45}$$

with $\lambda_n = -k^2 = -k_n^2$. The appropriate boundary conditions on $\psi_n$ are determined from the electric field. From (10.39) we have

$$\begin{aligned} E_\rho &= \frac{1}{i\omega\mu\varepsilon} \frac{\partial^2 A_z}{\partial\rho\partial z} - \frac{1}{\rho}\frac{\partial F_z}{\partial\phi}, \\ E_\phi &= \frac{1}{i\omega\mu\varepsilon} \frac{1}{\rho}\frac{\partial^2 A_z}{\partial\phi\partial z} + \frac{\partial F_z}{\partial\rho}, \\ E_z &= -i\omega A_z + \frac{1}{i\omega\mu\varepsilon}\frac{\partial^2 A_z}{\partial z^2}, \end{aligned} \tag{10.46}$$

with the boundary conditions $\mathbf{n} \times \mathbf{E} = \mathbf{0}$ on all three sides of the cavity. As with the rectangular coordinate scalar eigenvalue problem, the eigenvalue problem (10.45) is self-adjoint with real-valued, negative eigenvalues.

Expressing (10.45) as

$$\nabla^2 \psi_n = \frac{1}{\rho}\frac{\partial}{\partial\rho}\left(\rho\frac{\partial\psi_n}{\partial\rho}\right) + \frac{1}{\rho^2}\frac{\partial^2\psi_n}{\partial\phi^2} + \frac{\partial^2\psi_n}{\partial z^2} = \lambda_n\psi_n$$

and using separation of variables (see Section 5.4.1), for $E$-modes we obtain

$$A_z = \psi_{npq}^E(\mathbf{r}) = J_n(k_{np}\rho)\left\{\begin{array}{c}\sin n\phi \\ \cos n\phi\end{array}\right\}\cos\frac{q\pi z}{c}$$

with $n, q = 0, 1, 2, \ldots$ and $p = 1, 2, 3, \ldots$, where $k_{np}$ is such that $J_n(k_{np}a) = 0$. Therefore, $p$ indexes the various roots of $J_n(k_{np}a)$ for a given $n$. Eigenvalues are found as

$$\lambda_{nmp}^E = -k_{nmp}^2 = -\left[(k_{np})^2 + \left(\frac{q\pi}{c}\right)^2\right],$$

leading to the cavity resonances

$$\omega_{nmp}^E = \frac{1}{\sqrt{\mu\varepsilon}}\sqrt{(k_{np})^2 + \left(\frac{q\pi}{c}\right)^2}.$$

For $H$-modes we obtain

$$F_z = \psi_{npq}^H(\mathbf{r}) = J_n(k'_{np}\rho)\left\{\begin{array}{c}\sin n\phi \\ \cos n\phi\end{array}\right\}\sin\frac{q\pi z}{c}$$

with $n = 0, 1, 2, \ldots$ and $p, q = 1, 2, 3, \ldots$, and where $k'_{np}$ is such that $J'_n(k'_{np}a) = 0$. The prime indicates differentiation with respect to the argument, i.e., $J'_n(c\rho) = \partial J_n(c\rho)/\partial(c\rho)$. Eigenvalues are

$$\lambda_{nmp}^H = -k_{nmp}^2 = -\left[(k'_{np})^2 + \left(\frac{q\pi}{c}\right)^2\right],$$

leading to the cavity resonances

$$\omega_{nmp}^{H} = \frac{1}{\sqrt{\mu\varepsilon}} \sqrt{\left(k'_{np}\right)^2 + \left(\frac{q\pi}{c}\right)^2}.$$

As in the rectangular coordinate case, (10.46) can generate the solenoidal vector eigenfunctions **M** and **N** for a finite-length circular cylinder.

### Spherical Coordinate Cavities

Consider $\Omega \subset \mathbf{R}^3$ to represent the region interior to a sphere with $0 \leq r \leq a$, $0 \leq \phi \leq 2\pi$, and $0 \leq \theta \leq \pi$. In this case spherical cavity modes are conveniently decomposed into $E$-modes ($\mathrm{TM}^r$) and $H$-modes ($\mathrm{TE}^r$), with $E$-modes generated by

$$\mathbf{A} = \widehat{\mathbf{r}} A_r, \qquad \mathbf{F} = \mathbf{0},$$

and $H$-modes by

$$\mathbf{A} = \mathbf{0}, \qquad \mathbf{F} = \widehat{\mathbf{r}} F_r.$$

However, because $\left(\nabla^2 \mathbf{C}\right)_r \neq \nabla^2 C_r$, the analysis is not as straightforward as in the rectangular and cylindrical cases. In fact, it turns out that the Lorenz gauge $\phi_e = (i/(\omega\mu\varepsilon))\nabla \cdot \mathbf{A}$ $(\phi_m = (i/(\omega\mu))\nabla \cdot \mathbf{F})$ is not convenient for this problem. Therefore, rather than (10.38), which is obtained from the Lorenz gauge, we consider the more general (1.29) and (1.34). With $\mathbf{A} = \widehat{\mathbf{r}} A_r$ $(\mathbf{F} = \widehat{\mathbf{r}} F_r)$, and the choice $\phi_e = (i/(\omega\mu\varepsilon))\frac{\partial A_r}{\partial r}$ $(\phi_m = (i/(\omega\mu))\frac{\partial F_r}{\partial r})$ [84, pp. 267–268], the $\widehat{\theta}$- and $\widehat{\phi}$-components of (1.29) and (1.34) are identically satisfied, and the $\widehat{\mathbf{r}}$-components become

$$\left(\frac{\partial^2}{\partial r^2} + \frac{1}{r^2 \sin\theta} \frac{\partial}{\partial \theta}\left(\sin\theta \frac{\partial}{\partial r}\right) + \frac{1}{r^2 \sin^2\theta} \frac{\partial^2}{\partial \phi^2} + k^2\right) \left(\begin{array}{c} A_r \\ F_r \end{array}\right) = 0. \tag{10.47}$$

It is easy to see that the above are the same as

$$\left(\nabla^2 + k^2\right) \left(\begin{array}{c} \frac{A_r}{r} \\ \frac{F_r}{r} \end{array}\right) = 0.$$

Therefore, with $A_r = r\psi^E$ and $F_r = r\psi^H$, and letting $\psi$ represent either $\psi^E$ or $\psi^H$, we obtain

$$\left(\nabla^2 + k^2\right) \psi = 0. \tag{10.48}$$

The Helmholtz equation (10.48) is recognized as an eigenvalue equation involving $\psi = \psi_n$,

$$\nabla^2 \psi_n = \lambda_n \psi_n \tag{10.49}$$

with $\lambda_n = -k^2 = -k_n^2$. The appropriate boundary conditions on $\psi_n$ are determined from the electric field. From (1.27), (1.28), (1.32), and (1.33) we have

$$\mathbf{E} = -i\omega\mathbf{A} + \frac{1}{i\omega\mu\varepsilon}\nabla\frac{\partial A_r}{\partial r} - \nabla\times\mathbf{F},$$
$$\mathbf{H} = -i\omega\varepsilon\mathbf{F} + \frac{1}{i\omega\mu}\nabla\frac{\partial F_r}{\partial r} + \frac{1}{\mu}\nabla\times\mathbf{A}, \tag{10.50}$$

or, in component form for the electric field

$$E_r = -i\omega A_r + \frac{1}{i\omega\mu\varepsilon}\frac{\partial^2 A_r}{\partial r^2},$$
$$E_\theta = \frac{1}{i\omega\mu\varepsilon}\frac{1}{r}\frac{\partial^2 A_r}{\partial r\partial\theta} - \frac{1}{r\sin\theta}\frac{\partial F_r}{\partial\phi}, \tag{10.51}$$
$$E_\phi = \frac{1}{i\omega\mu\varepsilon}\frac{1}{r\sin\theta}\frac{\partial^2 A_r}{\partial r\partial\phi} + \frac{1}{r}\frac{\partial F_r}{\partial\theta},$$

and for the magnetic field

$$H_r = -i\omega\varepsilon F_r + \frac{1}{i\omega\mu}\frac{\partial^2 F_r}{\partial r^2},$$
$$H_\theta = \frac{1}{i\omega\mu}\frac{1}{r}\frac{\partial^2 F_r}{\partial r\partial\theta} + \frac{1}{r\mu\sin\theta}\frac{\partial A_r}{\partial\phi}, \tag{10.52}$$
$$H_\phi = \frac{1}{i\omega\mu}\frac{1}{r\sin\theta}\frac{\partial^2 F_r}{\partial r\partial\phi} - \frac{1}{r\mu}\frac{\partial A_r}{\partial\theta},$$

with the boundary conditions $\mathbf{n}\times\mathbf{E}|_\Gamma = \mathbf{0}$ on the spherical surface. As with the rectangular and cylindrical coordinate scalar eigenvalue problems, the eigenvalue problem (10.49) is self-adjoint in $\mathbf{L}^2(\Omega)$ with real-valued, negative eigenvalues.

Expressing (10.49) as

$$\nabla^2\psi_n = \frac{1}{r^2}\frac{\partial}{\partial r}\left(r^2\frac{\partial\psi_n}{\partial r}\right) + \frac{1}{r^2\sin\theta}\frac{\partial}{\partial\theta}\left(\sin\theta\frac{\partial\psi_n}{\partial\theta}\right) + \frac{1}{r^2\sin^2\theta}\frac{\partial^2\psi_n}{\partial\phi^2} = \lambda_n\psi_n$$

and using separation of variables, for $E$-modes we obtain

$$A_r = r\psi_{mnp}^E(\mathbf{r}) = \left(k'_{np}\rho\right)j_n\left(k'_{np}\rho\right)\left\{\begin{array}{c}\sin m\phi\\ \cos m\phi\end{array}\right\}P_n^m(\cos\theta)$$

with $m = 0, 1, 2, \ldots$ and $n, p = 1, 2, 3, \ldots$, where $k'_{np}$ is such that

$$j'_n\left(k'_{np}a\right) = 0.$$

Eigenvalues are found as

$$\lambda_{np}^E = -\left(k'_{np}\right)^2,$$

which are independent of $m$, leading to the cavity resonances

$$\omega_{np}^{E} = \frac{k'_{np}}{\sqrt{\mu\varepsilon}}.$$

For $H$-modes we obtain

$$F_r = r\psi_{mnp}^{H}(\mathbf{r}) = (k_{np}\rho)\, j_n(k_{np}\rho) \left\{ \begin{array}{c} \sin m\phi \\ \cos m\phi \end{array} \right\} P_n^m(\cos\theta)$$

with $m = 0, 1, 2, \ldots$ and $n, p = 1, 2, 3, \ldots$, where $k_{np}$ is such that

$$j_n(k_{np}a) = 0.$$

Eigenvalues are found as

$$\lambda_{np}^{H} = -(k_{np})^2,$$

which are independent of $m$, leading to the cavity resonances

$$\omega_{np}^{H} = \frac{k_{np}}{\sqrt{\mu\varepsilon}}.$$

As with the other cases, (10.51) can generate the solenoidal eigenfunctions $\mathbf{M}$ and $\mathbf{N}$ for a spherical cavity. Note also that while the eigenvalue will be real-valued, if material loss exists the resonance frequency will become complex.

# Bibliography for Part II


[1] Peterson, A.F., Ray, S.L., and Mittra, R. (1998). *Computational Methods for Electromagnetics*, New York: IEEE Press.

[2] Morita, N., Kumagai, N., and Mautz, J.R. (1990). *Integral Equation Methods for Electromagnetics*, Boston: Artech House.

[3] Jackson, J.D. (1975). *Classical Electrodynamics*, 2nd ed., New York: Wiley.

[4] Mikhlin, S.G. (1970). *Mathematical Physics, An Advanced Course*, Amsterdam: North-Holland.

[5] Kellogg, O.D. (1929). *Foundations of Potential Theory*, New York: Dover.

[6] Colton, D. (1988). *Partial Differential Equations*, New York: Random House/Birkhäuser.

[7] Stakgold, I. (1967). *Boundary Value Problems of Mathematical Physics*, Vol. II, New York: Macmillan.

[8] Dudley, D.G. (1994). *Mathematical Foundations for Electromagnetic Theory*, New York: IEEE Press.

[9] Morse, P.M. and Feshbach, H. (1953). *Methods of Theoretical Physics*, Vol. I, New York: McGraw-Hill.

[10] DiBenedetto, E. (1995). *Partial Differential Equations*, Boston: Birkhäuser.

[11] Polyanin, A.D. and Manzhirov, A.V. (1998). *Handbook of Integral Equations*, Boca Raton, FL: CRC Press.

[12] Shestopalov, Yu.V., Smirnov, Yu. G., and Chernokozhin, E.V. (2000). *Logarithmic Integral Equations in Electromagnetics*, Utrecht: VSP.

[13] Ramm, A.G. (1980). *Iterative Methods for Calculating Static Fields and Wave Scattering by Small Bodies*, New York: Springer-Verlag.

[14] Paul, C.R. (1992). *Introduction to Electromagnetic Compatibility*, New York: Wiley.

[15] Pozar, D.M. (1998). *Microwave Engineering*, 2nd ed., New York: Wiley.

[16] Collin, R.E. (1992). *Foundations for Microwave Engineering*, 2nd ed., New York: McGraw-Hill.

[17] Felson, L.B. and Marcuvitz, N. (1994). *Radiation and Scattering of Waves*, New York: IEEE Press.

[18] Van Bladel, J. (1991). *Singular Electromagnetic Fields and Sources*, Oxford: Clarendon Press.

[19] Baum, C.E., Liu, T.K., and Tesche, F.M. (1978). On the analysis of general multiconductor transmission-line networks, *Interaction Note 350,* Phillips Laboratory Note Series, Kirtland AFB.

[20] Tesche, F.M., Liu, T.K., Chang, S.K., and Giri, D.V. (1978). Field excitation of multiconductor transmission lines, *Interaction Note 351*, Phillips Laboratory Note Series, Kirtland AFB.

[21] Paul, C.R. (1976). Frequency response of multiconductor transmission lines illuminated by an electromagnetic field, *IEEE Trans. Electromag. Compatibility*, Vol. EMC-18, pp. 183–190, Nov.

[22] Hanson, G.W. and Yakovlev, A.B. (1998). An analysis of leaky-wave dispersion phenomena in the vicinity of cutoff using complex frequency plane singularities, *Radio Science*, Vol. 33, no. 4, pp. 803–820, July–Aug.

[23] Shevchenko, V.V. (1971). *Continuous Transitions in Open Waveguides*, Boulder: Golem Press.

[24] Kato, T. (1966). *Perturbation Theory for Linear Operators*, New York: Springer-Verlag.

[25] Friedman, B. (1956). *Principles and Techniques of Applied Mathematics*, New York: Dover.

[26] Papoulis, A. (1962). *The Fourier Integral and Its Applications*, New York: McGraw-Hill.

[27] Hanson, G.W. and Yakovlev, A.B. (1999). Investigation of mode interaction on planar dielectric waveguides with loss and gain, *Radio Science*, Vol. 34, no. 6, pp. 1349–1359, Nov.–Dec.

[28] Livernois, T.G. and Nyquist, D.P. (1987). Integral-equation formulation for scattering by dielectric discontinunities along open-boundary dielectric waveguides, *J. Opt. Soc. Am. A*, Vol. 4, no. 7, pp. 1289–1295.

[29] Kzadri, B. and Nyquist, D.P. (1992). TE wave excitation and scattering by obstacles in guiding and surround regions of asymmetric planar dielectric waveguide, *J. Electromagnetic Waves and Applications*, Vol. 6. no. 10, pp. 1353–1379.

[30] Uzunoglu, N.K. and Fikioris, J.G. (1982). Scattering from an inhomogeneity inside a dielectric-slab waveguide, *J. Opt. Soc. Am.*, Vol. 72, pp. 628–637, May.

[31] Cottis, P.G. and Uzunoglu, N.K. (1984). Analysis of longitudinal discontinunities in dielectric-slab waveguides, *J. Opt. Soc. Am. A.*, Vol. 1, pp. 206–215, Feb.

[32] Tsalamengas, J.L. and Fikioris, J.G. (1993). TM scattering by conducting strips right on the planar interface of a three-layered medium, *IEEE Trans. Antennas Propagat.*, Vol. 41, pp. 542–555, May.

[33] Nosich, A.I. (1999). The method of analytical regularization in wave scattering and eigenvalue problems: Foundations and review of solutions, *IEEE Antennas Propagat. Magazine*, Vol. 41, pp. 34–48, June.

[34] Butler, C.M. (1985). General solutions of the narrow strip (and slot) integral equations, *IEEE Trans. Antennas Propagat.*, Vol. AP-33, pp. 1085–1090, Oct.

[35] Chew, W.C. (1990). *Waves and Fields in Inhomogeneous Media*, New York: IEEE Press.

[36] Tai, C.T. (1994). *Dyadic Green Functions in Electromagnetic Theory*, 2nd ed., New York: IEEE Press.

[37] Bagby, J.S. and Nyquist, D.P. (1987). Dyadic Green's functions for integrated electronic and optical circuits, *IEEE Trans. Microwave Theory Tech.* Vol. MTT-35, pp. 206–210, Feb.

[38] Collin, R.E. (1991). *Field Theory of Guided Waves*, 2nd ed., New York: IEEE Press.

[39] Hanson, G.W. and Baum, C.E. (1998). Perturbation formula for the natural frequencies of an object in the presence of a layered medium, *Electromagnetics*, Vol.18, pp. 333–351.

[40] Abramowitz, M. and Stegun, I. (1965). *Handbook of Mathematical Functions*, New York: Dover.

[41] Lin, S.L. and Hanson, G.W. (2000). Efficient analysis of dielectric resonators immersed in inhomogeneous media, *IEEE Trans. Microwave Theory Tech.*, Vol. 48, no. 1, pp. 84–92, Jan.

[42] Gohberg, I., Goldberg, S., and Krupnik, N. (2000). *Traces and Determinants of Linear Operators*, Basel: Birkhäuser Verlag.

[43] Newman, E.H. and Forrai, D. (1987). Scattering from a microstrip patch, *IEEE Trans. Antennas Propagat.*, Vol. AP-35, pp. 245–251, March.

[44] Jones, D.S. (1994). *Methods in Electromagnetic Wave Propagation*, 2nd ed., New York: IEEE Press.

[45] Ilyinsky, A.S., Slepyan, G. Ya., and Slepyan, A. Ya., (1993). *Propagation, Scattering and Dissipation of Electromagnetic Waves*, London: Peregrinus.

[46] Prokhoda, I.G., Dmitryuk, S.G., and Morozov, V.M. (1985). *Tensor Green's Functions with Applications in Microwave Electrodynamics*, Dnepropetrovsk State University, Dnepropetrovsk, Ukraine (in Russian).

[47] Nosich, A.I. (1994). Radiation conditions, limiting absorption principle, and general relations in open waveguide scattering, *J. Electromagn. Waves Applications,* Vol. 8, no. 3, pp. 329–353.

[48] Hsiao, G.C. and Kleinman, R.E. (1997). Mathematical foundations for error estimation in numerical solutions of integral equations in electromagnetics, *IEEE Trans. Antennas Propagat.*, Vol. 45, no. 3, pp. 316–328.

[49] Li, L.W., Kooi, P.S., Leong, M.S., Yeo, T.S., and Ho, S.L. (1995). On the eigenfunction expansion of electromagnetic dyadic Green's functions in rectangular cavities and waveguides, *IEEE Trans. Microwave Theory Tech.,* Vol. 43, no. 3, pp. 700–702, Mar.

[50] Tai, C.T. and Rozenfeld, P. (1976). Different representations of dyadic Green's functions for a rectangular cavity, *IEEE Trans. Microwave Theory Tech.,* Vol. 24, no. 9, pp. 597–601, Sept.

[51] Mahon, J.P. (1990). An alternative representation for Green's functions used in rectangular waveguide slot analysis, *J. Electromagn. Waves Applications,* Vol. 4, no. 7, pp. 661–672.

[52] Rahmat-Samii, Y. (1975). On the question of computation of the dyadic Green's function at the source region in waveguides and cavities, *IEEE Trans. Microwave Theory Tech.,* Vol. 23, no. 9, pp. 762–765, Sept.

[53] Dmitryuk, S.G. and Petrusenko, I.V. (1994). The numerical analytical method of solving three-dimensional diffraction problems in the complex geometry domain, *Mathematical Methods in EM Theory Int. Conf.,* Kharkov, Ukraine, pp. 75–78.

[54] Petrusenko, I.V. and Dmitryuk, S.G. (1995). The method of partial overlapping regions for analysis the waveguide discontinuity and antenna problems, *URSI EM Theory Int. Symp.*, St. Petersburg, Russia, pp. 52–54.

[55] Gnilenko, A.B., Yakovlev, A.B., and Petrusenko, I.V., (1997). Generalized approach to modelling shielded printed circuit transmission lines, *IEE Proc.-Microwave Antennas Propagation,* Vol. 144, no. 2, pp. 103–110.

[56] Gnilenko, A.B. and Yakovlev, A.B. (1999). Electric dyadic Green functions for applications to shielded multilayered transmission line problems, *IEE Proc.-Microwave Antennas Propagation,* Vol. 146, no. 2, pp. 111–118, Apr.

[57] *Active and Quasi-Optical Arrays for Solid-State Power Combining* (1997). Edited by R. York and Z. Popovič, New York: John Wiley and Sons.

[58] *Active Antennas and Quasi-Optical Arrays* (1999). Edited by A. Mortazawi, T. Itoh, and J. Harvey, New York: IEEE Press.

[59] Khalil, A.I. and Steer, M.B. (1999). A generalized scattering matrix method using the method of moments for electromagnetic analysis of multilayered structures in waveguide, *IEEE Trans. Microwave Theory Tech.,* Vol. 47, no. 11, pp. 2151–2157, Nov.

[60] Yakovlev, A.B., Khalil, A.I., Hicks, C.W., Mortazawi, A., and Steer, M.B. (2000). The Generalized Scattering Matrix of closely spaced strip and slot layers in waveguide, *IEEE Trans. Microwave Theory Tech.,* Vol. 48, no. 1, pp. 126–137, Jan.

[61] Ortiz, S. and Mortazawi, A. (1999). A perpendicularly-fed patch array for quasi-optical power combining, in *IEEE MTT-S Intl. Microwave Symp. Dig.,* pp. 667–670, June.

[62] Ortiz, S. and Mortazawi, A. (1999). A perpendicular aperture-fed patch antenna for quasi-optical amplifier arrays, in *IEEE AP-S Int. Symp.,* pp. 2386–2389, July.

[63] Yakovlev, A.B., Ortiz, S., Ozkar, M., Mortazawi, A., and Steer, M.B. (2000). A waveguide-based aperture coupled patch amplifier array—Full-wave system analysis and experimental validation, *IEEE Trans. Microwave Theory Tech.,* Vol. 48, no. 12, pp. 2692–2699, Dec.

[64] Li, L.W., Kooi, P.S., Leong, M.S., and Yeo, T.S. (1994). Full-wave analysis of antenna radiation in a rectangular waveguide with discontinuity: Part I—Dyadic Green's function, in *ICCS Conf.,* Singapore, Vol. 2, pp. 459–463.

[65] Jin, H., Lin, W., and Lin, Y. (1994). Dyadic Green's functions for rectangular waveguide filled with longitudinally multilayered isotropic dielectric and their application, *IEE Proc.—Microwave Antennas Propagation,* Vol. 141, no. 6, pp. 504–508, Dec.

[66] Li, L.W., Kooi, P.S., Leong, M.S., Yeo, T.S., and Ho, S.L. (1995). Input impedance of a probe-excited semi-infinite rectangular waveguide with arbitrary multilayered loads: Part I - Dyadic Green's functions, *IEEE Trans. Microwave Theory Tech.,* Vol. 43, no. 7, pp. 1559–1566, July.

[67] Piefke, G. (1970). The tridimensional "Zwischenmedium" in the field theory, *Arch. Electr. Übertr.,* Vol. 24, p. 523.

[68] Piefke, G. (1973). *Feldtheorie,* B.I.-Wissenschaftsverlag.

[69] Reisdorf, F. and Schminke, W. (1973). Application of a "Zwischenmedium" to the radiation from a flanged waveguide, *Proc. IEE,* Vol. 120, pp. 739–740.

[70] Prokhoda, I.G. and Chumachenko, V.P. (1973). The method of partial intersecting domains for the investigation of waveguide-resonator systems having a complex shape, *Radiophysics and Quantum Electronics,* Vol. 16, no. 10, pp. 1219–1222.

[71] Petrusenko, I.V., Yakovlev, A.B., and Gnilenko, A.B., (1994). Method of partial overlapping regions for the analysis of diffraction problems, *IEE Proc.—Microwave Antennas Propagation,* Vol. 141, no. 3, pp. 196–198.

[72] Yakovlev, A.B. and Gnilenko, A.B. (1997). Analysis of microstrip discontinuities using the method of integral equations for overlapping regions, *IEE Proc.—Microwave Antennas Propagation,* Vol. 144, no. 6, pp. 449–457, Dec.

[73] Kantorovich, L.V. and Krilov, V.I. (1958). *Approximate Methods of Higher Analysis,* New York: Interscience.

[74] Iskander, M.F. and Hamid, M.A.K. (1974). Analysis of triangular waveguides of arbitrary dimensions, *Arch. Electr. Übertr.,* Vol. 28, pp. 455–461.

[75] Iskander, M.F. and Hamid, M.A.K. (1976). Scattering by a regular polygonal conducting cylinder, *Arch. Electr. Übertr.,* Vol. 30, pp. 403–408.

[76] Iskander, M.F. and Hamid, M.A.K. (1976). Scattering coefficients at a waveguide-horn junction, *Proc. IEE,* Vol.123, pp. 123–127.

[77] Iskander, M.F. and Hamid, M.A.K. (1977). Iterative solutions of waveguide discontinuity problems, *IEEE Trans. Microwave Theory Tech.*, Vol. 25, pp. 763–768.

[78] Gribanov, Yu.I. (1963). The coordinate spaces and infinite systems of linear equations. III, *Izv. Vuzov.—Mathematics,* Vol. 34, no. 3, pp. 27–39 (in Russian).

[79] Yakovlev, A.B. and Hanson, G.W. (1997). On the nature of critical points in leakage regimes of a conductor-backed coplanar strip line, *IEEE Trans. Microwave Theory Tech.*, Vol. 45, no. 1, pp. 87–94, Jan.

[80] Yakovlev, A.B. and Hanson, G.W. (1998). Analysis of mode coupling on guided-wave structures using Morse critical points, *IEEE Trans. Microwave Theory Tech.*, Vol. 46, no. 7, pp. 966–974, July.

[81] Yakovlev, A.B. and Hanson, G.W. (2000). Mode transformation and mode continuation regimes on waveguiding structures, *IEEE Trans. Microwave Theory Tech.*, Vol. 48, no. 1, pp. 67–75, Jan.

[82] Morse, P.M. and Feshbach, H. (1953). *Methods of Theoretical Physics*, Vol. II, New York: McGraw-Hill.

[83] Van Bladel, J. (1985). *Electromagnetic Fields*, Washington: Hemisphere.

[84] Harrington, R.F. (1961). *Time-Harmonic Electromagnetic Fields*, New York: McGraw-Hill.

[85] Mrozowski, M. (1997). *Guided Electromagnetic Waves, Properties and Analysis*, Somerset: Research Studies Press.

[86] Stakgold, I. (1999). *Green's Functions and Boundary Value Problems*, 2nd ed., New York: Wiley.

[87] Dennery, P. and Krzywicki, A. (1995). *Mathematics for Physicists*, New York: Dover.

[88] Sneddon, I.N. (1957). *Elements of Partial Differential Equations*, New York: McGraw-Hill.

[89] Dautray, R. and Lions, J.L. (1990). *Mathematical Analysis and Numerical Methods for Science and Technology*, Vol. 3, New York: Springer-Verlag.

[90] Zabreyko, P.P., Koshelev, A.I., Krasnosel'skii, M.A., Mikhlin, S.G., Rakovshchik, L.S., and Stet'senko, V.Ya. (1975). *Integral Equations—A Reference Text,* Noordhoff International Publishing: Leyden, The Netherlands.

# A

# Vector, Dyadic, and Integral Relations

## A.1 Vector Identities

$$\mathbf{A}\cdot\mathbf{B}=\mathbf{B}\cdot\mathbf{A}$$

$$\mathbf{A}\times\mathbf{B}=-\mathbf{B}\times\mathbf{A}$$

$$\mathbf{A}\cdot(\mathbf{B}\times\mathbf{C})=\mathbf{B}\cdot(\mathbf{C}\times\mathbf{A})=\mathbf{C}\cdot(\mathbf{A}\times\mathbf{B})$$

$$\mathbf{A}\times(\mathbf{B}\times\mathbf{C})=\mathbf{B}\,(\mathbf{A}\cdot\mathbf{C})-\mathbf{C}\,(\mathbf{A}\cdot\mathbf{B})$$

$$\nabla\,(\phi\psi)=\phi\nabla\psi+\psi\nabla\phi$$

$$\nabla\cdot(\psi\mathbf{A})=\mathbf{A}\cdot\nabla\psi+\psi\nabla\cdot\mathbf{A}$$

$$\nabla\times(\psi\mathbf{A})=\nabla\psi\times\mathbf{A}+\psi\nabla\times\mathbf{A}$$

$$\nabla\cdot(\mathbf{A}\times\mathbf{B})=\mathbf{B}\cdot\nabla\times\mathbf{A}-\mathbf{A}\cdot\nabla\times\mathbf{B}$$

$$\nabla\times(\mathbf{A}\times\mathbf{B})=(\mathbf{B}\cdot\nabla)\,\mathbf{A}-(\mathbf{A}\cdot\nabla)\,\mathbf{B}+\mathbf{A}\nabla\cdot\mathbf{B}-\mathbf{B}\nabla\cdot\mathbf{A}$$

$$\nabla\,(\mathbf{A}\cdot\mathbf{B})=(\mathbf{A}\cdot\nabla)\,\mathbf{B}+(\mathbf{B}\cdot\nabla)\,\mathbf{A}+\mathbf{A}\times\nabla\times\mathbf{B}+\mathbf{B}\times\nabla\times\mathbf{A}$$

$$\nabla\times\nabla\times\mathbf{A}=\nabla\nabla\cdot\mathbf{A}-\nabla^2\mathbf{A}$$

$$\nabla\cdot\nabla\psi=\nabla^2\psi$$

$$\nabla\times\nabla\psi=\mathbf{0}$$

$$\nabla\cdot\nabla\times\mathbf{A}=0$$

## A.2 Dyadic Identities

$$(\mathbf{AB}) \cdot (\mathbf{CD}) = \mathbf{A}\,(\mathbf{B} \cdot \mathbf{C})\,\mathbf{D}$$
$$(\mathbf{AB}) \times \mathbf{C} = \mathbf{A}\,(\mathbf{B} \times \mathbf{C})$$
$$\mathbf{A} \cdot (\underline{\mathbf{C}} \cdot \mathbf{B}) = (\mathbf{A} \cdot \underline{\mathbf{C}}) \cdot \mathbf{B} = \mathbf{A} \cdot \underline{\mathbf{C}} \cdot \mathbf{B}$$
$$(\mathbf{A} \cdot \underline{\mathbf{C}}) \times \mathbf{B} = \mathbf{A} \cdot (\underline{\mathbf{C}} \times \mathbf{B}) = \mathbf{A} \cdot \underline{\mathbf{C}} \times \mathbf{B}$$
$$(\mathbf{A} \times \underline{\mathbf{C}}) \cdot \mathbf{B} = \mathbf{A} \times (\underline{\mathbf{C}} \cdot \mathbf{B})$$
$$(\mathbf{A} \times \underline{\mathbf{C}}) \times \mathbf{B} = \mathbf{A} \times (\underline{\mathbf{C}} \times \mathbf{B}) = \mathbf{A} \times \underline{\mathbf{C}} \times \mathbf{B}$$
$$(\mathbf{A} \cdot \underline{\mathbf{C}}) \cdot \underline{\mathbf{D}} = \mathbf{A} \cdot (\underline{\mathbf{C}} \cdot \underline{\mathbf{D}}) = \mathbf{A} \cdot \underline{\mathbf{C}} \cdot \underline{\mathbf{D}}$$
$$(\underline{\mathbf{C}} \cdot \underline{\mathbf{D}}) \cdot \mathbf{A} = \underline{\mathbf{C}} \cdot (\underline{\mathbf{D}} \cdot \mathbf{A}) = \underline{\mathbf{C}} \cdot \underline{\mathbf{D}} \cdot \mathbf{A}$$
$$(\underline{\mathbf{C}} \cdot \underline{\mathbf{D}}) \times \mathbf{A} = \underline{\mathbf{C}} \cdot (\underline{\mathbf{D}} \times \mathbf{A}) = \underline{\mathbf{C}} \cdot \underline{\mathbf{D}} \times \mathbf{A}$$
$$(\mathbf{A} \times \underline{\mathbf{C}}) \cdot \underline{\mathbf{D}} = \mathbf{A} \times (\underline{\mathbf{C}} \cdot \underline{\mathbf{D}}) = \mathbf{A} \times \underline{\mathbf{C}} \cdot \underline{\mathbf{D}}$$
$$(\underline{\mathbf{C}} \times \mathbf{A}) \cdot \underline{\mathbf{D}} = \underline{\mathbf{C}} \cdot (\mathbf{A} \times \underline{\mathbf{D}})$$
$$\mathbf{A} \cdot (\mathbf{B} \times \underline{\mathbf{C}}) = -\mathbf{B} \cdot (\mathbf{A} \times \underline{\mathbf{C}}) = (\mathbf{A} \times \mathbf{B}) \cdot \underline{\mathbf{C}}$$
$$(\underline{\mathbf{C}} \times \mathbf{A}) \cdot \mathbf{B} = \underline{\mathbf{C}} \cdot \mathbf{A} \times \mathbf{B} = -(\underline{\mathbf{C}} \times \mathbf{B}) \cdot \mathbf{A}$$
$$\mathbf{A} \times (\mathbf{B} \times \underline{\mathbf{C}}) = \mathbf{B}\,(\mathbf{A} \cdot \underline{\mathbf{C}}) - (\mathbf{A} \cdot \mathbf{B})\,\underline{\mathbf{C}}$$
$$\mathbf{A} \cdot \underline{\mathbf{C}} = (\underline{\mathbf{C}})^{\top} \cdot \mathbf{A} = (\mathbf{A} \cdot \underline{\mathbf{C}})^{\top}$$
$$\left(\underline{\mathbf{C}}^{\top}\right)^{\top} = \underline{\mathbf{C}}$$
$$\mathbf{A} \times \underline{\mathbf{C}} = -\left[(\underline{\mathbf{C}})^{\top} \times \mathbf{A}\right]^{\top}$$
$$\mathbf{A} \cdot \underline{\mathbf{I}} = \underline{\mathbf{I}} \cdot \mathbf{A}$$
$$\mathbf{A} \times \underline{\mathbf{I}} = \underline{\mathbf{I}} \times \mathbf{A}$$
$$(\underline{\mathbf{I}} \times \mathbf{A}) \cdot \mathbf{B} = \mathbf{A} \cdot (\underline{\mathbf{I}} \times \mathbf{B}) = \mathbf{A} \times \mathbf{B}$$
$$(\underline{\mathbf{I}} \times \mathbf{A}) \cdot \underline{\mathbf{C}} = \mathbf{A} \times \underline{\mathbf{C}}$$
$$(\mathbf{A} \times \underline{\mathbf{I}}) \cdot \mathbf{B} = \mathbf{A} \times \mathbf{B}$$
$$(\mathbf{A}\,\mathbf{B}) \cdot \mathbf{C} = \mathbf{A}\,(\mathbf{B} \cdot \mathbf{C})$$
$$\nabla\,(\psi \mathbf{A}) = \psi \nabla \mathbf{A} + (\nabla \psi)\,\mathbf{A}$$
$$\nabla \cdot (\psi \underline{\mathbf{C}}) = (\nabla \psi) \cdot \underline{\mathbf{C}} + \psi \nabla \cdot \underline{\mathbf{C}}$$
$$\nabla \times (\psi \underline{\mathbf{C}}) = \nabla \psi \times \underline{\mathbf{C}} + \psi \nabla \times \underline{\mathbf{C}}$$
$$\nabla \times \nabla \times \underline{\mathbf{C}} = \nabla\,(\nabla \cdot \underline{\mathbf{C}}) - \nabla^2 \underline{\mathbf{C}}$$
$$\nabla \cdot (\psi\,\underline{\mathbf{I}}) = \nabla \psi$$
$$\nabla \cdot (\underline{\mathbf{I}} \times \mathbf{A}) = \nabla \times \mathbf{A}$$
$$\nabla \times (\psi\,\underline{\mathbf{I}}) = \nabla \psi \times \underline{\mathbf{I}}$$
$$\nabla \times \nabla \times (\psi\,\underline{\mathbf{I}}) = \nabla\,\nabla \psi - \underline{\mathbf{I}} \nabla^2 \psi$$
$$\nabla \cdot \nabla \mathbf{A} = \nabla^2 \mathbf{A}$$
$$\nabla \cdot \nabla \times \underline{\mathbf{C}} = \mathbf{0}$$

## A.3 Dyadic Analysis

Letting $x_i$ denote the Cartesian variables $(x, y, z)$, we represent a dyadic $\underline{\mathbf{B}}$ as

$$\underline{\mathbf{B}} = \sum_{i=1}^{3}\sum_{j=1}^{3} B_{i\,j}\,\widehat{\mathbf{x}}_i\,\widehat{\mathbf{x}}_j = \sum_{j=1}^{3} \mathbf{B}_j\,\widehat{\mathbf{x}}_j = \mathbf{B}_1\widehat{\mathbf{x}} + \mathbf{B}_2\widehat{\mathbf{y}} + \mathbf{B}_3\widehat{\mathbf{z}}$$

and

$$\underline{\mathbf{B}}^{\top} = \sum_{i=1}^{3}\sum_{j=1}^{3} B_{i\,j}\,\widehat{\mathbf{x}}_j\,\widehat{\mathbf{x}}_i = \sum_{j=1}^{3} \widehat{\mathbf{x}}_j\mathbf{B}_j = \widehat{\mathbf{x}}\mathbf{B}_1 + \widehat{\mathbf{y}}\mathbf{B}_2 + \widehat{\mathbf{z}}\mathbf{B}_3.$$

Then we have the following.

$$\begin{aligned}\mathbf{A}\cdot\underline{\mathbf{B}} &= \sum_{i=1}^{3}\sum_{j=1}^{3} A_i B_{i\,j}\,\widehat{\mathbf{x}}_j = \sum_{j=1}^{3} (\mathbf{A}\cdot\mathbf{B}_j)\,\widehat{\mathbf{x}}_j \\ &= (\mathbf{A}\cdot\mathbf{B}_1)\,\widehat{\mathbf{x}} + (\mathbf{A}\cdot\mathbf{B}_2)\,\widehat{\mathbf{y}} + (\mathbf{A}\cdot\mathbf{B}_3)\,\widehat{\mathbf{z}}\end{aligned}$$

$$\begin{aligned}\underline{\mathbf{B}}\cdot\mathbf{A} &= \sum_{i=1}^{3}\sum_{j=1}^{3} A_j B_{i\,j}\,\widehat{\mathbf{x}}_i \\ &= \sum_{j=1}^{3} \mathbf{B}_j\,(\widehat{\mathbf{x}}_j\cdot\mathbf{A}) = \mathbf{B}_1\,(\widehat{\mathbf{x}}\cdot\mathbf{A}) + (\widehat{\mathbf{y}}\cdot\mathbf{A})\,\widehat{\mathbf{y}} + (\widehat{\mathbf{z}}\cdot\mathbf{A})\,\widehat{\mathbf{z}}\end{aligned}$$

$$\mathbf{A}\times\underline{\mathbf{B}} = \sum_{j=1}^{3} (\mathbf{A}\times\mathbf{B}_j)\,\widehat{\mathbf{x}}_j = (\mathbf{A}\times\mathbf{B}_1)\,\widehat{\mathbf{x}} + (\mathbf{A}\times\mathbf{B}_2)\,\widehat{\mathbf{y}} + (\mathbf{A}\times\mathbf{B}_3)\,\widehat{\mathbf{z}}$$

$$\underline{\mathbf{B}}\times\mathbf{A} = \sum_{j=1}^{3} \mathbf{B}_j\,(\widehat{\mathbf{x}}_j\times\mathbf{A}) = \mathbf{B}_1\,(\widehat{\mathbf{x}}\times\mathbf{A}) + (\widehat{\mathbf{y}}\times\mathbf{A})\,\widehat{\mathbf{y}} + (\widehat{\mathbf{z}}\times\mathbf{A})\,\widehat{\mathbf{z}}$$

$$\begin{aligned}\nabla\cdot\underline{\mathbf{B}} &= \sum_{i=1}^{3}\sum_{j=1}^{3} \frac{\partial B_{i\,j}}{\partial x_i}\,\widehat{\mathbf{x}}_j \\ &= \sum_{j=1}^{3} (\nabla\cdot\mathbf{B}_j)\,\widehat{\mathbf{x}}_j = (\nabla\cdot\mathbf{B}_1)\,\widehat{\mathbf{x}} + (\nabla\cdot\mathbf{B}_2)\,\widehat{\mathbf{y}} + (\nabla\cdot\mathbf{B}_3)\,\widehat{\mathbf{z}}\end{aligned}$$

$$\begin{aligned}\nabla\times\underline{\mathbf{B}} &= \sum_{i=1}^{3}\sum_{j=1}^{3} (\nabla B_{i\,j}\times\widehat{\mathbf{x}}_i)\,\widehat{\mathbf{x}}_j = \sum_{j=1}^{3} (\nabla\times\mathbf{B}_j)\,\widehat{\mathbf{x}}_j \\ &= (\nabla\times\mathbf{B}_1)\,\widehat{\mathbf{x}} + (\nabla\times\mathbf{B}_2)\,\widehat{\mathbf{y}} + (\nabla\times\mathbf{B}_3)\,\widehat{\mathbf{z}}\end{aligned}$$

$$\nabla \mathbf{B} = \sum_{i=1}^{3}\sum_{j=1}^{3} \frac{\partial B_j}{\partial x_i} \widehat{\mathbf{x}}_i \widehat{\mathbf{x}}_j = \sum_{j=1}^{3} (\nabla B_j)\, \widehat{\mathbf{x}}_j = (\nabla B_1)\, \widehat{\mathbf{x}} + (\nabla B_2)\, \widehat{\mathbf{y}} + (\nabla B_3)\, \widehat{\mathbf{z}}$$

In Cartesian coordinates the identity dyadic is $\mathbf{I} = \widehat{\mathbf{x}}\widehat{\mathbf{x}} + \widehat{\mathbf{y}}\widehat{\mathbf{y}} + \widehat{\mathbf{z}}\widehat{\mathbf{z}}$, and

$$\nabla \cdot (\psi \underline{\mathbf{I}}) = \nabla \psi,$$
$$\nabla \times (\psi \underline{\mathbf{I}}) = \nabla \psi \times \underline{\mathbf{I}}.$$

## A.4 Integral Identities

In the following $V$ is a volume ($\subseteq \mathbf{R}^3$) bounded by closed surface $S$, and $\mathbf{n}$ is a unit normal vector that points outward from the surface $S$, with $\mathbf{n}\, dS = d\mathbf{S}$. In two dimensions $S$ is an open surface and $\mathbf{l}$ denotes the unit tangential vector to the edge (rim) of the open surface, with $\mathbf{m}$ a unit normal vector that points outward from the open surface $S$ with $\mathbf{m}$ following the right screw rule by turning $\mathbf{l}$. For an open surface $S$ the unit vector $\mathbf{n}$ is perpendicular to the rim of the surface, locally tangent to the open surface, with $\mathbf{n} = \mathbf{l} \times \mathbf{m}$. Note that we use bold for $\mathbf{n}, \mathbf{l}, \mathbf{m}$, but other unit vectors are indicated with a caret, e.g., $\widehat{\mathbf{x}}$, $\widehat{\mathbf{y}}$, $\widehat{\mathbf{z}}$. The dimensionality of the following integral theorems should be apparent from the given form.

Various mathematical conditions are placed on the functions appearing in the following theorems. For instance, the divergence theorem

$$\int_V \nabla \cdot \mathbf{A}\, dV = \oint_S \mathbf{n} \cdot \mathbf{A}\, dS$$

is valid for any regular region[1] $V$ with the vector function $\mathbf{A}$ continuous and piecewise continuously differentiable in $V$. The condition requiring that the function be piecewise continuously differentiable on the boundary surface $S$ may be removed as long as the left-side integral in the theorem converges. Similar conditions can be stated for the various Stokes' and gradient theorems, which in turn are used to derive the Green's theorems. While in many practical cases these conditions can be considerably weakened, the discussion of the dyadic Green's theorem in Section 1.3.5 illustrates that this is not always the case, depending on the order of the singularity involved.

In general, it is assumed that the regions in question are regular and that the involved functions are continuous and sufficiently continuously differentiable for the theorems to hold. The derivation of the standard theorems may be found in most calculus books, and the elevation to vector and vector-dyadic form is described in reference [3] of Chapter 1.

[1] A regular region is a bounded closed region whose boundary is a closed, sufficiently smooth surface. In this definition the boundary is part of the region. Alternatively, often one considers open regions $\Omega$, where $\widetilde{\Omega}$ is used to include the boundary.

**Vector Gauss or Divergence Theorem**

$$\int_V \nabla \cdot \mathbf{A}\, dV = \oint_S \mathbf{n} \cdot \mathbf{A}\, dS$$
$$\int_S \nabla \cdot \mathbf{A}\, dS = \oint_l \mathbf{n} \cdot \mathbf{A}\, dl$$

**Dyadic Gauss or Divergence Theorem**

$$\int_V \nabla \cdot \underline{\mathbf{A}}\, dV = \oint_S \mathbf{n} \cdot \underline{\mathbf{A}}\, dS$$

**Stokes' Theorem**

$$\int_S \mathbf{n} \cdot \nabla \times \mathbf{A}\, dS = \oint_l \mathbf{l} \cdot \mathbf{A}\, dl$$

**Dyadic Stokes' Theorem**

$$\int_S \mathbf{n} \cdot \nabla \times \underline{\mathbf{A}}\, dS = \oint_l \mathbf{l} \cdot \underline{\mathbf{A}}\, dl$$

**Vector Stokes' Theorem**

$$\int_V \nabla \times \mathbf{A}\, dV = \oint_S \mathbf{n} \times \mathbf{A}\, dS$$

**Scalar Stokes' Theorem**

$$\int_S \mathbf{n} \times \nabla\psi\, dS = \oint_l \mathbf{l}\psi\, dl$$

**Vector Gradient Theorem**

$$\int_V \nabla\psi\, dV = \oint_S \mathbf{n}\psi\, dS$$

**Dyadic Gradient Theorem**

$$\int_V \nabla\mathbf{A}\, dV = \oint_S \mathbf{n}\,\mathbf{A}\, dS$$

**Green's First Theorem**

$$\int_V \left[\psi_1 \nabla^2 \psi_2 + \nabla\psi_1 \cdot \nabla\psi_2\right] dV = \oint_S \psi_1 \nabla\psi_2 \cdot d\mathbf{S}$$
$$\int_S \left[\psi_1 \nabla^2 \psi_2 + \nabla\psi_1 \cdot \nabla\psi_2\right] dV = \oint_l \psi_1 \nabla\psi_2 \cdot d\mathbf{l}$$

### Green's Second Theorem

$$\int_V \left[\psi_1 \nabla^2 \psi_2 - \psi_2 \nabla^2 \psi_1\right] dV = \oint_S (\psi_1 \nabla \psi_2 - \psi_2 \nabla \psi_1) \cdot d\mathbf{S}$$

$$= \oint_S \left(\psi_1 \frac{\partial \psi_2}{\partial n} - \psi_2 \frac{\partial \psi_1}{\partial n}\right) dS$$

$$\int_S \left[\psi_1 \nabla^2 \psi_2 - \psi_2 \nabla^2 \psi_1\right] dS = \oint_l (\psi_1 \nabla \psi_2 - \psi_2 \nabla \psi_1) \cdot \mathbf{n}\, dl$$

$$= \oint_l \left(\psi_1 \frac{\partial \psi_2}{\partial n} - \psi_2 \frac{\partial \psi_1}{\partial n}\right) dl$$

### Vector Green's First Theorem

$$\int_V [(\nabla \times \mathbf{A}) \cdot (\nabla \times \mathbf{B}) - \mathbf{A} \cdot \nabla \times \nabla \times \mathbf{B}]\, dV = \oint_S \mathbf{n} \cdot (\mathbf{A} \times \nabla \times \mathbf{B})\, dS$$

### Vector Green's Second Theorem

$$\int_V [\nabla \times \nabla \times \mathbf{A} \cdot \mathbf{B} - \mathbf{A} \cdot \nabla \times \nabla \times \mathbf{B}]\, dV$$

$$= \oint_S [\mathbf{A} \times (\nabla \times \mathbf{B}) + (\nabla \times \mathbf{A}) \times \mathbf{B}] \cdot d\mathbf{S}$$

### Vector-Dyadic Green's First Theorem

$$\int_V [\nabla \times \mathbf{A} \cdot \nabla \times \underline{\mathbf{B}} - \mathbf{A} \cdot \nabla \times \nabla \times \underline{\mathbf{B}}]\, dV = \oint_S \mathbf{n} \cdot (\mathbf{A} \times \nabla \times \underline{\mathbf{B}})\, dS$$

### Vector-Dyadic Green's Second Theorem

$$\int_V (\nabla \times \nabla \times \mathbf{A}) \cdot \underline{\mathbf{B}} - \mathbf{A} \cdot (\nabla \times \nabla \times \underline{\mathbf{B}})\, dV$$

$$= \oint_S \mathbf{n} \cdot [\mathbf{A} \times \nabla \times \underline{\mathbf{B}} + (\nabla \times \mathbf{A}) \times \underline{\mathbf{B}}]\, dS$$

$$= \oint_S ((\mathbf{n} \times \mathbf{A}) \cdot (\nabla \times \underline{\mathbf{B}}) + (\mathbf{n} \times \nabla \times \mathbf{A}) \cdot \underline{\mathbf{B}})\, dS$$

which can be converted, using vector and dyadic identities, to

$$\int_V [(\nabla^2 \mathbf{A}) \cdot \underline{\mathbf{B}} - \mathbf{A} \cdot \nabla^2 \underline{\mathbf{B}}]\, dV = -\oint \{(\mathbf{n} \times \mathbf{A}) \cdot (\nabla \times \underline{\mathbf{B}}) + (\mathbf{n} \times \nabla \times \mathbf{A}) \cdot \underline{\mathbf{B}}$$

$$+ [\mathbf{n} \cdot \mathbf{A}\, \nabla \cdot \underline{\mathbf{B}} - \mathbf{n} \cdot \underline{\mathbf{B}}\, \nabla \cdot \mathbf{A}]\}\, dS.$$

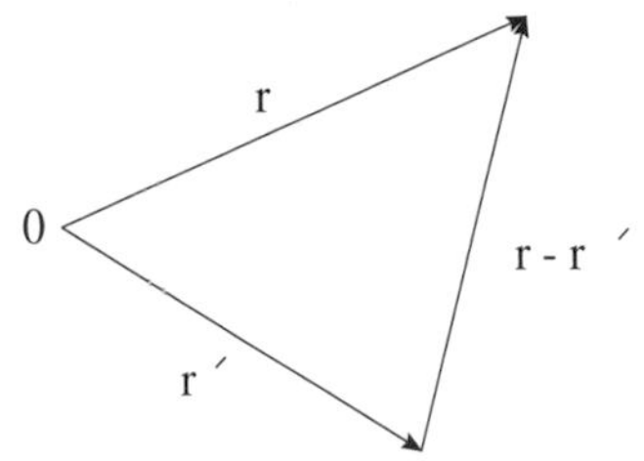

Figure A.1: Geometry of vectors $\mathbf{r}$ and $\mathbf{r}'$.

Note that Green's theorems can be considered as extensions of integration by parts to higher dimensions. Indeed, in one dimension we get

$$\int_a^b \left[\psi_1 \frac{d^2\psi_2}{dx^2} - \psi_2 \frac{d^2\psi_1}{dx^2}\right] dx = \left(\psi_1 \frac{d\psi_2}{dx} - \psi_2 \frac{d\psi_1}{dx}\right)\Bigg|_a^b .$$

# A.5 Useful Formulas Involving the Position Vector and Scalar Green's Functions

$\mathbf{r} = \sum_{i=1}^N \widehat{\mathbf{x}}_i x_i$

$\mathbf{R}(\mathbf{r},\mathbf{r}') = \mathbf{r} - \mathbf{r}'$, $\mathbf{R}(\mathbf{r}',\mathbf{r}) = \mathbf{r}' - \mathbf{r}$

$R(\mathbf{r},\mathbf{r}') = \left|\mathbf{R}(\mathbf{r},\mathbf{r}')\right| = |\mathbf{r}-\mathbf{r}'| = \left(\sum_{i=1}^N (x_i - x_i')^2\right)^{1/2} = R(\mathbf{r}',\mathbf{r})$

$\widehat{\mathbf{R}}(\mathbf{r},\mathbf{r}') = \frac{\mathbf{R}(\mathbf{r},\mathbf{r}')}{R(\mathbf{r},\mathbf{r}')} = -\widehat{\mathbf{R}}(\mathbf{r}',\mathbf{r})$

$\nabla R(\mathbf{r},\mathbf{r}') = \widehat{\mathbf{R}}(\mathbf{r},\mathbf{r}') = \nabla' R(\mathbf{r}',\mathbf{r}) = -\nabla' R(\mathbf{r},\mathbf{r}') = -\nabla R(\mathbf{r}',\mathbf{r})$

$\nabla \frac{1}{R(\mathbf{r},\mathbf{r}')} = -\frac{\nabla R(\mathbf{r},\mathbf{r}')}{R^2(\mathbf{r},\mathbf{r}')} = \nabla' \frac{1}{R(\mathbf{r}',\mathbf{r})} = -\nabla' \frac{1}{R(\mathbf{r},\mathbf{r}')} = -\nabla \frac{1}{R(\mathbf{r}',\mathbf{r})}$

$\frac{\partial}{\partial x_i} R(\mathbf{r},\mathbf{r}') = \frac{x_i - x_i'}{R(\mathbf{r},\mathbf{r}')} = \widehat{\mathbf{x}}_i \cdot \nabla R(\mathbf{r},\mathbf{r}') = -\frac{\partial}{\partial x_i'} R(\mathbf{r},\mathbf{r}')$

$\frac{\partial}{\partial x_i} \frac{1}{R(\mathbf{r},\mathbf{r}')} = -\frac{x_i - x_i'}{R^3(\mathbf{r},\mathbf{r}')} = -\frac{\partial}{\partial x_i'} \frac{1}{R(\mathbf{r},\mathbf{r}')}$

$\frac{\partial}{\partial n} \frac{1}{R(\mathbf{r},\mathbf{r}')} = \mathbf{n} \cdot \nabla \frac{1}{R(\mathbf{r},\mathbf{r}')} = \frac{-\mathbf{n}\cdot\widehat{\mathbf{R}}(\mathbf{r},\mathbf{r}')}{R^2(\mathbf{r},\mathbf{r}')} = \frac{\cos\left(\mathbf{r}'-\mathbf{r},\mathbf{n}\right)}{R^2(\mathbf{r},\mathbf{r}')}$

$\qquad = -\frac{\partial}{\partial n'} \frac{1}{R(\mathbf{r},\mathbf{r}')} = -\frac{\cos\left(\mathbf{r}-\mathbf{r}',\mathbf{n}\right)}{R^2(\mathbf{r},\mathbf{r}')}$

where $\cos(\mathbf{x},\mathbf{y})$ is the cos of the angle between the vectors $\mathbf{x}$ and $\mathbf{y}$. The vectors $\mathbf{r}$ and $\mathbf{r}'$ are depicted in Figure A.1.

Derivatives on the scalar Green's function

$$g(\mathbf{r},\mathbf{r}') = \frac{e^{-jk\,|\mathbf{r}-\mathbf{r}'|}}{4\pi\,|\mathbf{r}-\mathbf{r}'|} = g(\mathbf{r}',\mathbf{r})$$

follow from the above as

$$\nabla g(\mathbf{r},\mathbf{r}') = -g(\mathbf{r},\mathbf{r}')\left[jk + \frac{1}{R(\mathbf{r},\mathbf{r}')}\right]\nabla R(\mathbf{r},\mathbf{r}') = \nabla' g(\mathbf{r}',\mathbf{r})$$
$$= -\nabla' g(\mathbf{r},\mathbf{r}') = -\nabla g(\mathbf{r}',\mathbf{r})$$

$$\frac{\partial}{\partial x_i} g(\mathbf{r},\mathbf{r}') = \widehat{\mathbf{x}}_i \cdot \nabla g(\mathbf{r},\mathbf{r}') = -g(\mathbf{r},\mathbf{r}')\left[jk + \frac{1}{R(\mathbf{r},\mathbf{r}')}\right]\frac{x_i - x_i'}{R(\mathbf{r},\mathbf{r}')}$$
$$= -\frac{\partial}{\partial x_i'} g(\mathbf{r},\mathbf{r}')$$

## A.6 Scalar and Vector Differential Operators in the Three Principal Coordinate Systems

**Rectangular Coordinates** $(x, y, z)$

differential elements: $(dx, dy, dz)$

elemental volume: $dV = dx\,dy\,dz$

$\widehat{\mathbf{x}} = \widehat{\rho}\cos\phi - \widehat{\phi}\sin\phi = \widehat{\mathbf{r}}\cos\phi\sin\theta + \widehat{\theta}\cos\phi\cos\theta - \widehat{\phi}\sin\phi$

$\widehat{\mathbf{y}} = \widehat{\rho}\sin\phi + \widehat{\phi}\cos\phi = \widehat{\mathbf{r}}\sin\phi\sin\theta + \widehat{\theta}\sin\phi\cos\theta + \widehat{\phi}\cos\phi$

$\widehat{\mathbf{z}} = \widehat{\mathbf{r}}\cos\theta - \widehat{\theta}\sin\theta$

$\nabla f = \frac{\partial f}{\partial x}\widehat{\mathbf{x}} + \frac{\partial f}{\partial y}\widehat{\mathbf{y}} + \frac{\partial f}{\partial z}\widehat{\mathbf{z}}$

$\nabla \cdot \mathbf{A} = \frac{\partial A_x}{\partial x} + \frac{\partial A_y}{\partial y} + \frac{\partial A_z}{\partial z}$

$\nabla \times \mathbf{A} = \left(\frac{\partial A_z}{\partial y} - \frac{\partial A_y}{\partial z}\right)\widehat{\mathbf{x}} + \left(\frac{\partial A_x}{\partial z} - \frac{\partial A_z}{\partial x}\right)\widehat{\mathbf{y}} + \left(\frac{\partial A_y}{\partial x} - \frac{\partial A_x}{\partial y}\right)\widehat{\mathbf{z}}$

$\nabla^2 f = \frac{\partial^2 f}{\partial x^2} + \frac{\partial^2 f}{\partial y^2} + \frac{\partial^2 f}{\partial z^2}$

$\nabla^2 \mathbf{A} = \widehat{\mathbf{x}}\,\nabla^2 A_x + \widehat{\mathbf{y}}\,\nabla^2 A_y + \widehat{\mathbf{z}}\,\nabla^2 A_z$

**Cylindrical Coordinates** $(\rho, \phi, z)$

differential elements: $(d\rho, \rho\,d\phi,\, dz)$

differential volume: $dV = \rho\,d\rho\,d\phi\,dz$

$\widehat{\rho} = \widehat{\mathbf{x}}\cos\phi + \widehat{\mathbf{y}}\sin\phi = \widehat{\mathbf{r}}\sin\theta + \widehat{\theta}\cos\theta$

$\widehat{\phi} = -\widehat{\mathbf{x}}\sin\phi + \widehat{\mathbf{y}}\cos\phi$

$\widehat{\mathbf{z}} = \widehat{\mathbf{r}}\cos\theta - \widehat{\theta}\sin\theta$

$x = \rho\cos\phi, y = \rho\sin\phi, z = z$

$$\nabla f = \frac{\partial f}{\partial \rho}\widehat{\rho} + \frac{1}{\rho}\frac{\partial f}{\partial \phi}\widehat{\phi} + \frac{\partial f}{\partial z}\widehat{\mathbf{z}}$$

$$\nabla \cdot \mathbf{A} = \frac{1}{\rho}\frac{\partial}{\partial \rho}(\rho A_\rho) + \frac{1}{\rho}\frac{\partial A_\phi}{\partial \phi} + \frac{\partial A_z}{\partial z}$$

$$\nabla \times \mathbf{A} = \left(\frac{1}{\rho}\frac{\partial A_z}{\partial \phi} - \frac{\partial A_\phi}{\partial z}\right)\widehat{\rho} + \left(\frac{\partial A_\rho}{\partial z} - \frac{\partial A_z}{\partial \rho}\right)\widehat{\phi} + \left(\frac{1}{\rho}\frac{\partial}{\partial \rho}(\rho A_\phi) - \frac{1}{\rho}\frac{\partial A_\rho}{\partial \phi}\right)\widehat{\mathbf{z}}$$

$$\nabla^2 f = \frac{1}{\rho}\frac{\partial}{\partial \rho}\left(\rho\frac{\partial f}{\partial \rho}\right) + \frac{1}{\rho^2}\frac{\partial^2 f}{\partial \phi^2} + \frac{\partial^2 f}{\partial z^2}$$

$$\nabla^2 \mathbf{A} = \widehat{\rho}\left(\nabla^2 A_\rho - \frac{A_\rho}{\rho^2} - \frac{2}{\rho^2}\frac{\partial A_\phi}{\partial \phi}\right) + \widehat{\phi}\left(\nabla^2 A_\phi - \frac{A_\phi}{\rho^2} + \frac{2}{\rho^2}\frac{\partial A_\rho}{\partial \phi}\right) + \widehat{\mathbf{z}}\nabla^2 A_z$$

## Spherical Coordinates $(r, \theta, \phi)$

differential elements: $(dr, r\,d\theta, r\sin\theta\,d\phi)$

differential volume: $dV = r^2 \sin\theta\, dr\, d\theta\, d\phi$

$$\widehat{\mathbf{r}} = \widehat{\mathbf{x}}\sin\theta\cos\phi + \widehat{\mathbf{y}}\sin\theta\sin\phi + \widehat{\mathbf{z}}\cos\theta = \widehat{\rho}\sin\theta + \widehat{\mathbf{z}}\cos\theta$$

$$\widehat{\phi} = -\widehat{\mathbf{x}}\sin\phi + \widehat{\mathbf{y}}\cos\phi$$

$$\widehat{\theta} = \widehat{\mathbf{x}}\cos\theta\cos\phi + \widehat{\mathbf{y}}\cos\theta\sin\phi - \widehat{\mathbf{z}}\sin\theta = \widehat{\rho}\cos\theta - \widehat{\mathbf{z}}\sin\theta$$

$$x = r\sin\theta\cos\phi,\ y = r\sin\theta\sin\phi,\ z = r\cos\theta$$

$$\nabla f = \frac{\partial f}{\partial r}\widehat{\mathbf{r}} + \frac{1}{r}\frac{\partial f}{\partial \theta}\widehat{\theta} + \frac{1}{r\sin\theta}\frac{\partial f}{\partial \phi}\widehat{\phi}$$

$$\nabla \cdot \mathbf{A} = \frac{1}{r^2}\frac{\partial}{\partial r}\left(r^2 A_r\right) + \frac{1}{r\sin\theta}\frac{\partial}{\partial \theta}(A_\theta \sin\theta) + \frac{1}{r\sin\theta}\frac{\partial A_\phi}{\partial \phi}$$

$$\nabla \times \mathbf{A} = \frac{1}{r\sin\theta}\left(\frac{\partial}{\partial \theta}(A_\phi\sin\theta) - \frac{\partial A_\theta}{\partial \phi}\right)\widehat{\mathbf{r}} + \frac{1}{r}\left(\frac{1}{\sin\theta}\frac{\partial A_r}{\partial \phi} - \frac{\partial}{\partial r}(rA_\phi)\right)\widehat{\theta} + \frac{1}{r}\left(\frac{\partial}{\partial r}(rA_\theta) - \frac{\partial A_r}{\partial \theta}\right)\widehat{\phi}$$

$$\nabla^2 f = \frac{1}{r^2}\frac{\partial}{\partial r}\left(r^2\frac{\partial f}{\partial r}\right) + \frac{1}{r^2\sin\theta}\frac{\partial}{\partial \theta}\left(\sin\theta\frac{\partial f}{\partial \theta}\right) + \frac{1}{r^2\sin^2\theta}\frac{\partial^2 f}{\partial \phi^2}$$

$$\begin{aligned}\nabla^2 \mathbf{A} = \widehat{\mathbf{r}}&\left(\nabla^2 A_r - \frac{2A_r}{r^2} - \frac{2\cos\theta}{r^2}A_\theta - \frac{2}{r^2}\frac{\partial A_\theta}{\partial \theta}\right)\\ &+\widehat{\theta}\left(\nabla^2 A_\theta + \frac{2}{r^2}\frac{\partial A_r}{\partial \theta} - \frac{A_\theta}{r^2\sin^2\theta} - \frac{2\cos\theta}{r^2\sin^2\theta}\frac{\partial A_\phi}{\partial \phi}\right)\\ &+\widehat{\phi}\left(\nabla^2 A_\phi + \frac{2}{r^2\sin^2\theta}\frac{\partial A_r}{\partial \phi} - \frac{A_\phi}{r^2\sin^2\theta} + \frac{2\cos\theta}{r^2\sin^2\theta}\frac{\partial A_\theta}{\partial \phi}\right)\end{aligned}$$

# B

# Derivation of Second-Derivative Formula (1.59)

In this appendix we derive (1.59). Starting with

$$\phi(\mathbf{r}) = \int_V s(\mathbf{r}')\, g(\mathbf{r}, \mathbf{r}')\, dV'$$

and taking the first derivative yield

$$\begin{aligned}
\frac{\partial}{\partial x_i}\phi(\mathbf{r}) &= \frac{\partial}{\partial x_j} \lim_{\delta \to 0} \int_{V - V_\delta} s(\mathbf{r}')\, g(\mathbf{r}, \mathbf{r}')\, dV' \\
&= \lim_{\delta \to 0} \int_{V - V_\delta} s(\mathbf{r}') \frac{\partial}{\partial x_i} g(\mathbf{r}, \mathbf{r}')\, dV' \\
&= - \lim_{\delta \to 0} \int_{V - V_\delta} s(\mathbf{r}') \frac{\partial}{\partial x_i'} g(\mathbf{r}, \mathbf{r}')\, dV'
\end{aligned}$$

since we can bring the first derivative under the integral sign as described in Section 1.3.4. Note that $\mathbf{r} \in V_\delta$ with $V_\delta$ bounded by $S_\delta$, and $V_\delta \subset V$ as shown in Figure B.1.

Using

$$\int_V \left[ g_1 \frac{\partial g_2}{\partial x_i} + g_2 \frac{\partial g_1}{\partial x_i} \right] dV = \oint_S g_1\, g_2\, \widehat{\mathbf{x}}_i \cdot \mathbf{n}\, dS \tag{B.1}$$

(which can be derived using $\widehat{\mathbf{x}}\, g_1(x)\, g_2(x)$ as the vector function in the volume integral of the divergence theorem), which is valid because the point $\mathbf{r} = \mathbf{r}'$ is excluded from the integration, we obtain

$$\frac{\partial}{\partial x_i}\phi(\mathbf{r}) = - \lim_{\delta \to 0} \oint_{S + S_\delta} s(\mathbf{r}')\, g(\mathbf{r}, \mathbf{r}')\, \widehat{\mathbf{x}}_i \cdot \mathbf{n}\, dS' + \lim_{\delta \to 0} \int_{V - V_\delta} \frac{\partial\, s(\mathbf{r}')}{\partial x_i'}\, g(\mathbf{r}, \mathbf{r}')\, dV'.$$

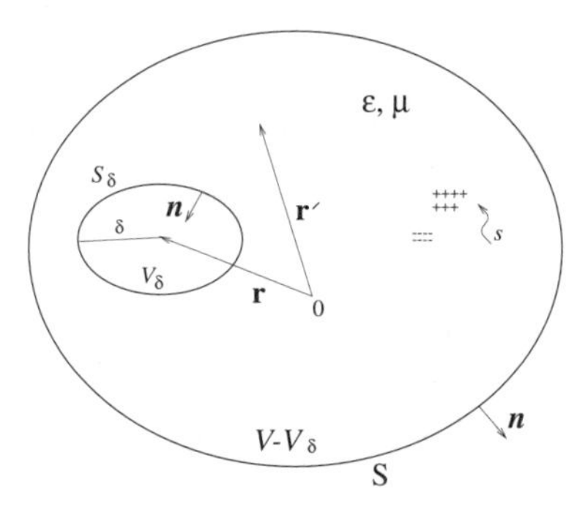


Figure B.1: Region $V$ containing sources $s$. Observation point $\mathbf{r}$ is excluded from $V$ by arbitrary exclusion volume $V_\delta$.

This becomes simply

$$\frac{\partial}{\partial x_i}\phi(\mathbf{r}) = -\oint_S s(\mathbf{r}')\, g(\mathbf{r},\mathbf{r}')\, \widehat{\mathbf{x}}_i \cdot \mathbf{n}\, dS' + \lim_{\delta\to 0}\int_{V-V_\delta} \frac{\partial\, s(\mathbf{r}')}{\partial x_i'}\, g(\mathbf{r},\mathbf{r}')\; dV' \tag{B.2}$$

because $\lim_{\delta\to 0}\oint_{S_\delta} s(\mathbf{r}')\, g(\mathbf{r},\mathbf{r}')\, \widehat{\mathbf{x}}_i \cdot \mathbf{n}\, dS' = 0$ since the integrand is $O(1/R)$ and the surface element is $O(R^2)$.

Each term in (B.2) may be differentiated under the integral sign (the volume integral looks like the potential caused by a source density $\partial\, s(\mathbf{r}')/\partial\, x_i'$ and the surface integral is proper with $\partial\, g/\partial\, x_i$ continuous on $S$), leading to

$$\begin{aligned}\frac{\partial^2}{\partial x_j\, \partial x_i}\phi(\mathbf{r}) = &-\oint_S s(\mathbf{r}')\, \frac{\partial}{\partial x_j} g(\mathbf{r},\mathbf{r}')\, \widehat{\mathbf{x}}_i \cdot \mathbf{n}\; dS' \\ &- \lim_{\delta\to 0}\int_{V-V_\delta} \frac{\partial\, s(\mathbf{r}')}{\partial x_i'}\,\frac{\partial}{\partial x_j'}\, g(\mathbf{r},\mathbf{r}')\; dV',\end{aligned} \tag{B.3}$$

where we have changed the derivative on $g$ in the volume integral to primed coordinates.[1] Subsequent to the substitution

$$\partial\, s(\mathbf{r}')/\partial\, x_i' = \partial\, \left(s(\mathbf{r}') - s(\mathbf{r})\right)/\partial\, x_i',$$

we use (B.1) to reexpress the volume term as

$$\begin{aligned}&\lim_{\delta\to 0}\oint_{S+S_\delta} [s(\mathbf{r}') - s(\mathbf{r})]\; \frac{\partial}{\partial x_j'}\, g(\mathbf{r},\mathbf{r}')\, \widehat{\mathbf{x}}_i \cdot \mathbf{n}\; dS' \\ &\quad - \lim_{\delta\to 0}\int_{V-V_\delta} [s(\mathbf{r}') - s(\mathbf{r})]\; \frac{\partial^2}{\partial x_i'\, \partial x_j'}\; g(\mathbf{r},\mathbf{r}')\; dV'\end{aligned}$$

[1] The expression (B.3) may be taken as an alternative formula for the second derivative and is presented in [45, p. 133].

and note again that $\lim_{\delta\to 0}\oint_{S_\delta}[s(\mathbf{r}') - s(\mathbf{r})]\,\frac{\partial}{\partial x_j'}\,g(\mathbf{r},\mathbf{r}')\,\widehat{\mathbf{x}}_i\cdot\mathbf{n}\,dS' = 0$ (for this to occur $s$ must be Hölder continuous as examined in more detail on p. 33). We therefore obtain

$$\begin{aligned}\frac{\partial^2}{\partial x_j\,\partial x_i}\phi(\mathbf{r}) = &-\oint_S s(\mathbf{r}')\,\frac{\partial}{\partial x_j}\,g(\mathbf{r},\mathbf{r}')\,\widehat{\mathbf{x}}_i\cdot\mathbf{n}\,dS' \\ &-\oint_S [s(\mathbf{r}') - s(\mathbf{r})]\,\frac{\partial}{\partial x_j'}\,g(\mathbf{r},\mathbf{r}')\,\widehat{\mathbf{x}}_i\cdot\mathbf{n}\,dS' \\ &+\lim_{\delta\to 0}\int_{V-V_\delta}[s(\mathbf{r}') - s(\mathbf{r})]\,\frac{\partial^2}{\partial x_i'\,\partial x_j'}\,g(\mathbf{r},\mathbf{r}')\,dV'.\end{aligned}$$

We can write the second term on the right side as two integrals since $g$ is not singular when $\mathbf{r}\notin S$, leading to

$$\begin{aligned}\frac{\partial^2}{\partial x_j\,\partial x_i}\lim_{\delta\to 0}\int_{V-V_\delta}s(\mathbf{r}')\,g(\mathbf{r},\mathbf{r}')\,dV' = &-s(\mathbf{r})\oint_S\frac{\partial}{\partial x_j}\,g(\mathbf{r},\mathbf{r}')\,\widehat{\mathbf{x}}_i\cdot\mathbf{n}\,dS' \\ &+\lim_{\delta\to 0}\int_{V-V_\delta}[s(\mathbf{r}') - s(\mathbf{r})]\,\frac{\partial^2}{\partial x_i'\,\partial x_j'}\,g(\mathbf{r},\mathbf{r}')\,dV',\end{aligned}\tag{B.4}$$

which is the desired equation (1.59). The derivation assumes $s(\mathbf{r})$ is continuously differentiable, but this requirement can be relaxed to Hölder-continuous functions [18, p. 249] by showing (1) that the right side of (B.4) exists, and (2) that it is, in fact, the derivative of $\phi$. In (B.4) $S$ is the surface that bounds $V$ and is clearly not related to the shape of $V_\delta$.

# C

# Gram–Schmidt Orthogonalization Procedure

The Gram–Schmidt procedure is a method of generating an orthonormal set of elements $\{u_1, u_2, \ldots, u_n\}$ from a linearly independent set of elements $\{x_1, x_2, \ldots, x_n\}$, where the set of elements may be finite or infinite. We'll work within a general inner product space $S$ with inner product $\langle x, y\rangle$ and norm $\|x\| = \langle x, x\rangle^{1/2}$. We define $v_1 = x_1$, and take the first element of the orthonormal set as

$$u_1 = \frac{v_1}{\|v_1\|} = \frac{x_1}{\|x_1\|}.$$

We next form $v_2 = x_2 - \langle x_2, u_1\rangle\, u_1$ and take

$$u_2 = \frac{v_2}{\|v_2\|} = \frac{x_2 - \langle x_2, u_1\rangle\, u_1}{\|x_2 - \langle x_2, u_1\rangle\, u_1\|}.$$

The interpretation of $v_2 = x_2 - \langle x_2, u_1\rangle\, u_1$ is that it is the quantity obtained by removing the projection of $x_2$ on $u_1$ from $x_2$ itself. We continue the process as $v_3 = x_3 - \langle x_3, u_2\rangle\, u_2 - \langle x_3, u_1\rangle\, u_1$ with $u_3 = v_3 / \|v_3\|$ such that

$$u_n = \frac{v_n}{\|v_n\|} = \frac{x_n - \sum_{k=1}^{n-1} \langle x_n, u_k\rangle\, u_k}{\left\| x_n - \sum_{k=1}^{n-1} \langle x_n, u_k\rangle\, u_k \right\|}.$$

It should be noted that the sets $\{x_1, x_2, \ldots, x_n\}$, $\{v_1, v_2, \ldots, v_n\}$, and $\{u_1, u_2, \ldots, u_n\}$ all generate the same linear subspace $M \subset S$.

# D

# Coefficients of Planar-Media Green's Dyadics

Dyadic Green's functions for planarly layered media are developed in Section 8.3. Here we provide the coefficients for any of the equivalent forms (8.190), (8.191), or (8.193). We assume the five-layer structure shown in Figure 8.12, where it is assumed that both the source point $\mathbf{r}'$ and the observation point $\mathbf{r}$ reside in layer three. In this case the coefficients are

$$R_t(x,x',\lambda) \tag{D.1}$$
$$= \frac{R_t^1(\lambda)e^{p_3(x-x')} + R_t^2(\lambda)e^{-p_3(x-x')} + R_t^3(\lambda)e^{p_3(x+x')} + R_t^4(\lambda)e^{-p_3(x+x')}}{Z^H(\lambda)},$$

$$R_n(x,x',\lambda)$$
$$= \frac{R_n^1(\lambda)e^{p_3(x-x')} + R_n^2(\lambda)e^{-p_3(x-x')} + R_n^3(\lambda)e^{p_3(x+x')} + R_n^4(\lambda)e^{-p_3(x+x')}}{Z^E(\lambda)},$$

$$R_c(x,x',\lambda)$$
$$= \frac{R_c^1(\lambda)e^{p_3(x-x')} + R_c^2(\lambda)e^{-p_3(x-x')} + R_c^3(\lambda)e^{p_3(x+x')} + R_c^4(\lambda)e^{-p_3(x+x')}}{Z^E(\lambda)Z^H(\lambda)},$$

which is the general form for any number of planar layers. For the five-layer structure considered here, we have

$$\begin{aligned}
Z^H(\lambda) &= g_{12}^{t+} g_{54}^{t+} e^{2p_3 d_2} - g_{12}^{t-} g_{54}^{t-}, \\
Z^E(\lambda) &= g_{12}^{n+} g_{54}^{n+} e^{2p_3 d_2} - g_{12}^{n-} g_{54}^{n-}, \\
R_t^1(\lambda) &= R_t^2(\lambda) = g_{12}^{t-} g_{54}^{t-}, \\
R_t^3(\lambda) &= g_{12}^{t+} g_{54}^{t-}, \qquad R_t^4(\lambda) = g_{12}^{t-} g_{54}^{t+} e^{2p_3 d_2}, \\
R_n^1(\lambda) &= R_n^2(\lambda) = g_{12}^{n-} g_{54}^{n-}, \\
R_n^3(\lambda) &= g_{12}^{n+} g_{54}^{n-}, \qquad R_n^4(\lambda) = g_{12}^{n-} g_{54}^{n+} e^{2p_3 d_2}, \\
R_c^1(\lambda) &= e^{2p_3 d_2} \left( a_7 g_{54}^{t+} g_{54}^{n-} - b_7 g_{12}^{n+} g_{12}^{t-} \right), \\
R_c^2(\lambda) &= e^{2p_3 d_2} \left( a_7 g_{54}^{n+} g_{54}^{t-} - b_7 g_{12}^{t+} g_{12}^{n-} \right), \\
R_c^3(\lambda) &= a_7 g_{54}^{t-} g_{54}^{n-} - b_7 e^{2p_3 d_2} g_{12}^{t+} g_{12}^{n+}, \\
R_c^4(\lambda) &= e^{2p_3 d_2} \left( a_7 g_{54}^{t+} g_{54}^{n+} e^{2p_3 d_2} - b_7 g_{12}^{t-} g_{12}^{n-} \right),
\end{aligned}$$

where

$$\begin{aligned}
a_7 &= \frac{1}{p_3} \left\{ F_2 \left[ \cosh p_2 d_1 - Z_{12}^n \sinh p_2 d_1 \right] + F_4 \right\}, \\
b_7 &= \frac{1}{p_3} \left\{ F_8 \left[ \cosh p_4 d_3 - Z_{54}^n \sinh p_4 d_3 \right] + F_6 \right\}, \\
F_2 &= \left( N_{12}^2 - 1 \right) T_{21}^t e^{-p_2 d_1} N_{32}^2 \left( 1 + Z_{12}^t \right), \\
F_4 &= \left( N_{23}^2 - 1 \right) \left( 1 + R_{12}^t e^{-2p_3 d_1} \right) N_{32}^2 \left( 1 + Z_{12}^t \right), \\
F_6 &= \left( N_{43}^2 - 1 \right) \left( 1 + R_{54}^t e^{-2p_4 d_3} \right) N_{34}^2 \left( 1 + Z_{54}^t \right), \\
F_8 &= \left( N_{54}^2 - 1 \right) T_{45}^t e^{-p_4 d_3} N_{34}^2 \left( 1 + Z_{54}^t \right),
\end{aligned}$$

and where

$$\begin{aligned}
g_{12}^{t\pm} &= 1 \pm \frac{p_2}{p_3} Z_{12}^t, \qquad g_{54}^{t\pm} = 1 \pm \frac{p_4}{p_3} Z_{54}^t, \\
g_{12}^{n\pm} &= 1 \pm \frac{p_2}{p_3 N_{23}^2} Z_{12}^n, \qquad g_{54}^{n\pm} = 1 \pm \frac{p_4}{p_3 N_{43}^2} Z_{54}^n, \\
Z_{12}^{t,n} &= \frac{e^{p_2 d_1} - e^{-p_2 d_1} R_{12}^{t,n}}{e^{p_2 d_1} + e^{-p_2 d_1} R_{12}^{t,n}}, \qquad Z_{54}^{t,n} = \frac{e^{p_4 d_3} - e^{-p_4 d_3} R_{54}^{t,n}}{e^{p_4 d_3} + e^{-p_4 d_3} R_{54}^{t,n}}, \\
R_{ij}^t &= \frac{p_j - p_i}{p_j + p_i}, \qquad R_{ij}^n = \frac{N_{ij}^2 p_j - p_i}{N_{ij}^2 p_j + p_i}, \qquad N_{ij}^2 = \frac{\varepsilon_i}{\varepsilon_j}.
\end{aligned}$$

As in Section 8.3, $p_j = \sqrt{\lambda^2 - k_j^2}$. Several special cases of particular interest follow.

### Homogeneous Parallel Plates

If $\varepsilon_1, \varepsilon_2, \varepsilon_4, \varepsilon_5 \to -i\infty$, $\varepsilon_3 = \varepsilon$, and $d_2 = a$, we have a parallel-plate structure, homogeneously filled with a medium characterized by $\varepsilon$ and $\mu_0$, with plate separation $a$, as shown in Figure 8.2. In this case the coefficients for the Green's dyadic are given by

$$R_t(x,x',\lambda) = \frac{e^{-pa}\cosh[p(x-x')] - \cosh[p(x+x'-a)]}{Z^H(\lambda)}, \tag{D.2}$$

$$R_n(x,x',\lambda) = \frac{e^{-pa}\cosh[p(x-x')] + \cosh[p(x+x'-a)]}{Z^E(\lambda)},$$

$$R_c(x,x',\lambda) = 0,$$

where

$$Z^E(\lambda) = Z^H(\lambda) = \sinh(pa)$$

and $p = \sqrt{\lambda^2 - k^2}$, such that the Hertzian-potential Green's dyadic (8.189) has the simple diagonal form

$$\begin{aligned} \underline{\mathbf{G}}_\pi &= \underline{\mathbf{G}}_\pi^p + \widehat{\mathbf{x}}\widehat{\mathbf{x}}G_n^s + (\widehat{\mathbf{y}}\widehat{\mathbf{y}}+\widehat{\mathbf{z}}\widehat{\mathbf{z}})\,G_t^s \\ &= \widehat{\mathbf{x}}\widehat{\mathbf{x}}\left(G_\pi^p + G_n^s\right) + (\widehat{\mathbf{y}}\widehat{\mathbf{y}}+\widehat{\mathbf{z}}\widehat{\mathbf{z}})\left(G_\pi^p + G_t^s\right). \end{aligned} \tag{D.3}$$

Note this implies that, because

$$\pi(\mathbf{r}) = \int_\Omega \underline{\mathbf{G}}_\pi(\mathbf{r},\mathbf{r}') \cdot \frac{\mathbf{J}_e(\mathbf{r}')}{i\omega\varepsilon}\, d\Omega',$$

then a current $\widehat{\alpha}J_\alpha$ maintains a parallel potential $\widehat{\alpha}\pi_\alpha$.

### Grounded Dielectric Layer

From Figure 8.12 with $\varepsilon_1 \to -i\infty$, and $\varepsilon_3 = \varepsilon_4 = \varepsilon_5 = \varepsilon_c$, we have a grounded dielectric characterized by $\varepsilon_2 = \varepsilon_f$, having thickness $d_1 = a$. This corresponds to the structure shown in Figure 8.6, with $\varepsilon_1 = \varepsilon_f$, $\varepsilon_2 = \varepsilon_c$, and $\mu_1 = \mu_2 = \mu_0$. The coefficients (D.1) become

$$R_t(x,x',\lambda) = \frac{p_c - p_f \coth p_f a}{Z^H} e^{-p_c(x+x')}, \tag{D.4}$$

$$R_n(x,x',\lambda) = \frac{N_{fc}^2 p_c - p_f \tanh p_f a}{Z^E} e^{-p_c(x+x')},$$

$$R_c(x,x',\lambda) = \frac{2\left(N_{fc}^2 - 1\right)p_c}{Z^H Z^E} e^{-p_c(x+x')},$$

where

$$Z^H(\lambda) = p_c + p_f \coth p_f\, a,$$
$$Z^E(\lambda) = N_{fc}^2 p_c + p_f \tanh p_f\, a,$$

with $N_{fc}^2 = \varepsilon_f/\varepsilon_c$, and $p_\alpha = \sqrt{\lambda^2 - k_\alpha^2}$, $\alpha = c, f$.

# E

# Additional Function Spaces

Some additional function spaces related to those described in Chapter 2 are presented in this appendix.

The Hölder space $\mathbf{H}^*_\beta$ is defined for a class of singular functions $y(t)$ as follows. Consider a function $y(t)$ defined and continuous everywhere on a closed, bounded interval $[a,b]$ except at a point $t_0 \in [a,b]$, where it is undefined. Assume that within a local neighborhood of $t_0$ the function $y(t)$ has the form

$$y(t) = f(t) \,/\, |t - t_0|^\beta$$

with $f(t) \in \mathbf{H}_{0,\alpha}(a,b)$ and $0 < \alpha \leq 1$. Then $y(t) \in \mathbf{H}^*_\beta((a,t_0) \cup (t_0,b))$, where $0 \leq \beta < 1$. In other words, if $x(t) \in \mathbf{H}_{0,\alpha}(a,b)$ at each point of the interval $[a,b]$, including point $t_0$, and it is differentiable everywhere on $[a,b]$ except maybe a point $t_0$, then $y(t) = dx(t)/dt \in \mathbf{H}^*_\beta((a,t_0) \cup (t_0,b))$ if in the vicinity of $t_0$ it can be represented in the form $y(t) = f(t)/|t - t_0|^\beta$.

If $t$ is a point from the local neighborhood of $t_0 \in [a,b]$ (i.e., $|t - t_0| = o(1)$ as $t \to t_0$), then $\mathbf{H}^*_{\beta_r} \subset \mathbf{H}^*_{\beta_p} \subset \mathbf{H}^*_\beta$ for $0 \leq \beta_r < \beta_p < \beta < 1$. Note that $\mathbf{H}^*_\beta((a,t_0) \cup (t_0,b)) \subset \mathbf{H}_{0,\alpha}(a,b)$ for any $0 < \alpha \leq 1$, $0 \leq \beta < 1$ and $\mathbf{H}_{0,\alpha}(a,b) \subset \mathbf{C}(a,b)$. Generalizing further, if

$$y(t) = f(t) \,/ \prod_{i=1}^{n} |t - t_i|^{\beta_i}$$

($n$ points $t_i$ where $y(t)$ is not defined but $x(t), f(t) \in \mathbf{H}_{\alpha_1 \ldots \alpha_n}(a,b)$), then $y(t) \in \mathbf{H}^*_{\beta_1 \ldots \beta_n}(I_n)$, where $I_n = [a,t_1) \cup \ldots \cup (t_i,t_{i+1}) \cup \ldots \cup (t_n,b]$ and $0 \leq \beta_i < 1$.

The one-dimensional Hölder space $\mathbf{H}^*_\beta((a,t_0) \cup (t_0,b))$ can be generalized for functions $y(\mathbf{t})$ of $n$ variables ($\mathbf{t} = (t_1,t_2,\ldots,t_n) \in \Omega$) such that

$y(\mathbf{t}) \in \mathbf{H}^*_\beta(\Omega/\Omega_0)$ if within a local neighborhood of $\mathbf{t_0} \in \Omega_0$ the function $y(\mathbf{t})$ can be represented in the form

$$y(\mathbf{t}) = f(\mathbf{t}) \,/\, |\mathbf{t} - \mathbf{t}_0|^\beta,$$

where $f(\mathbf{t}) \in \mathbf{H}_{0,\alpha}(\Omega)$ and $\beta$ $(0 \le \beta < 1)$ is a generalized constant.

Hölder spaces $\mathbf{H}^*_\beta$ and $\mathbf{H}_{0,\alpha}$ are associated with the behavior of the electromagnetic field components in the vicinity of conducting edges. Components longitudinal to the edge are defined within $\mathbf{H}_{0,\alpha}$, where they are described by $O(\varepsilon^\alpha)$ as $\varepsilon \to 0$ (e.g., $\alpha = 1/2$ for a sharp edge and $\alpha = 2/3$ for a rectangular edge), and components transverse to the edge are defined within $\mathbf{H}^*_\beta$ with the asymptotics $O(\varepsilon^{-\beta})$ as $\varepsilon \to 0$ (with $\beta = 1/2$ for a sharp edge and $\beta = 1/3$ for a rectangular edge). Note that electric and magnetic currents are defined conversely to the above in the vicinity of an edge (longitudinal current components are within $\mathbf{H}^*_\beta$, and the transverse ones are within $\mathbf{H}_{0,\alpha}$).

It is also useful to consider the Sobolev space $\mathbf{W}^k_p(\Omega)$ of noninteger order, where $\Omega \subset \mathbf{R}^n$ and $0 < k < 1$, which is defined for all $x(\mathbf{t}) \in \mathbf{L}^p(\Omega)$ such that the following norm is finite:

$$||x||^k_p = \left(||x||^p_p + \int_\Omega \int_\Omega \frac{|x(\mathbf{t}) - x(\mathbf{s})|^p}{|\mathbf{t} - \mathbf{s}|^{n+pk}}\, d\Omega_t\, d\Omega_s\right)^{1/p}.$$

$||x||_p$ is the $p$-norm in $\mathbf{L}^p(\Omega)$,

$$\|x\|_p = \left(\int_\Omega |x(\mathbf{t})|^p\, d\Omega\right)^{1/p}.$$

If $\mathbf{W}^k_p(\Omega)$ is a Hilbert space (e.g., $\mathbf{H}^k(\Omega) \equiv \mathbf{W}^k_2(\Omega)$), then the inner product is defined as

$$\langle x,y\rangle_k = \int_\Omega x(\mathbf{t})\overline{y(\mathbf{t})}\, d\Omega + \int_\Omega \int_\Omega \frac{[x(\mathbf{t}) - x(\mathbf{s})]\left[\overline{y(\mathbf{t})} - \overline{y(\mathbf{s})}\right]}{|\mathbf{t} - \mathbf{s}|^{n+2k}}\, d\Omega_t\, d\Omega_s.$$

In particular, in connection with boundary value problems of applied electromagnetics, it is important to introduce the Sobolev space $\mathbf{H}^{1/2}(\Gamma)$, which is used to characterize boundary values on $\Gamma$ of functions in $\mathbf{H}^1(\Omega)$ ($\Gamma \in \mathbf{R}^2$ is a smooth boundary of a bounded domain $\Omega \subset \mathbf{R}^3$). The space $\mathbf{H}^{1/2}(\Gamma)$ is a Hilbert space with the inner product [6]

$$\langle x, y\rangle_{1/2} = \int_\Gamma x(\mathbf{t})\overline{y(\mathbf{t})}\, d\Gamma + \int_\Gamma \int_\Gamma \frac{[x(\mathbf{t}) - x(\mathbf{s})]\left[\overline{y(\mathbf{t})} - \overline{y(\mathbf{s})}\right]}{|\mathbf{t} - \mathbf{s}|^3}\, d\Gamma_t\, d\Gamma_s.$$

The corresponding norm in $\mathbf{H}^{1/2}(\Gamma)$ is defined as $||x||_{1/2} = \langle x, x\rangle^{1/2}_{1/2}$. The dual of the space $\mathbf{H}^{1/2}(\Gamma)$ is the space of all normal derivatives to the boundary $\Gamma$, which is denoted as $\mathbf{H}^{-1/2}(\Gamma)$ with the norm [6] (see also a series of papers that appeared in the Russian literature, for example, [44]–[47]),

$$\left\|\frac{\partial x}{\partial n}\right\|_{-1/2} = \sup_{y \in H^{1/2}(\Gamma)} \frac{\left|\int_\Gamma \frac{\partial x(\mathbf{t})}{\partial n} y(\mathbf{t})\, d\Gamma\right|}{\|y\|_{1/2}}, \qquad y \neq 0.$$

We also have [6], [2]

$$\begin{aligned}\mathbf{H}^{-1/2}&(\mathrm{div}, \Gamma) \\ &= \{\mathbf{x}(\mathbf{t}) : \mathbf{x}(\mathbf{t}) \in \mathbf{H}^{-1/2}(\Gamma)^3,\ \mathbf{n}\cdot\mathbf{x}(\mathbf{t}) = 0,\ \nabla_\Gamma \cdot \mathbf{x}(\mathbf{t}) \in \mathbf{H}^{-1/2}(\Gamma)\}\end{aligned}$$

and

$$\begin{aligned}\mathbf{H}^{-1/2}(\mathrm{curl}, \Gamma) = &\{\mathbf{x}(\mathbf{t}) : \mathbf{x}(\mathbf{t}) \in \mathbf{H}^{-1/2}(\Gamma)^3,\ \mathbf{n}\cdot\mathbf{x}(\mathbf{t}) = 0, \\ &\nabla_\Gamma \cdot \mathbf{n} \times \mathbf{x}(\mathbf{t}) \in \mathbf{H}^{-1/2}(\Gamma)^3\},\end{aligned}$$

with norm defined as,

$$\|\mathbf{x}\|_{\mathbf{H}^{-1/2}(\mathrm{div},\Gamma)} = \left(\|\mathbf{x}\|_{-1/2}^2 + \|\nabla_\Gamma \cdot \mathbf{x}\|_{-1/2}^2\right)^{1/2}$$

and

$$\|\mathbf{x}\|_{\mathbf{H}^{-1/2}(\mathrm{curl},\Gamma)} = \left(\|\mathbf{x}\|_{-1/2}^2 + \|\nabla_\Gamma \cdot \mathbf{n} \times \mathbf{x}\|_{-1/2}^2\right)^{1/2},$$

where $\mathbf{n}$ is a normal to the boundary $\Gamma$ and $\nabla_\Gamma$ is the tangential del operator.

# Index


Adjoint, 144, 146, 149
    boundary conditions, 153
    formal, 154

Basis, 104, 105, 121, 123
    associated Legendre polynomials, 354
    Bessel function
        cylindrical, 347
        spherical, 357
    bi-orthonormal, 136, 335
    Chebyshev polynomial
        first-kind, 264, 361
        second-kind, 265, 361
    eigenfunction, 383
    exponential functions, 268, 354
    Hermite polynomials, 360
    Jordan, 253, 259
    Laguerre polynomial, 266, 267, 359
    Legendre polynomial, 353
    root system, 337
    spherical harmonics, 269, 354
    vector wavefunctions, 582
Bessel
    equation
        cylindrical, 343
        spherical, 355
    Fourier–Bessel expansion, 347, 357
    function
        cylindrical, 344
        spherical, 355
Born approximation, 52

Carleman's inversion formula, 398
Compact support, 65
Completeness relation, 284, 306, 319, 320, 327, 332, 336, 338, 382, 430, 431, 585
    associated Legendre polynomials, 354
    Bessel function
        cylindrical, 347, 348
        spherical, 357
    Chebyshev polynomial
        first-kind, 361
        second-kind, 362
    grounded dielectric, 476
    Hermite polynomials, 360
    impedance plane waveguide, 457
    improper eigenfunctions, 321
    Laguerre polynomials, 359
    Legendre polynomials, 353
    parallel-plate waveguide, 441
    sine functions, 416
    spherical harmonics, 355
    two-dimensional resonator, 445
Condition number, 137
Constitutive equations, 4
    complex, 11
Continuity, 88
    Hölder, 32, 89
    Lipschitz, 89
    piecewise, 89
    uniform, 88

Convergence, 90
absolute, 187
Cauchy, 91
energy, 168
mean, 93
mean-square, 93
pointwise, 92
strong, 114
uniform, 30, 92
weak, 114

Deficiency, 223
Depolarizing dyadic, 42, 44
Discontinunity conditions, 464
Dispersion equation, 465, 467
Domain, 130
Duality, 13, 43

Eigenbasis, 250, 253, 259, 260, 262, 269, 278
Eigenfunction, 425
associated, 284, 333, 337, 434, 464, 474, 481
basic, 258
Bessel equation
cylindrical, 346
spherical, 356
Fourier intergal operator, 360
generalized, 252, 269
grounded dielectric, 463
impedance plane, 449
improper, 318
parallel-plate waveguide, 435
short-circuit
electric, 579
magnetic, 580
vector, 581
Eigenfunction expansion, 303, 381
partial, 384, 429
Eigenspace, 222
Eigenvalue, 220
algebraic multiplicity, 222
geometric multiplicity, 222
Rayleigh quotient, 231
simple, 295
Eigenvalue problem
generalized, 220, 239
nonstandard, 220, 242
pseudo inner product, 241
standard, 220
Eigenvector, *see* Eigenfunction
associated, 253, 259
generalized, 252, 259
Expansion functions, 209
Exterior resonance, 245

Fredholm alternative, 205, 400

Generalized partial derivative, 69
Green's function, 20, 190, 428
adjoint, 192, 329
Dirichlet, 380, 385, 387
dyadic, 44, 148, 196, 199, 282, 504
cavity, 583
free-space, 38, 41, 162
grounded dielectric layer, 623
Hertzian potential, 491
parallel-plate, 499, 622
rectangular waveguide, 518, 553
Neumann, 374, 386, 387
scalar
free-space, 22, 162, 446
grounded dielectric, 471
impedance plane, 454
parallel-plate waveguide, 439
two-dimensional resonator, 444
Sturm–Liouville, 295
transmission line, 200
infinite, 416
resonator, 413
semi-infinite, 415
terminated, 410

Green's theorem, 23, 38
  generalized, 153

Harmonic function, 371
Helmholtz equation, 18

Impedance plane, 447
Inequality
  Bessel, 120
  Cauchy–Schwarz, 87
  Cauchy–Schwarz–Bunjakowsky, 112
  Hölder, 87, 109
  Minkowski, 87, 108
  triangle, 106
Inner product, 110
  energy, 167
  pseudo, 113
Integral equation, 391, 586
  domain, 48, 484
  electric field, 53, 486
  first-kind, 395, 487, 588
  magnetic field, 54, 174, 244, 246
  second-kind, 399, 400
  surface, 52
Inverse, *see* Operator
  right (left), 187
Isomorphism, 141, 249

Jordan canonical form, 252
Jump condition, 299

Kernel
  adjoint, 148, 149
  bounded, 139
  compact, 162, 399, 487
  degenerate, 172
  dyadic Green's function, 163
  Green's function, 140, 173
  Hilbert–Schmidt, 139, 172, 174, 296
  logarithmic, 174
  pseudo-adjoint, 162
  self-adjoint, 148, 159
  singular, 162
  weakly singular, 140

Laplace's equation, 368, 369
Leaky mode, 471
Legendre
  equation, 352
  equation, associated, 353
  Fourier–Legendre
    expansion, 353
  function, associated, 353
Linear functional, 143
Lorentz reciprocity theorem, 16
Lorenz gauge, 19

Manifold, 101
Matrix
  complex-symmetric, 163
  diagonalizable, 250
  equivalence class, 249
  Hermitian, 163
  similiar, 249
  simple, 250
  unitary, 249
Maxwell's equations, 4
Method of overlapping regions, 79, 560
Metric, *see* Space

Natural mode, 426
  grounded dielectric, 469
  impedance plane, 453
  parallel-plate waveguide, 437
  transmission line, 406
Norm, 106
  energy, 167
  matrix, 136
    absolute, 135
    column-sum, 136
    maximum, 137
    row-sum, 136
    Schur, 137
    spectral, 136
Null space, 131

Ohm's law, 10
Operator
- adjoint, 146, 153, 192, 230
- bounded, 131, 232
- bounded below, 185
- bounded below (above), 165
- Cauchy singular, 173
- closed, 180
- commutator, 157
- compact, 169, 234, 259, 377, 400, 485, 492, 561, 565, 567, 572
- compact, self-adjoint, 237, 258
- compact-normal, 237
- continuous, 132
- definite, 164
- densely defined, 130
- dissipative, 165, 236, 270
- Dunford integral representation, 277
- extension, 155
- Fredholm, 205
- Green's, 192, 282
- Hilbert–Schmidt, 139, 173
- Hilbert-transform, 140
- idempotent, 202
- inverse, 181, 233, 275
- Laplacian, 372
- linear, 130
- matrix, 135
- matrix representation, 249
- Neumann expansion, 187
- nonnegative, 236
- norm, 132
- norm convergence, 133, 187
- normal, 156, 238, 270
- nuclear, 270
- one-to-one, 182
- onto, 182
- positive, 164, 166
- positive-definite, 164
- projection, 202, 254, 272
- pseudo-adjoint, 161
- real, 154, 293
- resolvent, 223
- self-adjoint, 154, 235, 237, 239
- similar, 234
- simple, 271
- spectral representation, 254
- square-root, 165, 274
- Sturm–Liouville, 292
- symmetric, 155, 236
- unbounded, 132, 149
- unitary, 157, 238
- Volterra, 140
- weakly singular, 139, 140, 173

Operator equation, 278
- first-kind, 204, 264, 279
- second-kind, 204, 281
- well-posed, 186, 206

Operator-valued function, 220, 223, 245, 273, 285

Parseval's equality, 120
Piloting vector, 582
Poisson's equation, 368
Polynomial
- characteristic, 222
- Chebyshev
  - first-kind, 263, 360
  - second-kind, 265, 361
- Hermite, 359
- Laguerre, 266, 311, 359
- Legendre, 352

Potential
- double-layer, 394
- electric scalar, 18
- electric vector, 19
- Hertzian, 490
- magnetic scalar, 19
- magnetic vector, 18, 504
- single-layer, 394
- vector, 589

Poynting theorem, 14
Projection, 117

Radiation condition, 41, 56, 155, 161, 440, 480

Range, 131
Removable branch point, 318
Resolution of the identity, 254
Resonance frequency, 413, 595
Root system, 253, 259, 269, 333, 337

Scattering superposition, 298, 301, 380, 484
Set, 64
- closed, 96
- closure, 95
- compact, 96
- countable, 95
- dense, 94
- functions, 65
- numbers, 64
- resolvent, 224
- sequences, 73

Singular value, 259
Space
- Banach, 109
- bounded functions, 66
- compact metric, 96
- complete, 97, 109, 115
- continuous functions, 65
- dimension, 104
- direct sum, 103
- dual, 143
- Hilbert, 115
  - weighted, 293
- Hölder continuous functions, 70
- inner product, 110
- isomorphic, 141
- Lebesgue-integrable functions, 66
- linear, 100
- metric, 83
- normed, 105
- separable, 95
- Sobolev, 72

Span, 104
Spectral radius, 233
Spectrum
- approximate, 225
- bounded, 232
- continuous, 224, 314
- point, 224
- residual, 224, 238

Spherical harmonics, 354
Strictly diagonally dominant, 183
Subspace, 101
Surface wave, 470
Symmetric product, 113

Testing functions, 209
Theorem
- Hilbert–Schmidt, 257
- inverse mapping, 183
- orthogonal structure, 120
- projection, 120
- Riesz representation, 144
- Riesz–Fischer, 121
- spectral
  - finite-dimensional operator, 250
  - infinite-dimensional operator, 258
- spectral mapping, 275
- Steinberg, 244
- Weyl's, 311

Time-harmonic fields, 7
Transform, 432
- Fouier sine, 322
- Fourier cosine, 323
- Fourier exponential, 9, 325
- Fourier–Bessel (Hankel), 349

Transmission line, 403
- multiconductor, 417
- resonator, 412

Uniqueness theorem, 55

Vector wave equation, 17
Vector wavefunction, 581
Volume equivalence principle, 50, 483

Well posed problems, 186

Zwischenmedium, 560